ACOUSTICS

An Introduction to Its Physical Principles and Applications

ACOUSTICS

An Introduction to Its Physical Principles and Applications

Allan D. Pierce

1981 Edition: School of Mechanical Engineering
Georgia Institute of Technology

1989 Edition: Graduate Program in Acoustics
Pennsylvania State University

1994 Edition: Department of Aerospace
and Mechanical Engineering
Boston University

Published by the Acoustical Society of America
through the American Institute of Physics

CONTENTS

The more recent citations include most of the author's favorite references on acoustics; these are recommended reading for anyone who desires further elaboration on the subject matter. The author regrets that the pedagogical objectives of the book and the constraint that the book be of manageable length precluded the inclusion of some of the more important topics in modern acoustics (such as, for example, jet noise, acoustic emissions, cavitation, streaming, radiation pressure and levitation, combustion noise, parametric arrays, propagation through turbulence, sound-structural interaction, surface waves, and acoustical imaging). A consequence is that many works that the author esteems highly are not mentioned here. An introductory text with the objective of inculcating a deep understanding of the basic principles cannot, however, be encyclopedic and some hard decisions had to be made. The student should be able to proceed rapidly, once these basic principles are understood, toward any of the current frontiers of acoustics.

Along with the writings of Rayleigh and of other past contributors to the field, the style and content of this book have been influenced by the author's early teachers, Richard H. Duncan and Laszlo Tisza, and by his past associations with Albert Latter, Elisabeth Iliff, Charles A. Moo, S. H. Crandall, J. P. Den Hartog, Huw G. Davies, Y. K. Lin, T.-Y. Toong, Patrick Leehey, Richard Lyon, P. P. Lele, Joe W. Posey, Wayne A. Kinney, Warren Strahle, W. James Hadden, Jr., E.-A. Müller, W. Möhring, and F. Obermeier. The writing of the book has also been affected by conversations or correspondence with John Snowdon, Herbert S. Ribner, Dominic Maglieri, Lucio Maestrello, Richard K. Cook, R. Bruce Lindsay, Geoffrey Main, David T. Blackstock, K. Uno Ingard, David G. Crighton, Hugh G. Flynn, T. F. W. Embleton, Robert Waag, Robert E. Apfel, Robert W. Young, Jiri Tichy, Donald Lansing, M. C. Junger, H. M. Überall, C.-H. Chew, Edmund H. Brown, Prateen Desai, T. J. Lardner, Preston W. Smith, Jr., Michael Howe, Phillip A. Thompson, Joseph E. Piercy, Walter Soroka, Sigalia Dostrovsky, Wesley Cobb, Lawrence A. Crum, Henry E. Bass, Bill D. Cook, and Steven D. Pettyjohn. Thanks must also be expressed to the many students who pointed out weaknesses in the earlier class notes and who suggested improvements.

Although the writing of this book has extended over many years, the author's ideas concerning its substance crystallized during a year's sojourn (1976–1977) with the Max-Planck-Institut für Strömungsforschung in Göttingen. The Institute's research objectives and atmosphere were conducive to a sustained contemplation of the principles of acoustics, of their interconnections, and of their mechanical, thermodynamic, and mathematical foundations. The author is grateful to Professor E.-A. Müller and his colleagues for their hospitality and rapport and to the Alexander von Humboldt Foundation for the generous award that made the stay in Göttingen possible.

The author thanks the staff of the School of Mechanical Engineering at Georgia Tech for their forbearance throughout this long, seemingly interminable, project. The empathy and encouragement of S. Peter Kezios, the school's Director, is very much appreciated.

The author is also grateful to the library personnel who helped him in this endeavor; he especially thanks Robert Perrault for advice and for facilitating the procurement of rare bibliographic materials.

It was the author's extreme good fortune to have the collaboration of Rosie Atkins, an outstanding technical typist and manuscript stylist. Throughout several generations of manuscripts, Mrs. Atkins patiently and accurately interpreted and translated heavily scored, barely legible handscripts, laden with equations and symbols, into attractive and readable typescripts.

The author's largest debt of thanks is owed to his wife Penny and to his children, Jennifer and Bradford. Their loyalty, encouragement, cheerfulness, and willingness to sacrifice have contributed immeasurably to the successful completion of this book.

Allan D. Pierce

LIST OF SYMBOLS

a = radius of sphere, cylinder, or disk
= characteristic dimension of object
a_n = coefficient in modal expansion of pressure field
= zero of Airy function
a'_n = zero of derivative of Airy function
A = generic designation for amplitude factors
= cross-sectional area
= ray-tube area
$A_D(X)$ = diffraction integral, related to Fresnel integrals
A_s = absorbing power of room, metric sabins
$\mathrm{Ai}(\eta)$ = Airy function
$\mathcal{A}$ = aspect factor in echo-sounding equation
= age variable for accumulative nonlinear effects
b = frequency band
B = bias in estimate of statistical quantity
B_{pl} = plate-bending modulus
$B(\mathbf{x})$ = amplitude factor for waveform propagating along ray path
B/A = parameter of nonlinearity for compressible fluid
c = speed of sound
c_D, c_S = dilatational and shear elastic-wave speeds
c_p, c_v = specific-heat coefficients at constant pressure and constant volume
c_{pl} = phase velocity of flexural waves on a plate
c_T = isothermal sound speed = $c/\sqrt{\gamma}$
$c_{v\nu}$ = specific-heat contribution from internal vibrations of ν-type molecules

C = integration contour in complex plane
$C(X)$ = Fresnel (cosine) integral
$C_+(\Delta L)$ = decibel addition function, dB
C_A = acoustic compliance
$C_{bg}(\Delta L)$ = background correction function, dB
d = average number of excited degrees of freedom per molecule
$[D]$ = acoustic-mobility matrix
D/Dt = time derivative following flow = $\partial/\partial t + \mathbf{v} \cdot \nabla$
$D(\mathbf{e})$ = directional energy density, $J/(m^3 \cdot sr)$
$\hat{\mathbf{D}}$ = dipole-moment amplitude vector
$\mathcal{D}$ = rate of energy dissipation per unit volume
$\mathcal{D}_p(t - t')$ = autocovariance of acoustic pressure
e = base of natural logarithms, 2.71828 · · ·
= voltage
$\mathbf{e}$ = generic designation for unit (einheit) vector
$\mathbf{e}_x$ = unit vector in direction of increasing x
E = energy stored in a vibrating system
= Young's (elastic) modulus
= sound exposure, time integral of p^2
E_f = sound-exposure spectral density, $Pa^2 \cdot s/Hz$
E_K = kinetic energy
E_Q = estimate of quantity Q
$E(m)$ = complete elliptical integral of first kind
$\mathcal{E}$ = energy per unit volume
f = frequency, Hz
$f(X)$ = real part of diffraction integral $A_D(X)$
f_0 = geometric center of frequency band
f_1, f_2 = relaxation frequencies
= lower and upper limits of a frequency band
f_c = coincidence frequency
= cutoff frequency
$f_H(t)$ = Hilbert transform of function $f(t)$
f_r = resonance frequency
f_{Sch} = Schroeder cutoff frequency for room
$f_{ST}(\omega t)$ = sawtooth wave function
$f_{\nu 1}$ = fraction of ν-type molecules in first excited vibrational state
$\mathbf{f}_B$ = body force per unit volume
$\mathbf{f}_S$ = surface force per unit area
$F_W(\xi)$ = Whitham F function (sonic-boom theory)
$\mathbf{F}$ = force
g = acceleration due to gravity, 9.8 m/s^2

CHAPTER

ONE

THE WAVE THEORY OF SOUND

Acoustics is the science of sound, including its production, transmission, and effects.† In present usage, the term *sound* implies not only the phenomena in air responsible for the sensation of hearing but also whatever else is governed by analogous physical principles. Thus, disturbances with frequencies too low (*infrasound*) or too high (*ultrasound*) to be heard by a normal person are also regarded as sound. One may speak of underwater sound, sound in solids, or structure-borne sound. Acoustics is distinguished from optics in that sound is a mechanical, rather than an electromagnetic, wave motion.

The broad scope of acoustics as an area of interest and endeavor can be ascribed to a variety of reasons. First, there is the ubiquitous nature of mechanical radiation, generated by natural causes and by human activity. Then, there is the existence of the sensation of hearing, of the human vocal ability, of communication via sound, along with the variety of psychological influences sound has on those who hear it. Such areas as speech, music, sound recording and reproduction, telephony, sound reinforcement, audiology, architectural acoustics, and noise control have strong association with the sensation of hearing. That sound is a means of transmitting information, irrespective of our natural ability to hear, is also a significant factor, especially in underwater acoustics. A variety of applications, in basic research and in technology, exploit the fact that the transmission of sound is affected by, and consequently gives information concerning, the medium through which it passes and inter-

† Definitions in the present text conform to ANSI S1.1960 (R1976) American National Standard Acoustical Terminology, American National Standards Institute, Inc. (ANSI), New York, 1976. Selected symbols for physical quantities conform to ANSI/ASME Y10.11–1984, "Letter Symbols and Abbreviations for Quantities Used in Acoustics," American Society of Mechanical Engineers, New York, 1984.

vening bodies and inhomogeneities. The physical effects of sound on substances and bodies with which it interacts present other areas of concern and of technical application.

Some indication of the scope of acoustics and of the disciplines with which it is associated can be found in Fig. 1-1. The first annular ring depicts the traditional subdivisions of acoustics, and the outer ring names technical and artistic fields to which acoustics may be applied. (The chart is not intended to be complete, nor should any rigid interpretation be placed on the depicted proximity of any subdivision to a technical field. A detailed listing of acoustical topics can be found in the index classification scheme reprinted with the index of each volume of the *Journal of the Acoustical Society of America.*)

The present text, while intended as an introduction to acoustics, is concerned primarily with the physical principles underlying the discipline rather

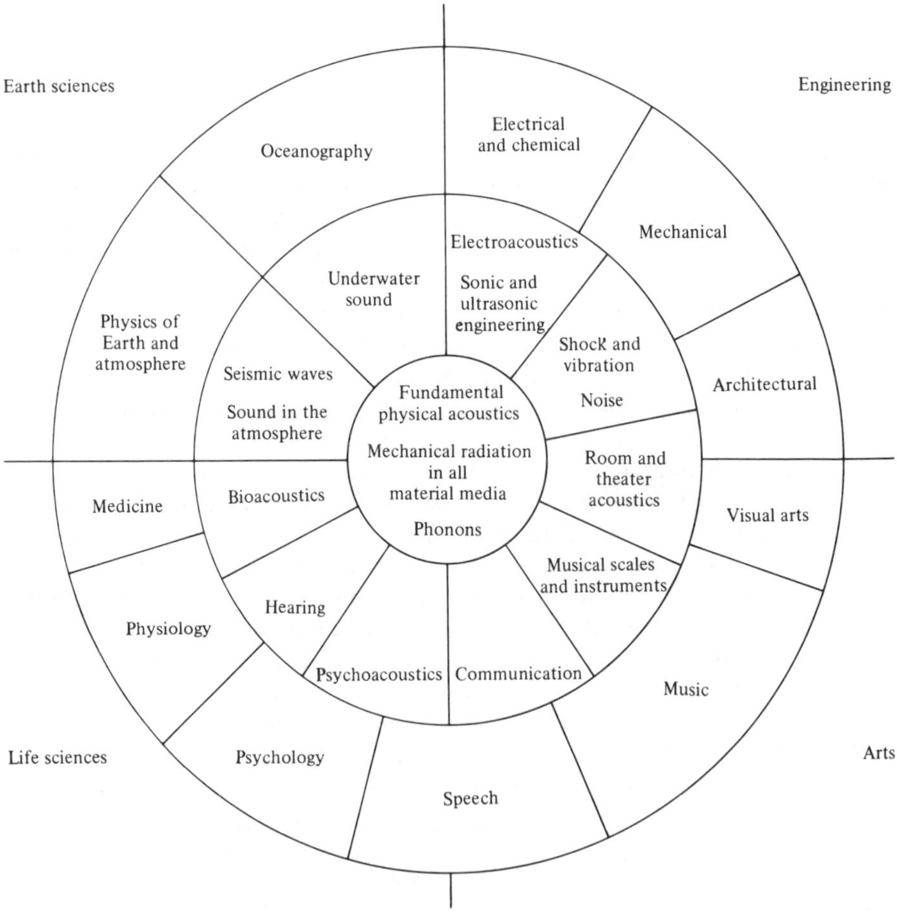

Figure 1-1 Circular chart illustrating the scope and ramifications of acoustics. [*Adapted from R. B. Lindsay, J. Acoust. Soc. Am.,* **36:***2242 (1964).*]

than with a summary of the current state of knowledge and technology in its many subfields. The general and specialized principles chosen for discussion are those which have found application in one or more of the following subfields: atmospheric acoustics, underwater acoustics, musical acoustics, ultrasonics, architectural acoustics, aeroacoustics, nonlinear acoustics, environmental acoustics, and noise control. For the most part, the selected subject matter is limited to sound in fluids, e.g., air and water.

We begin with a discussion of the wave theory of sound.

1-1 A LITTLE HISTORY

The speculation that sound is a wave phenomenon grew out of observations of water waves. The rudimentary notion of a *wave* is an oscillatory *disturbance* that moves away from some source and transports no discernable amount of matter over large distances of *propagation*. The possibility that sound exhibits analogous behavior was emphasized, for example, by the Greek philosopher Chrysippus (c. 240 B.C.), by the Roman architect and engineer Vetruvius (c. 25 B.C.), and by the Roman philosopher Boethius (A.D. 480–524). The wave interpretation was also consistent with Aristotle's (384–322 B.C.) statement† to the effect that air motion is generated by a source, "thrusting forward in like manner the adjoining air, so that the sound travels unaltered in quality as far as the disturbance of the air manages to reach."

A pertinent experimental result, inferred with reasonable conclusiveness by the early seventeenth century, with antecedents dating back to Pythagoras (c. 550 B.C.) and perhaps farther, is that the air motion generated by a vibrating body sounding a single musical note is also vibratory and of the same frequency as the body. The history of this is intertwined with the development of the laws for the natural frequencies of vibrating strings and of the physical interpretation of musical consonances.‡ Principal roles were played by Marin Mersenne (1588–1648), a French natural philosopher often referred to as the "father of acoustics," and by Galileo Galilei (1564–1642), whose *Mathematical Discourses Concerning Two New Sciences* (1638) contained§ the most lucid statement and discussion given up until then of the frequency equivalence.

Mersenne's description in his *Harmonie universelle* (1636) of the first absolute determination of the frequency of an audible tone (at 84 Hz) implies that he

† M. R. Cohen and I. E. Drabkin, *A Source Book in Greek Science*, Harvard University Press, Cambridge, Mass., 1948, pp. 289, 293–294, 307–308. Aristotle's statements on acoustics are also reprinted by R. B. Lindsay (ed.), *Acoustics: Historical and Philosophical Development*, Dowden, Hutchinson, and Ross, Stroudsburg, Penn., 1972, pp. 22–24. For a detailed account of the early history of acoustics, see F. V. Hunt, *Origins of Acoustics*, Yale University Press, New Haven, Conn., 1978. Hunt, p. 26, states that the above-cited aristotelian statement was probably written by Straton of Lampsacus (c. 340–269 B.C.).

‡ S. Dostrovsky, "Early Vibration Theory: Physics and Music in the Seventeenth Century," *Arch. Hist. Exact Sci.*, **14**:169–218 (1975).

§ The pertinent passages are reprinted in Lindsay, *Acoustics*, pp. 42–61, especially p. 48.

had already demonstrated that the absolute-frequency ratio of two vibrating strings, radiating a musical note and its octave, is as 1 : 2. The perceived harmony (consonance) of two such notes would be explained if the ratio of the air oscillation frequencies is also 1 : 2, which in turn is consistent with the source-air-motion-frequency-equivalence hypothesis.

The analogy with water waves was strengthened by the belief that air motion associated with musical sounds is oscillatory and by the observation that sound travels with a finite speed. Another matter of common knowledge was that sound bends around corners, which suggested diffraction, a phenomenon often observed in water waves. Also, Robert Boyle's (1660) classic experiment† on the sound radiation by a ticking watch in a partially evacuated glass vessel provided evidence that air is necessary, either for the production or transmission of sound.

The wave viewpoint was not unanimous, however. Gassendi‡ (a contemporary of Mersenne and Galileo), for example, argued that sound is due to a stream of "atoms" emitted by the sounding body; velocity of sound is speed of atoms; frequency is number emitted per unit time.

The apparent conflict§ between ray and wave theories played a major role in the history of the sister science optics, but the theory of sound developed almost from its beginning as a wave theory. When ray concepts were used to explain acoustic phenomena, as was done, for example, by Reynolds and Rayleigh‖ in the nineteenth century, they were regarded, either implicitly or explicitly, as mathematical approximations to a then well-developed wave theory; the successful incorporation of geometrical optics into a more comprehensive wave theory had demonstrated that viable approximate models of complicated wave phenomena could be expressed in terms of ray concepts. (This recognition has strongly influenced twentieth-century developments in architectural acoustics, underwater acoustics, and noise control.)

The mathematical theory of sound propagation began with Isaac Newton (1642–1727), whose *Principia*¶ (1686) included a mechanical interpretation of

† R. Boyle, *New Experiments, Physico-Mechanical, Touching the Spring of the Air,* 2d ed., 1662, Experiment 27, reprinted by Lindsay, pp. 68–73. Lindsay gives a modern interpretation of Boyle's experiment in "Transmission of Sound through Air at Low Pressure," *Am. J. Phys.,* **16:**371–377 (1948).

‡ R. B. Lindsay, "Pierre Gassendi and the Revival of Atomism in the Renaissance," *Am. J. Phys.,* **13:**235–242 (1945).

§ A. E. Shapiro, "Kinematic Optics: A Study of the Wave Theory of Light in the Seventeenth Century," *Arch. Hist. Exact Sci.,* **11:**134–266 (1973).

‖ O. Reynolds, "On the Refraction of Sound by the Atmosphere," *Proc. R. Soc. Lond.,* **22:** 531–548 (1874); J. W. Strutt, Baron Rayleigh, *The Theory of Sound,* vol. 2, 1878; 2d ed., 1896; reprinted by Dover, New York, 1945, secs. 286–290.

¶ There are several editions and translations. One generally available is the revision by F. Cajori of Andrew Motte's translation (1729), from Latin into English, of the third edition (1726): *Newton's Principia: Motte's Translation Revised,* University of California Press, Berkeley, 1934, reprinted 1947. Lindsay reprints passages from an 1848 edition of Motte's translation. Dostrovsky, "Early Vibration Theory," gives a detailed deciphering of Newton's analysis. The first such was given by Euler (1744).

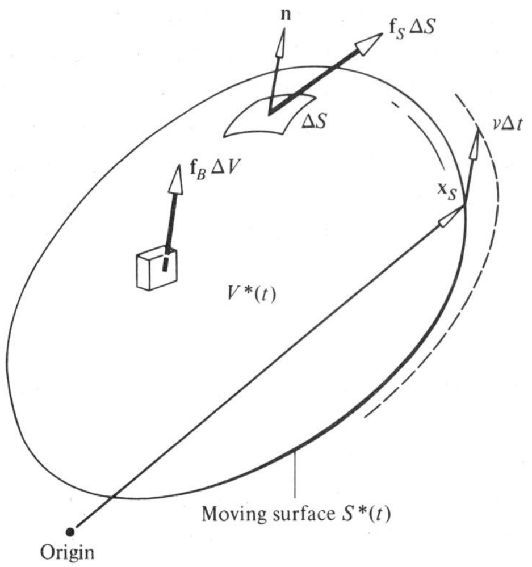

Figure 1-4 Forces acting on fluid particle occupying volume $V^*(t)$, each point on the surface of which moves with the local fluid velocity $\mathbf{v}(\mathbf{x}_S, t)$. Here $\mathbf{f}_S$ is surface force per unit area; $\mathbf{f}_B$ is body force per unit volume.

Although gravity is always present, it has negligible influence† on acoustic disturbances of all but extremely low frequencies, e.g., those of order or less than g/c, where g is acceleration due to gravity and c is the speed of sound; so, for simplicity, the body force term is here neglected at the outset. *Acoustic-gravity waves* (infrasonic waves with frequencies so low as to be strongly affected by gravity) have been a major topic of research during the past two decades but fall outside the scope of an introductory discussion.

The classical assumption regarding $\mathbf{f}_S$ is that it is directed normally into the surface S^*, that is,

$$\mathbf{f}_S = -\mathbf{n}p \tag{1-3.2}$$

with the magnitude p of this force per unit area identified as the *pressure*. The adoption of this relation, holding ideally for static equilibrium (*hydrostatics*), implies a neglect of *viscosity*. The lack of dependence of the pressure $p(\mathbf{x}, t)$ on the orientation of ΔS, that is, the direction of $\mathbf{n}$, may be regarded as a hypothesis but also follows‡ from a fundamental requirement that the net surface force divided by the mass of the fluid particle on which it acts should remain finite in the limit as the particle volume goes to zero. That $\mathbf{f}_S$ reverses direction when $\mathbf{n}$ reverses direction is consistent with Newton's third law.

† P. G. Bergmann, "The Wave Equation in a Medium with a Variable Index of Refraction," *J. Acoust. Soc. Am.*, **17**:329–333 (1946); N. A. Haskell, "Asymptotic Approximation for the Normal Modes in Sound Channel Wave Propagation," *J. Appl. Phys.*, **22**:157–168 (1951); C. O. Hines, "Atmospheric Gravity Waves: A New Toy for the Wave Theorist," *Radio Sci.*, **69D**:375–380 (1965); E. E. Gossard and W. H. Hooke, *Waves in the Atmosphere,* Elsevier, Amsterdam, 1975.

‡ H. Lamb, *Hydrodynamics,* 1879, 6th ed., 1932, reprinted by Dover, New York, 1945, pp. 1–2. The proof originated with A.-L. Cauchy, "On Pressure within a Fluid," 1827, reprinted in *Oeuvres complètes d'Augustin Cauchy,* ser. 2, vol. 7, Gauthier-Villars, Paris, 1889, pp. 37–39. (Here and throughout the balance of the book titles of articles cited are given in translation when the original is not in English.)

If p should be independent of position, the net surface force on a fluid particle integrates to zero, but otherwise it tends to be toward the direction of lower pressure. Mathematical substantiation of this comes from an application of Gauss' theorem to the surface integral of $-p\mathbf{n}$. The x component of this integral is of the form in Eq. (1-2.2) with $\mathbf{A}$ identified as $-p\mathbf{e}_x$. (Here $\mathbf{e}_x$ represents the unit vector in the direction of increasing x.) Since the divergence $\nabla \cdot (-p\mathbf{e}_x)$ is just $-\partial p/\partial x$, and since this is the x component of $-\nabla p$, Gauss' theorem implies

$$\iint_{S^*} \mathbf{f}_S \, dA = -\iiint_{V^*} \nabla p \, dV \tag{1-3.3}$$

when $\mathbf{f}_S = -p\mathbf{n}$, as in Eq. (2). Thus $-\nabla p$ is the equivalent force per unit volume due to pressure.

The time-rate-of-change-of-momentum term in Eq. (1) can similarly be expressed as a volume integral, without a time derivative operator outside the integral sign. A fluid particle is regarded as an aggregate of many "infinitesimal" fluid particles, each so small that the fluid velocity within it is everywhere nearly the same as the velocity of its center of mass. Since the mass of each fluid particle is constant, the time rate of change of momentum of a subparticle is $(\rho\Delta V^*)(d/dt)\mathbf{v}(\mathbf{x}_P(t),t)$, where $\mathbf{x}_P(t)$ is its position at time t. With help from the chain rule for differentiation, the acceleration factor becomes

$$\begin{aligned}\frac{d}{dt}\mathbf{v}(x_P(t), y_P(t), z_P(t), t) \\ &= \frac{\partial \mathbf{v}}{\partial t} + \frac{\partial \mathbf{v}}{\partial x}\frac{dx_P}{dt} + \frac{\partial \mathbf{v}}{\partial y}\frac{dy_P}{dt} + \frac{\partial \mathbf{v}}{\partial z}\frac{dz_P}{dt} \\ &= \frac{\partial \mathbf{v}}{\partial t} + (\mathbf{v}\cdot\nabla)\mathbf{v} = \frac{D\mathbf{v}}{Dt} \qquad (1\text{-}3.4)\end{aligned}$$

since $d\mathbf{x}_P/dt$ is just $\mathbf{v}(\mathbf{x}_P(t), t)$. (The operator $\partial/\partial t + \mathbf{v}\cdot\nabla$ is here abbreviated† D/Dt and represents the time rate of change as measured by someone moving with the fluid.) The resulting sum of infinitesimal masses times accelerations is equivalent to an integral, and so one obtains

$$\frac{d}{dt}\iiint_{V^*} \rho\mathbf{v} \, dV = \iiint_{V^*} \rho \frac{D\mathbf{v}}{Dt} \, dV \tag{1-3.5}$$

which represents an instance of *Reynolds' transport theorem.*‡

† This notation originated with G. G. Stokes, "On the Theories of the Internal Friction of Fluids in Motion, and of the Equilibrium and Motion of Elastic Fluids," *Trans. Camb. Phil. Soc.*, 8:287–319 (1845), especially sec. 5. Most of the article is reprinted in Lindsay, *Acoustics*, pp. 262–289.

‡ O. Reynolds, *Papers on Mathematical and Physical Subjects*, vol. 3, *The Sub-Mechanics of the Universe*, Cambridge University Press, London, 1903, secs. 13 and 14. A general statement of the transport theorem is

$$\frac{d}{dt}\iiint_{V^*} \rho f(\mathbf{x}, t) \, dV = \iiint_{V^*} \rho \frac{Df}{Dt} \, dV$$

where $f(\mathbf{x}, t)$ is an arbitrary function.

The equations discussed in the previous sections [mass conservation, Euler's equation, and the equation,† $p = p(\rho, s)$ with $s = s_0$, a constant] can be written in terms of the substitution (1) as

$$\frac{\partial}{\partial t}(\rho_0 + \rho') + \nabla \cdot [(\rho_0 + \rho')\mathbf{v}'] = 0 \qquad (1\text{-}5.2a)$$

$$(\rho_0 + \rho')\left(\frac{\partial}{\partial t} + \mathbf{v}' \cdot \nabla\right)\mathbf{v}' = -\nabla(p_0 + p') \qquad (1\text{-}5.2b)$$

$$p_0 + p' = p(\rho_0 + \rho', s_0) \qquad (1\text{-}5.2c)$$

Here $\mathbf{v}_0 = 0$; p_0 and ρ_0 are constants related by $p_0 = p(\rho_0, s_0)$. The terms in Eqs. (2*a*) and (2*b*) can be grouped into zero-order terms (all here identically zero), first-order [just one primed variable, for example, $\nabla \cdot (\rho_0\mathbf{v}')$], second-order [two primed variables, for example, $\nabla \cdot (\rho'\mathbf{v}')$], etc. In Eq. (2*c*), the grouping results from a Taylor-series expansion in ρ', that is,

$$p' = \left(\frac{\partial p}{\partial \rho}\right)_0 \rho' + \frac{1}{2}\left(\frac{\partial^2 p}{\partial \rho^2}\right)_0 (\rho')^2 + \cdots \qquad (1\text{-}5.2c')$$

where the indicated derivatives are evaluated at constant entropy and with density subsequently set to ρ_0.

The *linear approximation* (sometimes called the *acoustic approximation*) neglects second- and higher-order terms, so the *linear acoustic equations*‡ take the form

$$\frac{\partial \rho'}{\partial t} + \rho_0 \nabla \cdot \mathbf{v}' = 0 \qquad (1\text{-}5.3a)$$

$$\rho_0 \frac{\partial \mathbf{v}'}{\partial t} = -\nabla p' \qquad (1\text{-}5.3b)$$

$$p' = c^2 \rho' \qquad c^2 = \left(\frac{\partial p}{\partial \rho}\right)_0 \qquad (1\text{-}5.3c)$$

(Thermodynamic considerations require§ that c^2 always be positive.) For reasons made apparent in Sec. 1-7, c is referred to as the speed of sound.

† If the ambient state is inhomogeneous, $p = p(\rho, s_0)$ cannot be used and one falls back on $p = p(\rho, s)$, $Ds/Dt = 0$ as a starting point. If $p_0(\mathbf{x})$ and $\rho_0(\mathbf{x})$ are independent of t, these lead to

$$\frac{\partial p'}{\partial t} + \mathbf{v}' \cdot \nabla p_0 = c^2\left(\frac{\partial \rho'}{\partial t} + \mathbf{v}' \cdot \nabla \rho_0\right)$$

as the linear equation that replaces (3*c*).

‡ These particular equations (spatial or eulerian description, linearized, with p', ρ', $\mathbf{v}'$ as dependent variables, only with additional viscous terms) are given by Stokes ("Internal Friction of Fluids"). Equivalent formulations given by earlier authors differ from that above, either because of the use of the material description or because the authors chose to postpone the linearization to a later stage of the calculations, e.g., after the introduction of the velocity potential.

§ This is a special case of Le Châtelier's principle: "Experimental and Theoretical Research on Chemical Equilibrium," *Ann. Mines Carburants*, (8)**13**:157–380 (1888); L. D. Landau and E. M. Lifshitz, *Statistical Physics*, Addison-Wesley, Reading, Mass., 1959, pp. 32–66.

Some criteria for the validity of the linear approximation result from the requirement, for a representative solution, that each nonlinear term be almost everywhere and almost always much less than each of the dominant retained linear terms appearing in the same equation. A rough a priori estimate† of ratios of various terms ensues if one assigns a characteristic time T and a characteristic length L to the disturbance such that the order of magnitude of $\partial\psi'/\partial t$ (or $\partial\psi'/\partial x$) is $1/T$ (or $1/L$) times the order of magnitude of ψ' for any acoustic field quantity ψ'. This yields the related criteria

$$|p'| \ll \rho_0 \left(\frac{L}{T}\right)^2 \qquad |\mathbf{v}'| \ll \frac{L}{T} \qquad |\rho'| \ll \rho_0 \qquad \frac{|\rho'|}{\rho_0} \ll \frac{2c^2}{\rho_0|(\partial^2 p/\partial \rho^2)_0|} \tag{1-5.4}$$

For plane-wave propagation at constant frequency (discussed in Secs. 1-7 and 1-8) the identifications for T and L (period divided by 2π and wavelength divided by 2π) are such that L/T is c. Criteria based on this substitution, however, are not valid in the immediate vicinity of localized sources or in regions of wave focusing, since L can then be much smaller than cT. Also, even when the general criteria above are satisfied and nonlinear terms are all small, such terms can have an accumulative effect over large time intervals or large distances of propagation. For plane-wave propagation at constant frequency, these accumulative effects are significant when the ratio of propagation distance to wavelength becomes comparable to $\rho_0 c^2$ divided by a representative acoustic-pressure amplitude. There are in addition certain acoustic phenomena, e.g., acoustic streaming, that cannot be explained unless nonlinear effects are taken into account.

To the linear acoustic equations (3) can be added one for the temperature perturbation T'. From the thermodynamic relation $T = T(p, s)$, with $s = s_0$ constant, one has $T' = (\partial T/\partial p)_0 p'$ in the linear approximation. The coefficient can be reexpressed by means of thermodynamic identities‡ as $(\beta T/\rho c_p)_0$ in

† C. Eckart, "Vortices and Streams Caused by Sound Waves," *Phys. Rev.*, **73**:68–76 (1948).

‡ The stated relation follows from the mathematical identity

$$\left(\frac{\partial T}{\partial p}\right)_s = -\frac{(\partial s/\partial p)_T}{(\partial s/\partial T)_p}$$

and from the version of the second law of thermodynamics that states that

$$d\left(u - Ts + \frac{p}{\rho}\right) = -s\,dT + \frac{1}{\rho}\,dp$$

which implies the Maxwell relation

$$\left(\frac{\partial s}{\partial p}\right)_T = -\left(\frac{\partial}{\partial T}\frac{1}{\rho}\right)_p = +\rho^{-2}\left(\frac{\partial \rho}{\partial T}\right)_p$$

Thus

$$\left(\frac{\partial T}{\partial p}\right)_s = \frac{-\rho^{-1}(\partial\rho/\partial T)_p}{\rho T^{-1}[T(\partial s/\partial T)_p]} = \frac{\beta T}{\rho c_p}$$

terms of the coefficient of thermal (volume) expansion $\beta = -(1/\rho)(\partial\rho/\partial T)_p$ and the coefficient of specific heat at constant pressure $c_p = T(\partial s/\partial T)_p$. Thus one has

$$T' = \left(\frac{\beta T}{\rho c_p}\right)_0 p' \tag{1-5.5}$$

Typically, β is positive (distilled water near freezing temperature being an exception), and temperature peaks coincide with pressure peaks in a sound disturbance.

1-6 THE WAVE EQUATION

The *wave equation* results from the linear acoustic equations given above if one first uses (1-5.3*c*) to eliminate ρ' from the mass-conservation equation and then takes the time derivative of the resulting equation. If the order of time differentiation and the divergence operation† are interchanged in the second term, it then takes the form $\nabla\cdot(\rho_0\,\partial\mathbf{v}/\partial t)$, which is $-\nabla^2 p$ because of (3*b*). (Here we delete the primes on p' and $\mathbf{v}'$.) This sequence of steps yields

$$\nabla^2 p - \frac{1}{c^2}\frac{\partial^2 p}{\partial t^2} = 0 \tag{1-6.1}$$

where the operator ∇^2 is the laplacian sum of the second derivatives with respect to the three cartesian coordinates, i.e., the divergence of the gradient.

The one-dimensional version of this wave equation was first derived in 1747 by d'Alembert‡ for the case of the vibrating string. He subsequently recognized its possible applicability to sound in air but chose not to publish his derivation, presumably because of his strong reservations about the physical admissibility of its solutions. Euler (1747–1748, 1750) and Lagrange (1759) both treated the case of a sonorous line (see Fig. 1-5), a line of discrete masses connected by linear springs, and suggested its applicability to sound, although these early papers do not exhibit the wave equation per se. For reasons not completely

---

Here $(\partial s/\partial p)_T$ is an abbreviation for $\partial s(p, T)/\partial p$, etc. For more detailed discussions, see, for example, K. Wark, *Thermodynamics,* 3d ed., McGraw-Hill, New York, 1977, pp. 552–562; J. H. Keenan, *Thermodynamics,* M.I.T. Press, Cambridge, Mass., 1941, 1970, pp. 341–347; M. Tribus, *Thermostatics and Thermodynamics,* Van Nostrand, Princeton, N.J., 1961, pp. 243–256.

† The use of vector notation in the derivation of the wave equation was considered novel as recently as 1950. See, for example, W. J. Cunningham, "Application of Vector Analysis to the Wave Equation," *J. Acoust. Soc. Am.,* **22**:61 (1950); R. V. L. Hartley, "Note on the 'Application of Vector Analysis to the Wave Equation' ", ibid., 511.

‡ J.-le-Rond d'Alembert, "Investigation of the Curve Formed by a Vibrating String," 1747, trans. in Lindsay, *Acoustics,* pp. 119–130. For commentary, see Truesdell, "The Theory of Aerial Sound," p. XXXVII.

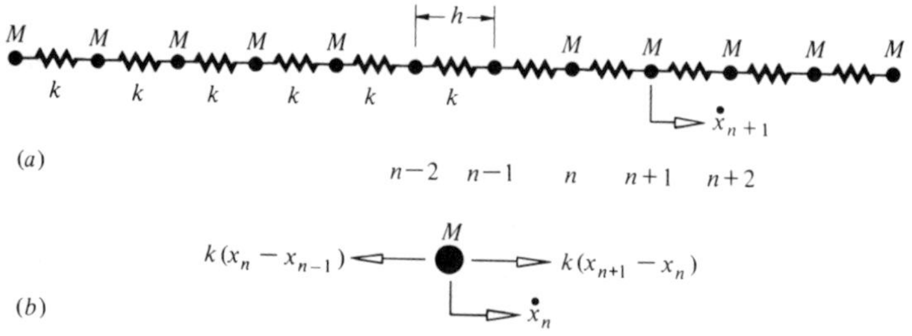

Figure 1-5 (*a*) Sonorous-line model used in early theories of sound propagation. A line of masses, each of mass M, separated at nominal intervals h and coupled by linear springs of spring constant k vibrates longitudinally. (*b*) Free-body diagram for the motion of the nth mass, corresponding to the equation $M\ddot{x}_n = k(x_{n+1} + x_{n-1} - 2x_n)$.

understood, Lagrange's analysis† was the catalyst that enabled Euler, within only a few days after first seeing Lagrange's paper, to develop the first theory of sound genuinely based on fluid-dynamic principles. The first derivation of the wave equation in one dimension for sound appeared in a paper submitted in 1759 by Euler; a derivation of the three-dimensional wave equation (with use of the material description) appeared in a second paper. Lagrange (1760, 1762) gave a subsequent derivation more nearly akin to that above, in which the linear approximation was made at an earlier stage.

This same wave equation occurs (although, generally also as an approximation) in a variety of other contexts: electromagnetic theory, gravity waves in shallow water, dilatational and shear elastic waves in solids, transverse vibrations in stretched membranes, Alfvén waves in magnetohydrodynamics, pressure surges in liquid-filled tubes with elastic walls, e.g., blood vessels, and electromagnetic transmission lines.

The derivation above was with acoustic pressure as the dependent field variable. The same equation (with change of dependent variable), however, holds for ρ', T', and $\nabla \cdot \mathbf{v}$, given the assumption that the ambient medium is homogeneous and quiescent. (The cartesian components of $\mathbf{v}$ also satisfy the wave equation if $\nabla \times \mathbf{v} = 0$.)

Two simple aspects of the wave equation may help one recall its form. First, since c has the units of velocity, ct has the units of length, so $(1/c^2)(\partial^2/\partial t^2)$ has the same units (1 over length squared) as ∇^2 and the equation

† J. L. Lagrange, "Researches on the Nature and Propagation of Sound," 1759, reprinted in *Oeuvres de Lagrange,* vol. 1, pp. 39–148; L. Euler, letter to J. L. Lagrange, dated Oct. 23, 1759; L. Euler, "On the Propagation of Sound," 1759, 1766, commentary by Truesdell, "Rational Fluid Mechanics, 1687–1765," pp. CXIX–CXXI; L. Euler, "Supplement to Researches on the Propagation of Sound," 1759, 1766, commentary by Truesdell, "Rational Fluid Mechanics, 1687–1765," pp. CXXII–CXXIII, "The Theory of Aerial Sound," pp. XLV–XLVII; J. L. Lagrange, "New Researches on the Nature and Propagation of Sound," 1760, 1762. Lindsay, *Acoustics,* gives translations of the second and third and of the introductory section of the first of these articles.

$0 < t < 2\pi/\omega$, then is zero again for $t > 2\pi/\omega$ (see Fig. 1-7), so that the waveform is a single cycle of a sinusoidal function. (Here ω and p_{pk} are positive constants.) At a given measurement site (coordinate s) there is no wave disturbance until $t = s/c$. Immediately before that time, the fluid particles just to the left of point s have an average velocity in the $+s$ direction, so the fluid starts to be compressed after the wave arrives and ρ' starts to increase with time. This compression in turn causes the pressure to increase. Since the pressure is temporarily larger to the left of s, the pressure gradient is in the $-s$ direction and fluid particles are accordingly accelerated in the $+s$ direction. This acceleration and compression continue until the pressure peak arrives (one-quarter of a period later). After this, the compression and overpressure start to diminish, although v, p, and ρ' are still positive. By the time the pressure node (one-half period after onset) arrives, the density is back to ambient, the fluid velocity has slowed to zero, and the net displacement of fluid particles to the right has reached its maximum value. However, the negative acceleration is still nonzero as there is a positive pressure gradient. Consequently, the fluid velocity goes negative, the density and pressure decrease to values below ambient, and the fluid is rarefacted. When the peak underpressure arrives, the fluid has attained its peak backward velocity. In the final quarter of the cycle, the acceleration is once again positive, the backward-moving fluid particles are slowed until, at the termination of the passage of the pulse, they are again motionless.

If the time integral of $f(t)$ is zero (as for the example discussed), the net displacement of the fluid particles is zero. The wave disturbance moved them temporarily to the right but then moved them back to their original positions.

One can infer (as originally hypothesized by Newton) that compression and rarefaction play an important role in sound propagation. In the example above, the disturbance is a moving region of compression followed by a moving region

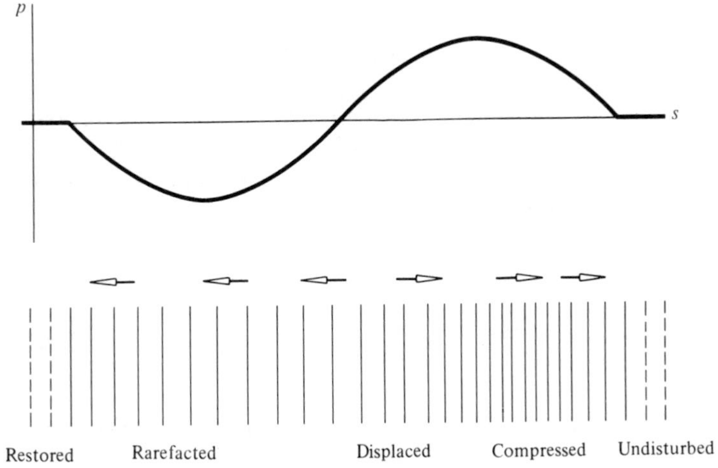

Figure 1-7 Fluid-particle positions during passage of one cycle of a sinusoidal plane traveling wave.

of rarefaction. Because of the presence of such density fluctuations, sound waves are *compressional waves*.

They are also *longitudinal waves* (as opposed to transverse waves) because the fluid velocity is parallel or antiparallel to the direction of propagation. This is a consequence of the vorticity's being zero. If $\mathbf{v}$ were of the form of a constant vector $\mathbf{V}$ times a scalar function of $t - \mathbf{n} \cdot \mathbf{x}/c$, the relation $\nabla \times \mathbf{v} = 0$ would require $\mathbf{n} \times \mathbf{V} = 0$, so $\mathbf{n}$ and the fluid velocity direction would have to be parallel or antiparallel.

The prediction of a zero net fluid displacement over a wave cycle demonstrates that it is the disturbance rather than the fluid itself that is moving with the sound speed. The disturbance may propagate over great distances, but the fluid particles themselves remain at all times close to their original positions.

1-8 WAVES OF CONSTANT FREQUENCY

An acoustic disturbance is of constant frequency if the field variables oscillate sinusoidally with time, such that (for the acoustic pressure p)

$$p = p_{\text{pk}} \cos (\omega t - \phi) = p_{\text{pk}} \sin (\omega t - \phi') = \text{Re}\, \hat{p} e^{-i\omega t} \tag{1-8.1}$$

where p_{pk} (the *amplitude* or *peak pressure*), ω (the *angular frequency*), $\hat{p}$ (the *complex pressure amplitude*), and ϕ (the *phase constant*) are independent of time t. (Re denotes "real part.") These three expressions above are equivalent, given the identifications

$$\phi' = \phi - \frac{\pi}{2} \qquad \hat{p} = p_{\text{pk}} e^{i\phi} \tag{1-8.2}$$

since

$$\sin\left(\alpha + \frac{\pi}{2}\right) = \cos \alpha \qquad e^{i\alpha} = \cos \alpha + i \sin \alpha \tag{1-8.3}$$

[The validity of the latter (Euler's formula) follows from a comparison of the power-series expansions of the two sides.]

The expressions in Eq. (1) oscillate between positive and negative values and repeat themselves whenever their arguments $\omega t - \phi$ or $\omega t - \phi'$ are changed by 2π. Thus the time per cycle (*period*) is $2\pi/\omega$, and the number of cycles per unit time (*frequency*) is

$$f = \frac{\omega}{2\pi} \tag{1-8.4}$$

The units of frequency are *hertz* (Hz), where 1 Hz equals† 1 cycle per second (or

† That the hertz is a superfluous unit has not escaped commentary. See, for example, H. M. Fitzpatrick, "The Hertz," *J. Acoust. Soc. Am.*, **42**:1098 (1967); R. W. Young, "On the Hertz," ibid.; M. Strasberg, "Name for Unit Radian Frequency" (the *avis*), *ibid.*, **41**:1367 (1967); F. Collins, "The Fitzpatrick Method," ibid., **43**:1460 (1968); L. G. Copley, "Angular Velocity," ibid.; H. M. Fitzpatrick, "Some Relevant Fundamentals," ibid., 1460–1461.

s^{-1}). The units of angular frequency (sometimes referred to simply as frequency without the qualifying adjective) are radians per second. Frequencies audible to a normal human ear are roughly between 20 and 20,000 Hz. As mentioned in Sec. 1-1, constant-frequency disturbances correspond to musical notes. A piano, for example, sounds a range of frequencies between 55 and 8360 Hz. Middle C corresponds to 262 Hz.

The complex-number representation in Eq. (1) is convenient† in theoretical studies; in particular, it replaces the amplitude and phase by a single complex number and condenses the writing of mathematical relations. One could take the time-dependent factor to be $e^{+i\omega t}$ instead of $e^{-i\omega t}$, but the latter is traditional‡ in wave-propagation studies and is advantageous for the description of traveling waves.

Although every wave disturbance, strictly speaking, has a beginning and an end and should therefore be regarded as a *transient*, some long-duration sounds can be idealized as being of constant frequency. [The terms "steady wave" and "continuous wave" (cw) are also used in the literature to denote the same property.] Also, even if not pure tones, persistent sounds may be superpositions of independently propagating constant-frequency disturbances. The mathematical apparatus of Fourier transforms, moreover, allows transients to be considered as a superposition of a continuous smear of constant-frequency components.

For disturbances like those described by Eq. (1), the *mean squared pressure* $(p^2)_{av}$ and *root-mean-squared* (rms) *pressure* p_{rms} are defined so that

$$(p^2)_{av} = \frac{1}{T}\int_{t_0}^{t_0+T} p^2\,dt = p_{rms}^2 \tag{1-8.5}$$

where T is either an integral number of half-wave periods or an interminably long time interval. Because of the trigonometric identity

$$\cos^2\alpha = \tfrac{1}{2} + \tfrac{1}{2}\cos 2\alpha \tag{1-8.6}$$

the square of $\cos(\omega t - \phi)$ oscillates about an average value of $\frac{1}{2}$ with a period of $1/2f$. Thus, Eqs. (1) and (5) lead to

$$(p^2)_{av} = \tfrac{1}{2}p_{pk}^2 = \tfrac{1}{2}|\hat{p}|^2 \tag{1-8.7}$$

Time Average of a Product

A related identity, stated here for future reference, concerns the time average of the product of two field quantities each oscillating with the same frequency

† This device was introduced into the acoustical literature by Rayleigh, *Theory of Sound*, vol. 1, sec. 104.

‡ The reasons for the choice are discussed by C. J. Bouwkamp: "A Contribution to the Theory of Acoustic Radiation," *Philips Res. Rep.*, **1**:251–277 (1946).

but not necessarily in phase. If one writes

$$X = \text{Re}\, \hat{X} e^{-i\omega t} \qquad Y = \text{Re}\, \hat{Y} e^{-i\omega t} \tag{1-8.8}$$

then

$$(XY)_{\text{av}} = \tfrac{1}{2}\, \text{Re}\, \hat{X}\hat{Y}^* \tag{1-8.9}$$

where $\hat{Y}^*$ is the complex conjugate of $\hat{Y}$. The derivation rests on the trigonometric identity [of which Eq. (6) is a special case]

$$\cos \alpha \cos \beta = \tfrac{1}{2} \cos (\alpha - \beta) + \tfrac{1}{2} \cos (\alpha + \beta) \tag{1-8.10}$$

If $\alpha = \omega t - \phi_X$, $\beta = \omega t - \phi_Y$, the second term averages out to zero while the first term has an average equal to $\frac{1}{2} \cos (\phi_Y - \phi_X)$. Since

$$|\hat{X}| \cdot |\hat{Y}| \cos (\phi_Y - \phi_X) = \text{Re}\, (|\hat{X}| \cdot |\hat{Y}| e^{\pm i(\phi_Y - \phi_X)})$$

relation (9) follows.

For sound in air, the lowest audible rms pressure amplitude is typically 2×10^{-5} Pa; a very loud sound would be one with $p_{\text{rms}} = 2$ Pa; one causing pain, with $p_{\text{rms}} = 60$ Pa, although these numbers vary with frequency and from individual to individual. (Here Pa is the unit symbol for the pascal, equal to 1 N/m^2.) In contrast, the ambient pressure at sea level is 10^5 Pa, so that the pressure amplitude in a sound wave is generally much less than p_0.

Field Equations for Spatially Dependent Complex Amplitudes

Since the field equations of Sec. 1-5 are (by design) linear and have time-independent coefficients, it is possible for the field variables to oscillate at each and every point with the same frequency. Thus ω may be considered independent of position. Equations governing the spatial dependences of the complex amplitudes can be developed by substituting expressions like Re $\hat{p}(\mathbf{x})e^{-i\omega t}$ into the linear acoustic equations. Because (1) the derivative (with respect to time or a spatial coordinate) commutes with the operation of taking the real part (so $\partial/\partial t \to -i\omega$), (2) the product of a real number with the real part of a complex number is the real part of the product, and (3) the sum of the real parts of several complex numbers is the real part of the sum, one obtains, for the mass-conservation equation,

$$\text{Re}\, [(-i\omega\hat{\rho} + \rho_0 \nabla \cdot \hat{\mathbf{v}})e^{-i\omega t}] = 0 \tag{1-8.11}$$

This will be satisfied if both the real and imaginary parts of the quantity in brackets are zero or, equivalently, if the quantity in parentheses is zero. That the latter should be zero follows since the above should be satisfied for all values of time (in particular, when $e^{-i\omega t}$ has the values 1 or $-i$).

Thus, one arrives at the prescription that the equations for the complex spatially dependent amplitudes can be obtained from the linear acoustic equations by (1) replacing the actual field variables by the corresponding amplitudes

relations

$$s' = \left(\frac{\partial s}{\partial \rho}\right)_{p,0}\left(\rho' - \frac{p}{c^2}\right) \qquad T' = \left(\frac{\partial T}{\partial \rho}\right)_{p,0}\left(\rho' - \frac{p}{c_T^2}\right) \qquad \left(\frac{\partial s}{\partial \rho}\right)_p = \left(\frac{\partial s}{\partial T}\right)_p\left(\frac{\partial T}{\partial \rho}\right)_p$$

along with the definition $T(\partial s/\partial T)_p$ for c_p.

A single wave equation for just one dependent variable is obtained from Eqs. (1) and (2) by taking the second time derivative of Eq. (2), commuting various operators and constants, and subsequently replacing $\partial^2\rho'/\partial t^2$ by $\nabla^2 p$ in accord with Eq. (1). Doing this gives

$$\frac{\partial}{\partial t}\left(\nabla^2 - \frac{1}{c^2}\frac{\partial^2}{\partial t^2}\right)p = \left(\frac{\kappa}{\rho c_p}\right)\nabla^2\left(\nabla^2 - \frac{1}{c_T^2}\frac{\partial^2}{\partial t^2}\right)p \tag{1-10.3}$$

which is the generalization of the wave equation when thermal conduction is taken into account.

The implications of Eq. (3) for plane-wave propagation at constant frequency ω can be explored with the substitution

$$p = \text{Re}\, Ae^{-i\omega t}e^{iks} \tag{1-10.4}$$

where A and k are independent of time and position. The same reasoning applies as in the derivation of Eqs. (1-8.12), so the "equation" for A results with the replacement of p by A in Eq. (3) and with the replacement of the differentiation operators $\partial/\partial t$ and ∇^2 by $-i\omega$ and $-k^2$. This gives a homogeneous linear algebraic equation whose solution for A is zero unless k is such that

$$\frac{k^2 - (\omega/c)^2}{k^2 - (\omega/c_T)^2} = \frac{\kappa}{\rho c_p}\frac{k^2}{i\omega} \tag{1-10.5}$$

(Any such relation between wave number k and angular frequency ω is termed a *dispersion relation*.) For fixed ω, this determines the values of k^2 such that plane-wave solutions are possible; the imaginary part of k corresponds to attenuation. Although this is a quadratic equation for k^2, we here limit our attention to the root closest in value to either $(\omega/c_T)^2$ or $(\omega/c)^2$.

If the adiabatic assumption is substantially better than the isothermal assumption, there is a root k^2 for which the right side of (5) has a magnitude much smaller than 1. In this case k^2 is approximately $(\omega/c)^2$ and the right side becomes $-i\omega/\omega_{\text{TC}}$ or $-if/f_{\text{TC}}$, where

$$\omega_{\text{TC}} = \frac{\rho c_p c^2}{\kappa} = 2\pi f_{\text{TC}} \tag{1-10.6}$$

is a characteristic number (units of s^{-1}) associated with thermal conduction (TC). From this, one can infer that the adiabatic approximation is valid if $\omega \ll \omega_{\text{TC}}$. In contrast, if $\omega \gg \omega_{\text{TC}}$, the propagation might be considered as isothermal (although in such circumstances the hitherto neglected viscosity would be expected to result in a high attenuation of sound). For angular frequencies between these limits, neither idealization is necessarily preferable, although nearly unattenuated propagation may still result if c_T and c are close to each other in value.

The frequencies of interest in acoustical studies are always much less than f_{TC}. For example, for air, $\rho c_p c^2 = \gamma^2 R p_0/(\gamma - 1)$ has the value 1.4×10^8 W/(m·s·K) at atmospheric pressure. The thermal conductivity varies from 2.4×10^{-2} to 2.7×10^{-2} W/(m·K) as the temperature ranges from 0 to 40 °C. Consequently, f_{TC} is of the order of 10^9 Hz. Also, for water, with the values given in the preceding section, $\rho c_p c^2 = 9 \times 10^{12}$ W/(m·s·K) at 10 °C and atmospheric pressure. The thermal conductivity varies from 0.56 to 0.60 W/(m·K) as the temperature ranges from 0 to 20 °C. Thus, for water, f_{TC} is of the order of 2×10^{12} Hz. In contrast, the highest known frequency in air detectable by animal life (bats and moths) is of the order of 1.5×10^5 Hz. Frequencies used in ultrasonic-propagation studies in water are typically less than 10^9 Hz; those used in underwater systems are typically less than 10^5 Hz.

The adiabatic approximation is better at lower frequencies than at higher frequencies because the heat production due to conduction is weaker when the wavelengths (varying inversely with frequency) are longer. For fixed amplitude A, the magnitude of the term $\kappa \nabla^2 T'$ in the linear version of the Fourier-Kirchoff equation (1-4.6) decreases with decreasing ω as ω^2; the term $\rho T_0 \, \partial s'/\partial t$ decreases as ω. Since the thermal-conduction term decreases more rapidly, the lower the frequency the more nearly valid the premise that the implication of the overall equation is $\partial s'/\partial t = 0$. (The often stated explanation, that oscillations in a sound wave are too rapid to allow appreciable conduction of heat, is wrong.)

1-11 ACOUSTIC ENERGY, INTENSITY, AND SOURCE POWER

Acoustic-Energy Corollary

The linear acoustic equations have a corollary (derived by Kirchhoff† in 1876) which resembles a statement of energy conservation for an acoustic field and which can be regarded as the acoustic counterpart of Poynting's theorem‡ for electromagnetic fields. To derive it, one takes the dot product of **v** with the

† G. Kirchhoff, *Vorlesungen über mathematische Physik: Mechanik,* 2d ed., Teubner, Leipzig, 1877, pp. 311, 336 (subsequently cited as *Mechanik*); Rayleigh, *The Theory of Sound,* vol. 2, sec. 295.

‡ Poynting's theorem is a corollary of Maxwell's equations; for electromagnetic fields in free space it takes the form of Eq. (2) with $w = \frac{1}{2}\varepsilon E^2 + \frac{1}{2}\mu H^2$ and $\mathbf{I} = \mathbf{E} \times \mathbf{H}$. The theorem was derived in integral form by J. Poynting in 1884 and again in the same year by O. Heaviside. For a full discussion, see J. A. Stratton, *Electromagnetic Theory,* McGraw-Hill, New York, 1941, pp. 131–133.

linear version of Euler's equation (with the deletion of the primes on p' and $\mathbf{v}'$), i.e.,

$$\mathbf{v} \cdot \left(\rho_0 \frac{\partial \mathbf{v}}{\partial t}\right) = -\mathbf{v} \cdot \nabla p = -\nabla \cdot (\mathbf{v}p) + p \nabla \cdot \mathbf{v}$$

$$= -\nabla \cdot (p\mathbf{v}) - p\rho_0^{-1} \frac{\partial \rho'}{\partial t} \tag{1-11.1}$$

Here the indicated mathematical steps follow from a vector identity and from the linear version of the mass-conservation equation. The term on the left can be alternately written as $(\partial/\partial t)(\frac{1}{2}\rho_0 v^2)$. Similarly, since $\rho' = p/c^2$, the expression $p\rho_0^{-1}\ \partial\rho'/\partial t$ can be written $(\partial/\partial t)(\frac{1}{2}p^2/\rho_0 c^2)$. Therefore, Eq. (1) can be reexpressed as†

$$\frac{\partial w}{\partial t} + \nabla \cdot \mathbf{I} = 0 \tag{1-11.2}$$

where

$$w = \tfrac{1}{2}\rho_0 v^2 + \frac{1}{2}\frac{p^2}{\rho_0 c^2} \qquad \mathbf{I} = p\mathbf{v} \tag{1-11.3}$$

The interpretation of (2) as a conservation law follows if we integrate it over an arbitrary fixed volume V within the fluid and reexpress the volume integral of $\nabla \cdot \mathbf{I}$ as a surface integral by means of Gauss' theorem. Doing this gives

$$\frac{d}{dt}\iiint_V w\, dV + \iint_S \mathbf{I} \cdot \mathbf{n}\, dA = 0 \tag{1-11.4}$$

where $\mathbf{n}$ is the unit normal vector pointing out of the surface S enclosing V. The form of this might be compared, for example, with the equation for conservation of mass, given in integral form by Eq. (1-2.1).

† Various generalizations (corresponding to alternate versions of the linear acoustic equations) are discussed in Chaps. 8 and 10. Another, of importance for very-low-frequency propagation in the atmosphere and oceans, results when ρ_0, p_0, and c are considered to be functions only of height z (or depth) under the influence of gravity, such that $dp_0/dz = -g\rho_0$. The linear acoustic equations with the gravitational-force term included lead to Eqs. (2) to (4), but w has an additional term

$$(\Delta w)_{\text{gravity}} = \tfrac{1}{2}\rho_0 \omega_{BV}^2 \xi_z^2$$

where

$$\omega_{BV} = \left(-\frac{g^2}{c^2} - \frac{g}{\rho_0}\frac{d\rho_0}{dz}\right)^{1/2} \qquad \omega_{BV}^2 > 0 \qquad \text{(stability)}$$

$$\xi_z = \frac{-s'}{ds_0/dz} \qquad v_z = \frac{\partial \xi_z}{\partial t}$$

are identified as the *Brunt-Vaissala frequency* and vertical particle displacement. For a derivation and discussion, see C. Eckart, *Hydrodynamics of Oceans and Atmospheres,* Pergamon, New York, 1960, pp. 53–60.

Energy Conservation in Fluids

The above corollary resembles† the energy-conservation law that can be derived from the original nonlinear fluid-dynamic equations [conservation of mass, Euler's equation, and $p = p(\rho, s)$ with s constant], i.e.,

$$\frac{\partial E}{\partial t} + \nabla \cdot (E\mathbf{v} + p\mathbf{v}) = 0 \qquad (1\text{-}11.5a)$$

$$\frac{d}{dt}\iiint_V E\, dV + \iint_S E\mathbf{v} \cdot \mathbf{n}\, dA + \iint_S p\mathbf{v} \cdot \mathbf{n}\, dA = 0 \qquad (1\text{-}11.5b)$$

$$E = \tfrac{1}{2}\rho v^2 + \rho U_P(\rho, s) \qquad U_P = \int_{1/\rho}^{1/\rho_0} p\,d\,\frac{1}{\rho} \qquad (1\text{-}11.6)$$

Here p is total pressure, E is energy per unit volume, and U_P is the potential energy per unit mass relative to the ambient state. [This last identification results from consideration of unit mass of fluid in a cylindrical vessel (cross-sectional area A) with a movable piston at its top. When the piston moves down a distance δh, the specific volume $1/\rho$ decreases by $A\ \delta h$. The work done by the force pA is $pA\ \delta h$, so $-p\ \delta(1/\rho)$ is the increase of potential energy.] In the integral form of the conservation law (5b), $E\mathbf{v} \cdot \mathbf{n}$ is energy convected out of the volume per unit surface area and time due to fluid motion; $p\mathbf{v} \cdot \mathbf{n}$ is rate of work done per unit area and by the fluid in V on its surroundings.

The resemblance mentioned above becomes apparent if E and $(E + p)\mathbf{v}$ are expanded to second order in $\rho - \rho_0$, $p - p_0$, and $\mathbf{v}$. To this order, one has

$$\rho U_P \approx \frac{p_0}{\rho_0}(\rho - \rho_0) + \frac{1}{2}\frac{c^2}{\rho_0}(\rho - \rho_0)^2 \qquad (1\text{-}11.7)$$

where $\rho - \rho_0$ can be replaced by its first-order equivalent $(p - p_0)/c^2$ in the second term. Thus one has

$$E \approx \tfrac{1}{2}\rho_0 v^2 + \left[\frac{p_0}{\rho_0}(\rho - \rho_0)\right] + \frac{1}{2}\frac{(p - p_0)^2}{\rho_0 c^2} \qquad (1\text{-}11.8a)$$

$$(E + p)\mathbf{v} \approx \left[\frac{p_0}{\rho_0}\rho\mathbf{v}\right] + (p - p_0)\mathbf{v} \qquad (1\text{-}11.8b)$$

† N. Andrejev, "On the Energy Expression in Acoustics," *J. Phys. (Moscow)*, **2**:305–312 (1940); J. J. Markham, "Second-Order Acoustic Fields: Energy Relations," *Phys. Rev.*, **86**:712–714 (1952); "Second-Order Acoustic Fields: Relations between Energy and Intensity," ibid., **89**:972–977 (1953); A. Schoch, "Remarks on the Concept of Acoustic Energy," *Acustica*, **3**:181–184 (1953); N. Andrejev, "Concerning Certain Second-Order Quantities in Acoustics," *Akust. Zh.*, **1**:2–11 (1955), trans. in *Sov. Phys.: Acoust.*, **1**:2–11 (1955). For a derivation of Eqs. (5), see Lamb, *Hydrodynamics*, sec. 10.

so that if $F(t)$ is periodic in time, and if I_r is averaged over an integral number of half periods, one has

$$I_{r,\text{av}} = \frac{(p^2)_{\text{av}}}{\rho c} \tag{1-12.7}$$

This is the same as the expression (1-11.10b) holding for a plane traveling wave; it is also consistent with the decrease of pressure amplitude as $1/r$ and with the decrease of time-averaged intensity as $1/r^2$.

For a constant-frequency disturbance, both p and v_r and consequently also $f(t)$ and $F(t)$ oscillate sinusoidally with time. One can write $f(t)$ as $|A| \cos (\omega t - \phi_A)$ or Re $Ae^{-i\omega t}$, where $A = |A|e^{i\phi_A}$. Then, since $F(t)$ is an oscillating function whose derivative is $f(t)$, it should be given by $\omega^{-1}|A| \sin (\omega t - \phi_A) = \text{Re}\,[(iA/\omega)e^{-i\omega t}]$. These expressions inserted into Eqs. (4b) and (5) yield

$$p = |A|r^{-1} \cos (\omega t - kr - \phi_A) = r^{-1}\,\text{Re}\, Ae^{-i\omega t}e^{ikr} \tag{1-12.8a}$$

$$\begin{aligned} \rho c v_r &= |A|r^{-1} \cos (\omega t - kr - \phi_A) + |A|k^{-1}r^{-2} \sin (\omega t - kr - \phi_A) \\ &= r^{-1}\,\text{Re}\left[\left(1 + \frac{i}{kr}\right) Ae^{-i\omega t}e^{ikr}\right] \end{aligned} \tag{1-12.8b}$$

where we use the abbreviation $k = \omega/c$. The second term in (8b) dominates if $kr \ll 1$; the first term if $kr \gg 1$. Since the time average of the cosine squared or of the sine squared is just $\frac{1}{2}$ while the time average of the cosine times the sine is zero, the following time averages result from the above relations:

$$I_{r,\text{av}} = \frac{|A|^2}{2\rho c r^2} \tag{1-12.9a}$$

$$\frac{1}{2}\frac{(p^2)_{\text{av}}}{\rho c^2} = \frac{|A|^2}{4\rho c^2 r^2} = \frac{I_{r,\text{av}}}{2c} \tag{1-12.9b}$$

$$\tfrac{1}{2}\rho(v_r^2)_{\text{av}} = \frac{|A|^2}{4\rho c^2 r^2}\left[1 + \frac{1}{(kr)^2}\right] \tag{1-12.9c}$$

for the intensity, potential energy density, and kinetic energy density. The average acoustic energy density w_{av} is the sum of the last two. In the limit $kr \ll 1$, the energy is predominantly kinetic, and the ratio $I_{r,\text{av}}$ to w_{av} is considerably less than the sound speed, but in the limit $kr \gg 1$ the intensity is cw and the potential and kinetic energy densities are the same.

Field at Large Distances from Source of Finite Extent

If the source is not spherically symmetric but is of limited size, the disturbance at large r locally resembles a plane wave propagating with speed c away from the source. Thus we can write $p \approx Bf(t - c^{-1}r, \theta, \phi)$ and $\mathbf{v} \approx p\mathbf{e}_r/\rho c$, where θ and ϕ denote the polar and azimuthal angles in spherical coordinates and B is some function slowly varying over distances (radial and transverse) comparable to a wavelength. To determine the general form of the dependence of B on r, θ, ϕ, let f be a sinusoidal function of time, so that the time-averaged intensity is

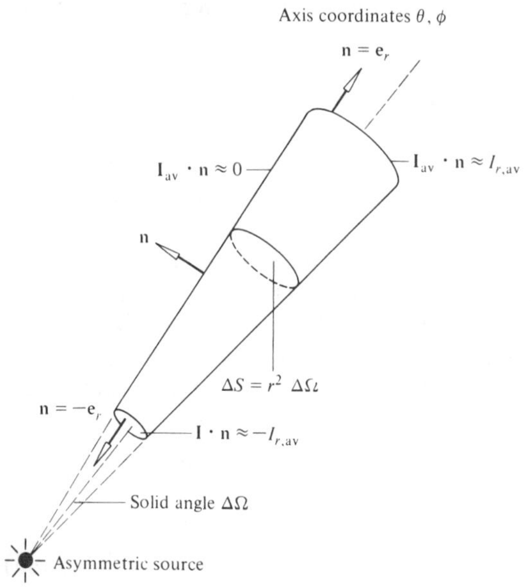

Figure 1-12 Segment of a cone of solid angle $\Delta\Omega$ with apex at central point in an asymmetric source. The indicated geometry is used to show that intensity along any radial line decreases as $1/r^2$ at large r from a finite-sized source.

$B^2(f^2)_{\text{av}}\mathbf{e}_r/\rho c$, with $(f^2)_{\text{av}}$ independent of r. The relation $\nabla \cdot \mathbf{I}_{\text{av}} = 0$ would then require, via Gauss' theorem (see Fig. 1-12), that the integral of $\mathbf{I}_{\text{av}} \cdot \mathbf{n}$ over any conical segment pointing radially away from the source vanish; so since the approximate $\mathbf{I}_{\text{av}}$ has only a radial component, the product $I_{r,\text{av}}\,\Delta S$ of intensity times cone cross-sectional area ΔS should be independent of radial distance r. But the area ΔS is $r^2\,\Delta\Omega$, where $\Delta\Omega$ is the solid angle subtended by the cone. This solid angle is constant along the cone, and $(f^2)_{\text{av}}$ and ρc are independent of r, so r^2B^2 is independent of r. Hence B varies inversely with r. Since any θ and ϕ dependence of B can be absorbed in the function f, we take B to be identically $1/r$.

The above reasoning leads to the following approximate expressions for the acoustic field at large distances from any source of finite extent:

$$p = \frac{1}{r} f(t - c^{-1}r, \theta, \phi) \qquad \mathbf{v} = \frac{p\mathbf{e}_r}{\rho c} \tag{1-12.10a}$$

$$\mathbf{I}_{\text{av}} = \frac{J(\theta, \phi)}{r^2}\,\mathbf{e}_r \qquad J(\theta, \phi) = \frac{1}{\rho c T}\int_{t_0}^{t_0+T} f^2(t, \theta, \phi)\,dt \tag{1-12.10b}$$

with T being a suitably chosen (very long or an integral number of half periods) averaging time. The first two expressions are not restricted to periodic signals, but the association of a time average with $\mathbf{I}$ normally implies that J should be independent of t_0.

The function $J(\theta, \phi)$ describes the *radiation pattern* of the source, acoustic power radiated per unit solid angle. The acoustic power radiated by the source is given by

$$\mathscr{P}_{\text{av}} = \iint_S \mathbf{I}_{\text{av}} \cdot \mathbf{n}_{\text{out}}\, dS = \int_0^{2\pi} \int_0^{\pi} J(\theta, \phi) \sin\theta \, d\theta \, d\phi \qquad (1\text{-}12.11)$$

since $r^2 \sin\theta \, d\theta \, d\phi$ is the differential element of area for a spherical surface ($\sin\theta \, d\theta \, d\phi$ is the differential of solid angle).

Equation (10*b*) indicates that the spherical spreading law is not restricted to spherically symmetric sources. The analysis assumes, however, an absence of reflections from external boundaries and ignores the absorption (loss of energy) of sound.

PROBLEMS

1-1 In an experiment pertaining to the anomalous effects of the atmosphere on sonic booms, B. A. Davy and D. T. Blackstock, *J. Acoust. Soc. Am.*, **49**:732–737 (1971), studied the propagation of transient acoustic pulses around and through a soap bubble filled with gaseous helium (monatomic with molecular weight 4). Verify from fundamental principles the authors' statement that the speed of sound in helium is about 1/0.34 times that in air.

1-2 Prove by any convenient method that the time rate of change of the volume $V^*(t)$ of a moving fluid particle is equal to the volume integral of the divergence of the fluid velocity.

1-3 Give an alternate derivation of the conservation-of-mass equation starting from the requirement that the mass in any moving fluid particle be constant.

1-4 Show that if gravity is taken into account, Euler's equation of motion for a fluid can be written as

$$\rho \frac{D\mathbf{v}}{Dt} = -\nabla p - g\rho \mathbf{e}_z$$

where g is the acceleration due to gravity and $\mathbf{e}_z$ is the unit vector in the vertical direction.

1-5 (*a*) Given an ideal gas for which $p = \rho R T$ with temperature-independent specific-heat coefficients c_p and c_v, where $\gamma = c_p/c_v$ and $c_p - c_v = R$, show that the entropy s per unit mass can be written as

$$s = s_0 + c_v \ln \frac{u}{u_0} - R \ln \frac{\rho}{\rho_0}$$

Here s_0 (a constant) is the specific entropy when the specific internal energy u and the density ρ have the values u_0 and ρ_0, respectively; u is defined so that it vanishes at $T = 0$.

(*b*) Derive an expression for the pressure p in terms of the specific entropy s and the density ρ. Compare your result with Eq. (1-4.2).

1-6 A common model for acoustic waves in inhomogeneous quiescent media is one in which gravity is neglected and p_0 is considered constant but ρ_0 and therefore also c vary with position (although not with time).

(*a*) Show that such a choice of ambient variables automatically satisfies the fluid-dynamic equations.

(*b*) Show that the linear acoustic equations for such a model can be written as

$$\frac{\partial p}{\partial t} + \rho_0 c^2 \nabla \cdot \mathbf{v} = 0 \qquad \rho_0 \frac{\partial \mathbf{v}}{\partial t} = -\nabla p$$

Is it necessarily still true that $p = \rho' c^2$?

(c) Show that the resulting wave equation for the acoustic pressure is

$$\rho_0 \nabla \cdot \left(\frac{1}{\rho_0} \nabla p \right) - \frac{1}{c^2} \frac{\partial^2 p}{\partial t^2} = 0$$

1-7 Consider vertical (z) propagation (no horizontal coordinate dependence) in an isothermal (c constant) quiescent ($\mathbf{v}_0 = 0$) atmosphere with gravity taken into account.

(a) Show that Euler's equation of motion as in Prob. 1-4 and the ideal-gas equation imply that p_0 and ρ_0 both decrease exponentially with height.

(b) Derive the linear acoustic equations for such a model and show in particular that they include the relation

$$\frac{\partial p'}{\partial t} + (\gamma - 1) g \rho_0 v_z = c^2 \frac{\partial \rho'}{\partial t}$$

(c) Show that the resulting one-dimensional wave equation for vertical propagation can be written in the form

$$\left[\frac{\partial^2}{\partial z^2} - \frac{1}{c^2} \left(\frac{\partial^2}{\partial t^2} + \omega_A^2 \right) \right] \frac{p}{\rho_0^{1/2}} = 0$$

where $\omega_A = (\gamma/2) g/c$ is a constant. [H. Lamb, *Proc. Lond. Math. Soc.*, **7**:122–141 (1908).]

1-8 Given that the vapor pressure of water at 30°C is 4.24×10^3 Pa, what is the speed of sound in air at 30°C when the relative humidity is 80 percent?

1-9 The acoustic pressure in a standing-wave pattern in an enclosed rectangular space in idealized cases may be of the form

$$p = A \cos \omega t \cos k_x x \cos k_y y \cos k_z z$$

where k_x, k_y, k_z are constants depending on the dimensions of the enclosure. What would the angular frequency ω have to be if this expression is to satisfy the wave equation?

1-10 Show that Reynolds' transport theorem and Euler's equation of motion (without gravity) lead for any given fluid particle to the angular-momentum conservation law

$$\frac{d}{dt} \iiint_{V^*} \rho \mathbf{x} \times \mathbf{v} \, dV = - \iint_{S^*} \mathbf{x} \times p \mathbf{n} \, dS$$

where $\mathbf{x}$ is a vector from a fixed point or from the center of mass of the fluid particle. *Hint:* You will need a number of vector identities and a version of Gauss' theorem that transforms the volume integral of the curl of a vector into a surface integral.

1-11 Starting from the relations $p = \rho R T$, $p \rho^{-\gamma} =$ const, for adiabatic disturbances in an ideal gas, show that the relation between temperature fluctuations and pressure fluctuations in a sound wave is given by $T'/T_0 = [(\gamma - 1)/\gamma] p'/p_0$.

1-12 (a) Verify that

$$p = A \cos \omega t \sin kx$$

is a solution of the one-dimensional wave equation provided that $\omega = ck$.

(b) Determine functions $f(t - c^{-1}x)$ and $g(t + c^{-1}x)$ such that their sum is equal to the expression above.

(c) What is the (x-component) fluid velocity associated with this acoustic pressure?

1-13 A longitudinal compressional wave of very long wavelength compared with h is propagating along the sonorous line sketched in Fig. 1-5. In terms of M, k, and h, what is the speed of such a wave in the limit $\lambda \gg h$? (L. Brillouin, *Wave Propagation in Periodic Structures*, Dover, New York, 1953, pp. 1–33.)

1-14 A transient plane wave propagates in the $+x$ direction through an initially undisturbed region. The acoustic pressure at a given point is zero for $t < 0$, is equal to $p_{\text{pk}} \sin \omega t$ for $0 < t < 2\pi/\omega$, and is equal to 0 for $t > 2\pi/\omega$. Give an expression in terms of p_{pk}, ω, ρ_0, and c for the peak displacement of any given fluid particle to the right.

1-15 The speed of sound in pure water is nominally about 1500 m/s; the mass per unit volume is 10^3 kg/m^3. A possible model for muddy water might be water with many small rigid particles (idealized as having the same density as water) suspended in it. Let f represent the fraction of any given volume normally occupied by such particles. In terms of f, what would you estimate for the velocity of sound in muddy water?

1-16 A plane sound wave propagating parallel to the ground has a waveform with one pronounced pressure peak. Microphone 1 at the origin receives this peak at time $t_1 = 0.0$ s; microphone 2 at $x = 1$ m, $y = 0$ receives it at time $t_2 = 0.00255$ s; microphone 3 at $x = 0$, $y = 1$ m receives it at time $t_3 = 0.00147$ s. What is the speed of the wave, and in what direction is it traveling?

1-17 If the oceans were isothermal and of constant salinity below a certain depth, how would the sound speed vary with further increase in depth?

1-18 The acoustic pressure in a standing wave within a narrow pipe closed at the end $x = 0$ and open at the end $x = L$ is

$$p = A \cos \frac{c\pi t}{2L} \cos \frac{\pi x}{2L}$$

What is the time-averaged energy density (in terms of A, c, L, and ρ_0) of this disturbance as a function of x?

1-19 A hypothetical instrument computes the rms pressure amplitude of an acoustic wave by averaging p^2 over a fixed time interval T and subsequently taking the square root. Given that the possible frequencies of the wave are greater than 1000 Hz, what is the smallest choice for T one should pick to ensure that the error in p_{rms} will not exceed 10 percent?

1-20 An initial-value problem for one-dimensional acoustic propagation in an unbounded space is posed when the values p_{in}, ρ'_{in}, $v_{x,\text{in}}$ (at $t = 0$) are specified for acoustic pressure, density, and fluid velocity as functions of x.

(*a*) Show that the general solution of the linear acoustic equations in one dimension for such an initial-value problem is

$$p = f(t - c^{-1}x) + g(t + c^{-1}x) \qquad \rho' = \frac{p}{c^2} + \left[\rho'_{\text{in}}(x) - \frac{p_{\text{in}}(x)}{c^2}\right]$$

$$v_x = \frac{1}{\rho c}[f(t - c^{-1}x) - g(t + c^{-1}x)]$$

where

$$2f(t - c^{-1}x) = p_{\text{in}}(x - ct) + \rho c v_{x,\text{in}}(x - ct)$$

$$2g(t + c^{-1}x) = p_{\text{in}}(x + ct) - \rho c v_{x,\text{in}}(x + ct)$$

(*b*) Given that, at $t = 0$, $p = A$ for $-L/2 < x < L/2$, while $p = 0$ for $x > L/2$ or for $x < -L/2$, sketch p, v_x, and ρ' versus x for $t = 3L/2c$. Assume that the initial values of ρ' and v_x are zero for all x.

(*c*) Derive expressions for the total acoustic kinetic and potential energies (densities integrated over x) per unit area transverse to the x axis at times $t = 0$ and $t = 3L/2c$ for the example above.

(*d*) After time $t = L/c$, the solution should exhibit less mass in the region $-L/2 < x < L/2$ than originally. What happened to this mass?

1-21 The rms acoustic pressure (in pascals) at a distance of 2 m from a small appliance suspended in an anechoic chamber filled with air is found to be $p_{\text{rms}} = 0.20|\cos \theta|$, where θ is the angle with respect to the vertical. Given that the acoustic disturbance at such a distance from the source locally resembles a plane wave propagating away from the source, what would you estimate for the sound power output of this appliance?

1-22 The acoustic pressure of an acoustic disturbance in a medium with ambient density ρ and sound speed c is given by

$$p = A \cos [\omega(t - c^{-1}x)] + B \sin [\omega(t - c^{-1}y)]$$

(*a*) Express p in the form Re $\hat{p}(\mathbf{x})e^{-i\omega t}$ and determine the complex pressure amplitude $\hat{p}(\mathbf{x})$.

(*b*) Derive expressions for the time-averaged acoustic energy density and acoustic intensity as functions of x and y.

(*c*) Verify by direct substitution that $\nabla \cdot \mathbf{I}_{\text{av}} = 0$.

1-23 Suppose the ambient density and sound speed vary with position $\mathbf{x}$ (as in Prob. 1-6), although the ambient pressure p_0 is constant. What modifications would this spatial variation require in the expressions given in the text for acoustic energy density and acoustic intensity?

1-24 The acoustic pressure in a spherically symmetric wave is given by

$$p = \frac{A}{r} \cos [\omega(t - c^{-1}r)]$$

where A is a constant. In terms of A, ω, c, ρ_0, and t, how much mass $\dot{m}$ passes per unit time out through a fixed spherical surface of radius R_0 in the limit $R_0 \ll c/\omega$? Assume that R_0 is larger than the radius of the source and that A is sufficiently small for nonlinear effects to be negligible.

1-25 Derive an explicit partial-differential equation for the radial component of the acoustic fluid velocity $v_r(r, t)$ in a spherically symmetric sound wave.

1-26 A spherically symmetric sound wave in water has an acoustic fluid velocity at a distance of $1/2\pi$ wavelengths from the source center given by

$$v_r(t) = (0.1)(2\pi) \sin \omega t \qquad \text{m/s}$$

(*a*) What is the acoustic-pressure amplitude at a distance of 10 wavelengths from the source center?

(*b*) If the wavelength is 0.1m, what is the average acoustic power output of the source?

1-27 A plane sound wave with frequency 2000 Hz is propagating through air along the axis of a duct of 0.1 m² cross-sectional area. What is the time average of the acoustic power transmitted by this wave if the fluid-velocity amplitude is 0.001 m/s?

1-28 A simple method of modifying the linear acoustic equations to simulate sound absorption introduced by Rayleigh (1877) is to add a term $\rho_0 \alpha \mathbf{v}$ to the left side of the linearized version of Euler's equation of motion. Here α is some positive constant with units of reciprocal time.

(*a*) What is the resulting form of the wave equation if such a term is taken into account?

(*b*) The energy-conservation corollary should be modified to

$$\frac{\partial w}{\partial t} + \nabla \cdot \mathbf{I} = -\mathscr{D}$$

where $\mathscr{D}$ is always nonnegative. Determine the expressions for w, $\mathbf{I}$, and $\mathscr{D}$.

(*c*) If plane waves of the form p = Re $Ae^{-i\omega t}e^{ikx}$ are to satisfy the wave equation derived in (*a*), what should the complex wave number k be?

1-29 An idealized sonic-boom pressure waveform (acoustic pressure versus time) is shown in the figure. Assume that such a wave is propagating freely through air (sound speed c, ambient density ρ) and derive an expression in terms of P, T, ρ, and c for the total acoustic energy carried across unit area normal to the wavefront during passage of the sonic boom.

1-30 Verify that the fluid-dynamic energy-conservation equation (1-11.5*a*) follows from the equation of mass conservation, from Euler's equation of motion, and from the assumption $p = p(\rho)$. Verify also that the expressions in Eqs. (1-11.8) are valid second-order approximations for E and $(E + p)\mathbf{v}$.

1-31 Show that if $\Phi(x, y, z, t)$ is a solution of the wave equation, then $\partial\Phi/\partial x$, $\partial^2\Phi/(\partial x \partial y)$, $\partial^2\Phi/\partial x^2$ are also solutions. If Φ is taken as $r^{-1}F(t - r/c)$, what forms do these solutions take when expressed in spherical coordinates?

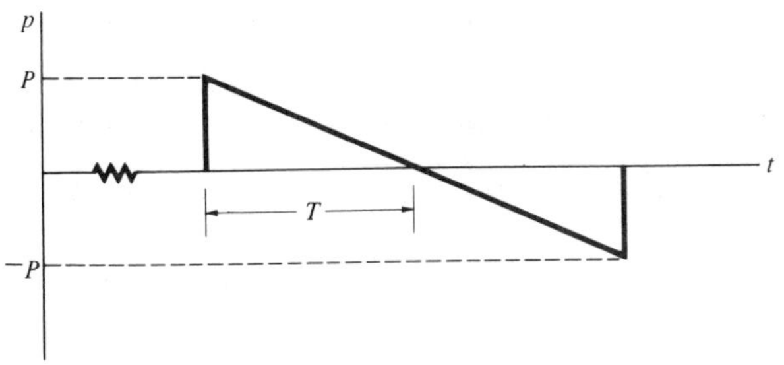

Problem 1-29 Sonic-boom pressure waveform.

1-32 (*a*) Derive an expression for $\nabla^2 p$ in spherical coordinates when p is a general function of r, θ, and ϕ.

(*b*) Show that one possible solution of the wave equation in spherical coordinates is

$$p = \text{Re}\,[Ae^{-i\omega t}(3\cos^2\theta - 1)(-k^2 - 3\,ikr^{-1} + 3r^{-2})r^{-1}e^{ikr}]$$

where A is an arbitrary complex constant. [If you have difficulty with part (*a*), consult the derivation outlined in Sec. 4-5.]

1-33 What is the time-averaged acoustic power output of an isolated source that generates the wave in Prob. 1-32?

1-34 Derive approximate two-term expressions in which each term is proportional to some power (not necessarily integer or positive) of ω/ω_{TC} for all of the roots of the dispersion relation (1-10.5) for complex wave number k in the limit $\omega/\omega_{TC} \ll 1$. Give a physical interpretation for each of the roots.

1-35 For a freely propagating plane acoustic wave of constant frequency, what is the relation between the time average of the square of the acoustic intensity and the square of the time average of the acoustic intensity?

1-36 Derive the relation $\nabla \cdot \mathbf{I}_{av} = 0$ with $\mathbf{I}_{av} = \frac{1}{2}\,\text{Re}\,\hat{p}^*\hat{\mathbf{v}}$ from Eqs. (1-8.12).

1-37 Sound is propagating through an ideal gas for which $p = \rho RT$, where R is a constant, but for which du/dT and the specific-heat ratio γ are functions of temperature. Prove that even though γ is not constant, one still has the sound speed given by $(\gamma RT)^{1/2}$ or by $(\gamma p/\rho)^{1/2}$.

1-38 Starting from the second law of thermodynamics and the definitions of c_p, β, and K_T, show that

$$\left(\frac{\partial c_p}{\partial p}\right)_T = -\frac{T}{\rho}\left[\beta^2 + \left(\frac{\partial \beta}{\partial T}\right)_p\right] \qquad \left(\frac{\partial \beta}{\partial p}\right)_T = \frac{1}{K_T^2}\left(\frac{\partial K_T}{\partial T}\right)_p$$

Are the coefficients in Eqs. (1-9.12) consistent with these identities?

1-39 A cylindrically symmetric (independent of z and azimuthal angle ϕ) wave is spreading out from a source extending along the z axis. From energy-conservation considerations, determine how the time average of the intensity pointing away from the source should vary with the radial distance $r = (x^2 + y^2)^{1/2}$. How is $I_{r,\text{av}}$ at a given value of r related to the average power $(d\mathcal{P}/dz)_{\text{av}}$ per unit length generated by the source?

1-40 A set of linear acoustic equations obtained by Stokes (1845), which include the effects of viscosity and apply to sound waves at points substantially removed from solid surfaces, can be taken as

$$\frac{\partial p}{\partial t} + \rho c^2\,\nabla \cdot \mathbf{v} = 0 \qquad \nabla \times \mathbf{v} = 0 \qquad \rho\frac{\partial \mathbf{v}}{\partial t} = -\nabla p + \tfrac{4}{3}\mu\,\nabla^2 \mathbf{v}$$

Here μ is the viscosity and may be considered constant.

(*a*) What are the corresponding partial-differential equations for the spatially dependent complex amplitudes $\hat{p}(\mathbf{x})$ and $\hat{\mathbf{v}}(\mathbf{x})$?

(*b*) Derive a single partial-differential equation for $p(\mathbf{x}, t)$ that does not involve $\mathbf{v}(\mathbf{x}, t)$.

(*c*) If one were to define a velocity potential Φ such that $\mathbf{v} = \nabla\Phi$, what would be an appropriate relation between p and Φ to replace the relation $p = -\rho\, \partial\Phi/\partial t$ used in the inviscid case?

(*d*) If $p(x, t) = \text{Re}\, Ae^{-i\omega t}e^{ikx}$, what relation should hold between k and ω? What are the real and imaginary parts of k (given that the real part is positive) to lowest order in ω?

1-41 (*a*) Show that for a homogeneous medium with constant ambient velocity $\mathbf{v}_0$, the linear acoustic equations take the form

$$\left(\frac{\partial}{\partial t} + \mathbf{v}_0 \cdot \nabla\right)p + \rho c^2\, \nabla \cdot \mathbf{v}' = 0 \qquad \rho\left[\frac{\partial \mathbf{v}'}{\partial t} + (\mathbf{v}_0 \cdot \nabla)\mathbf{v}'\right] = -\nabla p$$

(*b*) Show that the corresponding wave equation for p is

$$\nabla^2 p - \frac{1}{c^2}\left(\frac{\partial}{\partial t} + \mathbf{v}_0 \cdot \nabla\right)^2 p = 0$$

(*c*) If $\mathbf{v}_0 = v_0\mathbf{e}_x$ and if $p_{\text{NF}}(x, y, z, t)$ and $\mathbf{v}'_{\text{NF}}(x, y, z, t)$, where NF stands for no flow, are a solution of the equations when $v_0 = 0$, show that a solution when $v_0 \neq 0$ can be taken as $p_{\text{NF}}(x^*, y^*, z^*, t^*)$, $\mathbf{v}'_{\text{NF}}(x^*, y^*, z^*, t^*)$, where $x^* = x - v_0 t$, $y^* = y$, $z^* = z$, $t^* = t$. (This is known as a *galilean transformation.*) What is your interpretation of this result?

(*d*) Suppose one has a plane wave of the form $p = f(t - \mathbf{n} \cdot \mathbf{x}/v_{\text{ph}})$, where the phase velocity v_{ph} is some positive constant and $\mathbf{n}$ is the unit normal to surfaces of constant phase. What is v_{ph} in terms of c, v_0 and the angle θ between $\mathbf{n}$ and $\mathbf{v}_0$? Show that the corresponding expression for $\mathbf{v}'$ is $\mathbf{n}p/\rho c$ regardless of the directions of $\mathbf{n}$ and $\mathbf{v}_0$. *Hint:* Use the result of part (*b*).

(*e*) Verify that the energy corollary of the equations in (*a*) is

$$\frac{\partial w}{\partial t} + \nabla \cdot (\mathbf{v}_0 w + \mathbf{I}) = 0$$

where w and $\mathbf{I}$ are the expressions that apply for a medium at rest. Show that this leads to the prediction that

$$\mathbf{v}_w = \mathbf{v}_0 + \mathbf{n}c$$

is the velocity with which the energy is moving for a plane wave with unit vector $\mathbf{n}$ pointing normal to surfaces of constant phase. Give a simple interpretation of this result.

1-42 For a constant-frequency spherical wave propagating out from the origin, what is the ratio $(p^4)_{\text{av}}/(p^2)^2_{\text{av}}$? What is the ratio $(I_r^2)_{\text{av}}/(I_r)^2_{\text{av}}$? What would be the corresponding ratios for a plane wave?

1-43 A gas mixture is made up of equal parts (in terms of numbers of molecules) of O_2, NH_3, and CO_2 (a linear molecule). What would you estimate to be the specific heat ratio γ, gas constant R, and sound speed of this gas at 0°C?

1-44 For an acoustic disturbance of constant angular frequency ω, how is $[(\partial p/\partial t)^2]_{\text{av}}$ related to $(p^2)_{\text{av}}$? If the disturbance is a plane wave, how is $[(\nabla p)^2]_{\text{av}}$ related to $(p^2)_{\text{av}}$? How is $[(\partial p/\partial t)\, \nabla p]_{\text{av}}$ related to $\mathbf{I}_{\text{av}}$?

1-45 Two superimposed plane waves are propagating in the $+x$ and $-x$ directions such that

$$p = \text{Re}\, Ae^{-i\omega(t-x/c)} + \text{Re}\, Be^{-i\omega(t+x/c)}$$

What is the time average $I_{\text{av},x}$ of the net intensity in the $+x$ direction? How does $I_{\text{av},x}$ vary with x?

1-46 The acoustic pressure in a disturbance is of the form

$$p = \text{Re}\, Ae^{-i\omega(t-z/c)} + \text{Re}\, Br^{-1}e^{-i\omega(t-r/c)}$$

(1-11.3), only with $\mathbf{v}$ and p replaced by $\mathbf{v}_b$ and p_b. (It is not necessarily true that at any instant w is the sum of the w_b, even though $\mathbf{v}$ is always the sum of the $\mathbf{v}_b$ and p is always the sum of the p_b.)

2-2 PROPORTIONAL FREQUENCY BANDS

If the frequency scale is divided into contiguous bands, the bth band having lower frequency $f_1(b)$ and upper frequency $f_2(b)$, the partitioning is said to be into proportional frequency bands if $f_2(b)/f_1(b)$ is the same for each band. The *center frequency* f_0 of any such band is defined as the geometric mean $(f_1 f_2)^{1/2}$, which is always less than the arithmetic average $\frac{1}{2}(f_1 + f_2)$. The ratio of center frequencies of successive proportional bands is the same as f_2/f_1 for any one band; in addition, one has

$$\frac{f_0}{f_1} = \frac{f_2}{f_0} = \left(\frac{f_2}{f_1}\right)^{1/2} \tag{2-2.1}$$

An *octave band* is a band for which $f_2 = 2f_1$; a $\frac{1}{3}$-*octave band* is one for which $f_2 = 2^{1/3}f_1$; a $(1/N)$th-octave band is one for which $f_2 = 2^{1/N}f_1$. Three contiguous $\frac{1}{3}$-octave bands or N contiguous $(1/N)$th-octave bands are equivalent to an octave band. For example, the octave band (1000, 2000 Hz) is made up of the $\frac{1}{3}$-octave bands $(1000, 2^{1/3} \times 1000)$, $(2^{1/3} \times 1000, 2^{2/3} \times 1000)$ and $(2^{2/3} \times 1000, 2000)$. For a $(1/N)$th-octave band, Eq. (1) above shows that f_0 is $(1/2N)$th octave above f_1 and below f_2, so

$$f_1 = 2^{-1/2N}f_0 \qquad f_2 = 2^{1/2N}f_0 \tag{2-2.2}$$

Consequently, any proportional frequency band is defined by its center frequency and by N. An octave band ($N = 1$) with center frequency 1000 Hz, for example, would have $f_1 = 707$ Hz and $f_2 = 1414$ Hz.

Standard Frequencies and Bands

In some areas of acoustics (especially noise control) a standard compromised octave and $\frac{1}{3}$-octave frequency-partitioning scheme† uses the numerical accident that $2^{10/3} = 10.079$ is nearly 10. (Ten $\frac{1}{3}$-octaves are nearly a *decade*.) Since round numbers are convenient, center frequencies of the standard $\frac{1}{3}$-octave bands are chosen so that the $(b + 10)$th center frequency is 10 times the bth. Thus, given that 1000 Hz is the center frequency of a standard $\frac{1}{3}$-octave band, the scheme (see Table 2-1) is such that 1, 10, 100, 1000, 10,000 Hz, etc., are also standard $\frac{1}{3}$-octave-band f_0's. The other center frequencies are simple numerical approximations to the integer powers of $10^{1/10} = 1.25893$, these approximations

† ANSI S1.6-1967 (R1976), American National Standard Preferred Frequencies and Band Numbers for Acoustical Measurements, American National Standards Institute, New York, 1976.

Table 2-1 Center, lower, and upper frequencies for $\frac{1}{3}$-octave bands

Band no.	Frequency, Hz		
	Center	Lower	Upper
12	16†	14.0	18.0
13	20	18.0	22.4†
14	25	22.4†	28.0
15	31.5†	28.0	35.5
16	40	35.5	45†
17	50	45†	56
18	63†	56	71
19	80	71	90†
20	100	90†	112
21	125†	112	140
22	160	140	180†
23	200	180†	224
24	250†	224	280
25	315	280	355†
26	400	355†	450
27	500†	450	560
28	630	560	710†
29	800	710†	900
30	1,000†	900	1,120
31	1,250	1,120	1,400†
32	1,600	1,400†	1,800
33	2,000†	1,800	2,240
34	2,500	2,240	2,800†
35	3,150	2,800†	3,550
36	4,000†	3,550	4,500
37	5,000	4,500	5,600†
38	6,300	5,600†	7,100
39	8,000†	7,100	9,000
40	10,000	9,000	11,200†
41	12,500	11,200†	14,000
42	16,000†	14,000	18,000
43	20,000	18,000	22,400†
44	25,000	22,400†	28,000
45	31,500†	28,000	35,500

† Also an appropriate quantity for an octave band. The 1000-Hz octave band, for example, has lower and upper frequencies of 710 and 1400 Hz.

being

n	1	2	3	4	5	6	7	8	9
$10^{n/10} \approx$	1.25	1.6	2	2.5	3.15	4	5	6.3	8

Thus there are standard octave-band center frequencies at 16, 31.5, 63, 125,

250, 500, 1000, 2000, 4000, 8000, 16,000, and 31,500 Hz; a compromise has been made because $2 \times 16 \neq 31.5$ and $2 \times 63 \neq 125$. A rule of thumb is that successive $\frac{1}{3}$-octave-band center frequencies have ratios of 5 : 4. (The standard octave and $\frac{1}{3}$-octave-band center frequencies also serve as preferred frequencies for constant-frequency acoustical measurements.)

Equally Tempered Musical Scales

The concept of fixed frequency ratios (like those defining proportional frequency bands) also occurs in the theory of musical temperament. Certain instruments, e.g., the piano and stringed fretted instruments, once they are tuned, sound only a discrete set of notes. Temperament refers to the system by which these notes are systematically slightly mistuned (tempered) so that a larger variety of melodious combinations are possible.

When two notes are played together or in succession, the resulting sound is generally more harmonious to the ear when the corresponding frequencies are in simple ratios, and much music takes advantage of this fact. Classic musical intervals correspond to frequency ratios; particular intervals sounding especially harmonious are those with frequency ratios of 2 : 1 (octave), 3 : 2 (perfect fifth), 4 : 3 (perfect fourth), and 5 : 4 (major third). The terms, third, fourth, fifth, here refer to where the higher note falls in a musical scale (do, re, mi, fa, so, la, ti, do) when the lower note is the key note do. Such a scale is approximately realized by the notes C, D, E, F, G, A, B, C, represented by the white keys (starting with C as indicated in Fig. 2-1) on a piano keyboard. In *just intonation* (mathematically exact intervals) for a major key of C, the frequencies corresponding to D, E, F, G, A, B, and C are tuned to 9/8 (major interval), 5/4 (major third), 4/3 (fourth), 3/2 (fifth), 5/3 (sixth), 15/8 (seventh), and 2 (octave) times the frequency of the first C.

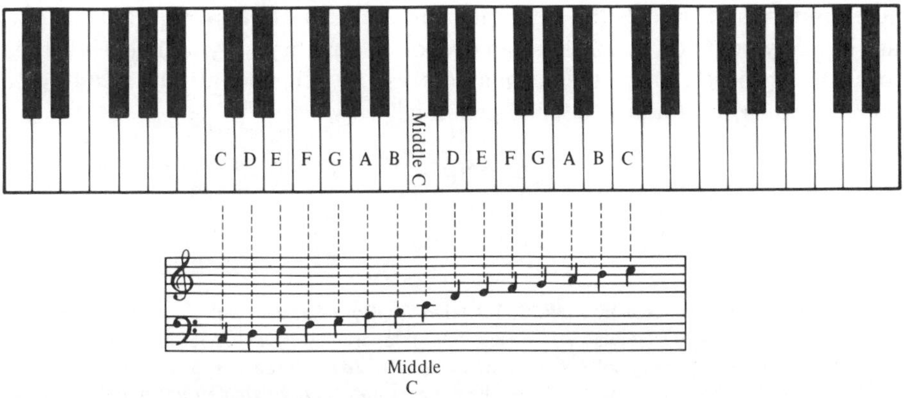

Figure 2-1 Segment of a piano keyboard showing letter designations of white keys and corresponding notes on the great staff (treble and bass clefs). *(From Beginning Piano Book for Older Students, Copyright © 1932, Clayton F. Summy Company. Used by permission. All rights reserved.)*

The option of playing all notes that can be reached by any succession of melodious intervals, e.g., fourths, fifths, and octaves, starting from a given keynote ideally requires a large number of notes within any given octave. The most common tuning system alleviating this problem is *equal temperament*† with a 12-note-per-octave scale in which successive notes are (1/12)-octave apart. An interval with a frequency ratio of $2^{1/12} = 1.0595$ is called a *half step*. Any two half steps approximate a major interval, any four a major third, any five a fourth, any seven a fifth, any nine a sixth, and any eleven a seventh. (Any twelve is exactly an octave.) Note that

$$2^{2/12} = 1.1225 \approx 9/8 = 1.1250 \qquad 2^{7/12} = 1.4893 \approx 3/2 = 1.5000$$
$$2^{4/12} = 1.2599 \approx 5/4 = 1.2500 \qquad 2^{9/12} = 1.6818 \approx 5/3 = 1.6667$$
$$2^{5/12} = 1.3348 \approx 4/3 = 1.3333 \qquad 2^{11/12} = 1.8877 \approx 15/8 = 1.8750$$

A piano keyboard has 7 white keys and 5 black keys (12 in all) per octave and can be tuned with such a scheme. Insofar as the human ear cannot perceive the discords caused by the deviations of the tempered ratios for fifths and fourths from their ideal values, the scheme is satisfactory, although to some trained listeners the discord in the major third is on the limit of unpleasantness. The scheme has the virtues of simplicity and of not requiring the instrument to be retuned whenever the key is changed. The interval G to the next higher D, for example, is as close to a perfect fifth as the interval from C to G.

2-3 LEVELS AND THE DECIBEL

Sound-Pressure Levels

Although sound-pressure amplitudes or rms pressures (corresponding to a given frequency component, a frequency band, or the acoustic pressure) can be measured in terms of pascals (or any other physical unit of pressure), it is customary in many contexts to measure and report a quantity varying linearly as the logarithm, base 10, of the mean squared pressure. This quantity is said to be a *sound-pressure level* and is defined generically by

$$L_p = 10 \log \frac{(p_s^2)_{\text{av}}}{p_{\text{ref}}^2} \qquad \text{(2-3.1)}$$

† This topic is discussed by J. W. S. Rayleigh, *Theory of Sound*, vol. 1, 1877; Dover, New York, 1945, secs. 15–20. See also A. J. Ellis "On Temperament," sec. A of appendix 20 to his translation (1885) of H. Helmholtz, *On the Sensations of Tone*, 2d ed., 1885; Dover, New York, 1954, pp. 430–441, 548. According to Ellis, the concept may have originated in China long before the time of Pythagoras (c. 540 B.C.). M. Mersenne, *Harmonie universelle*, 1636, however, was the first to give the correct frequency ratios for equal temperament. Although there is controversy whether J. S. Bach ever played on an instrument tuned according to equal temperament, his *Well-Tempered Clavier* (1722) had considerable influence on the use of the system.

logarithms of power. Thus 1 neper (Np) is roughly 1 bel (B). The exact relation is 1 Np = 2 log e B = 0.869 B. The transmission unit of the Bell System, identified as $\frac{1}{10}$ B, was given the name *decibel;* the sensation unit of the Bell System acousticians became the decibel. (The bel has rarely been used.) The subsequent widespread adoption† outside the Bell System of the decibel can be attributed to the inherent attractiveness of a logarithmic scale and to the prominence in the 1920s and 1930s of the Bell System's acoustical research staff. The choice of reference pressure (for sound in air) stems from the practice of plotting acoustical magnitudes in "units above auditory threshold"; note (from Fig. 2-2) that 0 dB is roughly the same as the auditory threshold in the midfrequency range.

Intensity and Power Levels

The decibel also occasionally describes average acoustic intensity and power. The intensity level L_I and the power level L_P are defined,‡ respectively, by

$$L_I = 10 \log \frac{|\mathbf{I}_{av}|}{I_{ref}} \qquad L_P = 10 \log \frac{\mathcal{P}_{av}}{\mathcal{P}_{ref}} \tag{2-3.5}$$

The preferred values for I_{ref} and $\mathcal{P}_{ref}$ are 10^{-12} W/m^2 and 10^{-12} W (1 picowatt), respectively. As with sound-pressure levels, one can also speak of intensity and power levels for a given frequency band.

In earlier literature, the term "intensity level" is occasionally used for sound-pressure level, but this is now discouraged because there is in general no simple relation between pressure and intensity and because acoustical standards assign a precise meaning to the term "intensity." (Intensity level is now rarely used.) However, for plane or spherical waves (see Secs. 1-11 and 1-12), $|\mathbf{I}_{av}|$ is $(p^2)_{av}/\rho c$, so in these cases

$$L_p = 10 \log \frac{|\mathbf{I}_{av}|}{p_{ref}^2/\rho c} \tag{2-3.6}$$

For air under normal conditions $\rho c \approx 400$ kg/(m$^2 \cdot$ s), and so $p_{ref}^2/\rho c \approx 10^{-12}$ W/m^2 when p_{ref} is taken as the preferred (for gases) value of 20 μPa. Consequently, for plane and spherical waves in air, sound-pressure level and intensity level are approximately the same.

† The first issue of the *Journal of the Acoustical Society of America* (1929) has perhaps the first article by someone outside the Bell System in which the term decibel is used in an acoustical context: V. O. Knudsen, "The Hearing of Speech in Auditoriums," *J. Acoust. Soc. Am.*, **1**:56–82 (1929). Knudsen defines the decibel on p. 58, n 4.

‡ ANSI S1.1-1960 (R1976), American National Standard Acoustical Terminology (1976); ANSI S1.21-1972, American National Standard Methods for the Determination of Sound Power Levels of Small Sources in Reverberation Rooms (1972), American National Standards Institute, New York.

2-4 FREQUENCY WEIGHTING AND FILTERS

Frequency Weighting Functions

In many contexts, a frequency-weighted mean squared pressure $(p^2)_{\mathrm{av},W}$ is used rather than the mean squared acoustic pressure $(p^2)_{\mathrm{av}}$. The weighted version is defined by a frequency-dependent *weighting function* $W(f)$ such that if $p(t)$ is a sum of discrete frequency components, then

$$(p^2)_{\mathrm{av},W} = \sum_n W(f_n)(p_n^2)_{\mathrm{av}} \tag{2-4.1}$$

If the $W(f_n)$ are all 1 (no weighting, or *flat response*), this reduces to Eq. (2-1.7). A decibel description of the weighting results with the substitution

$$W(f) = 10^{\Delta L_W(f)/10} \tag{2-4.2}$$

where $\Delta L_W(f)$ is the *relative response* (usually negative) in decibels. The weighted sound-pressure level results from Eq. (2-3.1) with $(p_s^2)_{\mathrm{av}}$ replaced by $(p^2)_{\mathrm{av},W}$; for a single-frequency waveform, the expressions above and the definition of a sound-pressure level imply that

$$L_{p,W} = L_p + \Delta L_W(f) \tag{2-4.3}$$

Three common weightings correspond to the A, B, and C relative response functions,† incorporated, for example, into commercially marketed *sound-level meters* (see Fig. 2-3). The A weighting is the most commonly used; the corresponding sound-pressure level is referred to as the *sound level* and denoted by L_{pA} (or L_A). This weighting was originally intended to be such that sounds of different frequencies giving the same decibel reading with A weighting would be equally loud. A sound having a higher sound level than a second sound (of different spectral content) would not always be louder, but it often is; from this standpoint, the sound level is an improvement over the unweighted sound-pressure level in that frequencies to which the human ear is less sensitive are weighted less than those to which the ear is more sensitive. Note that $\Delta L_A(f)$ is roughly the same as the negative of the threshold of audibility curve $L_{p,\min}(f)$ given in Fig. 2-2.

Sound-pressure levels associated with frequency bands can also be regarded as weighted sound-pressure levels. The mean squared pressure $(p_b^2)_{\mathrm{av}}$ associated with frequency band b results from Eq. (1) with $W(f) = 1$ $(\Delta L_W = 0)$ for frequencies within the band and $W(f) = 0$ $(\Delta L_W = -\infty)$ for frequencies outside the band. An octave-band sound-pressure level (OBSPL) is denoted by $L_{p,1/1}$ (or $L_{1/1}$), while a $\frac{1}{3}$-octave-band sound-pressure level (TOBSPL) is denoted by $L_{p,1/3}$ (or $L_{1/3}$). The first subscript corresponds to the physical quantity measured, but it is usually omitted for sound pressure.

† ANSI S1.4-1971 (R1976), American National Standard Specifications for Sound Level Meters, 1976.

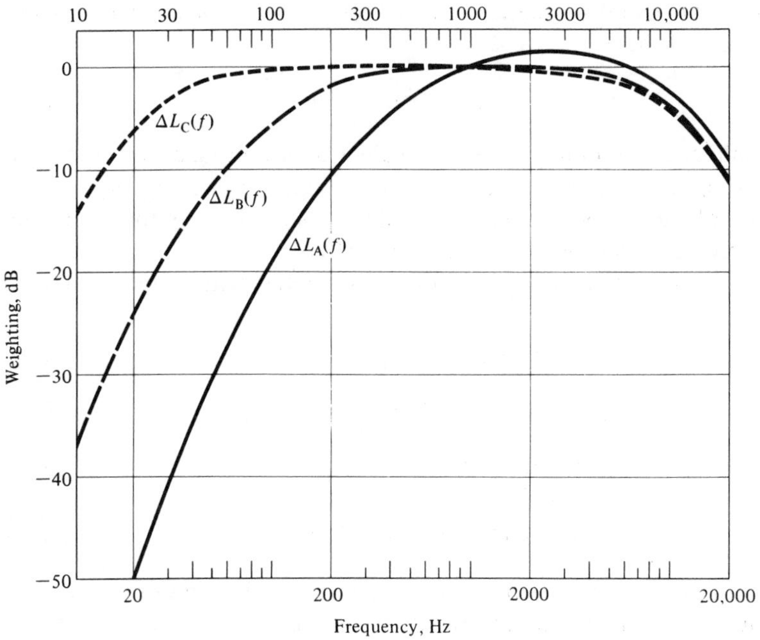

Figure 2-3 Relative response functions for A, B, and C weightings.

Linear Filters

Passing $p(t)$ through an appropriately designed filter, squaring the output, then averaging over time gives a measurement of $(p^2)_{av,W}$. The filter (see Fig. 2-4a) transforms $p(t)$ at its input terminal into $p_F(t) = \mathscr{L}\{p(t)\}$ at its output terminal,

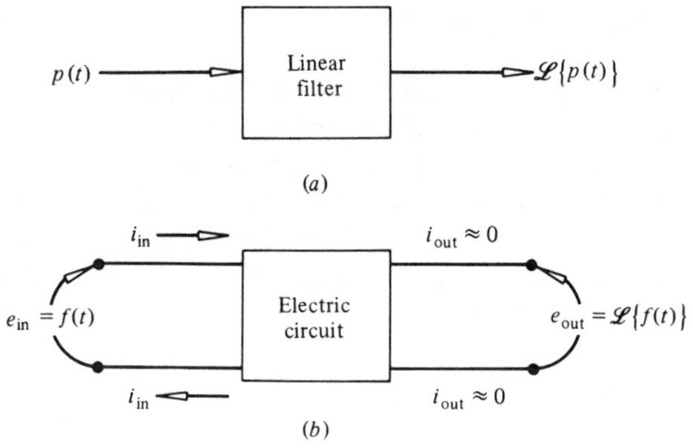

Figure 2-4 (a) Concept of a linear filter that transforms input into output function. (b) Electric-circuit representation; open-circuit voltage across output terminals is $\mathscr{L}\{f(t)\}$ when applied voltage across input is $f(t)$.

where $\mathscr{L}$ is a linear operator characteristic of the filter. The sequence of operations just described therefore yields

$$(p^2)_{\mathrm{av},W} = [(\mathscr{L}\{p\})^2]_{\mathrm{av}} = (p_F^2)_{\mathrm{av}} \tag{2-4.4}$$

A possible realization of a linear filter is an electric circuit (see Fig. 2-4*b*) with two wires leading in and two leading out. If the voltage across the input terminal is $f(t)$, the voltage across the output terminal when it is open (or terminated by an extremely high electric impedance) is $\mathscr{L}\{f(t)\}$.

Properties of the mathematical operator associated with a linear filter are such that

$$\mathscr{L}\{af(t)\} = a\mathscr{L}\{f(t)\} \tag{2-4.5a}$$

$$\mathscr{L}\{f_1(t) + f_2(t)\} = \mathscr{L}\{f_1(t)\} + \mathscr{L}\{f_2(t)\} \tag{2-4.5b}$$

$$\mathscr{L}\left\{\frac{d}{dt}f(t)\right\} = \frac{d}{dt}\mathscr{L}\{f(t)\} \tag{2-4.5c}$$

$$\mathrm{Re}\ \mathscr{L}\{f(t)\} = \mathscr{L}\{\mathrm{Re}\ f(t)\} \tag{2-4.5d}$$

Equation (5*c*) implies that the operation $\mathscr{L}$ is intrinsically time-invariant, and Eq. (5*d*) guarantees that the filtered function will be real if the input is real.

A corollary of the above relations is that for any angular frequency ω

$$\mathscr{L}\left\{\left(\frac{d^2}{dt^2} + \omega^2\right)f(t)\right\} = \left(\frac{d^2}{dt^2} + \omega^2\right)\mathscr{L}\{f(t)\} \tag{2-4.6}$$

Therefore if $f(t)$ is sinusoidal with angular frequency ω (such that the left side of the equation vanishes), $\mathscr{L}\{f(t)\}$ must satisfy the differential equation obtained by setting the right side to 0 and must therefore also be sinusoidal in time with the same angular frequency ω. Thus, if one writes Re $\hat{f}e^{-i\omega t}$ for $f(t)$, $\mathscr{L}\{f(t)\}$ must be of the general form

$$\mathscr{L}\{f(t)\} = \mathrm{Re}\ H(\omega)\hat{f}e^{-i\omega t} \tag{2-4.7}$$

where the *filter transfer function* $H(\omega)$ is a complex number independent of the amplitude $|\hat{f}|$ and phase of the input function but dependent on ω.

The considerations just stated plus the superposition property (5*b*) of a linear filter imply that if $p(t)$ is a multifrequency waveform of the general form of Eq. (2-1.1), the filtered waveform $p_F(t) = \mathscr{L}\{p(t)\}$ should be given by a similar expression with $\hat{p}_n$ replaced by $\hat{p}_{Fn} = H(\omega_n)\hat{p}_n$. Consequently, it follows from Eq. (2-1.7) that the mean square of $p_F(t)$ is

$$(p_F^2)_{\mathrm{av}} = \sum_n |H(\omega_n)|^2\,(p_n^2)_{\mathrm{av}} \tag{2-4.8}$$

A comparison of the above with Eq. (1) indicates that the frequency weighting function $W(f)$ is given by $|H(2\pi f)|^2$; since this is independent of the phase of

given by

$$[p^2(t)]_{\text{av}} = [p(t)p^*(t)]_{\text{av}} = \sum_{n=-\infty}^{\infty} \sum_{m=-\infty}^{\infty} \hat{q}_n \hat{q}_m^* (e^{-i(\omega_n - \omega_m)t})_{\text{av}}$$

$$= \sum_{n=-\infty}^{\infty} |\hat{q}_n|^2 = \sum_{n=0}^{\infty} (p_n^2)_{\text{av}} \qquad \text{for } p(t) \text{ real} \qquad (2\text{-}7.4)$$

Our previous deductions concerning multifrequency signals therefore apply to any $p(t)$, providing one restricts one's attention to a definite time segment and computes all averages with respect to this segment.

If one does not have a periodic waveform, a natural question is: Which numbers associated with the Fourier-series representation are insensitive to the choices of t_c and T? In this respect, many sounds of long duration are such that if $p(t)$ is passed through a filter designed to pass only frequencies (without alteration of amplitude) falling within some *passband* b, then long-term averages of the square of the filtered function will be insensitive to the duration and center of the time segment selected.† A sound satisfying this criterion may be called a *steady sound.* Given such a supposition (which can be checked by experiment), the Fourier coefficients should yield a meaningful estimate of $(p_b^2)_{\text{av}}$ for any given band provided T is sufficiently long. For bands with nonzero lower frequency, this supposition leads to

$$(p_b^2)_{\text{av}} = \lim_{T\to\infty} \sum_{n>0}^{(b)} 2 \left| \frac{1}{T} \int p(t) e^{i2\pi nt/T}\, dt \right|^2 \qquad (2\text{-}7.5)$$

where the sum extends over positive n such that $f_n = \omega_n/2\pi = n/T$ falls within the band. (As before, the limits of integration are $-T/2 + t_c$ and $T/2 + t_c$.) The number of terms included in the sum increases with increasing T and is approximately $T(\Delta f)_b$, where $(\Delta f)_b$ is the width of band b; the sum should be close to its limiting value when $T(\Delta f)_b \gg 1$.

Spectral Density

The band contribution $(p_b^2)_{\text{av}}$ can be regarded as being due to a continuous smear of frequency components; $(p_b^2)_{\text{av}}/(\Delta f)_b$ is then an average contribution per unit bandwidth to the mean squared acoustic pressure. Consequently, one conceives of a second limit in which the bandwidth becomes progressively smaller; the limit is the spectral density $p_f^2(f)$ of $p(t)$, that is,

$$p_f^2(f) = \lim_{(\Delta f)_b \to 0} \frac{(p_b^2)_{\text{av}}}{(\Delta f)_b} \qquad (2\text{-}7.6)$$

† See, for example, C. T. Morrow, "Averaging Time and Data Reduction Time for Random Vibration Spectra, I," *J. Acoust. Soc. Am.*, **30**:456–461 (1958).

f denoting the center frequency of the band. Thus, with this double-limit process (finite bandwidth, $T \to \infty$, then bandwidth $\to 0$, the order of taking limits being fixed), we have the concept of a spectral-density function $p_f^2(f)$, where

$$(p_b^2)_{av} = \int_{f_1}^{f_2} p_f^2(f)\, df \qquad (2\text{-}7.7)$$

gives the contribution to $(p^2)_{av}$ from a band of frequencies between f_1 and f_2.

Levels and Spectral Density

As discussed in previous sections for waveforms composed of a finite number of frequencies, one associates frequency-band sound-pressure levels (fixed frequency intervals, octaves, $\frac{1}{3}$ octaves, etc.) in decibels with any function $p(t)$ for which the concept of a spectral density is applicable (see Fig. 2-8). Levels of weighted sound pressure can be calculated by taking the weighted mean squared sound pressure as

$$(p^2)_{av,W} = \int_0^\infty W(f) p_f^2(f)\, df \approx \sum_b W(f_{0,b})(p_b^2)_{av} \qquad (2\text{-}7.8)$$

where $W(f)$ is the weighting function and $f_{0,b}$ is the center frequency for band b.

For a description of the spectral density in terms of decibels, the natural definition is that of the *sound-pressure spectrum level,*

$$L_{ps}(f) = 10 \log \frac{p_f^2(f)(\Delta f)_{ref}}{p_{ref}^2} \approx 10 \log \frac{(p_b^2)_{av}(\Delta f)_{ref}/(\Delta f)_b}{p_{ref}^2} \qquad (2\text{-}7.9)$$

where $(\Delta f)_{ref}$ is a reference bandwidth, usually taken as 1 Hz. In the second (approximate) expression, $(p_b^2)_{av}$ is the contribution to the mean squared pressure from a band of width $(\Delta f)_b$ centered at the frequency f.

White and Pink Noise

Two idealizations of the frequency dependence of the spectral density are $p_f^2(f)$ constant over the band of interest and $p_f^2(f)$ proportional to $1/f$. The first is called *white noise,* by analogy with white light, which is presumed composed uniformly of all optical frequencies. The second is called *pink noise* because the low frequencies are more prevalent. (Red light is lower-frequency light.)

White noise has the property that $(p_b^2)_{av}$ for any band is $(\Delta f)_b p_f^2$. Since $(\Delta f)_b = (2^{1/2N} - 2^{-1/2N}) f_0(b)$ for a $(1/N)$th-octave band, $(p_b^2)_{av}$ varies as the center frequency for proportional frequency bands. Thus the band sound-pressure levels for successive bands increase as

$$L_{b+1} - L_b = 10 \log \frac{f_0(b+1)}{f_0(b)} = \frac{1}{N} 10 \log 2 \approx \frac{3}{N} \qquad (2\text{-}7.10)$$

The difference is 3 dB for successive octave bands and 1 dB for successive $\frac{1}{3}$-octave bands.

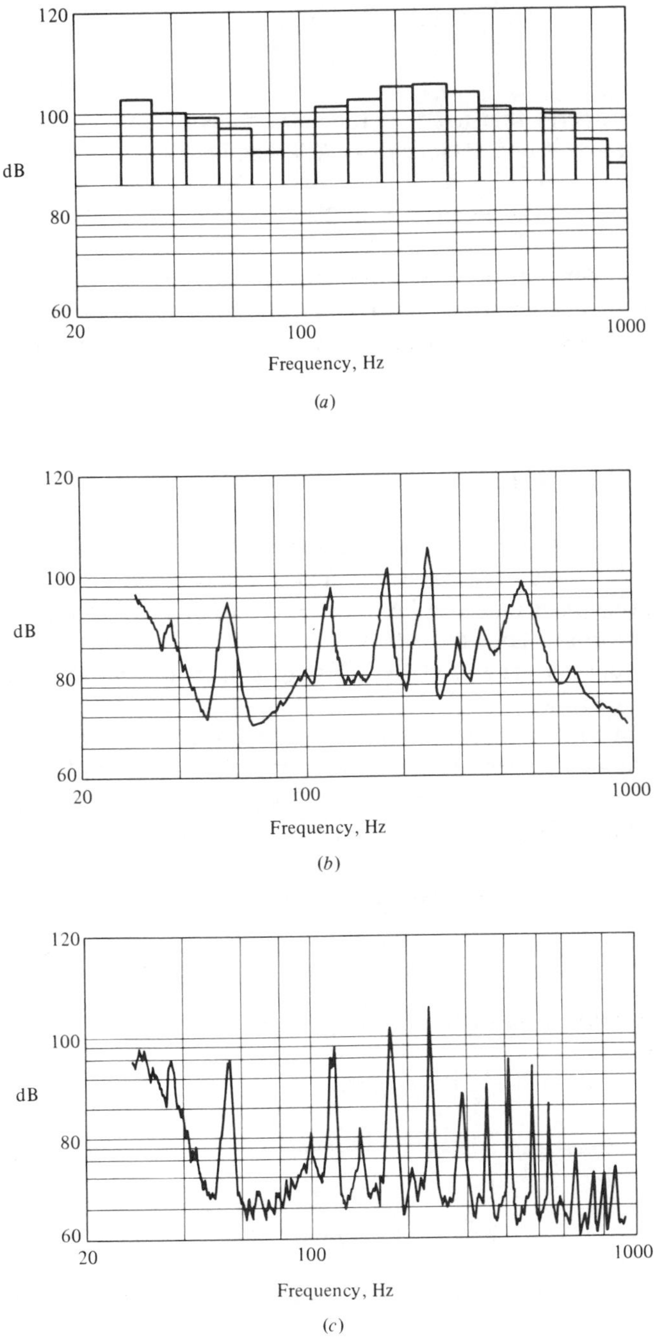

Figure 2-8 Dependence of refrigerator noise-spectrum analysis on bandwidth selection: (*a*) $\frac{1}{3}$-octave analysis for standard contiguous bands; (*b*) band sound-pressure level versus center frequency with bandwidth equal to 5 percent of center frequency; (*c*) band sound-pressure level versus center frequency with 2 Hz bandwidth. (*F. N. Fieldhouse, "Techniques for Identifying Sources of Noise and Vibration," Sound Vib.*, 4(12):16, 17, December 1970.)

Pink noise has the property that $(p_b^2)_{av}$ is the same for all $(1/N)$th-octave bands. This becomes evident if one sets $p_f^2(f) = K/f$, calculates

$$(p_b^2)_{av} = \int_{f_1}^{f_2} \frac{K}{f}\, df = K \ln 2^{1/N} \tag{2-7.11}$$

and notes that this is independent of center frequency. Thus, if one has pink noise over the range of, say, 31.5 to 31,500 Hz and the 500-Hz-octave-band sound-pressure level is 90 dB, then the 8000-Hz-octave-band sound-pressure level is also 90 dB.

2-8 TRANSIENT WAVEFORMS

A transient waveform is one where $p(t)$ is zero before some onset time and after some termination time. All waveforms are transients (there is always a beginning and an ending), although it may not be appropriate to consider them as such. Examples of waveforms whose transitory features may be an important consideration are sonic booms generated by supersonic aircraft and sounds generated by the impact of solids.

The frequency content of a transient waveform is described by its *Fourier transform* $\hat{p}(\omega)$. The basic concept follows from that of a Fourier series if one sets $t_c = 0$, $\hat{q}_n T = 2\pi\hat{p}(\omega_n)$ and then formally takes the limit as $T \rightarrow \infty$. The successive ω_n then become close together and can be considered as values of a continuous variable ω. The sum over n in Eq. (2-7.1) becomes an integral over $n = (T/2\pi)\omega$ or $T/2\pi$ times an integral over ω. The net result is

$$p(t) = \int_{-\infty}^{\infty} \hat{p}(\omega)e^{-i\omega t}\, d\omega \tag{2-8.1}$$

while the corresponding expression (2-7.3) for $\hat{q}_n$ gives

$$\hat{p}(\omega) = \frac{1}{2\pi}\int_{-\infty}^{\infty} p(t)e^{i\omega t}\, dt \tag{2-8.2}$$

[If $p(t)$ is real, then $\hat{p}(-\omega) = \hat{p}(\omega)^*$ and $|\hat{p}(-\omega)| = |\hat{p}(\omega)|$.] Similarly, Parseval's theorem, Eq. (2-7.4), in the same limit, gives

$$E = \int_{-\infty}^{\infty} |p(t)|^2\, dt = 2\pi \int_{-\infty}^{\infty} |\hat{p}(\omega)|^2\, d\omega \tag{2-8.3}$$

sometimes referred to as *Rayleigh's theorem*.† The indicated integral E is called the *sound exposure*.

† The generalization of Parseval's theorem to Fourier transforms was given by Rayleigh, "On the Character of the Complete Radiation at a Given Temperature," *Phil. Mag.*, (5)**27**:460–469 (1889). The common practice of referring to the generalization also as Parseval's theorem is followed throughout the present text.

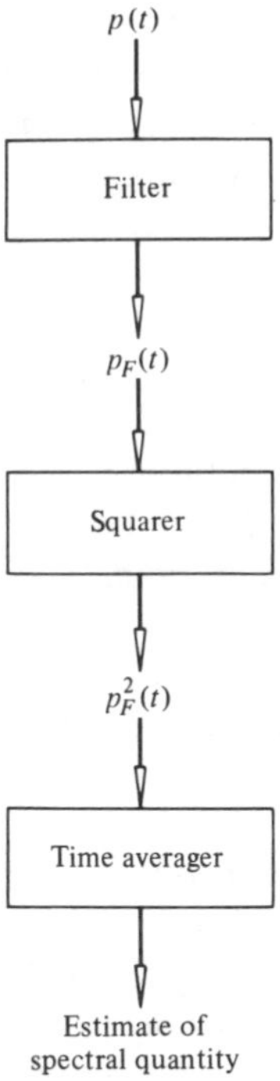

Figure 2-13 Sequence of operations forming basis for common analog method of spectral analysis.

2-11 BIAS AND VARIANCE

Although the expressions discussed in the previous sections for the mean squared band-filtered sound pressure $(p_b^2)_{av}$ and for the spectral density $p_f^2(f)$ involve taking one or more limits, in the real world we must work with just one or a limited number of data segments. Two questions should always be asked concerning data processing schemes for estimation of spectral quantities. First, if one were to repeat the same sequence of measurements and data processing a large number of times, would the numerical average of the individual estimates agree with the desired spectral quantity's actual value? If not, the estimating scheme has a *bias,* whose value is the difference between the average and the quantity's true value. Second, what is the mean squared deviation (*variance*) of the measured numbers from their average?

Perspective on the possible values of bias and variance can be obtained by consideration of a prototype analog† method (see Fig. 2-13) for measuring spectral quantities. The pressure signal passes continuously through a filter for which

† For a discussion of bias and variance associated with digital-computer estimation of spectral density from records of finite length, see R. B. Blackman and J. W. Tukey, *The Measurement of Power Spectra,* Dover, New York, 1958, pp. 11–25, 100–112. The above discussion of the analog case is similar to that given in Blackman and Tukey, pp. 25–28 and 112–116.

the magnitude of the frequency-response function squared (or frequency weighting function) is $W(f|Q)$, the dependence on frequency f being selected to facilitate the measurement of some spectral quantity Q. The filtered output is squared, and a weighted average over time, e.g., as by a measuring amplifier, is computed. If $t = 0$ is taken as the end of the averaging interval, the estimate E_Q for Q can be written

$$E_Q = \frac{1}{T}\int_{-\infty}^{0} A(t/T)p_F^2(t)\,dt \tag{2-11.1}$$

Here $p_F(t)$ is the output of the filter, and $A(t/T)$ is a weighting function characteristic† of the instrumentation, trailing off at large $-t$ (so the lower limit of integration is really finite), having a characteristic duration T, and being normalized such that its integral over t/T from $-\infty$ to 0 is 1. A possible $A(t/T)$ might be $e^{-|t|/T}$; the exact expression is not important in what follows, providing $A(t/T)$ is slowly varying with t over intervals of $1/f$, where f is a representative frequency of either the signal or of the filter's pass band; i.e., we assume $fT \gg 1$.

Let us first examine how the variance of the estimate E_Q depends on the functions $W(f|Q)$ and $A(t/T)$ and on the characteristic duration T. If the pressure signal is a stationary ergodic function, the ensemble average of E_Q is the (time-independent) ensemble average of p_F^2. The spectral density $p_{f,F}^2(f)$ of $p_F(t)$, according to Eq. (2-9.4), is $W(f|Q)$ times the spectral density $p_f^2(f)$ of the unfiltered signal. Because the average of the square of a function with zero mean is the integral over frequency of the corresponding spectral density, the ensemble average $\langle E_Q\rangle$ is given by the integral over f of $p_{f,F}^2(f)$, where it is assumed that the integrand goes to 0 as $f \to 0$.

The difference between a given estimate E_Q and its ensemble average results from Eq. (1) when $p_F^2(t)$ is replaced by $p_F^2 - \langle p_F^2\rangle$, the averaging brackets here implying an average over the ensemble. The variance is the expected square of the resulting integral expression. The product of the two integrals can be regarded as a double integral over t_1 and t_2, and so the variance becomes

$$\frac{1}{T^2}\int_{-\infty}^{0}\int A(t_1/T)A(t_2/T)L(t_1, t_2)\,dt_1\,dt_2 \tag{2-11.2}$$

$$L(t_1, t_2) = \langle[p_F^2(t_1) - \langle p_F^2\rangle][p_F^2(t_2) - \langle p_F^2\rangle]\rangle \approx 2[\mathscr{D}_{p,F}(t_1 - t_2)]^2 \tag{2-11.3}$$

Here the latter identification in terms of the autocovariance results (after some algebra) because the autocorrelation function and the autocovariance of $p_F(t)$

† Of some interest is what may be considered to be the characteristic averaging time of commercial *sound-level meters*. Taking the standard specifications [ANSI S1.4-1971 (R1976), p. 16] for such meters and assuming $A(t/T)$ is $\exp(-|t|/T)$, one can derive for the *fast dynamic characteristic* that $0 < T < 0.2$ s for type 1 instruments and $0 < T < 0.4$ s for type 2 and 3 instruments. For the *slow dynamic characteristic*, the corresponding ranges are $0.7 < T < 1.3$ and $0.5 < T < 1.7$ s.

2-24 The average acoustic-power output of a normal human voice is of the order of 50 μW [V. O. Knudsen, *J. Acoust. Soc. Am.*, **1**:56–82 (1929)]. How close must one be to a person in order to be assured that the received sound level is at least 70 dB?

2-25 If a wave is spreading cylindrically rather than spherically, by how many decibels does the sound-pressure level drop for each doubling of distance?

2-26 An approximate model for the statistical variations of a measured waveform sample of duration T is that the real and imaginary parts of all the Fourier components ($n \geq 0$) corresponding to a given frequency band are statistically independent and that $\langle \mathrm{Re}\, \hat{p}_n \rangle = 0$, $\langle (\mathrm{Re}\, \hat{p}_n)^2 \rangle = \sigma^2$, and $\langle (\mathrm{Re}\, \hat{p}_n)^4 \rangle = 3\sigma^4$, where σ^2 is independent of n and the same relations hold for ensemble averages of the powers of Im $\hat{p}_n$. From this model, what would you estimate to be the ratio of the variance to the square of the expected value for the segment's prediction of the mean squared value of the pressure signal's contribution from a frequency band of width Δf, where Δf is substantially larger than $1/T$?

2-27 A transient acoustic-pressure waveform, zero for $t < 0$, has the form $p_{\mathrm{pk}} \sin \omega t$ for $0 < t < 2\pi N/\omega$ and is thereafter zero, where N is an integer. Estimate how large N must be to ensure that at least 90 percent of the "energy" associated with the signal is carried by (angular) frequencies between 0.99ω and 1.01ω. Make whatever approximations seem appropriate.

2-28 Evaluate the integral

$$\int_0^1 (\sin^{-1} x)\delta(4x^2 - 3)\,dx$$

where $\delta(y)$ is the Dirac delta function.

2-29 In the usual equally tempered scale, the octave is divided into 12 parts, the choice of the number 12 being such that certain integer numbers of $\frac{1}{12}$-octave intervals correspond closely to frequency ratios of 3:2, 4:3, and 5:4. Is there any other choice between 12 and 24 for the number of intervals per octave that would accomplish the same purpose?

2-30 Suppose one took the definition of the spectral density $p_f^2(f)$ to be 4π times the Fourier transform of the autocovariance, as in Eq. (2-10.7). Show that this leads (with various assumptions that you should state) to the prediction that this spectral density is the same as would be obtained if one passed the signal through a filter of some narrow bandwidth Δf centered at f, took the time average of the square of the output, and divided the result by Δf.

2-31 Verify (with mathematical detail stating all pertinent assumptions) the assertion made in the legend of Fig. 2-10 that for a filtered signal made up of a sequence of discrete pulses the sum of successive peak values of the running time average of the square of the output is the contribution from frequencies within the filter's passband to the total time integral of the square of the original signal.

2-32 Nonlinear effects may distort an originally sinusoidal waveform into one of sawtooth shape, so that the time history of p at a given point would be approximately described by a periodic function $f(t) = f(t + T)$, where $f(t) = (P)(1 - 2t/T)$ for $0 < t < T$. For such a waveform, what fraction of the average value of p^2 is attributable to higher-order harmonics, i.e., frequencies other than $1/T$?

2-33 A generalization of Parseval's theorem for Fourier transforms is that, if $f(t)$ and $g(t)$ are two real functions having Fourier transforms $\hat{f}(\omega)$ and $\hat{g}(\omega)$, then

$$\int_{-\infty}^{\infty} f(t)g(t+\tau)\,dt = 2\pi \int_{-\infty}^{\infty} \hat{f}^*(\omega)\hat{g}(\omega)e^{i\omega\tau}\,d\omega$$

for any time shift τ. Give a proof of this, making use of the Dirac delta function.

2-34 Suppose that one has a stationary ergodic function $p(t)$, chooses a segment extending from $t = 0$ to $t = T$, and defines a function $g(t)$ as being equal to $p(t)$ for times within this interval and 0 outside this interval. The Fourier transform $\hat{g}(\omega)$ of $g(t)$ is then derived. How would one estimate the average spectral density p_f^2 of $p(t)$ over a band of frequencies (in hertz) extending from $100/T$ to $200/T$ from a knowledge of

$\hat{g}(\omega)$? How would you expect the average of $|\hat{g}(\omega)|^2$ over a fixed (narrow-band) frequency band to vary with the choice of T, in the limit of large T?

2-35 A hypothetical ideal filter is designed so that its transfer function $H(\omega)$ is $e^{i\omega\tau}$ for frequencies within an octave band consisting of angular frequencies between $2^{-1/2}\omega_0$ and $2^{1/2}\omega_0$. The function $H(\omega)$ is equal to zero for positive frequencies outside that band. [Recall that, for "negative" frequencies, $H(\omega)$ is defined such that $H(-\omega) = H^*(\omega)$.] Here τ is some relatively large delay time. What will the output of the filter be if the input signal equals 0 for $t < 0$ and equals $p_{\text{pk}}e^{-\alpha t}$ for $t > 0$? Give your result in the limit $\alpha \to 0$. What fraction of the "energy" of the output is concentrated within an interval of duration $20\pi/\omega_0$, that is, 10 periods, centered at time $t = \tau$?

2-36 A harmonic oscillator of mass m is acted upon by a time-varying force $F(t)$, and its motion is influenced by a spring with spring constant k and by a dashpot (constant b), such that its displacement $x(t)$ satisfies the differential equation

$$m\ddot{x} + b\dot{x} + kx = F(t)$$

The function $F(t)$ is a stationary ergodic time series characterized by a spectral density $F_f^2(f)$.

(*a*) What is the spectral density $v_f^2(f)$ of the velocity $v = \dot{x}$ of the oscillator?

(*b*) Assuming that $F_f^2(f)$ varies negligibly over a broad band of frequencies centered at the resonance frequency $[\omega_r = (k/m)^{1/2}]$ of the oscillator and that the oscillator is lightly damped $[b \ll (km)^{1/2}]$, derive a simple approximate expression for $(v^2)_{\text{av}}$. With what frequency would the oscillator appear to be predominantly vibrating?

2-37 Give an explicit proof that the operator $\oplus$ introduced in Sec. 2-5 to describe the addition of decibels satisfies the properties (2-5.3) and that Eqs. (2-5.4) and (2-5.5) ensure that

$$L_1 \oplus L_2 \oplus L_3 = 10 \log (10^{L_1/10} + 10^{L_2/10} + 10^{L_3/10})$$

What changes in these formulas would be necessitated if one chose to measure sound-pressure levels in nepers rather than decibels?

2-38 The autocovariance of a stationary ergodic time series must correspond to a spectral density that is nonnegative for all frequencies. Given this criterion, check whether each of the following is an admissible autocovariance ($a > 0$, $b > 0$):

(*a*) $\mathscr{D}_p(\tau) = e^{-a\tau^2}$

(*b*) $\mathscr{D}_p(\tau) = \dfrac{1}{1 + a\tau^2}$

(*c*) $\mathscr{D}_p(\tau) = (1 - b\tau^2)e^{-a\tau^2}$

2-39 A pressure signal is of the form of a sudden jump followed by a very slow exponential decrease, that is, $p(t) = 0$ if $t < 0$ and $p(t) = p_{\text{pk}}e^{-\alpha t}$ if $t > 0$. In the limit $\alpha \to 0$ determine an expression for the integrated octave-band sound-pressure level for an octave band centered at frequency f_0. By how many decibels does the integrated-band sound-pressure level differ for successive contiguous octave bands?

2-40 Suppose $p(t)$ is a function that goes to zero at least as fast as $e^{-a|t|}$ (for some positive value of a) when $t \to \infty$. We wish to know the asymptotic form of its Fourier transform $\hat{p}(\omega)$ without an explicit knowledge of $p(t)$.

(*a*) Show that if $p(t)$ has a positive discontinuity of Δp at $t = t_0$ and is otherwise continuous, then

$$\hat{p}(\omega) \to \frac{i\,\Delta p}{2\pi\omega}\, e^{i\omega t_0} \qquad \omega \to \infty$$

(*b*) Show that if $p(t)$ is everywhere continuous but $dp(t)/dt$ has a discontinuity of $\Delta\dot{p}$ at $t = t_0$ and is otherwise continuous, then

$$\hat{p}(\omega) \to \frac{-\Delta\dot{p}}{2\pi\omega^2}\, e^{i\omega t_0} \qquad \omega \to \infty$$

(Lighthill, *Fourier Analysis and Generalized Functions,* pp. 43, 46–57.)

2-41 Suppose the signals corresponding to the acoustic pressure and the three cartesian components of $\mathbf{v}$ are each passed through identical linear filters such that one obtains functions $p_F(\mathbf{x}, t)$ and $\mathbf{v}_F(\mathbf{x}, t)$.

(*a*) Show that p_F and $\mathbf{v}_F$ satisfy the same linear acoustic equations as the original unfiltered functions, that is, $\partial p_F/\partial t + \rho c^2 \nabla \cdot \mathbf{v}_F = 0$ and $\rho\, \partial \mathbf{v}_F/\partial t = -\nabla p_F$.

(*b*) Show that if an acoustic energy density w_F and intensity $\mathbf{I}_F$ are constructed according to Eqs. (1-11.3) from these filtered functions, the acoustic-energy corollary (1-11.2) will still be valid.

(*c*) Show in addition that this corollary holds for running time averages (rta), defined by

$$w_{F,\mathrm{rta}}(\mathbf{x}, t) = \int_{-\infty}^{t} A(t - t')w_F(t', \mathbf{x})\, dt'$$

with a function $A(t)$ whose integral from 0 to ∞ is 1, the function $A(t)$ being the same for the computation of both $w_{F,\mathrm{rta}}$ and $\mathbf{I}_{F,\mathrm{rta}}$.

CHAPTER

THREE

REFLECTION, TRANSMISSION, AND EXCITATION OF PLANE WAVES

When a sound wave strikes a surface (an interface between two substances), a reflected wave, or *echo,* results whose nature depends on the characteristics of the surface and of the adjoining substances. In some instances, one may be interested in the acoustic disturbance produced on the other side of the surface. A related topic is the generation of sound by a vibrating surface. Many acoustical phenomena involve such interactions of sound and surfaces, and we accordingly here examine the principles pertaining to them. For the most part, attention is restricted to situations where the plane-wave idealization is applicable, although certain concepts such as boundary conditions, causality, and specific acoustic impedance are introduced in more general terms.

3-1 BOUNDARY CONDITIONS AT IMPENETRABLE SURFACES

A vibrating or stationary surface S adjacent to a fluid imposes constraints, or *boundary conditions,* on the possible solutions of the fluid-dynamic equations. We here consider S to separate a solid material from a fluid, although much of the following discussion applies equally to an interface between two fluids, e.g., air and water. The surface S (see Fig. 3-1) is also regarded as smooth, so that, with any given (moving with the material in the solid) point $\mathbf{x}_S$ on S, we can associate a unit normal vector $\mathbf{n}_S$ pointing out of the solid into the fluid. One also associates with $\mathbf{x}_S$ a surface velocity $\mathbf{v}_S = d\mathbf{x}_S/dt$, representing the local average velocity of the solid particles near $\mathbf{x}_S$.

If the surface is *impenetrable* (not porous), a fluid particle adjacent to the surface S at a time t_0 must be adjacent to it at $t_0 + \Delta t$. During a short interval Δt,

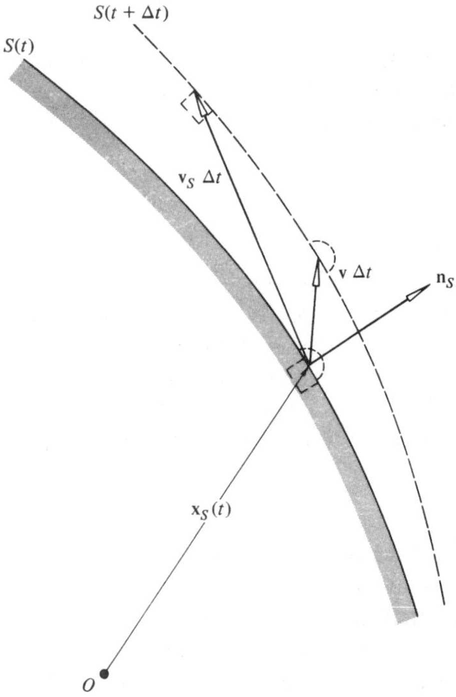

Figure 3-1 Idealized fluid-solid interface (surface S with unit normal $\mathbf{n}_S$). The position $\mathbf{x}_S(t)$ describes a material point in the solid; $\mathbf{v}_S(t)$ is its velocity; $\mathbf{v}(\mathbf{x}_S, t)$ is the velocity of a fluid particle adjacent to $\mathbf{x}_S(t)$ at time t.

the surface S moves normal to itself a distance $(\mathbf{v}_S\,\Delta t)\cdot\mathbf{n}_S = v_n\,\Delta t$, where $v_n = \mathbf{v}_S\cdot\mathbf{n}_S$ is the normal velocity of the surface. If one ignores viscosity or considers fluid particles that are close to, but not exactly at, the solid surface, e.g., just outside† a *viscous boundary layer*, the fluid may slip relative to the solid surface but nevertheless has the same normal displacement in time Δt as a solid particle in its immediate vicinity does. Otherwise, the fluid mass density would locally be anomalously very high or very small; both possibilities are implausible. Consequently, the normal component of the fluid velocity at the surface should be the same as that of the surface proper, so one has‡

$$\mathbf{v}\cdot\mathbf{n}_S = \mathbf{v}_S\cdot\mathbf{n}_S = v_n \tag{3-1.1}$$

at any point $\mathbf{x}_S$ on S.

† The thickness of the viscous boundary layer in typical cases of acoustical interest is of the order of $(2\mu/\rho\omega)^{1/2}$, where μ [$\sim 2\times 10^{-5}$ kg/(m · s) for air and $\sim 10^{-3}$ kg/(m · s) for water at 20°C] is the viscosity. This thickness is invariably much less than a wavelength for any frequency of interest. (Acoustic boundary layers are discussed in Sec. 10-4.)

‡ This condition may be recognized in early works by Euler, Lagrange, and Poisson. A statement similar in form to that in the text is given by G. G. Stokes, "On Some Cases of Fluid Motion," *Trans. Camb. Phil. Soc.*, **8:**105 (read May 29, 1843); *Mathematical and Physical Papers*, vol. 1, Cambridge University Press, Cambridge, 1880, pp. 17–68, especially p. 22.

Stationary Surfaces

If the surface S is stationary ($\mathbf{v}_S = 0$) though the fluid outside it may be moving, Eq. (1) reduces to $\mathbf{v} \cdot \mathbf{n}_S = 0$. If the linear acoustic equations (1-5.3) hold within the fluid, then Eq. (1) and the linear version of Euler's equation imply $\mathbf{n}_S \cdot \nabla p = 0$ on the surface.

Vibrating Surfaces

If the surface is vibrating, the application of Eq. (1) can be complicated because it applies at a moving rather than a fixed surface and because the unit normal $\mathbf{n}_S$ may be changing with time. However, if the surface-vibration amplitude is small compared with a representative acoustic wavelength and representative dimensions describing the surface, and if there is no ambient flow ($\mathbf{v}_0 = 0$), then it is consistent with the use of the linear acoustic equations to require instead that

$$\mathbf{v} \cdot \mathbf{n}_0 = \mathbf{v}_S \cdot \mathbf{n}_0 \tag{3-1.2}$$

hold at a nonmoving surface S_0 whose location is the average or nominal location of S. The unit vector $\mathbf{n}_0$ is normal to S_0 and therefore independent of time. The velocity $\mathbf{v}_S$ is the velocity (assumed small) of that point on the solid nominally at the same point on S_0. The premise is that the acoustic field within the fluid, predicted subject to specified normal component $\mathbf{v} \cdot \mathbf{n}_0$ of acoustic fluid velocity on a *fixed* surface, is very nearly the same as would be predicted if $\mathbf{v} \cdot \mathbf{n}_S$ were specified on the actual *moving* surface.

Example A rigid sphere of radius a rocks back and forth about an axle (Fig. 3-2) located a distance b from its center. The peak angular displacement is substantially less than $\pi/2$, so the motion of the center of the sphere is very nearly along a straight line. What boundary condition would one place on the linear acoustic equations to account for the presence of the oscillating sphere?

SOLUTION The axle is parallel to the y axis, with its center at $x = b$, $z = 0$. The angular velocity vector $\mathbf{\Omega}$ is accordingly in the y direction and can be denoted $\mathbf{\Omega} = \Omega(t)\mathbf{e}_y$. The velocity $\mathbf{v}_S$ of any point $\mathbf{x}_S$ of the surface is the vector cross product of angular velocity with a vector from any point on the axle to $\mathbf{x}_S$, so one has†

$$\mathbf{v}_S = \Omega\mathbf{e}_y \times (\mathbf{x}_S - b\mathbf{e}_x) = \Omega(\mathbf{e}_y \times \mathbf{x}_S) + \Omega b\mathbf{e}_z \tag{3-1.3}$$

† See, for example, S. H. Crandall, D. C. Karnopp, E. F. Kurtz, Jr., and D. C. Pridmore-Brown, *Dynamics of Mechanical and Electromechanical Systems*, McGraw-Hill, New York, 1968, pp. 61–78. The general relation

$$\frac{d}{dt}(\mathbf{x}_A - \mathbf{x}_B) = \mathbf{\Omega} \times (\mathbf{x}_A - \mathbf{x}_B)$$

for any two points fixed in a rigid body with angular velocity $\mathbf{\Omega}$ is sometimes referred to as *Euler's velocity equation* and stems from a 1776 paper by Euler.

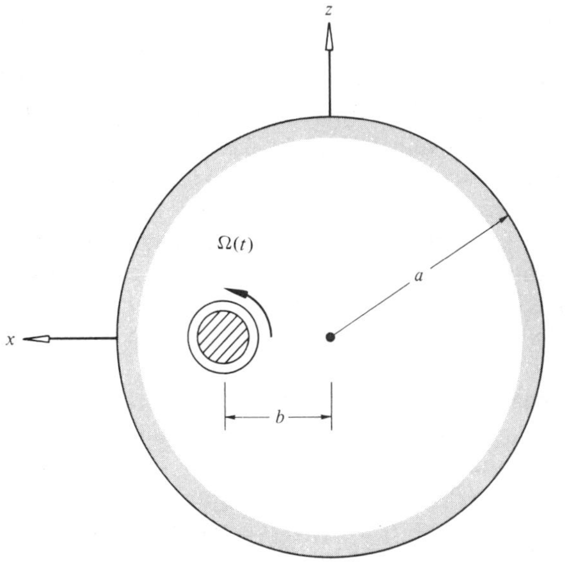

Figure 3-2 A rigid sphere of radius a pivoted about an axle displaced a distance b from its center. The angular velocity $\Omega(t)$ oscillates with a small amplitude, such that the sphere's center is always close to the origin.

To the approximation implied by Eq. (2), only an expression of first order in $\Omega(t)$ is desired, so the vector $\mathbf{x}_S$ in (3) can be replaced by the vector $a\mathbf{e}_r$. However, since the nominal boundary surface S_0 is a sphere of radius a centered at the origin, $\mathbf{n}_0$ is $\mathbf{e}_r$. Also, $\mathbf{e}_y \times \mathbf{e}_r$ is perpendicular to $\mathbf{e}_r$, so one obtains $(\mathbf{e}_y \times a\mathbf{e}_r) \cdot \mathbf{n}_0 = 0$ and

$$\mathbf{v}_S \cdot \mathbf{n}_0 = \Omega b \mathbf{e}_z \cdot \mathbf{n}_0 = \Omega b \cos\theta \tag{3-1.4}$$

where θ is the polar angle in spherical coordinates. This result is the same as would have been obtained if the sphere were translating without rotation back and forth in the z direction with a velocity $\mathbf{v}_C = \boldsymbol{\Omega} b\,\mathbf{e}_z$. The remaining motion, which is described by the term $\Omega(\mathbf{e}_y \times a\mathbf{e}_r)$ and which can be regarded as a rotation about the origin, gives no contribution to the acoustic boundary condition (2) because it describes a motion tangential to the surface.

The result (4) allows the boundary condition (2) to be taken as $v_r = \Omega b \cos\theta$ at $r = a$. Alternatively, since $\mathbf{e}_r \cdot \nabla p = \partial p/\partial r$, the radial component of the linear version of Euler's equation of motion would require $\partial p/\partial r$ to be $-\rho\dot{\Omega} b \cos\theta$ at $r = a$.

A generalization to this example is a moving rigid sphere of radius a whose center at time t is at $\mathbf{x}_C(t)$, where $|\mathbf{x}_C| \ll a$; the appropriate boundary condition is $v_r = \dot{\mathbf{x}}_C \cdot \mathbf{e}_r$ at $r = a$.

Continuity of Normal Component of Displacement

Boundary condition (2) raises conceptual difficulties when one seeks to understand phenomena in the near vicinity of the surface and moreover may be inappropriate† if there is an ambient flow. One way to resolve such difficulties is to regard the acoustic variables as functions‡ of x_0, y_0, z_0, t rather than x, y, z, t, where x_0, y_0, z_0 denote the cartesian coordinates a fluid particle would have had if there were no surface vibration or acoustic disturbance. Thus $v_x'(x_0, y_0, z_0, t)$ denotes the x component of acoustic fluid velocity for the fluid particle ordinarily at x_0, y_0, z_0 at that same time. Since $v_x'(x, y, z, t) - v_x'(x_0, y_0, z_0, t)$ and analogous differences are second order in acoustic amplitudes, the x_0, y_0, z_0 description necessitates no change in the linear equations of acoustics (with or without ambient flow). A vibrating impenetrable surface is then one whose mathematical description does not change with t when x_0, y_0, z_0, t are the independent variables. With the x_0, y_0, z_0, t description, all such surfaces formally appear stationary.

If there is an ambient flow past the surface, the appropriate principle replacing Eq. (2) is *continuity of normal displacement*. Consider a fluid particle P adjacent to the surface whose nominal location is $\mathbf{x}_0(P, t)$ and whose actual location is $\mathbf{x}(\mathbf{x}_0(P, t), t) = \mathbf{x}_0(P, t) + \Delta\boldsymbol{\xi}(P, t)$. A second fluid particle Q adjacent to the surface is selected such that $\mathbf{x}(\mathbf{x}_0(P, t), t) - \mathbf{x}_0(Q, t)$ is parallel to the unit normal $\mathbf{n}_0(P, t)$ to the ambient surface S_0 at $\mathbf{x}_0(P, t)$; that is, at time t particle P is on the same line extending out from the surface that passes through the nominal location of Q. The displacement of P from the nominal location of particle Q is $\Delta\xi_n(P, t)\mathbf{n}_0(P, t)$, where $\Delta\xi_n(P, t) \approx \Delta\xi_n(Q, t)$ is the normal displacement of the surface in the vicinity of particles P and Q at time t. Then, since $\mathbf{x} = \mathbf{x}_0 + \Delta\boldsymbol{\xi}$, one can write

$$\mathbf{x}_0(P, t) - \mathbf{x}_0(Q, t) + \Delta\boldsymbol{\xi}(\mathbf{x}_0(P, t), t) = \Delta\xi_n(P, t)\mathbf{n}_0(P, t) \tag{3-1.5}$$

Because the particles P and Q are close to each other for a typical small-amplitude acoustic disturbance, the difference $\mathbf{x}_0(P, t) - \mathbf{x}_0(Q, t)$ is nearly tangential to S_0, so $[\mathbf{x}_0(P, t) - \mathbf{x}_0(Q, t)] \cdot \mathbf{n}_0$ is much smaller than $\Delta\xi_n$ or $\Delta\boldsymbol{\xi} \cdot \mathbf{n}_0$. Consequently, to first order in acoustic amplitudes, Eq. (5) requires that the normal component of displacement of a fluid particle at the surface be the same as that of the adjacent element of surface. This condition, $\Delta\boldsymbol{\xi} \cdot \mathbf{n}_0 = \Delta\xi_n$, leads to Eq. (2) when there is no ambient flow, as can be demonstrated by a differentiation with respect to time.

3-2 PLANE-WAVE REFLECTION AT A FLAT RIGID SURFACE

An application of Eq. (1) in the previous section is the reflection of a plane wave from a flat rigid surface.§ The surface is taken as the $y = 0$ plane (see Fig. 3-3)

† An example when Eq. (2) is inappropriate is propagation across an interface (vortex sheet) between two fluids with different ambient fluid velocities. The proper boundary condition was pointed out by H. S. Ribner, "Reflection, Transmission, and Amplification of Sound by a Moving Medium," *J. Acoust. Soc. Am.*, **29**:435–441 (1957).

‡ C. Eckart, "Some Transformations of the Hydrodynamic Equations," *Phys. Fluids*, **6**:1037–1041 (1963); F. P. Bretherton and C. J. R. Garrett, "Wavetrains in Inhomogeneous Moving Media," *Proc. R. Soc. Lond.*, **A302**:529–554 (1969).

§ S. D. Poisson, "Memoir on the Theory of Sound," *J. Ec. Polytech.*, **7**:319–392 (April 1908), especially p. 351. The discussion in the present text derives in major part from that of George

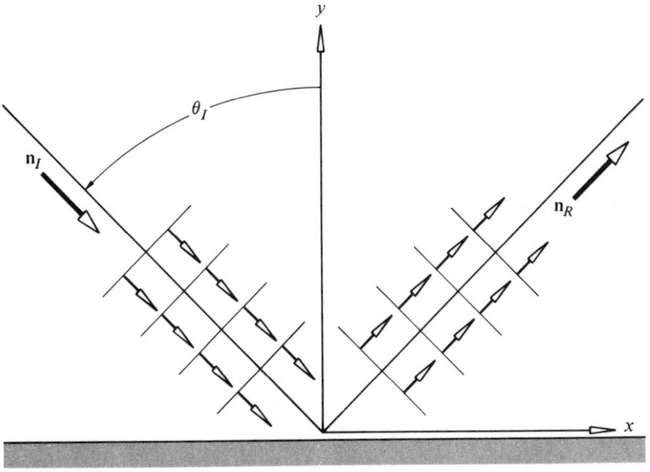

Figure 3-3 Reflection of a plane wave with angle of incidence θ_I at a flat rigid surface.

with the unit normal $\mathbf{n}_S$ as $\mathbf{e}_y$. The incident plane wave, in accord with Eqs. (1-7.7) and (1-7.8), can be written as

$$p_I = f(t - c^{-1}\mathbf{n}_I \cdot \mathbf{x}) \qquad \mathbf{v}_I = \frac{\mathbf{n}_I}{\rho c}\, p_I \tag{3-2.1}$$

The incident wave's direction of propagation (unit vector $\mathbf{n}_I$) can be considered to have no z component, so

$$\mathbf{n}_I = \mathbf{e}_x \sin\theta_I - \mathbf{e}_y \cos\theta_I \tag{3-2.2}$$

where θ_I, the *angle of incidence,* is the angle $\mathbf{n}_I$ makes with the unit vector $-\mathbf{e}_y$ pointing into the surface.

If the incident wave is a solution (throughout the spatial region of interest) of the linear acoustic equations (1-5.3) when the solid surface at $y = 0$ is not present, then the solution with the surface present, written as $p_I + p_R$, $\mathbf{v}_I + \mathbf{v}_R$, must be such that the pair p_R, $\mathbf{v}_R$ are themselves a solution of the linear acoustic equations. Moreover, the boundary condition $\mathbf{v} \cdot \mathbf{n}_S = 0$ at $y = 0$ requires $(\mathbf{v}_I + \mathbf{v}_R) \cdot \mathbf{e}_y = 0$ at $y = 0$.

In this particular case, the solution for the reflected wave is easily obtained from the alternate boundary condition, $\partial p/\partial y = 0$ at $y = 0$, which will be satisfied if

$$p_R(x, y, z, t) = p_I(x, -y, z, t) \tag{3-2.3}$$

Green, "On the Reflexion and Refraction of Sound," *Trans. Camb. Phil. Soc.*, 6:403–412 (1838), reprinted in R. P. Lindsay (ed.), *Acoustics: Historical and Philosophical Development,* Dowden, Hutchinson and Ross, Stroudsburg, Pa., 1972, pp. 231–241.

(This represents an example of the method of images.†) Here, for positive y, the quantity $p_I(x, -y, z, t)$ is the mirror extension of the acoustic pressure in the incident wave to negative values of y. If Eq. (3) is satisfied, the sum $p_I + p_R$ will be even in y and will therefore have zero y derivative at $y = 0$. Since $p_I(x, y, z, t)$ is given by Eq. (1), p_R becomes $f(t - c^{-1}\mathbf{n}_R \cdot \mathbf{x})$, where $\mathbf{n}_R$ differs from $\mathbf{n}_I$ in that its y (normal) component is of opposite sign; that is, $\mathbf{n}_R$ is $\mathbf{e}_x \sin \theta_I + \mathbf{e}_y \cos \theta_I$. That the angle between $\mathbf{n}_R$ and $\mathbf{e}_y$ is also θ_I is the *law of mirrors*: angle of incidence equals angle of reflection.

Because $f(t - c^{-1}\mathbf{n}_R \cdot \mathbf{x})$ describes a plane wave propagating in the direction $\mathbf{n}_R$, and because the fluid velocity in a plane traveling wave is $\mathbf{v} = \mathbf{n}p/\rho c$ [see Eq. (1-7.8)], one has

$$\mathbf{v}_R = \frac{\mathbf{n}_R}{\rho c} f(t - c^{-1}\mathbf{n}_R \cdot \mathbf{x}) = \frac{\mathbf{n}_R}{\rho c} p_R \tag{3-2.4}$$

which satisfies the boundary condition $(\mathbf{v}_R + \mathbf{v}_I) \cdot \mathbf{e}_y = 0$ at $y = 0$.

A consequence of the above solution is that, at $y = 0$, the acoustic pressure and the tangential component of the fluid velocity for the total wave disturbance are both exactly twice (or 10 log 4 ≈ 6 dB higher than) the corresponding quantities for the incident wave alone. If the incident wave is of constant frequency, then

$$\begin{aligned} p_I + p_R &= \mathrm{Re}[Ae^{-i\omega t}e^{ik_x x}(e^{-ik_y y} + e^{ik_y y})] \\ &= 2 \cos (ky \cos \theta_I) f(t - c^{-1}x \sin \theta_I) \end{aligned} \tag{3-2.5}$$

so the incident and reflected waves cancel whenever $ky \cos \theta_I$ is an odd multiple of $\pi/2$. (Here we use the abbreviations $k = \omega/c$, $k_x = k \sin \theta_I$, $k_y = k \cos \theta_I$.) Similarly, if the incident wave is a stationary ergodic time series with spectral density $p_{f,I}^2(f)$, the resulting acoustic pressure due to the combined incident and reflected waves will have a spectral density [see Eq. (2-9.4)]

$$p_f^2(f) = 4 \cos^2 \left(\frac{2\pi f}{c} y \cos \theta_I \right) p_{f,I}^2(f) \tag{3-2.6}$$

Consequently, if $p_{f,I}^2(f)$ is slowly varying over a frequency interval of width $\Delta f = c/(2y \cos \theta_I)$, then an average of $p_f^2(f)$ over an interval somewhat larger than Δf will be twice the corresponding average of $p_{f,I}^2(f)$. This leads to the rule of thumb that sound-pressure levels due to higher (and broad) frequency bands at points near (but not on) a rigid surface are 10 log 2 ≈ 3 dB higher than would be obtained if there were no reflection from the surface. The sound level exactly at the surface is 3 dB higher than at moderate distances from the surface.‡

† This dates back to Euler's "On the Propagation of Sound" (1759, 1766) and to his "More Detailed Enlightenment on the Generation and Propagation of Sound and on the Formation of Echos" (1765, 1767). The first paper is in Lindsay, *Acoustics*, pp. 136–154. The mathematical statement given in the text can be recognized in the previously cited paper by Poisson.

‡ E. W. Kellogg, "Estimating Room Errors in Loudspeaker Tests," *J. Acoust. Soc. Am.*, 4:56–62 (1932).

3-3 SPECIFIC ACOUSTIC IMPEDANCE

The concept of specific acoustic impedance leads to a boundary condition describing a surface, e.g., a porous wall, that is not necessarily impenetrable or rigid. To introduce the concept, we assume a linear relation (doubling one causes the other to double) between the acoustic pressure p and the inward normal component (into the surface and out of the fluid) $\mathbf{v} \cdot \mathbf{n}_{\text{in}}$ of the fluid velocity along a nonmoving surface S_0. If the surface vibrates under the influence of an acoustic disturbance, S_0 should represent the surface's nominal location, as described in Sec. 3-1.

If the properties of the environment on the other side of the surface S_0 are time-dependent, the existence of such a linear relation implies that different frequency components of p and $\mathbf{v} \cdot \mathbf{n}_{\text{in}} = v_{\text{in}}$ are uncoupled, so one need only specify the linear dependence for individual frequency components. For certain idealized situations, e.g., the reflection of a plane wave from a nominally flat surface of unlimited extent bounding a "wall" of uniform composition, the invariance of the overall model under translation parallel to the surface requires, moreover, that the ratio

$$\left(\frac{\hat{p}}{\hat{v}_{\text{in}}}\right)_{\text{on } S_0} = Z_s(\omega) = \rho c \zeta(\omega) \tag{3-3.1}$$

be independent of position along S_0. Here $\hat{p}$ is the complex amplitude of a single-frequency component of p (the latter being Re $\hat{p}e^{-i\omega t}$) at any given point on S_0, while $\hat{v}_{\text{in}}$ is the corresponding complex amplitude of the same frequency component of v_{in} at the same point. That a linear relation between $\hat{p}$ and $\hat{v}_{\text{in}}$ should be expressible in the above form is in accord with the expectation that when $\hat{p}$ vanishes, $\hat{v}_{\text{in}}$ should also, and conversely. The ratio $Z_s(\omega)$ is referred to as the *specific acoustic impedance* (or *unit area acoustic impedance*) of the surface S_0; the ratio $\zeta(\omega)$ of specific impedance $Z_s(\omega)$ to the *characteristic impedance* $Z_c = \rho c$ of the fluid is a convenient dimensionless quantity that simplifies writing mathematical relations. The real R_s and imaginary X_s parts of Z_s are the *specific acoustic resistance* and *reactance*, respectively. (In literature where the time dependence of oscillating quantities is described by $e^{j\omega t}$, where $j^2 = -1$, the reactance is the negative of what the definition adopted here would give.) Units of specific acoustic impedance are Pa · s/m or kg/(m² · s).

In mechanics, a ratio of a force amplitude to a velocity amplitude is referred to as an *impedance*. The term, although having an evident mechanical connotation (something impeding motion), was introduced first into electric-circuit theory as a ratio of voltage amplitude to current amplitude by Heaviside† in the late nineteenth century as a generalization of the concept of

† "Let us call the ratio of the impressed force to the current in a line when electrostatic induction is ignorable the Impedance of the line, from the verb impede. It seems as good a term as Resistance, from resist," O. Heaviside, "Electromagnetic Induction and Its Propagation," *Electrician (Lond.)*, **17:** July 23, 1886, pp. 212–213, reprinted in *Electrical Papers*, vol. 2, Copley, Boston, 1925, p. 64. Heaviside's definition has since been extended to imply the ratio of complex voltage amplitude to complex current amplitude.

electrical resistance for ac applications. Impedance was introduced into acoustics† by A. G. Webster in 1914 and independently in a context similar to that of Eq. (1) by Kennelly and Kurokawa in 1921. Since pressure is force per unit area, the ratio $\hat{p}/\hat{v}_{\text{in}}$ is an impedance per unit area or, since "specific" implies "per unit amount" (area in this instance), it is a specific impedance.‡

Plane Traveling Waves and Specific Acoustic Impedance

An instance to which Eq. (1) applies is a plane traveling wave, with $p = f(t - \mathbf{n}_I \cdot \mathbf{x}/c)$ and with $\mathbf{v} = \mathbf{n}_I p/\rho c$, propagating in a direction of incidence $\mathbf{n}_I$. If S_0 is a plane surface, and if a choice is made for the sense (toward which side) of $\mathbf{n}_{\text{in}}$, the impedance $Z_s(\omega)$ associated with S_0 in this context is $f/(\mathbf{n}_I \cdot \mathbf{n}_{\text{in}} f/\rho c)$ or

$$Z_s(\omega) = \frac{\rho c}{\mathbf{n}_I \cdot \mathbf{n}_{\text{in}}} = \frac{\rho c}{\cos \theta_I} \tag{3-3.2}$$

where θ_I is the angle between the propagation direction $\mathbf{n}_I$ and the inward normal $\mathbf{n}_{\text{in}}$. Although $Z_s(\omega)$ is independent of ω in this instance, it does depend on angle of incidence, so one could not consider Z_s to be an intrinsic property of the surface S_0. Another implication of this relation is that $\zeta(\omega)$ should be unity for a plane traveling wave passing at normal incidence through S_0.

Plane-Wave Reflection at a Surface with Finite Specific Impedance

The example (Sec. 3-2 and Fig. 3-3) of plane-wave reflection at a rigid surface can be generalized to reflection from a surface with finite specific impedance Z (possibly depending on the angle of incidence). (Here and in what follows the subscript s is omitted for brevity.) One takes the incident wave as given by Eqs. (3-2.1), with $\mathbf{n}_I$ as given by Eq. (3-2.2). The total disturbance consists of incident and reflected plane waves; the reflected wave pressure p_R, however, is $g(t - c^{-1}\mathbf{n}_R \cdot \mathbf{x})$, where the function $g(t)$ is not necessarily the same as the incident waveform $f(t)$.

If one considers $f(t)$ to be a superposition, e.g., Fourier series, of constant-frequency components, any one such component is of the form Re $\hat{f}e^{-i\omega t}$. The *pressure-amplitude reflection coefficient* $\mathcal{R}(\theta_I, \omega)$ is defined such that the quantity Re $\mathcal{R}(\theta_I, \omega)\hat{f}e^{-i\omega t}$ is the corresponding component of $g(t)$, so $\hat{g} = \mathcal{R}(\theta_I, \omega)\hat{f}$. Alternatively, if $f(t)$ and $g(t)$ are transient waveforms, $\mathcal{R}(\theta_I, \omega)$ is the ratio of the Fourier transform of $g(t)$ to that of $f(t)$. In either event, we can

† A. G. Webster, "Acoustical Impedance and the Theory of Horns and of the Phonograph," *Proc. Natl. Acad. Sci. (USA)*, **5**:275–282 (1919) (originally presented in 1914 at an American Physical Society Meeting); A. E. Kennelly and K. Kurokawa, "Acoustic Impedance and Its Measurement," *Proc. Am. Acad. Arts Sci.*, **61**:3–37 (1921).

‡ However, what is called acoustic impedance without the adjective "specific" has units of specific impedance divided by area rather than of specific impedance times area; see Sec. 7-2.

write

$$\hat{p} = \hat{f}e^{ik_x x}[e^{-ik_y y} + \mathcal{R}(\theta_I, \omega)e^{ik_y y}] \tag{3-3.3a}$$

$$\hat{v}_y = \frac{\cos\theta_I}{\rho c}\hat{f}e^{ik_x x}[-e^{-ik_y y} + \mathcal{R}(\theta_I, \omega)e^{ik_y y}] \tag{3-3.3b}$$

where $k_x = (\omega/c)\sin\theta_I$, $k_y = (\omega/c)\cos\theta_I$.

The boundary condition at $y = 0$ that $\hat{p}/\hat{v}_{\text{in}} = Z(\omega)$ leads in this case ($\hat{v}_{\text{in}} = -\hat{v}_y$) to

$$\frac{Z(\omega)\cos\theta_I}{\rho c} = \frac{1+\mathcal{R}(\theta_I,\omega)}{1-\mathcal{R}(\theta_I,\omega)} \qquad \mathcal{R}(\theta_I,\omega) = \frac{\zeta(\omega)\cos\theta_I - 1}{\zeta(\omega)\cos\theta_I + 1} \tag{3-3.4}$$

The magnitude of $\mathcal{R}$ is less than 1 if and only if the real part of Z is positive. Any surface having this property *absorbs* acoustic energy. The time-averaged acoustic power flowing into the surface per unit area of surface equals (for a single-frequency component)

$$(pv_{\text{in}})_{\text{av}} = \tfrac{1}{2}\,\text{Re}\,\hat{p}\hat{v}_{\text{in}}^* = \tfrac{1}{2}|\hat{v}_{\text{in}}|^2\,\text{Re}\,Z(\omega) \tag{3-3.5}$$

from Eq. (1) [and with the mathematical theorem of Eq. (1-8.9)]. The same quantity [with Eqs. (3) and $\hat{v}_{\text{in}} = -\hat{v}_y$ at $y = 0$] becomes

$$(pv_{\text{in}})_{\text{av}} = \frac{1}{2}\frac{\cos\theta_I}{\rho c}|\hat{f}|^2(1 - |\mathcal{R}|^2) \tag{3-3.6}$$

since the real part of $(1 + \mathcal{R})(1 - \mathcal{R}^*)$ is $1 - |\mathcal{R}|^2$. The surface absorbs energy if $\text{Re}\,Z(\omega) > 0$ or, equivalently, if $|\mathcal{R}| < 1$. This is so for a *passive surface* (one with no sound sources on its $-y$ side) that produces a reflected wave only when an incident wave is present.

The expression $\frac{1}{2}|\hat{f}|^2(\cos\theta_I)/\rho c$ gives the energy carried per unit time by the incident wave into the surface S_0 (per unit area of S_0), while the same quantity multiplied by $|\mathcal{R}|^2$ gives the energy carried away per unit time and area by the reflected wave. Thus, Eq. (6) yields the following principle: On a time-averaged basis, the acoustic energy incident equals the acoustic energy reflected plus the acoustic energy absorbed. The fraction absorbed is the *absorption coefficient* $\alpha(\theta_I, \omega)$; its value is here $(p\,v_{\text{in}})_{\text{av}}$ divided by $\frac{1}{2}|\hat{f}|^2(\cos\theta_I)/\rho c$ or, equivalently, is $1 - \rho_E$, where $\rho_E = |\mathcal{R}|^2$ (*energy reflection coefficient*) is the fraction of incident energy that is reflected.

If the pressure-amplitude reflection coefficient $\mathcal{R}$ is 1, then the expression for the reflected wave given above is such that $\hat{v}_y = 0$ at $y = 0$ and is the same as for reflection from a rigid surface. Since $\mathcal{R} = 1$ corresponds to $|Z| \to \infty$, the infinite specific-acoustic-impedance limit corresponds to a rigid surface. The limit $Z \to 0$ gives $\mathcal{R} = -1$ and requires $\hat{p} = 0$ on S_0 regardless of the value of $\hat{v}_{\text{in}}$, so, in this limit, the surface S_0 is said to be a *pressure-release surface*. [A circumstance discussed further below (Sec. 3-6) in which the latter idealization may be appropriate is when a wave propagating in water reflects from a water-air interface.]

Locally Reacting Surfaces

The pressure-amplitude reflection coefficient $\mathcal{R}$ varies with angle of incidence for surfaces not idealizable as rigid or as pressure-release surfaces, but in some cases, the specific acoustic impedance Z is very nearly independent of angle of incidence.† Such cases include, for example, surfaces of some typical thick and thin porous materials, surfaces of typical porous materials with air backing, with or without stiff impervious covering, with or without spaced supports. The premise would be that, if $Z(\omega)$ is computed from Eq. (4), given θ_I and realistic $\mathcal{R}(\theta_I, \omega)$, the result will be very nearly independent of θ_I for fixed frequency. The value of Z determined from $\mathcal{R}(\theta_I, \omega)$ when $\theta_I = 0$, termed the *normal-incidence surface impedance* (or the specific acoustic impedance of the surface for normal-incidence reflection), thus suffices to determine $\mathcal{R}(\theta_I, \omega)$ via Eq. (4) for any value of θ_I. A consequence is that if Z is finite, $\mathcal{R}(\theta_I, \omega)$ approaches -1 (as for a pressure-release surface) in the limit $\theta_I \to \pi/2$ (grazing incidence).

That Z should be independent of θ_I is consistent with the assumption that the value of v_{in} at a given point on S_0 depends on the acoustic pressure p at only the same point; i.e., pushing the surface at one point does not move it elsewhere. Thus, one can conceive of a *locally reacting surface* on which Eq. (1), $\hat{p} = Z\hat{v}_{\text{in}}$, holds at each and every point with fixed $Z(\omega)$ regardless of the nature of the acoustic field outside the surface. The model allows the possibility of the surface's being curved and, moreover, of Z's varying from point to point along the surface, e.g., a concrete-block wall partially covered with patches of corkboard.

The locally reacting model approximately accounts for passive wall vibrations caused by an external acoustic pressure. It can also approximately account for fluid being forced into, or sucked out of, the pores in the wall (leading to changes in normal fluid velocity on S_0) by pressure fluctuations outside the surface. It ignores the effect pressure at one point may have on fluid velocity at another point on the wall but has considerable advantage in simplicity over models that take explicit account of the mechanical properties of the wall.

Extensive measurements of the frequency dependence of the real and imaginary parts of $\zeta(\omega) = Z(\omega)/\rho c$ for commercial materials that might be idealized as locally reacting have been given by Beranek,‡ and an example from his paper is reproduced here (see Fig. 3-4). (Typically, such materials and backing combinations are stiffness-controlled at sufficiently low frequencies such that the specific acoustic reactance X is large and positive for small ω.) The locally reacting model is also commonly applied to ground surfaces§ (see Fig. 3-5).

† For a review, see P. M. Morse and R. H. Bolt, "Sound Waves in Rooms," *Rev. Mod. Phys.*, **16**:69–150 (1944).

‡ L. L. Beranek, "Acoustic Impedance of Commercial Materials and the Performance of Rectangular Rooms with One Treated Surface," *J. Acoust. Soc. Am.*, **12**:14–23 (1940).

§ T. F. W. Embleton, J. E. Piercy, and N. Olson, "Outdoor Sound Propagation over Ground of Finite Impedance," *J. Acoust. Soc. Am.*, **59**:267–277 (1976); J. E. Piercy, T. F. W. Embleton,

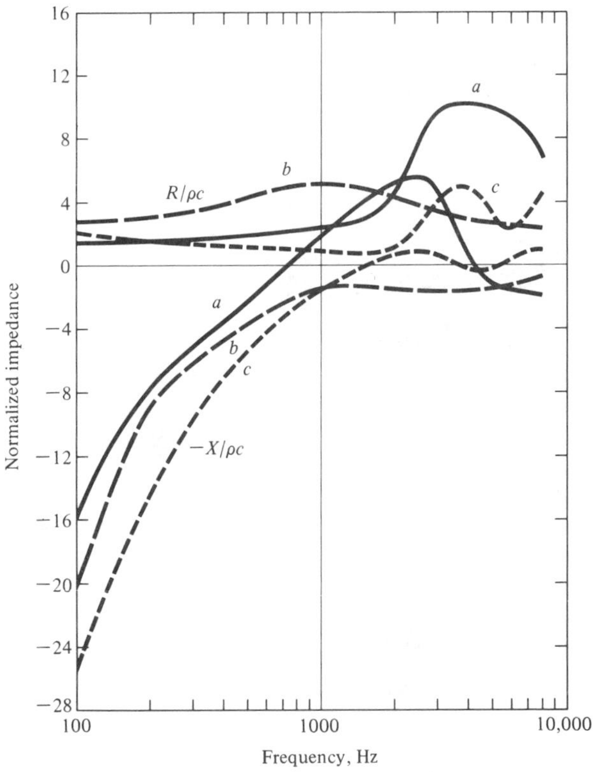

Figure 3-4 Specific acoustic impedance Z of small samples with a rigid wall backing. Plotted are $R/\rho c$ and $-X/\rho c$, where $Z = R + iX$. (*a*) Celotex C-4, 3.2 cm thickness. (*b*) Johns-Manville Permoacoustic, 2.5 cm thickness. (*c*) Johns-Manville Acoustex, 2.2 cm thickness. [*L. L. Beranek, J. Acoust. Soc. Am.*, **12**:*14 (1940)*.]

Theory of the Impedance Tube

Values of $Z(\omega)$ are frequently deduced† from the standing-wave pattern resulting outside a surface when a plane wave is incident upon it. The incident and reflected waves propagate along a cylindrical tube (*impedance tube*) with the

---

and L. C. Sutherland, "Review of Noise Propagation in the Atmosphere," ibid., **61**:1403–1418 (1977); P. J. Dickinson and P. E. Doak, "Measurements of the Normal Acoustic Impedance of Ground Surfaces," *J. Sound Vib.*, **13**:309–322 (1970).

† Detailed specifications for conducting such measurements are given in the ASTM standard C384-58, Impedance and Absorption of Acoustical Materials by the Tube Method, American Society for Testing and Materials, Philadelphia, 1958. The method dates back to J. Tuma (1902), F. Weisbach (1910), Hawley Taylor (1913), and E. T. Paris (1927). The earlier references are cited in E. T. Paris, "On the Stationary Wave Method of Measuring Sound-Absorption at Normal Incidence," *Proc. Phys. Soc. (Lond.)*, **39**:269–295 (1927). An early explicit use of the method to determine impedance rather than absorption coefficient was in W. M. Hall, "An Acoustic Transmission Line for Impedance Measurement," *J. Acoust. Soc. Am.*, **11**:140–146 (1939).

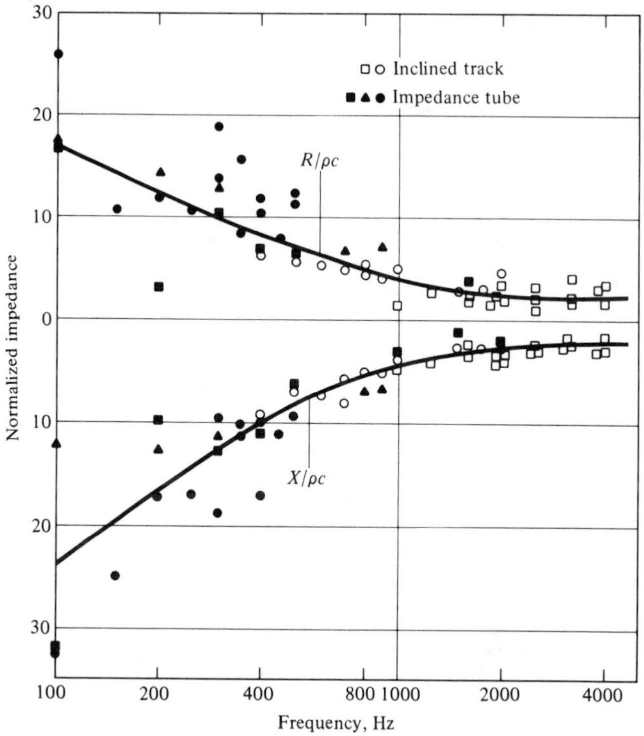

Figure 3-5 Real and imaginary components of the specific acoustic impedance of different samples of grass-covered ground from two sites in Ottawa. The agreement of the inclined-track data (derived from reflection at two different angles of oblique incidence) with impedance-tube data for normal incidence supports the locally reacting hypothesis. [*T. F. W. Embleton, J. E. Piercy, and N. Olson, J. Acoust. Soc. Am.,* **59**:*272* (*1976*).]

sample surface at one end (see Fig. 3-6). The mean squared amplitude of the total acoustic pressure, in accord with Eq. (3*a*), varies with y as

$$(p^2)_{\text{av}} = \tfrac{1}{2}|\hat{f}|^2|1 + \mathcal{R}e^{i2ky}|^2$$
$$= \tfrac{1}{2}|\hat{f}|^2[1 + |\mathcal{R}|^2 + 2|\mathcal{R}|\cos(2ky + \delta_R)] \tag{3-3.7}$$

where δ_R is the phase of $\mathcal{R}$. Thus, $(p^2)_{\text{av}}$ has a maximum of $\tfrac{1}{2}|\hat{f}|^2(1 + |\mathcal{R}|)^2$ whenever $2ky + \delta_R$ is an even multiple of π (so successive maxima are $\tfrac{1}{2}$ wavelength apart); it has its minimum value of $\tfrac{1}{2}|\hat{f}|^2(1 - |\mathcal{R}|)^2$ whenever $2ky + \delta_R$ is an odd multiple of π (so successive minima are also $\tfrac{1}{2}$ wavelength apart). It follows that the ratio s^2 of maximum to minimum values is given by

$$s^2 = \frac{(p^2)_{\text{av,max}}}{(p^2)_{\text{av,min}}} = \frac{(1 + |\mathcal{R}|)^2}{(1 - |\mathcal{R}|)^2} \tag{3-3.8}$$

and that
$$\delta_R = -2ky_{\max,1} + 2m\pi = -2ky_{\min,1} + (2n + 1)\pi \tag{3-3.9}$$

Here $y_{\max,1}$ is the smallest distance y from the surface at which $(p^2)_{\text{av}}$ attains a maximum; $y_{\min,1}$ is the smallest distance at which it attains a minimum. The

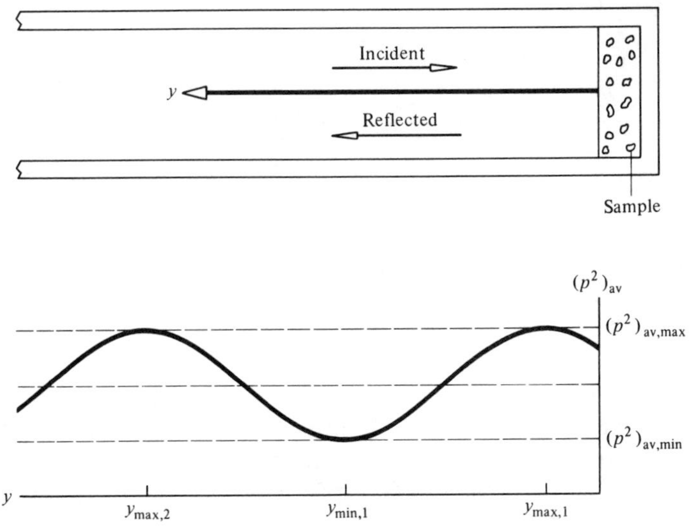

Figure 3-6 Theory of the impedance tube. The incident wave undergoes amplitude change and phase shift when reflected by sample. The resulting interference and reinforcement of reflected and incident waves causes $(p^2)_{av}$ along the tube to have successive maxima and minima whose ratios and locations determine Z.

quantities n and m are arbitrary integers whose values are immaterial insofar as the determination of the real and imaginary parts of the reflection coefficient $\mathcal{R}$ is concerned.

Once $\mathcal{R} = |\mathcal{R}|e^{i\delta_R}$ has been determined from the above equations, the normal-incidence surface impedance can be determined from Eq. (4) (with θ_I set to 0). Thus, for example, if $s^2 = 4$ and $y_{max,1} = \lambda/8$, one has $|\mathcal{R}| = \frac{1}{3}$ and $\delta_R = -\pi/2$, so $\zeta(\omega)$ is $(1 - i/3)/(1 + i/3)$ or $0.8 - 0.6i$.

The plane-wave absorption coefficient α (equal to $1 - |\mathcal{R}|^2$) is found from Eq. (8) to be $4s/(s + 1)^2$. The same relations suffice to determine $|\mathcal{R}|$ and α when the wave pattern results from partial reflection of an obliquely incident (θ_I not 0) plane wave.† In the determination of δ_R from Eqs. (9), however, k should be replaced by $k \cos \theta_I$.

3-4 RADIATION OF SOUND BY A VIBRATING PISTON WITHIN A TUBE

Some key concepts associated with the generation of sound by vibrating bodies are exemplified by the model‡ of a piston (see Fig. 3-7) that fits snugly inside a

† L. Cremer, "Determination of the Degree of Absorption in the Case of Oblique Sound Incidence with the Help of Standing Waves," *Elektr. Nachrichtentech.*, **10**:302–315 (1933).

‡ This example was considered by S. D. Poisson, "Memoir on the Movement of an Elastic Fluid through a Cylindrical Tube, and on the Theory of Wind Instruments," *Mem. Acad. Sci. Paris*, **2**:305–402 (1819). It is also discussed by Rayleigh, *The Theory of Sound*, vol. 2, Dover, 1945, secs. 255–259.

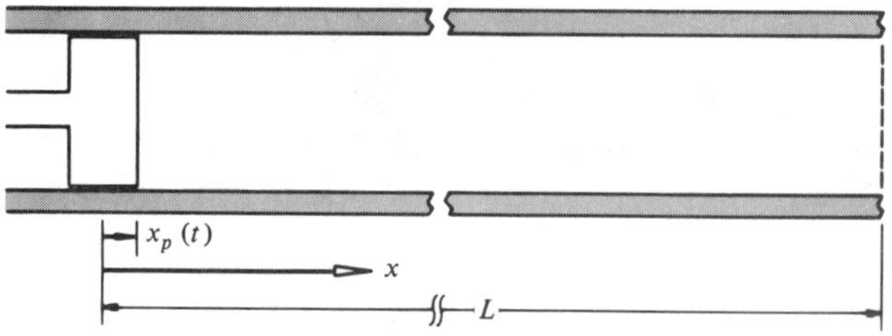

Figure 3-7 Vibrating piston at one end of a rigid-walled tube. The face of the piston at $x_p(t)$ oscillates about $x = 0$.

hollow rigid tube of cross-sectional area A filled with fluid; the piston oscillates back and forth due to some external cause, making sound waves that propagate in the fluid. The $+x$ face of the piston is flat and transverse to the (x) tube axis; the cross section of the tube is independent of x, so the acoustic field in the tube is independent of the other coordinates y and z. (We neglect viscosity and thermal conductivity.)

Inside the tube on the $+x$ side of the piston, the acoustic field variables, satisfying Eqs. (1-5.3), can be taken to be of the form (1-7.4) and (1-7.6) as a superposition of left- and right-traveling plane waves, i.e.,

$$\begin{Bmatrix} p/\rho c \\ v_x \end{Bmatrix} = U(t - c^{-1}x) \pm W(t + c^{-1}x) \tag{3-4.1}$$

where the functions U and W remain to be determined. If the $+x$ face of the piston is oscillating with small amplitude about $x = 0$ so that its position is given by $x_p(t)$, the (approximate) boundary condition (3-1.2) gives $U(t) - W(t)$ for dx_p/dt.

Causality

Other relations relevant to the determination of U and W come from considerations of causality; e.g., the piston's oscillations cause the sound field. If the piston does not start to oscillate until $t = 0$, and if the tube is of length L, the expressions on the right side of Eqs. (1) should be 0 if one has both t less than 0 and x between 0 and L, so $U(\tau) = 0$ if $\tau < 0$ and $W(\tau) = 0$ if $\tau < L/c$. If the far end of the tube is passive, one expects, moreover (causality again), that no disturbance will originate at that end until the wave generated by the piston reaches it. Since $U(t - c^{-1}x)$ does not become nonzero at $x = L$ until $t = L/c$, one accordingly does not expect $W(t + c^{-1}x)$ to become nonzero until $t + c^{-1}x$ exceeds $L/c + c^{-1}L$ or $2L/c$, so $W(\tau)$ is 0 if $\tau < 2L/c$.

The analysis just given allows one to take

$$\frac{p}{\rho c} = v_x = v_p(t - c^{-1}x) \tag{3-4.2}$$

for values of x between 0 and L and for times t up to $(2L - x)/c$, that is, until the echo from the far end of the tube first comes back to x. Here $v_p(t) = dx_p/dt$ is the velocity of the piston at time t, so $v_p(t - x/c)$ is the velocity of the piston at a *retarded time* $t - x/c$ which is x/c earlier than the time at which the acoustic disturbance is currently being sensed at x.

Equations (2) will still describe the acoustic field in the tube at later times if the echo from the far end is weak compared with the primary wave generated by the piston. Attenuating mechanisms (discussed in Chap. 10 of the present text) may cause the amplitude of the generated wave to decrease exponentially as $e^{-\alpha x}$ with increasing propagation distance x, where α is a positive frequency-dependent quantity. If L is sufficiently large to ensure that $e^{-\alpha L} \ll 1$ for all frequencies of interest in the generated wave, the echo will be negligible. Moreover, if $\alpha\lambda/2\pi \ll 1$, and if one limits one's attention to (not large) values of x such that $e^{-\alpha x}$ is not appreciably different from 1, although $e^{-2\alpha L} \ll 1$, Eqs. (2) may still give an adequate description of the acoustic field, even for times larger than $2L/c$. Thus, the concept of an infinitely long tube, while an idealization, applies if there is a small amount of attenuation in a long tube.

A common technique for *anechoic* (without echo) termination is to design the tube and its lining so that the attenuation per unit length increases slowly (to avoid partial reflection) but steadily from a small value near the source end to a large value at the far end such that

$$\exp\left(-2\int_0^L \alpha\, dx\right) \ll 1 \tag{3-4.3}$$

The use of wedges of absorbing material on the walls of *anechoic chambers*† (rooms without echoes) is based on a similar principle.

Tube with Rigid End; Resonance

If the attenuation within the tube is idealized as zero, and if the far end of the tube is a rigid plane reflector, the incident wave of Eqs. (2) upon reflection at $x = L$ gives rise to a similar wave traveling in the $-x$ direction; the pressure in this wave must be $\rho c v_p(t - (2L - x)/c)$ for the sum of the two pressure terms to be symmetric about $x = L$. [This is an application of the *method of images;* replacing $L - x$ by $-(L - x)$ is the same as replacing x by $2L - x$.] This reflected wave in turn reflects at $x = 0$, giving rise to a wave with acoustic pressure $\rho c v_p(t - (2L + x)/c)$, so the fluid-velocity contributions at $x = 0$ from the second and third terms cancel each other. [The solution is such that the first

† L. L. Beranek and H. P. Sleeper, Jr., "Design and Construction of Anechoic Sound Chambers," *J. Acoust. Soc. Am.*, **18**:140–150 (1946); W. Koidan, G. R. Hruska, and M. A. Pickett, "Wedge Design for National Bureau of Standards Anechoic Chamber," ibid., **52**:1071–1076 (1972).

term alone satisfies the boundary condition at $x = 0$ of $v_x = v_p(t)$, so the sum of all successive terms must give a contribution to v_x that vanishes at $x = 0$.] If one extends the reasoning just described,† whereby each reflected wave successively generates another reflected wave at the opposite end of the tube, the net result is

$$v_x = v_p\left(t - \frac{x}{c}\right) - v_p\left(t - \frac{2L - x}{c}\right) + v_p\left(t - \frac{2L + x}{c}\right) - v_p\left(t - \frac{4L - x}{c}\right) + v_p\left(t + \frac{4L + x}{c}\right) - \cdots \quad (3\text{-}4.4)$$

so, with reference to Eqs. (1), one identifies

$$U\left(t - \frac{x}{c}\right) = \sum_{n=0}^{\infty} v_p\left(t - \frac{x}{c} - \frac{2nL}{c}\right) \quad (3\text{-}4.5a)$$

$$W\left(t + \frac{x}{c}\right) = \sum_{m=1}^{\infty} v_p\left(t + \frac{x}{c} - \frac{2mL}{c}\right) \quad (3\text{-}4.5b)$$

Note that the lower limits, $n = 0$ and $m = 1$, on the two sums are different; $n = 0$ corresponds to the primary wave. The various terms in the above sums do not become nonzero until t is sufficiently large for their arguments to be positive, so there are only a finite number of nonzero terms in the sum for any finite value of t.

If the end at $x = L$ is a *pressure-release surface,* instead of a rigid surface, the same analysis applies except that additional factors of $(-1)^n$ and $(-1)^m$ should multiply the terms of Eqs. (5*a*) and (5*b*). The pressure-release surface is an approximate boundary condition for a narrow (diameter small compared to wavelength) open-ended tube protruding into an unbounded space; a classic application is the upper end of an organ pipe.‡

Resonance arises when the successive echoes reinforce the pressure on the piston face. Suppose the piston velocity is 0 up to $t = 0$ and thereafter is periodic with a period equal to the round-trip time $2L/c$. Then $v_p(t - 2nL/c)$ is equal to $v_p(t)$ if $n < ct/2L$ or is equal to 0 if $n > ct/2L$; so one has, from Eqs. (1)

† The application of the method of images to account for multiple reflections of plane waves in tubes is described by L. Euler in his "On the Propagation of Sound," 1766; trans. in Lindsay, *Acoustics,* pp. 136–154.

‡ Daniel Bernoulli, "Physical, Mechanical, and Analytical Researches on Sound and on the Tones of Differently Constructed Organ Pipes," 1762; J. L. Lagrange, "New Researches on the Nature and the Propagation of Sound," 1762; L. Euler, "More Detailed Enlightenment on the Generation and Propagation of Sound and on the Formation of Echoes," 1767. A synopsis of these papers is given by C. A. Truesdell, "The Theory of Aerial Sound, 1687–1788," in *Leonhardi Euleri Opera Omnia,* ser. 2, vol. 13, Orell Füssli, Lausanne, 1955, pp. LI–LXIII. (The validity of this boundary condition is discussed in Sec. 7-6.)

and (5),

$$(p)_{x=0} = \rho c[1 + 2N(t)]v_p(t) \tag{3-4.6}$$

where $N(t)$ is the largest integer less than $ct/2L$ or, equivalently, the total number of echoes returned to the piston within time t. For such periodic motion of the piston, the pressure at $x = 0$ is always in phase with the velocity and moreover has an amplitude increasing stepwise in time, so the acoustic power output pv_xA of the piston tends on the average to increase linearly with time. Thus, the acoustic energy (equal to the time integral of the input power) stored in the tube by time $t = 2L(N + 1)/c$ is

$$E = 2\rho AL(v_p^2)_{\text{av}} \sum_{n=0}^{N} (1 + 2n) \tag{3-4.7}$$

Since the indicated sum on n is $(N + 1)^2$ or $(ct/2L)^2$, the acoustic energy tends to increase quadratically with time.† Both the acoustic power output by the source and the stored energy increase without bound unless some account is taken of dissipative processes.

Because a function with period $2L/mc$ (with m a positive integer) automatically repeats itself at intervals of $2L/c$, the above analysis holds if the repetition period of $v_p(t)$ is $2L/mc$, so if $v_p(t)$ is a sinusoidal function of time, the frequencies f_m (in hertz) at which resonance will occur are $f_m = mc/2L$ for $m = 1, 2, 3, \ldots$. The lowest resonant frequency (corresponding to $m = 1$) is when $L = \lambda/2$. If the end at $x = L$ is a pressure-release surface (approximately the case for a narrow hollow tube protruding into an open space), the resonance criterion is that $(-1)^n v_p(t - 2nL/c)$ equal $v_p(t)$ for all $n < ct/2L$. This will be so if $v_p(t)$ is oscillating sinusoidally at resonance frequencies $f_m = (m + \frac{1}{2})(c/2L)$ for $m = 0, 1, 2, 3, \ldots$. This follows because $\sin[(2\pi f_m)(t - 2nL/c)]$ is $\sin[2\pi f_m t - (2m + 1)n\pi]$ or $(-1)^{(2m+1)n} \sin 2\pi f_m t$. This in turn reduces to $(-1)^n \sin 2\pi f_m t$. The resonance frequencies $mc/2L$ for the tube with two rigid ends do not occur when one end is a pressure-release surface since contributions to the pressure at the piston from successive echoes cancel each other when the piston is driven at such frequencies.

Constant-Frequency Oscillations

Any damping mechanism attenuates transients, so that if a source is set into motion with a periodic vibration, the acoustic field variables eventually oscillate with the same repetition period. We demonstrate this for the example just

† This is analogous to the result for an undamped harmonic oscillator driven at its resonance frequency, whereby the particular solution describing motion starting from rest has an amplitude increasing linearly with time. See, for example, L. Meirovitch, *Elements of Vibration Analysis*, McGraw-Hill, New York, 1975, pp. 45–46.

discussed of an oscillating piston in a tube. The velocity $v_p(t)$ is taken to be 0 for $t < 0$ and to be $V_0 \cos \omega t$ for $t > 0$, where the angular frequency ω is not necessarily an integral multiple of $\pi c/L$.

If any weak damping mechanism is taken into account, one can expect the solution given by Eqs. (1) and (5) to be qualitatively correct, except that terms corresponding to very high order echoes may have suffered a large attenuation and phase shift. For larger values of n, an appropriate replacement† of terms such as $v_p(t \pm x/c - 2nL/c)$ in Eqs. (5) is $e^{-\beta n} v_p(t \pm x/c - 2nL/c + n\,\Delta\phi)$ where β and $\Delta\phi$ are small constants but βn and $n\,\Delta\phi$ are not necessarily small. The premise here is that the net attenuation and phase shift suffered during successive round trips are the same. With such a substitution, $U(t - x/c)$ in Eq. (5a) becomes

$$U_{\text{damp}}(t, x) = \text{Re}\left(V_0 e^{-i\omega t} e^{ikx} \sum_{n=0}^{N} \psi^n\right) \tag{3-4.8}$$

where we use the abbreviation $\psi = e^{i2kL}e^{-\beta}e^{-i\Delta\phi}$ and where N is the largest integer less than $(ct - x)/2L$.

The sum over n in the above is $(1 - \psi^{N+1})/(1 - \psi)$, which is nearly $1/(1 - \psi)$ in the limit $e^{-\beta N} \ll 1$ or, equivalently, when $t \gg 2L/c\beta$. Also, unless kL is very close to a multiple of π, the factor $1/(1 - \psi)$ for smaller values of β and $\Delta\phi$ is essentially the same as would be obtained if β and $\Delta\phi$ were set to 0. Thus, in the limit of large t, $U_{\text{damp}}(t, x)$ reduces to

$$V_0 \,\text{Re}\, \frac{e^{-i\omega t}e^{ikx}}{1 - e^{i2kL}} = \frac{V_0}{2} \frac{\sin \omega(t - x/c + L/c)}{\sin kL} \tag{3-4.9}$$

Similarly, the analogous version with damping included of the sum in Eq. (5b) has a limit given by the above but with $-x/c$ replaced by $+x/c$. Consequently, with the aid of the trigonometric identity for $\sin(A + B)$, one has

$$\begin{Bmatrix} p/\rho c \\ v_x \end{Bmatrix} \approx V_0 \frac{\begin{matrix}\sin \\ \cos\end{matrix}(\omega t) \begin{matrix}\cos \\ \sin\end{matrix}(k(L - x))}{\sin kL} \tag{3-4.10}$$

for the asymptotic (steady-state) solution. The expression for v_x reduces to $V_0 \cos(\omega t)$ at $x = 0$ in accord with the boundary condition $v_x = v_p(t)$ at $x = 0$.

The steady-state solution, while not appreciably affected in mathematical form by the presence of damping, depends on the existence of damping for its eventual asymptotic emergence as the dominant response to a periodic excitation.‡ Since p and v_x are everywhere 90° out of phase in this asymptotic solu-

† If the only attenuation mechanism were viscous drag at the tube walls, approximate values for β and $\Delta\phi$ would be $2L(\omega\mu/8\rho c^2)^{1/2}L_P/A$ and its negative. Here L_P is the perimeter of the tube and μ the viscosity (see Sec. 10-5).

‡ This assertion is commonly proved in texts on mechanical vibrations or electric-circuit theory for a spring-mass-dashpot system or an RLC circuit. See, for example, J. P. Den Hartog, *Mechanical Vibrations,* 4th ed., McGraw-Hill, New York, 1956, p. 54.

tion, the actual acoustic power supplied to the tube by the oscillating piston, once the steady-state field is realized, is small if the damping is weak.

Resonance is manifested by Eqs. (10) because p and v_x become singular when $\sin kL$ is 0. If damping is taken into account, the acoustic amplitudes at such frequencies (where kL is a multiple of π) will be large but not singular. A prediction of the actual magnification can be made by carrying through the derivation leading to Eq. (9) without approximating $1/(1 - \psi)$ by $1/(1 - e^{i2kL})$.

An implication of Eqs. (10) is that at any frequency near a resonance frequency $f_m = mc/2L$ (where $k_m = \pi m/L$), p is $P \sin 2\pi ft \cos (m\pi x/L)$ approximately, where P is independent of x and t. This, however, for given P and with $f = f_m$, corresponds to a solution with constant frequency of the linear acoustic equations that could exist within the tube if both ends were closed by rigid planes, so that $\partial p/\partial x = 0$ at both $x = 0$ and $x = L$. This is accordingly a *natural acoustic motion of constant frequency,* which in the absence of damping does not require a source for its maintenance. Such natural constant-frequency disturbances are referred to as *modes* and occur only for certain discrete frequencies (the $f_m = mc/2L$ in this instance) termed *natural frequencies.* The analysis illustrates two general principles: (1) the resonance frequencies are the same as the natural frequencies, and (2) the spatial dependence of the acoustic field when driven at a frequency close to a resonance frequency is nearly the same as that of the corresponding natural mode.

The resonance frequencies and associated mode shapes are found by assuming $e^{-i\omega t}$ time dependence at the outset and then solving the eigenvalue problem posed by the Helmholtz equation and the appropriate boundary conditions at $x = 0$ and $x = L$. For example, if the end at $x = 0$ is rigid and that at $x = L$ is a pressure-release surface, one has

$$\frac{d^2\hat{p}}{dx^2} + k^2\hat{p} = 0 \tag{3-4.11a}$$

$$\frac{d\hat{p}}{dx} = 0 \quad \text{at } x = 0 \qquad \hat{p} = 0 \quad \text{at } x = L \tag{3-4.11b}$$

The differential equation and the $x = 0$ boundary condition are satisfied if $\hat{p}(x) = P \cos kx$, where P is any constant. Only for certain discrete values (*eigenvalues*) of k can a nontrivial ($\hat{p}$ not identically 0) solution be found that satisfies both boundary conditions; the k_m are such that $\cos k_m x = 0$ at $x = L$, so $k_m L$ should be an odd multiple of $\pi/2$. Since $k_m = 2\pi f_m/c$, one accordingly concludes that $f_m = (c/4L)(2m + 1)$, where m is an integer. The *mode shapes* (*eigenfunctions*) are given by $P_m \cos [(2m + 1)\pi x/2L]$. There are $m + 1$ *pressure nodes* (including that at $x = L$) representing values of x at which $\hat{p}(x) = 0$ and $m - 1$ *pressure antinodes* (including that at $x = 0$) at which $d\hat{p}/dx = 0$.

Tube with Impedance Boundary Condition at End

The steady-state acoustic field generated by a piston with velocity $V_0 \cos \omega t$ can be derived directly by taking $U(t) = \text{Re } ae^{-i\omega t}$ and $W(t) = \text{Re } be^{-i\omega t}$

in Eqs. (1) and subsequently choosing the constants a and b such that the boundary conditions at the ends of the tube are met. The derivation† is carried through here with the end at $x = L$ characterized by a specific acoustic impedance Z. We write Eqs. (1) in the form

$$\begin{Bmatrix} p/\rho c \\ v_x \end{Bmatrix} = \text{Re}\,[e^{-i\omega t}(ae^{ikx} \pm be^{-ikx})] \tag{3-4.12}$$

Then, since $\hat{v}_x = V_0$ at $x = 0$ and since $\hat{p}/\hat{v}_x = Z$ at $x = L$, one has

$$a - b = V_0 \qquad (Z - \rho c)ae^{ikL} = (Z + \rho c)be^{-ikL} \tag{3-4.13}$$

Thus, with some algebra, it follows that

$$\hat{p} = \rho c V_0 \frac{Z \cos k(L - x) - i\rho c \sin k(L - x)}{\rho c \cos kL - iZ \sin kL} \tag{3-4.14}$$

Note that Re $\hat{p}e^{-i\omega t}$ reduces to the expression in Eqs. (10) (for tube with rigid end) in the limit of large $|Z/\rho c|$. Also, the above expression is identical to that appropriate to the normal-incidence ($\theta_I = 0$) reflection of a plane wave from a wall with impedance Z. One can obtain Eq. (14) from Eq. (3-3.3a) by replacing the symbols θ_I, y, and $\hat{v}_y$ by 0, $L - x$, and $-\hat{v}_x$ and choosing the $\hat{f}$ in (3-3.3a) so that $\hat{v}_x = V_0$ at $y = L$; that is, $\hat{f} = ae^{ikL}$ is the complex amplitude of the net incident wave on the far end ($x = L$) of the tube. The amplitude-reflection coefficient $\mathcal{R}$ [given, according to Eq. (3-3.4), by $(Z - \rho c)/(Z + \rho c)$] is the same as be^{-ikL}/ae^{ikL}.

The Q of a Resonance

The above solution exemplifies the behavior of an acoustic system driven near a resonance frequency. For simplicity, we consider the case when the end at $x = L$ is "nearly rigid," so $|Z| \gg \rho c$; we accordingly anticipate resonant behavior near any angular frequency $\omega_n^0 = n\pi c/L$ with n an integer. If both numerator and denominator in Eq. (14) are divided by $Z \cos kL$ and terms of higher than first order in either $\rho c/Z$ or $\omega - \omega_n^0$ are discarded in the denominator and terms of higher than zero order are discarded in the numerator, the result (with some algebra) is

$$\hat{p} \approx \frac{2Q_n}{k_n^0 L} \rho c V_0 \frac{\cos k_n^0 x}{1 - i2Q_n\,\Delta\omega/\omega_n^0} \tag{3-4.15}$$

with $$Q_n = \frac{k_n^0 L(R^2 + X^2)}{2\rho c R} \qquad \Delta\omega = \omega - \omega_n^0 + \frac{(\rho c^2/L)X}{X^2 + R^2} \tag{3-4.16}$$

Here R and X are the real and imaginary parts of Z and are evaluated at ω_n^0. The approximate expression for $\hat{v}_x$ is similar to that of Eq. (15), but $\rho c V_0$ should be

† E. T. Paris, "On Resonance in Pipes Stopped with Imperfect Reflectors," *Phil. Mag.*, 4:907–917 (1927).

replaced by V_0 and $\cos k_n^0 x$ should be replaced by $i \sin k_n^0 x$. Note that Eq. (15) is not valid near points where $\cos k_n^0 x = 0$ (nominal locations of *pressure nodes*), while the equation for $\hat{v}_x$ is not valid near points where $\sin k_n^0 x = 0$ (nominal locations of *antinodes*). Given the previously stated assumption that $|Z| \gg \rho c$, both $Q_n/k_n^0 L$ and Q_n are much larger than 1.

A principal implication of Eq. (15) is that for any fixed value of x (other than a pressure node) and for fixed piston velocity amplitude V_0, one has, for variable but small $\Delta\omega$ (see Fig. 3-8),

$$(p^2)_{\text{av}} \simeq \frac{(p^2)_{\text{av,max}}}{1 + (2Q_n \, \Delta\omega/\omega_n)^2} \tag{3-4.17}$$

where $(p^2)_{\text{av,max}}$ is the maximum mean squared pressure for frequencies in the vicinity of ω_n^0. The frequency at which the maximum is obtained is that at which $\Delta\omega = 0$, that is, approximately at ω_n^0. The above indicates that $(p^2)_{\text{av}}$ drops to one-half of its resonant value and the sound-pressure level drops by 3 dB when $|\Delta\omega| = \omega_n/2Q_n$ or $|\Delta f| = f_n/2Q_n$. The quantity f_n/Q_n is accordingly the frequency width Δf of the resonance peak measured between its *half-power points,* i.e., where $(p^2)_{\text{av}} = \frac{1}{2}(p^2)_{\text{av,max}}$. For such resonance peaks, a *quality factor* Q can

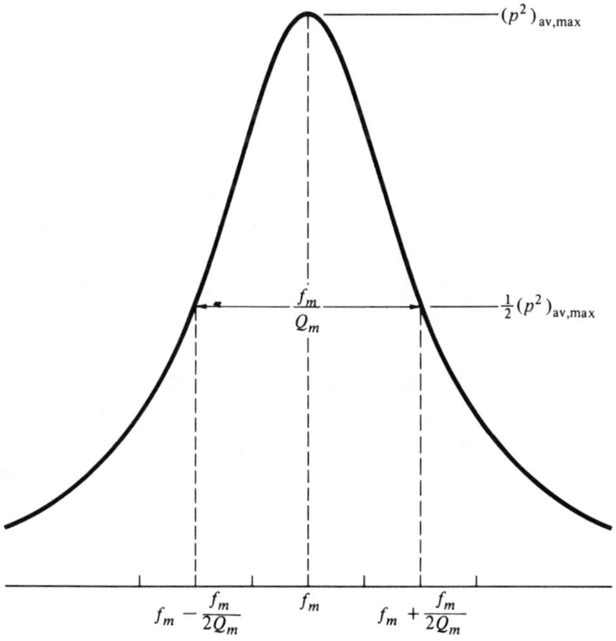

Figure 3-8 Sketch of a resonance peak in the frequency response of a system driven at constant frequency. Plotted is $(p^2)_{\text{av}}$ at a typical point for frequencies near the mth resonance frequency f_m. Peak drops to one-half maximum value at $f_m \pm f_m/2Q_m$, where Q_m is the quality factor for the resonance.

be defined† as $f_n/\Delta f$, resonance frequency divided by bandwidth between half-power points. Thus, the Q in the example above is the Q_n given by Eq. (16).

An alternate definition of the Q associated with a resonance is the energy within the system divided by the average energy lost per radian when the system is vibrating at a resonance frequency. In the steady state, the average energy loss per unit time is the same as the average power $\mathcal{P}_{\text{av}}$ supplied by the source. A radian corresponds to a time increment of $1/\omega$, so the energy loss per radian is $\mathcal{P}_{\text{av}}/\omega$. Thus, if ω_n is a resonant frequency, one should have

$$Q_n = \omega_n \frac{E_{\text{av}}}{\mathcal{P}_{\text{av}}} \qquad \text{(3-4.18)}$$

We here show that Eq. (18) is consistent with Eq. (16) for the example discussed in the preceding paragraphs. The average energy per unit volume is $\frac{1}{4}|\hat{p}|^2/\rho c^2 + \frac{1}{4}\rho|\hat{v}_x|^2$, where $\hat{p}$ is given by Eq. (15) with $\Delta\omega = 0$ and $\hat{v}_x$ is given as described in the discussion following Eq. (15). This yields $(Q_n/k_n^0 L)^2\,(\rho V_0^2)AL$ for the time-averaged acoustic energy E_{av} within the tube. The time-averaged power is $\frac{1}{2}$ Re $(\hat{p}\,\hat{v}_x^*)A$ evaluated at any value of x. Although the approximate expression (15) and its counterpart for $\hat{v}_x$ indicate that $\hat{p}$ and $\hat{v}_x$ are 90° out of phase, this is not exactly the case and $\mathcal{P}_{\text{av}}$ is not zero; it is only small. A good approximation for $\mathcal{P}_{\text{av}}$ results from using $\hat{v}_x = V_0$ at $x = 0$, so $\mathcal{P}_{\text{av}}$ is $\frac{1}{2}V_0 A$ times Re $\hat{p}$ at $x = 0$, or $(Q_n/k_n^0 L)\rho c V_0^2 A$, or $ck_n^0 E_{\text{av}}/Q_n$, where $\hat{p}$ is taken from Eq. (15). Since $ck_n^0 \approx \omega_n$, Eq. (18) results.

3-5 SOUND RADIATION BY TRAVELING FLEXURAL WAVES

As a second example of plane-wave generation‡ by a vibrating solid, we consider a wall consisting of a large plate (idealized as infinite) whose right face nominally is flush with the $y = 0$ plane but which is undergoing transverse vibrations (see Fig. 3-9). Thus, a given point on the plate's face has y coordinate $\eta(x, z, t)$, which if positive, represents the displacement of that portion of the plate to the right (toward $y > 0$).

A given displacement field $\eta(x, z, t)$ can be represented via a triple Fourier transform as a superposition of traveling transverse waves, and since the acoustic disturbance due to the overall vibration is a superposition of acoustic waves caused by the individual transverse waves, it is sufficient (as an initial step for analysis) to limit one's attention to a single traveling wave. Thus, we consider

† The definitions given here are consistent with those given in *IEEE Standard Dictionary of Electrical and Electronics Terms*, Wiley-Interscience, New York, 1972, p. 453.

‡ J. Brillouin, "Problems of Radiation in the Acoustics of Buildings," *Acustica*, **2**:65–76 (1952). The method of analysis dates back to Green, "On the Reflexion and Refraction of Sound," 1838. The closely related problem of radiation by flexural waves, periodic along axis, on a transversely oscillating cylinder of infinite length was analyzed by A. Kalähne, "The Wave Motion about a Transversely Vibrating String in an Unbounded Fluid," *Ann. Phys.*, (4)**45**:657–705 (1914).

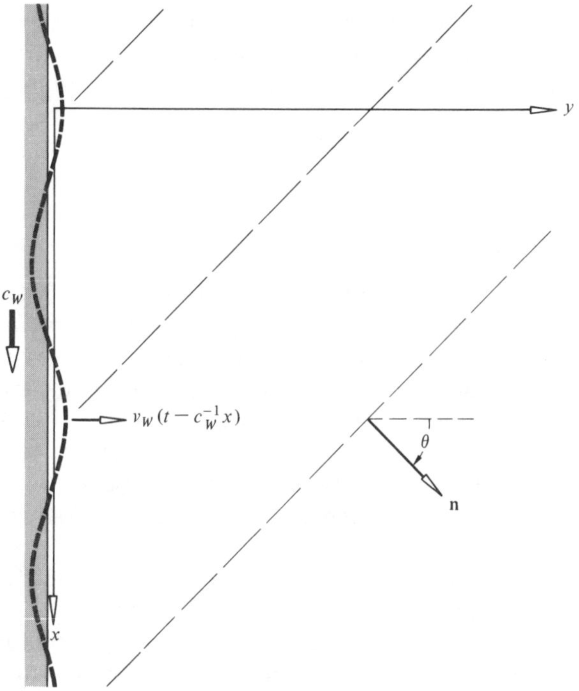

Figure 3-9 Sound radiation by flexural wave moving along a wall with supersonic speed c_W. Wall moves in y direction with velocity $v_W(t - c_W^{-1}x)$ and generates plane waves propagating at angle θ. If the flexural-wave speed is subsonic, the disturbance (for constant-frequency excitation) dies out exponentially with y.

the special case when η is such that $\partial\eta/\partial t = v_W(t - c_W^{-1}x)$, where $|v_W| \ll c$. Here the wall (or plate) normal vibrational velocity v_W (W being an abbreviation for wall) is independent of z and depends on t and x only through the combination $t - c_W^{-1}x$; the function $v_W(t - c_W^{-1}x)$ represents a transverse wave (flexural wave) moving in the $+x$ direction with the *flexural-wave speed* c_W and without change of form. (If the flexural wave is a natural wave motion of the plate, the only such wave moving without change of form is one of constant frequency, but this restriction need not be taken into account at present.) As described below, the nature of the acoustic disturbance in the fluid depends critically on whether the flexural wave is moving at supersonic ($c_W > c$) or subsonic ($c_W < c$) speed.

Sound Generated by Supersonic Flexural Waves

If $c_W > c$, the steady-state solution of the linear acoustic equations satisfying the boundary condition $v_y = v_W(t - c_W^{-1}x)$ at $y = 0$, corresponding to the notion (causality again) that the sound is actually caused by the vibrating surface, and neglecting reflections from distant walls or surfaces on the far $+y$ side of the plate, is a plane wave. To demonstrate this, we consider a plane traveling-wave solution (propagating at any angle θ with the y axis) of the linear acoustic equations of the form $p = f(t - \mathbf{n} \cdot \mathbf{x}/c)$, $\mathbf{v} = \mathbf{n}p/\rho c$. The z independence of the boundary conditions suggests that $\mathbf{n}$ has no z component, so we set $\mathbf{n} = n_x\mathbf{e}_x + n_y\mathbf{e}_y$, where $n_x = \sin\theta$ and $n_y = \cos\theta$ are the x and y components of $\mathbf{n}$. Then the boundary condition $v_y = v_W(t - c_W^{-1}x)$ at $y = 0$ is satisfied by

$$\frac{c}{n_x} = \frac{c}{\sin\theta} = c_W \qquad f(t) = \frac{\rho c}{n_y} v_W(t) \tag{3-5.1}$$

Trace-Velocity Matching Principle

If any function, such as $f(t - \mathbf{n}\cdot\mathbf{x}/c)$ above, depends on t and x in the combination $t - v_{\text{tr}}^{-1}x$, where v_{tr} is some constant, one says that v_{tr} is the *trace velocity* corresponding to the x direction. If a line of microphones or sensors were placed parallel to the x axis so that each had the same y and z coordinates, the relation between the signals received by the various sensors could be interpreted as if the disturbance were moving in the x direction with speed v_{tr} (see Fig. 3-10). The actual disturbance might in reality be moving at an angle with the x axis and, if it is a plane wave, its speed in the direction of propagation will be less than v_{tr}.

The trace-velocity matching principle† states that, under steady-state circumstances, the trace velocity of effect equals the trace velocity of the cause. If a disturbance has t and x dependence only in the combination $t - v_{\text{tr}}^{-1}x$, and if this causes other disturbances, they should also depend on t and x in the same combination. This presumes that the governing equations are unchanged if one changes the time origin and the spatial origin such that $t \to t + \Delta t$, $x \to x + v_{\text{tr}}\,\Delta t$ for arbitrary Δt; that is, the governing equations and boundary conditions must have an invariance under time and x-direction translations. In the present example, this is guaranteed because the linear acoustic equations are the same regardless of the choice of time and spatial origins and because the interface between the vibrating solid and the fluid is nominally flat and parallel to the x axis. The cause (the wall vibrations) has trace velocity c_W along the x direction, so the trace velocity c/n_x of the effect (the radiated sound wave) must also be c_W.

Outgoing versus Incoming Waves

In the solution represented by Eq. (1), there are two possible choices for n_y. Since $\mathbf{n}$ is a unit vector, one has $n_x^2 + n_y^2 = 1$, and thus $n_y = \pm[1 - (c/c_W)^2]^{1/2}$. The plus sign, leading to a plane wave propagating obliquely away from the plate, is a plausible choice since it agrees with the notion that a wave should

† An early explicit stating of this is given by Rayleigh, *The Theory of Sound*, vol. 2, sec. 270. The term "trace matching" is also used to denote the related phenomenon by which matching the trace velocity of an incident wave with the propagation velocity of a free wave in a wall tends to reduce the transmission loss of a wall (L. Cremer, M. Heckl, and E. E. Ungar, *Structure-Borne Sound*, Springer-Verlag, New York, 1973, p. 409).

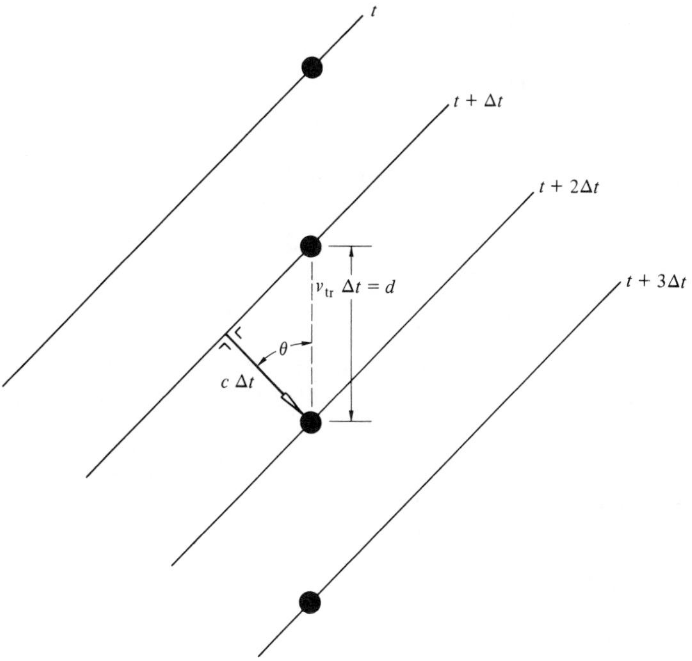

Figure 3-10 Plane-wave passage past linear array of microphones. Trace velocity v_{tr} is distance d between microphones divided by time lapse Δt for reception of given wave feature. The sketch indicates that $v_{tr} = c/(\cos\theta)$.

propagate away from rather than toward its source. There do exist,† among other physical categories of wave propagation, counterexamples to this notion, but here the choice of the plus sign also leads to an I_y that is everywhere positive. Thus, if we want a solution in which acoustic energy (as well as the wave itself) propagates away from the source, $n_y > 0$ is required. Two other methods of substantiating this choice may also be mentioned. First, one can solve a modified version of the linear acoustic equations in which a damping mechanism‡ (causing internal loss of acoustic energy) is introduced. It is sufficient to consider v_W as a sinusoidal function of its argument and to take the acoustic variables as being the real parts of complex spatially dependent amplitudes times $e^{-i\omega t}$. Then, although the source is not explicitly considered to be bounded in duration and spatial extent (with the steady-state idealization of a

† For counterexamples, see S. H. Crandall, "Negative Group Velocities in Continuous Structures," *J. Appl. Mech.*, **24**:622–623 (1957); H. Lamb, "On Group Velocity," *Proc. Lond. Math. Soc.*, (2)**1**:473–479 (1903–1904).

‡ J. W. S. Rayleigh, "On Progressive Waves," *Proc. Lond. Math. Soc.*, **9**:21–26 (1877); reprinted as an appendix to vol. 1 of the Dover edition of *The Theory of Sound;* H. Lamb, *Hydrodynamics,* 6th ed., 1932, Dover, New York, 1945, pp. 399, 413.

traveling flexural wave), the wave far from the plate should die out in amplitude with large y. One discards a possible wave that grows with increasing distance as being unphysical and then examines the resulting solution in the limit as the damping goes to zero. This results in just the $n_y > 0$ wave. A second method is to solve a transient problem in which the plate is completely at rest at an early time t_0 and then starts (gradually growing in amplitude) after that time to vibrate so that $\partial\eta/\partial t$ is of the form $v_W(t - c_W^{-1}x)$. The wave field is required initially to be zero everywhere, and it evolves gradually after the source has been turned on. At late times, the acoustic field in the vicinity of the vibrating portions of the plate resembles the physically realistic steady-state solution. The procedure just described can be formally carried through by Fourier transform techniques; the asymptotic steady-state solution at finite y (the transient radiates away) is the same as what results from the considerations previously mentioned.

The solution for acoustic waves generated by supersonic ($c_W > c$) flexural waves moving along a plate can be summarized as

$$p = \rho c_W v_x = \frac{\rho c}{n_y} v_W(t - \mathbf{n} \cdot \mathbf{x}/c) \tag{3-5.2a}$$

$$v_y = v_W\left(t - \mathbf{n} \cdot \frac{\mathbf{x}}{c}\right) \qquad n_x = \frac{c}{c_W} \qquad n_y = \left[1 - \left(\frac{c}{c_W}\right)^2\right]^{1/2} \tag{3-5.2b}$$

The intensity in the acoustic field is $p\mathbf{v}$, or $p^2\mathbf{n}/\rho c$ since the disturbance is a plane traveling wave. With p as given above, one accordingly has

$$\mathbf{I} = \frac{\rho c}{n_y^2}\left[v_W\left(t - \mathbf{n} \cdot \frac{\mathbf{x}}{c}\right)\right]^2 \mathbf{n} \tag{3-5.3}$$

The energy radiated per unit time by the vibrating plate per unit area of its surface is $pv_y = I_y$, evaluated at $y = 0$, or $(\rho c/n_y)v_W^2$, where v_W is evaluated at $t - c_W^{-1}x$. In the limit $c_W \to \infty$, v_W is independent of x, and the plate is moving back and forth as a unit, so the solution reduces to that of the example discussed previously of a piston in a long tube. However, when c_W decreases to near the sound speed c in the fluid, $n_y \to 0$ and p, $\mathbf{I}$, and the radiated acoustic power per unit area become large. The infinite limit cannot be realized because, among other reasons, the generation of acoustic energy must result in a decrease of the vibrational energy in the plate.

Acoustic Disturbances Created by Subsonic Flexural Waves

When $c_W < c$ (subsonic flexural wave), the plane-wave solution described above is inapplicable because it would require n_y to be imaginary, but the trace-velocity matching principle still applies. If one limits oneself to flexural waves of constant frequency (a building block for more general cases) such that

$v_W(t - c_W^{-1}x)$ is of the form $V_0 \cos(\omega t - \omega c_W^{-1}x)$, the boundary condition at the plate is satisfied if one sets

$$v_y = V_0 \cos(\omega t - \omega c_W^{-1}x)F(y) \tag{3-5.4}$$

where $F(y)$ is 1 at $y = 0$. This above expression, representing a cartesian component of **v**, should satisfy the wave equation and (since the latter is separable in a cartesian coordinate system) one finds that it does, provided $F(y)$ satisfies the ordinary differential equation

$$\frac{d^2F}{dy^2} - \left(\frac{\omega}{c}\right)^2 \beta^2 F = 0; \qquad \text{where } \beta = \left[\left(\frac{c}{c_W}\right)^2 - 1\right]^{1/2} \tag{3-5.5}$$

This equation has linearly independent solutions that grow or die out exponentially with increasing y. Since the medium is here idealized as being unbounded on the right ($+y$ side), we discard the former as unphysical and consequently obtain $e^{-(\omega/c)\beta y}$ for $F(y)$.

The acoustic pressure is found from expression (4) for v_y, in conjunction with the trace-velocity matching principle, and from the y component of Euler's equation of motion. (We rule out any term not having the same y dependence as v_y, since such a term that satisfied the conditions just stated would not also satisfy the wave equation.) The x component of **v** is similarly found from the expression for p, from the trace-velocity matching principle, and from the x component of Euler's equation of motion. In this manner, one obtains a wave field of the form

$$p = \rho c_W v_x = -\rho c V_0 \beta^{-1} \sin(\omega t - \omega c_W^{-1}x)e^{-(\omega/c)\beta y} \tag{3-5.6a}$$

$$v_y = V_0 \cos(\omega t - \omega c_W^{-1}x)e^{-(\omega/c)\beta y} \tag{3-5.6b}$$

Such a wave disturbance of constant frequency, propagating in one direction but decaying exponentially in another, is an *inhomogeneous plane wave*.†

The acoustic-energy implications of the above solution are

$$w_{\text{av}} = \tfrac{1}{2}\rho V_0^2 \left(\frac{c/c_W}{\beta}\right)^2 e^{-2(\omega/c)\beta y} \tag{3-5.7a}$$

$$I_{x,\text{av}} = c_W w_{\text{av}} \qquad I_{y,\text{av}} = 0 \tag{3-5.7b}$$

where w is the acoustic energy per unit volume given by Eq. (1-11.3); the time averages here are over an integral number of half periods. Here $I_{y,\text{av}}$ is zero because the y component of fluid velocity is 90° out of phase with the acoustic pressure, so the time average of their product is zero. The acoustic energy in the fluid associated with the presence of the flexural wave stays close to the plate, as evidenced by the factor $e^{-2(\omega/c)\beta y}$, and moves as a unit parallel to the plate in the $+x$ direction with speed c_W.

† L. M. Brekhovskikh, *Waves in Layered Media*, Academic, New York, 1960, pp. 4–6.

The Coincidence Frequency

The prediction that the flexural wave radiates sound only if $c_W > c$ applies to the idealized case where the plate is of infinite extent and the flexural wave continues indefinitely, but the model's predictions have approximate validity when a plate of finite size large in terms of flexural and acoustic wavelengths is vibrating. The enhanced radiation when c_W is near c can be demonstrated† by suspending a large metal plate by strings and causing it to vibrate by means of an electromagnetic shaker attached to the plate. If the shaker is oscillating at fixed frequency $f = \omega/2\pi$, the vibration over the surface of the plate for higher frequencies can be considered for the most part (except near the shaker and near the plate edges) as a superposition of freely propagating plane flexural waves traveling in various directions, each with speed (phase velocity) c_W. The speed c_W is proportional to $\omega^{1/2}$ for a thin plate, the theoretical relation‡ being

$$c_W = c_{\mathrm{pl}} = K^{1/4}\omega^{1/2} \qquad K = \frac{Eh^2}{12\rho_S(1 - \nu^2)} \tag{3-5.8}$$

where E = Young's modulus
h = plate thickness
ρ_S = mass in plate per unit volume
ν = Poisson's ratio

For an aluminum ($E = 72 \times 10^9$ Pa, $\rho_S = 2.7 \times 10^3$ kg/m^3, $\nu = 0.34$) plate of 0.5 cm thickness, for example, K is 63 N · m^3/kg; thus, for a frequency of 1000 Hz ($\omega = 6283$ rad/s) one has $c_W = 220$ m/s. Each of the superimposed plane flexural waves contributes independently to the radiated sound amplitudes in accord with the linear nature of the boundary conditions. Consequently, at typical points outside the plate, there should be a noticeable increase in the received sound when the shaker frequency goes from somewhat below to somewhat above the coincidence frequency§ at which $c_W = c$. This coincidence frequency f_c is $c^2/2\pi K^{1/2}$, that is, of the order of 2.3 kHz for the 0.5-cm-thick aluminum-plate example just cited. Providing the plate dimensions are large compared with c/f_c (about 15 cm for the example), the effect is quite observable. [One should design the demonstration so that the averaged (over surface of plate) squared vibrational velocity caused by the shaker does not vary substantially with frequency.]

† L. Wittig, "Random Vibration of Point Driven Strings and Plates," Ph.D. thesis, Department of Mechanical Engineering, Massachusetts Institute of Technology, Cambridge, Mass., 1971; S. H. Crandall and L. Wittig, "Chladni Patterns for Random Vibrations of a Plate," in G. Hermann and N. Perrone (eds.), *Dynamic Response of Structures*, Pergamon, New York, 1972, pp. 55–72.

‡ L. Cremer, M. Heckl, and E. E. Ungar, *Structure-Borne Sound*, Springer-Verlag, New York, 1973, pp. 95–101; Rayleigh, *The Theory of Sound*, vol. 1, secs. 214–217; Y. C. Fung, *Foundations of Solid Mechanics*, Prentice-Hall, New York, 1965, pp. 456–463.

§ The concept originated with L. Cremer, "Theory of the Sound Blockage of Thin Walls in the Case of Oblique Incidence," *Akust. Z.*, 7:81–104 (1942). The definition of coincidence frequency given in the text is that of M. C. Junger and D. Feit, *Sound, Structures, and Their Interaction*, M.I.T. Press, Cambridge, Mass., 1972, pp. 158–159.

Specific Radiation Impedance

For a body vibrating at fixed frequency, the ratio of complex pressure amplitude $\hat{p}$ to the outward component $\hat{v}_{\text{out}}$ of the acoustic-fluid-velocity complex amplitude is the *local specific radiation impedance* Z_{rad} of the surface.† Thus,

$$Z_{\text{rad}} = \left(\frac{\hat{p}}{\hat{v}_{\text{out}}}\right)_{\text{on } S_0} \qquad \text{where } \hat{v}_{\text{out}} = \hat{\mathbf{v}}_S \cdot \mathbf{n}_{\text{out}} \tag{3-5.9}$$

where $\mathbf{v}_S$ is the surface velocity of the body, and $\mathbf{n}_{\text{out}}$ is the unit normal to the surface pointing into the fluid. In general, Z_{rad} varies from point to point along the surface and with frequency. It also depends on the environment of the body; e.g., the specific radiation impedance of the example of the vibrating piston in a tube, discussed previously, depends on the impedance at the far end ($x = L$) and on the length L of the tube. In addition, the specific radiation impedance at any given point depends on the relative phasing and amplitudes of vibration at points all over the vibrating body. In the example just discussed of sound generated by flexural waves on a plate, one finds from Eqs. (2) and (6) that

$$Z_{\text{rad}} = \begin{cases} \rho c \left[1 - \left(\dfrac{c}{c_w}\right)^2\right]^{-1/2} = \dfrac{\rho c}{\cos\theta} & c < c_W \quad \text{(3-5.10}a\text{)} \\[2ex] -i\rho c \left[\left(\dfrac{c}{c_W}\right)^2 - 1\right]^{-1/2} = -\dfrac{i\rho c}{\beta} & c > c_W \quad \text{(3-5.10}b\text{)} \end{cases}$$

i.e., it depends critically on the flexural-wave speed and changes from purely resistive (Z_{rad} real) to purely reactive (Z_{rad} imaginary) when the flexural-wave speed drops below the sound speed in the fluid.

The concept of specific radiation impedance is useful in the prediction of the effects of the surrounding fluid on the vibration of a solid.‡ (This is of substantial importance when a body is vibrating under water and of less importance when it is vibrating in air.) In addition, it is useful in the analysis of the efficiency with which a vibration can generate sound. If the outward component of the acoustic fluid velocity is known along the surface, its complex amplitude $\hat{v}_{\text{out}}$ and the radiation impedance Z_{rad} give a prediction of the time average of the acoustic power generated per unit area of the solid's surface:

$$(\mathbf{I} \cdot \mathbf{n}_{\text{out}})_{\text{av}} = \tfrac{1}{2}|\hat{v}_{\text{out}}|^2 \operatorname{Re} Z_{\text{rad}} \tag{3-5.11}$$

The acoustic power radiated by the body is the area integral of this expression over the ambient surface S_0 of the vibrating body. [See Eq. (1-11.14).]

† The term "radiation impedance" without the adjective "specific" is often used for the complex-amplitude ratio of the net reaction force exerted by acoustic pressure on a radiating body to a surface-averaged outward component of velocity. See, for example, P. M. Morse, *Vibration and Sound,* McGraw-Hill, New York, 1948, p. 237; L. L. Beranek, *Acoustics,* McGraw-Hill, New York, 1954, pp. 116–128.

‡ See, for example, Junger and Feit, *Sound, Structures, and Their Interaction,* pp. 163–165; G. Kurtze and R. H. Bolt, "On the Interaction between Plate Bending Waves and Their Radiation Load," *Acustica,* **9**:238–242 (1959).

3-6 REFLECTION AND TRANSMISSION AT AN INTERFACE BETWEEN TWO FLUIDS

The concepts of trace velocity, specific radiation impedance, and the trace-velocity matching principle apply to the example† of a plane wave incident on an interface between two fluids (see Fig. 3-11). The incident wave (henceforth indicated by subscript I in place of I) propagates through a medium ($y < 0$) with sound speed c_{I} and ambient density ρ_{I} in the direction $\mathbf{n}_{\mathrm{I}} = \mathbf{e}_x \sin\theta_{\mathrm{I}} + \mathbf{e}_y \cos\theta_{\mathrm{I}}$ toward an interface separating the first medium from a second medium (c_{II}, ρ_{II}, with $y > 0$). The interface nominally coincides with the $y = 0$ plane but oscillates and flexes because of the acoustical disturbance. (The two fluids are presumed not to mix.)

The analysis in this and succeeding sections regarding the transmission and reflection of plane waves at one or more parallel interfaces applies also when one or more of the considered substances is an elastic solid, providing one limits one's attention to normal incidence ($\theta_{\mathrm{I}} = 0$), considers only longitudinal waves, and replaces acoustic pressure p in the solid by $-\sigma_{yy}$, the negative of the normal stress acting on surfaces perpendicular to the direction of propagation. The sound speed in the solid is interpreted as c_D, the dilatational elastic-wave speed; it and the shear-wave speed c_S are given by

$$c_D = \left[\frac{E(1-\nu)}{(1+\nu)(1-2\nu)\rho}\right]^{1/2} \qquad c_S = \left[\frac{E}{2(1+\nu)\rho}\right]^{1/2} \tag{3-6.1}$$

where E = elastic modulus
ν = Poisson's ratio
ρ = mass per unit volume

A brief list of values of c_D, ρ, and other pertinent properties for common solid materials is given in Table 3-1. The restriction to normal incidence is necessary because a longitudinal wave striking an interface obliquely will also excite shear (transverse) waves‡ within a solid. The ensuing analysis, however, is written as

† The discussion in the text is similar to that of Green, "On the Reflexion and Refraction of Sound," 1838. A treatment of sound reflection and refraction earlier than that of Green had been given by S. D. Poisson, "Memoir on the Movement of Two Superimposed Elastic Fluids," *Mem. Acad. Sci. Paris,* **10**:317–404 (1831). Poisson dealt with the normal-incidence case earlier in his "Memoir on the Movement of an Elastic Fluid through a Cylindrical Tube." The optical counterpart of the reflection-refraction problem had been considered in terms of a mechanical model of light waves by Fresnel in 1823.

‡ Insofar as the reflected wave is concerned, the analysis in Sec. 3-3, leading to Eqs. (3-3.3) to (3-3.6), is applicable for oblique plane-wave reflection from a solid. If the wave is incident from a fluid onto a homogeneous isotropic elastic solid half space, the appropriate identification [replacing Eq. (4)] for the specific acoustic impedance of the reflecting surface is

$$Z_{\mathrm{II}} = \rho_{\mathrm{II}} c_D \left\{ \frac{[1 - 2(c_S/v_{\mathrm{tr}})^2]^2}{[1 - (c_D/v_{\mathrm{tr}})^2]^{1/2}} + 4\frac{c_S}{c_D}\left(\frac{c_S}{v_{\mathrm{tr}}}\right)^2 \left[1 - \left(\frac{c_S}{v_{\mathrm{tr}}}\right)^2\right]^{1/2} \right\}$$

where each radical is understood to have a phase of $\pi/2$ when its argument is negative. An elastic

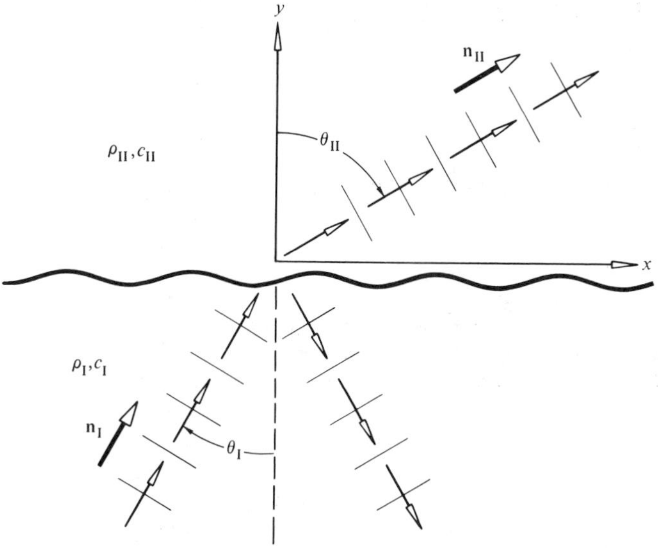

Figure 3-11 Plane-wave reflection and refraction at an interface between two fluids. Refracted wave (direction $\mathbf{n}_{\mathrm{II}}$) is generated in fluid II if $c_{\mathrm{I}}/(\sin\theta_{\mathrm{I}}) > c_{\mathrm{II}}$.

if both materials were ideal fluids and makes no a priori restriction to normal incidence.

The trace velocity v_{tr} of the incident wave along the x axis, i.e., along the $y = 0$ plane, is c_{I} divided by the x component of $\mathbf{n}_{\mathrm{I}}$, or $c_{\mathrm{I}}/\sin\theta_{\mathrm{I}}$. Whatever disturbance is generated within the second fluid must have the same trace velocity. For the reflected wave, this leads again to the law of mirrors (angle of incidence equals angle of reflection), and the reflected wave is a plane wave propagating in the direction $\mathbf{e}_x \sin\theta_{\mathrm{I}} - \mathbf{e}_y \cos\theta_{\mathrm{I}}$, that is, similar to that of $\mathbf{n}_{\mathrm{I}}$ except that the y component has changed sign.

If the trace velocity is supersonic with respect to the second medium ($v_{\mathrm{tr}} > c_{\mathrm{II}}$), the analysis above of the radiation of sound by a supersonic flexural wave traveling along a plate [leading to Eqs. (3-5.2)] suggests that the disturbance in the second fluid will be a plane wave propagating away from the interface. The propagation direction (unit vector $\mathbf{n}_{\mathrm{II}}$ making angle θ_{II} with the y

---

solid is such that $c_D^2 > 2c_S^2$, so Z_{II} is imaginary and $|\mathcal{R}_{\mathrm{I,II}}| = 1$ if $v_{\mathrm{tr}} < c_S$. There is a value of v_{tr} (the Rayleigh wave speed) somewhat less than c_S for which Z_{II} is identically zero and for which $\mathcal{R}_{\mathrm{I,II}} = -1$; but in cases when $\rho_{\mathrm{II}} \gg \rho_{\mathrm{I}}$, $c_D \gg c_{\mathrm{I}}$, the range of incidence angles where $|Z_{\mathrm{II}}|$ is comparable or smaller than $|Z_{\mathrm{I}}|$ is very small and typically $|Z_{\mathrm{II}}| \gg |Z_{\mathrm{I}}|$, so $\mathcal{R}_{\mathrm{I,II}} \approx 1$ and the half space can be idealized as rigid. A derivation of the above is given by Brekhovskikh, *Waves in Layered Media*, pp. 30–31. Brekhovskikh's $Z_1 \cos^2 2\gamma_1 + Z_t \sin^2 2\gamma_1$ in his eq. (4.25) is the same as our Z_{II} with the identifications $Z_1 = \rho_{\mathrm{II}} c_D/[1 - (c_D/v_{\mathrm{tr}})^2]^{1/2}$, $Z_t = \rho_{\mathrm{II}} c_S/[1 - (c_S/v_{\mathrm{tr}})^2]^{1/2}$, $\cos\gamma_1 = [1 - (c_S/v_{\mathrm{tr}})^2]^{1/2}$, $\sin\gamma_1 = c_S/v_{\mathrm{tr}}$.

Table 3-1 Representative mechanical and thermal properties of common solid materials at room temperature

Material	Composition	Density ρ, 10^3 kg/m^3	Dilatational wave speed c_D, 10^3 m/s	Elastic modulus E, 10^{10} N/m^2	Poisson's ratio ν	Thermal conductivity κ, W/(m · K)	Specific heat c_p, 10^3 J/(kg · K)
Aluminum	Pure and alloy	2.7–2.9	6.4	6.8–7.9	0.32–0.34	234	0.96
Brass	60–70% Cu, 40–30% Zn	8.4–8.5	4.7	10.0–11.0	0.33–0.36	146	0.37
Copper		8.9	5.0	11–13	0.33–0.36	385	0.39
Iron, cast	2.7–3.6% C	7.0–7.3	5.0	9–15	0.21–0.30	52	0.42
Lead		11.3	2.0	1.4	0.40–0.45	35	0.13
Steel	Carbon and low alloy	7.7–7.9	5.9	19–22	0.26–0.29	45	0.42
Stainless steel	18% Cr, 8% Ni	7.6–7.9	5.8	19–21	0.30	15	0.46
Titanium	Pure and alloy	4.5	6.1	10.6–11.4	0.34	8	0.54
Glass	Various	2.4–3.9	4.0–6.4	5.0–7.9	0.21–0.27	1.0	0.50–0.83
Methyl methacrylate		1.2	1.8–2.2	0.24–0.35	0.35		
Polyethylene		0.91	2.0	0.014–0.076	0.45		
Rubber		1.0–1.3		0.00008–0.0004	0.50	0.14–0.16	1.1–2.0

Source: S. H. Crandall, N. C. Dahl, and T. J. Lardner (eds.), *An Introduction to the Mechanics of Solids,* 2d ed., McGraw-Hill, New York, 1972, p. 286; T. Baumeister (ed.), *Standard Handbook for Mechanical Engineers,* 7th ed., McGraw-Hill, New York, 1967, pp. 4-11, 4-92, 4-95, 5-6, 6-7, 6-10; D. E. Gray (coord. ed.), *American Institute of Physics Handbook,* 3d ed., McGraw-Hill, New York, 1972, pp. 3-101, 3-104, 4-106, 4-154, 4-155.

axis) of this *transmitted wave* has a trace velocity in the x direction along the interface of $c_{\text{II}}/(\sin\theta_{\text{II}})$. The trace-velocity matching principle requires this be the same as the trace velocity of the incident wave, so one has† (*Snell's law*)

$$c_{\text{I}}^{-1}\sin\theta_{\text{I}} = c_{\text{II}}^{-1}\sin\theta_{\text{II}} = \frac{1}{v_{\text{tr}}} \tag{3-6.2}$$

This phenomenon, whereby propagation direction changes on passage into a medium with different sound speed, is known as *refraction*.

Internal boundary conditions coupling the solutions of the wave equation in the two fluids are the continuity of normal particle velocity and of total pressure at the actual (deformed) interface. The former leads to the approximate requirement that the normal component of displacement be continuous at the nominal interface location or, in the absence of ambient flow, that v_y be continuous at $y = 0$. The requirement of pressure continuity assumes no mass transport across the interface and neglects surface tension; under such circumstances it is the fluid-dynamic counterpart of Newton's third law. Since the ambient pressure is constant (with the neglect of gravity), and since the acoustic pressure changes negligibly over distances comparable to a particle displacement, the appropriate approximate boundary condition is the continuity of acoustic pressure at the nominal interface location.

[With gravity taken into account and with y denoting the vertical direction, however, the requirement, that acoustic pressure be continuous at $y = 0$, must be modified‡ to

$$p'(x, 0^-, z, t) - \rho_{\text{I}} g\eta = p'(x, 0^+, z, t) - \rho_{\text{II}} g\eta \tag{3-6.3}$$

where $\eta = \Delta\boldsymbol{\xi}\cdot\mathbf{e}_y$ at interface = normal (y-direction) displacement of interface

g = acceleration due to gravity

$p'(x, 0^-, z, t)$ = acoustic pressure in fluid I extrapolated to $y = 0$

This results because the total pressure in, say, medium II at $y = \eta$ is $[p_0(\eta) + p'(x, \eta, z, t)]_{\text{II}}$. Then, since η is small and $(dp_0/dy)_{\text{II}} = -g\rho_{\text{II}}$ (hydrostatic relation), the total pressure is equal to approximately $p_0(0) - g\rho_{\text{II}}\eta + p'_{\text{II}}$, where p'_{II} denotes the acoustic part of the pressure just above the interface in medium II.]

† The hypothesis that $(\sin\theta_{\text{I}})/(\sin\theta_{\text{II}})$ is independent of θ_{I} in the case of optical radiation was advocated with supporting (although incorrect) mathematical reasoning by Descartes in his *Dioptics* (Leyden, 1637), but it is believed that Descartes learned about this experimental fact from a manuscript (no longer in existence) circulated c. 1621 by Willebrord Snell (1591–1626). The earliest discovery of this law of sines was by Thomas Harriott (c. 1560–1621). [J. W. Shirley, "Early Experimental Determination of Snell's Law," *Am. J. Phys.*, **19**:507–508 (1951); W. B. Joyce and A. Joyce, "Descartes, Newton, and Snell's Law," *J. Op. Soc. Am.*, **66**:1–8 (1976).]

‡ F. Press and D. G. Harkrider, "Propagation of Acoustic-Gravity Waves in the Atmosphere," *J. Geophys. Res.*, **67**:3889–3908 (1962).

The disturbance in medium II is equivalent to what would be produced by a traveling [with trace velocity $c_I/(\sin\theta_I)$] flexural wave moving along the interface, so if medium II is unbounded, the ratio $\hat{p}/\hat{v}_y$ at the interface (which is continuous since $\hat{p}$ and $\hat{v}_y$ are continuous) is given by the radiation impedance of Eqs. (3-5.10) with ρc replaced by $\rho_{II}c_{II}$ and c_W replaced by $c_I/(\sin\theta_I)$; that is, $\hat{p}/\hat{v}_y = Z_{II}$ at $y = 0$, where we use the abbreviation

$$Z_{II} = \begin{cases} \dfrac{\rho_{II}c_{II}}{\cos\theta_{II}} & \text{if } \sin\theta_I < \dfrac{c_I}{c_{II}} \qquad (3\text{-}6.4a) \\[2ex] -\dfrac{i\rho_{II}c_{II}}{\beta_{II}} & \text{if } \sin\theta_I > \dfrac{c_I}{c_{II}} \qquad (3\text{-}6.4b) \end{cases}$$

where
$$\cos^2\theta_{II} = -\beta_{II}^2 = 1 - \left(\frac{c_{II}}{c_I}\right)^2 \sin^2\theta_I \tag{3-6.5}$$

For the respective cases in Eqs. (4*a*) and (4*b*), $\cos\theta_{II}$ and β_{II} are understood to be positive.

Since $\hat{p}/\hat{v}_y$ is continuous across the interface, Z_{II} is also the specific acoustic impedance at $y = 0$. The analysis given previously of plane-wave reflection from a surface of fixed impedance is therefore applicable here. In particular, the acoustic field variables in the region $y < 0$ are given by Eqs. (3-3.3) (providing one replaces y and $\hat{v}_y$ there by their negatives to take into account the difference between the choices of coordinate systems). The pressure-amplitude reflection coefficient $\mathscr{R}$ is identified from Eq. (3-3.4) as

$$\mathscr{R}_{I,II} = \frac{Z_{II} - Z_I}{Z_{II} + Z_I} \tag{3-6.6}$$

where, by analogy to Eqs. (4), we define $Z_I = \rho_I c_I/(\cos\theta_I)$. This reflection coefficient has the significance that if

$$\hat{p}_I = \hat{f}e^{i(\omega/c_I)\mathbf{n}_I\cdot\mathbf{x}} \tag{3-6.7}$$

is the complex pressure amplitude of the incident wave, the corresponding quantity for the reflected wave $\hat{p}_R$ is Eq. (7) multiplied by $\mathscr{R}_{I,II}$ with $\mathbf{n}_I$ replaced by $\mathbf{n}_R$ in the exponent. The analogous expression for the complex pressure amplitude in the second medium is of the form of a constant $\mathscr{T}_{I,II}$ times Eq. (7) with $\mathbf{n}_I/c_I$ replaced by $\mathbf{n}_{II}/c_{II}$ in the exponent if $\sin\theta_I < c_I/c_{II}$. For the other possibility, when $\sin\theta_I > c_I/c_{II}$, the transmitted wave is of the form [see Eqs. (3-5.6)]

$$\hat{p}_T = \mathscr{T}_{I,II}\hat{f}e^{i(\omega/c_I)(\sin\theta_I)x}e^{-(\omega/c_{II})\beta_{II}y} \tag{3-6.8}$$

In either event, $\hat{v}_y = \hat{p}/Z_{II}$ and $\hat{v}_x = \hat{p}/\rho_{II}v_{tr}$ throughout the second medium. Also, the continuity of the pressure at the interface requires that the transmission coefficient $\mathscr{T}_{I,II}$ be $1 + \mathscr{R}_{I,II}$ or $2Z_{II}/(Z_{II} + Z_I)$.

In the constant-frequency case, the energy per unit time and per unit area of interface (averaged over an integral number of half cycles) carried in toward the interface by the incident wave and carried out from the interface by the reflected and transmitted waves can be identified, respectively, as

$$\left(\frac{d\mathcal{P}}{dA}\right)_{\text{av},\text{I}} = \tfrac{1}{2}|\hat{f}|^2/Z_\text{I} \qquad \left(\frac{d\mathcal{P}}{dA}\right)_{\text{av},R} = |\mathcal{R}_{\text{I,II}}|^2\left(\frac{d\mathcal{P}}{dA}\right)_{\text{av},\text{I}}$$

$$\left(\frac{d\mathcal{P}}{dA}\right)_{\text{av},T} = \left(\frac{d\mathcal{P}}{dA}\right)_{\text{av},\text{I}} - \left(\frac{d\mathcal{P}}{dA}\right)_{\text{av},R} \tag{3-6.9}$$

These follow from such considerations as those giving Eqs. (3-3.5) and (3-3.6); the latter is in accord with the conservation of acoustic energy.

Water-Air Interfaces

A plane sound wave incident from a medium with a higher sound speed onto an interface separating it from a medium with lower sound speed is reflected as if the interface had a real specific acoustic impedance given by Eqs. (4*a*) and (5). If $c_\text{II} \ll c_\text{I}$, it is a good approximation to replace $\cos\theta_\text{II}$ in Eq. (4*a*) by 1, giving $Z_\text{II} \simeq \rho_\text{II}c_\text{II}$ independent of angle of incidence θ_I, so the surface is locally reacting. If, in addition, $\rho_\text{II}c_\text{II} \ll \rho_\text{I}c_\text{I}$, then, insofar as the prediction of the reflected wave is concerned, it is also a good approximation to consider Z_II as identically zero, so the surface is idealized as a pressure-release surface.

The above considerations apply in particular to underwater sound reflection from the water's surface (a water-air interface), since $(c_\text{air}/c_\text{water})^2 \simeq 0.05$ and $(\rho c)_\text{air}/(\rho c)_\text{water} \simeq 0.0003$.

Transient Reflection

If the incident waveform is not of constant frequency but is described by $f(t - \mathbf{n}_\text{I} \cdot \mathbf{x}/c_\text{I})$ for the acoustic pressure, then providing $c_\text{I} > c_\text{II}$ or θ_I is less than the *critical angle* $\sin^{-1}(c_\text{I}/c_\text{II})$, the reflected and transmitted waveforms are similar to that of the incident waveform:

$$p_R = \mathcal{R}_{\text{I,II}} f\left(t - \mathbf{n}_R \cdot \frac{\mathbf{x}}{c_\text{I}}\right) \tag{3-6.10a}$$

$$p_T = \mathcal{T}_{\text{I,II}} f\left(t - \mathbf{n}_\text{II} \cdot \frac{\mathbf{x}}{c_\text{II}}\right) \tag{3-6.10b}$$

These follow from the inverse Fourier transforms of the previously described expressions for $\hat{p}_R$ and $\hat{p}_T$ for the constant-frequency case when one recognizes, for the circumstances just described, that the reflection and transmission coefficients are real and frequency-independent. (Note that $|\mathcal{R}_{\text{I,II}}| \le 1$ but $|\mathcal{T}_{\text{I,II}}|$ can be larger than 1.)

However, if the second medium should have a sound speed greater than the first, the reflected waveform will no longer be a constant times the incident waveform when θ_I is greater than the critical angle, i.e., the θ_I giving a θ_II equal to $\pi/2$ from Snell's law, although one still has $p_R = g(t - \mathbf{n}_R \cdot \mathbf{x}/c_\text{I})$, where the

Fourier transform of $g(t)$ is related† to that of $f(t)$ by $\hat{g}(\omega) = \mathscr{R}_{\mathrm{I,II}}\hat{f}(\omega)$ for positive real ω. In this circumstance, Eqs. (4*b*) and (6) require that $\mathscr{R}_{\mathrm{I,II}}$ have a magnitude equal to 1 but be complex, so it may be written (for $\omega > 0$) as $\exp(-i\phi_{\mathrm{I,II}})$ where

$$\phi_{\mathrm{I,II}} = 2\tan^{-1}\frac{\rho_{\mathrm{I}}c_{\mathrm{I}}/(\cos\theta_{\mathrm{I}})}{\rho_{\mathrm{II}}c_{\mathrm{II}}/\beta_{\mathrm{II}}} \tag{3-6.11}$$

is an angle between 0 and π. Since $g(t)$ should be a real function, $\hat{g}(-\omega)$ equals $\hat{g}(\omega)^*$, and since $\hat{f}(\omega)$ also has the same property, $\mathscr{R}_{\mathrm{I,II}}$ for negative real ω should be the complex conjugate of that for $\omega > 0$. The Fourier integral relations (2-8.1) and (2-8.2) accordingly give

$$g(t) = (\cos\phi_{\mathrm{I,II}})f(t) + (\sin\phi_{\mathrm{I,II}})f_H(t) \tag{3-6.12}$$

where $$f_H(t) = -\frac{1}{\pi}\,\mathrm{Re}\left(\int_0^{\infty} e^{-i\omega t}\, i\int_{-\infty}^{\infty} e^{i\omega t'}f(t')\,dt'\,d\omega\right)$$

The order of integration in the above can be interchanged after insertion of a factor $e^{-\omega\tau}$ [similar to what is done in Eq. (2-8.5)], with the understanding that one should eventually take the limit as $\tau \to 0$. In this manner, one finds

$$f_H(t) = \lim_{\tau\to 0}\left[\frac{1}{\pi}\int_{-\infty}^{\infty} f(t')\,\frac{t'-t}{\tau^2 + (t'-t)^2}\,dt'\right]$$

If τ is extremely small and $f(t')$ is continuous, then, since the fractional quantity is odd in $t' - t$, the contribution to the integral over t' from $t - \varepsilon$ to $t + \varepsilon$ is negligible (ε being taken as, say, some large but fixed integer times τ). Outside this range of t', the fractional quantity is very nearly $1/(t' - t)$, so the limit above is equivalent to

$$f_H(t) = \frac{1}{\pi}\,\mathrm{Pr}\int_{-\infty}^{\infty}\left[\frac{f(t')}{t'-t}\right]dt' \tag{3-6.13}$$

where Pr (denoting *principal value*) is an abbreviation for what is implied by the above discussion; i.e., one performs the integration omitting an interval of width 2ε centered at the singularity and takes the limit as $\varepsilon \to 0$. In the mathematical-physics literature $f_H(t)$ is called the *Hilbert transform*‡ of $f(t)$. Three examples are shown in Fig. 3-12.

An apparent paradox presented by Eqs. (12) and (13) is that $f_H(t)$ and therefore $g(t)$ may be nonzero at times arbitrarily long before $f(t)$ first becomes nonzero. Thus, a person in the first medium hears a portion (*precursor*) of the

† A. B. Arons and D. R. Yennie, "Phase Distortion of Acoustic Pulses Obliquely Reflected from a Medium of Higher Sound Velocity," *J. Acoust. Soc. Am.*, 22:231–237 (1950); B. F. Cron and A. H. Nuttall, "Phase Distortion of a Pulse Caused by Bottom Reflection," ibid., 37:486–492 (1965).

‡ P. M. Morse and H. Feshbach, *Methods of Theoretical Physics*, vol. 1, McGraw-Hill, New York, 1953, p. 372.

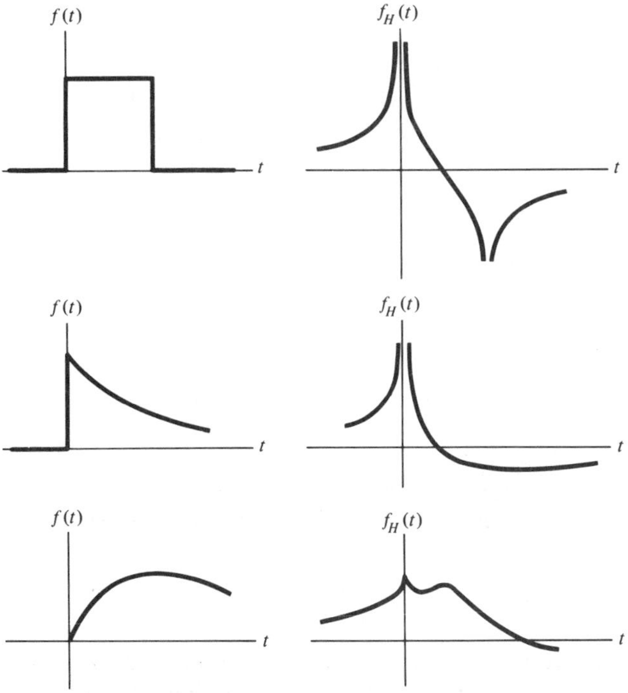

Figure 3-12 Three simple pulse shapes and their Hilbert transforms. [*D. Sachs and A. Silbiger, J. Acoust. Soc. Am.*, **49**:*835 (1971)*.]

echo before he hears the direct wave. This, however, is not a violation of causality, since the solution just described is for a steady-state circumstance for which the incident wave has been impinging on the interface (although, at large negative values of x) at all times in the remote past. Since the solution described requires, in particular, that c_{II} be greater than c_I, it is possible for acoustic energy to arrive earlier at the listener location via a faster path that takes advantage of the higher sound speed in the second medium.

3-7 MULTILAYER TRANSMISSION AND REFLECTION

The foregoing analysis can be extended to plane-wave transmission through any number of fluid layers of different density and sound speed† (see Fig. 3-13). The trace-velocity matching principle applies for each layer, so $\hat{p}$ throughout has a common x-dependent factor of $\exp [i(\omega/c_I)(\sin \theta_I)x]$. In any given layer,

† Rayleigh, *The Theory of Sound,* vol. 2, sec. 271; R. W. Boyle and W. F. Rawlinson, "Passage of Sound through Contiguous Media," *Trans. R. Soc. Can.*, (3)**22**:55–68 (1928).

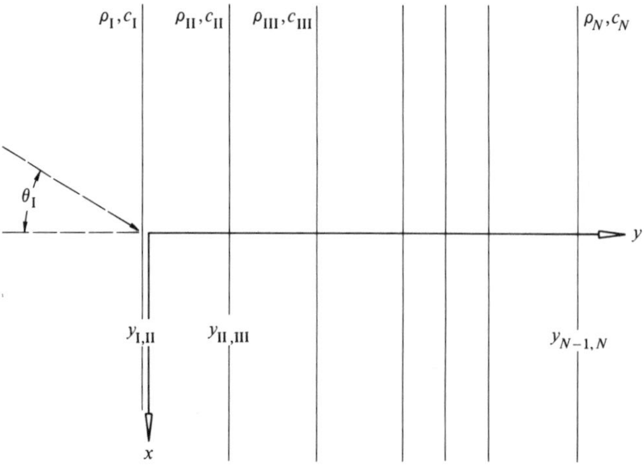

Figure 3-13 Plane-wave transmission through a sequence of nominally parallel fluid layers with differing densities and sound speeds; $y_{N-1,N}$ gives the y coordinate of the interface between the $(N - 1)$th and Nth layers.

the disturbance is a superposition of two obliquely propagating plane waves if $c < c_I/(\sin \theta_I)$ or of exponentially growing and decaying (with y) inhomogeneous plane waves if $c > c_I/(\sin \theta_I)$. The internal boundary conditions, continuity of $\hat{p}$ and $\hat{v}_y$, allow one to define a y-dependent specific impedance $Z_{\text{local}}(y)$ as the local ratio of $\hat{p}$ to $\hat{v}_y$, which is continuous across interfaces. Within each layer, one can define an intrinsic specific impedance (Z_I, Z_{II} for the first and second layers, etc.) such that, say, Z_{II} is given by Eqs. (3-6.4) with Z_{III}, Z_{IV} defined analogously.

A technique† for analyzing such multilayer transmission-reflection problems is based on an intermediate determination of Z_{local} at the interface $y_{I,II}$ between the first and second layers. Once $Z_{\text{local}}(y_{I,II})$ is determined, the reflection coefficient is given [by analogy with Eq. (3-6.6)] by

$$\mathscr{R} = \frac{Z_{\text{local}}(y_{I,II}) - Z_I}{Z_{\text{local}}(y_{I,II}) + Z_I} \tag{3-7.1}$$

and the fractions of incident energy reflected and transmitted are $|\mathscr{R}|^2$ and $1 - |\mathscr{R}|^2$. [The latter follows from Eq. (3-3.6) and from the relation $\nabla \cdot \mathbf{I}_{av} = 0$. Since translational symmetry transverse to the y axis requires $\partial I_{x,av}/\partial x = 0$, the relation $\nabla \cdot \mathbf{I}_{av} = 0$ implies that $(pv_y)_{av}$ is independent of y. The average energy transmitted past the I,II interface per unit time and area transverse to the y axis equals that transmitted into the last layer.]

To determine $Z_{\text{local}}(y_{I,II})$, one begins with the "known" local specific impedance at the last (largest y) interface $y_{N-1,N}$. This may be some specified specific

† Brekhovskikh, *Waves in Layered Media*, pp. 56–61.

acoustic impedance of a surface, or if the last layer is idealized as unbounded, it is the intrinsic specific impedance Z_N. To find the local specific impedance at the interface between the $(N-2)$th and $(N-1)$th layers, one makes use of an *impedance-translation theorem* (proved below), which states that, within any homogeneous layer, with intrinsic specific impedance Z_{int}, the local specific impedance $Z_{\text{local}}(y-L)$ at $y-L$ is related to that at y by

$$Z_{\text{local}}(y-L) = Z_{\text{int}} \frac{Z_{\text{local}}(y)\cos KL - iZ_{\text{int}}\sin KL}{Z_{\text{int}}\cos KL - iZ_{\text{local}}(y)\sin KL} \tag{3-7.2}$$

[which can be considered a generalization of Eq. (3-4.14)]. Here we abbreviate

$$K = \frac{\omega\rho}{Z_{\text{int}}} = \frac{\omega}{c}\begin{cases}\cos\theta & c<c_I/(\sin\theta_I) \\ i\beta & c>c_I/(\sin\theta_I)\end{cases} \tag{3-7.3}$$

where ρ and c are the ambient density and sound speed of the layer and $\cos\theta$ and β are determined as in Eqs. (3-6.5). (Recall that for any ϕ, $\cos i\phi$ and $\sin i\phi$ are $\cosh\phi$ and $i\sinh\phi$, respectively.)

To prove this impedance-translation theorem, note that, within such a layer, the general solution of the linear acoustic equations, given $e^{-i\omega t}$ time dependence and $e^{i(\omega/v_{\text{tr}})x}$ dependence [with $v_{\text{tr}} = c_I/(\sin\theta_I)$] on coordinate x, is

$$\begin{Bmatrix}\hat{p} \\ Z_{\text{int}}\hat{v}_y\end{Bmatrix} = e^{i(\omega/v_{\text{tr}})x}(Ae^{iKy} \pm Be^{-iKy})$$

where A and B are constants. The quantity $\hat{p}/\hat{v}_y$ at y or $y-L$ gives $Z_{\text{local}}(y)$ or $Z_{\text{local}}(y-L)$, respectively. Solution of the first such equation for Be^{-iKy}/Ae^{iKy} and substitution of that ratio into the second equation yields Eq. (2).

The impedance-translation equation, plus the continuity of Z_{local} across layer interfaces, allows one to successively work back, layer by layer, from $Z_{\text{int}}(y_{N-1,N})$ to $Z_{\text{int}}(y_{\text{I,II}})$. As an illustration, consider three layers, one intervening layer of thickness L sandwiched between two semi-infinite half spaces $(c_{\text{I}}, \rho_{\text{I}})$ and $(c_{\text{III}}, \rho_{\text{III}})$. As long as $c_{\text{III}} < c_{\text{I}}/(\sin\theta_{\text{I}})$, there will be a transmitted plane wave in region III propagating (in accord with the trace-velocity matching principle and Snell's law) at an angle θ_{III} with respect to the y axis, where $(\sin\theta_{\text{III}})/c_{\text{III}}$ is $(\sin\theta_{\text{I}})/c_{\text{I}}$. The local specific impedance at the $+y$ side of layer II (and throughout layer III) is Z_{III}. The local specific impedance at the (I,II) interface results from Eq. (2) with $Z_{\text{local}}(y)$ and Z_{int} identified as Z_{III} and Z_{II}, respectively, so the reflection coefficient becomes

$$\mathscr{R} = \frac{(Z_{\text{II}}Z_{\text{III}} - Z_{\text{I}}Z_{\text{II}})\cos K_{\text{II}}L - i(Z_{\text{II}}^2 - Z_{\text{I}}Z_{\text{III}})\sin K_{\text{II}}L}{(Z_{\text{II}}Z_{\text{III}} + Z_{\text{I}}Z_{\text{II}})\cos K_{\text{II}}L - i(Z_{\text{II}}^2 + Z_{\text{I}}Z_{\text{III}})\sin K_{\text{II}}L} \tag{3-7.4}$$

If Z_{II} is real $(c_{\text{II}} < c_{\text{I}}/\sin\theta_{\text{I}})$, this reflection coefficient [as well as the local specific impedance $Z_{\text{local}}(y_{\text{I,II}})$] is periodic in layer thickness L with a repetition length π/K_{II}. It is also periodic in frequency.

One of the implications of Eq. (4) is that $|\mathscr{R}| = 1$ whenever Z_{III} is purely imaginary $[c_{\text{III}} > c_{\text{I}}/(\sin\theta_{\text{I}})]$, regardless of the properties of the intervening layer. In general, $|\mathscr{R}| = 1$ if the sound speed in the last layer exceeds the trace velocity, for any number of intervening layers, provided the last layer is idealized as a half space (unbounded at large y). This must be so because $\hat{p}$ and $\hat{v}_y$ are 90° out of phase in the last layer; the time average of power transmitted is zero.

Another implication of the above expression is that $\mathcal{R}$ may be identically zero under circumstances other than the trivial one where $Z_I = Z_{II} = Z_{III}$. For example, if the angle of incidence and the layer properties are such that $Z_{II}^2 = Z_I Z_{III}$, then $\mathcal{R}$ will be zero if $K_{II}L$ is an odd multiple of $\pi/2$ (such that its cosine is zero). A special case would be $\theta_I = 0$ (in which case the analysis also applies to longitudinal elastic-wave transmission through solid slabs). Then, if one wants perfect transmission without reflection into medium III from a source in medium I, a *transmission plate*† made of *buffer material* is placed between the two substances; this buffer material should have (or approximate) the property

$$\rho_{II}c_{II} = (\rho_I c_I \rho_{III} c_{III})^{1/2} \tag{3-7.5}$$

The thickness of the layer would be selected so that $(\omega/c_{II})L = \pi/2$ or, for fixed frequency $f = \omega/2\pi$, so that $L = \frac{1}{4}(c_{II}/f)$ is a quarter of the sound wavelength at that frequency in the buffer material.

If the properties of medium III are the same as those of medium I (so one has a layer of foreign material in an otherwise homogeneous medium), Eq. (4) reduces (with $Z_{III} = Z_I$ and after dividing numerator and denominator by $Z_{II}Z_I$) to

$$\mathcal{R} = \frac{-i(r - r^{-1}) \sin K_{II}L}{2 \cos K_{II}L - i(r + r^{-1}) \sin K_{II}L} \tag{3-7.6}$$

with the abbreviation $r = Z_{II}/Z_I$. The fraction of incident energy transmitted is $1 - |\mathcal{R}|^2$, and since both the incident wave and the transmitted wave (on the far side of the intervening layer) are plane waves propagating in the same direction through the same medium, the mean squared pressures have the ratio $1 - |\mathcal{R}|^2$. After some algebra one therefore obtains

$$\frac{(p_T^2)_{av}}{(p_I^2)_{av}} = \frac{1}{1 + \frac{1}{4}(r - r^{-1})^2 \sin^2 K_{II}L} \tag{3-7.7}$$

Because $(r - r^{-1}) \sin K_{II}L$ is real regardless of the sign of Z_{II}^2, the above relation holds (recall that $i^4 = 1$) also when Z_{II} is imaginary. Note that $(p_T^2)_{av}/(p_I^2)_{av} \leq 1$ and that it equals 1 (perfect transmission) when $K_{II}L$ is a multiple of π.

3-8 TRANSMISSION THROUGH THIN SOLID SLABS, PLATES, AND BLANKETS

Transmission Loss

For circumstances, like those described in the last part of the preceding section, when a sound wave is incident on an intervening slab of material (not necessar-

† P. J. Ernst, "Ultrasonic Lenses and Transmission Plates," *J. Sci. Instrum.*, 22:238–243 (1945).

ily a fluid layer), one defines a *sound-power transmission coefficient* τ as the fraction of the incident sound power transmitted to the far side of the slab. If the incident wave is a plane wave, and if the slab (or partition) has properties unchanging with displacements parallel to its faces, the transmitted wave will be a plane wave propagating in the same direction as the incident wave. One can accordingly argue, as in the discussion preceding Eq. (3-7.7), that the fraction of incident power transmitted is the same as the quotient of the mean squares of transmitted and incident acoustic pressures. Consequently, the *plane-wave sound-power transmission coefficient* $\tau(\theta_I, \omega)$ (corresponding to angle of incidence θ_I and angular frequency ω) for such circumstances becomes $(p_T^2)_{av}/(p_I^2)_{av}$. The *transmission loss* R_{TL} (in decibels) is defined in general in terms of the transmitted fraction τ of incident power as 10 log $(1/\tau)$ and thus, for the plane-wave constant-frequency case, the *plane-wave transmission loss* equals

$$R_{TL} = L_{p,I} - L_{p,T} \tag{3-8.1}$$

where $L_{p,I}$ and $L_{p,T}$ are the sound-pressure levels for the incident and transmitted plane waves.

Slab Specific Impedance

The analysis of transmission loss simplifies for the case (see Fig. 3-14) of an intervening slab, i.e., a layer of different material, whose properties are such that $v_{front} = v_{back}$, where v_{front} denotes the normal component of the fluid velocity (in the direction from front toward back) at the front of the slab and v_{back} denotes the analogous quantity on the opposite side of the slab. (Which side one wishes to designate as the front is arbitrary, but in a subsequent discussion we take the side from which the incident wave is coming as the front side.) The

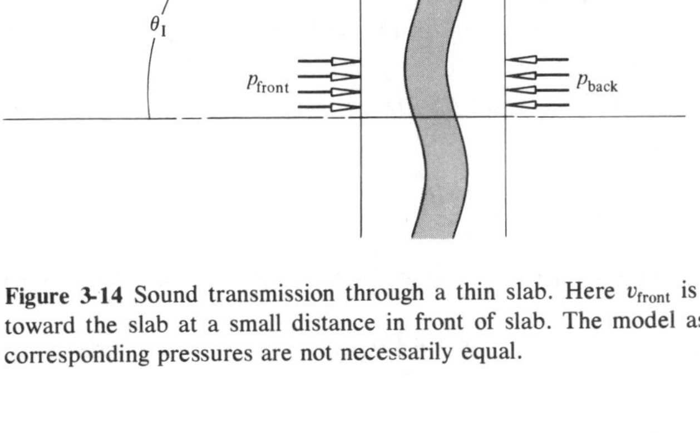

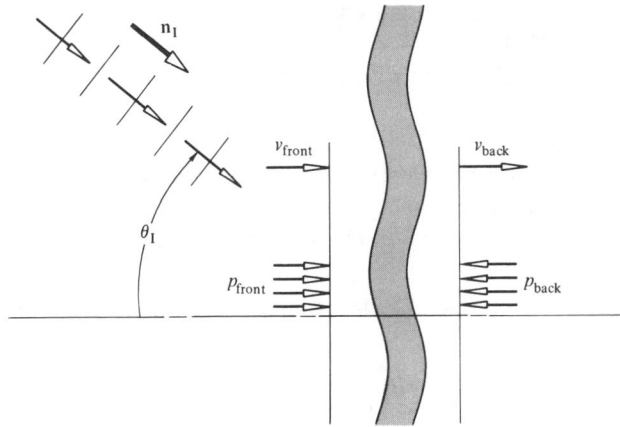

Figure 3-14 Sound transmission through a thin slab. Here v_{front} is the fluid-velocity component toward the slab at a small distance in front of slab. The model assumes that $v_{front} = v_{back}$, but corresponding pressures are not necessarily equal.

assumption that the two velocities are nearly equal is appropriate if the time for an acoustic disturbance to propagate across the slab is substantially less than one-quarter of a wave period and if the ratio of the characteristic impedance of the material in the slab to the local specific acoustic impedance at the back of the slab is large compared to 2π times the ratio of the thickness of the slab to a wavelength. For solid walls of typical thicknesses, with air on both sides, such is invariably the case at audible frequencies.

If the slab is porous, so that there is a net flow of fluid through it, the transverse velocity of the solid material in the slab may not be the same as v_{front} or v_{back}, but $v_{\text{front}} = v_{\text{back}}$ nevertheless may be a good approximation if the pore volume per unit slab area is substantially less than $\frac{1}{4}$ wavelength. (This follows from conservation-of-mass considerations and from the assumption that density fluctuations of fluid within the pores are not markedly different from those on either side of the slab. If there is flow through the pores, then, on a microscopic scale, the fluid velocity just at the surface will vary substantially over distances comparable to pore sizes and pore spacings, but such variations smooth out for regions only slightly removed from the slab surface. The quantities v_{front} and v_{back} can be considered as local averages over small areas parallel to the slab faces.)

Given this equivalence of fluid velocities on opposite sides of the slab, one can define a slab specific impedance $Z_{\text{sl}}(v_{\text{tr}}, \omega)$ such that

$$\hat{p}_{\text{front}} - \hat{p}_{\text{back}} = Z_{\text{sl}}(v_{\text{tr}}, \omega)\hat{v}_{\text{front}} = Z_{\text{sl}}(v_{\text{tr}}, \omega)\hat{v}_{\text{back}} \tag{3-8.2}$$

Here $\hat{p}_{\text{front}}$ and $\hat{p}_{\text{back}}$ represent the complex acoustic-pressure amplitudes at the front and back sides of the slab; v_{tr} is the common, parallel to slab face, trace velocity of the acoustic disturbances on the two sides of the slab, each appropriate complex acoustic amplitude having the common factor $\exp(i\omega x/v_{\text{tr}})$ for its x dependence. An additional assumption implied in this definition is that the slab's dynamics are governed by linear equations.

Dividing both sides of Eq. (2) by $\hat{v}_{\text{front}} = \hat{v}_{\text{back}}$ and making use of the definition of local specific impedance as ratio of $\hat{p}$ to $\hat{v}_y$ yields

$$Z_{\text{local}}(y_{\text{front}}) = Z_{\text{local}}(y_{\text{back}}) + Z_{\text{sl}} \tag{3-8.3}$$

(This is analogous to the result that the electric impedance of two circuit elements in series is the sum of the impedances of the two elements.)

If the incident acoustic wave impinges on the slab from the front side, the pressure-amplitude reflection coefficient $\mathcal{R}$ is given by Eq. (3-7.1) with $y_{\text{I,II}}$ identified as y_{front}; also, Eq. (3-3.3*b*) requires that $\hat{v}_{\text{front}}$ be $(\hat{p}_{\text{I}}/Z_{\text{I}})(1 - \mathcal{R})$. The relation (3) therefore yields

$$\hat{v}_{\text{front}} = \frac{2\,\hat{p}_{\text{I}}}{2Z_{\text{I}} + Z_{\text{sl}}} \tag{3-8.4}$$

Here $\hat{p}_{\text{I}}$ denotes the complex amplitude of the incident wave's acoustic pressure at the front of the slab; Z_{I} is $\rho c/(\cos\theta_{\text{I}})$.

Since pressure and the y component of fluid velocity on the back side of the slab are related as for a plane wave propagating at angle θ_{I} with the y axis, at the back side of the slab one has $\hat{p}_T = Z_{\mathrm{I}}\hat{v}_{\mathrm{back}}$. Thus, Eq. (4) leads to $2Z_{\mathrm{I}}/(2Z_{\mathrm{I}} + Z_{\mathrm{sl}})$ for the pressure-amplitude transmission coefficient. The square of the magnitude of this is the plane-wave sound-power transmission coefficient, so the transmission loss, from Eq. (1), becomes

$$R_{\mathrm{TL}} = 10 \log \left(\left| 1 + \frac{1}{2} \frac{Z_{\mathrm{sl}}}{\rho c} \cos \theta_{\mathrm{I}} \right|^2 \right) \tag{3-8.5}$$

with the insertion of $(\cos \theta_{\mathrm{I}})/\rho c$ for $1/Z_{\mathrm{I}}$.

The energy theorem for the circumstances just described can be derived with appropriate identifications from Eqs. (3-3.5) and (3-3.6), i.e.,

$$\left(\frac{d\mathscr{P}}{dA}\right)_{\mathrm{av},T} = \left(\frac{d\mathscr{P}}{dA}\right)_{\mathrm{av,I}} - \left(\frac{d\mathscr{P}}{dA}\right)_{\mathrm{av},R} - \left(\frac{d\mathscr{P}}{dA}\right)_{\mathrm{av},d} \tag{3-8.6}$$

where

$$\left(\frac{d\mathscr{P}}{dA}\right)_{\mathrm{av},T} = \tfrac{1}{2}|\hat{v}_{\mathrm{front}}|^2 \,\mathrm{Re}\, Z_{\mathrm{local}}(y_{\mathrm{back}}) \tag{3-8.7a}$$

$$\left(\frac{d\mathscr{P}}{dA}\right)_{\mathrm{av},d} = \tfrac{1}{2}|\hat{v}_{\mathrm{front}}|^2 \,\mathrm{Re}\, Z_{\mathrm{sl}} \tag{3-8.7b}$$

represent the power transmitted per unit face area and the rate at which energy is dissipated per unit area within the slab. The latter follows because the average rate at which work is done on the slab is $\frac{1}{2}$ Re $[(\hat{p}_{\mathrm{front}} - \hat{p}_{\mathrm{back}})\hat{v}^*_{\mathrm{front}}]$. If Z_{sl} is purely imaginary, there is no energy dissipation and Eq. (6) reverts to a strict conservation-of-energy statement.

Oblique-Incidence Mass Law

A simple model† of a slab or a plate useful in the discussion and interpretation of acoustic transmission phenomena is the *perfectly limp plate,* whose specific impedance comes solely from the inertia of its mass. The model is such that, if m_{pl} is the plate mass per unit area, then, from Newton's second law,

$$m_{\mathrm{pl}} \frac{\partial v_{\mathrm{pl}}}{\partial t} = p_{\mathrm{front}} - p_{\mathrm{back}} \tag{3-8.8}$$

Also, given that the plate is not porous, the boundary condition (3-1.2) requires that $\hat{v}_{\mathrm{pl}} = \hat{v}_{\mathrm{front}} = \hat{v}_{\mathrm{back}}$, so from Eq. (2) and the prescription $\partial/\partial t \to -i\omega$ one identifies $Z_{\mathrm{sl}} = -i\omega m_{\mathrm{pl}}$. The transmission loss of Eq. (5) accordingly becomes (limp-wall mass-law transmission loss)‡

$$R_{\mathrm{TL}} = 10 \log \left[1 + \left(\frac{\omega m_{\mathrm{pl}}}{2\rho c}\right)^2 \cos^2 \theta_{\mathrm{I}} \right] \tag{3-8.9}$$

† L. Cremer, "Theory of the Sound Blockage of Thin Walls in the Case of Oblique Incidence," *Akust. Z.*, 7:81–104 (1942).

‡ I. L. Ver and C. I. Holmer, "Interaction of Sound Waves with Solid Structures," in L. L. Beranek (ed.), *Noise and Vibration Control,* McGraw-Hill, New York, 1971, pp. 270–361.

where we recognize $2\rho c/\omega$ as the mass per unit area of a slab of thickness λ/π filled with fluid of density ρ.

For a slab of solid material in air for frequencies in the audible range it is invariably true that $2\pi m_{\mathrm{pl}}/\rho\lambda \gg 1$. [For example, for a $\frac{1}{2}$-cm-thick aluminum plate and a frequency of 340 Hz, one has $m_{\mathrm{pl}} \approx 13$ kg/m^2 and $\lambda \approx 1$ m, and (with $\rho = 1.2$ kg/m^3) the ratio $2\pi m_{\mathrm{pl}}/\rho\lambda$ is of the order of 70.] Given this assertion and providing θ_{I} is not close to grazing incidence (so $\cos\theta_{\mathrm{I}}$ is not too small), the 1 in the argument of the logarithm in Eq. (9) is negligible. In this limit, doubling the plate mass m_{pl} or frequency f increases R_{TL} by $10 \log 4 \approx 6$ dB.

The oblique-incidence mass law also follows from the expression (3-7.7) for the sound-transmission coefficient of an intervening fluid layer in the limit $|K_{\mathrm{II}}L| \ll 1$ and $|Z_{\mathrm{II}}| \gg |Z_{\mathrm{I}}|$ ($\rho_{\mathrm{II}}c_{\mathrm{II}} \gg \rho_{\mathrm{I}}c_{\mathrm{I}}$). Then one can neglect r^{-1} in the expression $r - r^{-1}$ and approximate $\sin K_{\mathrm{II}}L$ by $K_{\mathrm{II}}L$. Since $rK_{\mathrm{II}}L$ is $(Z_{\mathrm{II}}/Z_{\mathrm{I}})(\omega\rho_{\mathrm{II}}/Z_{\mathrm{II}})L$, which in turn is $\omega(\rho_{\mathrm{II}}L/\rho_{\mathrm{I}}c_{\mathrm{I}}) \cos\theta_{\mathrm{I}}$, while $\rho_{\mathrm{II}}L = m_{\mathrm{pl}}$ is the slab mass per unit area, the quantity 10 times the logarithm, base 10, of the right side of Eq. (3-7.7) in the limit described is the same as the R_{TL} of Eq. (9) above.

Transmission through Euler-Bernoulli Plates

The springlike resistance of a thin plate to bending can be approximately taken into consideration by replacing Eq. (8) by the *Euler-Bernoulli plate equation*†

$$m_{\mathrm{pl}} \frac{\partial^2 \xi_{\mathrm{pl}}}{\partial t^2} = p_{\mathrm{front}} - p_{\mathrm{back}} - B_{\mathrm{pl}} \left(\frac{\partial^2}{\partial x^2} + \frac{\partial^2}{\partial z^2} \right)^2 \xi_{\mathrm{pl}} \tag{3-8.10}$$

where ξ_{pl} is the normal displacement of the plate (positive if in y direction), so $v_{\mathrm{pl}} = \partial\xi_{\mathrm{pl}}/\partial t$. The quantity B_{pl} is the plate bending modulus (proportionality factor between torque per unit length and curvature for cylindrical bending), given, according to the theory of elasticity, for a homogeneous isotropic plate by $Eh^3/[12(1 - \nu^2)]$. Here E is elastic modulus, h is plate thickness, and ν is Poisson's ratio.

agating without dependence on z and with a trace velocity v_{tr} along the x axis, the prescription $\partial/\partial t \to -i\omega$, $\partial/\partial x \to i\omega/v_{\mathrm{tr}}$, $\partial/\partial z \to 0$ converts Eq. (10) into

† So called because it is based on the same general principles as the Euler-Bernoulli model of a beam, which dates back to papers published by James Bernoulli (1705), Daniel (James's nephew) Bernoulli, (1741–1743, published 1751), and L. Euler (1779, 1782). The theory of thin plates is due to S. D. Poisson, "Memoir on Elastic Surfaces," 1814, "Memoir on the Equilibrium and Movement of Elastic Bodies," 1820; Sophie Germain, "Researches on the Theory of Elastic Surfaces," 1821; and G. Kirchhoff, "On the Equilibrium and the Motion of an Elastic Plate," 1850. Summaries and bibliographical data for all these works are given by I. Todhunter and K. Pearson, *A History of the Theory of Elasticity and of the Strength of Materials*, vol. 1, 1866, reprinted by Dover, New York, 1960, pp. 10–13, 30–32, 50–56, 147–160, 208–276; vol. 2, pt. 2, 1893, reprinted 1960, pp. 39–48. For a modern derivation of the thin-plate equation see, for example, C.-T. Wang, *Applied Elasticity*, McGraw-Hill, 1953, pp. 276–280.

an algebraic equation relating complex amplitudes. Consequently, the slab specific impedance is identified, with reference to Eq. (2), as

$$Z_{\text{sl}} = -i\omega m_{\text{pl}} \left[1 - \left(\frac{c_{\text{pl}}}{v_{\text{tr}}}\right)^4\right] \tag{3-8.11}$$

where c_{pl}, abbreviated for $\omega^{1/2}(B_{\text{pl}}/m_{\text{pl}})^{1/4}$, is the same as in Eq. (3-5.8) and represents the natural-phase velocity (so called because it is associated with the speed of lines of constant phase) for traveling waves with straight wavefronts of angular frequency ω on a plate. If v_{tr} should equal c_{pl}, one could have a disturbance propagating along the plate without any external influence; i.e., Eq. (10) can then be satisfied with $p_{\text{front}} - p_{\text{back}} = 0$ but with ξ_{pl} of the form of a constant-frequency plane traveling wave that is not identically zero.

The oblique-incidence transmission loss for the Euler-Bernoulli plate model is as given by Eq. (9) but with the prescription

$$m_{\text{pl}} \rightarrow m_{\text{pl}} \left[1 - \left(\frac{c_{\text{pl}}}{v_{\text{tr}}}\right)^4\right] = m_{\text{pl}} \left[1 - \left(\frac{f}{f_c}\right)^2 \sin^4 \theta_{\text{I}}\right] \tag{3-8.12}$$

where, in the latter expression, v_{tr} has been identified as $c/(\sin \theta_{\text{I}})$ and where the variation of c_{pl} as the square root of the frequency f has been used to express $c_{\text{pl}} = (f/f_c)^{1/2}c$, f_c being the coincidence frequency at which $c_{\text{pl}} = c$. As is described in the discussion following Eq. (3-5.8), f_c should equal $c^2/2\pi K^{1/2}$ with K equaling $B_{\text{pl}}/m_{\text{pl}}$.

If $f \ll f_c/(\sin^2 \theta_{\text{I}})$, the factor in brackets in Eq. (12) is nearly 1; then the transmission loss is unaffected by plate stiffness and is the same as that predicted by the mass-law equation. However, if $f = f_c/(\sin^2 \theta_{\text{I}})$ [or, equivalently, if $\theta_{\text{I}} = \sin^{-1} (f_c/f)^{1/2}$ or if $v_{\text{tr}} = c_{\text{pl}}$], the transmission loss predicted by Eq. (9) with the substitution (12) is identically 0. It is also zero in the limit of zero frequency. Thus, when considered as a function of frequency, the transmission loss must have a maximum somewhere between 0 and $f_c/(\sin^2 \theta_{\text{I}})$. The maximum coincides with that of $2\pi f m_{\text{pl}}[1 - (f/f_c)^2 \sin^4 \theta_{\text{I}}]$ and is accordingly at $f_c/(3^{1/2} \sin^2 \theta_{\text{I}})$, that is, smaller by a factor of $1/(3^{1/2}) = 0.58$ than the frequency at which perfect transmission occurs.

Internal energy losses within solids are frequently taken into account with the replacement† of the elastic modulus E by $(1 - i\eta)E$ [or, equivalently, of B_{pl} by $(1 - i\eta)B_{\text{pl}}$ in the case of a plate] in relations involving complex amplitudes. Here η is a real quantity termed the *loss factor* (Table 3-2), which can be measured for a given plate by a variety of methods‡ and which in general varies with frequency. It should not strictly be considered a material constant as it is strongly affected, in the case of metals, for example, by such processes as cold

† A. Schoch, "On the Asymptotic Behavior of Forced Plate Vibrations at High Frequencies," *Akust. Z.*, **2**:113–128 (1937).

‡ L. Cremer, M. Heckl, and E. E. Ungar, *Structure-Borne Sound*, Springer-Verlag, New York, 1973, pp. 189–205.

rolling, heat treatment, and irradiation.† Typical values for metals range from 10^{-4} (aluminum) to 10^{-2} (lead). A plate of laminar construction or one covered with a viscoelastic layer has a composite loss factor that can be estimated if one knows the dynamical properties of the individual layers.‡

The substitution of a complex plate bending modulus $(1 - i\eta)B_{\text{pl}}$ into Eq. (11) leads to

$$Z_{\text{sl}} = \omega\eta m_{\text{pl}}\left(\frac{f}{f_c}\right)^2 \sin^4\theta_{\text{I}} - i\omega m_{\text{pl}}\left[1 - \left(\frac{f}{f_c}\right)^2 \sin^4\theta_{\text{I}}\right] \tag{3-8.13}$$

for the slab specific impedance of a lossy plate. The transmission loss of Eq. (5) derived from this when $\eta \ll 1$ is close to that for $\eta = 0$ except in the vicinity of the frequency $f_c/(\sin^2\theta_{\text{I}})$, where the lossless-plate theory would predict a zero transmission loss. The modified theory gives instead

$$R_{\text{TL}} = 20 \log \left(1 + \frac{1}{2}\frac{\omega\eta m_{\text{pl}}}{\rho c}\cos\theta_{\text{I}}\right)$$

at this frequency.

Transmission through Porous Blankets§

The simplest model of a porous slab is a blanket whose resistance to flow is described by the *specific flow resistance* R_f, defined so that the transverse fluid velocity on either side relative to the velocity v_{bl} of the blanket is given for steady flow by

$$v_{\text{front}} - v_{\text{bl}} = v_{\text{back}} - v_{\text{bl}} = \frac{1}{R_f}(p_{\text{front}} - p_{\text{back}}) \tag{3-8.14}$$

This is a fluid-dynamic analog to Ohm's law of electric resistance. For a homogeneous material of fixed density, R_f is proportional to the blanket thickness; the specific flow resistance per unit thickness is the *flow resistivity* (Table 3-3). The quantity R_f can be determined from a steady-flow experiment in which the blanket is held fixed and fluid is forced to flow through it at a set rate with the pressure measured on both sides of the blanket. The application of Eq. (14) to situations in which v_{front} or v_{bl} may be oscillating with time is consistent with the assumption that R_f is independent of frequency.

The determination of transmission loss can be carried through with various

† C. Zener, *Elasticity and Anelasticity of Metals,* University of Chicago Press, Chicago, 1948, pp. 41–59, 94–95, 115–121.

‡ A review citing principal references is given by E. E. Ungar, "Damping of Panels," in Beranek, *Noise and Vibration Control,* pp. 434–475.

§ L. L. Beranek, "Acoustical Properties of Homogeneous, Isotropic Rigid Tiles and Flexible Blankets," *J. Acoust. Soc. Am.,* **19**:556–568 (1947); R. H. Nichols, Jr., "Flow-Resistance Characteristics of Fibrous Acoustical Materials," ibid., **19**:866–871 (1947); ASTM C522-69, Standard Method of Test for Airflow Resistance of Acoustical Materials, American Society for Testing and Materials, Philadelphia.

Table 3-2 Typical loss factors (flexural) at audio frequencies for common materials

Material	Loss factor η	Material	Loss factor η
Aluminum	10^{-4}	Magnesium	10^{-4}
Brass, bronze	$<10^{-3}$	Masonry blocks	$5–7 \times 10^{-3}$
Brick	$1–2 \times 10^{-2}$	Oak, fir	$0.8–1 \times 10^{-2}$
Concrete:		Plaster	5×10^{-3}
Light	1.5×10^{-2}	Plexiglass, Lucite	$2–4 \times 10^{-2}$
Porous	1.5×10^{-2}	Plywood	$1–1.3 \times 10^{-2}$
Dense	$1–5 \times 10^{-2}$	Sand, dry	0.6–0.12
Copper	2×10^{-3}	Steel, iron	$1–6 \times 10^{-4}$
Cork	0.13–0.17	Tin	2×10^{-3}
Glass	$0.6–2 \times 10^{-3}$	Wood fiberboard	$1–3 \times 10^{-2}$
Gypsum board	$0.6–3 \times 10^{-2}$	Zinc	3×10^{-4}
Lead	$0.5–2 \times 10^{-3}$		

Source: E. E. Ungar, "Damping of Panels," in L. L. Beranek (ed.), *Noise and Vibration Control,* McGraw-Hill, New York, 1971, p. 453.

idealizations of how the blanket is supported. A particular case would be that when the blanket is hanging freely. If the blanket is perfectly limp, Eq. (8) applies but with m_{pl} and v_{pl} replaced by m_{bl} and v_{bl}, the change of subscript implying that we are concerned with a blanket. Equations (14) and (8) together then give an equation of the form of Eq. (2) in which the slab specific impedance

Table 3-3 Flow resistivity of porous materials of various densities

Material	Density, kg/m³	Flow resistivity, 10^3 N · s/m⁴
Fiberglas AA	11.2	58
	7.4	34
Fiberglas H-33	41.6	29
Rock wool (Johns-Manville Stonefelt, type M)	54.1	28
	42.6	31
Kaowool Blanket B (Babcock and Wilcox)	50	65
Wood fiber	32.2	39
Ultralite no. 200 (Gustin Bacon Co.)	20.0	7
	100.0	90
Ultrafine no. 1001 (Certain-teed)	40	30
Acoustiform-Mat Ceiling Board (Celotex)	160	70
Thermafiber insulating blanket (U.S. Gypsum)	30	3.5

Source: L. L. Beranek, *J. Acoust. Soc. Am.,* **19:**556–568 (1947); D. A. Bies, "Acoustical Properties of Porous Materials," in L. L. Beranek (ed.), *Noise and Vibration Control,* McGraw-Hill, New York, 1971, pp. 250–251.

for the blanket is consequently identified as

$$Z_{\text{sl}} = \left[\frac{-1}{i\omega m_{\text{bl}}} + \frac{1}{R_f}\right]^{-1} \tag{3-8.15}$$

This, in terms of an electric-circuit analogy, consists of impedances $-i\omega m_{\text{bl}}$ and R_f in parallel. The expression for transmission loss, resulting from a substitution of Eq. (15) into Eq. (5), is cumbersome, but if $\omega \ll R_f/m_{\text{bl}}$, it reduces to the mass-law transmission loss. In the other limit of $\omega \gg R_f/m_{\text{bl}}$, it reduces to the transmission loss for an immobile blanket, i.e.,

$$R_{\text{TL}} = 10 \log \left(\left| 1 + \frac{1}{2}\frac{R_f}{\rho c}\cos\theta_{\text{I}}\right|^2\right) \tag{3-8.16}$$

which is independent of frequency.

PROBLEMS

3-1 A solid sphere of radius a is rotating with uniform angular velocity ω about an axle displaced a slight distance b from its center, where $b \ll a$ and $b \ll c/\omega$. Here c is the sound speed in the surrounding fluid. Let the rotational axis lie along the z axis and let the sphere's center lie in the $z = 0$ plane, so that with an appropriate choice of time origin the sphere's center at time t is at $x_C = b \cos \omega t$, $y_C = b \sin \omega t$. In terms of spherical coordinates (r, θ, ϕ), where $z = r \cos \theta$, $x = r \sin \theta \cos \phi$, what boundary condition would be imposed on the acoustic fluid velocity on a sphere of radius a centered at the origin to enable one to predict the resulting acoustic field approximately?

3-2 A broadband plane wave at an angle of incidence of 45° and propagating through air with sound speed 340 m/s is reflected from a rigid surface. Over the octave band centered at 500 Hz the incident sound has nearly constant spectral density, and the sound level corresponding to this band for the incident wave alone is 80 dB (re 20 μPa). Determine and plot as a function of distance from the wall the octave-band sound-pressure level for the same band that results because of the sound reflection. Beyond what minimum distance can one assume that the octave-band level is within ±0.5 dB of 83 dB? How does this answer change if one considers instead an octave band centered at 250 or 1000 Hz?

3-3 An interface between two fluids nominally lies on the $y = 0$ plane. In the absence of an acoustic disturbance, the fluid in the region $y > 0$ is moving with a velocity v_0 in the x direction while that in the region $y < 0$ is motionless. The sound speeds and ambient densities in the regions $y < 0$ and $y > 0$ are c_{I}, ρ_{I} and c_{II}, ρ_{II}, respectively. A plane wave of angular frequency ω is incident from the $y < 0$ side of the interface with a propagation direction characterized by a unit vector $\mathbf{n}_{\text{I}} = \mathbf{e}_x \sin \theta_{\text{I}} + \mathbf{e}_y \cos \theta_{\text{I}}$. Show that one of the appropriate linear acoustic boundary conditions at the interface is

$$(v_y)_0^{(-)} = \frac{(v_y)_0^{(+)}}{1 - (v_0/c_{\text{I}}) \sin \theta_{\text{I}}}$$

where $(v_y)_0^{(-,+)}$ denote the y components of the acoustic fluid velocity on the two sides of the interface.

3-4 The acoustic pressure (incident wave plus reflected wave) just outside a specimen of sound-absorbing material (interface coinciding with $y = 0$ plane) when an incident wave of frequency $f = \omega/2\pi$ is propagating toward it at an angle of incidence of 45° is

$$p = A \cos\left[\omega\left(t - \frac{x - y}{2^{1/2}c}\right)\right] - 0.5\, A \sin\left[\omega\left(t - \frac{x + y}{2^{1/2}c}\right)\right]$$

where A is the amplitude of the incident wave and y is the distance from the interface. What is the specific acoustic impedance of this interface in units of ρc? If the material is locally reacting, what will the absorption coefficient for reflection with the same frequency at normal incidence be?

3-5 A particular type of acoustic tile is locally reacting and for a frequency of 200 Hz has a normal-incidence specific impedance of $1000 + i2000$ kg/(m$^2 \cdot$ s). A plane wave in air of 200-Hz sound with a sound-pressure level of 70 dB in the absence of reflection is incident on the tile at an angle of θ_I.

(*a*) How close must θ_I be to grazing incidence for the resulting sound-pressure level just at the surface of the tile to be less than 67 dB?

(*b*) Determine and plot the absorption coefficient as a function of θ_I.

3-6 Suppose that one knew at the outset that a particular interface was locally reacting and had determined, for a given frequency, the absorption coefficient versus angle of incidence θ_I. Would it be possible to determine the specific acoustic impedance of the surface from these data? If so, give instructions and a numerical example for a possible data-analysis scheme. [F. V. Hunt, *J. Acoust. Soc. Am.*, **10**:216–217 (1939); L. L. Beranek, *ibid.*, **12**:14–23 (1940).]

3-7 A plane wave is incident at an angle of incidence θ_I on a reflecting surface of unknown specific acoustic impedance. The net acoustic pressure at a point just outside the surface is measured and found to be $B\cos(\omega t - \psi)$; at the same point in the absence of reflection it would be $A\cos\omega t$. In terms of ρ, c, A, B, ω, θ_I, and ψ, determine an expression for the specific acoustic impedance of the surface. [U. Ingard and R. H. Bolt, *J. Acoust. Soc. Am.*, **23**:509–516 (1951).]

3-8 A long circular duct (length idealized as infinite) of radius a whose axis coincides with the x axis is filled with fluid of ambient density ρ and sound speed c. At $x = 0$ the duct has stretched across it a thin membrane. The dynamics of the membrane are such that in circumstances of interest it can be modeled as a thin rigid piston of effective mass m_{eff} whose displacement x_p (equal to the membrane's displacement averaged over the cross-sectional area) is resisted by a force proportional to x_p, the proportionality factor (spring constant) being k_{eff}. Thus, the membrane's displacement satisfies the differential equation

$$m_{\text{eff}}\ddot{x}_p + k_{\text{eff}}x_p = \pi a^2(p_{\text{front}} - p_{\text{back}})$$

If a plane wave of angular frequency ω is incident on the membrane from the $-x$ side, what fraction of the incident power will be transmitted to the air on the $+x$ side of the membrane?

3-9 The membrane of Prob. 3-8 is displaced a distance $x_p = x_p^0$ and released from rest at time $t = 0$. Before that time there is no acoustic disturbance in the tube. Given the idealization that the only cause of vibrational-energy loss of the membrane is the radiation of sound, determine x_p as a function of time. What is the net acoustic energy radiated by the membrane in the $+x$ direction in the limit of large t? Under what circumstances will the pressure variation be nonoscillatory?

3-10 A piston at one end of a tube (cross-sectional area 0.01 m^2) whose length is exactly one-fourth wavelength at a frequency of 1000 Hz is oscillating with a displacement amplitude of 0.0001 m and with a frequency 1000 Hz + Δf, where Δf is much smaller than 1000 Hz. The apparent specific impedance at the other end of the tube is $\rho c(0.02 - i.006)$ where $\rho = 1.2$ kg/m^3 and $c = 340$ m/s. For what value of Δf is the average acoustic power generated by the oscillating piston a maximum? What is the quality factor Q for the resonance?

3-11 Two fluids with sound speeds and densities (c_{I}, ρ_{I}) and (c_{II}, ρ_{II}), respectively, are separated by a plane interface. In one experiment, a plane wave at angle of incidence θ_{I} (less than the critical angle) is incident on the interface from the first fluid and a plane wave propagating at angle θ_{II} with the interface normal is generated in the second fluid, while in a second experiment a plane wave is incident on the interface from the second fluid at an angle of incidence θ_{II}. Prove that the fractions of incident power transmitted are the same for the two experiments and the fractions of incident power reflected are also the same.

3-12 A plastic transmission plate is to be designed to allow perfect transmission (without reflection) of normal-incidence plane waves from water ($\rho = 1000$ kg/m^3, $c = 1500$ m/s) into steel ($\rho = 7700$ kg/m^3, $c = 6100$ m/s). The frequency of interest is 20,000 Hz, and the available plastics all have a density of 1500 kg/m³. What should the sound speed in the plastic and the plate's thickness be? (A minimal thickness is desired.) Suppose the same plate is used for transmission of the same frequency, also at normal incidence, from steel into water. What fraction of the incident power will be transmitted?

3-13 If a fluid occupying the region $y > 0$ is bounded by a locally reacting surface of finite specific impedance, it is sometimes possible to have an acoustic disturbance (*surface wave*) with an acoustic pressure of the form

$$p = \text{Re}\; Pe^{-\alpha y}e^{ikx}e^{-i\omega t}$$

where, for a given real angular frequency ω, the quantities P, α, and k are complex constants, the real parts of α and k being positive and P being arbitrary but nonzero. As an example, take $Z = \rho c(100 + i200)$ and determine expressions for α and k. In terms of P, x, and y, what are the time-averaged y and x components of the acoustic intensity in the fluid? What is the time-averaged energy loss per unit surface area and per unit time of the surface wave? How do the answers change if the specific impedance is $Z = \rho c(100 - i200)$?

3-14 A porous blanket of mass per unit area high enough not to move under the influence of acoustic disturbances of interest is suspended a distance L in front of a flat rigid wall. The flow resistance of the blanket is R_f. A plane wave of angular frequency ω (wave number $k = \omega/c$) is incident normally on the blanket. Determine an expression for the specific acoustic impedance on a surface just in front of the blanket. What fraction of the incident power is absorbed? For given k and R_f, what choice of L gives maximum absorption? What would the absorption coefficient be in the latter case?

3-15 A piston at the $x = 0$ end of a tube of length L is set into motion at time $t = 0$ with a velocity V_0 for $0 < t < L/c$, $-V_0$ for $L/c < t < 2L/c$, V_0 for $2L/c < t < 3L/c$, $-V_0$ for $3L/c < t < 4L/c$, etc. The far end of the tube is presumed rigid, and loss mechanisms within the tube are of negligible significance. Determine and sketch the acoustic pressure at the piston face as a function of t for t up to $10L/c$. Also determine and sketch the instantaneous acoustic power output of the piston over the same interval of time. How much acoustic energy is in the tube by time $10L/c$?

3-16 For the idealized model (no viscosity) discussed in the text for reflection of obliquely incident plane waves at an interface between two fluids, is the tangential component of acoustic fluid velocity continuous across the interface? Is the ambient density times tangential acoustic fluid velocity continuous? Is the velocity potential continuous?

3-17 Following the Alaskan earthquake of March 28, 1964, Rayleigh waves traveling at a velocity of the order of 10 times the speed of sound in air passed across the United States. At Boulder, Colorado, the resulting infrasonic pressure oscillation near the ground was at an amplitude of 2 Pa and a period of 25 s. Estimate the amplitude of the transverse velocity of the ground motion. What was the time-averaged intensity of the resulting acoustic wave? Assuming that all the radiated energy propagated to ionospheric heights without reflection or refraction, what would the fluid-velocity amplitude have been at an altitude where the ambient density is 10^{-8} that at the earth's surface? [R. K. Cook, "Radiation of Sound by Earthquakes," pap. K19 in D. E. Commins (ed.), *5ᵉ Congr. Int. Acoust.*, G. Thone, Liège, 1965, vol. 1b.]

3-18 A sheet of porous material is suspended in air at a distance of $\frac{1}{4}$ wavelength in front of a rigid wall. For the frequency of interest, the mass of the sheet is high enough not to move significantly under the influence of a sound wave. When a constant-frequency plane wave is normally incident on the sheet, a microphone just in front of it registers an acoustic pressure with a rms amplitude of 0.3 Pa, while a microphone behind it at the wall surface registers a rms amplitude of 0.2 Pa. What is the specific flow resistance of the sheet? What fraction of the incident sound power is absorbed? What would the transmission loss of the same sheet be if the wall were not present?

3-19 Sound waves in air are incident at an angle of 45° on a 0.5-cm-thick sheet of steel.

(*a*) At what frequency would you expect perfect transmission to occur (with the neglect of internal losses)?

(*b*) At what lower frequency does the transmission loss have a maximum?

(*c*) At what frequency above that of part (*a*) does the transmission loss first exceed that of part (*b*)?

(*d*) Discuss the general dependence of the ratio of the frequencies of parts (*c*) and (*b*) on the elastic modulus E and Poisson's ratio of the material in the plate, the thickness of the plate, the angle of incidence, and the mass per unit volume of the material in the plate. (Use the thin-plate model for the sheet.)

3-20 It is planned to construct a sound barrier by suspending two identical lead sheets at a distance d apart. Assuming that all the sound arrives at normal incidence, is there some optimal nonzero choice for d (in terms of sound wavelengths) that will give a maximum transmission loss? If so, by how many decibels would the resulting transmission loss exceed that of a single sheet of twice the mass per unit area? Express your answers in terms of $X = \omega m_{pl}/\rho c$.

3-21 A subsonic flexural wave with phase speed $c/3$ and angular frequency ω is propagating along the surface of a plate immersed in a fluid of ambient density ρ. Discuss the acoustically induced trajectories of fluid particles moving with the local fluid velocity, nominally located at a distance h from the plate. Are they circles, ellipses, or straight lines? When a given particle is at a point on its trajectory that is closest to the plate surface, what is the phase of the plate's transverse displacement at the nearest point on the plate?

3-22 A plane wave of angular frequency ω is incident normally on a slab of foreign material (assumed lossless) of width d. Let ρ_{I} and c_{I} denote ambient density and sound speed of the material on both sides of the slab and let ρ_{II} and c_{II} denote the analogous quantities for the slab itself.

(*a*) Let $\hat{R}_{\mathrm{I}}$ be the complex amplitude of the reflected pressure wave just at the near surface of the slab and let $\hat{T}_{\mathrm{III}}$ be the complex amplitude of the transmitted pressure wave just at the far surface of the slab; show that

$$\frac{\hat{R}_{\mathrm{I}}}{\hat{T}_{\mathrm{III}}} = \frac{i}{2}\left[\frac{(\rho c)_{\mathrm{I}}}{(\rho c)_{\mathrm{II}}} - \frac{(\rho c)_{\mathrm{II}}}{(\rho c)_{\mathrm{I}}}\right] \sin \frac{\omega d}{c_{\mathrm{II}}}$$

(*b*) Suppose two identical transducers (which generate and receive sound) are placed on opposite sides of the slab at distances L and $L + \Delta L$, respectively, from the nearer side of the slab and are caused to oscillate in phase but with different amplitudes for a short time less than $2L/c_{\mathrm{II}}$ but larger than several $2\pi/\omega$. The net received plane wave at the farther transducer is found to have negligible amplitude throughout most of its time of reception for some choice of ΔL and for some ratio B/A of the two amplitudes of the incident pressure pulses. How are B/A and ΔL related to ω, d, and the acoustical properties of the two materials? Could one determine c_{II} and ρ_{II} from such an experiment? [H. J. McSkimin, *J. Acoust. Soc. Am.*, **23**:429–434 (1951).]

3-23 In Prob. 3-22, suppose that Euler's equation of motion does not hold within the slab proper; instead, for waves going in the $+x$ and $-x$ directions, suppose that the complex amplitude $\hat{v}_x$ of fluid velocity is related to the corresponding amplitude $\hat{\sigma}_{xx}$ of the normal component of stress by

$$Z_{\mathrm{II}}(\omega)\hat{v}_x = \mp\hat{\sigma}_{xx}$$

where $Z_{\mathrm{II}}(\omega)$ is some complex number depending on frequency. Suppose that $\hat{\sigma}_{xx}$ varies with distance x through the slab as

$$\hat{\sigma}_{xx} = Ae^{ik_{\mathrm{II}}x} + Be^{-ik_{\mathrm{II}}x}$$

where $k_{\mathrm{II}}(\omega)$ is another complex number, the two terms here corresponding to waves traveling in the $+x$ and $-x$ direction, respectively. Take $\hat{\sigma}_{xx} = -\hat{p}$ (Newton's third law) to hold at the two faces of the slab and discuss how the experiment described above should be modified and how the results should be interpreted in order to obtain information concerning Z_{II} and k_{II}. Is it appropriate to assume that $Z_{\mathrm{II}} = \rho_{\mathrm{II}}\omega/k_{\mathrm{II}}$?

3-24 A sonic boom with acoustic-pressure waveform given by $f(t - \mathbf{n}_{\mathrm{I}} \cdot \mathbf{x}/c_{\mathrm{I}})$, where $f(t)$ is as sketched in Prob. 1-29, is incident from air onto an air-water interface at an angle of incidence of 45°. Discuss the general characteristics of the signature of the pressure signal received at a depth h

below the interface. Neglect viscosity and nonlinear effects. [R. K. Cook, *J. Acoust. Soc. Am.*, **47**:1430–1436 (1970); J. C. Cook, T. Goforth, and R. K. Cook, ibid., **51**:729–741 (1972).]

3-25 Determine the natural frequencies and the corresponding eigenfunctions describing the x dependence of acoustic pressure for a narrow tube extending from $x = 0$ to $x = L$ with both ends open. Take the boundary condition at each open end to be $p = 0$.

3-26 A piston at the $x = 0$ end of a tube (cross-sectional area A) of length L is oscillating with a velocity amplitude V_0. The specific acoustic impedance at the other end ($x = L$) is $\varepsilon\rho c$, where ε is a small positive real number much less than 1. Give approximate expressions for the lowest resonance frequency, the Q of this resonance, and the peak time-averaged acoustic power output of the piston for frequencies in the vicinity of this resonance.

3-27 A stretched membrane nominally lies in the xz plane and is surrounded on both sides by a fluid of ambient density ρ and sound speed c. The flexural vibrations of the membrane are governed by the partial differential equation

$$\sigma \frac{\partial^2 \eta}{\partial t^2} - T\left(\frac{\partial^2 \eta}{\partial x^2} + \frac{\partial^2 \eta}{\partial z^2}\right) = p(x, 0^-, z, t) - p(x, 0^+, z, t)$$

where σ = mass per unit area of membrane
T = tension per unit length to which the membrane is stretched
η = transverse displacement of membrane in $+y$ direction

The two pressures correspond to the $y < 0$ and $y > 0$ sides. It is here assumed that $(T/\sigma)^{1/2} \ll c$. Suppose one has a sinusoidal wave

$$\eta = A \cos\left[\omega\left(t - \frac{x}{c_W}\right)\right]$$

traveling in the x direction. What is the speed c_W in terms of ω, σ, T, ρ, and c?

3-28 A plane wave is incident on a continuously stratified medium for which c and ρ are functions of y; the acoustic pressure is of the form $p = \text{Re}\ \hat{p}(y)e^{-i\omega(t-x/v_{\text{tr}})}$, and v_y is given by an analogous expression, where v_{tr} is some given trace velocity. Show that the local specific acoustic impedance $Z_{\text{local}}(y) = \hat{p}(y)/\hat{v}_y(y)$ satisfies the differential equation

$$-\frac{dZ_{\text{local}}}{dy} = -i\omega\rho + \frac{i\omega}{\rho}(c^{-2} - v_{\text{tr}}^{-2})Z_{\text{local}}^2$$

Discuss how, with appropriate approximations, one can derive the limp-wall oblique-incidence mass-law transmission loss from this equation.

CHAPTER

FOUR

RADIATION FROM VIBRATING BODIES

Attention in the present chapter is directed toward models of sound generation and propagation for which the resulting phenomena are more conveniently described in terms of spherical coordinates than cartesian coordinates. We begin with the fundamental examples of sound radiation from radially and transversely oscillating spheres and subsequently show that they can be used as building blocks for analyses of sound radiation in less idealized circumstances.

4-1 RADIALLY OSCILLATING SPHERE

The prototype of an omnidirectional source is a sphere† (see Fig. 4-1) centered at the origin whose radius oscillates about some nominal value a with velocity $v_S(t)$. Given that the external medium is unbounded, the acoustic field is spherically symmetric, and so Eqs. (1-12.4) apply. With $F(t - r/c)/\rho c$ replaced by an equivalent "to be determined" function $\psi(t - r/c + a/c)$, these equations become

$$v_r = \frac{\dot{\psi}}{r} + \frac{c\psi}{r^2} \qquad p = \frac{\rho c \dot{\psi}}{r} \tag{4-1.1}$$

† G. G. Stokes, "On the Communication of Vibration from a Vibrating Body to a Surrounding Gas," *Phil. Trans. R. Soc. Lond.*, **158**:447–463 (1868); A. E. H. Love, "Some Illustrations of Modes of Decay of Vibratory Motions," *Proc. Lond. Math. Soc.*, (2) **2**:88–113 (1905); J. Brillouin, "Transient Radiation of Sound Sources and Related Problems," *Ann. Telecommun.*, **5**:160–172, 179–194 (1950).

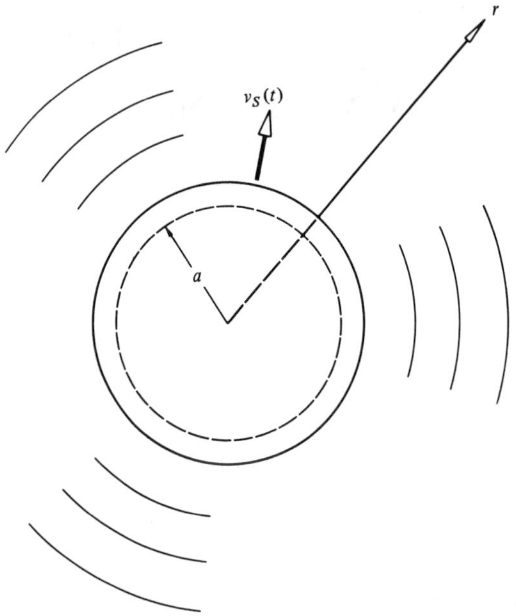

Figure 4-1 Sound generation by a radially oscillating sphere with radial surface velocity $v_S(t)$.

where $\dot{\psi}$ denotes the derivative of ψ with respect to its argument (here understood to be $t - r/c + a/c$). The boundary condition, $v_r(a, t) = v_S(t)$, resulting from Eq. (3-1.2), therefore requires

$$a^{-1} \frac{d}{dt} \psi(t) + ca^{-2}\psi(t) = v_S(t) \tag{4-1.2}$$

which integrates to

$$\psi(t) = a \int_{-\infty}^{t} e^{-(c/a)(t-\tau)} v_S(\tau)\, d\tau \tag{4-1.3}$$

with the requirement that $\psi(t)$ be zero before $v_S(t)$ first becomes nonzero. The above expression for $\psi(t)$ and Eqs. (1) for $v_r(r, t)$ and $p(r, t)$ describe the *transient solution* for the acoustic field radiated by the sphere.

The constant-frequency solution, resulting when v_S has been oscillating for a long time with an angular frequency ω, can be derived directly from Eqs. (1-12.8); the constant A appearing there is identified from the requirement that the $\hat{v}_r$ in Eq. (1-12.8*b*) be $\hat{v}_S$ at $r = a$, where $\hat{v}_S$ is the complex amplitude of $v_S(t)$. Thus, Eqs. (1-12.8) yield

$$\frac{\hat{p}}{\rho c} = \frac{\hat{v}_r}{1 + i/kr} = \frac{-ika^2\hat{v}_S}{r(1 - ika)} e^{ik(r-a)} \tag{4-1.4}$$

These also result from the transient solution for a sphere that starts oscillating at time t_0 in the limit when the retarded time $t - r/c + a/c$ (minus the time t_0) is large compared with the time $c/2\pi a$ for a disturbance to travel around the perimeter of the sphere. (Here $t - r/c + a/c$ is the time the wave currently being received left the surface.)

The time-averaged intensity $I_{r,\text{av}}$ of a spherically symmetric wave is $|\hat{p}|^2/2\rho c$, in accord with (4) and with the relation $I_{r,\text{av}} = \frac{1}{2}\,\text{Re}\,\hat{p}\hat{v}_r^*$. Consequently, the time-averaged power $\mathcal{P}_{\text{av}}$ radiated by the radially oscillating sphere is $4\pi r^2|\hat{p}|^2/2\rho c$, so, from (4) one has

$$\mathcal{P}_{\text{av}} = \frac{(ka)^2}{1 + (ka)^2}\,\rho c (v_S^2)_{\text{av}}(4\pi a^2) \tag{4-1.5}$$

In the limit $ka \gg 1$, the power radiated per unit surface area of the sphere is $\rho c (v_S^2)_{\text{av}}$ (the same as for radiation to one side from a plate vibrating without flexure with a velocity amplitude $|\hat{v}_S|$).

Low-Frequency Approximation

If $v_S(t)$ changes slowly over times of the order of a/c, a suitable approximation to $p(r, t)$ results from a neglect of the first term in Eq. (2), such that $\psi(t) = (a^2/c)v_S(t)$. Also, it is consistent to ignore the distinction between $v_S(t - r/c + a/c)$ and $v_S(t - r/c)$ when $\psi(t - r/c + a/c)$ is inserted into Eqs. (1). In this manner, the acoustic pressure reduces to

$$p(r, t) = \frac{\rho}{4\pi r}\left(\frac{dQ_S}{dt}\right)_{t \to t - r/c} \tag{4-1.6}$$

where $Q_S(t) = 4\pi a^2 v_S$ (surface area of sphere times radial velocity) is the time derivative of the volume enclosed by the source and is referred to as the *source-strength function*. The result, moreover, is a good approximation† even if, over any interval of time, the radius may change by an increment comparable to, or larger than, its original value, providing $Q_S(t)$ is the instantaneous derivative of the actual volume enclosed by the sphere. It is required that the velocity of the surface always be substantially less than c and that the surface acceleration be substantially less than c^2 divided by the sphere radius.

The constant-frequency version of Eq. (6) also results from Eq. (4) if one neglects the term $-ika$ in the factor $(1 - ika)^{-1}$ and approximates e^{-ika} by 1. The equivalence is evident if one notes that dQ_S/dt evaluated at $t - r/c$ is equal to $\text{Re}\,(-i\omega\hat{Q}_S e^{ikr}e^{-i\omega t})$, where $\hat{Q}_S$ is $4\pi a^2\hat{v}_S$. The prescription for incorporation of a time shift, $t \to t - r/c$, is to multiply the complex amplitude by a factor of e^{ikr}.

† This was recognized and applied by M. Strasberg, "Gas Bubbles as Sources of Sound in Liquids," *J. Acoust. Soc. Am.*, **28**:20–26 (1956). A rigorous justification is given by P. A. Frost and E. Y. Harper, "Acoustic Radiation from Surfaces Oscillating at Large Amplitude and Small Mach Number," ibid., **58**:318–325 (1975).

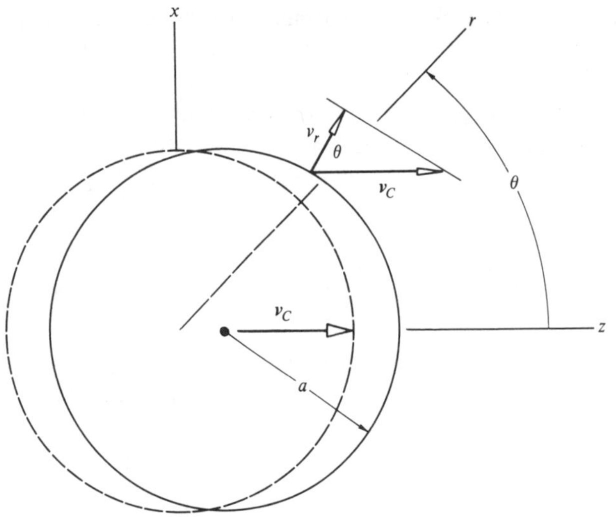

Figure 4-2 Sound generation by a transversely oscillating rigid sphere of radius a. The center of the sphere moves back and forth along the z axis with velocity $\mathbf{v}_C(t)$.

4-2 TRANSVERSELY OSCILLATING RIGID SPHERE

A rigid sphere (see Fig. 4-2) whose center is oscillating back and forth along the z axis about the origin is the simplest model† of a source whose volume does not change with time. The appropriate boundary condition, deduced from Eq. (3-1.2), at the nominal location of the sphere's surface is

$$v_r(a, \theta, t) = \mathbf{v}_C(t) \cdot \mathbf{e}_r = v_C(t) \cos \theta \tag{4-2.1}$$

where $\mathbf{v}_C(t) = v_C(t)\mathbf{e}_z$ is the velocity of the sphere's center. To construct a solution of the linear acoustic equations satisfying this boundary condition, we note that (1) the derivative with respect to z of any solution of the wave equation is also a solution and (2) a known solution is $1/r$ times any function of $t - r/c$, so

$$\Phi = \frac{\partial}{\partial z}\left[\frac{1}{r}\psi\left(t - \frac{r}{c} + \frac{a}{c}\right)\right] \tag{4-2.2}$$

is a possible candidate for the velocity potential. Here the differentiation is carried out at fixed x and y, and so, since $r^2 = x^2 + y^2 + z^2$, one has $\partial r/\partial z = z/r$ or $\cos \theta$. Thus, the operator $\partial/\partial z$ can be replaced by $(\cos \theta)\, \partial/\partial r$ in the above. Because Eq. (2) and the expression in Eq. (1) both depend on θ only through the

† S. D. Poisson, "On the Simultaneous Movement of a Pendulum and of the Surrounding Air," *Mem. Acad. Sci., Paris,* **11**:521–582 (1832); Stokes, "On the Communication of Vibration," 1868.

multiplicative factor cos θ, a function ψ can be found that ensures that $v_r = \partial\Phi/\partial r$ will reduce to $v_C(t)$ cos θ when $r = a$.

The ordinary differential equation that $\psi(t)$ must satisfy so that $\partial\Phi/\partial r$ will equal $v_C(t)$ cos θ at $r = a$ results from Eq. (2) if one recognizes that the first and second derivatives with respect to r of $\psi(t - r/c + a/c)$ are $-(1/c)\dot{\psi}(t)$ and $(1/c)^2\ddot{\psi}(t)$, respectively, at $r = a$. The resulting substitutions into the boundary-condition equation then yield

$$\ddot{\psi}(t) + 2\,\frac{c}{a}\,\dot{\psi}(t) + 2\left(\frac{c}{a}\right)^2\psi(t) = c^2 a v_C(t) \tag{4-2.3}$$

Such an inhomogeneous linear second-order ordinary differential equation with constant coefficients can be solved† as a superposition of indicial responses, but we here limit ourselves to the steady-state case, such that $v_C(t)$ equals Re $\hat{v}_C e^{-i\omega t}$. The prescription $\partial/\partial t \to -i\omega$ (discussed in Sec. 1-8) converts Eq. (3) into an algebraic equation for the complex amplitude associated with ψ, the solution of which leads to

$$\psi(t) = \text{Re}\, A e^{ika} e^{-i\omega t} \tag{4-2.4}$$

with the abbreviation

$$A = \frac{\hat{v}_C a^3 e^{-ika}}{2 - (ka)^2 - 2ika} = \frac{\hat{v}_C}{[(d^2/dr^2)(r^{-1}e^{ikr})]_{r=a}} \tag{4-2.5}$$

The quantity $\psi(t - r/c + a/c)$ is obtained by inserting an additional factor of $e^{ik(r-a)}$ in Eq. (4). With this insertion, with subsequent substitution of $\psi(t - r/c + a/c)$ into Eq. (2), and with the relations $p = -\rho\partial\Phi/\partial t$ and $\mathbf{v} = \nabla\Phi$, the spatially dependent amplitudes of the field quantities are identified as

$$\hat{\Phi} = A \cos\theta\, \frac{d}{dr}\frac{e^{ikr}}{r} \tag{4-2.6a}$$

$$\hat{p} = i\omega\rho\hat{\Phi} \qquad \hat{v}_r = \frac{\partial\hat{\Phi}}{\partial r} \qquad \hat{v}_\theta = r^{-1}\frac{\partial\hat{\Phi}}{\partial\theta} \tag{4-2.6b}$$

In the above expressions, the operation by d/dr and d^2/dr^2 leads to multiplicative factors of $d/dr \to ik - 1/r$ and $d^2/dr^2 \to -k^2 + 2r^{-2} - 2ikr^{-1}$. Thus, in the *far field*, where $kr \gg 1$, one has

$$\hat{p} \approx -\frac{\omega^2\rho}{c} A \cos\theta\,\frac{e^{ikr}}{r} \approx \rho c \hat{v}_r \tag{4-2.7}$$

† See, for example, K. N. Tong, *Theory of Mechanical Vibration*, Wiley, New York, 1960, pp. 31–37. The differential equation can also be solved by the *method of variation of parameters* described, for example, by C. R. Wylie, Jr., *Advanced Engineering Mathematics*, McGraw-Hill, New York, 1951, pp. 41–44. Transient solutions and their implications for sound radiated by a transversely accelerating sphere are reviewed by A. Akay and T. H. Hodgson, "Sound Radiation from an Accelerated or Decelerated Sphere," *J. Acoust. Soc. Am.*, **63**:313–318 (1978). The earliest such solutions, for spheres suddenly accelerated from rest to a uniform velocity and to a sinusoidally oscillating velocity, are given by G. Kirchhoff, *Mechanik*, 2d ed., Teubner, Leipzig, 1877, pp. 317–321.

Because $|\hat{v}_\theta|$ decreases at large r as r^{-2} rather than r^{-1}, it is negligible. The time-averaged intensity $(pv_r)_{av}$ derived from this far-field approximation is

$$I_{r,av} = \frac{(ka)^4}{4 + (ka)^4} \rho c[(\mathbf{v}_C \cdot \mathbf{e}_r)^2]_{av} \left(\frac{a}{r}\right)^2 \tag{4-2.8}$$

The same expression, moreover, holds for all points outside the sphere. The time average $(pv_\theta)_{av}$ vanishes identically since $\hat{p}$ and $\hat{v}_\theta$ are 90° out of phase. Because of the cos θ factor in the acoustic pressure and of the $\cos^2\theta$ factor in the acoustic intensity, there is no sound at right angles to the direction of the sphere's translation; the sound is most intense in the directions $\theta = 0$ or 180°, where the sphere's motion is directly toward or away from the listener.

The time average of the acoustic power emitted is the integral of $I_{r,av}$ over the surface of a sphere, or

$$\mathcal{P}_{av} = \left[\frac{(ka)^4}{4 + (ka)^4}\right] \rho c \frac{(v_C^2)_{av}}{3} 4\pi a^2 \tag{4-2.9}$$

Here $4\pi a^2$ is the surface area of the oscillating sphere; $\frac{1}{3}(v_C^2)_{av}$ is the surface average of $[(\mathbf{v}_C \cdot \mathbf{n})^2]_{av}$. In the large ka limit (when the factor in brackets becomes 1), each element of the sphere's surface radiates sound as if it were a segment of a very large flat surface vibrating perpendicularly to itself with velocity $\mathbf{v}_C \cdot \mathbf{n}$. In the opposite limit, where $ka \ll 1$, $\mathcal{P}_{av}$ is smaller than its high-frequency limit by a factor of $(ka)^4/4$, while the corresponding factor in the same limit for a radially oscillating sphere [see Eq. (4-1.5)] is $(ka)^2$. Since, in this low ka limit, $(ka)^2 \gg \frac{1}{4}(ka)^4$, the radially oscillating sphere is a much more efficient radiator of sound at low frequencies than the transversely oscillating sphere, given that the surface-averaged mean squared normal velocities are of comparable magnitude.

Force Exerted by Transversely Oscillating Sphere

The net force exerted on the fluid by the sphere, in accord with Newton's third law, is the surface integral of $p(a, \theta, t)\mathbf{e}_r$. Symmetry requires that this force have only a z component, so one has

$$\mathbf{F}(t) = F_z(t)\mathbf{e}_z = \mathbf{e}_z a^2 \int_0^{2\pi} \int_0^{\pi} p(a, \theta, t) \cos\theta \sin\theta \, d\theta \, d\phi \tag{4-2.10}$$

The complex amplitude associated with this force is consequently found, from Eqs. (6), to be

$$\hat{F}_z = i\omega\rho A(ika - 1)\tfrac{4}{3}\pi e^{ika} \tag{4-2.11}$$

where A is the constant in Eq. (5). The time-averaged acoustic power $\mathcal{P}_{av}$ transmitted to the fluid must be $\frac{1}{2}$ Re $\hat{F}_z \hat{v}_C^*$ and, in accord with the acoustic-energy conservation theorem, this leads to the same result as Eq. (9).

Small-ka Approximation

In the low-frequency limit, when $v_C(t)$ is oscillating at frequencies such that $ka \ll 1$, the appropriate approximation to Eq. (3) results when the first two terms on the left side are neglected, so $\psi(t) = \frac{1}{2}a^3 v_C(t)$; it is also consistent in this limit to approximate $\psi(t - r/c + a/c)$ by $\psi(t - r/c)$. Consequently, the velocity potential in Eq. (2) approximates to

$$\Phi = \tfrac{1}{2}a^3 \cos\theta \, \frac{\partial}{\partial r}\left[\frac{1}{r} v_C\left(t - \frac{r}{c}\right)\right] \tag{4-2.12}$$

The corresponding approximation for p, resulting from the relation $p = -\rho\, \partial\Phi/\partial t$, yields

$$p = \frac{1}{2}\frac{\rho a^3}{c}\frac{1}{r}\mathbf{e}_r \cdot \left[\left(\frac{\partial}{\partial t} + \frac{c}{r}\right)\dot{\mathbf{v}}_C\left(t - \frac{r}{c}\right)\right] \tag{4-2.13}$$

Also, in this low-frequency or $ka \ll 1$ approximation, the force amplitude $\hat{F}_z$ given by Eq. (11) reduces, in lowest nonzero order, to $-i\omega\rho A\frac{4}{3}\pi$, while A, from Eq. (5), reduces to $\hat{v}_C a^3/2$. The time-dependent force $\mathbf{F}(t) = \mathbf{e}_z \operatorname{Re} \hat{F}_z e^{-i\omega t}$ consequently appears in this approximation as†

$$\mathbf{F}(t) = \tfrac{1}{2}m_d \dot{\mathbf{v}}_C(t) \tag{4-2.14}$$

where $m_d = \frac{4}{3}\pi a^3 \rho$ is the mass displaced by the sphere. This resembles Newton's second law, force equals mass times acceleration, with an apparent *entrained mass* equal to $m_d/2$. However, this approximate $F_z(t)$ is 90° out of phase with $v_C(t)$ and is consequently inadequate for a nonzero estimate of the time-averaged acoustic power $\mathcal{P}_{\text{av}} = \frac{1}{2} \operatorname{Re} \hat{F}_z \hat{v}_C^*$. The lowest-order approximation for the resistive part (that in phase with $\hat{v}_C$) of the complex amplitude, derived from Eq. (11), is

$$(\hat{F}_z)_{\text{resist}} \approx \frac{(ka)^4 \rho c \pi a^2 \hat{v}_C}{3} \tag{4-2.15}$$

and this suffices to reproduce the $\mathcal{P}_{\text{av}}$ in Eq. (9) to lowest nonzero order in ka.

4-3 MONOPOLES AND GREEN'S FUNCTIONS

Concept of a Point Source

Any spherically symmetric source of sound of angular frequency ω in an unbounded fluid gives rise to an outgoing spherically symmetric wave, the

† P. M. Morse, *Vibration and Sound,* 2d ed., McGraw-Hill, New York, 1948, p. 319. For incompressible flow, this dates back to George Green, "On the Vibrations of Pendulums in Fluid Media," 1833, reprinted in N. M. Ferrers (ed.), *Mathematical Papers of the Late George Green,* Macmillan, London, 1871, pp. 315–324, and to G. G. Stokes, "On Some Cases of Fluid Motion," 1843, reprinted in G. G. Stokes, *Mathematical and Physical Papers,* vol. 1, Cambridge University Press, Cambridge, 1880, pp. 2–68. A modern derivation is given by C.-H. Yih, *Fluid Mechanics,* McGraw-Hill, New York, 1969, pp. 99–108.

complex velocity-potential amplitude, the complex pressure amplitude, and time-averaged power output of which [see Eqs. (1-12.8a) and (1-12.9a)] are representable in the form

$$\hat{\Phi} = -\hat{Q}_S \frac{e^{ikR}}{4\pi R} \qquad \hat{p} = \hat{S}\frac{e^{ikR}}{R} \qquad \mathcal{P}_{\text{av}} = \frac{2\pi|\hat{S}|^2}{\rho c} \tag{4-3.1}$$

where the *source-strength amplitude* $\hat{Q}_S = -4\pi\hat{S}/i\omega\rho$ is a constant and $R = |\mathbf{x} - \mathbf{x}_S|$ is radial distance from the center of the source (at $\mathbf{x}_S$). The constant $\hat{S}$ is here referred to as the *monopole amplitude*. One possible realization of a source of such a wave would be the radially oscillating sphere discussed in Sec. 4-1, in which case $\hat{S}$ results from the coefficient of $r^{-1}e^{ikr}$ in Eq. (4-1.4). One can consider a hypothetical limiting case for which a becomes progressively smaller but $\hat{v}_S$ becomes simultaneously larger, such that $\hat{S} \approx -i\omega\rho a^2\hat{v}_S$ remains constant. The sphere is then idealized as a point. Although an extremely small source of sufficiently large strength to generate audible sound at appreciable distances would in actuality require consideration of nonlinear terms, the concept of a *point source*† (or *acoustic monopole*) generating waves governed by the linear acoustic equations is a convenient extrapolation consistent with the general framework of linear acoustic theory. Typically, any small source, with time-varying mass of fluid in any small volume enclosing it, has all the attributes of a point source, providing the dimensions of the source are small compared with a wavelength and the discussion of the sound field is restricted to radial distances greater than several body diameters. (This is discussed in Sec. 4-7.)

The field of Eq. (1) satisfies the Helmholtz equation (1-8.13) everywhere except at the source; it is a limiting form (as $\varepsilon \to 0$) of some particular solution of the inhomogeneous equation‡ (see Fig. 4-3)

$$(\nabla^2 + k^2)\hat{p}_\varepsilon = -4\pi\hat{S}\,\Delta_\varepsilon(R) \tag{4-3.2}$$

where the right side has an R-dependent factor $\Delta_\varepsilon(R)$.

Insight into the possible choices for $\Delta_\varepsilon(R)$ such that $\hat{S}R^{-1}e^{ikR}$ will be a solution at finite R results after integration of both sides of Eq. (2) over the volume of a sphere of radius R_0 centered at $\mathbf{x}_S$. Since $\nabla^2\hat{p}_\varepsilon$ is the divergence of $\nabla\hat{p}_\varepsilon$, the resulting first term becomes a surface integral. If $\hat{p}_\varepsilon$ is spherically symmetric (which would follow from symmetry if there were no external boundaries and which is approximately true if R_0 is sufficiently small compared with the distance to the nearest boundary), then the angular integration in each term results in a factor 4π, representing the total solid angle about a point. In this

† The concept, which is analogous to those of a point mass and of a point charge, was introduced into acoustics by H. Helmholtz, "Theory of Air Oscillations in Tubes with Open Ends," *J. Reine Angew. Math.*, **57**:1–72 (1860).

‡ The inhomogeneous Helmholtz equation for $k = 0$ is the mathematical equivalent of *Poisson's equation*, $\nabla^2 V = -4\pi G\rho$, originally introduced as a relation between gravitational potential and mass density by Poisson in 1813.

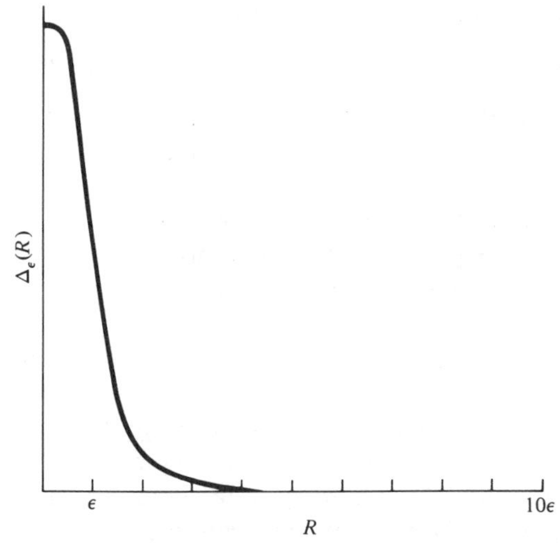

Figure 4-3 Possible form of a function $\Delta_\varepsilon(R)$ that is concentrated where $R < \varepsilon$, negligibly small for $R > 10\varepsilon$. As explained in the text, the integral of $4\pi R^2\,\Delta_\varepsilon(R)$ over R should be 1.

manner, one obtains

$$4\pi R_0^2 \left(\frac{\partial \hat{p}_\varepsilon}{\partial R}\right)_{R_0} + 4\pi k^2 \int_0^{R_0} \hat{p}_\varepsilon R^2\, dR = -4\pi \hat{S} \iiint \Delta_\varepsilon(R)\, dV \qquad (4\text{-}3.3)$$

Suppose $\Delta_\varepsilon(R)$ is concentrated within the region $R < \varepsilon$ and is negligible for R greater than, say, 10ε. If one takes $R_0 = 10\varepsilon$, the integral on the right side is approximately the same as if carried over all space. Also, if Δ_ε is to be such that $\hat{p}$ in Eq. (1) is a solution for $R \geq R_0$, then, for sufficiently small ε, the overall volume integral of $\Delta_\varepsilon(R)$ must be such that (3) is satisfied if $\hat{p}_\varepsilon$ is replaced by $\hat{S}R^{-1}e^{ikR}$. This insertion and subsequent evaluation of the indicated derivative and integral lead to a value for the left side that is identically $-4\pi\hat{S}$ regardless of the value of R_0. The right side of Eq. (3) must have the same value, so the appropriate identification of $\Delta_\varepsilon(R)$ is any function whose volume integral is 1 and which is of appreciable magnitude only for values of R less than, say, 10ε. Such a function would be $\delta_\varepsilon(x - x_S)\delta_\varepsilon(y - y_S)\delta_\varepsilon(z - z_S)$, where $\delta_\varepsilon(x)$, defined by Eq. (2-8.7), is an element in the sequence describing the Dirac delta function. This identification leads to the generalized function relation†

$$(\nabla^2 + k^2)\hat{p} = -4\pi\hat{S}\,\delta(\mathbf{x} - \mathbf{x}_S) = -4\pi\hat{S}\,\delta(x - x_S)\,\delta(y - y_S)\,\delta(z - z_S) \qquad (4\text{-}3.4)$$

The indicated product expression defining $\delta(\mathbf{x} - \mathbf{x}_S)$ implies that this Dirac delta

† A delta function on the right side of the wave equation to denote the presence of a point source was used as early as 1937 in their theory of Cherenkov radiation by I. Frank and I. Tamm, *CR Dokl. Acad. Sci. URSS*, **14**:109–114 (1937). Its widespread use today was undoubtedly considerably influenced by the chapter on Green's functions in P. M. Morse and H. Feshbach, *Methods of Theoretical Physics*, vol. 1, McGraw-Hill, New York, 1953, pp. 791–895.

function with vector argument must be such that for any function $\psi(\mathbf{x})$

$$\iiint \psi(\mathbf{x})\delta(\mathbf{x} - \mathbf{x}_S)\, dV = \psi(\mathbf{x}_S) \tag{4-3.5}$$

The strict interpretation of Eq. (4) is that $\hat{p}$ should be the limit as $\varepsilon \to 0$ of the solution $\hat{p}_\varepsilon$ of Eq. (2), but for most purposes the process of taking such a limit need not be considered explicitly.

Another interpretation of Eq. (4) is that $\hat{p}$ should be a solution of the homogeneous equation except in the near neighborhood (of vanishing volume) of $\mathbf{x}_S$ and that near $\mathbf{x}_S$ it should become singular as R^{-1} in such a way that

$$\hat{p} = \frac{\hat{S}}{R} + \hat{S}\, f(x, y, z) \tag{4-3.6}$$

where $f(x, y, z)$ is bounded at $\mathbf{x}_S$. To prove this assertion, one recognizes that any solution of the inhomogeneous differential equation can be represented as any particular solution plus some solution of the homogeneous equation. The particular solution $\hat{S}R^{-1}e^{ikR}$ approaches $\hat{S}/R$ plus bounded terms as $R \to 0$, and the solution of the homogeneous equation is bounded; so Eq. (6) results. This interpretation applies in particular when the propagation of sound away from the source is altered by the presence of bounding surfaces, e.g., a source above the ground. Note also that Eq. (6) is equivalent to the condition $R\hat{p} \approx \hat{S}$ near $\mathbf{x}_S$.

Point Mass Source

A differential equation analogous to (4) results from the linear acoustic equations when a point-mass-source term is added to the linear version of the mass-conservation equation; i.e., one replaces the zero on the right side of Eq. (1-5.3a) by $\dot{m}_S(t)\,\delta(\mathbf{x} - \mathbf{x}_S)$, where $\dot{m}_S(t)$ is the rate at which mass is added (negative if extracted) to the fluid existing outside some small fixed region enclosing the source (see Fig. 4-4). Alternately, one can interpret $\dot{m}_S$ as ρQ_S, where $Q_S(t)$ is the integral of $\mathbf{v} \cdot \mathbf{n}$ over a small surface enclosing the source and accordingly represents the time rate of change of the volume excluded from the fluid by the source.

If the derivation outlined in Sec. 1-6 of the wave equation is carried through with the mass-conservation equation modified by the inclusion of a point-mass-source term, the result is the inhomogeneous wave equation

$$\nabla^2 p - \frac{1}{c^2}\frac{\partial^2 p}{\partial t^2} = -\ddot{m}_S(t)\,\delta(\mathbf{x} - \mathbf{x}_S) = -\rho\dot{Q}_S(t)\,\delta(\mathbf{x} - \mathbf{x}_S) \tag{4-3.7}$$

The solution appropriate to an unbounded fluid can be developed from the solution $\hat{S}R^{-1}e^{ikR}$ of Eq. (4) and from the superposition principle. The quantities $\hat{p}$ and $\hat{S}$ can be interpreted as the Fourier transforms of p and $\ddot{m}_S/4\pi$, so Eq. (4) follows from (7). The product of e^{ikR} times the Fourier transform of $\ddot{m}(t)$, however, is the Fourier transform of $\ddot{m}(t - R/c)$. The Fourier integral theorem consequently gives

$$p = (4\pi R)^{-1}\ddot{m}_S\left(t - \frac{R}{c}\right) \tag{4-3.8}$$

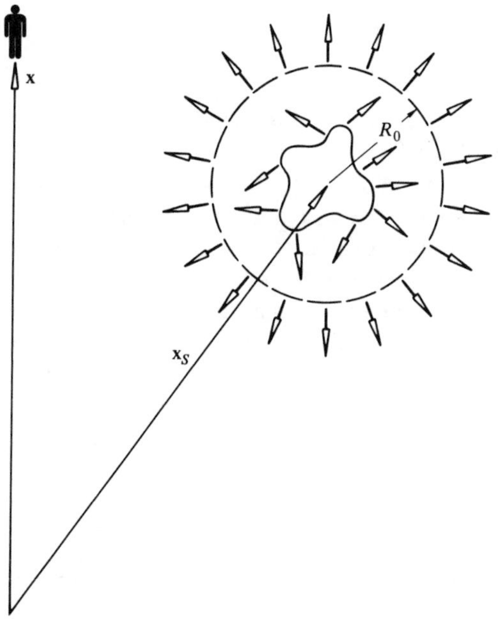

Figure 4-4 Sketch supporting idealization of a small source with time-varying volume as a point-mass source. The fluid flow in a small sphere (radius R_0) surrounding the source is approximately such that the rate of mass flow through the sphere's surface is ρ times the time derivative of the volume enclosed by the source.

which is equivalent to Eq. (4-1.6), previously derived for the radially oscillating sphere in the limit $ka \ll 1$.

If boundaries are to be taken into account, an appropriate solution of the homogeneous equation should be added. Regardless of what such solution is added, one can argue [in a manner similar to that leading to Eq. (6)] that the presence of the delta function on the right side of the wave equation is equivalent to the requirement that in the vicinity of $\mathbf{x}_S$

$$p \approx \frac{\ddot{m}_S(t)}{4\pi R} + f(x, y, z, t) \tag{4-3.9}$$

where $f(x, y, z, t)$ is bounded in magnitude.

Green's Functions

The solution of Eq. (4) with $\hat{S} = 1$, satisfying whatever boundary conditions (presumed passive) are imposed by the presence of external surfaces or causality considerations, is the *Green's function*† $G_k(\mathbf{x}|\mathbf{x}_S)$, the first argument denoting the location of the listener and the second the location of the source. Thus,

† The name derives from George Green's use of analogous functions in connection with Laplace's equation to derive solutions of electrostatic and magnetostatic boundary value problems. (G. Green, *An Essay on the Application of Mathematical Analysis to the Theories of Electricity and Magnetism,* Nottingham, 1828, pp. 10–13.)

$G_k(\mathbf{x}|\mathbf{x}_S)$ satisfies the inhomogeneous equation

$$(\nabla^2 + k^2)G_k(\mathbf{x}|\mathbf{x}_S) = -4\pi\delta(\mathbf{x} - \mathbf{x}_S) \tag{4-3.10}$$

and if the medium external to the source is unbounded, G_k is identified from Eq. (1) as the *free-space Green's function* $R^{-1}e^{ikR}$.

A universal property of Green's functions is the *reciprocity relation* $G_k(\mathbf{x}|\mathbf{x}_S) = G_k(\mathbf{x}_S|\mathbf{x})$; that is, G_k is unchanged if source and listener locations are interchanged. The free-space Green's function satisfies this trivially; the proof of reciprocity for more general circumstances is deferred to Sec. 4-9.

The superposition principle allows the Green's function to be used in the construction of solutions corresponding to several point sources (see Fig. 4-5). Thus, if one has N point sources, the complex acoustic-pressure amplitude should satisfy the Helmholtz equation with a sum of source terms, $-4\pi\hat{S}_n\delta(\mathbf{x} - \mathbf{x}_n)$, on the right side; the appropriate solution resulting from Eq. (10) is

$$\hat{p} = \sum_{n=1}^{N} \hat{S}_n G_k(\mathbf{x}|\mathbf{x}_n) \tag{4-3.11}$$

Similarly, for a continuous smear of sources where $\hat{s}(\mathbf{x})$ denotes the monopole-amplitude distribution per unit volume, one has

$$\nabla^2\hat{p} + k^2\hat{p} = -4\pi\hat{s}(\mathbf{x}) = -4\pi\iiint\hat{s}(\mathbf{x}_S)\,\delta(\mathbf{x} - \mathbf{x}_S)\,dV_S \tag{4-3.12}$$

$$\hat{p} = \iiint G_k(\mathbf{x}|\mathbf{x}_S)\hat{s}(\mathbf{x}_S)\,dV_S \tag{4-3.13}$$

where the integration extends over the source volume.

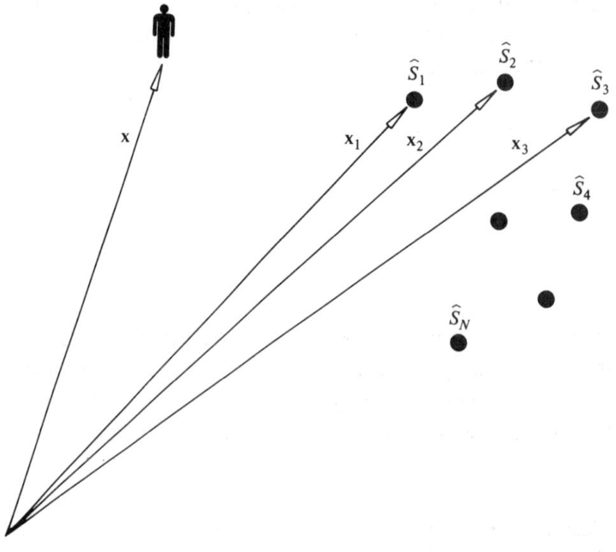

Figure 4-5 Nomenclature for discussion of sound radiation from N point sources. Here $\hat{S}_n$ and $\mathbf{x}_n$ denote monopole amplitude and location of the nth point source.

The Green's function $G(\mathbf{x}, t|\mathbf{x}_S, t_S)$ (corresponding to a unit point impulsive source) for the wave equation satisfies

$$\left(\nabla^2 - \frac{1}{c^2}\frac{\partial^2}{\partial t^2}\right) G(\mathbf{x}, t|\mathbf{x}_S, t_S) = -4\pi\delta(t - t_S)\,\delta(\mathbf{x} - \mathbf{x}_S) \qquad \text{(4-3.14)}$$

and from causality considerations should be zero if $t < t_S$. The solution when the external medium is unbounded results from Eqs. (7) and (8) with $\ddot{m}(t) \to 4\pi\delta(t - t_S)$; that is,

$$G(\mathbf{x}, t|\mathbf{x}_S, t_S) = \frac{\delta(t - t_S - R/c)}{R} \qquad \text{(4-3.15)}$$

The function satisfying Eq. (14) can be used to develop a solution for a distributed transient source of the inhomogeneous wave equation, where a source term $-4\pi s(\mathbf{x}, t)$ is on the right side. The source function $s(\mathbf{x}, t)$ is written as a time and volume integral (the differential of integration being $dt_S\, dV_S$) in a manner analogous to that depicted in Eq. (12). The superposition principle and Eq. (14) then yield

$$p = \iiiint G(\mathbf{x}, t|\mathbf{x}_S, t_S) s(\mathbf{x}_S, t_S)\, dV_s\, dt_S \qquad \text{(4-3.16)}$$

When the Green's function is given by Eq. (15), the t_S integration can be done using the property (2-8.9) of the delta function and one accordingly obtains

$$p = \iiint \frac{s(\mathbf{x}_S, t - R/c)}{R}\, dV_S \qquad \text{(4-3.17)}$$

The retarded time $t - R/c$ in the argument of s implies that the contribution from each portion of the source travels to the listener with the sound speed.

4-4 DIPOLES AND QUADRUPOLES

Dipoles

The superposition of fields of two or more monopoles located at different points gives a possible acoustic field because of the linearity of the basic equations. One can conceive, in particular, of two point sources (see Fig. 4-6) of opposite monopole amplitudes $\hat{S}$ and $-\hat{S}$, that is, 180° out of phase with each other, and located a distance d apart at $\mathbf{x}_S + \mathbf{d}/2$ and $\mathbf{x}_S - \mathbf{d}/2$. [If the monopoles are both radially oscillating spheres of nominal radius a, then a should be substantially less than d so that the acoustic-pressure field in the vicinity of either source will be dominated by a $1/R$ term, as required in Eq. (4-3.6).]

A point dipole corresponds to the limit in which d becomes small enough to ensure that $kd \ll 1$. In this limit and given $|\mathbf{x} - \mathbf{x}_S| \gg d$, $G_k(\mathbf{x}|\mathbf{x}_S \pm \mathbf{d}/2)$ can be approximated with a truncated Taylor series as $G_k(\mathbf{x}|\mathbf{x}_S) \pm (\mathbf{d}/2) \cdot \nabla_S G_k(\mathbf{x}|\mathbf{x}_S)$, where the operator ∇_S denotes the gradient with respect to the source coordi-

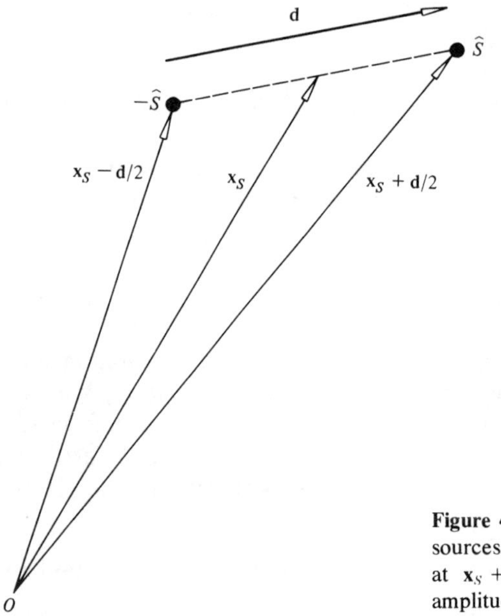

Figure 4-6 Acoustic dipole modeled by two point sources of monopole amplitudes $\hat{S}$ and $-\hat{S}$ located at $\mathbf{x}_S + \mathbf{d}/2$ and $\mathbf{x}_S - \mathbf{d}/2$. The dipole-moment amplitude vector $\hat{\mathbf{D}}$ is $\hat{S}\mathbf{d}$.

nates. Thus, the superimposed pressure field becomes

$$\hat{p} = \hat{\mathbf{D}} \cdot \nabla_S G_k(\mathbf{x}|\mathbf{x}_S) \tag{4-4.1}$$

where the complex amplitude $\hat{\mathbf{D}}$ (*dipole-moment amplitude vector*) replaces $\hat{S}\mathbf{d}$. Since $G_k(\mathbf{x}|\mathbf{x}_S)$ satisfies Eq. (4-3.10), the differential equation that (1) must satisfy is

$$\nabla^2\hat{p} + k^2\hat{p} = (\hat{\mathbf{D}} \cdot \nabla_S)[-4\pi\delta(\mathbf{x} - \mathbf{x}_S)] = 4\pi\hat{\mathbf{D}} \cdot \nabla\delta(\mathbf{x} - \mathbf{x}_S) \tag{4-4.2}$$

If the fluid surrounding the dipole is unbounded, the function $G_k(\mathbf{x}|\mathbf{x}_S)$ is $R^{-1}e^{ikR}$. Since $\nabla_S f = (df/dR)\nabla_S R$ for any function $f(R)$ of R and since $\nabla_S R = (\mathbf{x}_S - \mathbf{x})/R$, the acoustic field (1) for a dipole in an unbounded fluid becomes

$$\hat{p} = -\hat{\mathbf{D}} \cdot \mathbf{e}_R \frac{d}{dR}\frac{e^{ikR}}{R} = -\nabla \cdot (\hat{\mathbf{D}}R^{-1}e^{ikR}) \tag{4-4.3}$$

Here $\mathbf{e}_R = (\mathbf{x} - \mathbf{x}_S)/R$ is the unit vector pointing radially outward from the dipole center toward the observation point.

Point Force in a Fluid

The model of a point time-varying concentrated force† $\mathbf{F}(t)$ applied at a point $\mathbf{x}_S$ within a fluid furnishes another instance of the generation of a dipole field. Such

† J. W. S. Rayleigh, *The Theory of Sound,* vol. 2, 2d ed., 1896, reprinted by Dover, New York, 1945, sec. 375.

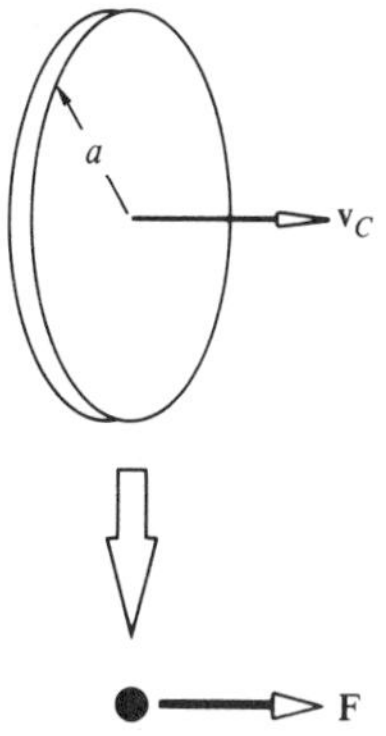

Figure 4-7 Transversely oscillating thin disk of radius a (where $ka \ll 1$) as a possible physical realization of a point force applied to a fluid. As discussed in Sec. 4-8, the apparent equivalent force $\mathbf{F}(t)$ is $\frac{8}{3}\rho a^3 \dot{\mathbf{v}}_C$, where $\dot{\mathbf{v}}_C$ is the transverse acceleration of the disk.

a model can be approximately realized by a very thin rigid disk of radius a (see Fig. 4-7) oscillating transverse to its face, with $\mathbf{F}(t)$ identified as the net force exerted by the disk on the adjacent fluid. [The value $\mathbf{F}(t) = \frac{8}{3}\rho a^3 \dot{\mathbf{v}}_C$ is derived in Sec. 4-8.] The presence of the force is taken into account by the inclusion of a term $\mathbf{F}(t)\,\delta(\mathbf{x} - \mathbf{x}_S)$ on the right side of the linear version of Euler's equation of motion for a fluid.

The corresponding inhomogeneous wave equation is derived by taking the divergence of both sides of the Euler equation with the source term included and subsequently replacing $\rho\nabla \cdot \mathbf{v}$ by $-\partial\rho'/\partial t$, in accord with the conservation of mass equation, then replacing $\partial\rho'/\partial t$ by $c^{-2}\,\partial p/\partial t$. In this manner, one obtains

$$\nabla^2 p - \frac{1}{c^2}\frac{\partial^2 p}{\partial t^2} = \nabla \cdot [\mathbf{F}(t)\,\delta(\mathbf{x} - \mathbf{x}_S)] = -\mathbf{F}(t) \cdot \nabla_S \delta(\mathbf{x} - \mathbf{x}_S) \qquad (4\text{-}4.4)$$

Consequently, for the constant-frequency case, an equation of the same form as Eq. (1) results, but with $4\pi\hat{\mathbf{D}}$ replaced by $\hat{\mathbf{F}}$. The solution when the fluid is unbounded is given by Eq. (3) with $\hat{\mathbf{D}}$ replaced by $\hat{\mathbf{F}}/4\pi$. Therefore, by the same process by which Eq. (4-3.8) was derived, one can identify the transient solution as

$$p = \frac{1}{4\pi}\mathbf{e}_R \cdot \left(\frac{1}{R} + \frac{1}{c}\frac{\partial}{\partial t}\right)\frac{\mathbf{F}(t - R/c)}{R} \qquad (4\text{-}4.5)$$

Quadrupoles

The simplest conceptual realization of a quadrupole is two closely spaced dipoles† (see Fig. 4-8) with equal but opposite dipole-moment amplitude vectors.

† The definition here of quadrupole radiation is the same as that of M. J. Lighthill, "On Sound Generated Aerodynamically, I: General Theory," *Proc. R. Soc. Lond.*, **A211**:564–587 (1952). The term is sometimes used to denote the portion of a field whose amplitude falls off with r as r^{-2} or to denote the portion expressible in terms of second-order spherical harmonics, but the proper definition is for a field resembling that corresponding to a limiting case of four closely spaced point monopoles whose aggregate source strength and dipole moment vanish. In the case of solutions of Laplace's equation $\nabla^2\Phi = 0$, a quadrupole field in an unbounded space also has the properties mentioned above.

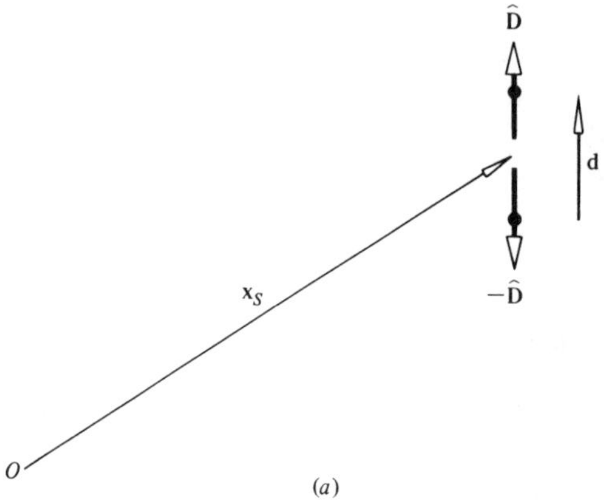

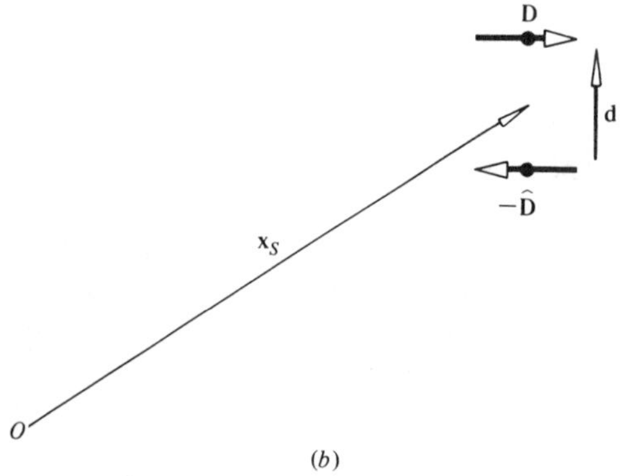

Figure 4-8 Possible models of acoustic quadrupoles: (*a*) longitudinal quadrupole; (*b*) lateral quadrupole. The general model discussed in the text consists of two dipoles with dipole-moment amplitude vectors $\hat{\mathbf{D}}$ and $-\hat{\mathbf{D}}$ at $\mathbf{x}_S + \mathbf{d}/2$ and $\mathbf{x}_S - \mathbf{d}/2$.

Such a model would give, from Eq. (1), a superposition of the dipole fields $\pm\mathbf{D} \cdot \nabla_S G_k(\mathbf{x}|\mathbf{x}_S \pm \mathbf{d}/2)$; the sum, in the limit of $d \ll R$, $kd \ll 1$, approximates to

$$\hat{p} = (\hat{\mathbf{D}} \cdot \nabla_S)(\mathbf{d} \cdot \nabla_S)G_k(\mathbf{x}|\mathbf{x}_S) \tag{4-4.6}$$

If the medium is unbounded, the Green's function is $R^{-1}e^{ikR}$ and since $\nabla_S = -\nabla$ when applied to a function of $\mathbf{x} - \mathbf{x}_S$, one has

$$\hat{p} = (\hat{\mathbf{D}} \cdot \nabla)(\mathbf{d} \cdot \nabla)\frac{e^{ikR}}{R} = \sum_{\mu,\nu=1}^{3} \hat{Q}_{\mu\nu} \frac{\partial^2}{\partial x_\mu\, \partial x_\nu}\frac{e^{ikR}}{R} \tag{4-4.7}$$

where we write $\hat{Q}_{\mu\nu} = \hat{D}_\mu d_\nu$. (One can also define $\hat{Q}_{\mu\nu}$ as the average of $\hat{D}_\mu d_\nu$ and $\hat{D}_\nu d_\mu$.)

Since $\hat{\mathbf{D}}$ and $\mathbf{d}$ are vectors whose directions are arbitrary and since $\partial^2/(\partial x \partial y)$ is the same as $\partial^2/(\partial y \partial x)$, the above implies that any quadrupole field in an unbounded space is a linear combination of six functions corresponding to the differential operators $\partial^2/\partial x^2$, $\partial^2/\partial y^2$, $\partial^2/\partial z^2$, $\partial^2/(\partial x \partial y)$, $\partial^2/(\partial x \partial z)$, and $\partial^2/(\partial y \partial z)$ applied to $R^{-1}e^{ikR}$. Of these, there are two basic types: a *longitudinal quadrupole,* for which $\hat{\mathbf{D}}$ and $\mathbf{d}$ are parallel, and a *lateral quadrupole,* for which they are perpendicular.

The field of an axial quadrupole aligned along the z axis is given, according to Eq. (7), by

$$\hat{p} = \hat{Q}_{zz} \left[(1 - 3\cos^2\theta)\left(\frac{ik}{R} - \frac{1}{R^2} + \frac{k^2}{3}\right) - \frac{k^2}{3}\right] \frac{e^{ikR}}{R} \tag{4-4.8}$$

where θ is the angle between $\mathbf{e}_R$ and the z direction, so $\cos\theta$ is $(z - z_S)/R$. Similarly, for a lateral quadrupole with $\hat{\mathbf{D}}$ in the x direction and with $\mathbf{d}$ in the y direction, one finds

$$\hat{p} = \hat{Q}_{xy} \frac{(x - x_S)(y - y_S)}{R^2} (-k^2 - 3ikR^{-1} + 3R^{-2}) \frac{e^{ikR}}{R} \tag{4-4.9}$$

Since the intensity in the far field ($kR \gg 1$) is radial, and since its time average equals $\frac{1}{2}|\hat{p}|^2/\rho c$, Eqs. (8) and (9) yield

$$I_{r,\text{av}} = \begin{cases} \dfrac{(k^4 \cos^4\theta)|\hat{Q}_{zz}|^2}{2\rho c R^2} & \text{longitudinal} \quad (4\text{-}4.10a) \\[2ex] \dfrac{k^4 \sin^4\theta \cos^2\phi \sin^2\phi}{2\rho c R^2} |\hat{Q}_{xy}|^2 & \text{lateral} \quad (4\text{-}4.10b) \end{cases}$$

The radiation patterns in the two cases vary with θ and ϕ as $\cos^4\theta$ and as $\sin^4\theta \cos^2\phi \sin^2\phi$ (see Fig. 4-9). The total acoustic power outputs (time average) found by integrating the appropriate expression for $I_{r,\text{av}}$ over the surface of a sphere of radius R are $\pi k^4/\rho c$ times $\frac{2}{5}|\hat{Q}_{zz}|^2$ and $\frac{2}{15}|\hat{Q}_{xy}|^2$, since the area averages of $\cos^4\theta$ and $\sin^4\theta \cos^2\phi \sin^2\phi$ are $\frac{1}{5}$ and $\frac{1}{15}$.

Multipole Expansions

A number (*array*) of monopole sources of the same frequency gives rise to a composite acoustic field whose complex acoustic-pressure amplitude is of the

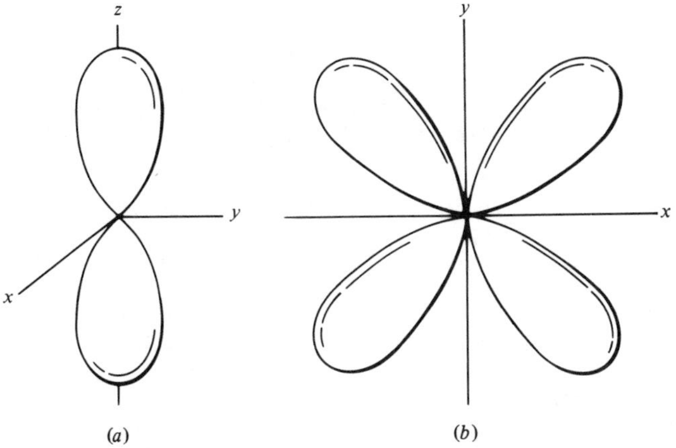

Figure 4-9 Radiation patterns of (*a*) a longitudinal quadrupole and (*b*) a lateral quadrupole. Here distance from the origin to a point on a sketched surface is proportional to the magnitude of the acoustic intensity in the same direction.

form of Eq. (4-3.11); we assume in what follows that the external medium is unbounded, so that the contribution to the sum from the nth source is $\hat{S}_n R_n^{-1} e^{ikR_n}$, where $R_n = |\mathbf{x} - \mathbf{x}_n|$. If the sources are clustered in the vicinity of the origin within a volume of radius d, where $kd \ll 1$, an expansion of $R_n^{-1} e^{ikR_n}$ in a multiple power series in the source coordinates should be rapidly convergent at $r \gg d$, so we replace†

$$R_n^{-1} e^{ikR_n} = [\exp(-\mathbf{x}_n \cdot \nabla)]\, (r^{-1} e^{ikr}) \tag{4-4.11a}$$

$$\exp(-\mathbf{x}_n \cdot \nabla) = 1 - \mathbf{x}_n \cdot \nabla + \frac{1}{2!}(\mathbf{x}_n \cdot \nabla)(\mathbf{x}_n \cdot \nabla) - \cdots \tag{4-4.11b}$$

The sum over sources then becomes

$$\hat{p} = \hat{S} r^{-1} e^{ikr} - \hat{\mathbf{D}} \cdot \nabla(r^{-1} e^{ikr}) + \sum_{\mu,\nu} \hat{Q}_{\mu\nu} \frac{\partial^2}{\partial x_\mu\, \partial x_\nu} r^{-1} e^{ikr} + \cdots \tag{4-4.12}$$

where we use the abbreviations

$$\hat{S} = \sum_n \hat{S}_n \qquad \hat{\mathbf{D}} = \sum_n \mathbf{x}_n \hat{S}_n \qquad \hat{Q}_{\mu\nu} = \frac{1}{2} \sum_n x_{n\mu} x_{n\nu} \hat{S}_n \tag{4-4.13}$$

Thus, the acoustic field formally appears as a monopole field plus a dipole field plus a quadrupole field, etc.‡

Given $kd \ll 1$, the field is generally well approximated by that of a single point monopole. An exception is when the sum of the $\hat{S}_n$ vanishes, either by design or because of symmetry. Then the dipole term would dominate, and the

† The derivation proceeds from

$$f(x - \varepsilon) = f(x) - \varepsilon \frac{d}{dx} f(x) + \frac{1}{2!} \varepsilon^2 \frac{d^2}{dx^2} f(x) - \cdots$$

$$= f(x) - (\varepsilon \mathbf{e}_x \cdot \nabla) f(x) + \frac{1}{2!} (\varepsilon \mathbf{e}_x \cdot \nabla)^2 f(x) - \cdots$$

If one has a function of $\mathbf{x} - \mathbf{x}_S$, the coordinate system can be temporarily oriented so that one of the axes points in the direction of $-\mathbf{x}_S$. The above then applies if the components of $\mathbf{x} - \mathbf{x}_S$ perpendicular to $\mathbf{x}_S$ are held constant, with the result

$$f(\mathbf{x} - \mathbf{x}_S) = f(\mathbf{x}) - (\mathbf{x}_S \cdot \nabla) f(\mathbf{x}) + \frac{1}{2!} (\mathbf{x}_S \cdot \nabla)^2 f(\mathbf{x}) - \cdots$$

For a fuller explanation, see R. Courant, *Differential and Integral Calculus,* vol. 2, Wiley-Interscience, Glasgow, 1936, pp. 80–81.

‡ The theory of a multipole expansion of a static field described by a potential satisfying Laplace's equation originated with J. C. Maxwell, *A Treatise on Electricity and Magnetism,* vol. 1, Oxford University Press, Oxford, 1873, pp. 157–178; the extension to the dynamic case for electromagnetic fields is due to H. A. Lorentz, "Extension of the Maxwell Theory, Theory of Electrons: State of the Field If the Exciting Charge Lies in an Infinitely Small Space," in A. Sommerfeld (ed.), *Encyklopädie der mathematischen Wissenshaften,* vol. 5, pt. 2, no. 1, 1904, reprinted by Teuber, Leipzig, 1922, pp. 177–178. A concise statement of the theory for the acoustical case is given by P. E. Doak, "Multipole Analysis of Acoustic Radiation," pap. K56 in D. E. Commins (ed.), *5e Congr. Int. Acoust.,* G. Thone, Liège, 1965, vol. 1b.

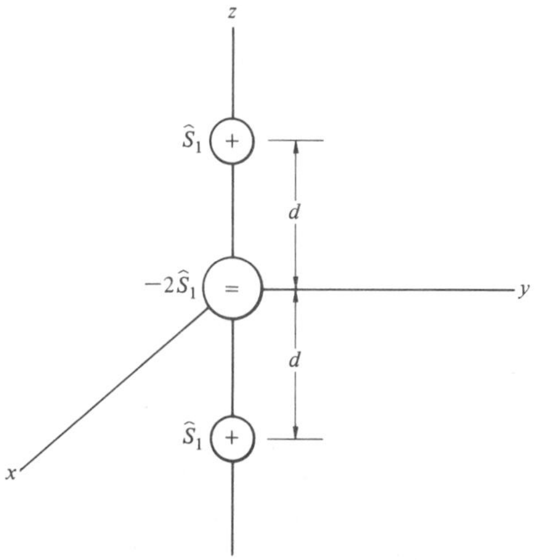

Figure 4-10 Example of sound radiation from three point sources lying on the z axis at $z = d$, $z = 0$, and $z = -d$ with monopole amplitudes $\hat{S}_1$, $-2\hat{S}$, and $\hat{S}_1$. The field for $kd \ll 1$ is that of a longitudinal quadrupole.

far-field pressure would have an amplitude diminished by a factor of the order of kd from that nominally expected. If $\hat{S}$ is zero, $\hat{\mathbf{D}}$, as computed by Eq. (13), should be independent of the choice of coordinate origin.

When both the monopole amplitude and dipole-moment-amplitude vector vanish, the quadrupole term ordinarily dominates. In such a case, the far-field pressure and the acoustic power output are decreased by factors of the order of $(kd)^2$ and $(kd)^4$ from what would nominally be expected.

Example Suppose three point sources (see Fig. 4-10) lie on the z axis at $z = d$, $z = 0$, and $z = -d$, with monopole amplitudes of $\hat{S}_1$, $-2\hat{S}_1$, and $\hat{S}_1$, respectively. The total monopole amplitude is zero; the dipole-moment-amplitude vector is also zero. The only nonzero quadrupole component is $\hat{Q}_{zz} = d^2\hat{S}_1$, so the acoustic field is that of a longitudinal quadrupole and the net acoustic power output, resulting from Eq. (10a), is $\frac{2}{5}\pi(kd)^4|\hat{S}_1|^2/\rho c$. If the phase of the center source is reversed, so that all three are in phase, the field will be that of a monopole with monopole amplitude $4\hat{S}_1$ and the acoustic power output will be $32\pi|\hat{S}_1|^2/\rho c$, larger by a factor of $80/(kd)^4$.

4-5 UNIQUENESS OF SOLUTIONS OF ACOUSTIC BOUNDARY-VALUE PROBLEMS

Many physical phenomena in acoustics are modeled as *boundary-value problems,* whereby some features of the acoustic field are specified on bounding surfaces or throughout a spatial region at an initial instant. Using this information, one seeks to predict the acoustic field at other points and at other times. Such problems need not be solved explicitly by mathematical analysis or numerical computation; answers to major questions can be obtained by direct ex-

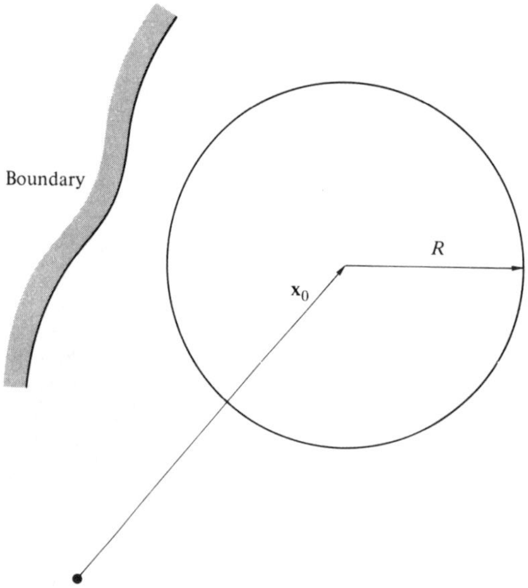

Figure 4-11 Geometry for discussion of Poisson's theorem, which relates the acoustic pressure at $\mathbf{x}_0$ at time t to the value and the time and spatial derivatives of the acoustic pressure at time $t - R/c$ averaged over the surface of a sphere of radius R centered at $\mathbf{x}_0$.

perimental measurement, by similitude analysis of the governing equations, or possibly by experimentation on an analogous physical system that can be modeled, with a suitable translation of symbols, by the same equations. It is desirable (especially from the latter standpoint when one is planning experiments) to know just how many initial data or boundary data are required for a unique prediction.

Poisson's Theorem and Its Implications

Causality is often incorporated, either explicitly or implicitly, in posing acoustic boundary-value problems. To characterize the wave caused by a source, one must require the wave to be absent before the source is first turned on. The earliest time at which such a wave disturbance appears at a distant point is delayed by the minimum time of propagation at the sound speed c from source to listener. This property of acoustic fields results with some generality from a relationship derived originally by Poisson.†

Suppose the acoustic pressure $p(\mathbf{x}, t)$ satisfies the wave equation in some region. We let $\mathbf{x}_0$ be any point in the region and consider a hypothetical sphere of radius R centered at the point $\mathbf{x}_0$ (see Fig. 4-11). A restriction on R is that during times $t_0 - R/c$ to t_0 the spherical region must be entirely within the fluid. Let $\bar{p}(\mathbf{x}_0, R, t)$ be the average (*spherical mean*) of $p(\mathbf{x}_0 + \mathbf{n}R, t)$ over the spherical

† The version of the proof given here is due to J. Liouville, "On Two Memoirs by Poisson," *J. Math. Pures Appl.*, (2)**1**:1–6 (1856). Poisson's original proof appeared in "Memoir on the Integration of Some Partial Differential Equations and, in Particular, That of the General Equation of Movement of Elastic Fluids," *Mem. Acad. Sci. Paris*, **3**:121–176 (1818).

surface, i.e.,

$$\bar{p}(\mathbf{x}_0, R, t) = \frac{1}{4\pi R^2}\iint p(\mathbf{x}_0 + \mathbf{n}R, t)\, dS \tag{4-5.1}$$

where $\mathbf{n}$ is the surface's outward unit normal vector. Then Poisson's relationship (derived further below) is

$$p(\mathbf{x}_0, t_0) = \left[\left(\frac{\partial}{\partial R} + \frac{1}{c}\frac{\partial}{\partial t}\right) R\bar{p}(\mathbf{x}_0, R, t)\right]_{t\to t_0 - R/c} \tag{4-5.2}$$

This implies that if one knew p, $\mathbf{n}\cdot\nabla p$, and $\partial p/\partial t$ at all points on the surface at time $t_0 - R/c$, this information would be sufficient to determine $p(\mathbf{x}_0, t_0)$ at a time R/c later. (The relation also holds if one replaces c by $-c$.)

To demonstrate Eq. (2) it is sufficient to choose the coordinate system so that $\mathbf{x}_0$ is at the origin and to use spherical coordinates (r, θ, ϕ). Since p satisfies the wave equation, one has (with r set to R)

$$\lim_{\varepsilon\to 0}\frac{1}{4\pi}\int_0^{2\pi}\int_\varepsilon^{\pi-\varepsilon}\left(\nabla^2 p - \frac{1}{c^2}\frac{\partial^2 p}{\partial t^2}\right)\sin\theta\, d\theta\, d\phi = 0$$

Here, in terms of spherical coordinates, the laplacian† of p is

† Spherical coordinates constitute an *orthogonal curvilinear coordinate system* (discussed in general here for future reference). If ξ_1, ξ_2, ξ_3 are properly ordered coordinates, the unit vectors $\mathbf{a}_i = \nabla\xi_i/|\nabla\xi_i|$ must form a right-handed set such that $\mathbf{a}_1\cdot\mathbf{a}_2 = 0$, $\mathbf{a}_1\times\mathbf{a}_2 = \mathbf{a}_3$, etc. The incremental-displacement vector $d\mathbf{x}$ can be written

$$d\mathbf{x} = h_1\, d\xi_1\, \mathbf{a}_1 + h_2\, d\xi_2\, \mathbf{a}_2 + h_3\, d\xi_3\, \mathbf{a}_3 \tag{i}$$

where

$$h_i = \left[\sum_j \left(\frac{\partial x_j}{\partial \xi_i}\right)^2\right]^{1/2} \tag{ii}$$

represents distance associated with unit change in ξ_i. In terms of the h_i, the expressions for the gradient, divergence, laplacian, and the $\mathbf{a}_i$ are

$$\nabla p = \sum_{i=1}^{3}\mathbf{a}_i\frac{1}{h_i}\frac{\partial p}{\partial \xi_i} \tag{iii}$$

$$\nabla\cdot\mathbf{v} = \frac{1}{h_1h_2h_3}\left(\frac{\partial}{\partial\xi_1}h_2h_3v_1 + \frac{\partial}{\partial\xi_2}h_3h_1v_2 + \frac{\partial}{\partial\xi_3}h_1h_2v_3\right) \tag{iv}$$

$$\nabla^2 p = \frac{1}{h_1h_2h_3}\left[\frac{\partial}{\partial\xi_1}\left(\frac{h_2h_3}{h_1}\frac{\partial p}{\partial\xi_1}\right) + \frac{\partial}{\partial\xi_2}\left(\frac{h_3h_1}{h_2}\frac{\partial p}{\partial\xi_2}\right) + \frac{\partial}{\partial\xi_3}\left(\frac{h_1h_2}{h_3}\frac{\partial p}{\partial\xi_3}\right)\right] \tag{v}$$

$$\mathbf{a}_i = \sum_j \frac{1}{h_i}\frac{\partial x_j}{\partial\xi_i}\mathbf{e}_j \tag{vi}$$

where $\mathbf{e}_1,\mathbf{e}_2,\mathbf{e}_3$ are unit vectors in the x_1,x_2,x_3 directions and $v_1 = \mathbf{v}\cdot\mathbf{a}_1$. For spherical coordinates r,θ,ϕ with $x = r\sin\theta\cos\phi$, $y = r\sin\theta\sin\phi$, $z = r\cos\theta$, one finds from (ii) that $h_r = 1$, $h_\theta = r$, $h_\phi = r\sin\theta$, so Eq. (3) results from (v). These details are discussed in almost any text on vector analysis and in many texts on mathematical techniques, electromagnetic theory, and fluid mechanics. See, for example, I. S. Sokolnikoff and R. M. Redheffer, *Mathematics of Physics and Modern Engineering*, 2d ed., McGraw-Hill, New York, 1966, pp. 416–417. Expression (v) is due to G. Lamé, "On the Laws of Equilibrium of the Fluid Ether," *J. Ec. Polytech.*, **14**:191–288 (1834).

$$\nabla^2 p = \frac{1}{r}\frac{\partial^2}{\partial r^2} rp + \frac{1}{r^2 \sin\theta}\frac{\partial}{\partial\theta}\left(\sin\theta \frac{\partial p}{\partial\theta}\right) + \frac{1}{r^2 \sin^2\theta}\frac{\partial^2}{\partial\phi^2} p \qquad (4\text{-}5.3)$$

but the second and third terms give no contribution to the above average over solid angles since $\partial p/\partial\theta$ is finite at $\theta = 0$ and $\theta = \pi$ and p is periodic in ϕ with period 2π. The angular averaging operation can be carried out on p first [giving the spherical mean $\bar{p}(0, R, t)$] for the remaining two terms in the integrand because R and t are independent of θ and ϕ. Consequently, $\bar{p}(0, R, t)$ satisfies the wave equation (1-12.2) for a spherically symmetric wave.

If one now defines

$$F(R, t) = \frac{\partial}{\partial R} R\bar{p} + \frac{1}{c}\frac{\partial}{\partial t} R\bar{p}$$

this wave equation can be written

$$\left(\frac{\partial}{\partial R} - \frac{1}{c}\frac{\partial}{\partial t}\right) F(R, t) = 0$$

which has the general solution $f(t + R/c)$ for $F(R, t)$. However, if one takes the above definition for $F(R, t)$ in the limit $R \to 0$ (given that $\partial\bar{p}/\partial R$ and $\partial\bar{p}/\partial t$ remain finite), one must identify $f(t)$ as $\bar{p}(0, 0, t)$ or, equivalently, as $p(0, t)$; so one has $p(0, t + R/c) = F(R, t)$. Substituting $p(0, t + R/c)$ for $F(R, t)$ into the above differential equation and setting $t = t_0 - R/c$, we obtain Eq. (2), thereby verifying the theorem.

A simple consequence of Poisson's theorem is that if, at some time t_1, both $p(\mathbf{x}, t_1)$ and $\partial p(\mathbf{x}, t_1)/\partial t_1$ are identically zero within a sphere of radius R_0 centered at $\mathbf{x}_0$, then $p(\mathbf{x}_0, t)$ must remain zero up until time $t_1 + R_0/c$. Hence wave disturbances (with the neglect of nonlinear terms and ambient flow) cannot move faster than the speed of sound. If initially the acoustic field in some bounded or partially bounded space is zero, and if the walls are set in vibration at time t_{init}, the earliest time one can expect a wave disturbance at a given point is $t_{\text{init}} + R_{\min}/c$, where $R_{\min}$ is the minimum distance from that point to the boundary.

The above reasoning leads to *Huygens' construction*† (see Fig. 4-12) for determination of time of onset of a wave disturbance. The surface (wavefront) separating disturbed and undisturbed regions moves into the undisturbed region with speed c.

Closed Regions

We here consider the question of uniqueness when the region of interest (Fig. 4-13) is enclosed by surfaces on which the normal component $\mathbf{v} \cdot \mathbf{n}_S$ of the

† Huygens' exposition on the principles underlying such a construction is in his *Traité de la lumière*, Leyden, 1678. For a detailed summary and relevant history, see E. Mach, *The Principles of Physical Optics*, 1926, reprinted by Dover, New York, 1954, pp. 255–271. The modern viewpoint on Huygens' principle is described by B. B. Baker and E. T. Copson, *The Mathematical Theory of Huygens' Principle*, Oxford, 1950, pp. 1–3.

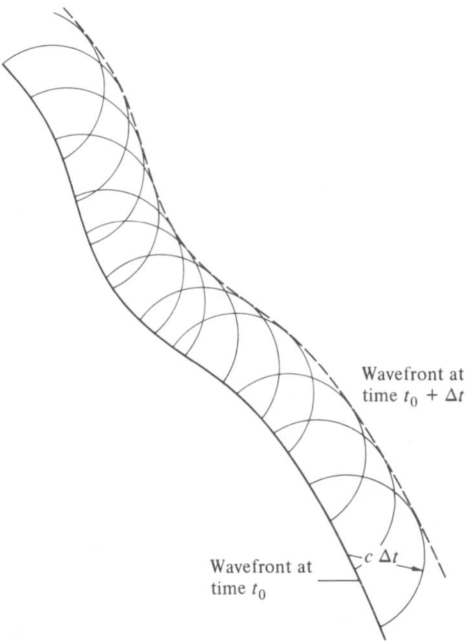

Figure 4-12 Huygens' construction of a wavefront at time $t_0 + \Delta t$ from wavefront at time t_0. The new wavefront is the envelope of spheres of radius $c\ \Delta t$ centered at points on the old wavefront.

acoustic fluid velocity is specified as a function of time. If the acoustic field within the enclosure is zero before the walls begin to vibrate, the subsequent acoustic field is unique. A proof† results if one assumes that there are two such fields and then demonstrates that their difference is zero. This difference satisfies the same (zero) initial conditions and the same homogeneous partial differential equations, but satisfies the requirement that $\Delta\mathbf{v} \cdot \mathbf{n}_S = 0$ at all boundary surfaces. The energy theorem of Eq. (1-11.2) applies, with p replaced by Δp and $\mathbf{v}$ replaced by $\Delta\mathbf{v}$. The integral version of the latter for the total volume V takes the form

$$\frac{\partial}{\partial t} \iiint \left[\frac{(\Delta p)^2}{2\rho c^2} + \frac{\rho(\Delta\mathbf{v})^2}{2} \right] dV = \iint \Delta p\ \Delta\mathbf{v} \cdot \mathbf{n}_S\, dS \qquad \text{(4-5.4)}$$

where the surface's unit normal $\mathbf{n}_S$ points into V.

Since $\Delta\mathbf{v} \cdot \mathbf{n}_S = 0$ at every point on the surface, the volume integral in Eq.

† The general method of proving uniqueness with energy integrals dates back to C. F. Gauss, "General Theorems Concerning the Attracting and Repelling Forces That Vary with the Inverse Square of Distance," Leipzig, 1840, reprinted in *Carl Friedrich Gauss Werke*, vol. 5, Königlichen Gesellschaft der Wissenschaften, Göttingen, 1877, pp. 197–242, especially pp. 226–237. The generalization to the wave equation is due to G. Kirchhoff, *Mechanik*, 2d ed., Teubner, Leipzig, 1877, pp. 311, 336. For a modern discussion with pertinent twentieth-century references, see R. Courant, *Methods of Mathematical Physics*, vol. 2, *Partial Differential Equations*, Interscience, New York, 1962, pp. 642–647.

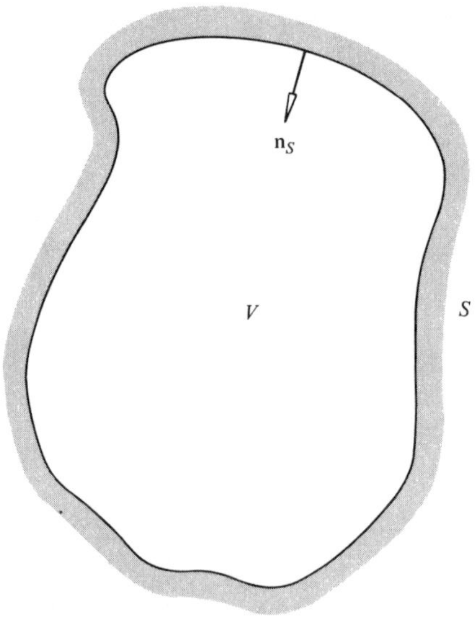

Figure 4-13 Geometry for discussion of the uniqueness of solutions of the wave equation for a closed region consisting of a volume V with bounding surface S. Here $\mathbf{n}_S$ is the unit normal to S pointing out of the surface into the fluid.

(4) must be independent of time. The initial values of Δp and $\Delta \mathbf{v}$, however, are zero, so the volume integral must be zero for all time. The only way such an integral can be zero is for its integrand to vanish. Hence, $\Delta \mathbf{v}$ and Δp are zero at all points in V for all times. Thus, the two solutions of the boundary-value problem must be the same, and uniqueness follows.

Uniqueness can be demonstrated similarly when p rather than $\mathbf{v} \cdot \mathbf{n}_S$ is specified at each point on the boundary, given that p and $\mathbf{v}$ are initially specified everywhere. Also, one could specify the problem by giving one or the other, p or $\mathbf{v} \cdot \mathbf{n}_S$, at each point on the bounding surfaces. One cannot arbitrarily specify both along the boundary, since use of either one or the other might lead to different solutions. Nevertheless, if the problem is to be physically meaningful, the boundary data taken in a single experiment must be consistent with the mathematical model, so it should not in principle make any difference what subset of boundary data is used in the prediction of p and $\mathbf{v}$ at interior points. Also, there is here an implication for the possible design of acoustic systems. Given the broad assumptions that lead to the linear acoustic equations (1-5.3) and the boundary condition (3-1.2), one cannot independently control surface pressures and normal velocities.

Uniqueness and Open Regions

The above conclusions apply even when the fluid's spatial extent is unbounded in certain directions (see Fig. 4-14). One limits one's attention to a finite region partly enclosed by solid surfaces and partly enclosed by a hypothetical surface that lies within the fluid. This latter surface is taken to be far enough removed from the cause of the sound, e.g., some vibrating solid surface, to ensure that,

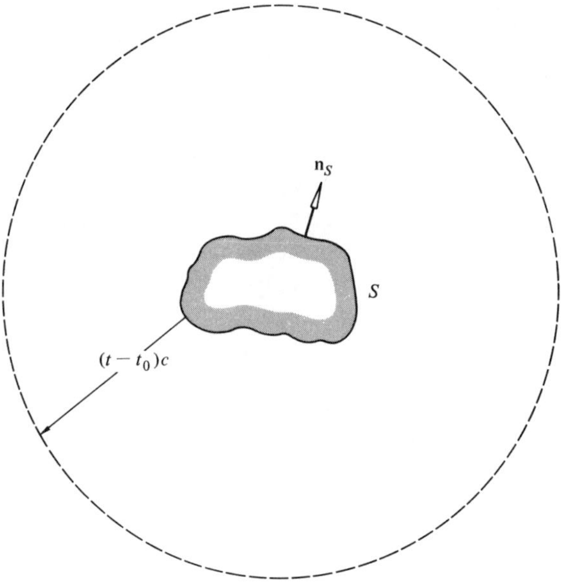

Figure 4-14 Conceptual device used for proof of uniqueness of transient solutions of the wave equation for an open region. The outer surface is at least a distance $(t - t_0)c$ from any point on the inner boundary; t_0 is the time of source excitation.

for all times of interest, the wave disturbance has not yet reached it. The existence of such a surface is guaranteed by Poisson's theorem. One chooses it to be at least a distance $(c)(t - t_0)$, t_0 being time of initial source excitation, from any active surface. Then p and $\mathbf{v}$ are zero on the surface. Consequently, if one postulates two solutions, each initially zero, and specifies that they must both satisfy the same boundary conditions (specified values of either p or $\mathbf{v}$ for all times up to t at each point on S), Eq. (4) again results and leads to the conclusion that Δp and $\Delta \mathbf{v}$ must be zero up to time t. The solution is unique up to time t, but since t is arbitrary, the solution is unique for all time.

Sommerfeld's Radiation Condition

The boundary condition that the acoustic field vanish at points farther than $(t - t_0)c$ from the source is awkward to apply in analytical studies. Often used instead is the *Sommerfeld radiation condition,*† which states that (in spherical

† A. Sommerfeld, "The Green's Function of the Oscillation Equation," *Jahresber. Dtsch. Math. Ver.*, **21**:309–353 (1912). Sommerfeld's *Ausstrahlungsbedingung* appears on p. 331. For later statements of radiation conditions (and proofs of uniqueness) see K. Rellich, "On the Asymptotic Behavior of Solutions of $\nabla^2 u + ku = 0$ in Infinite Regions," ibid., **53**:57–64 (1943); F. V. Atkinson, "On Sommerfeld's Radiation Condition," *Phil. Mag.*, (7)**40**:645–651 (1949); C. H. Wilcox, "A Generalization of Theorems of Rellich and Atkinson," *Proc. Am. Math. Soc.*, **7**:271–276 (1956); R. Leis, "On the Neumann Boundary Value Problem for the Helmholtz Oscillation Equation," *Arch. Ration. Mech. Anal.*, **2**:101–113 (1958); C. H. Wilcox, "Spherical Means and Radiation Conditions," ibid., **3**:133–148 (1959).

coordinates)

$$\lim_{r\to\infty}\left[r\left(\frac{\partial p}{\partial r}+\frac{1}{c}\frac{\partial p}{\partial t}\right)\right]=0 \qquad \lim_{r\to\infty}\left[r\left(\frac{\partial \hat{p}}{\partial r}-ik\hat{p}\right)\right]=0 \tag{4-5.5}$$

(The constant-frequency version results from the first equation with the prescription $\partial/\partial t \to -i\omega$.) This can be derived, given that all the bodies generating or perturbing the acoustic field are within a finite region centered at the origin. At sufficiently large distance r, the acoustic field varies more strongly with radial displacements than with displacements perpendicular to the radial direction, and $\nabla^2 p$ is approximately $r^{-1}\partial^2(rp)/\partial r^2$; one therefore concludes that p at large r is of the form of Eq. (1-12.3), where the functions f and g depend on the angular coordinates θ, ϕ, in addition to r and t. The function $g(t + r/c, \theta, \phi)$ is argued to be zero from causality considerations, so one is left with just the f term, the error being of the order of $1/r^2$ times another function of $t - r/c$, θ, and ϕ. Consequently, one obtains the radiation condition (5) above.

An equivalent statement of the Sommerfeld radiation condition is

$$\lim_{r\to\infty}[r(p - \rho c v_r)] = 0 \tag{4-5.6}$$

which results because the wave disturbance locally resembles a plane wave ($\mathbf{v} \approx \mathbf{n}p/\rho c$) propagating in the radial direction at large r. This version leads to the identification of ρc as the apparent specific acoustic impedance $Z = \hat{p}/\hat{v}_r$ associated with a sphere of radius r in the limit of large r.

With condition (6) imposed, the boundary-value problem for sound radiation from a collection of vibrating solids all of finite extent and on the surface of each of which either p or $\mathbf{v} \cdot \mathbf{n}_S$ is prescribed (but not both) must also have a unique solution. If one assumes that there are two solutions, then Eq. (4) holds. If V is taken to be finite and bounded by a sphere of large radius r [not necessarily greater than $(t - t_0)c$], the right side is not a priori zero but reduces, because of Eq. (6), to the nonpositive quantity

$$-\iint_{S_r} \rho c(\Delta v_r)^2\, dS$$

where the integration extends over the sphere of radius r (on which $\mathbf{n}_S$ is $-\mathbf{e}_r$). The time integral of the above cannot be positive, so the volume integral in (4) is either 0 or negative at any given instant. It cannot, however, be negative, so it must be zero. One concludes that $\Delta\mathbf{v}$ and Δp are zero throughout the volume V and, in particular, that Δv_r is zero on the outer sphere. The solution is therefore unique.

Uniqueness of Constant-Frequency Fields

A constant-frequency acoustic field (or the Fourier transforms of acoustic variables in a transient disturbance) is uniquely specified in a closed volume V when $\hat{p}$, $\hat{\mathbf{v}} \cdot \mathbf{n}_S$, or $Z = \hat{p}/(-\hat{\mathbf{v}} \cdot \mathbf{n}_S)$ (but only one of the three at any point) is given at each and every point on the confining surface S, providing that, on one portion of S, it is Z (rather than $\hat{p}$ or $\hat{\mathbf{v}} \cdot \mathbf{n}_S$) that is specified and on this surface Re $Z > 0$ and $|Z|$ is finite. The proof results from the corollary $\nabla \cdot (\text{Re } \hat{p}^*\hat{\mathbf{v}}) = 0$ of the steady-state field equations (1-8.12). If one has two solutions, the differences $\Delta\hat{p}$ and $\Delta\hat{\mathbf{v}}$ must also satisfy this divergence relation; the integral of such a relation over V, in conjunction with Gauss' theorem, requires a zero value for the integral of Re $(\Delta\hat{p}^* \, \Delta\hat{\mathbf{v}} \cdot \mathbf{n}_S)$ over the surface confining the volume V. If both solutions are required to satisfy boundary conditions with either $\hat{p}$ or $\hat{\mathbf{v}} \cdot \mathbf{n}_S$ (but not both) prescribed on various portions of S, then $\Delta\hat{p}$ or $\Delta\hat{\mathbf{v}} \cdot \mathbf{n}_S$, respectively, will vanish on those portions. On the remaining portions, the specific impedance $Z = \hat{p}/(-\hat{\mathbf{v}} \cdot \mathbf{n}_S)$ is prescribed, so $\Delta\hat{p} = -Z\Delta\hat{\mathbf{v}} \cdot \mathbf{n}_S$ and the requirement for a zero value of the surface integral reduces to

$$\iint \text{Re } Z|\Delta\hat{\mathbf{v}} \cdot \mathbf{n}_S|^2 \, dS = 0 \tag{4-5.7}$$

where the integral extends over just those surfaces on which an impedance boundary condition is prescribed. Equation (7) results in the conclusion that on any surface of finite specific impedance over which Re $Z > 0$ one must have $\Delta\hat{\mathbf{v}} \cdot \mathbf{n}_S = 0$. The relation $\Delta\hat{p} = -Z\Delta\hat{\mathbf{v}} \cdot \mathbf{n}_S$ then requires $\Delta\hat{p} = 0$ on the same portion of surface.

The above analysis indicates that both $\Delta\hat{p}$ and its normal derivative vanish on some finite surface. Because $\Delta\hat{p}$ must satisfy the Helmholtz equation (1-8.13), each and every higher derivative of $\Delta\hat{p}$ is zero on this surface. [For example, if the surface lies on the $z = 0$ plane, $\Delta\hat{p}$ and $\partial\Delta\hat{p}/\partial z$ are zero for a finite range of x and y. Within this range, $\partial\Delta\hat{p}/\partial x$, $\partial^2\Delta\hat{p}/\partial x^2$, etc., are zero because $\Delta\hat{p}$ is constantly zero. Similarly, $\partial^2\Delta\hat{p}/(\partial x \, \partial z)$ is zero because $\partial\Delta\hat{p}/\partial z$ is constantly zero. The Helmholtz equation then predicts that $\partial^2\Delta\hat{p}/\partial z^2$ will be

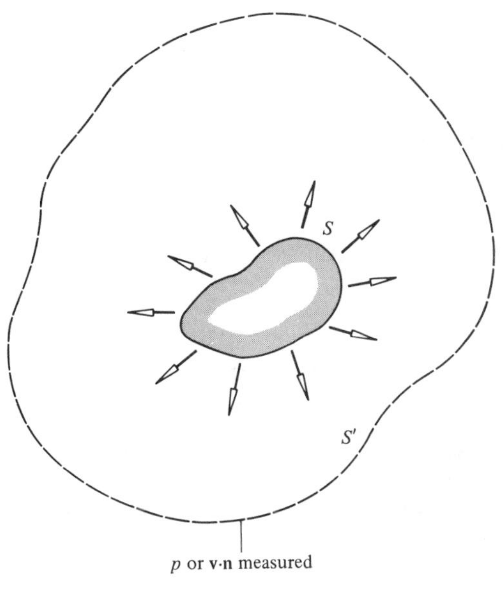

Figure 4-15 An implication of the uniqueness theorem: the acoustic field outside any surface S' enclosing the source can be determined from the knowledge of either p or $\mathbf{v} \cdot \mathbf{n}$ on S'.

zero on the surface. Zero values for the higher derivatives result because $\partial\Delta\hat{p}/\partial z$, $\partial^2\Delta\hat{p}/\partial z^2$, etc., also satisfy the Helmholtz equation.]

Since $\Delta\hat{p}$ and all its derivatives vanish on a portion of S, the prediction of $\Delta\hat{p}$ for points away from that surface based on a Taylor-series expansion is zero. This then leads to the conclusion that the solution is unique.

The analysis just given implies that sufficient boundary conditions for constant-frequency radiation from a finite-sized vibrating body (or an assemblage of vibrating bodies) in an *open space* result from specification of $\hat{\mathbf{v}} \cdot \mathbf{n}_S$, $\hat{p}$, or Z at each point on the body and from specification of the Sommerfeld radiation condition (6) on a large sphere surrounding the body. The latter device formally makes the open region appear to be a finite volume V; because ρc is real and positive, there is a portion (the outer sphere) of the confining surface on which Re $Z > 0$; the solution is therefore unique.

Since predictions of acoustic fields can be modeled as boundary-value problems, one has considerable latitude in the selection of what data might be taken in the near field of a source to predict the field at moderate to large distances. Over any surface (see Fig. 4-15) enclosing the source one can measure either $\hat{p}$ or $\hat{\mathbf{v}} \cdot \mathbf{n}$. The source need not be a vibrating solid, and the surface on which near-field measurements are made need not be the surface of the source, but it is required that the acoustic field equations hold outside the surface of measurement. Because such data lead (although, possibly with the aid of a large computer) to a unique prediction of the field outside the surface, any such prediction of $\hat{p}$ or $\hat{\mathbf{v}}$ at a distant point is the same as would be obtained from any other valid choice of near-field data.

4-6 THE KIRCHHOFF-HELMHOLTZ INTEGRAL THEOREM

Discussions of sound radiation are often facilitated by a mathematical theorem† due to Kirchhoff and Helmholtz, derived here for an isolated vibrating body (or for a fixed surface enclosing a source) in an otherwise unbounded fluid; each point on the surface S of the body vibrates with the same angular frequency ω.

The derivation begins with the vector identity‡

$$G(\nabla^2 + k^2)\hat{p} - \hat{p}(\nabla^2 + k^2)G = \nabla \cdot (G\,\nabla\hat{p} - \hat{p}\,\nabla G) \qquad (4\text{-}6.1)$$

† Helmholtz, "Theory of Air Oscillations . . . ," especially pp. 22–25; G. Kirchhoff, "Toward a Theory of Light Rays," *Ann. Phys. Chem.*, **18**:663–695 (1883), especially pp. 666–669. A frequently cited modern derivation is that of J. A. Stratton, *Electromagnetic Theory*, McGraw-Hill, New York, 1941, pp. 424–428. The basic mathematical ideas were used in the case of Laplace's and Poisson's equations by Green, *Essay on the Application of Mathematical Analysis*, 1828.

‡ *Green's theorem* can be derived from this by integrating both sides over a fixed volume, then converting the integral on the right to a surface integral by means of Gauss' theorem. Green, *Essay on the Application of Mathematical Analysis*, 1828.

where G is any function of position. Both sides are integrated over a volume V consisting of all points outside S that are within some large sphere of radius R centered at the origin. The contribution from the first term on the left is zero because $(\nabla^2 + k^2)\hat{p} = 0$ within V. Gauss' theorem transforms the volume integration over the right side into a surface integral; there are contributions from the inner surface S and from the outer sphere. The integration accordingly yields

$$-\iiint \hat{p}(\nabla^2 + k^2)G\, dV = -\iint_S (G\,\nabla\hat{p} - \hat{p}\,\nabla G)\cdot \mathbf{n}_S\, dS + I_R \qquad (4\text{-}6.2)$$

where

$$I_R = R^2 \int_0^{2\pi}\int_0^{\pi} \left(G\,\frac{\partial \hat{p}}{\partial R} - \hat{p}\,\frac{\partial G}{\partial R}\right) \sin\theta\, d\theta\, d\phi \qquad (4\text{-}6.3)$$

is the surface integral over the outer sphere. The minus sign appears in front of the first term on the right of (2) because $\mathbf{n}_S$ is here understood to point out of the surface S into the external volume.

We stipulate that G is a Green's function $G_k(\mathbf{x}|\mathbf{x}_0)$ that throughout V satisfies the inhomogeneous Helmholtz equation (4-3.10). This stipulation causes the left side of Eq. (2) to be $4\pi\hat{p}(\mathbf{x}_0)$ (given that $\mathbf{x}_0$ is in V) because of the integral property of the Dirac delta function. Moreover, if G is required to satisfy the Sommerfeld radiation condition, if $|G|$ goes to zero at least as fast as $1/R$ at large R, and if $\hat{p}$ has the same properties (which must be true for the actual solution), I_R vanishes in the limit of large R. Because the remaining terms in (2) are independent of the choice for R, one must conclude that I_R is identically zero for any sphere containing the surface and the point $\mathbf{x}_0$. Thus, for $\mathbf{x}_0$ exterior to S, Eq. (2) reduces to

$$\hat{p}(\mathbf{x}_0) = -\frac{1}{4\pi}\iint (G\,\nabla\hat{p} - \hat{p}\,\nabla G)\cdot \mathbf{n}_S\, dS \qquad (4\text{-}6.4)$$

where the integration extends over the vibrating surface only. (If $\mathbf{x}_0$ were within the interior of S, a similar equation would result but with the left side replaced by zero.)

One has some latitude in the selection of the Green's function G. One could choose it, for example, so that G or $\nabla G\cdot\mathbf{n}_S$ vanishes on the surface S, and then one of the two terms in the integrand of (4) would drop out and one would need only know (besides G) $\hat{p}$ or $\nabla\hat{p}\cdot\mathbf{n}_S$, respectively, to evaluate $\hat{p}(\mathbf{x}_0)$. However, the simplest explicit choice for G is the free-space Green's function $R^{-1}e^{ikR}$; we here make this choice to obtain the Kirchhoff-Helmholtz integral theorem.

One may note, from Eq. (1-8.12), that $\nabla\hat{p}\cdot\mathbf{n}_S = i\omega\rho\hat{v}_n$, and also that

$$\nabla G = \frac{\mathbf{x} - \mathbf{x}_0}{R^3}(ikR - 1)e^{ikR} \qquad (4\text{-}6.5)$$

The transient version of Eq. (4) can consequently be identified with the prescriptions that $i\omega \rightarrow -\partial/\partial t$ and that a factor e^{ikR} multiplying $e^{-i\omega t}$ is equivalent to shifting t to $t - R/c$. Thus, with the symbol change $\mathbf{x} \to \mathbf{x}_S$, $\mathbf{x}_0 \to \mathbf{x}$ we obtain

$$p(\mathbf{x}, t) = \frac{\rho}{4\pi} \iint \frac{\dot{v}_n(\mathbf{x}_S, t - R/c)}{R}\, dS + \frac{1}{4\pi c} \iint \mathbf{e}_R \cdot \mathbf{n}_S \left(\frac{\partial}{\partial t} + \frac{c}{R} \right) \frac{p(\mathbf{x}_S, t - R/c)}{R}\, dS \qquad (4\text{-}6.6)$$

where here we write $R = |\mathbf{x} - \mathbf{x}_S|$ and $\mathbf{e}_R = (\mathbf{x} - \mathbf{x}_S)/R$. The symbol $\mathbf{x}_S$ here denotes a point on the surface of the body; $\mathbf{x}$ denotes a point outside the body. [The derivation of Eq. (4) led to a representation of the listener location by the symbol $\mathbf{x}_0$, but since the choice of symbols to denote position is only a matter of definition, one can make the substitutions $\mathbf{x} \to \mathbf{x}_S$ and $\mathbf{x}_0 \to \mathbf{x}$.] The constant-frequency version of the Kirchhoff-Helmholtz integral theorem is recovered if one replaces $\dot{v}_n(\mathbf{x}_S, t - R/c)$ by $-i\omega\hat{v}_n(\mathbf{x}_S)e^{ikR}$, $\partial/\partial t \to -i\omega$, etc.

Result (6) holds if $\mathbf{x}$ is any point outside S and, in particular, if $\mathbf{x} = \mathbf{x}_S' + \mathbf{n}_S'\delta$ is a point displaced a slight distance δ from a point $\mathbf{x}_S'$ on the surface. In the limit as δ becomes zero, the integrands become singular, but the right side of (6) remains finite and approaches the sum of the principal values (i.e., omit a small patch of radius ϵ centered at $\mathbf{x}_S'$ and take the limit as $\epsilon \to 0$) of the integrals plus $\frac{1}{2}p(\mathbf{x}, t)$. Alternately, one can regard the right side of (6) as yielding $\frac{1}{2}p(\mathbf{x}, t)$ rather than $p(\mathbf{x}, t)$ when $\mathbf{x}$ is on the surface. If $\mathbf{x}$ is inside the surface, the right side should yield zero.

Equation (6) or its constant-frequency counterpart is not a solution of an acoustic boundary-value problem since, as discussed in the previous section, one cannot specify both p and v_n independently on the surface. Instead, it is a corollary of the governing partial-differential equation and of the Sommerfeld radiation condition. If $\hat{v}_n$, for example, is specified on S, the solution of the acoustic boundary-value problem will have to be such that it gives values of $\hat{p}(\mathbf{x}_S)$ on the surface S satisfying the $\mathbf{x} \to \mathbf{x}_S$ version (as described above) of Eq. (6). Such an equation can be regarded as an integral equation for $\hat{p}(\mathbf{x}_S)$ and, indeed, the numerical solution of this integral equation is a common first step for prediction of the acoustic field of a vibrating object.†

Multipole Expansions of the Kirchhoff-Helmholtz Integral

The integral theorem leads to convenient expressions for the coefficients in the multipole expansion of a small vibrating body.‡ Let us assume that the body is

† Solution of the integral equation is not unique for certain discrete frequencies, but can be made unique if one specifies that the Kirchhoff-Helmholtz integral vanish for all $\mathbf{x}$ within the surface. [H. A. Schenck, "Improved Integral Formulation for Acoustic Radiation Problems," *J. Acoust. Soc. Am.*, **44**:41–58 (1968); L. G. Copley, "Fundamental Results Concerning Integral Representations in Acoustic Radiation," ibid., **44**:28–32 (1968); P. H. Rogers, "Formal Solution of the Surface Helmholtz Integral Equation at a Nondegenerate Characteristic Frequency," ibid., **54**:1662–1666 (1973).]

‡ H. L. Oestreicher, "Representation of the Field of an Acoustic Source as a Series of Multipole Fields," *J. Acoust. Soc. Am.*, **29**:1219–1222 (1957), **30**:481 (1958).

confined to the vicinity of the origin and that any dimension a characterizing the body's size satisfies the criterion $ka \ll 1$, where $k = \omega/c$ and ω is any angular frequency characterizing the surface vibrations. For simplicity, we here use the transient expression (6); the constant-frequency result can be determined with the prescription that the retarded time $t - R/c$ in the argument of a function corresponds to the presence of a factor of e^{ikR} in the complex amplitude and with the replacement of $\partial/\partial t$ by $-i\omega$.

The derivation of an appropriate multipole expansion is similar to that of Eq. (4-4.12). One replaces $p(\mathbf{x}_S, t - R/c)/R$ by the expansion resulting from the application of the operator $\exp(-\mathbf{x}_S\cdot\nabla)$ to $p(\mathbf{x}_S, t - r/c)/r$, where the $\exp(-\mathbf{x}_S\cdot\nabla) = 1 - \mathbf{x}_S\cdot\nabla + \cdots$ is the expansion operator in Eq. (4-4.11b). A similar expansion replaces $\dot{v}_n(\mathbf{x}_S, t - R/c)/R$. Note also that the operator $\mathbf{e}_R(\partial/\partial t + c/R)$ applied to $p(\mathbf{x}_S, t - R/c)/R$ is equivalent to $-c\nabla$ applied to the same function. Thus, Eq. (6) becomes

$$p(\mathbf{x}, t) = \frac{\rho}{4\pi}\iint [\exp(-\mathbf{x}_S\cdot\nabla)]\,\frac{\dot{v}_n(\mathbf{x}_S, t - r/c)}{r}\,dS - \frac{1}{4\pi}\iint [\exp(-\mathbf{x}_S\cdot\nabla)](\mathbf{n}_S\cdot\nabla)\,\frac{p(\mathbf{x}_S, t - r/c)}{r}\,dS \tag{4-6.7}$$

A rearrangement of terms and application of differential calculus identities subsequently yields the multipole expansion

$$p = \frac{S(t - r/c)}{r} - \nabla\cdot\frac{\mathbf{D}(t - r/c)}{r} + \sum_{\mu,\nu=1}^{3}\frac{\partial^2}{\partial x_\mu\,\partial x_\nu}\,\frac{Q_{\mu\nu}(t - r/c)}{r} + \cdots \tag{4-6.8}$$

where

$$S(t) = \frac{\rho}{4\pi}\iint \dot{v}_n(\mathbf{x}_S, t)\,dS = \frac{\rho}{4\pi}\,\dot{Q}_S(t) \tag{4-6.9a}$$

$$\mathbf{D}(t) = \frac{1}{4\pi}\iint [\rho\mathbf{x}_S\dot{v}_n(\mathbf{x}_S, t) + \mathbf{n}_S p(\mathbf{x}_S, t)]\,dS \tag{4-6.9b}$$

$$Q_{\mu\nu}(t) = \frac{1}{8\pi}\iint [\rho x_{S\mu}x_{S\nu}\dot{v}_n(\mathbf{x}_S, t) + (x_{S\mu}n_\nu + x_{S\nu}n_\mu)p(\mathbf{x}_S, t)]\,dS \tag{4-6.9c}$$

are identified as the monopole function, the dipole-moment vector, and the $\mu\nu$th quadrupole component, respectively. Definition (9c) is such that $Q_{\mu\nu} = Q_{\nu\mu}$. In Eq. (9a), $Q_S(t)$ is the instantaneous time derivative of the volume enclosed by a surface that moves with the same normal velocity as the fluid just outside the reference surface S and is consequently identified as the source strength.

4-7 SOUND RADIATION FROM SMALL VIBRATING BODIES

We have seen (Secs. 4-1 and 4-2) that simple expressions result for the sound radiation from spherical bodies undergoing radial or transverse oscillations in the limit $ka \ll 1$. Similar expressions, appropriate for sound at large distances from small vibrating bodies of arbitrary shape, are derived here. The analysis also gives some insight into the nature of acoustic fields near such bodies.

For vibrations of a given angular frequency ω or, alternately, of a given value of $k = \omega/c$, the boundary-value problem for radiation from an isolated vibrating body is posed by the Helmholtz equation (1-8.13), by a specification of the normal component $\hat{\mathbf{v}}_S \cdot \mathbf{n} = \hat{v}_n$ of the complex amplitude of the outward-normal component of the body's surface velocity, and by the Sommerfeld radiation condition. The boundary condition (3-1.2) implies that at the surface, $\mathbf{n} \cdot \nabla \hat{p}$ should be $ik\rho c\hat{v}_n$. The resulting boundary-value problem, in accordance with the remarks in Sec. 4-5, should have a unique solution.

An approximate solution scheme results from consideration of a sequence of problems in which the frequency and therefore k varies continuously from problem to problem but for which the complex surface velocity amplitude $\hat{v}_n$ at a given point on the surface is held fixed. If we let a be a representative length characterizing the dimensions of the body, then ka is a dimensionless parameter distinguishing various problems in the overall set. Two possible expansions of $\hat{p}$ in terms of ka would be *an inner expansion* in which r/a is kept fixed and *an outer expansion* in which kr is kept fixed. Such expansions exist as simple power series in ka for the known solutions [see Eqs. (4-1.4), (4-2.5), and (4-2.6)] for a radially oscillating sphere and for a transversely oscillating rigid sphere, so one can proceed with some hope of finding such expansions for more general classes of vibrating bodies. The leading term in the inner expansion should be at most of order ka; that in the outer expansion should be at most of order $(ka)^2$. Thus, one can write the inner expansion as

$$\hat{p} = \sum_{n=1}^{N} \hat{p}_{\text{in},n} + R_N^{\text{in}} \tag{4-7.1a}$$

where $\hat{p}_{\text{in},n}$ is of the form

$$\hat{p}_{\text{in},n} = i\rho c\hat{v}_{\text{typ}}(ka)^n F_n\left(\frac{r}{a}, \theta, \phi\right) \tag{4-7.1b}$$

with the dimensionless functions $F_n(r/a, \theta, \phi)$ for $n = 1, 2, \ldots,$ yet to be determined. Here $\hat{v}_{\text{typ}}$ is some typical value of the $\hat{v}_n$; the quantity R_N^{in} is the remainder. Similarly, the outer expansion can be written

$$\hat{p} = \sum_{n=2}^{N} \hat{p}_{\text{out},n} + R_N^{\text{out}} \tag{4-7.2a}$$

with

$$\hat{p}_{\text{out},n} = i\rho c\hat{v}_{\text{typ}}(ka)^n G_n(kr, \theta, \phi) \tag{4-7.2b}$$

These are (at worst) *asymptotic expansions* in the sense that, for given ϵ, r/a, ϕ, θ, and N, there is some finite value δ such that if $ka < \delta$, the remainder R_N^{in} in the inner expansion has absolute value less than $\epsilon(ka)^{N+1}$, even though, for fixed ka, the quantity $|R_N^{\text{in}}|$ may not go to zero when N becomes large without limit.

The *method of matched asymptotic expansions*† as applied to the general boundary-value problem posed above is a scheme whereby the $\hat{p}_{\text{in},n}$ and $\hat{p}_{\text{out},n}$ can be determined in a systematic fashion from the following requirements:

1. Both Eqs. (1*a*) and (2*a*) represent solutions of the Helmholtz equation.
2. The inner expansion (1*a*) must satisfy the inner boundary condition.
3. The outer-expansion terms must satisfy the Sommerfeld radiation condition.
4. The first few terms in both expansions describe the same function in a hypothetical range where $a \ll r \ll 1/k$.

Requirement 1 applied to the inner expansion is satisfied when Eqs. (1) are substituted into the Helmholtz equation and when the resulting coefficients of different powers of ka are equated to zero. Similarly, requirement 2 is satisfied if the inner expansion is substituted into the inner boundary condition and if this is required to be identically satisfied for arbitrary ka. In this manner, the following sequence of (incompletely posed) boundary-value problems results:

$$\nabla^2 \hat{p}_{\text{in},1} = 0 \qquad \text{with } \mathbf{n} \cdot \nabla \hat{p}_{\text{in},1} = i\omega\rho\hat{v}_n \text{ on } S \tag{4-7.3a}$$

$$\nabla^2 \hat{p}_{\text{in},2} = 0 \qquad \text{with } \mathbf{n} \cdot \nabla \hat{p}_{\text{in},2} = 0 \text{ on } S \tag{4-7.3b}$$

$$\nabla^2 \hat{p}_{\text{in},3} = -k^2 \hat{p}_{\text{in},1} \qquad \text{with } \mathbf{n} \cdot \nabla \hat{p}_{\text{in},3} = 0 \text{ on } S \tag{4-7.3c}$$

The form of the outer expansion can be derived from the constant-frequency version of the multipole expansion, Eq. (4-6.8), of the Kirchhoff-Helmholtz integral. For the evaluation of coefficients depending on surface pressure, we use the inner expansion, Eq. (1*a*), for $\hat{p}$. Thus, one can consider $\hat{\mathbf{D}}$ and the $\hat{Q}_{\mu\nu}$ as being expanded in a power series in ka, that is, $\mathbf{D} = \hat{\mathbf{D}}_1 + \hat{\mathbf{D}}_2 + \cdots$, etc., where $\hat{\mathbf{D}}_1$ results from Eq. (4-6.9*b*) with $\dot{v}_n$ replaced by $-i\omega\hat{v}_n$ and with p replaced by $\hat{p}_{\text{in},n}$ and where

$$4\pi\hat{\mathbf{D}}_n = \iint \mathbf{n}_S \hat{p}_{\text{in},n}\, dS \qquad n \geq 2 \tag{4-7.4}$$

The $\hat{Q}_{\mu\nu,n}$ are defined analogously with reference to Eq. (4-6.9*c*).

One can establish from Eq. (4-6.9*a*) that $\hat{S}$ is of the form $-i\rho c\hat{v}_{\text{typ}}ka^2$ times a dimensionless quantity independent of ka. Similarly, from Eqs. (1*b*) and (4-6.9*b*), one establishes that each $\hat{\mathbf{D}}_n$ is of the form of $-i\rho c\hat{v}_{\text{typ}}a^2(ka)^n$ times such a dimensionless quantity. Each of the $\hat{Q}_{\mu\nu,n}$ is of the form of $-i\rho c\hat{v}_{\text{typ}}a^3(ka)^n$ times

† Texts discussing the method of matched asymptotic expansions are A. H. Nayfeh, *Perturbation Methods,* Wiley-Interscience, New York, 1973, pp. 111–154; J. D. Cole, *Perturbation Methods in Applied Mathematics,* Blaisdell, Waltham, Mass., 1968, pp. 11–78, 129–162; M. Van Dyke, *Perturbation Methods in Fluid Mechanics,* Academic, New York, 1964, pp. 77–97. A general review of the method as applied to acoustics is given by M. B. Lesser and D. G. Crighton, "Physical Acoustics and the Method of Matched Asymptotic Expansions," in W. P. Mason (ed.), *Physical Acoustics,* vol. 11, Academic, New York, 1976, pp. 69–149. The modern development of the method was inaugurated by S. Kaplun, P. A. Lagerstrom, and J. D. Cole in articles published c. 1955. The basic concept that the near field of a small vibrating body is approximately the same as if the fluid were incompressible can be discerned in papers by Rayleigh published in 1871 (Rayleigh scattering) and 1897.

a dimensionless quantity independent of ka. Analogous considerations hold for the coefficients arising from higher-order terms in the multipole expansion. Consequently, a comparison of the ka dependence for fixed kr of the various order (in ka) terms in the multipole expansion with those in the outer expansion results in the identifications

$$\hat{p}_{\text{out},2} = \hat{S}\,\frac{e^{ikr}}{r} \tag{4-7.5a}$$

$$\hat{p}_{\text{out},3} = -\hat{\mathbf{D}}_1 \cdot \nabla\,\frac{e^{ikr}}{r} \tag{4-7.5b}$$

$$\hat{p}_{\text{out},4} = \left(-\hat{\mathbf{D}}_2 \cdot \nabla\,\frac{e^{ikr}}{r}\right) + \sum_{\mu,\nu=1}^{3} Q_{\mu\nu,1}\,\frac{\partial^2}{\partial x_\mu\,\partial x_\nu}\,\frac{e^{ikr}}{r} \tag{4-7.5c}$$

[Below it is demonstrated that the dipole term (in parentheses) of Eq. (5c) is zero.]

The determination of boundary conditions for the asymptotic behavior at large r of the inner expansion functions $\hat{p}_{\text{in},n}$ is accomplished with the help of a general matching condition that both expansions represent the same function at intermediate distances r, where $a \ll r \ll 1/k$, so that the inner expansion's form at large r/a should resemble the outer expansion's form at small ka. The latter can be derived by expanding each of the e^{ikr} appearing in Eqs. (5) in a power series in kr

$$e^{ikr} = \sum_{m=0}^{\infty} \frac{(ika)^m (r/a)^m}{m!} \tag{4-7.6}$$

so that one has, for example, that the mth term in the expansion of $[\partial^2/(\partial x_\mu\,\partial x_\nu]$ $(r^{-1}e^{ikr})$ is $(ka)^m/a^3$ times a dimensionless function of r/a, θ, and ϕ. Thus, since $\hat{Q}_{\mu\nu,1}$ is $-i\rho c\hat{v}_{\text{typ}}ka^4$ times a dimensionless quantity independent of ka, the product of $\hat{Q}_{\mu\nu,1}$ and the mth term in the kr expansion varies with ka for fixed r/a as $(ka)^{m+1}$. Consequently, such a term gives information concerning $\hat{p}_{\text{in},n}$ for $n = m + 1$ at large r/a. In such a manner, one establishes that, in the limit of large r/a,

$$\hat{p}_{\text{in},1} \to \frac{\hat{S}}{r} - \hat{\mathbf{D}}_1 \cdot \nabla\,\frac{1}{r} + \sum_{\mu,\nu=1}^{3} \hat{Q}_{\mu\nu,1}\,\frac{\partial^2}{\partial x_\mu\,\partial x_\nu}\,\frac{1}{r} - \cdots \tag{4-7.7a}$$

$$\hat{p}_{\text{in},2} \to ik\hat{S} - \hat{\mathbf{D}}_2 \cdot \nabla\,\frac{1}{r} + \sum_{\mu,\nu=1}^{3} \hat{Q}_{\mu\nu,2}\,\frac{\partial^2}{\partial x_\mu\,\partial x_\nu}\,\frac{1}{r} - \cdots \tag{4-7.7b}$$

$$\begin{aligned}\hat{p}_{\text{in},3} \to {} & -\frac{1}{2}k^2\left(\hat{S}r - \hat{\mathbf{D}}_1 \cdot \nabla r + \sum_{\mu,\nu=1}^{3} \hat{Q}_{\mu\nu,1}\,\frac{\partial^2 r}{\partial x_\mu\,\partial x_\nu} - \cdots\right) \\ & - \hat{\mathbf{D}}_3 \cdot \nabla\,\frac{1}{r} + \sum_{\mu,\nu=1}^{3} \hat{Q}_{\mu\nu,3}\,\frac{\partial^2}{\partial x_\mu\,\partial x_\nu}\,\frac{1}{r} - \cdots\end{aligned} \tag{4-7.7c}$$

That Eqs. (7) are consistent with Eqs. (3) follows since the individual terms in Eqs. (7a) and (7b) are solutions of Laplace's equation $\nabla^2\psi = 0$. Also, since $\nabla^2 r = 2/r$ [see Eq. (4-5.3)], Eqs. (7a) and (7c) are such that $\nabla^2\hat{p}_{\text{in},3} = -k^2\hat{p}_{\text{in},1}$. The function $\hat{p}_{\text{in},1}$ is uniquely determined by Eq. (3a) and by the requirement, derived from (7a), that it go to zero at large r at least as fast as $1/r$. That the asymptotic expansion of $\hat{p}_{\text{in},1}$ should be given by Eq. (7a), where the coefficients $\hat{S}$, $\hat{\mathbf{D}}_1$, $\hat{Q}_{\mu\nu,1}$ are given by the constant-frequency versions of Eqs. (4-6.9) with $\hat{p} \rightarrow \hat{p}_{\text{in},1}$, follows from the $k = 0$ analog of the multipole expansion of the Kirchhoff-Helmholtz integral.

The only way the asymptotic expansion (7b) can be consistent with the boundary condition in Eq. (3b) that $\nabla\hat{p}_{\text{in},2} \cdot \mathbf{n} = 0$ on the vibrating surface is for one to have $\hat{p}_{\text{in},2} = ik\hat{S}$ identically. Equation (4) then yields the relation $\hat{\mathbf{D}}_2 = 0$. The $\hat{Q}_{\mu\nu,2}$ calculated from Eq. (4-6.9c) (with the v_n term omitted and p replaced by $ik\hat{S}$) are zero unless $\mu = \nu$. The third term in Eq. (7b) vanishes nevertheless because all three of the $\hat{Q}_{\mu\nu,2}$ are equal and because $\nabla^2(1/r) = 0$. Analogous considerations apply to the higher-order terms.

We now summarize the results of the preceding analysis, explicitly taking into account the time dependence using the prescription $-i\omega \rightarrow \partial/\partial t$ and using the correspondence of the factor e^{ikr} to the time shift $t \rightarrow t - r/c$. Equations (3a) and (3b) imply that up to second order in ka the acoustic pressure at distances $r \ll 1/k$ satisfies Laplace's equation (which results for *incompressible potential flow*),

$$\nabla^2 p_{\text{in}}(\mathbf{x}, t) = 0 \tag{4-7.8a}$$

with the boundary condition

$$\mathbf{n} \cdot \nabla p_{\text{in}}(\mathbf{x}_S, t) = -\rho \frac{\partial}{\partial t} v_n(\mathbf{x}_S, t) \tag{4-7.8b}$$

at points $\mathbf{x}_S$ on the surface of the vibrating body. At large r/a the inner solution approaches

$$p_{\text{in}}(\mathbf{x}, t) \rightarrow \left[\frac{S(t)}{r} - \mathbf{D}_1(t) \cdot \nabla \frac{1}{r} + \sum_{\mu,\nu=1}^{3} Q_{\mu\nu,1}(t) \frac{\partial^2}{\partial x_\mu \, \partial x_\nu} \frac{1}{r} - \cdots \right] - \frac{\dot{S}(t)}{c} \tag{4-7.9}$$

where $S(t)$, $\mathbf{D}_1(t)$, and the $Q_{\mu\nu,1}(t)$ are as given by Eqs. (4-6.9), the latter two with p replaced by p_{in}. The quantity in brackets corresponds to first order in ka (for fixed r/a), and the last term corresponds to second order. Equation (9) imposes an outer boundary condition on p_{in}, that it plus $\dot{S}/c$ go to zero at least as fast as $1/r$. In conjunction with Eqs. (8), this specifies $p_{\text{in}}(\mathbf{x}, t)$, $\mathbf{D}_1(t)$, and the $Q_{\mu\nu,1}(t)$ uniquely.

Another implication of the analysis is that the acoustic-pressure field at

$r \gg a$ is given up to fourth order in ka (for fixed kr) by†

$$p_{\text{out}} = \frac{S(t - r/c)}{r} - \nabla \cdot \frac{\mathbf{D}_1(t - r/c)}{r} + \sum_{\mu,\nu=1}^{3} \frac{\partial^2}{\partial x_\mu \, \partial x_\nu} \frac{Q_{\mu\nu,1}(t - r/c)}{r} \tag{4-7.10}$$

where the monopole, dipole, and quadrupole terms correspond, respectively, to second, third, and fourth order in ka for fixed kr. This satisfies the wave equation and matches Eq. (9).

The monopole term in Eq. (10) is the same [see Eq. (4-1.6)] as derived for the radially oscillating sphere in the limit $ka \ll 1$. The implication here, however, is that this should be a good approximation for sound radiation at distances $r \gg a$ from any small vibrating body whose volume changes with time. This confirms the assertion that any sufficiently small source with time-varying volume can be considered as a point monopole source regardless of the shape of the body.

Instances when the monopole term might be insufficient to explain radiation from a small vibrating body are when the body is moving very nearly as a rigid body or it is a vibrating plate or shell whose thickness changes negligibly. For the latter case, $v_n \mathbf{x}_S$ is equal and opposite on opposite sides of the shell, so the integral over the first term vanishes in Eq. (4-6.9*b*). Since the surface integral over $p\mathbf{n}_S$ is the net force $\mathbf{F}(t)$ exerted by the body on the surrounding fluid, one identifies the leading term in the acoustic-pressure field at $r \gg a$ as being the same as Eq. (4-4.5), derived for a point force applied to a fluid.

For a rigid body, one can in general write (see Sec. 3-1) v_n as $\mathbf{n} \cdot (\mathbf{v}_C + \boldsymbol{\Omega} \times \mathbf{x}_S)$ where $\mathbf{v}_C(t)$ is the velocity of the body's geometric center (taken as the origin) and $\boldsymbol{\Omega}(t)$ is the body's angular velocity. In such a case, an application of Gauss' theorem converts the surface integral of $\mathbf{x}_S \dot{v}_n$ to the volume integral

$$\iint \mathbf{x}_S \dot{v}_n \, dS = \iiint [\dot{\mathbf{v}}_C + \dot{\boldsymbol{\Omega}} \times \mathbf{x} + \nabla \cdot (\dot{\boldsymbol{\Omega}} \times \mathbf{x})] \, dV = \rho^{-1} m_d \dot{\mathbf{v}}_C \tag{4-7.11}$$

The second equality results because the choice of geometric center as origin forces the volume integral of $\mathbf{x}$ to be zero and because $\nabla \cdot (\dot{\boldsymbol{\Omega}} \times \mathbf{x}) = 0$. (Here m_d is the mass of fluid displaced by the body.) Consequently, the dipole-moment vector $\mathbf{D}_1(t)$ is $1/4\pi$ times $\mathbf{F}_1(t) + m_d \, d\mathbf{v}_C/dt$, and the leading term in the associated pressure field at $r \gg a$ is‡

† A brief derivation of the first two terms here (taken individually) is given by L. D. Landau and E. M. Lifshitz, *Fluid Mechanics*, Addison-Wesley, Reading, Mass., 1959, pp. 280–281. Although Landau and Lifshitz do not use the full liturgy of what is now called the *method of matched asymptotic expansions*, their approach employs the same concepts.

‡ An alternate derivation applicable when $\mathbf{F}_1$ and $\mathbf{v}_C$ are parallel, e.g., because of symmetry, dates back to H. Lamb, *The Dynamical Theory of Sound*, 2d ed., 1925, reprinted by Dover, New York, 1960, pp. 240–241. A general statement, developed by H. M. Fitzpatrick and M. Strasberg, c. 1957, is summarized by Strasberg, "Radiation from Unbaffled Bodies of Arbitrary Shape at Low Frequencies," *J. Acoust. Soc. Am.*, **34**:520–521 (1962).

$$p_{\text{dipole}} = -\frac{1}{4\pi}\nabla\cdot\left\{\frac{1}{r}\left[\mathbf{F}_1\left(t-\frac{r}{c}\right) + m_d\dot{\mathbf{v}}_C\left(t-\frac{r}{c}\right)\right]\right\} \qquad (4\text{-}7.12)$$

This is consistent with the result (4-2.13) for radiation from a transversely oscillating sphere in the limit $ka \ll 1$. Since, for that special case, $\mathbf{F}_1(t) = \frac{1}{2}m_d d\mathbf{v}_C/dt$, the sum $\mathbf{F}_1 + m_d\mathbf{v}_C$ is $\frac{3}{2}m_d a\mathbf{v}_C/dt$ and, with $m_d = \frac{4}{3}\pi a^{3r}$ and a vector identity, the above reduces to Eq. (4-2.13).

Instances where a vibrating body would radiate predominantly as a quadrupole would be when (1) the intrinsic symmetry of the body and of the vibration is such that $\mathbf{F}_1(t)$ must be identically zero and (2) either $\dot{\mathbf{v}}_c$ is identically zero throughout the motion or the vibrating body can be modeled as a thin shell. As an example, consider the rigid body in Fig. 4–16, whose nominal position is such that its surface is even in x and y, the body undergoing rocking oscillations about the z axis passing through its geometric center. The symmetry of the body and of the motion require that v_n be antisymmetric in x and y, so the normal derivative of p at the surface is also antisymmetric in x and y. Since the wave equation and the radiation condition are unchanged if $x \rightarrow -x$ or if $y \rightarrow -y$, the solution of the resulting boundary-value problem must conform to the symmetry properties of the boundary conditions, so p is odd in both x and y. This automatically rules out monopole and dipole fields. The symmetry requires that the lowest-order (in some ka) outer solution for fixed kr be a lateral quadrupole field of the form

$$p_Q = 2\,\frac{\partial^2}{\partial x\,\partial y}\,\frac{Q_{xy,1}(t-r/c)}{r} \qquad (4\text{-}7.13)$$

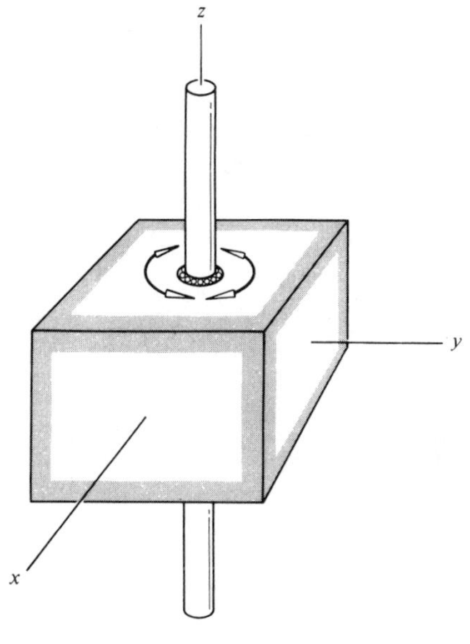

Figure 4-16 Example of a quadrupole radiator: a symmetric rigid body undergoing rocking motion about its geometric center. (If the cross-section is a square, the radiation is octupole.)

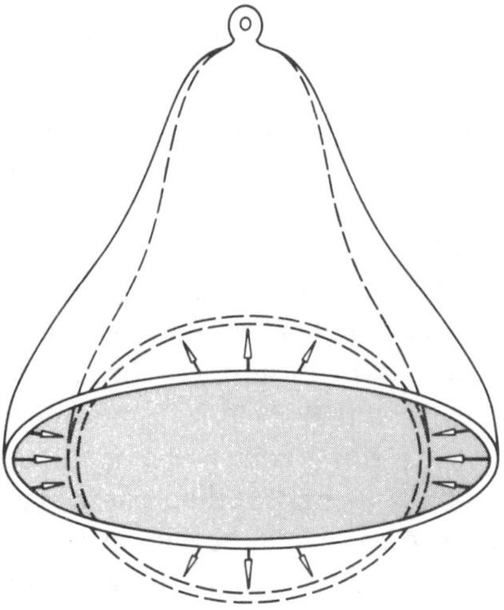

Figure 4-17 A symmetric bell vibrating in a mode that produces quadrupole radiation.

Another example of quadrupole radiation is a *vibrating bell*† (see Fig. 4-17). When the bell is vibrating with constant frequency in any one of its natural vibration modes, the bell's circular symmetry requires the normal velocity $\hat{v}_n$ to be periodic in azimuthal angle ϕ with period $2\pi/N$, where N is an integer. Any breathing mode with $N = 0$ typically corresponds to a frequency far above the audible range; one mode with $N = 1$ is a simple pendulum oscillation (caused by gravity) and corresponds to an infrasonic frequency; other $N = 1$ modes involve flexing (as in transverse vibration of a beam) of the bell without changing the circular shapes of its cross sections and correspond to ultrasonic frequencies. Since the pressure radiated by any mode has the same ϕ dependence as $\hat{v}_n$, one concludes that the monopole and dipole terms vanish identically for any vibration corresponding to audible frequencies. The only vibrational modes giving rise to quadrupole radiation are those corresponding to $N = 2$, and if the bell's symmetry axis is the z axis, the only nonzero quadrupole components are $Q_{xy} = Q_{yx}$, Q_{xx} and Q_{yy}. Symmetry also requires $Q_{xx} = -Q_{yy}$. Thus, the radiated acoustic pressure at $r \gg a$ is given predominantly by an expression of the form

$$p_Q = 2\,\frac{\partial^2}{\partial x\,\partial y}\,\frac{Q_{xy,1}(t - r/c)}{r} + \left(\frac{\partial^2}{\partial x^2} - \frac{\partial^2}{\partial y^2}\right)\frac{Q_{xx,1}(t - r/c)}{r} \qquad \text{(4-7.14)}$$

† The first mathematical discussion of note of sound radiation by bells is that of Stokes, "On the Communication of Vibration," 1868, who modeled the bell as a sphere. His identification of the radiation as quadrupole is implicit in his choice of the spherical harmonic of second order to describe "the principal vibration for a sphere vibrating in the manner of a bell." J. W. S. Rayleigh, *The Theory of Sound*, 2d ed., Dover, New York, 1945, vol. 2, sec. 324, quotes the relevant passages from Stokes' paper verbatim. An extensive discussion of vibrations of bells and of their acoustic radiation is given by Rayleigh, "On Bells," *Phil. Mag.*, (5)**29**:1–17 (1890).

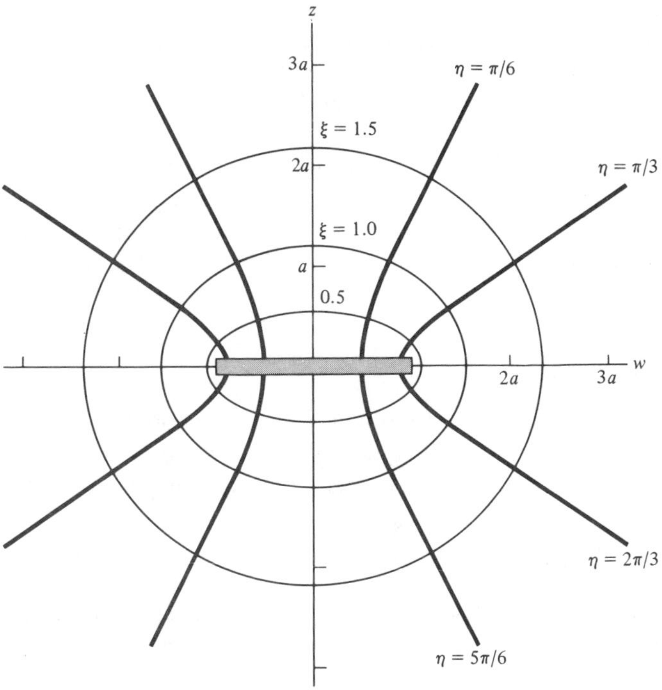

Figure 4-18 Oblate-spheroidal coordinates used in analysis of radiation from a vibrating disk. The limiting surface $\xi = 0$ coincides with the disk's nominal location.

An implication† of this equation is that there should be no sound along the z axis ($x = 0$, $y = 0$).

4-8 RADIATION FROM A CIRCULAR DISK

As an application of the analytical technique described in the previous section, we here consider a small circular disk‡ (see Fig. 4-18) of radius a oscillating parallel to its axis with velocity $v_C(t)$. Such an example furnishes a model for sound radiation from an unbaffled loudspeaker and leads to a prediction of

† J. W. S. Rayleigh, "Acoustical Observations I," *Phil. Mag.*, (5)**3**:456–464 (1877).

‡ The problem of radiation by a vibrating disk is closely related to that of diffraction by a disk, so that solution for one leads to solution of the other. This is discussed by F. M. Wiener, "On the Relation between the Sound Fields Radiated and Diffracted by Plane Obstacles," *J. Acoust. Soc. Am.*, **23**:697–700 (1951). The solution of the latter problem in the small ka limit is due to Rayleigh, "On the Passage of Waves through Apertures in Plane Screens, and Allied Problems," *Phil. Mag.*, (5)**43**:259–272 (1897). The low-frequency result for the oscillating disk was explicitly stated by Lamb, *Dynamical Theory of Sound*, p. 241.

acoustic power substantially less than what would be obtained if the loudspeaker were mounted in a baffle. If the disk nominally lies in the xy plane with its center at the origin, the inner boundary condition is that $v_z = v_C(t)$ for $w < a$, where $w = (x^2 + y^2)^{1/2}$, and for z both slightly greater and slightly less than 0.

The boundary-value problem for determination of the inner field can be posed in terms of a velocity potential $\Phi_{\text{in}}(\mathbf{x}, t)$, whose gradient is $\mathbf{v}_{\text{in}}$ and which is such that $p_{\text{in}} = -\rho\, \partial\Phi_{\text{in}}/\partial t$. It follows from Eqs. (4-7.8) that Φ_{in} should satisfy Laplace's equation and the boundary condition $\partial\Phi_{\text{in}}/\partial z = v_C(t)$ for $w < a$ and for $z = 0^+$ and $z = 0^-$.

Oblate-Spheroidal Coordinates

The natural coordinates for the problem are oblate-spheroidal coordinates† (ξ, η, ϕ) where $w = a \cosh \xi \sin \eta$, $z = a \sinh \xi \cos \eta$, $x = w \cos \phi$, $y = w \sin \phi$ with $\xi \geq 0$, $0 < \eta < \pi$, and $0 < \phi < 2\pi$. A surface of constant ξ is given by

$$\frac{w^2}{a^2 \cosh^2 \xi} + \frac{z^2}{a^2 \sinh^2 \xi} = 1 \tag{4-8.1}$$

and represents an oblate spheroid formed by rotation of an ellipse (distance $2a$ between its foci, major semidiameter $a \cosh \xi$, minor semidiameter $a \sinh \xi$) about its minor axis, which coincides with the z axis. The disk is a degenerate member of this family and corresponds to the surface $\xi = 0$.

In oblate-spheroidal coordinates, Laplace's equation takes the general form‡

$$\nabla^2\Phi_{\text{in}} = \frac{1}{a^2(\cosh^2 \xi - \sin^2 \eta)} \left[\frac{1}{\cosh \xi} \frac{\partial}{\partial \xi} \left(\cosh \xi \frac{\partial \Phi_{\text{in}}}{\partial \xi} \right) + \frac{1}{\sin \eta} \frac{\partial}{\partial \eta} \left(\sin \eta \frac{\partial \Phi_{\text{in}}}{\partial \eta} \right) \right] + \frac{1}{a^2 \cosh^2 \xi \sin^2 \eta} \frac{\partial^2 \Phi_{\text{in}}}{\partial \phi^2} = 0 \tag{4-8.2}$$

and the component of $\nabla\Phi_{\text{in}}$ pointing in the direction of increasing ξ and perpendicular to a surface of constant ξ is given in general by

$$\nabla\Phi_{\text{in}} \cdot \mathbf{e}_\xi = \frac{1}{a(\cosh^2 \xi - \sin^2 \eta)^{1/2}} \frac{\partial \Phi_{\text{in}}}{\partial \xi} \tag{4-8.3}$$

† H. Lamb, *Hydrodynamics*, 6th ed., 1932, reprinted by Dover, New York, 1945, sec. 107, pp. 142–143. Our a is Lamb's k, our ξ is Lamb's η, our η is Lamb's θ, our ϕ is Lamb's ω.

‡ The general statements on p. 173n. apply to oblate-spheroidal coordinates with the identifications $\xi_1, \xi_2, \xi_3 \rightarrow \xi, \eta, \phi$. Thus one has

$$h_\xi = h_\eta = a\,(\cosh^2 \xi - \sin^2 \eta)^{1/2} \qquad h_\phi = a \cosh \xi \sin \eta$$

such that $h_\xi\, d\xi$ is incremental displacement associated with $\xi \rightarrow \xi + d\xi$, etc. Equations (2) to (4) follow from Eqs. (v), (iii), and (vi) in the footnote with the substitutions just described.

Here

$$\mathbf{e}_\xi = \frac{1}{(\cosh^2 \xi - \sin^2 \eta)^{1/2}} [\sinh \xi \sin \eta \, (\mathbf{e}_x \sin \phi + \mathbf{e}_y \cos \phi) + \cosh \xi \cos \eta \, \mathbf{e}_z] \quad (4\text{-}8.4)$$

is the unit vector in the direction of increasing ξ. Thus, on the surface of the disk ($\xi = 0$), $\mathbf{e}_\xi$ is $+\mathbf{e}_z$ if $\cos \eta > 0$ ($z = 0^+$) and $-\mathbf{e}_z$ if $\cos \eta < 0$ ($z = 0^-$), so $\mathbf{e}_\xi$ is the unit outward-normal vector $\mathbf{n}$ to a flat disk when $\xi = 0$.

Solutions of Laplace's Equation

For future reference, we here digress to list three particular solutions (corresponding to monopole, dipole, and quadrupole fields) of Eq. (2):

$$F_0(\xi) \qquad \cos \eta \, F_1(\xi) \qquad \cos \eta \sin \eta \sin \phi \, F_2^1(\xi) \quad (4\text{-}8.5)$$

Their substitution into Laplace's equation shows that each of the functions $F_n^m(\xi)$ (no superscript if $m = 0$) must satisfy the ordinary differential equation†

$$\frac{1}{\cosh \xi} \frac{d}{d\xi} \left(\cosh \xi \frac{dF_n^m}{d\xi} \right) + \left[\frac{m^2}{\cosh^2 \xi} - n(n+1) \right] F_n^m = 0 \quad (4\text{-}8.6)$$

The equation for $F_0(\xi)$ is relatively simple, the solution being any constant times the indefinite integral of $1/(\cosh \xi)$. If we require $F_0(\xi) \to 0$ as $\xi \to \infty$, the resulting integration leads to

$$F_0(\xi) = \sin^{-1}\left[\frac{1}{\cosh \xi}\right] \quad (4\text{-}8.7a)$$

The two differential equations for $F_1(\xi)$ and $F_2^1(\xi)$ have the respective properties, which can be verified by substitution, that they have particular solutions

$$F_1(\xi) = \frac{d}{d\xi}[(\cosh \xi) F_0(\xi)] \quad (4\text{-}8.7b)$$

$$F_2^1(\xi) = \frac{d}{d\xi}\left[(\sinh \xi) F_1 + \tfrac{1}{3}(\cosh \xi) \frac{dF_1}{d\xi}\right] \quad (4\text{-}8.7c)$$

† This is related to the differential equation satisfied by the associated Legendre functions. The function $F_n^m(\xi)$ is a constant times $Q_n^m(i \sinh \xi)$, that is, an associated Legendre function of the second kind with imaginary argument. Definitions and properties of the Legendre functions are given in M. Abramowitz and I. A. Stegun (eds.), *Handbook of Mathematical Functions,* Dover, New York, 1965, pp. 331–341. Our choice for $F_0(\xi)$ is $iQ_0(i \sinh \xi)$. The expressions for F_1 and F_2^1 follow from eqs. (8.5.3) and (8.6.7) in the *Handbook*. For derivations, see E. T. Whittaker and G. N. Watson, *A Course of Modern Analysis,* 4th ed., Cambridge, 1927, pp. 318, 324.

Here the $F_0(\xi)$ and $F_1(\xi)$ on the right sides are any particular solutions of the $n = 0$, $m = 0$ and $n = 1$, $m = 0$ equations. Thus, with $F_0(\xi)$ as given above, one has

$$F_1(\xi) = \sinh \xi \sin^{-1}\left(\frac{1}{\cosh \xi}\right) - 1 \tag{4-8.7b'}$$

$$F_2^1(\xi) = 3 \sinh \xi \cosh \xi \sin^{-1}\left(\frac{1}{\cosh \xi}\right) - 3 \cosh \xi + \frac{1}{\cosh \xi} \tag{4-8.7c'}$$

Both go to zero as $\xi \to \infty$.

For the boundary-value problem of the transversely oscillating disk, the requirement that $\nabla\Phi_{\text{in}} \cdot \mathbf{n}$ equal v_C or $-v_C$ if $z = 0^+$ or $z = 0^-$ is satisfied if one requires $\nabla\Phi \cdot \mathbf{e}_\xi = v_C$ when $\xi = 0$ or, from (3) above, if $\partial\Phi/\partial\xi = v_C a \cosh \eta$ when $\xi = 0$. This suggests that one look for a solution of Laplace's equation of the form $\cos \eta\, F(\xi)$, where $dF(\xi)/d\xi = v_C a$ at $\xi = 0$ and $F(\xi) \to 0$ as $\xi \to \infty$. The function $F(\xi)$ is identified from Eqs. (5) and ($7b'$) as $(2av_C/\pi)F_1(\xi)$, so the velocity potential of the inner field is given by†

$$\Phi_{\text{in}} = \frac{2av_C}{\pi} \cos \eta \left[\sinh \xi \sin^{-1}\left(\frac{1}{\cosh \xi}\right) - 1\right] \tag{4-8.8}$$

Determination of the Outer Solution

The inner-field potential function is such that, on the two faces of the disk ($\xi = 0$, $\cos \eta = \pm(1 - (w/a)^2]^{1/2}$), one has‡

$$\Phi_{\text{in}} = \mp \frac{2v_C a}{\pi}\left[1 - \left(\frac{w}{a}\right)^2\right]^{1/2} \qquad w < a \tag{4-8.9}$$

† Lamb, *Hydrodynamics,* sec. 108, p. 144. Our Eq. (8) follows from Lamb's expression (3) for his ϕ (which is the negative of our Φ_{in}) with $\mu \to \cos \eta$, $\zeta \to \sinh \xi$, $\zeta_0 \to 0$, $e \to 1$, $\sin^{-1} e \to \pi/2$, $\varepsilon \to a$, $U \to v_C$. The mathematical identity $\cot^{-1}(\sinh \xi) = \sin^{-1}(1/\cosh \xi)$ has also been used. The solution is due to E. Heine, "Concerning Some Problems that Lead to Partial Differential Equations," *J. Reine Angew. Math.,* **26:**185–216 (1843).

‡ The prediction in Eq. (9) gives infinite tangential velocity at the edge of the disk, so if the convection term $\rho\mathbf{v} \cdot \nabla\mathbf{v}$ [equal to $\nabla(\rho v^2/2)$ for irrotational flow] is taken into account, the pressure at the edge will also be infinite when the plate is moving with constant speed. The ideal-fluid solution is unrealistic for the steady-motion case, the actual flow developing a wake behind the disk and eddies being generated at the edges that are swept downstream with the fluid. In the acoustical case, however, the disk is not moving with steady velocity but is oscillating back and forth with a small velocity amplitude. The theoretical prediction is not valid within a distance of the order of $(2\mu/\rho\omega)^{1/2}$ from the edge of the plate (where μ is the viscosity of the fluid), but this length is much smaller than a and the potential-flow solution gives a prediction that is on the whole reasonably accurate. For a discussion with accompanying photographs for the related problem of nominally steady flow past a strip (with a disclaimer in regard to the acoustical case) see A. Sommerfeld, *Mechanics of Deformable Bodies,* 2d ed., 1947, Academic, New York, 1950, pp. 207–215. The feeble influence of viscosity on flows associated with oscillatory motion is explained by Lamb, *Hydrodynamics,* pp. 619–623, 654–657.

while Φ_{in} is identically 0 for $w > a$ on the plane $z = 0$ ($\eta = \pi/2$). In the limit of large ξ, $\sinh \xi \sin^{-1}(1/\cosh \xi)$ approaches $1 - \frac{4}{3}e^{-2\xi}$ and $e^{\xi} \to 2r/a$, where r is the radial (spherical coordinates) distance from the origin. Consequently, for $r \gg a$, one has

$$\Phi_{\text{in}} \to \frac{2v_C a^3}{3\pi} \frac{\partial}{\partial z} \frac{1}{r} \tag{4-8.10}$$

which is characteristic of the potential for the incompressible-flow field of a dipole. The pressure in the far field corresponding to this is $-\rho\, \partial\Phi_{\text{in}}/\partial t$, so, with reference to Eq. (4-7.9), one identifies the dipole-moment vector as

$$\mathbf{D}_1(t) = \frac{2\rho\dot{\mathbf{v}}_C(t)a^3}{3\pi} = \frac{\mathbf{F}_1(t)}{4\pi} \tag{4-8.11}$$

The matching procedure corresponding to Eq. (4-7.9) allows us to identify the acoustic pressure at distances $r \gg a$ (outer solution) as $-\nabla \cdot [\mathbf{D}_1(t - r/c)/r]$.

For a transversely oscillating sphere of radius a_1, the quantity $m_d\, dv_C/dt + F_z$ is $(\frac{3}{2})(\frac{4}{3})\pi a_1^3 \rho\, dv_C/dt$ while for the disk it is $\frac{8}{3}\rho a^3\, dv_C/dt$; one can therefore conclude that the far field of a transversely oscillating disk (with $ka \ll 1$) is equivalent to that radiated by a transversely oscillating sphere of radius $a_1 = (4/3\pi)^{1/3}a = 0.7515a$.

4-9 RECIPROCITY IN ACOUSTICS

Reciprocity† refers to situations for which a magnitude associated with an "effect" at a point is unchanged when the locations of "cause" and "point of observation" are interchanged.

Reciprocity in Vibrating Systems

As an example, consider the mechanical system in Fig. 4-19 consisting of three coupled masses that move because of applied forces F_1, F_2, and F_3. The motion is influenced by a spring with spring constant k_2 and by dashpots (constants c_2 and c_3). If x_1, x_2, x_3 denote the displacements of the corresponding masses, the coupled equations of motion (derived from mechanical principles) can be written in matrix form as

$$\begin{bmatrix} D_{11} & -k_2 & -c_3 \dfrac{d}{dt} \\ -k_2 & D_{22} & -c_2 \dfrac{d}{dt} \\ -c_3 \dfrac{d}{dt} & -c_2 \dfrac{d}{dt} & D_{33} \end{bmatrix} \begin{bmatrix} x_1 \\ x_2 \\ x_3 \end{bmatrix} = \begin{bmatrix} F_1 \\ F_2 \\ F_3 \end{bmatrix} \tag{4-9.1}$$

† The concept dates back to Helmholtz, "Theory of Air Oscillations in Tubes with Open Ends," 1860, and to J. C. Maxwell, "On the Calculations of the Equilibrium and Stiffness of Frames," *Phil. Mag.*, (4)**27**:294–299 (1864).

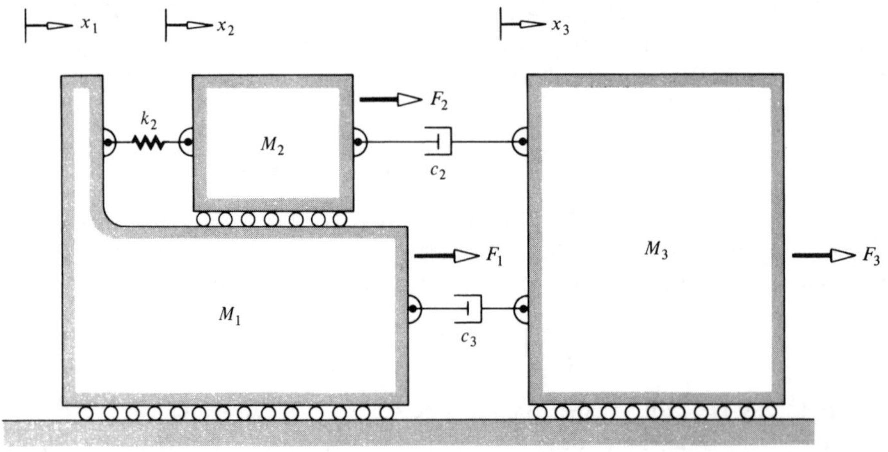

Figure 4-19 A mechanical system that satisfies the reciprocity principle.

where $D_{11} = M_1\, d^2/dt^2 + c_3\, d/dt + k_2$, etc., are linear operators. The pertinent property of the matrix is its symmetry about the diagonal. Thus, if each force is oscillating with angular frequency ω, such that $F_1 = \text{Re}\, \hat{F}_1 e^{-i\omega t}$, the corresponding algebraic equations for the complex amplitudes (with the prescription $d/dt \to -i\omega$) of the velocities u_1, u_2, u_3 (dx_1/dt, dx_2/dt, dx_3/dt), written as

$$\sum_{j=1}^{3} Y_{ij}\hat{u}_j = \hat{F}_i \qquad i = 1, 2, 3 \tag{4-9.2}$$

are such that the mobility matrix $[Y]$ is also symmetric†; that is, $Y_{ij} = Y_{ji}$.

$$Y_{ij} = Y_{ji}.$$

The solution of Eqs. (2) for the $\hat{u}_i$ takes the form

$$\hat{u}_i = \sum_j Z_{ij}\hat{F}_j \tag{4-9.3}$$

where the coefficients Z_{ij} are elements of the matrix $[Z]$ representing the inverse

† This was first demonstrated by J. W. S. Rayleigh, "Some General Theorems relating to Vibrations," *Proc. Lond. Math. Soc.*, **4**:357–368 (1873); *Theory of Sound*, vol. 1, pp. 91–104, 150–157. The symmetry is because a dissipation function $D(\dot{x}_1, \dot{x}_2, \ldots)$ exists such that Lagrange's equations for a conservative linear system can be extended to give

$$\frac{d}{dt}\frac{\partial T}{\partial \dot{x}_i} + \frac{\partial D}{\partial \dot{x}_i} + \frac{\partial V}{\partial x_i} = F_i$$

where the kinetic-energy function T and potential-energy function V are quadratic in the $\dot{x}_i$ and the x_i, respectively. The generalized force F_i is such that $F_i\, \delta x_i$ represents the work done on the system during an admissible variation δx_i. The proof is also given by E. T. Whittaker, *A Treatise on the Analytical Dynamics of Particles and Rigid Bodies*, 4th ed., Cambridge University Press, London, 1937, pp. 230–232.

of $[Y]$. This mechanical-impedance matrix $[Z]$ is also symmetric ($Z_{ij} = Z_{ji}$) because the inverse of a symmetric matrix must also be symmetric. Consequently, if a force with complex amplitude $\hat{F}_a$ is applied to mass M_i, no other active forces being applied, the velocity amplitude $\hat{u}_j$ of mass M_j ($j \neq i$) is the same as would be obtained for the velocity amplitude of mass M_i if the force $\hat{F}_a$ were applied to M_j. This is a statement of the principle of reciprocity.

Another statement of the reciprocity principle comes from a consideration of two separate experiments in which the impressed forces are given by $\hat{F}_{1a}$, $\hat{F}_{2a}$, $\hat{F}_{3a}$ and $\hat{F}_{1b}$, $\hat{F}_{2b}$, $\hat{F}_{3b}$, respectively. Let $\hat{u}_{1a}$, $\hat{u}_{2a}$, $\hat{u}_{3a}$ and $\hat{u}_{1b}$, $\hat{u}_{2b}$, $\hat{u}_{3b}$ denote the corresponding velocity amplitudes for the two experiments. Then one can demonstrate that

$$\sum_i (\hat{F}_{ia}\hat{u}_{ib} - \hat{F}_{ib}\hat{u}_{ia}) = 0 \tag{4-9.4}$$

The proof follows from either Eq. (2) or Eq. (3).

The above results [$Y_{ij} = Y_{ji}$, $Z_{ij} = Z_{ji}$, and Eq. (4)] apply to any lumped-parameter vibrational system undergoing small-amplitude oscillations of constant frequency. Analogous results apply to electric circuits.† Reciprocity does not depend on the system's being nondissipative and is thus not directly related to any requirement of energy conservation.

Reciprocity and the Linear Acoustic Equations

The linear acoustic equations derived in Chap. 1 require (given a nonmoving time-independent ambient medium) that the complex amplitudes $\hat{p}(\mathbf{x})$ and $\hat{\mathbf{v}}(\mathbf{x})$ for a constant-frequency disturbance satisfy

$$-i\omega\hat{p} + \rho c^2 \nabla \cdot \hat{\mathbf{v}} = 0 \qquad -i\omega\rho\hat{\mathbf{v}} + \nabla\hat{p} = 0 \tag{4-9.5}$$

These also apply if ρ and c are position-dependent, given that p_0 is constant; in what follows, we allow for this possibility.‡ Suppose one has two sets of solutions, $\hat{p}_a$, $\hat{\mathbf{v}}_a$ and $\hat{p}_b$, $\hat{\mathbf{v}}_b$, of the above equations. Then, the following statement (leading to a reciprocity principle) is in general true:

$$\nabla \cdot (\hat{p}_a\hat{\mathbf{v}}_b - \hat{p}_b\hat{\mathbf{v}}_a) = 0 \tag{4-9.6}$$

The proof is as follows:

$$\begin{aligned}
\nabla \cdot (p_a\hat{\mathbf{v}}_b) &= p_a\nabla \cdot \hat{\mathbf{v}}_b + \hat{\mathbf{v}}_b \cdot (\nabla\hat{p}_a) = \hat{p}_a\left(\frac{i\omega}{\rho c^2}\hat{p}_b\right) + \hat{\mathbf{v}}_b \cdot (i\omega\rho\,\hat{\mathbf{v}}_a) \\
&= \hat{p}_b\left(\frac{i\omega}{\rho c^2}\hat{p}_a\right) + \hat{\mathbf{v}}_a \cdot (i\omega\rho\,\hat{\mathbf{v}}_b) \\
&= \hat{p}_b\,\nabla \cdot \hat{\mathbf{v}}_a + \hat{\mathbf{v}}_a \cdot \nabla\hat{p}_b = \nabla \cdot (\hat{p}_b\hat{\mathbf{v}}_a)
\end{aligned}$$

where the successive steps follow from Eqs. (5) and from vector identities.

† See, for example, H. H. Skilling, *Electrical Engineering Circuits*, Wiley, New York, 1957, pp. 303–304, 331–332.

‡ The proof of the acoustic-reciprocity principle for an inhomogeneous medium is due to L. M. Lyamshev, "A Question in Connection with the Principle of Reciprocity in Acoustics," *Sov. Phys. Dokl.*, **4**:405-409 (1959).

Integration of Eq. (6) over a volume V and application of Gauss' theorem yields

$$\iint \hat{\mathbf{v}}_b \cdot \mathbf{n}_{\text{in}} \hat{p}_a \, dS - \iint \hat{\mathbf{v}}_a \cdot \mathbf{n}_{\text{in}} \hat{p}_b \, dS = 0 \tag{4-9.7}$$

where $\mathbf{n}_{\text{in}} = -\mathbf{n}_{\text{out}}$ is the unit normal pointing into the volume V. This is analogous to Eq. (4); $\hat{p}_a(\mathbf{x}_S)\, dS$ is the force applied in the a experiment to the volume by the external environment on a surface element of area dS centered at $\mathbf{x}_S$; $\hat{\mathbf{v}}_a(\mathbf{x}_S) \cdot \mathbf{n}_{\text{in}}$ is the corresponding velocity at $\mathbf{x}_S$ in the direction of the impressed force.

Interchange of Source and Listener

To prove the version of the acoustic-reciprocity theorem that involves interchange of listener and source positions, we let $\hat{p}_a(x)$, $\hat{\mathbf{v}}_a(x)$ be the field caused by a point source at $\mathbf{x}_1$ with source strength amplitude $\hat{Q}_a$, such that Re $\hat{Q}_a e^{-i\omega t}$ represents the time rate of volume efflux from the source. Then the first of Eqs. (5) is modified to

$$-i\omega\hat{p}_a + \rho c^2 \nabla \cdot \hat{\mathbf{v}}_a = \rho c^2 \hat{Q}_a \delta(\mathbf{x} - \mathbf{x}_1) \tag{4-9.8}$$

Similarly, let $\hat{p}_b(\mathbf{x})$, $\hat{v}_b(\mathbf{x})$ describe the field caused by a point source $\hat{Q}_b$ at $\mathbf{x}_2$. Then a derivation analogous to that leading to Eq. (6) yields

$$\nabla \cdot (\hat{p}_a \hat{\mathbf{v}}_b - \hat{p}_b \hat{\mathbf{v}}_a) = \hat{p}_a \hat{Q}_b \delta(\mathbf{x} - \mathbf{x}_2) - \hat{p}_b \hat{Q}_a \delta(\mathbf{x} - \mathbf{x}_1) \tag{4-9.9}$$

On the boundaries of the volume of interest it is assumed that conditions such as $\hat{p} = 0$, or $\hat{\mathbf{v}} \cdot \mathbf{n}_{\text{out}} = 0$, or $\hat{p}/\hat{\mathbf{v}} \cdot \mathbf{n}_{\text{out}} = Z(\mathbf{x}_S)$, or the Sommerfield radiation condition are prescribed. Both the a and b fields satisfy the same boundary conditions. Consequently, if one integrates both sides of Eq. (9) over the volume, the surface integral resulting from the divergence on the left side is zero,† so one is left with

$$\frac{\hat{p}_a(\mathbf{x}_2)}{\hat{Q}_a} = \frac{\hat{p}_b(\mathbf{x}_1)}{\hat{Q}_b} \tag{4-9.10}$$

The ratio of pressure amplitude to source strength remains the same if locations of source and listener are interchanged.

† The recognition that the reciprocity principle for point sources applies when portions of the boundary are locally reacting is due to E. Skudrzyk, *Die Grundlagen der Akustik*, Springer, Vienna, 1954, p. 380. Lyamshev, "Principle of Reciprocity," 1959, showed that the principle applies if the medium has within it elastic bodies, e.g., plates, shells, or membranes. However, J. H. Janssen, "A Note on Reciprocity in Linear Passive Accoustical Systems," *Acustica*, **8**:76–78 (1958), has shown that reciprocity is violated if the medium has within it a porous material described by equations of motion like those devised by C. Zwikker and C. W. Kosten, *Sound Absorbing Materials*, Elsevier, Amsterdam, 1949, chap. 3. Whether this is a necessary consequence of the material properties or an artifact of the model remains to be determined.

Reciprocity and Green's Functions

For a homogeneous medium, $\rho(\mathbf{x}_2) = \rho(\mathbf{x}_1)$, and the ratio $\hat{p}_a(\mathbf{x}_2)/\hat{Q}_a$ is $-i\omega\rho/4\pi$ times the Green's function $G_k(\mathbf{x}_2|\mathbf{x}_1)$ (see Sec. 4-3). Thus, Eq. (10) implies that

$$G_k(\mathbf{x}_2|\mathbf{x}_1) = G_k(\mathbf{x}_1|\mathbf{x}_2) \tag{4-9.11}$$

which can be regarded as the reciprocity principle for Green's functions corresponding to point-source solutions of the Helmholtz equation. This holds trivially for the free-space Green's function $R^{-1}e^{ikR}$, where $R = |\mathbf{x}_2 - \mathbf{x}_1|$, but the analysis above shows that it has considerable general applicability.†

Example A barrier extending to some height h is to be erected between a noise source and a region where quiet is desired. One side of the barrier is to be treated with special sound-absorbing material; the other side is to be left untreated. On which side should the treatment be applied?

SOLUTION Given that the source radiates very nearly as a point source, that the surfaces are locally reacting, and that the source and possible listeners are symmetrically located on opposite sides of the barrier, the answer, according to the principle of reciprocity, is that it makes no difference which side is treated.

4-10 TRANSDUCERS AND RECIPROCITY

A transducer is any device that changes one form of energy into another; loudspeakers and microphones are examples of *electroacoustic transducers*. A model of a linear electroacoustic transducer‡ can be taken as a "black box" (Fig. 4-20) embedded in a fluid with two wires at one end which carry a current i into and out of the transducer and across which the voltage is e. On the other

† A reciprocity relation when the source is a dipole rather than a monopole is derived by J. W. S. Rayleigh, "On the Application of the Principle of Reciprocity to Acoustics," *Proc. R. Soc. Lond.*, **25**:118–122 (1876); *Theory of Sound*, vol. 2, sec. 294. A well-known case (also discussed by Rayleigh) where reciprocity is not applicable is when the medium has an ambient motion. For example, if the wind velocity increases with height, sound is always heard better downwind than upwind. Reciprocity still applies, however, if the ambient flow direction is reversed at each point when source and listener locations are interchanged. From a strictly mathematical standpoint, reciprocity of the Green's function follows if the governing boundary-value problem (partial-differential equations and boundary conditions) is self-adjoint. Analogous considerations hold for the set of Green's functions corresponding to a system of equations. If the problem is not self-adjoint, a reciprocity principle can be derived relating the Green's functions to those corresponding to the adjoint system. For a full discussion, see C. Lanczos, *Linear Differential Operators*, Van Nostrand, London, 1961, pp. 239–244.

‡ For a general account of the mathematical description and properties of transducers, see F. Hunt, *Electroacoustics*, Harvard University Press, Cambridge, Mass., 1954, especially pp. 92–94, 103–109.

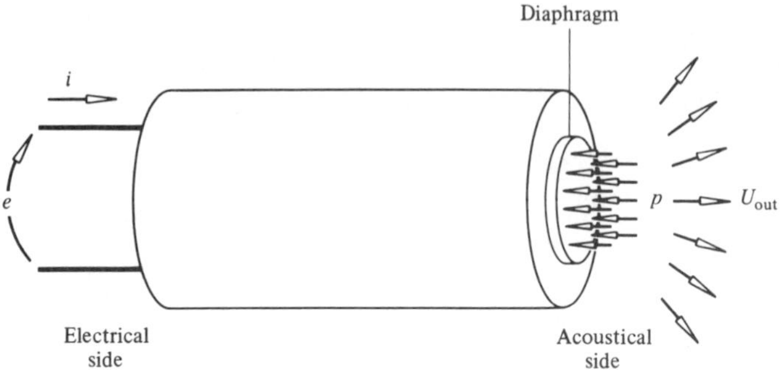

Figure 4-20 Sketch of an idealized transducer. Voltage e is across wires on electric side; current i flows through transducer. Pressure p on acoustical side acts on a diaphragm, whose vibration causes a volume velocity U_{out} flowing out from the transducer.

side is a movable surface whose motion is characterized by a *volume velocity* U representing the time rate of change of the volume enclosed by the surface or, equivalently, the area integral over the transducer surface of its outward-normal velocity. This surface is acted upon by some perturbation pressure p. If the pressure is nonuniform over the surface of the transducer, the value of p we use is a weighted surface average, the weighting being such that, for this p, $-pU$ is the net mechanical-power input to the transducer. The product ei represents the net electric-power input, so with such identifications we refer to $-p$ and U or to e and i as conjugate variables; $-p$ and e are *generalized forces;* U and i are *generalized velocities*.

When all variables e, i, $-p$, and U are oscillating with the same angular frequency ω, the physical properties of a linear transducer impose two algebraic relations† between the complex amplitudes $\hat{e}$, $\hat{\imath}$, $-\hat{p}$, and $\hat{U}$; we write them as

$$\begin{bmatrix} \hat{e} \\ -\hat{p} \end{bmatrix} = \begin{bmatrix} Z_{ec} & T_{ea} \\ T_{ae} & Z_a \end{bmatrix} \begin{bmatrix} \hat{\imath} \\ \hat{U} \end{bmatrix} \tag{4-10.1}$$

The matrix element Z_{ec} is the *clamped electrical impedance* ($\hat{e}/\hat{\imath}$ when $\hat{U}$ is zero), while Z_a is the *open-circuit acoustic impedance* ($-\hat{p}/\hat{U}_{out}$ when $\hat{\imath}$ is 0). (The term "acoustic impedance" is discussed in detail in Sec. 7-2.) The values of the matrix elements can be derived from fundamental principles if one has a detailed model of the transducer. Alternately, they can be obtained by experiment. The physical principles governing typical designs‡ result in either

† This was first recognized by H. Poincaré, "Study of Telephonic Reception," *Eclairage Electr.*, **50**:221–372 (1907). Writing the equations in terms of mechanical impedances as well as electric impedances is due to R. L. Wegel, "Theory of Magneto-Mechanical Systems as Applied to Telephone Receivers and Similar Structures," *J. Am. Inst. Electr. Eng.*, **40**:791–802 (1921).

‡ Reciprocity theorems for electroacoustic transducers date back to W. Schottky, "The Law of Low-Frequency Reception in Acoustics and Electroacoustics" *Z. Phys.*, **36**:689–736 (1926). For a general discussion and detailed proofs, see L. L. Foldy and H. Primakoff, "A General Theory of

$T_{ea} = T_{ae}$ or $T_{ea} = -T_{ae}$, although this is not invariably the case (transducers having the property $|T_{ea}| = |T_{ae}|$ are called *reciprocal transducers*). When this is so, the generalized velocity at one side of the transducer resulting from an application of a generalized force on the other side has the same direct proportionality to this force as when locations of generalized force and generalized velocity are interchanged, i.e.,

$$\left|\frac{\hat{U}}{\hat{e}}\right|_{\hat{p}=0} = \left|\frac{\hat{\imath}}{\hat{p}}\right|_{\hat{e}=0} \tag{4-10.2}$$

This is a reciprocity principle analogous to those discussed previously for mechanical and acoustical systems.

In general, one makes a distinction between the portion of $\hat{p}$ due to external causes, e.g., another sound source, and that caused by the motion of the surface, which causes a local motion of the surrounding fluid and which radiates sound to the far field. For a given environment, the latter portion $\hat{p}_{\text{rad}}$ is directly proportional to $\hat{U}$, so we write $\hat{p}_{\text{rad}}/\hat{U} = Z_{a,\text{rad}}$, this serving to define the *acoustic radiation impedance* of the transducer. With this definition, the second of the two algebraic equations implied by (1) can be rewritten

$$-\hat{p}_{\text{ext}} - Z_{a,\text{rad}}\hat{U} = T_{ae}\hat{\imath} + Z_a\hat{U} \tag{4-10.3}$$

This takes a form similar to the original equation if the second term on the left is transferred to the right and if $Z_a + Z_{a,\text{rad}}$ is abbreviated as Z_a'. Consequently, Eq. (1) also holds with the substitutions, $\hat{p} \to \hat{p}_{\text{ext}}$, $Z_a \to Z_a'$. If the transducer is a reciprocal transducer, Eq. (2) remains valid when $\hat{p}$ is replaced by $\hat{p}_{\text{ext}}$.

A transducer is acting as a *loudspeaker* when $\hat{p}_{\text{ext}} = 0$. In this case, its performance is characterized by the ratio $\hat{U}/\hat{\imath}$ (with $\hat{p}_{\text{ext}} = 0$). The transducer equation (with the substitutions described above) gives this ratio as $-T_{ae}/Z_a'$. If the loudspeaker dimensions are small compared with a wavelength and if the loudspeaker is located in an open space, it radiates as a monopole; the monopole amplitude is identified from Eq. (4-6.9*a*) as $-i\omega\rho\hat{U}/4\pi$. The far-field pressure amplitude is $(\hat{S}/r)e^{ikr}$, so one has

$$\hat{p}(r) = \left(\frac{\hat{U}}{\hat{\imath}}\right)_{p_{\text{ext}}=0}\left(\frac{-i\omega\rho}{4\pi r}\,e^{ikr}\right)(\hat{\imath}) \tag{4-10.4}$$

for the acoustic pressure amplitude in the far field.

If the transducer is acting as a *microphone*, the ideal operation is such that negligible current passes through the transducer; $\hat{e}$ will then vary in direct proportion to the external pressure $\hat{p}_{\text{ext}}$, the proportionality factor derived from Eq. (1) being $-T_{ea}/Z_a'$ (with the substitutions described previously). The mag-

---

Passive Linear Electroacoustic Transducers and the Electroacoustic Reciprocity Theorem, I and II," *J. Acoust. Soc. Am.*, **17**:109–120 (1945); **19**:50–58 (1947). That transducers are not necessarily reciprocal was demonstrated in 1942 by E. M. McMillan; the analysis is given in his "Violation of the Reciprocity Theorem in Linear Passive Electromechanical Systems," ibid., **18**:344–347 (1946).

nitude of this factor is the *microphone response* M (open-circuit voltage response to pressure in sound field).

If the transducer is a reciprocal transducer such that $|T_{ea}| = |T_{ae}|$, Eqs. (1) (with $\hat{p} \to \hat{p}_{\text{ext}}$) and (4) lead to *Schottky's law of low-frequency reception*

$$M = \left|\frac{\hat{e}}{\hat{p}_{\text{ext}}}\right|_{\hat{\imath}=0} = \frac{4\pi r}{\omega\rho}\left|\frac{\hat{p}(r)}{\hat{\imath}}\right|_{\hat{p}_{\text{ext}}=0} \tag{4-10.5}$$

which is completely independent of the constants of the transducer.

An application of (5) is in the *calibration of microphones*.† Suppose one wants to determine the *microphone response* M_A of microphone A. One has in the laboratory a loudspeaker C and a reciprocal transducer B, neither of which are necessarily calibrated. In a first experiment (see Fig. 4-21) the loudspeaker C is turned on, transducer B is placed a distance d from the loudspeaker, and its open-circuit voltage $|\hat{e}_B|_{\text{E1}}$ (caused by the pressure from the loudspeaker) is measured. Here E1 denotes experiment 1. In the second experiment, transducer B is removed and microphone A is placed in the identical position, the loudspeaker's input voltage being unchanged. Then the open-circuit voltage $|\hat{e}_A|_{\text{E2}}$ is measured. It is expected that $\hat{p}_{\text{ext}}$ will be the same in the two experiments, so

$$\frac{|\hat{e}_A|_{\text{E2}}}{|\hat{e}_B|_{\text{E1}}} = \frac{M_A}{M_B} \tag{4-10.6}$$

The third experiment is with the loudspeaker C replaced by the reciprocal transducer B and with microphone A left in the same position as in experiment 2. The transducer B is driven as a loudspeaker and its input current $|\hat{\imath}_B|_{\text{E3}}$ is measured. One also measures the open-circuit voltage $|\hat{e}_A|_{\text{E3}}$ induced in microphone A by the sound from transducer B. According to (5), the external pressure at microphone A in this experiment should be given by

$$|\hat{p}_{\text{ext},A}|_{\text{E3}} = \frac{\omega\rho}{4\pi d} M_B |\hat{\imath}_B|_{\text{E3}} = \frac{|\hat{e}_A|_{\text{E3}}}{M_A} \tag{4-10.7}$$

where the second equality results from the definition of M_A. Elimination of M_B (which is not necessarily known) from Eqs. (6) and (7) and subsequent solution

† The use of reciprocity in calibration of microphones was suggested by S. Ballantine in 1929 but was not used until 1940, when R. K. Cook and W. R. MacLean independently invented the absolute-calibration method and Cook demonstrated its practicality. [S. Ballantine, "Reciprocity in Electromagnetic, Mechanical, Acoustical, and Interconnected Systems," *Proc. Inst. Radio Eng.*, **17**:929–951 (1929); R. K. Cook, "Absolute Pressure Calibrations of Microphones," *J. Res. Nat. Bur. Stand.*, **25**:489–505 (1940); W. R. MacLean, "Absolute Measurement of Sound without a Primary Standard," *J. Acoust. Soc. Am.*, **12**:140–146 (1940).] A general review and historical account is given by H. B. Miller, "Acoustical Measurements and Instrumentation," ibid., **61**:274–282 (1977). The free-field method (due to MacLean) discussed in the text is less commonly used than the pressure-chamber method. Detailed calibration methods are described in ANSI S1.10-1966 (R1976), American National Standard Method for the Calibration of Microphones, American National Standards Institute, New York, 1976.

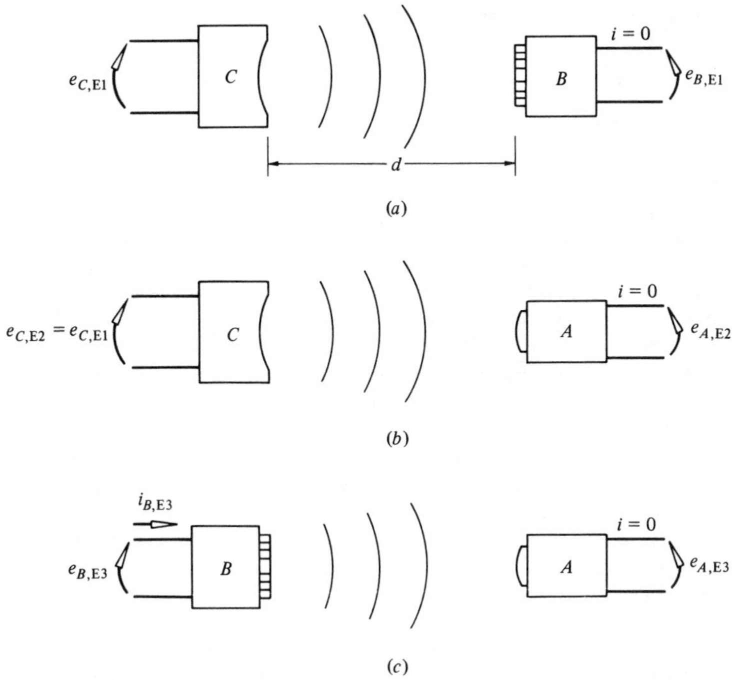

Figure 4-21 Free-field method for absolute calibration of a microphone A by use of a loudspeaker C and a reciprocal transducer B. Successive experiments E1, E2, E3 are sketched in (a), (b), and (c).

of the resulting equation for M_A then yields

$$M_A = \left(\frac{4\pi d}{\omega\rho}\right)^{\frac{1}{2}} \left(\frac{|\hat{e}_A|_{E3}|\hat{e}_A|_{E2}}{|\hat{e}_B|_{E1}|\hat{i}_B|_{E3}}\right)^{\frac{1}{2}} \tag{4-10.8}$$

Thus one has a measurement of the microphone response M_A without ever explicitly measuring a pressure.

PROBLEMS

4-1 A spherical body immersed in a compressible fluid has constant radius a up until time $t_0 = a/c$ and then suddenly begins to expand so that the radial velocity at the surface is V_0 for $t > a/c$, where $V_0 \ll c$. Determine the acoustic pressure and sketch p versus t for fixed r. (Limit your analysis to when $t - r/c \ll a/V_0$ and use an approximate boundary condition at $r = a$.) Show that the net acoustic energy imparted to the fluid is approximately $4\pi a^3 \rho V_0^2$. What fraction of this energy propagates to the far field? What happens to the rest of the energy? [M. C. Junger, *J. Acoust. Soc. Am.*, **40:**1025–1030 (1966).]

4-2 Show that the transient solution of the differential equation (4-2.3) is

$$e^{ct/a}\psi(t) = ca^2 \int_{-\infty}^{t} \sin\left[\frac{c(t-\tau)}{a}\right] v_C(\tau) e^{c\tau/a}\, d\tau$$

Show that for a sphere suddenly (at $t = 0$) accelerated from rest to constant speed v_C the above integral gives

$$\psi(t) = \begin{cases} 0 & t < 0 \\ \dfrac{v_C a^3}{2}\left[1 - e^{-ct/a}\left(\cos\dfrac{ct}{a} + \sin\dfrac{ct}{a}\right)\right] & t > 0 \end{cases}$$

4-3 Use the result of Prob. 4-2 in a discussion of sound radiation from an impulsively accelerated sphere of radius a whose translational velocity is 0 before $t = 0$ and equal to a constant value v_C for $t > 0$. Determine an explicit expression for the acoustic pressure during the early history of wave disturbances at radial distances $r \gg a$. Show that the pressure has a sudden jump at the onset of the pulse and determine the magnitude of this jump. Sketch a typical pressure waveform and indicate how one can determine a and v_C from it when these quantities are not known a priori. [M. C. Junger and W. Thompson, Jr., *J. Acoust. Soc. Am.*, **38**:978–986 (1965).]

4-4 The center of a rigid sphere of radius a is moving along a circular path of radius b with constant angular velocity Ω, where $\Omega b \ll c$, $\Omega a \ll c$. Determine an expression for the acoustic pressure in the far field resulting from this motion. What is the time-averaged acoustic power radiated?

4-5 A rigid sphere of radius a is oscillating back and forth along the z axis about the origin with angular frequency ω such that its center moves with velocity $\hat{v}_C \mathbf{e}_z \cos \omega t$, where $\hat{v}_C$ is a constant. Determine expressions for the time averages of the net acoustic kinetic energy and potential energy contained within a large sphere of radius r (centered at the origin) and verify that the difference of the two approaches a nonzero constant in the limit of large r. Determine this constant and give an interpretation of its magnitude for the case $\omega a/c \ll 1$ in terms of the related incompressible-flow problem. [J. E. Jones (Lennard-Jones), *Proc. Lond. Math. Soc.*, (2)**20**:347–364 (1922).]

4-6 Give explicit expressions for the inner and outer expansions (in powers of ka with a/r or kr held fixed) for the example of a radially oscillating sphere. Discuss the order of magnitude of successive terms for the cases $ka = 0.01$ and $kr = 0.1$ for both expansions. Show explicitly that the outer expansion of the inner expansion is the same as the inner expansion of the outer expansion for at least the first three terms.

4-7 An accelerometer mounted on the surface of a radially oscillating sphere of nominal radius a indicates that the acceleration is composed of a very large number of frequencies such that the mean squared acceleration associated with any finite frequency band of width Δf is $a_f^2 \Delta f$, where a_f^2 (spectral density of acceleration) is nearly constant over the range of 250 to 2000 Hz. Determine an expression for the spectral density of the received acoustic pressure at arbitrary radius r from the sphere for the same range of frequencies. Given that $ka \ll 1$ for all the frequencies of interest, by how many decibels would the octave-band sound-pressure levels corresponding to two successive octave bands be expected to differ?

4-8 Answer the questions in Prob. 4-7 for a rigid sphere undergoing transverse oscillations along the z axis, the accelerometer being mounted at a point on the sphere corresponding to $\theta = 0$.

4-9 Two point sources of monopole amplitudes $\hat{S}$ and $-\hat{S}$, both radiating at angular frequency ω, are located a distance d apart, where kd is not necessarily small. Determine expressions for the far-field acoustic pressure and the time-averaged net acoustic power radiated by this combination of sources. For what values of kd is the acoustic power within 10 percent of what would be predicted for a dipole with dipole-moment amplitude $\hat{S}d$? Beyond what value of kd can one be assured that the radiated power is within 10 percent of that corresponding to the sum of what would be radiated by each source in the absence of the other source? How do you reconcile your results with the prediction (Sec. 1-11) that the power output by a collection of sources is the sum of the powers output by the individual sources?

4-10 *Acoustic similitude.* Show that the complex amplitude $\hat{p}$ of acoustic pressure in a sound field radiated by a body of characteristic dimension a vibrating with angular frequency ω has the general and asymptotic forms

$$\hat{p} \approx \rho c \hat{v}_{\text{typ}} F\left(\frac{\mathbf{x}}{a}, ka\right) \rightarrow \rho c \hat{v}_{\text{typ}} M(\theta, \phi, ka) \frac{a}{r} e^{ikr}$$

while the time average of the net acoustic power radiated by the body is of the form

$$\mathcal{P}_{av} = \rho c \left[\iint (v_n^2)_{av}\, dA\right] Q(ka)$$

Here the functions $F(\mathbf{x}/a, ka)$, $M(\theta, \phi, ka)$, and $Q(ka)$ should be dimensionless and should in general depend on the shape of the body and on the relative amplitudes and phases of the normal velocity on the surface of the body; $\hat{v}_{typ}$ is a complex amplitude of the normal velocity at a typical point on the surface; $\iint (v_n^2)_{av}\, dA$ is the integral of the mean squared normal velocity over the body's surface. Show also that, in the limit of small ka, the functions $M(\theta, \phi, ka)$ and $Q(ka)$ vary with ka as ka and $(ka)^2$, respectively, for monopole radiation; as $(ka)^2$ and $(ka)^4$, respectively, for dipole radiation; and as $(ka)^3$ and $(ka)^6$, respectively, for quadrupole radiation.

4-11 A small vibrating body ($ka \ll 1$) radiates primarily as a quadrupole into an unbounded fluid. Assuming that the surface vibrations are unaffected by the surrounding fluid, show that the time-averaged acoustic power output varies with the ambient density and sound speed of the fluid as ρ/c^5. Suppose that the power output is $\mathcal{P}_{av,0}$ when the surrounding fluid is air at a pressure of 10^5 Pa and a temperature of 20°C. What is the power output when the pressure is pumped down to 10^3 Pa with the temperature held constant? Suppose, after the pumping down, hydrogen (a diatomic gas with molecular weight 2) is added to the fluid until the pressure once again is 10^5 Pa (the temperature still being held constant). What is the resulting sound power output of the body in this air-hydrogen mixture? [G. G. Stokes, *Phil. Trans. R. Soc. Lond.*, **158**:447–463 (1868).]

4-12 A sphere of nominal radius a, nominally centered at the origin, is simultaneously undergoing radial and transverse oscillations such that its centerpoint has velocity $\hat{v}_C \mathbf{e}_z \cos \omega t$ and its instantaneous radius is $a + (\hat{v}_S/\omega) \sin \omega t$. Determine an expression for the complex amplitude of the acoustic pressure at an arbitrary point outside the sphere. Determine the net time-averaged acoustic power output of the body and show that the contributions from radial and transverse oscillations are additive. Given that $ka = 0.1$, what would the ratio $|\hat{v}_C/\hat{v}_S|$ have to be for the two contributions to be equal? Is your result consistent with the assertion that any body with time-varying volume tends to radiate primarily as a monopole in the limit $ka \ll 1$?

4-13 (The following exercise is intended to demonstrate that the near-field pressure of a vibrating body may possibly be predicted from an incompressible-flow model even when ka is comparable to 1.) A spherical body of nominal radius a is undergoing quadrupole-type contortions such that the normal velocity at the surface is given by $V_0 \sin^2 \theta \cos \phi \sin \phi \cos \omega t$. Determine the ratio of the complex pressure amplitude at the surface to what would be obtained if the surrounding fluid were incompressible and plot the real and imaginary parts of this ratio versus ka. Up to what value of ka is the real part within 25 percent of its low-frequency limit? Up to what value is the imaginary part less than 25 percent of the real part?

4-14 Verify that an explicit substitution into the Kirchhoff-Helmholtz integral formula of the surface pressures and normal velocities for a radially oscillating sphere leads to the expression for the acoustic pressure outside the sphere derived in Sec. 4-1. For simplicity, limit your comparison to the constant-frequency case and to points where $r \gg a$, $kr \gg 1$, but do not necessarily assume that ka is small.

4-15 Carry through the exercise described in Prob. 4-14 for the example of a transversely oscillating sphere.

4-16 One possible scheme to determine the acoustic power output of a vibrating body is to measure p and v_n simultaneously on the surface, compute the time average of their product, and integrate the result over the surface area. Suppose this method is tried for a transversely oscillating sphere vibrating such that $ka = 0.1$. To what accuracy would the relative phase between v_n and p have to be measured at each point in order to guarantee an accuracy of 10 percent in the acoustic power estimate? Would one expect less stringent instrumentation requirements if the measurements were made instead on a sphere whose radius were such that $kr = 1$?

4-17 Show that it is possible for three longitudinal quadrupoles to be mutually oriented so that the resulting acoustic field is completely spherically symmetric. How would the acoustic power output of the combination of the three quadrupoles compare with what would be expected for the sum of the three acoustic powers associated with each quadrupole when radiating alone?

4-18 In a large unbounded space, a sphere of fluid of radius a is suddenly heated, e.g., by nuclear irradiation, to a temperature increment ΔT above the ambient temperature T_0, such that, at $t = 0$, the sphere has pressure $p_0 + \Delta p$ but is of ambient density and the fluid within it has not yet begun to move. Assuming that the linear acoustic approximation is valid, what is the time dependence of acoustic pressure p at an arbitrary radius $r > a$? Give a sketch of your result.

4-19 A rectangular solid, a by $1.5a$ by $2a$, is centered at the origin with each of its six faces nominally perpendicular to the corresponding coordinate axis; it is undergoing rotational oscillations (angular frequency ω) about the z axis. Determine an approximate expression for the dependence on r, θ, ϕ (spherical coordinates) of the acoustic pressure at distances $r \gg a$. If the amplitude of the pressure oscillations at a distance corresponding to $kr = 10$ at a point on the $x = y$ line is p_{10}, what would you estimate as the total time-averaged acoustic power output (in terms of p_{10}, ω, c, ρ) of this sound source? Assume $ka \ll 1$. How would you expect p_{10} and the power output to vary if the frequency were doubled but the peak angle of rotation of the solid were kept constant?

4-20 The acoustic pressure on the surface of a vibrating sphere of radius a is measured and found to be given by

$$p = A \cos \omega t \cos \theta$$

where A and ω are constant and θ is the polar angle in spherical coordinates. What would you estimate as the time-averaged acoustic power generated by this source in terms of A, ω, ρ, and c?

4-21 A sphere of radius a and mass M is suspended from a fixed point in an otherwise open space by a spring with spring constant k_{sp}, such that the tether point lies on the z axis and the sphere's center is nominally at the origin. The sphere is displaced a distance z_0 ($\ll a$) and released from rest. Discuss the subsequent motion of the sphere assuming $M \gg \frac{4}{3}\pi\rho a^3$. How long will it be before 90 percent of the potential energy initially stored in the spring is radiated away as sound? (Neglect viscosity.)

4-22 A cubical loudspeaker enclosure, dimensions a on each edge, has four loudspeakers of radius b centrally placed in each of its four sides (but not on the top and bottom). The enclosure is suspended in a large open space. If only one loudspeaker is excited, the average acoustical power output is $\mathcal{P}_{av,1}$. What would the power output be if all four are excited with the same amplitude and all four are in phase? (Assume $ka \ll 1$.) If each loudspeaker moved as a rigid disk of area A and with velocity amplitude V_0 and angular frequency ω, what would you estimate for $\mathcal{P}_{av,1}$? Discuss the nature of the radiation when the loudspeakers 2, 3, and 4 (numbered counterclockwise looking down from the top) have phases of 90, 180, and 270° relative to the first loudspeaker.

4-23 When a small loudspeaker that radiates as a monopole in an open space is placed in the corner of a room, the sound-pressure level in the center of the room is 100 dB. The loudspeaker is then moved to the center of the room, and the vibrational amplitude of its moving face is increased by a factor of 2. What would you expect for the sound-pressure level in the corner of the room (old loudspeaker position)?

4-24 A sound source located at point A in a building gives rise to a sound level outside the building 100 m away at point B of 75 dB. It is known that the sound leaves through an open window. In a second experiment, it is found that a second sound source located a large distance away from the building (in the same relative direction as B) causes a sound level inside the building of 60 dB at point A. The sound level at the same distance from the source along an unobstructed path is 65 dB. Estimate the acoustic power output the first sound source, i.e., in the building, would have if it were radiating into an open space. Assume both sources to be nominally omnidirectional and to have dimensions small compared with a wavelength. Both sources have the same frequency content.

4-25 A small body of unspecified shape is oscillating with angular frequency ω in a fluid with sound speed c and ambient density ρ. Any representative dimension a of the body is much less than c/ω. At distances r, where $r \gg a$ and $r \ll c/\omega$, the pressure perturbation caused by the body's oscillations is found to be given approximately by

$$p \approx \frac{Kx}{r^3} \cos \omega t$$

where K is a constant. Estimate the time-averaged acoustic power that this body radiates to the far field in terms of K, ρ, c, and ω.

4-26 A rigid square plate of dimensions a on a side is oscillating back and forth along the z axis normal to its face, such that its center has velocity $V_0 \cos \omega t$. Assume that $(\omega/c)a \ll 1$, $V_0/\omega \ll a$. The value of V_0 is not measured, but it is known to be the same in two successive experiments. In experiment 1, the ambient density is ρ, the sound speed is c, and the angular frequency ω_1. In experiment 2, the ambient density is pumped down to $10^{-3}\rho$, the fluid is heated so that its sound speed becomes $2c$, and the frequency is increased to $2\omega_1$. In the first experiment, the acoustic pressure is measured on the z axis at a distance c/ω_1 from the plate and is found to be given by $K_1 \cos \omega_1 t$. Give an expression for the acoustic pressure at radial distances $r \gg a$ (but r not necessarily large compared with c/ω_1) for the second experiment. Express your result in terms of the parameters c, ω_1, ρ, K_1 as well as the spherical coordinates r and θ.

4-27 Two identical reciprocal transducers are separated a distance of 4.8 m in an unbounded fluid (sound speed 340 m/s, ambient density 1.2 kg/m^3). One transducer is used as a loudspeaker, the other as a microphone. When an oscillating current of rms amplitude 10^{-2} A is input to the first transducer, it is found that an oscillating voltage of rms amplitude 1 V is induced in the open circuit of the second transducer. The frequency is 200 Hz. What is the rms acoustic pressure incident on the moving face of the second transducer?

4-28 The disk described in Sec. 4-8 is undergoing rocking oscillations about the diameter lying along the x axis, such that a point on the disk with a given y coordinate has velocity $v_z = \Omega y$. Here Ω is the time-varying angular velocity of the disk. Show that the acoustic-pressure field at large distances from the disk is given in the small ka approximation by

$$p = 2\frac{\partial^2}{\partial y \partial z}\frac{Q_{yz,1}(t - r/c)}{r} \qquad Q_{yz,1}(t) = \frac{2\rho\dot{\Omega}(t)a^5}{45\pi}$$

4-29 The circumstances of Prob. 4-28 are altered so that the disk is undergoing rocking oscillations about the line $y = \Delta$. Show that the resulting pressure on the front face ($z = 0^+$) of the disk is approximately

$$p \approx \frac{4}{3\pi}\rho\dot{\Omega}(t)(a^2 - w^2)^{1/2}(y - \tfrac{3}{2}\Delta)$$

and that the acoustic pressure in the far field is

$$p \approx \frac{4\rho a^5}{45\pi}\left(\frac{\partial}{\partial y} + \frac{15}{2}\frac{\Delta}{a^2}\right)\frac{\partial}{\partial z}\frac{\dot{\Omega}(t - r/c)}{r}$$

4-30 Devise any linear circuit having as elements at least one resistor, two inductors, and a capacitor and demonstrate that reciprocity holds in the sense that the complex amplitude of the current flowing through the second inductor caused by a specified voltage imposed in series with the first inductor is the same as when the voltage is imposed in series with the second inductor and the measured current is that flowing through the first inductor.

4-31 Give an alternate derivation of the reciprocity relation Eq. (4-9.10) starting from Eq. (4-9.7) with a volume bounded externally by the fluid's natural boundaries and internally by two tiny spheres centered at $\mathbf{x}_1$ and $\mathbf{x}_2$. Boundary conditions on the inner sphere centered at $\mathbf{x}_1$ should be such that, for the a field, the net volume flowing per unit time out through the sphere has complex amplitude $\hat{Q}_a$ in the limit of vanishing sphere radius while, for the b field, the corresponding limit is zero.

CHAPTER

FIVE

RADIATION FROM SOURCES NEAR AND ON SOLID SURFACES

The present chapter begins with a discussion of the effects of nearby solid surfaces on the radiation of sound and then continues with the closely related topic of radiation from a planar surface when a portion of it is vibrating. This topic serves to introduce and illustrate concepts helpful in understanding the influence of baffles on sound sources, the radiation from extended bodies, the transition from near field to far field, and common phenomena associated with the diffraction of sound.

5-1 SOURCES NEAR PLANE RIGID BOUNDARIES

The sound field radiated by a source is often appreciably affected by a neighboring surface. If this surface (referred to here as a *wall*) is idealized as rigid, planar, and of infinite extent, only simple considerations are required to take its presence into account.

Image Sources

The conceptual device commonly used is an *image source* (see Fig. 5-1) such that the original boundary-value problem of source plus wall is replaced by one with two sources (original source and image source) but no wall. The image source is the mirror image in all respects of the original source. Thus, if the wall corresponds to the plane $z = 0$ and if (x_S, y_S, z_S) is a point on the surface of the original source, $(x_S, y_S, -z_S)$ must be a point on the surface of the corresponding image source. If the velocity at a point on the source's surface has cartesian

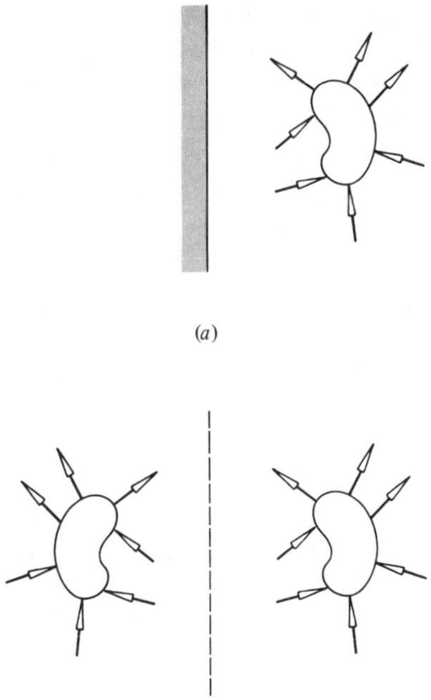

(a)

(b)

Figure 5-1 Concept of an image source. The original boundary-value problem (*a*) of a vibrating body outside a rigid plane surface is equivalent to the boundary-value problem (*b*) of radiation from source and image source in an unbounded medium.

components (v_1, v_2, v_3), the velocity at the corresponding point on the image source must have components $(v_1, v_2, -v_3)$.

The mirror symmetry of the boundary-value problem of two sources and no wall requires the z component of the fluid velocity to vanish on the plane $z = 0$. This is the condition imposed by the presence of the wall in the original boundary-value problem with source and wall, so the solution to the problem with source and image source but no wall satisfies the fluid-dynamic equations and the boundary conditions appropriate to the original problem. Our uniqueness theorems of Sec. 4-5 require the two solutions to be identical in the region $z > 0$.

Remarks concerning Acoustic Power and Spherical Spreading

Symmetry requires that one-half of the power radiated by a source and its image in an open space be transmitted to the source side of the symmetry plane. Consequently, the total power (radiating into the region $z > 0$) emitted by a source near a wall is half what would be radiated by the isolated source-image combination (no wall) in all directions.

At radial distances large compared with source-image separation, source dimensions, and a wavelength, the (far-field) acoustic pressure is of the form†

† F. A. Fischer, "Directionality and Radiation Intensity of Acoustic Ray Groups in the Vicinity of a Reflecting Plane Surface," *Elektr. Nachrichtentech.*, **10**:19–24 (1933).

$f(t - r/c, \theta, \phi)/r$, the radial component of acoustic fluid velocity is $p/\rho c$, and the radial component of time-averaged intensity is $(p^2)_{av}/\rho c$. The intensity and mean squared pressure at such large distances decrease as $1/r^2$ with increasing radial distance r (fixed θ and ϕ), so our conclusions concerning spherical spreading for an isolated source apply equally well for a source near a plane rigid wall, given r sufficiently large. If $J(\theta, \phi)/r^2$ gives the far-field intensity, the acoustic power $\mathcal{P}_{av,W}$ radiated into the region $z > 0$ by a source near a wall is

$$\mathcal{P}_{av,W} = \int_0^{\pi/2} \int_0^{2\pi} J(\theta, \phi)\, d\phi \sin\theta\, d\theta \tag{5-1.1}$$

Note that we integrate over a hemisphere rather than a sphere; θ ranges from 0 to $\pi/2$ rather than from 0 to π.

Cases When More than One Wall Is Present

For a source between two parallel rigid walls, one needs an infinite array of images (see Fig. 5-2a). There are, first, the two images corresponding to reflections of the source through the two walls, then images of the images corre-

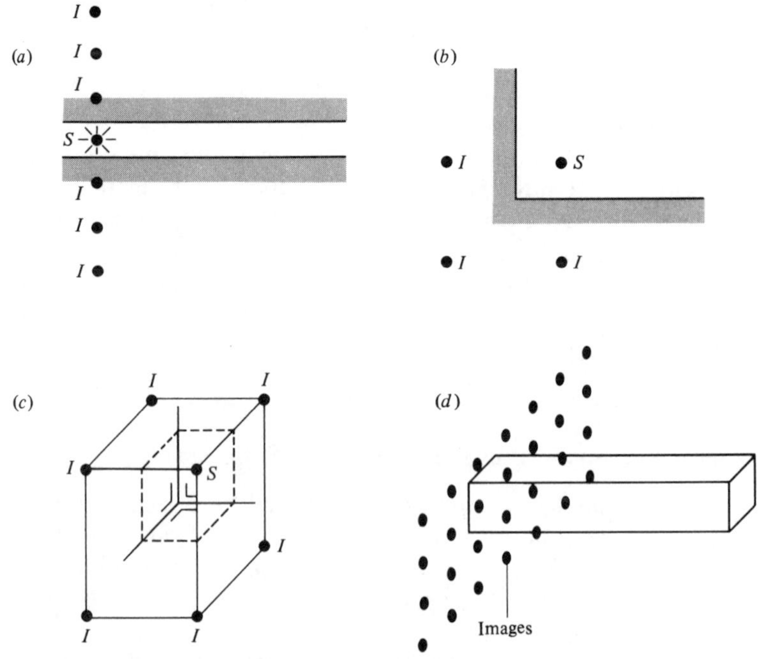

Figure 5-2 Situations in which more than one image source is required to satisfy the boundary conditions: (a) source between plane parallel walls; (b) source near where two perpendicular walls meet; (c) source near intersection of three mutually perpendicular surfaces; (d) source in a rectangular duct.

sponding to reflections of the images through the opposite walls, then images of these images, etc. The total array of sources has a repetition distance of twice the distance between walls. (This is what one sees in a room with mirrors on two parallel walls.)

The array of sources is not confined to a region of limited spatial extent, so our previous discussion concerning spherical spreading does not apply. Energy-conservation considerations imply instead, at large cylindrical radial distance w, that the integral over z between walls of the time-averaged radial component of intensity should fall off with w as $1/w$ for fixed azimuthal angle ϕ. In general, the z component of intensity will not be negligible compared with the radial component, and one cannot assume that the plane-wave relation $p = \rho c v_w$ holds at large w.

The method of images also applies when a source is near two rigid walls meeting at right angles (see Fig. 5-2*b*); three image sources are required in the equivalent boundary-value problem. If the source is near the corner of three walls at right angles to each other, one obtains an equivalent boundary-value problem by adding seven image sources (Fig. 5-2*c*). Since the source and images are confined to a region of limited spatial extent, deductions analogous to those for the single-wall case can be made concerning spherical spreading at large distances from the source.

A more complicated example (Fig. 5-2*d*) is a source in an infinitely long rectangular duct with rigid walls. In this case, there is a twofold infinity of image sources, all lying in a plane transverse to the duct. For a source in a six-sided rectangular room with rigid walls, there is a threefold infinity of image sources arrayed in a three-dimensional rectangular lattice.

Dependence of Acoustic Far Field and Net Acoustic Power Output on Distance from a Wall

For a source of characteristic dimension a vibrating at angular frequency $\omega = ck$ and located a nominal distance z_S from a single flat rigid wall (at $z = 0$), the far-field pressure and the net acoustic power output depend on kz_S and a/z_S. A principal assumption is that the state of vibration of the body is independent of z_S. (This is a good approximation for a solid body vibrating in air.) In the limit $a/z_S \ll 1$, that is, where distance from the wall is great compared with a body dimension, the total acoustic field is well approximated by the superposition of those fields resulting from separate consideration of the source and image. Thus, if the far-field acoustic pressure due to the source alone (no wall) is $\hat{f}(\theta, \phi)R_S^{-1}e^{ikR_S}$, where R_S is distance from the source's nominal location, the combination of the source and image has a far-field pressure (with $R_S \gg a$, $R_I \gg a$) given by

$$\hat{p} \approx \hat{f}(\theta, \phi)\frac{e^{ikR_S}}{R_S} + \hat{f}(\pi - \theta, \phi)\frac{e^{ikR_I}}{R_I} \tag{5-1.2}$$

where R_I is distance from the image source. At distances $r \gg z_S$, one has

$R_S \approx r - z_S \cos\theta$ and $R_I \approx r + z_S \cos\theta$, so the above reduces to

$$\hat{p} \approx \frac{e^{ikr}}{r} [e^{-ikz_S \cos\theta} \hat{f}(\theta, \phi) + e^{ikz_S \cos\theta} \hat{f}(\pi - \theta, \phi)] \tag{5-1.3}$$

From this one derives the time-averaged acoustic intensity $\frac{1}{2}|\hat{p}|^2/\rho c$. The average acoustic power output results from Eq. (1) with $J(\theta, \phi) = r^2 I_{r,\text{av}}$. Taking $\hat{p}$ as given by Eq. (3), changing the θ integration variable to $\theta' = \pi - \theta$ in appropriate terms, then replacing the symbol θ' by θ, we find

$$\mathcal{P}_{\text{av},W} = \mathcal{P}_{\text{av,ff}} + \Delta\mathcal{P}_{\text{av}} \tag{5-1.4}$$

where
$$\mathcal{P}_{\text{av,ff}} = \frac{1}{2\rho c} \int_0^{2\pi} \int_0^{\pi} |\hat{f}(\theta, \phi)|^2 \sin\theta \, d\theta \, d\phi \tag{5-1.5}$$

$$\Delta\mathcal{P}_{\text{av}} = \frac{1}{2\rho c} \text{Re} \left[\int_0^{2\pi} \int_0^{\pi} e^{i2kz_S \cos\theta} \hat{f}(\pi - \theta, \phi) \hat{f}^*(\theta, \phi) \sin\theta \, d\theta \, d\phi \right] \tag{5-1.6}$$

Here $\mathcal{P}_{\text{av,ff}}$ is the free-field power output (wall not present), and $\Delta\mathcal{P}_{\text{av}}$ is the power increment (possibly negative) caused by the presence of the wall.

If the far-field radiation of the source when isolated is spherically symmetric (as for a monopole), $\hat{f}(\theta, \phi)$ is the monopole amplitude $\hat{S}$ and the above expressions reduce to†

$$\hat{p} \approx 2\hat{S} \frac{e^{ikr}}{r} \cos(kz_S \cos\theta) \qquad r \gg z_S \tag{5-1.7a}$$

$$\mathcal{P}_{\text{av},W} = \mathcal{P}_{\text{av,ff}} \left[1 + \frac{\sin 2kz_S}{2kz_S} \right] \tag{5-1.7b}$$

When $kz_S \ll 1$, the acoustic pressure in the far field is doubled, the intensity increases by a factor of 4, and the power increases by a factor of 2. (Recall that the power is going only into the region $z > 0$.) When $2kz_S = 4.49$ ($z_S = 0.358\lambda$), the power output has its minimum value of $0.783\,\mathcal{P}_{\text{av,ff}}$; it oscillates about $\mathcal{P}_{\text{av,ff}}$ at larger z_S and is within 5 percent of the free-field value for $2kz_S > 20$ ($z_S > 1.59\lambda$).

If the radiation pattern for the source alone resembles that of a dipole perpendicular to the wall, $\hat{f}(\theta, \phi)$ is $-ik\hat{D}_z \cos\theta$ and $\hat{f}(\pi - \theta, \phi)$ is the negative of $\hat{f}(\theta, \phi)$; Eq. (3) therefore yields

$$\hat{p} = -2 \sin(kz_S \cos\theta) k\hat{D}_z \cos\theta \frac{e^{ikr}}{r} \tag{5-1.8a}$$

where $\hat{D}_z$ is the source's dipole-moment amplitude. The field for $kz_S \ll 1$ is consequently that of a longitudinal quadrupole with quadrupole-moment amplitude $2z_S\hat{D}_z$. The power output for arbitrary kz_S is given (with $\eta = 2kz_S$), according to Eqs. (5) and (6), by

$$\mathcal{P}_{\text{av},W} = \mathcal{P}_{\text{av,ff}}(1 - 6\eta^{-2} \cos\eta - 3\eta^{-1} \sin\eta + 6\eta^{-3} \sin\eta) \tag{5-1.8b}$$

† U. Ingard and G. Lamb, Jr., "Effect of a Reflecting Plane on the Power Output of Sound Sources," *J. Acoust. Soc. Am.*, **29**:743–744 (1957).

The quantity in parentheses reduces to $\frac{3}{10}\eta^2$ and to $1 - 3\eta^{-1} \sin \eta$ in the limits $\eta \ll 1$ and $\eta \gg 1$. Although the source's acoustic power vanishes when the source is at the wall, it is within 5 percent of the free-field value when $\eta > 60$ ($z_S > 4.77\lambda$).

One concludes from the above examples and from a study of Eq. (6) that $\Delta\mathscr{P}_{\text{av}}$ can be regarded as 0 if kz_S is sufficiently large. Since the real and imaginary parts of $\exp\,(i2kz_S \cos \theta)$ oscillate rapidly with θ if kz_S is large, in the limit of very large kz_S the overall integrand is an oscillatory function, the integrals over whose peaks tend to cancel integrals over troughs. Just how far the source must be from the surface before the limit is nearly realized depends on the complexity of the source.

5-2 SOURCES MOUNTED ON WALLS: THE RAYLEIGH INTEGRAL; FRESNEL-KIRCHHOFF THEORY OF DIFFRACTION BY AN APERTURE

A model for a source with a *baffle,* e.g., a loudspeaker on one side of a large enclosure, is that in which a limited portion of a surface has prescribed normal velocity, the remainder of the surface being idealized as rigid. The surface is here taken as the $z = 0$ plane, and the region on the $+z$ side of the surface is idealized as unbounded (see Fig. 5-3).

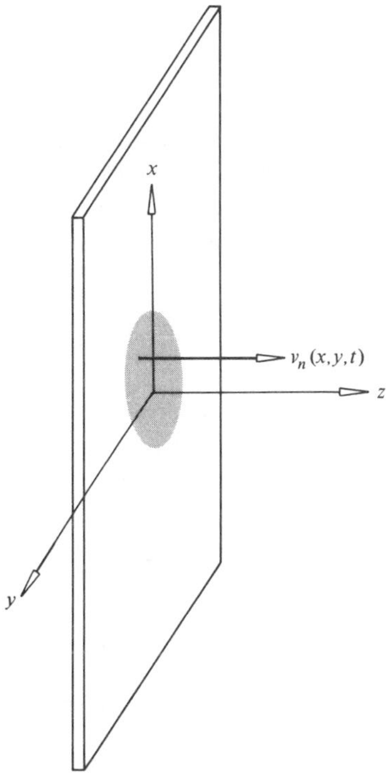

Figure 5-3 Nomenclature for description of radiation from a nominally flat and rigid surface ($z = 0$ plane), a limited portion of which is vibrating with normal velocity $v_n(x, y, t)$.

An expression for the acoustic pressure outside the surface can be extracted from the Kirchhoff-Helmholtz integral theorem, Eq. (4-6.6), with the aid of the method of images. The boundary-value problem, with nonzero $v_n(x, y, t)$ specified on some area of the $z = 0$ plane and otherwise zero, is equivalent to that of radiation from a thin disk of time-varying thickness in an unbounded medium. The normal velocity v_n for given x and y on the two sides of the disk has the same value, i.e., both sides are either moving outward simultaneously or moving inward simultaneously, so that the resulting z symmetry requires p, v_x, and v_y to be even in z but v_z to be odd in z. Consequently, the integrals in Eq. (4-6.6) over the surface pressure give equal and opposite contributions, and the net contribution from surface pressure to the Kirchhoff-Helmholtz integral is zero. (The distance R from listener position to either of any two surface points on opposite sides of the disk has the same value since the disk is infinitesimally thin.) The integrals over the surface-normal velocity from the front and back surfaces of the disk give equal contributions, so one need integrate only over the front face providing the resulting expression is multiplied by 2.

The result of the reasoning just outlined is that the Kirchhoff-Helmholtz integral reduces to the *Rayleigh integral*†

$$p(\mathbf{x}, t) = \frac{\rho}{2\pi} \iint \frac{\dot{v}_n(x_S, y_S, t - R/c)}{R}\, dx_S\, dy_S \tag{5-2.1}$$

where R^2 is $z^2 + (x - x_S)^2 + (y - y_S)^2$. This is equivalent to the field generated by a continuous smear of monopole sources distributed on the $z = 0$ plane; that is, p satisfies the inhomogeneous wave equation

$$\nabla^2 p - \frac{1}{c^2}\frac{\partial^2 p}{\partial t^2} = -2\rho \dot{v}_n(x, y, t)\,\delta(z) \tag{5-2.2}$$

in an unbounded space. The apparent mass added to the fluid per unit surface area has a time derivative equal to $2\rho v_n(x, y, t)$; the volume excluded from the fluid per unit area of the $z = 0$ plane by the source has a time derivative equal to $2v_n(x, y, t)$. The factor of 2 appears because both sides of the disk are moving outward with velocity v_n.

Green's-Function Derivation of Rayleigh Integral

An alternate derivation‡ of Eq. (1) results for the constant-frequency case from the Green's-function formulation in Sec. 4-6. One can rephrase Eq. (4-6.4) for the problem under consideration here as

$$\hat{p}(\mathbf{x}) = \frac{1}{4\pi} \iint [\hat{p}(\mathbf{x}_S)\, \nabla_S G_k(\mathbf{x}_S|\mathbf{x}) - G_k(\mathbf{x}_S|\mathbf{x})\, \nabla_S \hat{p}(\mathbf{x}_S)]_{z_S=0} \cdot \mathbf{e}_z\, dx_S\, dy_S \tag{5-2.3}$$

† J. W. S. Rayleigh, *The Theory of Sound,* vol. 2, 2d ed., 1896, reprinted by Dover, New York, 1945, sec. 278.

‡ A. Sommerfeld, "The Freely Vibrating Piston Membrane," *Ann. Phys.*, (5)**42**:389–420 (1943).

where $G_k(\mathbf{x}_S|\mathbf{x})$ is a Green's function for the Helmholtz equation, which we choose to be that corresponding to a point source outside a rigid flat surface. It can be derived by the method of images and is

$$G_k(\mathbf{x}_S|\mathbf{x}) = R_1^{-1}e^{ikR_1} + R_2^{-1}e^{ikR_2} \tag{5-2.4}$$

where $$R_{1,2} = [(x_S - x)^2 + (y_S - y)^2 + (z_S \mp z)^2]^{\frac{1}{2}} \tag{5-2.5}$$

[Here $G_k(\mathbf{x}_S|\mathbf{x}) = G_k(\mathbf{x}|\mathbf{x}_S)$, in accord with the *principle of reciprocity* discussed in Sec. 4-9.] The Green's function of Eq. (4) has the property that $\nabla_S G_k(\mathbf{x}_S|\mathbf{x}) \cdot \mathbf{e}_z$ vanishes at $z_S = 0$, so the first term in Eq. (3) drops out. In regard to the second term, $G_k(\mathbf{x}_S|\mathbf{x})$ at $z_S = 0$ is $2R^{-1}e^{ikR}$. Also $\nabla_S\hat{p}(\mathbf{x}_S) \cdot \mathbf{e}_z$ at $z_S = 0$ is $i\omega\rho\hat{v}_n(x_S, y_S)$, in accord with Euler's equation of motion, so Eq. (3) reduces to

$$\hat{p}(\mathbf{x}) = \frac{-i\omega\rho}{2\pi}\iint \hat{v}_n(x_S, y_S)R^{-1}e^{ikR}\,dx_S\,dy_S \tag{5-2.6}$$

which is recognized as the constant-frequency form (involving complex amplitudes) of Eq. (1).

Fresnel-Kirchhoff Theory of Diffraction

There is a resemblance between the Rayleigh integral in Eq. (6) and what results from the Fresnel-Kirchhoff theory of diffraction† by an aperture A in a screen (Fig. 5-4). If a wave disturbance, e.g., plane wave or diverging spherical wave, is incident from the $-z$ side of the screen on the aperture, the classic assumptions (expressed in terms of acoustic quantities) of Kirchhoff would be that insofar as the evaluation of the pressure on the $+z$ side is concerned, the $\hat{p}(x_S, y_S)$ and $\hat{v}_n(x_S, y_S)$ in the Kirchhoff-Helmholtz integral can be taken as $\hat{p}_i(x_S, y_S)$ and $\hat{\mathbf{v}}_i(x_S, y_S) \cdot \mathbf{e}_z$ (i for incident) within the aperture and as zero at points on the screen surface outside the aperture. This would then give ($z > 0$)

$$\hat{p}(\mathbf{x}) = \frac{1}{4\pi}\iint_A [-\hat{p}_i(\mathbf{x}_S)(ik - R^{-1})\mathbf{e}_R - i\omega\rho\hat{\mathbf{v}}_i(\mathbf{x}_S)] \cdot \mathbf{e}_z R^{-1}e^{ikR}\,dx_S\,dy_S \tag{5-2.7}$$

where the integral extends only over the aperture. At distances large compared with a wavelength, the quantity R^{-1} is neglected compared with ik. Furthermore, if the incident wave is a plane wave with propagation direction $\mathbf{n}_i$, then $p_i = \rho c \mathbf{v}_i \cdot \mathbf{n}_i$ and $\mathbf{v}_i \cdot \mathbf{e}_z = (\mathbf{v}_i \cdot \mathbf{n}_i)\mathbf{n}_i \cdot \mathbf{e}_z$, so Eq. (7) would reduce to

$$\hat{p}(\mathbf{x}) = \frac{-i\omega\rho}{2\pi}\iint_A \left[\frac{1}{2}\left(1 + \frac{\mathbf{e}_R \cdot \mathbf{e}_z}{\mathbf{n}_i \cdot \mathbf{e}_z}\right)\right] \hat{\mathbf{v}}_i(x_S, y_S) \cdot \mathbf{e}_z R^{-1}e^{ikR}\,dx_S\,dy_S \tag{5-2.8}$$

† M. Born and E. Wolf, *Principles of Optics*, 4th ed., Pergamon, Oxford, 1970, pp. 378–381. Pertinent original references are A. Fresnel, "On the Diffraction of Light; Examination of the Colored Fringes Existing in the Shadow of an Illuminated Body," *Ann. Chim. Phys.*, (2)**1**:239–281 (1816); G. G. Stokes, "On the Dynamical Theory of Diffraction," *Trans. Camb. Phil. Soc.*, **9**:1 (1849), reprinted in Stokes, *Mathematical and Physical Papers*, vol. 2, Cambridge University Press, Cambridge, 1883, pp. 243–328; G. Kirchhoff, "On the Theory of Light Rays," *Ann. Phys. Chem.*, **18**:663–695 (1883).

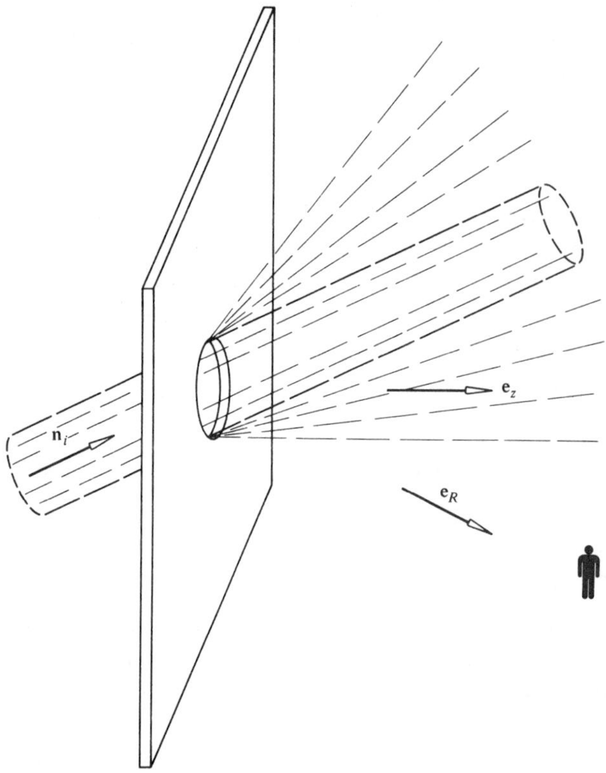

Figure 5-4 Unit vectors $\mathbf{n}_i$, $\mathbf{e}_R$, $\mathbf{e}_z$ used in the Fresnel-Kirchhoff approximation for diffraction by an aperture in a thin screen.

An equivalent version (with the assumptions described above) results when $\hat{\mathbf{v}}_i$ is replaced by $\hat{p}_i\mathbf{n}_i/\rho c$.

Equation (8) can be compared with Eq. (6). The two agree if $\hat{v}_n$ is interpreted as $\hat{\mathbf{v}}_i \cdot \mathbf{e}_z$ and if the location of the observation point $\mathbf{x}$ is far enough distant to make $\mathbf{e}_R$ approximately constant and nearly equal to $\mathbf{n}_i$ for all straight lines connecting points on the aperture with $\mathbf{x}$. If the Kirchhoff assumption that $\hat{\mathbf{v}} \cdot \mathbf{e}_z = \hat{\mathbf{v}}_i \cdot \mathbf{e}_z$ on the aperture is accepted, expression (8) would have to be erroneous unless $\mathbf{e}_R \cdot \mathbf{e}_z/\mathbf{n}_i \cdot \mathbf{e}_z$ is identically 1, since Eq. (6) represents the exact solution when $\hat{v}_n$ is known over the plane $z = 0$ and since $\hat{v}_n$ must be zero on the plane at points outside the aperture.

The Fresnel-Kirchhoff theory of diffraction is intrinsically a high-frequency approximation; it gives incorrect results when the aperture dimensions are much smaller than a wavelength.† Furthermore, even if such dimensions are

† H. A. Bethe, "Theory of Diffraction by Small Holes," *Phys. Rev.*, **66:**163–182 (1944); R. D. Spence, "A Note on the Kirchhoff Approximation in Diffraction Theory," *J. Acoust. Soc. Am.*, **21:**98–100 (1949).

large and one uses the theory to predict fields at only those distances which are large compared with a wavelength, the predictions may be in substantial error at large angular deviations from the direction $\mathbf{n}_i$. Nevertheless, the theory is satisfactory for explaining small-angle high-frequency diffraction phenomena and has an advantage in simplicity compared with rigorous theories of diffraction. It is extensively used in optics; applications to acoustics are limited because many of the diffraction phenomena of interest either involve dimensions small compared with a wavelength or require an understanding of diffraction through large angles.

5-3 LOW-FREQUENCY RADIATION FROM SOURCES MOUNTED ON WALLS

Insight into the implications of the Rayleigh integral can be obtained from examination of limiting cases. If the region in which $\hat{v}_n$ is nonzero is confined to a distance a from the origin, and if $ka \ll 1$, the concepts of matched asymptotic expansions discussed in Sec. 4-7 are applicable. The near-field pressure satisfies Laplace's equation and has a complex amplitude found from Eq. (5-2.6) with e^{ikR} replaced by $1 + ikR$

$$\hat{p}_{\text{in}}(\mathbf{x}) = \frac{-i\omega\rho}{2\pi} \iint \hat{v}_n(x_S, y_S) R^{-1}\, dx_S\, dy_S + \frac{\rho c k^2}{2\pi} \hat{Q}_S \qquad (5\text{-}3.1)$$

where $\hat{Q}_S$ is the surface integral of $\hat{v}_n(x_S, y_S)$ and represents the complex amplitude of the rate of volume flow out from the source.

The acoustic-pressure amplitude at distance $r \gg a$ is given by the multipole expansion that matches Eq. (1); to fourth order in ka, one has

$$\hat{p}_{\text{out}}(\mathbf{x}) = \hat{S}\, \frac{e^{ikr}}{r} - \left(\hat{D}_x \frac{\partial}{\partial x} + \hat{D}_y \frac{\partial}{\partial y} \right) \frac{e^{ikr}}{r} + \left(\hat{Q}_{xx} \frac{\partial^2}{\partial x^2} + 2\hat{Q}_{xy} \frac{\partial^2}{\partial x\, \partial y} + \hat{Q}_{yy} \frac{\partial^2}{\partial y^2} \right) \frac{e^{ikr}}{r} \qquad (5\text{-}3.2)$$

where $\hat{S}$ is $-(i\omega\rho/2\pi)\hat{Q}_S$, while $\hat{D}_x$ and $\hat{Q}_{xy}$ are given by $-(i\omega\rho/2\pi)$ times the area integrals of $x_S \hat{v}_n$ (for $\hat{D}_x$) and of $\frac{1}{2}x_S y_S \hat{v}_n$ (for $\hat{Q}_{xy}$). The leading term in Eq. (2), with the time dependence explicitly inserted, gives the prediction

$$p_{\text{out}}(\mathbf{x}, t) = \frac{\rho}{2\pi r} \dot{Q}_S \left(t - \frac{r}{c} \right) \qquad (5\text{-}3.3)$$

which describes a radially symmetric spherical wave. This is the same as the expression (4-1.6) for monopole radiation from a vibrating body of time-varying volume if we replace $\dot{Q}_S$ by $2\dot{Q}_S$; the factor of 2 results because of the image source.

The above solution indicates the substantial effect a baffle has on sound radiation. If a circular disk of radius a is vibrating with constant frequency

($ka \ll 1$) transverse to its face in an open space, it radiates primarily as a dipole and the acoustic power output to one side (see Sec. 4-8) is given by $(16/27\pi)^2\rho c(ka)^4(\pi a)^2(v_n^2)_{av}/2$. However, if the disk is baffled by placing it in an aperture of the same size in a large screen, the radiation is primarily as a monopole and the power output to one side is $\rho c(ka)^2(\pi a^2)(v_n^2)_{av}/2$. Insofar as $ka \ll 1$, the second case corresponds to a much greater power output.

Pressure on Vibrating Circular Piston at Low Frequencies

For a vibrating circular piston of radius a mounted in a rigid wall (an idealization of a baffled loudspeaker), the pressure amplitude at the wall ($z = 0$), given $ka \ll 1$, can be determined from Eq. (1) with $\hat{v}_n$ set equal to a constant over the surface of the piston; one then has

$$(\hat{p}_{\text{in}})_{z=0} = \frac{\rho c}{\pi}\,\hat{v}_n\left[-ika\,\psi\left(\frac{w}{a}\right) + \frac{\pi}{2}(ka)^2\right] \tag{5-3.4}$$

where

$$2\psi\left(\frac{w}{a}\right) = a^{-1}\iint (R^{-1})_{z=0}\,dx_S\,dy_S \tag{5-3.5}$$

Because of the cylindrical symmetry and because of its lack of dimensionality, (5) is a function only of w/a, where w is the distance of the point (x, y) from the center of the piston.

To evaluate $\psi(w/a)$ it is sufficient to let $y = 0$, $x = -w$. Then one can use a cylindrical coordinate system in which $x_S = -w + \xi a\cos\phi$, $y_S = \xi a\sin\phi$, such that ξa is the radial distance (cylindrical coordinates) from the point $(-w, 0)$. The differential area element is then $a^2\xi\,d\xi\,d\phi$ and, moreover, $(R)_{z=0}$ is $a\xi$, so $2\psi(w/a) = \iint d\xi d\phi$ with appropriate integration limits. With the abbreviations $\eta = w/a$, $\zeta = (1 - \eta^2\sin^2\phi)^{1/2}$, and $\phi_m = \sin^{-1}(1/\eta)$ we find that the disk occupies the region $0 < \xi < \eta\cos\phi + \zeta$, $0 < \phi < 2\pi$, for $\eta < 1$, and the region $\eta\cos\phi - \zeta < \xi < \eta\cos\phi + \zeta$, $-\phi_m < \phi < \phi_m$, for $\eta > 1$. Consequently, one has

$$2\psi(\eta) = \iint d\xi\,d\phi = \begin{cases} \displaystyle\int_0^{2\pi}(\eta\cos\phi + \zeta)\,d\phi & \eta < 1 \quad (5\text{-}3.6a) \\ \displaystyle 2\int_{-\phi_m}^{\phi_m}(1 - \eta^2\sin^2\phi)^{1/2}\,d\phi & \eta > 1 \quad (5\text{-}3.6b) \end{cases}$$

The second expression can be cast into a more convenient form if one changes the integration variable to $u = \sin^{-1}(\eta\sin\phi)$, such that u is $\pi/2$ when $\phi = \phi_m$ and such that $\zeta\,d\phi/du$ is the sum of $-(\eta - \eta^{-1})(1 - \eta^{-2}\sin^2 u)^{-1/2}$ and $\eta(1 - \eta^{-2}\sin^2 u)^{1/2}$.

The integral over $\cos\phi$ from 0 to 2π in the $\eta < 1$ expression in Eqs. (6) vanishes, and the integral over ζ from 0 to 2π is 4 times the integral from 0 to

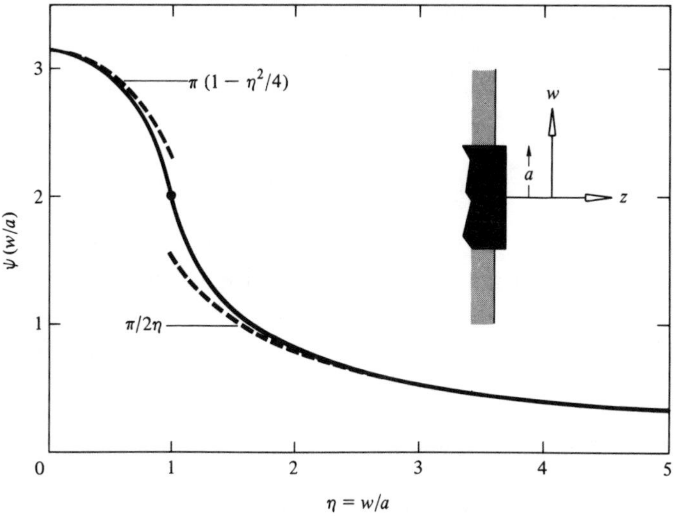

Figure 5-5 Plot of function $\psi(w/a)$ describing the relative magnitude of the acoustic pressure [with complex amplitude $-(ika/\pi)\rho c\hat{v}_n\psi(w/a)$] at radius $w = \eta a$ outside ($z = 0^+$) a wall in which a piston of radius a is oscillating with very low frequency ($ka \ll 1$).

$\pi/2$; the indicated integrations reduce in this manner to†

$$\psi(\eta) = \begin{cases} 2E(\eta^2) & \eta < 1 \quad (5\text{-}3.7a) \\ 2\eta E\left(\dfrac{1}{\eta^2}\right) - 2(\eta - \eta^{-1})K\left(\dfrac{1}{\eta^2}\right) & \eta > 1 \quad (5\text{-}3.7b) \end{cases}$$

Here we abbreviate

$$\begin{Bmatrix} E(m) \\ K(m) \end{Bmatrix} = \int_0^{\pi/2} (1 - m \sin^2 \phi)^{\pm 1/2}\, d\phi \tag{5-3.8}$$

for the *complete elliptical integrals*‡ of the first and second kinds, respectively. [Both $K(m)$ and $E(m)$ are $\pi/2$ at $m = 0$; as $m \to 1$, $K(m) \to \frac{1}{2} \ln [16/(1 - m)]$ and $E(m) \to 1$.] The function $\psi(\eta)$ (see Fig. 5-5) has the value of π at $\eta = 0$, decreases monotonically to 2 at $\eta = 1$, and further decreases for $\eta > 1$ to an asymptotic form $\psi(\eta) \to \pi/2\eta$ at large η. This latter behavior is consistent with the requirement that $\hat{p}_{\text{in}}$ match the expression in Eq. (2) for $a \ll w \ll 1/k$.

† H. Lamb, "On the Vibrations of an Elastic Plate in Contact with Water," *Proc. R. Soc. Lond.*, **A98:**205–216 (1920). A general result holding for arbitrary *ka* was later derived by N. W. McLachlan, "The Acoustic and Inertia Pressure at Any Point on a Vibrating Circular Disk," *Phil. Mag.*, (7)**14:**1012–1025 (1932).

‡ L. M. Milne-Thomson, "Elliptical Integrals," in M. Abramowitz and I. Stegun (eds.), *Handbook of Mathematical Functions,* Dover, New York, 1965, pp. 590–592, 608–611.

Force Exerted by the Slowly Oscillating Baffled Piston

The complex amplitude of the force exerted by the piston on the fluid outside the wall is the integral of $(\hat{p}_{\text{in}})_{z=0}$ over the area of the piston. In this respect, note that

$$\int_0^{2\pi}\int_0^a \psi\left(\frac{w}{a}\right) w\, dw\, d\phi = 4\pi a^2 \int_0^1 \left[\int_0^{\pi/2} (1 - \eta^2 \sin^2 u)^{1/2}\, du\right] \eta\, d\eta$$

A change of integration order allows the η integration to be performed; the resulting integrand for the u integration is subsequently recognized as the derivative of $\frac{1}{3}[\tan(u/2) + \sin u]$. Consequently, the above expression is $\frac{8}{3}\pi a^2$. Equation (4) therefore gives† the force exerted by the piston on the fluid to second order in ka as

$$\hat{F}_z = (\rho c \hat{v}_n)\pi a^2 \left[-ika\frac{8}{3\pi} + \frac{(ka)^2}{2}\right] \tag{5-3.9}$$

or with the time dependence explicitly inserted,

$$F_z(t) = \rho\pi a^2 \frac{8a}{3\pi}\, \dot{v}_n(t) - \frac{\rho\pi a^4}{2c}\, \ddot{v}_n(t) \tag{5-3.10}$$

The leading term, from the viewpoint of Newton's second law, indicates that the fluid entrained by the piston has an apparent mass of $\rho\pi a^2(8a/3\pi)$, corresponding to the fluid in a cylinder of area πa^2 and length $8a/3\pi$.

5-4 RADIATION IMPEDANCE OF BAFFLED-PISTON RADIATORS

The ratio of the force amplitude $\hat{F}_z$ to the normal velocity amplitude $\hat{v}_n$ for a baffled piston (with $\hat{v}_n$ constant over the piston's area) is the piston's *mechanical radiation impedance* (here denoted by $Z_{m,\text{rad}}$) and is the area integral of the specific radiation impedance $\hat{p}/\hat{v}_n$. Thus, from Eq. (5-2.6), one has

$$Z_{m,\text{rad}} = \frac{-i\omega\rho}{2\pi}\iiiint R^{-1}e^{ikR}\, dx_S\, dy_S\, dx\, dy \tag{5-4.1}$$

where R is $[(x - x_S)^2 + (y - y_S)^2]^{1/2}$ and the limits are such that (x_S, y_S) and (x, y) are within the area A of the piston. The ratio $(\hat{F}_z/A)/\hat{v}_n A = Z_{m,\text{rad}}/A^2$ is the *acoustic radiation impedance* $Z_{a,\text{rad}}$. The quadruple integral in Eq. (1) is known as the *Helmholtz integral*.

Electroacoustic Significance of Radiation Impedance

This parameter $Z_{m,\text{rad}}$ is of importance in transducer design because it describes the influence of the environment on transducer performance. In particular, it is

† J. W. S. Rayleigh, "On the Theory of Resonance," *Phil. Trans. R. Soc. Lond.*, **161:**77–118 (1870).

required for the evaluation of the transducer's electroacoustic efficiency. For a linear electroacoustic transducer operating at constant angular frequency ω, Eq. (4-10.1) relates the complex amplitudes (see Fig. 4-20) $\hat{e}$ and $-\hat{F}_z/A$ to the complex amplitudes $\hat{\imath}$ and $\hat{U} = \hat{v}_n A$. [The $\hat{p}$ in Eq. (4-10.1) is the complex amplitude of an average pressure p, the averaging being such that $-pU$ is the power input to the transducer by the external fluid. Since, for the rigid piston, this power is $-F_z v_n$, and since U is $v_n A$, we replace $\hat{p}$ by $\hat{F}_z/A$.] If the transducer constants Z_{ec}, T_{ea}, T_{ae}, and Z_a are known, the additional knowledge of the radiation impedance $Z_{m,\text{rad}}$ allows a prediction of the ratios $\hat{v}_n/\hat{e}$ and $\hat{\imath}/\hat{e}$ when the transducer is operated as a loudspeaker; i.e.,

$$A\hat{v}_n = \frac{T_{ae}}{T_{ae}T_{ea} - Z_{ec}Z'_a} \qquad \hat{\imath} = \frac{-Z'_a}{T_{ae}T_{ea} - Z_{ec}Z'_a} \tag{5-4.2}$$

where Z'_a abbreviates $Z_a + Z_{m,\text{rad}}/A^2$. These relations, given the applied voltage $\hat{e}$, determine the electric power $\frac{1}{2}$ Re $\hat{e}\hat{\imath}^*$ supplied and the acoustic power output $\frac{1}{2}|\hat{v}_n|^2$ Re $Z_{m,\text{rad}}$. The ratio of the latter to the former is the *electroacoustic efficiency* η, given in terms of the symbols introduced above by

$$\eta = \frac{|T_{ae}|^2 \operatorname{Re} Z_{m,\text{rad}}/A^2}{T_{ae}T_{ea} - Z_{ec}Z'_a} \tag{5-4.3}$$

Evaluation of Radiation Impedance for a Baffled Circular Piston

The fourfold integration in Eq. (1) reduces† to tabulated functions of a single variable for a circular piston of radius a with a series of mathematical manipulations. Because of the symmetry in interchange of x and y with x_S and y_S, it is sufficient to restrict the integration range so that $(x_S^2 + y_S^2)^{1/2} \le (x^2 + y^2)^{1/2}$ and subsequently to multiply the result by 2. For the x_S, y_S, integration, one uses a coordinate system centered at the point (x, y) and rotated so that the center of the disk lies at $x'_S = w$, $y'_S = 0$, where $w = (x^2 + y^2)^{1/2}$, and introduces cylindrical coordinates R, ϕ_S, such that $x'_S = R\cos\phi_S$ and $y'_S = R\sin\phi_S$. The region $(x_S^2 + y_S^2)^{1/2} < w$ then comprises points where $-\pi/2 < \phi_S < \pi/2$ and $0 < R < 2w\cos\phi_S$. In this manner, one obtains

$$Z_{m,\text{rad}} = \frac{-i\omega\rho}{\pi}\int_0^{2\pi} d\phi \int_0^a w\,dw \int_{-\pi/2}^{\pi/2} d\phi_S \int_0^{2w\cos\phi_S} e^{ikR}\,dR \tag{5-4.4}$$

The ϕ integration gives a factor of 2π; the last two integrations yield

$$\frac{1}{\pi}\int_{-\pi/2}^{\pi/2} d\phi_S \int_0^{2w\cos\phi_S} e^{ikR}\,dR = \frac{1}{\pi ik}\int_{-\pi/2}^{\pi/2} e^{i2kw\cos\phi_S}\,d\phi_S - \frac{1}{ik}$$
$$= \frac{1}{ik}[J_0(2kw) + i\mathbf{H}_0(2kw) - 1] \tag{5-4.5}$$

† Rayleigh, *The Theory of Sound*, vol. 2, sec. 302.

where

$$J_0(\eta) = \frac{2}{\pi}\int_0^{\pi/2} \cos(\eta \cos\phi_S)\, d\phi_S \qquad \mathbf{H}_0(\eta) = \frac{2}{\pi}\int_0^{\pi/2} \sin(\eta\cos\phi_S)\, d\phi_S \tag{5-4.6}$$

are the *Bessel function* and the *Struve function*† of zero order (see Table 5-1). The functions $J_0(\eta)$ and $\mathbf{H}_0(\eta)$ have the properties‡

$$\int_0^\eta J_0(\eta)\eta\, d\eta = \eta J_1(\eta) = -\eta \frac{d}{d\eta} J_0(\eta) \tag{5-4.7a}$$

$$\int_0^\eta \mathbf{H}_0(\eta)\eta\, d\eta = \eta \mathbf{H}_1(\eta) = \eta\left[\frac{2}{\pi} - \frac{d}{d\eta}\mathbf{H}_0(\eta)\right] \tag{5-4.7b}$$

where $J_1(\eta)$ and $\mathbf{H}_1(\eta)$ are the Bessel function and the Struve function of first order. These relations permit an evaluation of the remaining integration over w in Eq. (4); the net result for the mechanical radiation impedance is

$$Z_{m,\text{rad}} = \rho c \pi a^2[R_1(2ka) - iX_1(2ka)] \tag{5-4.8}$$

with (see Fig. 5-6)

$$R_1(2ka) = 1 - \frac{2J_1(2ka)}{2ka} \qquad X_1(2ka) = \frac{2\mathbf{H}_1(2ka)}{2ka} \tag{5-4.9}$$

For small values of the argument η, a power-series expansion and a term-

† The Bessel function $J_n(\eta)$ and the Struve function $\mathbf{H}_n(\eta)$ for positive integer order n can be considered to be defined by the integrals

$$\begin{Bmatrix} J_n(\eta) \\ \mathbf{H}_n(\eta) \end{Bmatrix} = \frac{2(2n+1)\eta^n}{[(2n+1)(2n-1)\cdots 3\cdot 1]\pi}\int_0^{\pi/2} \begin{Bmatrix} \cos \\ \sin \end{Bmatrix}(\eta\cos\phi)(\sin\phi)^{2n}\, d\phi$$

For a full discussion, see G. N. Watson, *A Treatise on the Theory of Bessel Functions*, 2d ed., Cambridge University Press, London, 1966, pp. 24–25, 328–338. The expression for $J_n(\eta)$ is known as *Poisson's integral* for the Bessel function. The boldface symbol $\mathbf{H}_n(\eta)$ for the Struve function is traditional and should not be construed as denoting a vector.

‡ For the Struve functions, the identity (7*b*) follows from

$$\begin{aligned} 1 &= \int_0^{\pi/2} \frac{\partial}{\partial\phi}[\sin\phi\cos(\eta\cos\phi)]\, d\phi \\ &= \int_0^{\pi/2}\left\{\frac{\partial}{\partial\eta}[\sin(\eta\cos\phi)] + \eta\sin^2\phi\sin(\eta\cos\phi)\right\} d\phi \qquad \text{(i)} \\ &= \int_0^{\pi/2}\left\{\frac{\partial}{\partial\eta}\left[\eta\frac{\partial}{\partial\eta}\sin(\eta\cos\phi)\right] + \eta\sin(\eta\cos\phi)\right\} d\phi \qquad \text{(ii)} \end{aligned}$$

Equation (i) leads to $1 = (\pi/2)(d\mathbf{H}_0/d\eta + \mathbf{H}_1)$, while (ii) leads to $1 = (\pi/2)[(d/d\eta)(\eta\, d\mathbf{H}_0/d\eta) + \eta\mathbf{H}_0]$. Since $\eta\, d\mathbf{H}_0/d\eta = 0$ at $\eta = 0$, the integral from 0 to η of the latter yields $\eta = (\pi/2)(\eta\, d\mathbf{H}_0/d\eta + \text{L})$, where L is the left side of (7*b*). The derivation of (7*a*) for the Bessel functions proceeds in an analogous manner from

$$0 = \eta\int_0^{\pi/2}\frac{\partial}{\partial\phi}[\sin\phi\cos\phi\cos(\eta\cos\phi)]\, d\phi$$

Table 5-1 Bessel and Struve functions of orders 0 and 1

η	$J_0(\eta)$	$J_1(\eta)$	$\mathbf{H}_0(\eta)$	$\mathbf{H}_1(\eta)$
0	1.00	0.00	0.00	0.00
0.5	0.94	0.24	0.31	0.05
1.0	0.77	0.44	0.57	0.20
1.5	0.51	0.56	0.74	0.41
2.0	0.22	0.58	0.79	0.65
2.5	−0.05†	0.50	0.73	0.86
3.0	−0.26	0.34	0.57	1.02
3.5	−0.38	0.14	0.36	1.09
4.0	−0.40	−0.07†	0.14	1.07
4.5	−0.32	−0.23	−0.06†	0.97
5.0	−0.18	−0.33	−0.19	0.81
5.5	−0.01	−0.34	−0.23	0.63
6.0	+0.15†	−0.28	−0.18	0.48
6.5	0.26	−0.15	−0.08	0.38
7.0	0.30	−0.00	+0.06†	0.35
7.5	0.27	+0.14†	0.20	0.39
8.0	0.17	0.23	0.30	0.49
8.5	0.04	0.27	0.34	0.62
9.0	−0.09†	0.25	0.32	0.75
9.5	−0.19	0.16	0.24	0.85
10.0	−0.25	0.04	0.12	0.89

† Zeros of $J_0(\eta)$ are 2.405, 5.520, 8.654; zeros of $J_1(\eta)$ are 3.832, 7.016, 10.173; zeros of $\mathbf{H}_0(\eta)$ are 4.323, 6.780, 10.481.

by-term integration of Eqs. (6) and (7) yields

$$J_1(\eta) = \frac{\eta/2}{(1!)^2} - \frac{2(\eta/2)^3}{(2!)^2} + \frac{3(\eta/2)^5}{(3!)^2} - \cdots \tag{5-4.10a}$$

$$\mathbf{H}_1(\eta) = \frac{2}{\pi}\left(\frac{\eta^2}{1^2\cdot 3} - \frac{\eta^4}{1^2\cdot 3^2\cdot 5} + \frac{\eta^6}{1^2\cdot 3^2\cdot 5^2\cdot 7} - \cdots\right) \tag{5-4.10b}$$

so, for small values of $2ka$, the *piston impedance functions* $R_1(2ka)$ and $X_1(2ka)$ are given by

$$R_1(2ka) = \frac{(2ka)^2}{4\cdot 2} - \frac{(2ka)^4}{6\cdot 4^2\cdot 2} + \frac{(2ka)^6}{8\cdot 6^2\cdot 4^2\cdot 2} - \cdots \tag{5-4.11a}$$

$$X_1(2ka) = \frac{(4/\pi)(2ka)}{3} - \frac{(4/\pi)(2ka)^3}{5\cdot 3^2} + \frac{(4/\pi)(2ka)^5}{7\cdot 5^2\cdot 3^2} + \cdots \tag{5-4.11b}$$

Both series are absolutely convergent but slow to converge when $2ka$ is substantially larger than 1. Note that these are consistent with Eq. (5-3.10) in the limit $2ka \ll 1$.

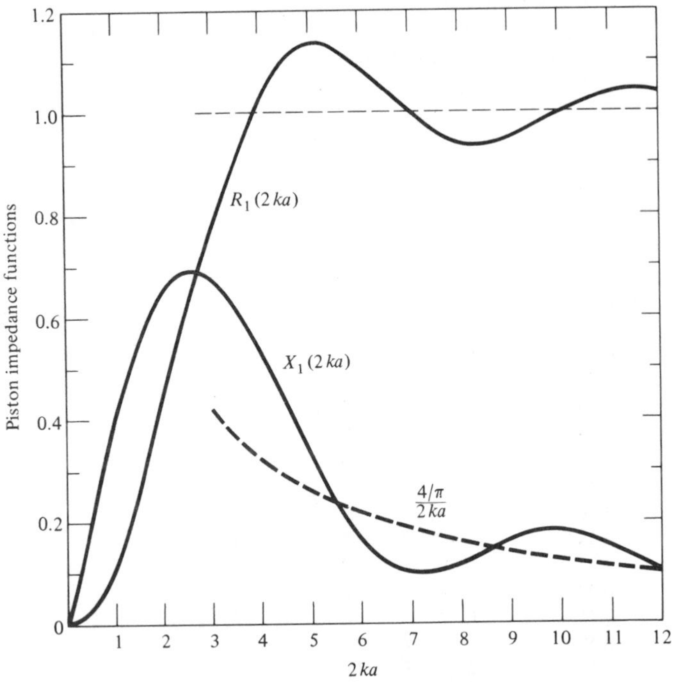

Figure 5-6 Piston impedance functions $R_1(2ka)$ and $X_1(2ka)$ for a circular piston of radius a mounted in a rigid planar baffle. These functions are such that the mechanical radiation impedance of the piston is $\rho c\pi a^2(R_1 - iX_1)$.

In the other limit, when $2ka \gg 1$, one uses the asymptotic expressions†

$$J_1(\eta) \to \left(\frac{2}{\pi\eta}\right)^{1/2} \cos\left(\eta - \frac{3\pi}{4}\right) \tag{5-4.12a}$$

$$\mathbf{H}_1(\eta) \to \frac{2}{\pi} + \left(\frac{2}{\pi\eta}\right)^{1/2} \sin\left(\eta - \frac{3\pi}{4}\right) \tag{5-4.12b}$$

to obtain

$$R_1(2ka) \to 1 - \frac{(8/\pi)^{1/2}\cos(2ka - 3\pi/4)}{(2ka)^{3/2}} \tag{5-4.13a}$$

$$X_1(2ka) \to \frac{4/\pi}{2ka} + \frac{(8/\pi)^{1/2}\sin(2ka - 3\pi/4)}{(2ka)^{3/2}} \tag{5-4.13b}$$

† To derive the asymptotic expression for $\mathbf{H}_1(\eta)$, we write the integrand in Eq. (6) for $\mathbf{H}_0(\eta)$ as the real part of $i \exp(-i\eta \cos \phi)$ and interchange the order of taking the real part and of integrating. The integration path is then deformed to one going from 0 to $\pi/2 + i\infty$ plus one going from $\pi/2 + i\infty$ to $\pi/2$. For the first segment, the variable of integration is changed to s, so that $\cos\phi = 1 - is^2$ and s goes from 0 to $+\infty$ along the path. In the second segment, one lets $\xi = \text{Im}\,\phi$ be the integration variable. Doing all this yields

$$\mathbf{H}_0(\eta) = \left(\frac{2}{\pi}\right) 2^{1/2}\, \text{Re}\left[e^{-i(\eta-\pi/4)} \int_0^\infty \frac{e^{-\eta s^2}\, ds}{(1 - is^2/2)^{1/2}}\right] + \frac{2}{\pi}\int_0^\infty e^{-\eta \sinh \xi}\, d\xi$$

The limiting expressions of 1 and $(4/\pi)/2ka$ are approached in an oscillatory manner, the amplitude decreasing as $(2ka)^{-3/2}$ with increasing ka. The limiting value of $\rho c \pi a^2$ for $Z_{m,\text{rad}}$ is what would be expected if the acoustic disturbance near $z = 0$ over the major portion of the piston were the same as in a plane wave emanating from an unbounded wall vibrating without flexure.

5-5 FAR-FIELD RADIATION FROM LOCALIZED WALL VIBRATIONS

When the wall area undergoing constant-frequency vibrations is confined to a distance a from the origin, a characteristic far field is realized at points where the radial distance r is much larger than either a or ka^2. In this event, a suitable approximation for the Rayleigh integral (5-2.6) results when R is replaced† by $r - \mathbf{x}_S \cdot \mathbf{e}_r$ in the exponent and by r in the denominator, so that $R^{-1}e^{ikR}$ becomes $r^{-1}e^{ikr} \exp(-ik\mathbf{x}_S \cdot \mathbf{e}_r)$.

In this limit of large r Eq. (5-2.6) is reduced to the form of an outgoing spherical wave with nonuniform directivity, i.e.,

$$\hat{p} \approx f(\theta, \phi) r^{-1} e^{ikr} \tag{5-5.1}$$

where we abbreviate

$$\begin{aligned} f(\theta, \phi) &= \frac{-i\omega\rho}{2\pi} \iint \hat{v}_n(x_S, y_S) e^{-ik\mathbf{x}_S \cdot \mathbf{e}_r} \, dx_S \, dy_S \\ &= \frac{-i\omega\rho}{2\pi} g(k \sin\theta \cos\phi, k \sin\theta \sin\phi) \end{aligned} \tag{5-5.2}$$

with

$$g(\xi, \eta) = \iint \hat{v}_n(x_S, y_S) e^{-i\xi x_S} e^{-i\eta y_S} \, dx_S \, dy_S \tag{5-5.3}$$

representing the two-dimensional Fourier transform‡ of $\hat{v}_n(x_S, y_S)$.

where the phase of the radical is understood to be between 0 and $-\pi/4$. For large η one can approximate $(1 - is^2/2)^{1/2}$ by 1 and sinh ξ by ξ without appreciably changing the value of either integral, the resulting approximate integrals being then readily performed, so one obtains

$$\mathbf{H}_0(\eta) \to \frac{2}{\pi\eta} + \left(\frac{2}{\pi\eta}\right)^{1/2} \cos\left(\eta - \frac{3\pi}{4}\right)$$

From (7*b*), one has $\mathbf{H}_1(\eta) = (2/\pi) - d\mathbf{H}_0/d\eta$; using the above and keeping only terms of order $\eta^{-1/2}$, we obtain (12*b*). The derivation of (12*a*) proceeds in an analogous manner from Eq. (6) except that one takes the imaginary part of $i \exp(-i\eta \cos\phi)$. The asymptotic expression for $J_1(\eta)$ is obtained from that of $J_0(\eta)$ with the identity $J_1(\eta) = -dJ_0(\eta)/d\eta$.

† In the analogous Fresnel-Kirchhoff theory of diffraction by an aperture (Sec. 5-2), the diffraction is said to be *Fraunhofer diffraction* when the R in e^{ikR} can be replaced by $r - \mathbf{x}_S \cdot \mathbf{e}_r$. Points at which this approximation is satisfactory are said to lie in the *Fraunhofer region*. Similarly the terms *Fresnel diffraction* and *Fresnel region* are used when the quadratic terms (but not the higher-order terms) in the expression $R \approx r - \mathbf{x}_S \cdot \mathbf{e}_r + \frac{1}{2}[x_S^2 + y_S^2 - (\mathbf{x}_S \cdot \mathbf{e}_r)^2]/r$ affect the value of the integral. See Born and Wolf, *Principles of Optics*, p. 383.

‡ R. C. Jones, "On the Theory of the Directional Patterns of Continuous Source Distributions on a Plane Surface," *J. Acoust. Soc. Am.*, **16**:147–171 (1945).

For a *circular piston*, where $\hat{v}_n$ is constant up to radius a and thereafter zero, the integral in Eq. (2) leads (after a change of integration variables to cylindrical coordinates u, ϕ_S, where $x_S = u \cos \phi_S$, $y_S = u \sin \phi_S$) to

$$f(\theta, \phi) = -i\omega\rho\hat{v}_n \int_0^a \left(\frac{1}{2\pi} \int_0^{2\pi} e^{-iku \sin\theta \cos(\phi - \phi_S)}\, d\phi_S \right) u\, du$$

The periodicity of the integrand allows the integration on ϕ_S to be replaced by one on $\phi_S - \phi$ from 0 to 2π. Since the exponential is symmetrical in $\phi_S - \phi$, it can be replaced by the cosine of its argument. With this replacement, the integrations from 0 to $\pi/2$, $\pi/2$ to π, π to $3\pi/2$, and $3\pi/2$ to π yield identical values, so the quantity in parentheses is $2/\pi$ times the integral from 0 to $\pi/2$ over

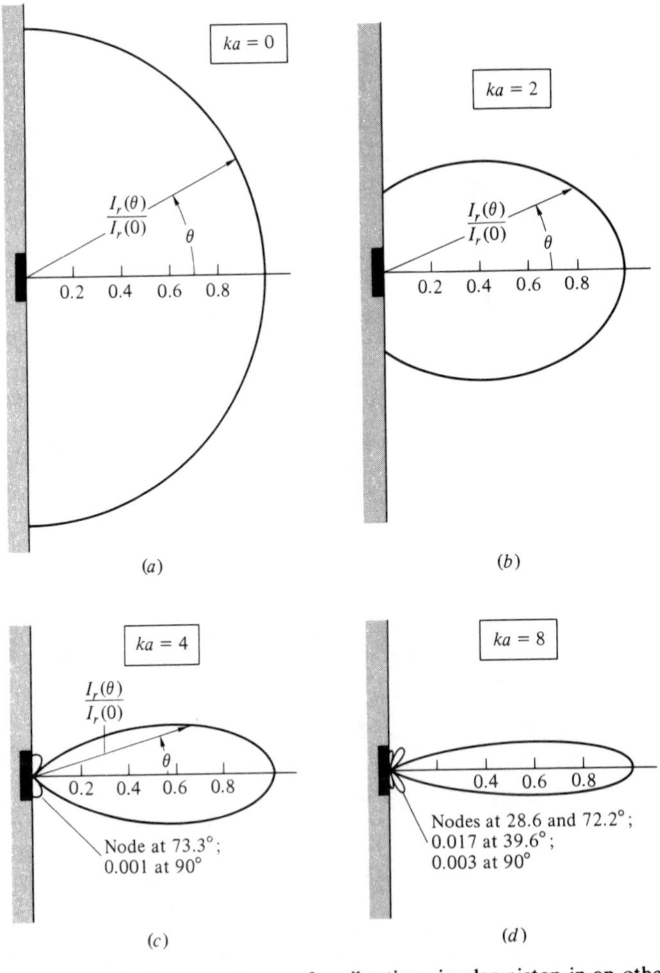

Figure 5-7 Radiation patterns of a vibrating circular piston in an otherwise rigid wall for various values of ka. The quantity plotted is $I_r(\theta)/I_r(0)$, where $I_r(\theta)$ is the time-averaged intensity as a function of polar angle θ and $I_r(0)$ is the intensity at $\theta = 0$. (*a*) $ka = 0$; (*b*) $ka = 2$; (*c*) $ka = 4$; (*d*) $ka = 8$.

cos $[ku \sin\theta \cos(\phi_S - \phi)]$, the integration variable being $\phi_S - \phi$. This quantity is subsequently recognized, from Eq. (5-4.6), as $J_0(ku \sin\theta)$. Consequently, $f(\theta, \phi)$ reduces† to

$$f(\theta) = \frac{-i\omega\rho\hat{v}_n}{k^2 \sin^2\theta} \int_0^{ka\sin\theta} J_0(\eta)\eta \, d\eta = -i \frac{\rho c \hat{v}_n k a^2}{2} \frac{2J_1(ka\sin\theta)}{ka\sin\theta} \qquad (5\text{-}5.4)$$

The Bessel function of first order in the latter expression results from Eq. (5-4.7a). We have here deleted ϕ as an argument of $f(\theta)$, since the result, because of the circular symmetry, is independent of ϕ.

The Bessel function $J_1(\eta)$ is $\eta/2$ for small η [see Eq. (5-4.10a)], while, for large η, it has the asymptotic form given in Eq. (5-4.12a). The first three zeros are at $\eta = 3.832$, 7.016, and 10.173; the nth zero in the limit of large n is asymptotically $(n + \frac{1}{4})\pi$. Consequently, the factor $2J_1(ka\sin\theta)/(ka\sin\theta)$, considered as a function of θ, is 1 at $\theta = 0$ and has one zero between 1 and $\pi/2$ if $3.832 < ka < 7.016$, two zeros if $7.016 < ka < 10.173$, three zeros if $10.173 < ka < 13.32$, etc. Note that the far-field value of $\hat{p}$ at $\theta = 0$ is the same as the leading term in the low-frequency ($ka \ll 1$) outer expansion (5-3.2).

The far-field intensity corresponding to Eqs. (1) and (4) is

$$I_{r,\text{av}} = \frac{|f(\theta)|^2}{2\rho c r^2} = (I_{r,\text{av}})_{\theta=0} \left[\frac{2J_1(ka\sin\theta)}{ka\sin\theta}\right]^2 \qquad (5\text{-}5.5)$$

so the radiation pattern (see Fig. 5-7) given by $r^2 I_{r,\text{av}}$ when plotted versus θ exhibits, for $ka > 3.83$, a central lobe centered at $\theta = 0$ that is bounded at $\theta = \pm\sin^{-1}(3.83/ka)$, plus one or more side lobes.

The acoustic power output $\mathcal{P}_{\text{av}}$ by the vibrating baffled piston is the surface integral over a hemisphere ($0 < \theta < \pi/2$) of large radius r of $I_{r,\text{av}}$. The acoustic-energy corollary requires $\mathcal{P}_{\text{av}}$ to be the same as the integral of $\frac{1}{2}\text{Re}\,\hat{p}\,\hat{v}_n^*$ over the front face of the piston or to be $\frac{1}{2}|\hat{v}_n|^2 \,\text{Re}\, Z_{m,\text{rad}}$, where $Z_{m,\text{rad}}$ is the radiation impedance. Consequently, the function $R_1(2ka)$ appearing in Eqs. (5-4.8) and (5-4.9) should be the same as

$$R_1(2ka) = \frac{(ka)^2}{2} \int_0^{\pi/2} \left[\frac{2J_1(ka\sin\theta)}{ka\sin\theta}\right]^2 \sin\theta \, d\theta \qquad (5\text{-}5.6)$$

and, indeed, a substitution of the power-series expansion (5-4.10a) of $J_1(\eta)$ into the above reproduces Eq. (5-4.11a).

5-6 TRANSIENT SOLUTION FOR BAFFLED CIRCULAR PISTON

We here discuss the transient radiation‡ from a baffled piston [radius a, centered at the origin, $v_n = v_n(t)$ on the piston face, 0 on the remainder of the wall] that

† N. W. McLachlan, "Pressure Distribution in a Fluid due to the Axial Vibration of a Rigid Disc," *Proc. R. Soc. Lond.*, **A122**:604–609 (1928). For Fraunhofer diffraction by a circular aperture, the formula was first derived, although in a somewhat different form, by G. B. Airy, *Trans. Camb. Phil. Soc.*, **5**:283 (1835).

‡ J. W. Miles, "Transient Loading of a Baffled Piston," *J. Acoust. Soc. Am.*, **25**:200–203 (1953); F. Oberhettinger, "Transient Solutions of the Baffled Piston Problem," *J. Res. Nat. Bur.*

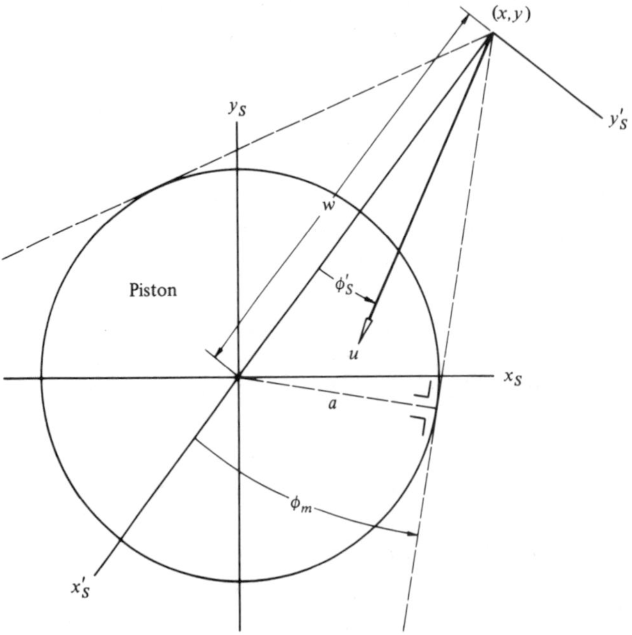

Figure 5-8 Coordinate systems for derivation of the transient acoustic field of a circular piston in a rigid baffle. The coordinate system (x'_S, y'_S) is centered at the projection $(x, y, 0)$ of the listener position on the piston plane and oriented so that the piston center is at $x'_S = w$, $y'_S = 0$. The polar coordinates u and ϕ_S are such that $x'_S = u \cos \phi_S$, $y'_S = u \sin \phi_S$.

results immediately following switch-on. To transform the double integral in Eq. (5-2.1) into a single integral, one first changes the coordinate system (x_S, y_S) to one centered at the point $(x, y, 0)$ and rotated so that the center of the piston is at $x'_S = w$, $y'_S = 0$, where $w = (x^2 + y^2)^{1/2}$. The integration variables are taken as u and ϕ'_S, where $x'_S = u \cos \phi'_S$, $y'_S = u \sin \phi'_S$, so that $R = (u^2 + z^2)^{1/2}$ and the differential area element $dx_S\, dy_S$ becomes $u\, du\, d\phi'_S$ (see Fig. 5-8). Points on the perimeter of the piston then correspond to values of u and ϕ'_S such that

$$u^2 + w^2 - 2uw \cos \phi'_S = a^2 \tag{5-6.1}$$

For $w < a$ (listener location within cylinder extending outward from the piston face), the values of u corresponding to points within the piston area range from 0 to $a + w$, and for u within these limits ϕ'_S ranges from $-\pi$ to π for $0 < u < a - w$, but for $a - w < u < a + w$ it ranges from $-\phi_m$ to ϕ_m, where, from Eq. (1), we define

$$\phi_m(u) = \cos^{-1} \frac{u^2 + w^2 - a^2}{2wu} \tag{5-6.2}$$

Stand., **65B:** 1–6 (1961). The derivation in the text is similar to that of P. R. Stepanishen, "Transient Radiation from Pistons in an Infinite Planar Baffle," *J. Acoust. Soc. Am.*, **49:** 1628–1638 (1971).

to be such that it lies between 0 and π. For $w < a$, ϕ_m decreases monotonically from π to 0 when u ranges from $a - w$ to $a + w$.

For $w > a$ (listener outside the piston's projection), the integration variable u ranges from $w - a$ to $w + a$, and for u fixed ϕ_S ranges from $-\phi_m$ to ϕ_m, where ϕ_m is still as given by Eq. (2). In this case, however, ϕ_m increases from 0 (at $u = w - a$) up to a maximum of $\sin^{-1}(a/w)$ [occurring when $u = (w^2 - a^2)^{1/2}$] and thereafter decreases, reaching 0 at $u = w + a$.

Since $v_n(t - R/c)$ is independent of ϕ_S', the ϕ_S' integration in Eq. (5-2.1) (with the changes in integration variables described above) can be done directly, with the result

$$p = -\rho c H(a - w) \int_0^{a-w} \frac{d}{du}\left[v_n\left(t - \frac{R}{c}\right)\right] du - \frac{\rho c}{\pi} \int_{|a-w|}^{a+w} \phi_m(u) \frac{d}{du}\left[v_n\left(t - \frac{R}{c}\right)\right] du \tag{5-6.3}$$

because $(d/du)[v_n(t - R/c)] = -(1/c)\dot{v}_n(t - R/c)u/R$. Here $H(a - w)$ is the Heaviside unit step function (1 if $w < a$, 0 if $w > a$). Note that the first integral is $v_n(t - R_s/c) - v_n(t - z/c)$, where $R_s = [(a - w)^2 + z^2]^{1/2}$ is the smallest distance from the listener to the perimeter of the piston.

An alternate version (used in subsequent sections) of Eq. (3) results after an integration by parts of the second term, such that

$$p = \rho c H(a - w) v_n\left(t - \frac{z}{c}\right) + \frac{\rho c}{\pi} \int_{|a-w|}^{a+w} \frac{d\phi_m}{du} v_n\left(t - \frac{R}{c}\right) du \tag{5-6.4}$$

In addition, we make a further change of integration variable to ψ, where

$$u^2 = w^2 + a^2 + 2wa \sin \psi \tag{5-6.5}$$

such that ψ ranges from $-\pi/2$ to $\pi/2$ as u ranges from $|a - w|$ to $a + w$. Also, it follows from (2) and the definition of ψ that

$$\frac{d\phi_m}{du} du = -au^{-2}(a + w \sin \psi)\, d\psi \tag{5-6.6}$$

Consequently Eq. (4) yields†

$$p = \rho c H(a - w) v_n\left(t - \frac{z}{c}\right) - \frac{\rho c}{\pi} \int_{-\pi/2}^{\pi/2} \frac{a(a + w \sin \psi)}{w^2 + a^2 + 2wa \sin \psi} v_n\left(t - \frac{R}{c}\right) d\psi \tag{5-6.7}$$

where, in terms of ψ, the distance R is now $(w^2 + a^2 + z^2 + 2wa \sin \psi)^{1/2}$.

Yet another version (used directly below) results from the change of integration variable in the second integral in Eq. (3) to $\tau = t - R/c$ such that

$$u = [c^2(t - \tau)^2 - z^2]^{1/2} \qquad \frac{d}{du} v_n\left(t - \frac{R}{c}\right) du = \dot{v}_n(\tau)\, d\tau \tag{5-6.8}$$

† A. Schoch, "Considerations in Regard to the Sound Field of a Piston Diaphragm," *Akust. Z.*, **6**:318–326 (1941).

Consequently, one obtains

$$p = \rho c H(a - w)\left[v_n\left(t - \frac{z}{c}\right) - v_n\left(t - \frac{R_s}{c}\right)\right] + \frac{\rho c}{\pi}\int_{t-R_l/c}^{t-R_s/c} \dot{v}_n(\tau)\phi_m(u)\, d\tau \qquad (5\text{-}6.9)$$

with R_l and R_s representing the largest and smallest distances, $[(a \pm w)^2 + z^2]^{1/2}$, from the listener position to the perimeter of the piston.

Equation (9) is frequently used with a numerical integration of the second term to determine the transient field of the baffled circular piston when v_n is a given function. The overall expression can be rewritten as

$$p = \int_{-\infty}^{t} \dot{v}_n(\tau) p_{\text{us}}(\mathbf{x}, t - \tau)\, d\tau \qquad (5\text{-}6.10)$$

where $p_{\text{us}}(\mathbf{x}, t)$ is the *unit step response,* acoustic pressure resulting at the listener location at time t when v_n is zero before $t = 0$ and thereafter has value 1. The expression for $p_{\text{us}}(\mathbf{x}, t)$ results from Eq. (9) if one sets $v_n(t) = H(t)$, so $\dot{v}_n(t) = \delta(t)$, such that (see Fig. 5-9)

$$p_{\text{us}}(\mathbf{x}, t) = \begin{cases} 0 & t < \dfrac{z}{c} \\ 0 & w > a,\ \dfrac{z}{c} < t < \dfrac{R_s}{c} \\ \rho c & w < a,\ \dfrac{z}{c} < t < \dfrac{R_s}{c} \\ \dfrac{\rho c}{\pi} \cos^{-1} \dfrac{c^2t^2 - z^2 + w^2 - a^2}{2w(c^2t^2 - z^2)^{1/2}} & \dfrac{R_s}{c} < t < \dfrac{R_l}{c} \\ 0 & t > \dfrac{R_l}{c} \end{cases} \qquad (5\text{-}6.11)$$

This multiplied by V_0 gives the field radiated by a piston that is suddenly accelerated to velocity V_0 at time $t = 0$. Its implication for this case is that, for $w < a$, the received acoustic-pressure pulse begins abruptly with a jump to a value $\rho c V_0$ at $t = z/c$, stays constant until $t = R_s/c$, and then decreases monotonically, reaching 0 at $t = R_l/c$, and staying 0 thereafter. For $w > a$, p stays 0 up until time R_s/c and increases from 0 following onset up to a maximum value of $(\rho c V_0/\pi) \sin^{-1} (a/w)$ [achieved when $t = (w^2 + z^2 - a^2)^{1/2}/c$] and thereafter decreases, reaching 0 (and remaining 0 thereafter) at $t = R_l/c$.

The various arrival times characterizing the field radiated by the piston in the idealized situation just described are consistent with Poisson's theorem and Huygens' construction and can be derived from simple considerations. If the listener lies in the projection of the piston's area, the earliest arrival time is z/c and the arrival should be the same as for radiation from a piston of infinite area up until the first arrival from the perimeter of the piston, occurring at time R_s/c. At points outside the piston's projection, the first wave to arrive must come from the nearest point on the piston perimeter, so it arrives at time R_s/c. Since

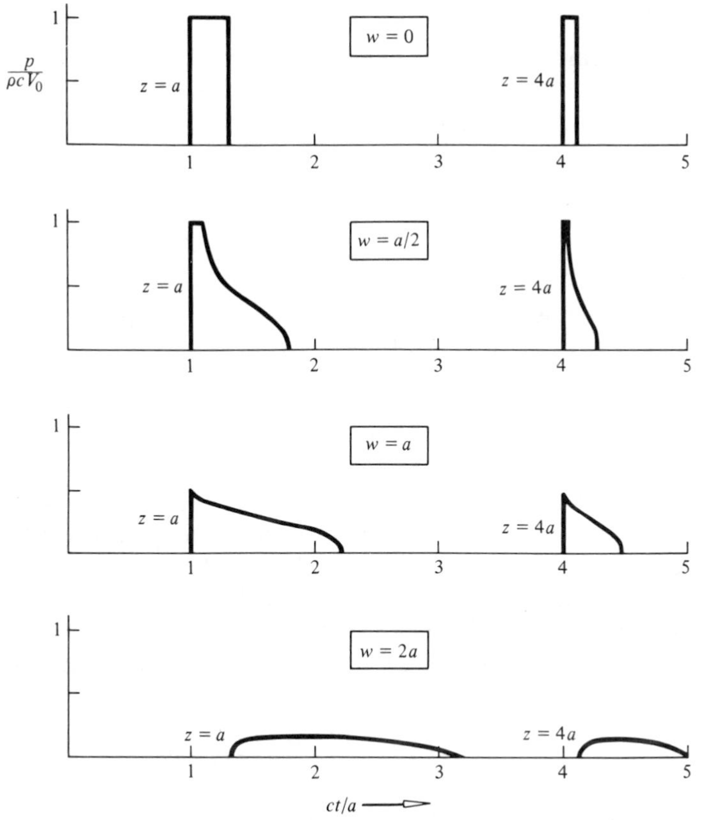

Figure 5-9 Transient acoustic-pressure waveforms at $z = a$ and $z = 4a$ caused by an impulsively accelerated circular piston in an otherwise rigid wall. The piston is motionless before $t = 0$ and thereafter has constant velocity V_0. To take advantage of the model's intrinsic similitude, $p/\rho c V_0$ is plotted versus ct/a for fixed values of w/a and z/a.

the Rayleigh integral gives no contribution from points at which $\dot{v}_n$ is zero, the last arrival in both cases must come from the farthest point on the perimeter of the piston and arrives at time R_l/c.

5-7 FIELD ON AND NEAR THE SYMMETRY AXIS

The expressions derived in the previous section demonstrate that the field of an oscillating baffled circular piston is not necessarily easy to describe at intermediate radial distances. However, a simple expression results for the field along the symmetry axis ($x = 0$, $y = 0$). This expression follows trivially from Eq. (5-6.9) if w is set to zero, so that $R_s = R_l$, but inasmuch as the steps leading to that equation are somewhat intricate, an alternate derivation for the special case $w = 0$ is given here.

Field on Symmetry Axis

The derivation proceeds from the Rayleigh integral (5-2.1) with x and y set to 0 and with the integration variables x_S, y_S replaced by cylindrical coordinates w_S, ϕ_S, where $x_S = w_S \cos \phi_S$ and $y_S = w_S \sin \phi_S$. Thus we have

$$p(0, 0, z, t) = \frac{\rho}{2\pi} \int_0^{2\pi} \int_0^a \frac{\dot{v}_n(t - R/c)}{R} w_S \, dw_S \, d\phi_S \tag{5-7.1}$$

where $R^2 = z^2 + w_S^2$.

The ϕ_S integration yields 2π; the w_S integration can be replaced by one over R, such that $R^{-1}w_S \, dw_S$ becomes dR and the integration limits become z and $(z^2 + a^2)^{1/2}$. Since $\dot{v}_n(t - R/c)$ is $-c(\partial/\partial R)[v_n(t - R/c)]$, we accordingly obtain†

$$p = \rho c \left[v_n\left(t - \frac{z}{c}\right) - v_n\left(t - \frac{(z^2 + a^2)^{1/2}}{c}\right) \right] \tag{5-7.2}$$

This can be regarded as the superposition of two waves, one propagating from the center of the piston and the other (with a minus sign prefixed) propagating from the edge of the piston ($w_S = a$).

When the piston is oscillating with constant angular frequency ω, the two terms in Eq. (2) may cancel for certain values of z. With the prescription that the complex amplitude of $v_n(t - \tau)$ is $\hat{v}_n e^{i\omega\tau}$, Eq. (2) yields, after some algebra, the expression

$$\hat{p} = -2i\rho c \hat{v}_n \exp \left\{ \frac{ik[z + (z^2 + a^2)^{1/2}]}{2} \right\} \sin \left[\frac{k(z^2 + a^2)^{1/2} - kz}{2} \right] \tag{5-7.3}$$

This (see Fig. 5-10) is zero whenever $k(z^2 + a^2)^{1/2}$ differs from kz by a multiple of 2π or when

$$kz = \frac{(ka)^2 - (2n\pi)^2}{4n\pi} \tag{5-7.4}$$

where n is any positive integer less than $ka/2\pi$. Thus, if $ka/2\pi$ is between 5 and 6, there would be five pressure nodes along the z axis. Moreover, if ka should be

† H. Backhaus and F. Trendelenberg, "On the Unidirectional Beaming of Piston Diaphragms," *Z. Tech. Phys.*, **7**:630–635 (1926). The analogous result for diffraction by a circular aperture dates back to Fresnel, "On the Diffraction of Light . . .," 1816, and to A. Schuster, "Elementary Treatment of Problems on the Diffraction of Light," *Phil. Mag.*, (5)**31**:77–86 (1891). The result is related to Poisson's famous prediction (originally intended to debunk Fresnel's theory of diffraction but shortly thereafter experimentally confirmed by Arago) that there should be a bright spot in the shadow of a circular disk along the axis of the disk. If the Fresnel-Kirchhoff integral with $\mathbf{e}_R \cdot \mathbf{e}_z = \mathbf{n}_i \cdot \mathbf{e}_z = 1$ in Eq. (5-2.8) is used with $\hat{\mathbf{v}}_i \cdot \mathbf{e}_z = \hat{\mathbf{v}}_n$ for $w_S > a$, 0 for $w_S < a$, and with a small attenuation factor inserted to make the integral convergent, one obtains (*Babinet's principle*) an expression equal to the original incident plane wave minus what would be predicted for the problem of diffraction by a circular aperture of the same size. This difference for points on the symmetry axis, according to Eq. (2), is $\rho c v_n(t - (z^2 + a^2)^{1/2}/c)$, which has exactly the same amplitude as that of the incident acoustic-pressure wave. For a historical account, see E. Mach, *The Principles of Physical Optics: An Historical and Philosophical Treatment*, 1926, reprinted by Dover, New York, 1954, pp. 285–286.

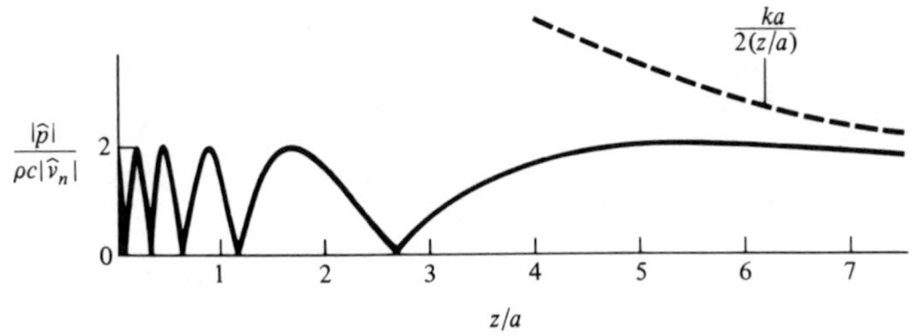

Figure 5-10 Variation along symmetry axis of acoustic-pressure amplitude $|\hat{p}|$ with distance z (units of a) from center of oscillating circular piston of radius a. Plot of $|\hat{p}|/|\rho c\hat{v}_n|$ versus z/a is for $ka/2\pi = 5.5$.

an integer multiple of 2π, one of these nodes (largest n) is on the face of the piston at $z = 0$, $w = 0$. There are one or more nodes only if $ka > 2\pi$.

The existence of such nodes is a consequence of the circular symmetry of the piston; they would not be expected for a piston of irregular shape. Beyond the farthest node ($n = 1$), the pressure amplitude $|\hat{p}|$ rises to one additional maximum of $|2\rho c\hat{v}_n|$ at $kz = [(ka)^2 - \pi^2]/2\pi$ and thereafter decreases. In the limit $z \gg a$, one has $(z^2 + a^2)^{1/2} \approx z + \frac{1}{2}a^2/z$ and if, moreover, $z \gg ka^2$, Eq. (3) above reduces to

$$\hat{p} \to -\frac{i}{2}(ka^2)\rho c\hat{v}_n \frac{e^{ikz}}{z} \tag{5-7.5}$$

which has the characteristic form for spherical spreading and is the same as would be predicted for a piston vibrating at low frequencies. [See Eq. (5-3.2).] The reason for the latter behavior is that, if one is directly in front of a piston (not necessarily circular) and sufficiently far from it, the phases e^{ikR} of contributions from various points on the piston are all nearly the same. The criterion for the leading term in Eq. (5-3.2) to hold is that the path lengths from any two points on the piston to the listener differ by a quantity considerably less than a wavelength.

Field near Symmetry Axis

To study the field when w is not identically zero but merely small compared with a, we make use of Eq. (5-6.7). Within the integrand of the second term, it is a good approximation to set $w = 0$ everywhere except in the time delay R/c; the latter is approximated by a power-series expansion in w truncated to first order, such that $R \approx (a^2 + z^2)^{1/2} + wa(a^2 + z^2)^{-1/2} \sin\psi$. With these approximations, the ψ integration for the determination of the complex amplitude requires the evaluation of

$$\int_{-\pi/2}^{\pi/2} \exp\left[\frac{ikwa}{(a^2 + z^2)^{1/2}} \sin\psi\right] d\psi = 2\int_0^{\pi/2} \cos\left[\frac{kwa}{(a^2 + z^2)^{1/2}} \sin\psi\right] d\psi \tag{5-7.6}$$

This, however, is recognized from Eq. (5-4.6), after a change of integration variable to $\pi/2 - \psi$, as $\pi J_0(kwa/(z^2 + a^2)^{1/2})$. Consequently, the constant-frequency version of Eq. (5-6.7), for $w/a \ll 1$, becomes†

$$\hat{p} = \rho c \hat{v}_n \left[e^{ikz} - e^{ik(z^2+a^2)^{1/2}} J_0\left(\frac{kwa}{(z^2 + a^2)^{1/2}}\right)\right] \tag{5-7.7}$$

Since $J_0(0) = 1$, the above expression for $\hat{p}$ reduces to Eq. (2) when $w = 0$. However, since‡

$$J_0(\eta) \to \left(\frac{2}{\pi\eta}\right)^{1/2} \cos\left(\eta - \frac{\pi}{4}\right) \tag{5-7.8}$$

for $\eta \gg 1$, the second term in Eq. (7) is small compared with the first when $kw \gg [1 + (z/a)^2]^{1/2}$. This could be so even for $w \ll a$ if $ka \gg 1$. For example, if $ka = 100$ and $z = a$, the criterion would be met for $kw = 10$ or $w = a/10$. One concludes that if $ka \gg 1$, the field is approximately a plane wave at points where $a \gg w \gg (z^2 + a^2)^{1/2}/ka$. Such a region exists for $z \ll ka^2$.

5-8 TRANSITION TO THE FAR FIELD

If $ka \gg 1$, the field of a vibrating baffled piston persists as a collimated beam of radius a for distances up to the order of ka^2 from the piston with some anomalous behavior due to symmetry (as discussed in the previous section) near the beam's axis and with some deterioration at the edge of the beam. To describe the latter behavior and the transition to the far field, we return to expression (5-6.7). Our interest here is in circumstances for which $kR_l - kR_s$ is substantially larger than 1, so that the real and imaginary parts of the integrand in the second term undergo a large number of oscillations over the range of integration. The integrals over adjacent peaks and troughs tend to cancel each other, the exceptions being those near $\psi = -\pi/2$ and $\psi = \pi/2$, where the derivative of the phase with respect to ψ vanishes. To take advantage of this, we change the variable of integration to $\xi = \sin\psi$ (so R^2 becomes $z^2 + w^2 + a^2 + 2wa\xi$) and then deform the path of integration going from $\xi = -1$ to 1 to the contour $C = C_1 + C_2$ sketched in Fig. 5-11. The variable of integration for the C_1 contour is changed to u_1, so that

$$kR = kR_s + iu_1^2 \qquad 2k^2wa(\xi + 1) = 2ikR_s u_1^2 - u_1^4$$

The first equation defines u_1 in terms of ξ; the second results from squaring both sides of the first. Note that $\exp ikR$ dies out exponentially with increasing u_1 if the contour C_1 is specified so that u_1 is real and positive all along C_1. Similarly, the variable of integration for the integration along contour C_2 is taken as u_2,

† Schoch, "Considerations . . . ," 1941.

‡ The derivation of this asymptotic expression proceeds as outlined on p. 225*n*; the result is due to Poisson (1823). For a general derivation that includes higher-order terms, see Watson, *Treatise on the Theory of Bessel Functions*, pp. 196–198.

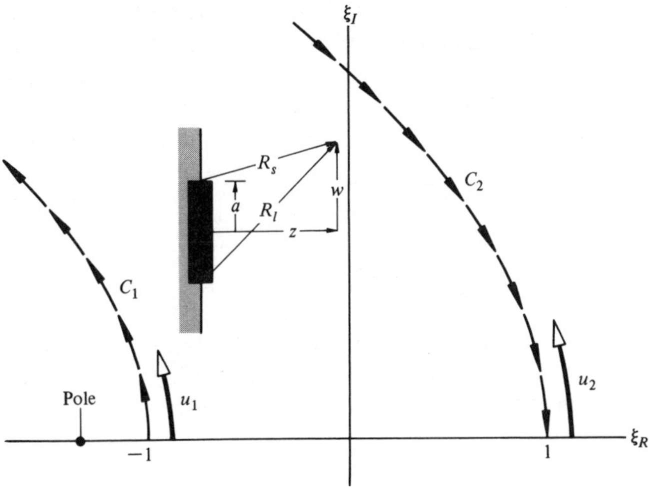

Figure 5-11 Deformed integration contour in the complex ξ plane for evaluation of the acoustic-pressure field from a vibrating circular piston in the limit $ka \gg 1$, $kR_l - kR_s \gg 1$. The original integration path was from $\xi = -1$ to $\xi = +1$ along the real axis. The contour C_1 is the parabola $2(\xi_R + 1) = -(wa\xi_I/R_s)^2$. Contour C_2 is defined analogously.

where

$$kR = kR_l + iu_2^2 \qquad 2k^2wa(\xi - 1) = 2ikR_lu_2^2 - u_2^4$$

and C_2 is specified such that u_2 is real and positive along C_2. (The integral over the arc at infinity connecting C_1 and C_2 vanishes for w not identically zero.)

With the substitutions just described, Eq. (5-6.7) leads to the expression

$$\hat{p} = \rho c\hat{v}_n H(a - w)e^{ikz} - \frac{\rho c\hat{v}_n}{\pi} e^{i(kR_s + \pi/4)} \int_0^\infty e^{-u_1^2} \phi_1(u_1)\, du_1 - \frac{\rho c\hat{v}_n}{\pi} e^{i(kR_l - \pi/4)} \int_0^\infty e^{-u_2^2} \phi_2(u_2)\, du_2 \tag{5-8.1}$$

where

$$\phi_{1,2}(u) = \frac{2[2k^2a(a \mp w) + G_{1,2}](kR_{s,l} + iu^2)}{[k^2(a \mp w)^2 + G_{1,2}](4k^2wa \mp G_{1,2})^{1/2}(2kR_{s,l} + iu^2)^{1/2}} \tag{5-8.2}$$

with the abbreviation

$$G_{1,2}(u) = 2ikR_{s,l}u^2 - u^4 \tag{5-8.3}$$

The phases of the radicals in the integrands are here understood to be 0 when $u = 0$ and to vary continuously with increasing u when u is real.

To obtain approximate expressions for the above integrals that elucidate the phenomena occurring at intermediate values of z near the edges of the original beam (i.e., near $w = a$) emanating from a piston of large ka, we limit our attention here to circumstances in which $ka \gg 1$ and $1/k \ll z \ll ka^2$,

$w > a/2$. For these circumstances, such quantities as kR_s, kR_l, kwa/R_l, kwa/R_s, and $k(w + a)a/R_l$ are all large compared with 1. Since the integrands in Eq. (1) are concentrated near $u_1 = 0$ and $u_2 = 0$, respectively, one can approximate the quantities $\phi_1(u_1)$ and $\phi_2(u_2)$ by setting u_1^2 or u_2^2 to zero in any factor whose magnitude is large compared with 1. In this manner, we obtain

$$\phi_1(u) \approx \left(\frac{2R_s}{kwa}\right)^{1/2} \frac{ka(a - w) + iR_s u^2}{k(a - w)^2 + 2iR_s u^2} \tag{5-8.4}$$

$$\phi_2(u) \approx \frac{a}{a + w}\left(\frac{2R_l}{kwa}\right)^{1/2} \tag{5-8.5}$$

(Note that in the former expression we allow for the possibility of $a - w$ being close to zero.) To facilitate the evaluation of the corresponding integral, we rewrite the above approximate expression for $\phi_1(u)$ in the form

$$\phi_1(u) \approx \left(\frac{R_s}{2kwa}\right)^{1/2} + \frac{a + w}{4(wa)^{1/2}}\left[\frac{1}{(\pi/2)^{1/2}X + e^{-i\pi/4}\,u} + \frac{1}{(\pi/2)^{1/2}X - e^{-i\pi/4}\,u}\right] \tag{5-8.6}$$

where we use the abbreviation

$$X = \left(\frac{k}{\pi R_s}\right)^{1/2}(a - w) \tag{5-8.7}$$

In regard to the insertion of these expressions for ϕ_1 and ϕ_2 into Eq. (1), note that the integral from 0 to ∞ of exp $(-u^2)$ is $\frac{1}{2}\pi^{1/2}$ and that the integral arising from the second term in the brackets in Eq. (6) can be rewritten after a change of integration variable, $u \to -u$, in the same form as the integral arising from the first term but with integration limits of $-\infty$ and 0. Consequently, one obtains†

$$\frac{\hat{p}}{\rho c\hat{v}_n} = H(a - w)e^{ikz} - \left(\frac{R_s}{8\pi kwa}\right)^{1/2} e^{i(kR_s+\pi/4)} - \frac{2a}{a + w}\left(\frac{R_l}{8\pi kwa}\right)^{1/2} e^{i(kR_l-\pi/4)} - \frac{a + w}{(8wa)^{1/2}} A_D(X)e^{i(kR_s+\pi/4)} \tag{5-8.8}$$

† The limiting case of $a \to \infty$, $w - a$ finite and abbreviated by x, corresponds to the case when the $x < 0$ portion of the plane $z = 0$ is vibrating with constant amplitude and phase and the $x > 0$ portion is motionless. This limit applied to (8) gives

$$\frac{\hat{p}}{\rho c\hat{v}_n} = H(-x)e^{ikz} - 2^{-1/2}A_D(X)\exp\left\{i\left[k(x^2 + 2^2)^{1/2} + \frac{\pi}{4}\right]\right\} \tag{i}$$

with $X = -\{k/[\pi(x^2 + z^2)^{1/2}]\}^{1/2}x$. This, with $z \gg |x|$, reduces to

$$\frac{\hat{p}}{\rho c\hat{v}_n} = e^{ikz}[H(-x) - 2^{-1/2}e^{i\pi/4}A_D(X)e^{i(\pi/2)X^2}] = e^{ikz}2^{-1/2}e^{-i\pi/4}\int_{-X}^{\infty} e^{i(\pi/2)t^2}\,dt \tag{ii}$$

The mathematical steps leading to (ii) are explained later in the present section. This in the limit considered is the same as the classical result for Fresnel diffraction of a plane wave by a straight edge in the Fresnel-Kirchhoff theory. See Born and Wolf, *Principles of Optics*, pp. 433–434.

where $A_D(X)$ is the *diffraction integral*† given by

$$A_D(X) = \frac{1}{\pi 2^{1/2}} \int_{-\infty}^{\infty} \frac{e^{-u^2}\, du}{(\pi/2)^{1/2} X - e^{-i\pi/4}\, u} \tag{5-8.9}$$

$$= f(X) - ig(X) \tag{5-8.9a}$$

the latter serving to define the *auxiliary Fresnel functions*‡ $f(X)$ and $g(X)$, which represent the real and negative imaginary parts of $A_D(X)$.

Properties of the Diffraction Integral

The diffraction integral $A_D(X)$ has the properties of being odd in X but discontinuous at $X = 0$ and of being related to the *Fresnel integrals*

$$C(X) = \int_0^X \cos\left(\frac{\pi}{2} t^2\right) dt \qquad S(X) = \int_0^X \sin\left(\frac{\pi}{2} t^2\right) dt \tag{5-8.10}$$

by the relation

$$A_D(X) = \frac{1-i}{2} e^{-i(\pi/2)X^2} \{\operatorname{sign}(X) - (1-i)[C(X) + iS(X)]\} \tag{5-8.11}$$

[This equivalence is demonstrated for $X > 0$ by replacing (a mathematical identity)

$$\frac{1}{\zeta - e^{-i\pi/4} u} = e^{-i\pi/4} \int_0^{\infty} \exp\,[i(\zeta e^{i\pi/4} - u)s]\, ds$$

in Eq. (9) with $\zeta = (\pi/2)^{1/2} X$, interchanging the order of s and u integrations, and subsequently writing the total exponent as

$$-u^2 + i(\zeta e^{i\pi/4} - u)s = -i\zeta^2 - y^2 - \left(u + \frac{is}{2}\right)^2$$

with $y = s/2 + e^{-i\pi/4}\zeta$. The integral over u of $e^{-(u+is/2)^2}$ yields $\pi^{1/2}$. The integral over s of e^{-y^2} is changed to an integral over y from $e^{-i\pi/4}\zeta$ to ∞, which in turn is broken into an integral from 0 to ∞ (which evaluates to $\pi^{1/2}$) minus an integral from 0 to $e^{-i\pi/4}\zeta$. In the latter integral, the variable of integration is changed to t, where $y = (\pi/2)^{1/2} t e^{-i\pi/4}$, such that the t integration limits become 0 and X. The cited result then follows from Euler's formula (1-8.3), from Eqs. (10), and from the recognition that $e^{\pm i\pi/4}$ is $(1 \pm i)/2^{1/2}$.]

Behavior of $A_D(X)$ at large and small values of $|X|$ is determined, respectively, by (1) expanding the integrand in Eq. (9) in an inverse power series in X, then integrating term by term, and (2) expanding the integrands in Eqs. (10) in a

† So called here because it is a ubiquitous feature of any asymptotic solution of the wave equation when the boundary involves a sharp edge. Born and Wolf, *Principles of Optics,* p. 428, use the term to refer, with some multiplicative factors, to the integral of e^{ikR} over the aperture.

‡ W. Gautschi, "Error Function and Fresnel Integrals," in Abramowitz and Stegun (eds.), *Handbook of Mathematical Functions,* pp. 297–302, 323–324. Note that our $A_D(X)$ is $(1 - i)/2$ times the $w(z)$ in Gautschi's eq. (7.1.4) with $z = (\pi/2)^{1/2} X e^{i\pi/4}$, so our (9*a*), giving $iA_D(X) = [(1 + i)/2]w(z)$ as $g(X) + if(X)$, is consistent with Gautschi's (7.3.23) and (7.3.24).

power series in $(\pi/2)t^2$, then integrating term by term, subsequently substituting the results plus a power-series expansion of $\exp[-i(\pi/2)X^2]$ into Eq. (11). In this manner, the large X limit yields

$$f(X) \to \frac{1}{\pi X} - \frac{3}{\pi^3 X^5} + \cdots \tag{5-8.12a}$$

$$g(X) \to \frac{1}{\pi^2 X^3} - \frac{15}{\pi^4 X^7} + \cdots \tag{5-8.12b}$$

while the small X limit yields

$$f(X) = \text{sign}(X)\left(\frac{1}{2} - \frac{\pi}{4}X^2 + \frac{\pi}{3}|X|^3 - \cdots\right) \tag{5-8.13a}$$

$$g(X) = \text{sign}(X)\left(\frac{1}{2} - |X| + \frac{\pi}{4}X^2 - \cdots\right) \tag{5-8.13b}$$

The plots in Fig. 5-12 of $f(X)$ and $g(X)$ along with the leading terms in their asymptotic expressions indicate that, for most purposes, the asymptotic expressions are sufficient for $|X| > 2$.

Field Near Edge of Main Beam

If w is very close to a, that is, a listener at a point on a hypothetical cylinder projecting out from the piston's perimeter, the parameter X is vanishingly small and, in accord with Eqs. (9a) and (13), $A_D(X)$ is $(1 - i)/2$ if $X = 0^+$ $(w = a - 0^+)$ and $-(1 - i)/2$ if $X = 0^-$ $(w = a + 0^+)$, so the last term (with the

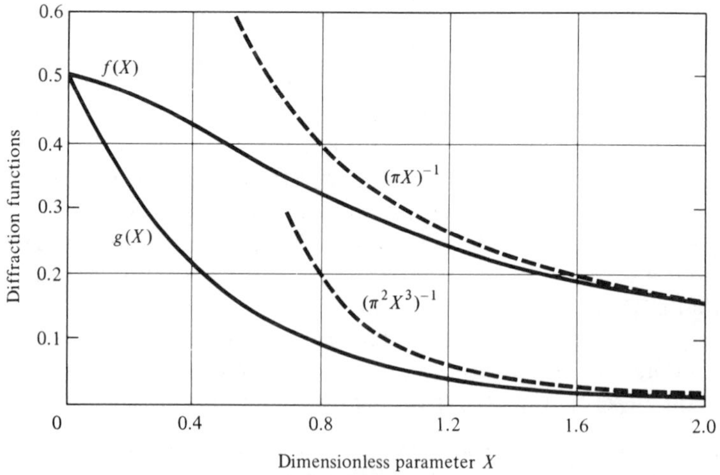

Figure 5-12 Auxiliary Fresnel functions $f(X)$ and $g(X)$ versus their argument X, representing the real and negative imaginary parts of the diffraction integral $A_D(X)$ (an odd function of X). The leading terms in the asymptotic expressions for $f(X)$ and $g(X)$ are also shown. [*A. D. Pierce, J. Acoust. Soc. Am.*, **55**:*946 (1974)*.]

minus sign) in Eq. (8) is $-\frac{1}{2}e^{ikz}$ sign $(a - w)$. Regardless of which direction the limit is approached from, the sum of the first and fourth terms gives $\frac{1}{2}e^{ikz}$ at $w = a$, so the right side in Eq. (8) is continuous at $w = a$ (as it should be). The complete expression at $w = a$ consequently reduces to

$$\frac{\hat{p}}{\rho c \hat{v}_n} \approx \tfrac{1}{2}e^{ikz}\left[1 - e^{i\pi/4}\left(\frac{z}{2\pi k a^2}\right)^{1/2}\right] - \frac{(z^2 + 4a^2)^{1/4}}{(8\pi k a^2)^{1/2}} e^{-i\pi/4} e^{ik(z^2+4a^2)^{1/2}} \tag{5-8.14}$$

The range of values of z for which the above is valid can be assessed with reference to the exact expression [derived from Eqs. (5-6.3) or (5-6.7)] for $\hat{p}/\rho c \hat{v}_n$ when $w = a$, that is,

$$\frac{\hat{p}}{\rho c \hat{v}_n} = \tfrac{1}{2}e^{ikz} - \frac{1}{\pi}\int_0^{\pi/2} e^{ik[z^2 + (2a)^2 \sin^2\phi]^{1/2}}\, d\phi \tag{5-8.15}$$

For $z = 0$, this has the value† [see Eq. (5-4.6)]

$$\left(\frac{\hat{p}}{\rho c \hat{v}_n}\right)_{z=0} = \tfrac{1}{2}[1 - J_0(2ka) - i\mathbf{H}_0(2ka)] \tag{5-8.16}$$

If $ka \gg 1$, both the Bessel function and the Struve function are small compared with 1 and the right side here is close to $\frac{1}{2}$.

In general, the second term in Eq. (15) is of small magnitude until z reaches values comparable to ka^2, in which case the appropriate approximate form [derived after replacing the radical in the exponent by its truncated binomial expansion $z + (2a^2/z)\sin^2\phi$] is

$$\frac{\hat{p}}{\rho c \hat{v}_n} \approx \tfrac{1}{2}e^{ikz}\left[1 - e^{ika^2/z} J_0\left(\frac{ka^2}{z}\right)\right] \tag{5-8.17}$$

which may be compared with Eq. (5-7.7). When the argument of the Bessel function is small compared with 1, Eq. (17) reduces to Eq. (5-7.5) (as it should), but it is equivalent to Eq. (14) [with $(z^2 + 4a^2)^{1/2}$ replaced by $z + 2a^2/z$ in the latter] in the limit when the Bessel function can be replaced by the leading term in its asymptotic expansion, e.g., when ka^2/z is of the order of 1 or greater. Consequently, one can conclude that, near $w = a$, Eq. (14) gives a good description of the pressure field up to $z = ka^2$. In addition, since the terms other than $\frac{1}{2}e^{ikz}$ in both Eqs. (14) and (17) are of minor significance unless z becomes comparable to ka^2, Eq. (14) is also a good approximation (for w near a) when z is close to the plane of the piston.

Characteristic Single-Edge Diffraction Pattern

In the range of values of z where both z and $(z^2 + 4a^2)^{1/2}$ are small compared with $8\pi ka^2$, given that $|w - a| \ll a$, the second and third terms in Eq. (8) are of

† A. G. Warren, "A Note on the Acoustic Pressure and Velocity Relations on a Circular Disc and in a Circular Orifice," *Proc. Phys. Soc.* (*Lond.*), **40**:296–299 (1928). Warren omits all details; an explicit derivation is given by McLachlan, "The Acoustic and Inertia Pressure . . . ," *Phil. Mag.*, (7)**14**:1012–1025 (1932).

smaller magnitude than the first and fourth, so insight into the phenomena occurring near the edge of the primary sound beam results from the neglect of these two terms. (The stated criteria would apply, for example, if $ka = 100$ and if $z/a < 100$.) To the same order of approximation, one can set $(a + w)/(8wa)^{1/2} = 1/\sqrt{2}$ in the coefficient preceding $A_D(X)$; one can also set R_s equal to $z + (w - a)^2/2z$ in the exponential factor e^{ikR_s} and equal to z in the argument of X. Thus, Eq. (8) reduces to

$$\hat{p} = \rho c \hat{v}_n e^{ikz} \left[H(X) - \frac{e^{i\pi/4}}{2^{1/2}} A_D(X) e^{i(\pi/2)X^2} \right] \tag{5-8.18}$$

$$= \rho c \hat{v}_n e^{ikz} \left(2^{-1/2} e^{-i\pi/4} \int_{-X}^{\infty} e^{i(\pi/2)t^2} \, dt \right) \tag{5-8.18a}$$

with X now approximated to $(k/\pi z)^{1/2}(a - w)$. Here we have also replaced the $a - w$ in the argument of the Heaviside unit step function by X, since the latter has the same sign as $a - w$. Note that the overall function is continuous in X (as it should be) since, near $X = 0$, the second term (without the minus sign) is $\frac{1}{2}$ if $X = 0^+$ and $-\frac{1}{2}$ if $X = 0^-$.

An implication of the above approximate expression for $\hat{p}$ is that the spatial and frequency dependence of the mean squared pressure is contained in a single dimensionless parameter X, that is,

$$\frac{(p^2)_{\text{av}}}{(\rho c)^2 (v_n^2)_{\text{av}}} = \left| H(X) - \frac{e^{i\pi/4}}{2^{1/2}} A_D(X) e^{i(\pi/2)X^2} \right|^2 = \frac{1}{2} \left| \int_{-X}^{\infty} e^{i(\pi/2)t^2} \, dt \right|^2 \tag{5-8.19}$$

$$= \tfrac{1}{2}\{[\tfrac{1}{2} + C(X)]^2 + [\tfrac{1}{2} + S(X)]^2\} \tag{5-8.19a}$$

$$= \tfrac{1}{2}\{[f(X)]^2 + [g(X)]^2\} \qquad X < 0 \; (w > a) \tag{5-8.19b}$$

This function, plotted in Fig. 5-13, occurs also in the theory of diffraction by edges and may accordingly be called the *characteristic single-edge diffraction pattern*. It decreases monotonically with increasing negative X, asymptotically approaching $1/2\pi^2X^2$; at $X = 0$ it has the value $\frac{1}{4}$, while at large positive X it approaches

$$\frac{(p^2)_{\text{av}}}{(\rho c)^2 (v_n^2)_{\text{av}}} \to 1 - \frac{2^{1/2} \cos [(\pi/2)X^2 + \pi/4]}{\pi X} \qquad w < a \; (X > 0) \tag{5-8.20}$$

i.e., it oscillates† about 1 with an amplitude that decreases with increasing X.

The latter approximate expression exhibits local pressure minima whenever $(\pi/2)X^2 + \pi/4$ is a multiple of 2π, that is, when (with $\lambda = 2\pi/k$)

$$a - w \approx (2\lambda z)^{1/2}(n - \tfrac{1}{8})^{1/2} \tag{5-8.21}$$

The positions of the local pressure maxima are given by an analogous expression, but with the number $\frac{1}{8}$ replaced by $\frac{5}{8}$. With increasing $a - w$ (decreasing w) or, equivalently, with increasing n, these maxima and minima become progres-

† Photographs resulting from exposure of a photographic plate to an ultrasonic beam radiating from a baffled piston exhibit such interference rings in a vivid manner. [J. T. Dehn, "Interference Patterns in the Near Field of a Circular Piston," *J. Acoust. Soc. Am.*, **32**:1692–1696 (1960).]

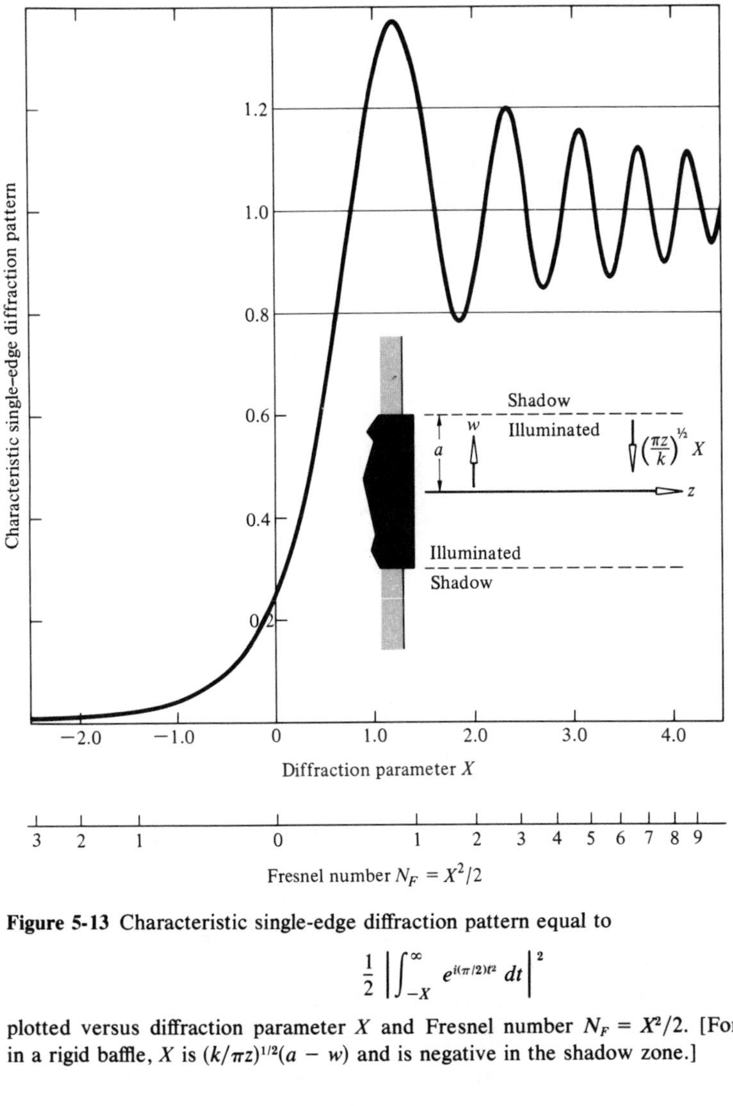

Figure 5-13 Characteristic single-edge diffraction pattern equal to

$$\frac{1}{2}\left|\int_{-X}^{\infty} e^{i(\pi/2)t^2}\,dt\right|^2$$

plotted versus diffraction parameter X and Fresnel number $N_F = X^2/2$. [For a circular piston in a rigid baffle, X is $(k/\pi z)^{1/2}(a - w)$ and is negative in the shadow zone.]

sively closer together. With increasing distance z from the piston, the overall pattern spreads out; the radial distance between the nth and $(n + 1)$th maxima increases with z as $z^{1/2}$.

Similarly, if $w > a$, the radial distance $w'(z)$ at which $(p^2)_{\text{av}}$ first drops below some set fraction ε (assumed substantially less than one-fourth) of the nominal plane-wave value $(\rho c)^2(v_n^2)_{\text{av}}$ tends to increase with z, the quantity $w'(z) - a$ being approximately $(\lambda z/\varepsilon)^{1/2}/2\pi$. If the so-defined $w'(z)$ is taken as a measure of the radius of the broadened beam, the axial distance at which the beam radius has increased by 2 wavelengths is 4 times that at which it has increased by 1 wavelength and the beam therefore broadens at a slower rate

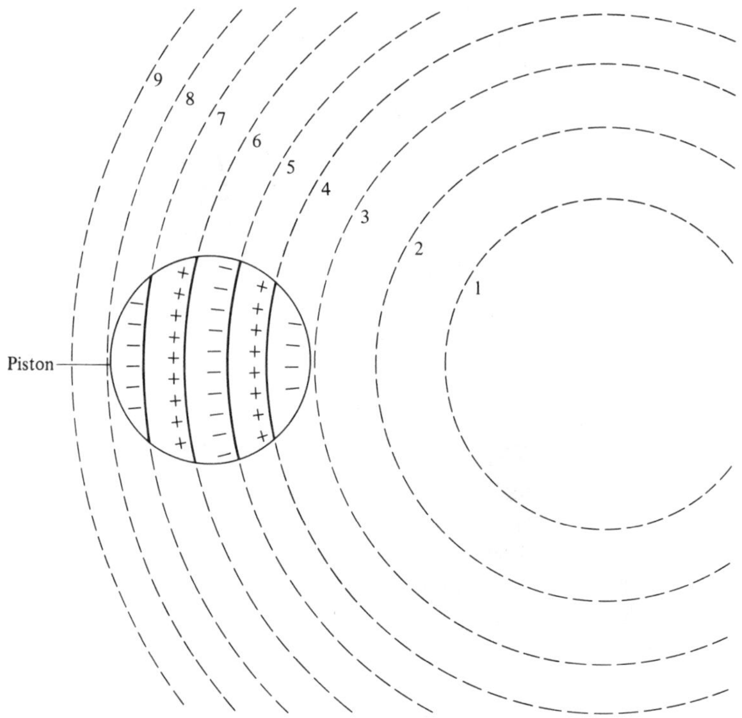

Figure 5-14 Fresnel zones on a circular piston. Example plotted is for $ka = 20$, $w_L/a = 6$, $z_L/a = 4$, where a is piston radius and w_L and z_L are cylindrical coordinates of listener.

with increasing z. However, the heights and depths of particular maxima or minima do not vary with z in the approximation considered here.

The successive minima and maxima within the beam near $w = a$ can be interpreted as partial interference and reinforcement of a plane wave coming from the face of the piston with phase kz and a wave coming from the nearest point on the perimeter of the piston with phase $kR_s + \pi + \delta$, where δ varies with position but is between 0 and $\pi/4$ (asymptotically $\pi/4$). Thus one has

$$N_F = \frac{R_s - z}{\lambda/2} = \begin{cases} (2n - 1) - \delta/\pi & \text{for reinforcement} \quad (5\text{-}8.22a) \\ 2n - \delta/\pi & \text{for partial cancellation} \quad (5\text{-}8.22b) \end{cases}$$

The left side, representing the difference between the path length from the edge and the direct path length in units of half wavelengths, is the *Fresnel number* N_F. Since $R_s - z$ is $(w - a)^2/2z$ in the approximation considered here, the parameter X is $(2N_F)^{1/2}$.

The term "Fresnel number" derives from the concept of *Fresnel zones*† (see Fig. 5-14). The set of all points on the surface at radial distance R (from the

† A. Sommerfeld, *Optics*, Academic, New York, 1950, pp. 218–220; Born and Wolf, *Principles of Optics*, pp. 371–375; F. W. Sears, *Optics*, 3rd ed., Addison-Wesley, Reading, Mass., 1949, pp. 245–251.

listener) between z and $z + \lambda/2$ is said to lie in the first Fresnel zone; those for which R lies between $z + \lambda/2$ and $z + \lambda$ lie in the second Fresnel zone, etc. The Rayleigh integral (5-2.6) can be interpreted as a sum over contributions from the various Fresnel zones that overlap the active face of the vibrating piston. Phase variations of wavelets that originate from points on the same Fresnel zone are relatively minor, while wavelets originating from two adjacent zones tend (on the average) to partially cancel each other. The Fresnel number in Eq. (5-8.22) can be identified as the number of Fresnel zones that separate the projection of the listener point on the $z = 0$ plane from the nearest point on the piston's perimeter. A unit change in Fresnel number corresponds to the addition of the contribution from another Fresnel zone to the Rayleigh integral, which partially cancels the contribution from the previously added zone. This qualitatively explains why the distance from a maximum to the next minimum or from a minimum to the next maximum corresponds asymptotically to a unit change in N_F. However, no special significance should be attached to integer values of N_F.

Since the approximate expression Eq. (18) depends on the radius a of the piston only through the distance $w - a$, it and all the intervening remarks apply to the radiation from uniformly vibrating baffled pistons that are not necessarily of circular shape. One can interpret $w - a$ as transverse distance from the listener position to the nearest point on the outward projection of the piston's perimeter. The solution's validity is primarily limited to points near the nominal edge of the beam; the restrictions described previously apply if a is taken as a characteristic dimension of the piston.

Field far Outside the Central Beam

To describe the pressure field at points at a moderate distance from the edge of the central beam, yet for circumstances in which the inequalities assumed at the beginning of the present section are valid, one can approximate $A_D(X)$ in Eq. (8) by its asymptotic limit $1/\pi X$ with X given by Eq. (7). (This presumes that $w - a$ is sufficiently large to ensure that $|X| \geq 2$.) For such circumstances, the first term in (8) vanishes, and the second and fourth combine into one similar to the third but with $a - w$ replacing $a + w$. In this limit, the acoustic disturbance resembles the sum of two waves, coming from the nearest and farthest points, respectively, on the piston's perimeter. These waves set up an interference and reinforcement pattern; local minima in $(p^2)_{\text{av}}$ occur when

$$\frac{R_l - R_s}{\lambda/2} \approx 2n + \tfrac{1}{2} \qquad (5\text{-}8.23)$$

where n is an integer less than $2a/\lambda - \frac{1}{4}$. (Note that the maximum possible value of $R_l - R_s$ is $2a$.)

For the considered range of z for which the approximation described above is valid, the maxima in this interference pattern are substantially lower in

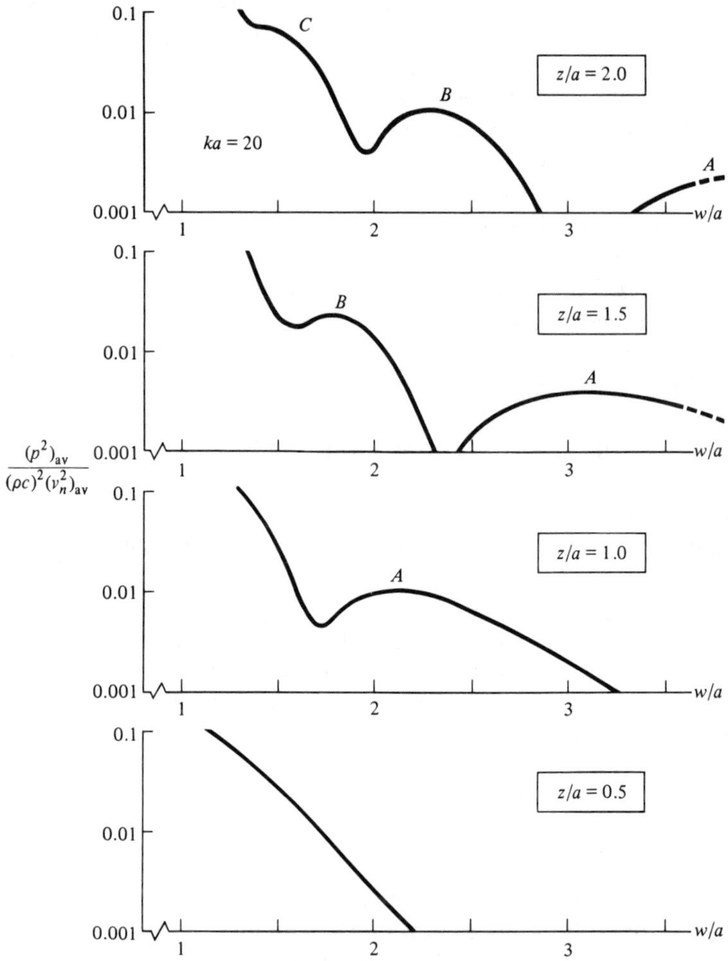

Figure 5-15 The development with increasing axial distance z of side lobes A, B, and C in the radiation pattern of a circular piston (radius a) vibrating at a frequency such that $ka = 20$. The quantity plotted is $(p^2)_{av}$ in units of the nominal average value $(\rho c)^2(v_n^2)_{av}$ expected for plane-wave propagation in the central beam; w is the radial distance in cylindrical coordinates from the axis of the piston. The computations are based on Eq. (5-8.8).

magnitude than those found in the central beam ($w < a$). The first discernible minimum, for z fixed and for $w > a$, corresponds to a value of n for which the cylindrical radial distance w satisfying Eq. (23) is somewhat greater than a, so the minima corresponding to lower integer values of n are not present until z has increased to some threshold value, depending on n. Typical patterns† are shown in Fig. 5-15.

† The analysis in the present section is largely due to Schoch, "Considerations . . . ," 1941. For a comparable but mathematically dissimilar discussion of the field of a circular plane piston in the $ka \gg 1$ limit, see P. H. Rogers and A. O. Williams, Jr., "Acoustic Field of a Circular Plane Piston

The partial cancellation at a minimum becomes nearly complete at radial distances r sufficiently large to ensure that $R_s/R_l \approx 1$, $(w - a)/(w + a) \approx 1$. In this limit one can set $R_l \approx R_s \approx r$ and $w - a \approx w + a \approx w$ in the coefficients of the exponentials. However, to account for phase variations over a hemisphere of fixed r, one should retain the first-order corrections to R_l and R_s in the exponentials; that is, $R_{l,s} \approx r \pm a \sin \theta$. In this manner, one finds that Eq. (8) reduces to what is given by Eqs. (5-5.1) and (5-5.4) but with the Bessel function replaced by its asymptotic expression (5-4.10a). Consequently, Eq. (8) matches the far-field expression in the limit $w \gg a$ (as it should).

PROBLEMS

5-1 At the time a small airplane passes at 150 m altitude over point A on the ground (see sketch), the sound level at A is 100 dB. Estimate the sound level received at the same time at a point B (150 m from A) on the intersection of an isolated building with the ground.

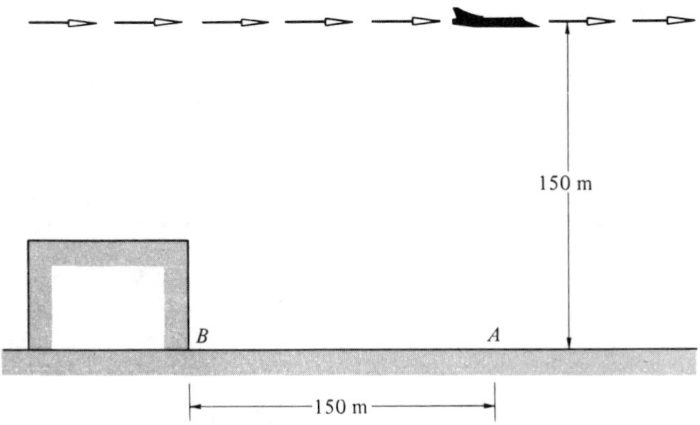

Problem 5-1

5-2 Verify that the method of images applies for a source near a planar pressure-release surface if the image source's surface motion is appropriately chosen. What is the Green's function for a unit-monopole-amplitude point source near a pressure-release surface? Show that the field approaches that of a dipole when a monopole source is sufficiently close to a pressure-release surface.

5-3 An acoustic monopole is near the corner of a large room. Take the floor as the $z = 0$ plane and the two neighboring walls as the $x = 0$ and $y = 0$ planes; let the source be at the point (d, d, d) and let the power output the source would have in an unbounded space be $\mathcal{P}_{av,ff}$. Assuming that the surfaces are perfectly rigid, determine and plot the resulting acoustic power as a function of kd. Beyond what value of kd can one assume the acoustic power output to be within 10 percent of $\mathcal{P}_{av,ff}$? [J. Tickner, *J. Sound Vib.*, **36**:133–145 (1974).]

---

in Limits of Short Wavelength or Large Radius," *J. Acoust. Soc. Am.*, **52**:865–870 (1972). Some detailed computational results for the intermediate range of ka are displayed by H. Stenzel, *Leitfaden zur Berechnung von Schallvorgängen*, Springer, Berlin, 1939, pp. 75–79; they are also given by S. N. Rschevkin, *A Course of Lectures on the Theory of Sound*, Pergamon, Oxford, 1963, pp. 441–443.

5-4 The space ($x > 0$, $y > 0$, $z > 0$) is bounded by three rigid planes at $x = 0$, $y = 0$, and $z = 0$.

(*a*) Derive an expression for the Green's function $G_k(\mathbf{x}|\mathbf{x}_0)$ for the Helmholtz equation that satisfies the appropriate boundary conditions and verify that $G_k(\mathbf{x}|\mathbf{x}_0) = G_k(\mathbf{x}_0|\mathbf{x})$.

(*b*) When $|\mathbf{x}_0|$ is a large distance from the corner but $\mathbf{x}$ is much closer, show that this Green's function assumes the approximate form

$$G_k(\mathbf{x}|\mathbf{x}_0) = F(k\mathbf{x}, \mathbf{e}_i)r_0^{-1}e^{ikr_0}$$

and determine the function $F(k\mathbf{x}, \mathbf{e}_i)$. Do not necessarily assume $kr \gg 1$. Here $r_0 = |\mathbf{x}_0|$ and $\mathbf{e}_i = -\mathbf{x}_0/r_0$ is the unit vector pointing from source to corner.

(*c*) How does this result apply when a plane wave rather than a wave from a point source is incident on the corner?

5-5 An underwater monopole source with angular frequency $\omega = ck$ is at depth z_S below the water's surface (a pressure-release surface) and is at a distance x_S from a large rigid surface occupying the $x = 0$ plane. Otherwise the region occupied by the water is unbounded.

(*a*) Determine the Green's function $G_k(x, y, z|x_S, y_S, z_S)$ for the Helmholtz equation that satisfies the boundary conditions appropriate to this problem and verify that the Green's function satisfies the reciprocity condition.

(*b*) Determine the far-field radiation pattern of the source at distances $|\mathbf{x}| \gg |\mathbf{x}_S|$ when $k|\mathbf{x}| \gg 1$.

(*c*) Determine the time-averaged acoustic power of the source and discuss the limiting cases of $kx_S \to 0$ and $kz_S \to 0$.

5-6 Two loudspeakers of area A are mounted on a large rigid wall ($z = 0$) with their centers at $x = -l/2$, $y = 0$, and $x = l/2$, $y = 0$. Both loudspeakers have the same velocity amplitude $|\hat{v}_n|$, but they are 90° out of phase. Determine the time-averaged far-field acoustic intensity and power output of this two-loudspeaker system. Consider the dimensions of the loudspeakers to be small compared with a wavelength or with l but carry through the derivation for arbitrary kl. (The analysis is simpler if the polar axis of the spherical coordinate system is selected so that the resulting field is independent of ϕ.)

5-7 Four small loudspeakers (labeled 1, 2, 3, 4) are mounted at $(-l/2, l/2)$, $(l/2, l/2)$, $(l/2, -l/2)$, and $(-l/2, -l/2)$ on a rigid wall occupying the $z = 0$ plane. The separation distance l is large compared with a loudspeaker radius a but small compared with a wavelength of the radiated sound. Determine the power radiated out from the wall by this system to lowest nonzero order in kl when each loudspeaker oscillates with velocity amplitude $|\hat{v}_n|$ for the following possible phase selections: (*a*) all loudspeakers in phase; (*b*) speakers 1 and 2 in phase but 180° out of phase with 3 and 4; (*c*) speakers 1 and 3 in phase but 180° out of phase with 2 and 4.

5-8 A rigid circular diaphragm of mass $m = 0.015$ kg and radius 0.15 m moves inside a cylindrical cavity whose mouth has a very large baffle. The diaphragm is separated from the inner end of the cavity by an elastic material that behaves like a spring with a spring constant of 2000 N/m. A sinusoidally varying force with a frequency of 330 Hz causes the diaphragm to vibrate and to radiate 0.5 W of acoustic power.

(*a*) What is the velocity amplitude of the diaphragm?

(*b*) What force amplitude is required to produce this power? [Take $\rho c = 400$ kg/(m² · s) and $c = 350$ m/s.]

5-9 A square piston, dimensions a on each side, is mounted in a rigid wall ($z = 0$) and vibrates with angular frequency ω and velocity amplitude $|\hat{v}_n|$.

(*a*) Derive an expression for the far-field intensity for arbitrary ka.

(*b*) For $ka = 2\pi$, plot the ratio of intensity at polar angle θ to that at $\theta = 0$ versus θ for fixed azimuthal angle ϕ when $\phi = 0°$ and when $\phi = 45°$. Also plot the analogous ratio for fixed θ versus ϕ when $\theta = 90°$.

(*c*) Determine the smallest value of ka for which the far-field radiation pattern has a nodal direction. Take the piston as occupying the region $-a/2 < x < a/2$, $-a/2 < y < a/2$ in the $z = 0$ plane and let $x = r \sin\theta \cos\phi$, $y = r \sin\theta \sin\phi$, $z = r\cos\theta$.

5-10 A small baffled loudspeaker driven by a transducer and oscillating at 1000 Hz frequency with rms velocity of 1 m/s causes the sound in air at a radial distance of 10 m to have a rms acoustic pressure of 0.1 Pa. The electroacoustic transducer (with baffled loudspeaker included) is such that when it acts as a loudspeaker, a voltage Re $1.0e^{-i\omega t}$ V causes an area-averaged loudspeaker velocity of Re $(1 - i)e^{-i\omega t}$ m/s and a current of Re $(1 - i)e^{-i\omega t}$ A. What is the electroacoustic efficiency of this system?

5-11 An annular piston with inner radius a and outer radius $\frac{4}{3}a$ is mounted on a wall so that the inner area, $0 < w < a$, does not move, while the piston, $a < w < \frac{4}{3}a$, oscillates with velocity amplitude $|\hat{v}_n|$ and angular frequency ω.

(*a*) What is the smallest nonzero value of ω at which the acoustic pressure just in front of the center point (0, 0, 0) is zero?

(*b*) If ω is systematically varied, what is the maximum acoustic-pressure amplitude one can expect at any given point on the symmetry axis?

5-12 A *zone plate* is constructed to enhance the acoustic-pressure amplitude at a point on the symmetry axis 10 wavelengths from the center of a baffled circular piston oscillating at angular frequency ω. The radius of the piston is such that, at this frequency and for the cited listener point, it corresponds to the outer edge of the fifth Fresnel zone. The piston is oscillating with velocity amplitude $|\hat{v}_n|$, but the zone plate blocks out the second and fourth zones so that only zones 1, 3, and 5 contribute to the radiated field. What is the acoustic-pressure amplitude at the chosen listener point?

5-13 A rigid sphere of radius a moves back and forth with small displacement amplitude and angular frequency ck in a circular hole of the same radius in a large rigid screen.

(*a*) Given that $ka \ll 1$ and that the velocity amplitude of the sphere is $|\hat{v}_C|$, determine the acoustic power radiated to one side of the screen.

(*b*) How does your result compare with what would be expected without the screen?

5-14 A baffled circular piston of radius a begins to vibrate at time $t = 0$ such that $v_n(t) = 0$ for $t < 0$, $v_n(t) = |\hat{v}_n| \sin \omega t$ for $t > 0$. Plot the acoustic pressure versus time at a point on the symmetry axis at a distance $3\pi c/\omega$ from the piston center when $\omega = 4\pi c/a$.

5-15 (*a*) Show that the method of images applies for a point source within the interior region of a wedge formed by two rigid walls that intersect at an angle of π/n, where n is a positive integer.

(*b*) Determine the location of all necessary images of a source at a point described by cylindrical coordinates w_S, ϕ_S, z_S within a wedge-shaped region formed by the planes $\phi = 0$ and $\phi = \pi/3$.

(*c*) Give an expression for the Green's function that satisfies boundary conditions appropriate to the circumstances of (*b*).

(*d*) How much enhancement in acoustic-power output relative to that expected in a free-field environment is obtained in the limit $w_S \to 0$?

5-16 Verify that the expressions in Eqs. (5-4.9) and (5-5.6) for $R_1(2ka)$ are equivalent.

5-17 Determine a definite-integral expression for the acoustic power radiated by the baffled square piston of Prob. 5-9 and show that its average approximates to $(\rho c)(ka)^2 a^2 (v_n^2)_{av}/2\pi$ for $ka \ll 1$ and to $\rho c a^2 (v_n^2)_{av}$ for $ka \gg 1$.

5-18 For the low-frequency limit, when the acoustic field near an oscillating baffled circular piston can be described as incompressible flow, determine the component v_w of the fluid velocity that corresponds to flow radially away from the symmetry axis for points on the piston ($z = 0$). Plot your result in a suitable dimensionless form versus w/a.

5-19 A limiting case of interest is when the $x < 0$ half of the $z = 0$ plane has normal velocity Re $\hat{v}_n e^{-i\omega t}$ while the other half remains rigid.

(*a*) Prove that the complex acoustic-pressure amplitude $\hat{p}$ along the plane $x = 0$ is $\frac{1}{2}\rho c \hat{v}_n e^{ikz}$.

(*b*) Show that $\hat{p}$ is given by the expression on page 236*n* in the limit $kz \gg 1$. Give a derivation that proceeds from the Rayleigh integral without the artifice of extracting the $ka \gg 1$ limit from the result for a circular piston.

5-20 (*a*) Show for the circumstances for which Eq. (5-8.18) is applicable that the radial component (cylindrical coordinates) v_w of the fluid velocity at $w = a$ has a complex amplitude $\hat{v}_w$ equal to $[(1 + i)/2]\hat{v}_n(\pi kz)^{-1/2}e^{ikz}$.

(*b*) Use this result to estimate to what distance z the primary beam (occupying the cylinder of radius a) propagates before the acoustic power transported within it drops by 50 percent of its value near the piston surface. (Assume $ka \gg 1$.)

5-21 Show that the quadruple Helmholtz integral in Eq. (5-4.1) (whose value determines the piston's radiation impedance) can be reduced to evaluation of the double integral

$$\oint\oint \mathrm{Ein}(-ikR)\, d\mathbf{l} \cdot d\mathbf{l}_S$$

where $d\mathbf{l}$ and $d\mathbf{l}_S$ are differential line elements, the two integrations proceeding around the perimeter of the piston. Here

$$\mathrm{Ein}(\eta) = \int_0^\eta \frac{1 - e^{-t}}{t}\, dt$$

is the exponential integral. Do not necessarily assume that the piston is circular. [O. A. Lindemann, "Transformation of the Helmholtz Integral into a Line Integral," *J. Acoust. Soc. Am.*, **40**:914–915 (1966).]

5-22 (*a*) Show that, in the limit of small ka, where a is a characteristic piston dimension, the result in Prob. 5-21 reduces to the evaluation of

$$\oint\oint R\, d\mathbf{l} \cdot d\mathbf{l}_S$$

(*b*) Hence show that the mechanical radiation impedance of a baffled rectangular piston of dimensions a by b is given in the limit of $ka \ll 1$, $kb \ll 1$, by

$$Z_{m,\mathrm{rad}} = -i\frac{\rho c}{2\pi} k(ab)^{3/2} f\left(\frac{a}{b}\right) + \frac{\rho c}{2\pi} k^2(ab)^2$$

where

$$f(\zeta) = 2\zeta^{1/2}\sinh^{-1}\zeta^{-1} + 2\zeta^{-1/2}\sinh^{-1}\zeta + \tfrac{2}{3}\zeta^{3/2} + \tfrac{2}{3}\zeta^{-3/2} - \tfrac{2}{3}(\zeta + \zeta^{-1})^{3/2}$$

[O. A. Lindemann, "Radiation Impedance of a Rectangular Piston at Very Low Frequencies," *J. Acoust. Soc. Am.*, **44**:1738–1739 (1968).]

5-23 A point source of monopole amplitude $\hat{S}$ and oscillating at angular frequency ω is at $(0, 0, z_S)$ between two parallel rigid walls, $z = 0$ and $z = h$.

(*a*) Show that the image sources have z coordinates $2nh \pm z_S$, where the integer n is positive, negative, or zero.

(*b*) Show that the complex amplitude of the acoustic pressure can be alternately written as

$$\hat{p} = \hat{S}\int_{-\infty}^{\infty} \frac{e^{ik(\zeta^2+w^2)^{1/2}}}{(\zeta^2 + w^2)^{1/2}} \sum_{n=-\infty}^{\infty} [\delta(\zeta - z + z_S + 2nh) + \delta(\zeta - z - z_S + 2nh)]\, d\zeta$$

$$= \frac{\hat{S}}{h}\int_{-\infty}^{\infty} \frac{e^{ik(\zeta^2+w^2)^{1/2}}}{(\zeta^2 + w^2)^{1/2}} \left[\sum_{n=0}^{\infty} \varepsilon_n \cos\frac{n\pi z_S}{h} \cos\frac{n\pi(z - \zeta)}{h}\right] d\zeta$$

$$= \frac{\hat{S}}{h}\sum_{n=0}^{\infty} \varepsilon_n \cos\frac{n\pi z_S}{h} \cos\frac{n\pi z}{h} \int_{-\infty}^{\infty} \frac{e^{ik(\zeta^2+w^2)^{1/2}}}{(\zeta^2 + w^2)^{1/2}} \cos\frac{n\pi\zeta}{h}\, d\zeta$$

where

$$\varepsilon_n = \begin{cases} 1 & \text{for } n = 0 \\ 2 & \text{for } n \ge 1 \end{cases}$$

(*c*) Express the above definite integral as k times a function of $[k^2 - (n\pi/h)^2]^{1/2}w$ and show that the result is proportional to what is defined as the Hankel function in standard reference texts.

5-24 (*a*) Verify that the complex acoustic-pressure amplitude at the perimeter of an oscillating baffled circular piston is given by Eq. (5-8.16).

(*b*) Show that the result is compatible with Eqs. (5-3.4) and (5-3.7) in the limit $ka \ll 1$.

5-25 (*a*) Determine an expression for the time-averaged axial component $I_{z,\mathrm{av}}$ of the acoustic intensity along the symmetry axis of a baffled circular piston oscillating at constant frequency.

(*b*) What is the corresponding limiting value ($w \to 0$) of $w^{-1}I_{w,\mathrm{av}}$ along the symmetry axis? (Here w denotes the radial distance in cylindrical coordinates.)

(*c*) Sketch the energy flow lines (lines everywhere tangential to **I**) in the vicinity of the symmetry axis for $ka = 6\pi$. Indicate the direction of energy flow with arrows.

5-26 A highly directional acoustic radiator is to be designed using a baffled circular piston. The sound-pressure level in the far field at angles greater than 10° should be at least 10 dB less than that at the same radial distance along the symmetry axis. What is the minimum value of ka to accomplish this objective?

CHAPTER

SIX

ROOM ACOUSTICS

The sound in a room consists of that coming directly from the source plus sound reflected or scattered (see Fig. 6-1) by the walls and by objects in the room. Sound having undergone one or more reflections is called *reverberant* sound because it corresponds for an impulsive source to a series of echoes. If the direct wave predominates almost everywhere, the room is *anechoic* (without echoes); rooms so designed† are *anechoic chambers*. A *reverberation chamber* is a room designed‡ so that the reverberant field predominates overwhelmingly.

The bulk of the present chapter is concerned with sound in reverberant rooms. Many of the concepts introduced here, e.g., room absorption, reverberation time, random-incidence absorption coefficients, and random wave fields, have implications extending beyond room acoustic applications and correspond to analogous concepts in such diverse areas as the propagation of sound in the ocean, the vibrations of large complex bodies, the radiation of sound by such bodies, and the propagation of sound within and out of ducts.

† J. Duda, "Basic Design Considerations for Anechoic Chambers," *Noise Control Eng.*, **9**:60–67 (1977); W. Koidan and G. R. Hruska, "Acoustical Properties of the National Bureau of Standards Anechoic Chamber," *J. Acoust. Soc. Am.*, **64**:508–516 (1978).

‡ Standard design criteria are set forth in American National Standard Methods for the Determination of Sound Power Levels of Small Sources in Reverberation Rooms, ANSI S1.21-1972, American National Standards Institute, New York, 1972. See also the discussion by W. K. Blake and L. J. Maja, "Chamber for Reverberant Acoustic Power Measurements in Air and in Water," *J. Acoust. Soc. Am.*, **57**:380–384 (1975).

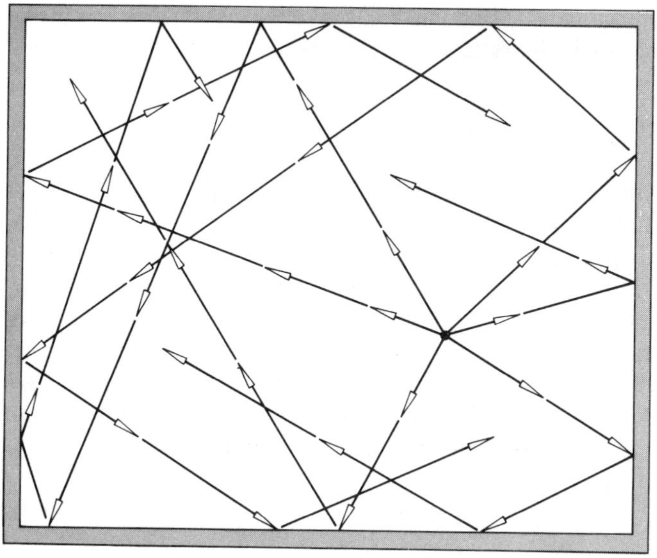

Figure 6-1 Sketch of ray paths from a source in a reverberant room.

6-1 THE SABINE-FRANKLIN-JAEGER THEORY OF REVERBERANT ROOMS

An appropriate idealization (discovered by W. C. Sabine† at the turn of the century) is that the sound "fills" a reverberant room in such a way that the average energy per unit volume in any region is nearly the same as in any other region. The corresponding mathematical model (*reverberant-field model*) that Sabine deduced from a series of ingenious experiments has a relation to the full-wave model (wave equation plus boundary conditions) of classical acoustics similar to that of radiative heat transfer to electromagnetic theory or of kinetic theory to classical mechanics. It applies best to "large" rooms whose characteristic dimensions are substantially larger than a typical wavelength and to "live" (as opposed to "dead") rooms, for which the time determined by the ratio of the total propagating energy within the room to the time rate at which energy is being lost from the room (absorbed or transmitted out) is considerably

† W. C. Sabine, "Architectural Acoustics," *Eng. Rec.*, **38:**520–522 (1898); "Architectural Acoustics," ibid., **41:**349–351, 376–379, 400–402, 426–427, 450–451, 477–478, 503–505 (1900); both the 1898 paper and the series of 1900 are also printed in *Am. Archit. Build. News*, **62:**71–73 (1898), ibid., **68:**3–5, 19–22, 35–37, 43–45, 59–61, 75–76, 83–84 (1900). All except that of 1898 are reprinted in W. C. Sabine, *Collected Papers on Acoustics*, Dover, New York, 1964. Historical sidelights are given by L. L. Beranek: "The Notebooks of Wallace C. Sabine," *J. Acoust. Soc. Am.*, **61:**629–639 (1977).

larger than the time required for a sound wave to travel across a representative dimension of the room. (Other limitations are discussed in Secs. 6-3 and 6-6.)

Energy Conservation Equation for Rooms

The basic concepts involved in the Sabine model are best explained within the context of the principle of conservation of acoustic energy. The portion of the field associated with a given frequency band can be defined, even for nonsteady fields, in terms of functions $p_b(\mathbf{x}, t)$ and $\mathbf{v}_b(\mathbf{x}, t)$ that correspond to the instantaneous outputs when $p(\mathbf{x}, t)$ and $\mathbf{v}(\mathbf{x}, t)$ are passed through frequency filters. These filtered field variables also satisfy the linear acoustics equations (see Prob. 2-41), and so the derivation of Eq. (1-11.2) is still applicable. After an integration over the interior volume V of the room, the analogous differential equation involving p_b and $\mathbf{v}_b$ yields the energy-conservation relation

$$\frac{d}{dt}\iiint w_b\, dV = \mathcal{P}_b - \mathcal{P}_{b,d} \qquad (6\text{-}1.1)$$

where w_b is the acoustic energy density given by (1-11.3) with p_b and $\mathbf{v}_b$ replacing p and $\mathbf{v}$. Here $\mathcal{P}_b$ is the net acoustic power associated with the frequency band of interest supplied by sources in the room. The power dissipated $\mathcal{P}_{b,d}$ is the power within the same frequency band leaving the room through its bounding surfaces and is defined as a surface integral of $p_b\mathbf{v}_b \cdot \mathbf{n}_{\text{out}}$. The dissipation within the interior of the room proper is usually not significant, except at higher frequencies, but Eq. (1) (with a broader interpretation of $\mathcal{P}_{b,d}$) can still be used when one wants to take this into account (see Sec. 10-8).

Equation (1), holding at every instant, is also true (Prob. 2-41) if w_b, $\mathcal{P}_b$, and $\mathcal{P}_{b,d}$ are replaced by running time averages, $\bar{w}_b$, $\bar{\mathcal{P}}_b$, and $\bar{\mathcal{P}}_{b,d}$. One can also argue that if the effective duration of the averaging interval is sufficiently long, these running time averages are additive functions for nonoverlapping bands. For example, the function $\bar{w}_b$ for the band 1000 to 2000 Hz should equal the sum of those corresponding to the bands 1000 to 1500 Hz and 1500 to 2000 Hz.

Spatial Uniformity

The principal assumption on which the Sabine model is based is that over the major portion of the interior space of the room, the *local spatial average* of $\bar{w}_b$ is independent of position. (A local spatial average is here understood to be an average over a volume with dimensions substantially larger than a representative acoustic wavelength but substantially smaller than those of the room as a whole.) This assumption may not be valid near a source and may also not be true near protruding obstacles, but one can limit the volume of consideration to whatever portion V' of V the assumption applies. It must nevertheless be assumed that only a small fraction of V is excluded.

This spatial uniformity requires the presence of the walls for its existence and maintenance. If a source is suddenly turned on, the time interval within which such a uniformity is established can be estimated as the time lapse until the hundredth reflected wave arrives. For a rectangular room with nearly rigid walls, the various reflected waves can be considered as coming from a rectangular array of image sources (see Sec. 5-1); in the extended space there is one image source per volume V, so the first 100 images lie within a radius of the order of $(3/4\pi)^{1/3}(100)^{1/3}V^{1/3} = 2.9V^{1/3}$. This suggests that an average spatial uniformity is well established within a time interval of the order of $3l/c$, where l is a representative dimension of the room. For l equal to, say, 10 m and with $c = 340$ m/s, this gives a time interval of 0.1 s.

The Sabine model regards all acoustic fields with the same average energy density $\bar{w}$ as equivalent insofar as a field's statistical properties are concerned. (Here and in what follows $\bar{w}$ represents the local spatial average of the running time average; the subscript b is omitted, and no additional symbolism is used to imply spatial averaging. Also, in accord with the remarks above, $\bar{w}$ is assumed independent of position.)

A consequence of the statistical-equivalence assumption is that $\bar{\mathcal{P}}_d$ depends on the reverberant field in the room only through $\bar{w}$. Furthermore, because the boundary conditions at surfaces bounding V are governed by linear equations relating the primary acoustic field variables p and $\mathbf{v}$, this relationship should be a direct proportionality. (Both $\bar{w}$ and $\bar{\mathcal{P}}_d$ increase by the factor K^2 when the field variables are each increased by a factor K.) The proportionality constant is a property of the room as a whole, independent of the nature and position of the source but possibly dependent on frequency.

The proportionality just described can be written

$$\bar{\mathcal{P}}_d = \frac{c}{4} A_s \bar{w} \qquad (6\text{-}1.2)$$

where c is the speed of sound and A_s is a frequency-dependent room property having units of area that can be considered to be defined by this equation. For reasons explained below, A_s is referred to as the *equivalent area of open windows* or the *absorbing power* of the room and is said to have the units of *metric sabins,* the term *sabin* identifying the context in which it is used. (The unit *sabin* without the adjective, refers to the area A_s in square feet, although Sabine used metric units in his first papers.)

With the substitution of Eq. (2) for $\bar{\mathcal{P}}_d$, the running time average of the energy-conservation law (1) is reduced to the differential equation†

$$V\frac{d\bar{w}}{dt} + \frac{c}{4}A_s\bar{w} = \bar{\mathcal{P}} \qquad (6\text{-}1.3)$$

† G. Jaeger, "Toward a Theory of Reverberation," *Sitzungsber. Kais. Akad. Wiss. (Vienna), Math. Naturwiss. Kl.*, sec. IIa, **120:**613–634 (1911).

Reverberation Time

After the sudden extinction of a source in a reverberant room, the running time average of sound pressure squared, as indicated by a sound-level meter with the "fast" response, for example, may fluctuate somewhat erratically (Fig. 6-2), but the gross tendency resembles an exponential decay, similar to that experienced by the volume average $\bar{w}$ of energy density. The latter behavior results from an integration of Eq. (3) with $\bar{\mathcal{P}}$ set to zero, i.e.,

$$\bar{w}(t) = \bar{w}_{\text{init}} e^{-t/\tau} \qquad \tau = \frac{4V}{cA_s} \tag{6-1.4}$$

The so-defined *characteristic decay time* τ has units of seconds per half neper, since whenever the amplitude of the primary acoustic variables decreases by a factor of e^{-1} or by 1 neper (Np), the energy density (a bilinear quantity) decreases by a factor of e^{-2}.

The usual descriptor for the exponential decay of reverberant sound is the time T_{60} required for the spatial average of the energy density to drop by a factor of 10^6 (60 dB). This *reverberation time* T_{60} is such that when $t = T_{60}$ in Eq. (4), $\bar{w}_{\text{init}}/\bar{w}$ is 10^6; therefore, T_{60} is $(6 \ln 10)\tau = 13.82\tau$. Because $\bar{w}$ is proportional to $\overline{p^2}$ (a relation $\bar{w} = \overline{p^2}/\rho c^2$ is derived below), and because a decrease of $\overline{p^2}$ by a factor of 10^6 corresponds to a decrease in sound level by 60 dB, T_{60} has the units of seconds per 60 dB; its relation to τ expresses the equivalence of 60 dB to 13.82 Np/2.

Sabine's Equation

One of Sabine's principal contributions to room acoustics was the experimental discovery that for an empty room of volume V the reverberation time T_{60} is

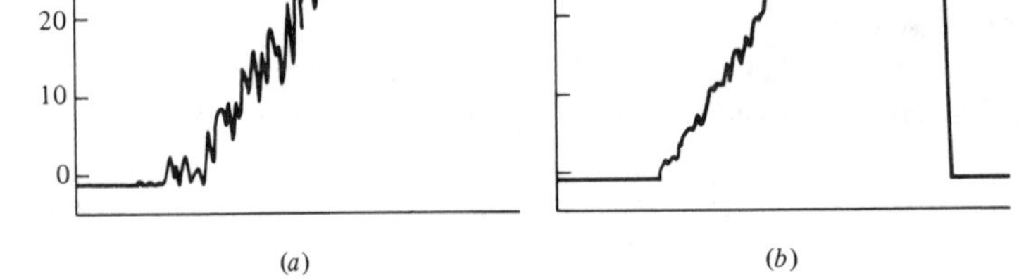

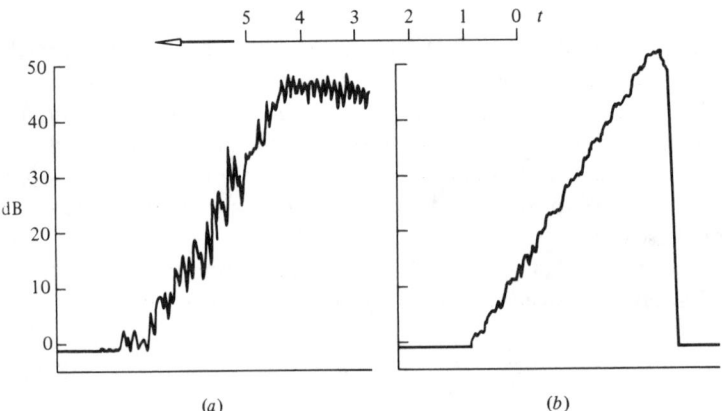

Figure 6-2 Reverberant decay of running time average of square of acoustic pressure as displayed by a high-speed level recorder. (*a*) Sudden turnoff of a narrow-band source (1000 ± 50 Hz) and (*b*) firing a pistol shot (600 to 1200 Hz). *(W. Furrer, Room and Building Acoustics and Noise Abatement, Butterworths, London, 1964, p. 89.)*

predictable from the relation† (in SI units)

$$T_{60} = \frac{0.161V}{\sum_i \alpha_i A_i} \tag{6-1.5}$$

Here the sum extends over all the distinct portions of the total surface area of the room, each element of area A_i characterized by an *absorption coefficient* α_i determined from measurements of T_{60} with various mixtures of wall coverings and from the requirement that α_i be 1 for an open window. The model presumes that α_i is an intrinsic property of the wall material (depending also on frequency), independent of the source, source location, magnitude (given that it is sufficiently large), and location of area A_i and of the coverings on other portions of the bounding surfaces. Sabine's experimental data indicated that Eq. (5) can predict reverberation times for specific cases using values of the α_i derived from previous measurements of reverberation times in different circumstances. Typical numbers measured by Sabine with a source of 512 Hz frequency for the absorption coefficient α were wood sheathing (hard pine), 0.061; plaster on wood lath, 0.034; plaster on wire lath, 0.033; glass, single thickness, 0.027; plaster on tile, 0.025; brick set in Portland cement, 0.025; seat cushions, 0.80; carpeting, 0.20; oriental rugs, extra heavy, 0.29; linoleum, loose on floor, 0.12. (Table 6-1 lists absorption coefficients extracted from more recent literature.)

The extension to Sabine's derivation of Eq. (5) that successfully predicts the numerical coefficient is due to W. S. Franklin;‡ a derivation similar in basic concept but explicitly related to the wave theory of sound is given below.

Diffuse Sound Fields

To demonstrate the equivalence of Eqs. (4) and (5) when A_s is as defined by Eq. (2), it is sufficient to limit one's consideration to the constant-frequency case. Within the interior of a reverberant room, the field is regarded as a superposition of freely propagating plane waves, no two of which are traveling in the same direction (see Fig. 6-3*a*), so for the complex amplitudes we write

$$\hat{p} = \sum_q \hat{p}_q e^{ik\mathbf{n}_q \cdot \mathbf{x}} \qquad \rho c\hat{\mathbf{v}} = \sum_q \mathbf{n}_q \hat{p}_q e^{ik\mathbf{n}_q \cdot \mathbf{x}} \tag{6-1.6}$$

The time average of the energy density associated with this field, expressed using Eqs. (1-11.3) and (1-8.9), involves a double sum over indices q and q', but the process of taking a local spatial average causes the cross terms ($q \neq q'$) to tend to average out. The spatial average of $\exp\,[ik(\mathbf{n}_q - \mathbf{n}_{q'}) \cdot \mathbf{x}]$ is nearly zero

† Various slightly different experimentally determined values for the numerical coefficient are mentioned in Sabine's writings; 0.164 s/m is, for example, given in a 1906 paper (*Collected Papers on Acoustics*, p. 103). The value 0.161 is predicted by theory when the room temperature is 18.3°C (65°F); 0.164 corresponds to 9.4°C (49°F).

‡ W. S. Franklin, "Derivation of Equation of Decaying Sound in a Room and Definition of Open Window Equivalent of Absorbing Power," *Phys. Rev.*, **16**:372–374 (1903).

Table 6-1 Representative absorption coefficients of surfaces

Material	Absorption coefficient α					
	125 Hz	250 Hz	500 Hz	1000 Hz	2000 Hz	4000 Hz
Brick, unglazed	0.03	0.03	0.03	0.04	0.05	0.07
Plaster, gypsum or lime, on brick	0.01	0.02	0.02	0.03	0.04	0.05
On concrete block	0.12	0.09	0.07	0.05	0.05	0.04
Concrete block, coarse	0.36	0.44	0.31	0.29	0.39	0.25
Painted	0.10	0.05	0.06	0.07	0.09	0.08
Plywood, 1-cm-thick paneling	0.28	0.22	0.17	0.09	0.10	0.11
Cork, 2.5 cm thick with airspace behind	0.14	0.25	0.40	0.25	0.34	0.21
Glass, typical window	0.35	0.25	0.18	0.12	0.07	0.04
Drapery, lightweight, flat on wall	0.03	0.04	0.11	0.17	0.24	0.35
Heavyweight, draped to half area	0.14	0.35	0.55	0.72	0.70	0.65
Floor, concrete	0.01	0.01	0.02	0.02	0.02	0.02
Linoleum on	0.02	0.03	0.03	0.03	0.03	0.02
Heavy carpet on	0.02	0.06	0.14	0.37	0.66	0.65
Wood	0.15	0.11	0.10	0.07	0.06	0.07
Ceiling, gypsum board	0.29	0.10	0.05	0.04	0.07	0.09
Plastered	0.14	0.10	0.06	0.05	0.04	0.03
Plywood, 1 cm thick	0.28	0.22	0.17	0.09	0.10	0.11
Suspended acoustical tile, 2 cm thick	0.76	0.93	0.83	0.99	0.99	0.94
Gravel, loose and moist, 10 cm thick	0.25	0.60	0.65	0.70	0.75	0.80
Grass, 5 cm high	0.11	0.26	0.60	0.69	0.92	0.99
Rough soil	0.15	0.25	0.40	0.55	0.60	0.60
Water surface, as in a pool	0.01	0.01	0.01	0.02	0.02	0.03

Source: M. D. Egan, *Concepts in Architectural Acoustics*, McGraw-Hill, 1972, pp. 32–34.

for a sufficiently large averaging volume. Moreover, the spatial averages of the cross terms should have a variety of magnitudes; either sign is equally likely for terms having a given magnitude, so the total sum of such terms should be small. The terms for which $q = q'$, however, are positive and must be retained. With the neglect of cross terms, the time average of the energy density reduces to the sum of the time averages of its constituent plane waves [see Eq. (1-11.11a)], so one obtains

$$\bar{w} = \frac{1}{2\rho c^2} \sum_q |\hat{p}_q|^2 \approx \frac{1}{\rho c^2} \overline{p^2} \tag{6-1.7}$$

which is analogous to Parseval's theorem (see Secs. 2-1 and 2-7).

The portion $\bar{w}_{\Delta\Omega}$ of the average energy density propagating with directions lying within a cone of solid angle $\Delta\Omega$ is that part of the sum in Eq. (7) for which

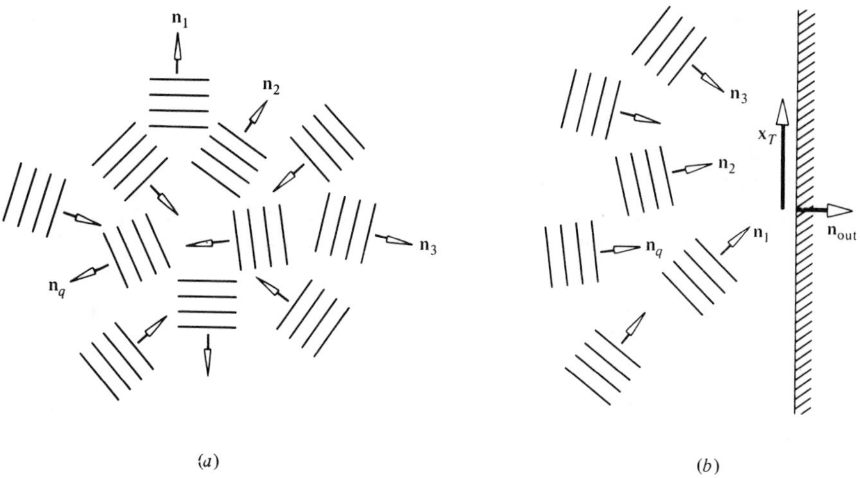

Figure 6-3 (*a*) Reverberant field represented as a superposition of traveling plane waves. (*b*) Waves incident on a surface adjacent to a reverberant field.

$\mathbf{n}_q$ lies in $\Delta\Omega$. One can conceive of a *directional energy density* $D(\mathbf{e})$ as the quasi limit as $\Delta\Omega$ becomes small of $\bar{w}_{\Delta\Omega}/\Delta\Omega$, where $\Delta\Omega$ is the solid angle centered on the direction $\mathbf{e}$. This $D(\mathbf{e})$ (energy per unit volume and per unit solid angle of propagation direction†) must accordingly be such that its integral over all directions, 4π sr (steradians), is $\bar{w}$.

A field satisfying the criterion that $D(\mathbf{e})$ be independent of $\mathbf{e}$, so $D(\mathbf{e}) = \bar{w}/4\pi$, is a perfectly *diffuse field.* Near an absorbing surface (especially at an open window), the field departs from this ideal, but nevertheless $D(\mathbf{e})$ for directions pointing into the surface (out of the room) is representative of the acoustic state within the interior of the room and should therefore be nearly $\bar{w}/4\pi$, where $\bar{w}$ is the room's average energy density.

The above considerations allow one to describe the energy lost at any large flat (or nearly flat) portion of the room's bounding surface. If many plane waves are simultaneously incident on such a wall (Fig. 6-3*b*), the individual waves reflect independently and the principle of superposition can be used in conjunction with the theory of plane-wave reflection described in Sec. 3-3. Such an analysis requires that the time average of the rate at which energy is absorbed (not reflected) by the surface per unit area be

$$\frac{1}{2\rho c}\,\mathrm{Re}\sum_{q,r}{}'\;\hat{p}_q\hat{p}_r^*(1+\mathscr{R}_q)(1-\mathscr{R}_r^*)e^{ik(\mathbf{n}_q-\mathbf{n}_r)\cdot\mathbf{x}_T}\,\mathbf{n}_r\cdot\mathbf{n}_{\text{out}}$$

† In the theory of radiative heat transfer, an *intensity of radiation I* is defined as the energy emitted by a surface per unit area of surface per unit time and per unit solid angle of propagation direction. The analog of the directional energy density defined in the text can be identified for volumes just outside such a surface and for directions pointing obliquely away from it as I/c, where c is the speed at which the energy propagates. See, for example, F. Kreith, *Principles of Heat Transfer,* 3d ed., Intext, New York, 1973, p. 229.

where $\mathscr{R}_q$ = pressure-amplitude reflection coefficient corresponding to incidence direction $\mathbf{n}_q$

$\mathbf{x}_T$ = displacement vector tangential to surface

$\mathbf{n}_{\text{out}}$ = unit vector pointing out of room

The prime implies that the sum is restricted to incident waves, such that $\mathbf{n}_r$ points obliquely toward the wall.

If the surface portion is sufficiently large, one can replace the above expression by its average over surface area. For reasons similar to those given in the derivation of Eq. (7), the surface-area averages of the cross terms are small and tend to average out. Consequently, one is left with just the area averages of the terms for which $q = r$, for which the exponential factor is 1, and for which $\hat{p}_q\hat{p}_r^* = |\hat{p}_q|^2$ is real. Moreover, the real part of $(1 + \mathscr{R}_q)(1 - \mathscr{R}_q^*)$ is the absorption coefficient $\alpha(\mathbf{n}_q)$ for a plane wave incident in the $\mathbf{n}_q$ direction. The resulting expression is therefore

$$\frac{d\bar{\mathscr{P}}_d}{dA} = \frac{1}{2\rho c}\sum_q \alpha(\mathbf{n}_q)|\hat{p}_q|^2\mathbf{n}_q \cdot \mathbf{n}_{\text{out}} \tag{6-1.8}$$

To eliminate explicit reference to the amplitudes $|\hat{p}_q|$ of individual plane waves, the above sum is arranged into a double sum, first over terms for which $\mathbf{n}_q$ lies within solid angle $\Delta\Omega$, then over solid angles. If an individual solid-angle element is sufficiently small, the factors $\alpha(\mathbf{n}_q)$ and $\mathbf{n}_q \cdot \mathbf{n}_{\text{out}}$ for all the constituent terms can be approximated with $\mathbf{n}_q$ replaced by the solid angle's central direction, unit vector $\mathbf{e}$. Furthermore, the partial sum of the $|\hat{p}_q|^2$, corresponding to $\mathbf{n}_q$ lying within this small range of solid angle, can be recognized from Eq. (7) as $2\rho c^2$ times $D(\mathbf{e})\,\Delta\Omega$. The sum over solid-angle elements goes into an integral over solid angle, so Eq. (8) yields

$$\frac{d\bar{\mathscr{P}}_d}{dA} = c\iint' \alpha(\mathbf{e})D(\mathbf{e})\mathbf{e} \cdot \mathbf{n}_{\text{out}}\,d\Omega = \frac{c}{4}\,\alpha_{\text{ri}}\bar{w} \tag{6-1.9}$$

where the integral extends over just those directions for which $\mathbf{e} \cdot \mathbf{n}_{\text{out}} \geq 0$. The second equality follows from the perfectly diffuse idealization, $D = \bar{w}/4\pi$, and with the definition

$$\alpha_{\text{ri}} = \frac{1}{\pi}\iint' \alpha(\mathbf{e})\mathbf{e} \cdot \mathbf{n}_{\text{out}}\,d\Omega \tag{6-1.10}$$

for the *random incidence absorption coefficient* α_{ri}.

Equation (10) describes a weighted average of plane-wave absorption coefficients because when $\alpha(\mathbf{e})$ is constant, the right side integrates to $\alpha(\mathbf{e})$. This is verified if one chooses a coordinate system such that $\mathbf{n}_{\text{out}}$ is in the z direction and if one uses the spherical coordinates θ, ϕ to describe directions, so that $\mathbf{e} \cdot \mathbf{n}_{\text{out}} = \cos\theta$ and $d\Omega = \sin\theta\,d\theta\,d\phi$; the integration limits are $(0, \pi/2)$ and $(0, 2\pi)$ for θ and ϕ. Ordinarily, $\alpha(\theta, \phi)$ is independent of ϕ, so Eq. (10) reduces to

$$\alpha_{\text{ri}} = 2\int_0^{\pi/2} \alpha(\theta)\cos\theta\,\sin\theta\,d\theta \tag{6-1.11}$$

Equivalent Area of Open Windows

For an open window of sufficiently large area, one would expect $\alpha(\theta)$ to be 1 regardless of angle of incidence, so α_{ri} would also be 1. Thus, the average absorption coefficient α for a given surface of area ΔA can alternately be defined in the manner originally chosen by Sabine as the ratio of $\Delta\bar{\mathcal{P}}_d/\Delta A$ to that expected for an open window. The latter is identified from Eq. (9), with $\alpha = 1$, as $(c/4)\bar{w}$. (In what follows the subscript ri is omitted.)

Sabine's definition allows a broader conception† of absorption coefficient transcending some of the limitations of the derivation. The average rate of dissipation $\Delta\bar{\mathcal{P}}_d$ by any portion of the walls or by any object in the room can be written as $(c/4)\bar{w}\ \Delta A_s$, where ΔA_s is the equivalent area of open windows yielding the same $\Delta\bar{\mathcal{P}}_d$. The sum of all such $\Delta\bar{\mathcal{P}}_d$ gives Eq. (2), so A_s is the equivalent area of open windows for the room as a whole.

If all such contributions come from surfaces for which it is meaningful to associate an absorption coefficient, A_s becomes the sum of the $\alpha_i A_i$. The reverberation time $T_{60} = (6 \ln 10)\tau$, where τ is given by Eq. (4), becomes

$$T_{60} = \frac{(24 \ln 10)V}{c \sum_i \alpha_i A_i} = \frac{55.3V}{cA_s} \qquad (6\text{-}1.12)$$

The first version, which has been referred to as the *Sabine-Franklin reverberation time*,‡ reduces to Eq. (5) when $c = 342$ m/s (corresponding to a temperature of 18.3°C or 65°F).

Absorbing Power of Objects and Persons

To account for objects or people in a room, one adds the appropriate increment ΔA_s for each object to the absorbing power A_s. The following examples show how ΔA_s can be determined.

Example 1 A room of volume V has reverberation times of $T_{60,\text{I}}$ or $T_{60,\text{II}}$ when a person is not or is present in the room. The total A_s for each case is determined from the second version of Eq. (12), and the increment ΔA_s due to the person's presence is the difference, i.e.,

$$\Delta A_s = \frac{(24 \ln 10)V}{c}\left(\frac{1}{T_{60,\text{II}}} - \frac{1}{T_{60,\text{I}}}\right) \qquad (6\text{-}1.13)$$

† That the absorption coefficient defined by Eq. (10) is not necessarily the same as what is required to yield the reverberation time via Eq. (5) is discussed at some length by T. F. W. Embleton, "Sound in Large Rooms," in L. L. Beranek (ed.), *Noise and Vibration Control*, McGraw-Hill, New York, 1971, pp. 219–244.

‡ W. B. Joyce, "Sabine's Reverberation Time and Ergodic Auditoriums," *J. Acoust. Soc. Am.*, **58**:643–655 (1975).

Example 2 An area ΔA of the room in Example 1 nominally having absorption coefficient α_0 is covered by an oil painting, and the reverberation time decreases to $T_{60,\mathrm{III}}$. To determine the ΔA_s associated with the painting, one follows the analysis of Example 1 but recognizes that the painting replaces a wall portion having absorbing power $\alpha_0\,\Delta A$. The difference of the two A_s's is the ΔA_s intrinsically due to the painting minus $\alpha_0\,\Delta A$. Consequently, the painting's ΔA_s is

$$\Delta A_s = \alpha_0\,\Delta A + \frac{(24 \ln 10)V}{c}\left(\frac{1}{T_{60,\mathrm{III}}} - \frac{1}{T_{60,\mathrm{I}}}\right) \tag{6-1.14}$$

In such a manner, Sabine determined that the absorbing-power increment associated with an isolated man is of the order of 0.48 metric sabin at 512 Hz. (For a woman dressed in the style of 1900, it was 0.54 metric sabin.) For oil paintings with an area of the order of 1 m^2, he found the average absorption coefficient $\Delta A_s/\Delta A$ (where ΔA included the frames) to be 0.28.

A chief premise in typical applications is that the absorbing-power increment associated with an object is intrinsic to that object. It should be the same for every room, regardless of position and orientation of the object, regardless of the position of the source and of other objects, and regardless of the room's construction. However, even if the diffuse-field idealization is appropriate in the bulk of the room, the premise is poor if two such objects are close together or if a number obtained when the object was suspended in the center of the room is to be used when the object is resting on the floor.

Such exceptions are generally recognizable as such. For example, if one wishes to estimate the incremental absorbing power of an audience in an auditorium,† one refers to data not for isolated persons but for other audiences seated on the same type of chairs with the same seating density (see Table 6-2). The premise would be that the average increment per person is the same for both audiences.

6-2 SOME MODIFICATIONS

The Sabine-Franklin-Jaeger model introduced in the preceding section rests on restrictive assumptions and holds at best only in an averaged sense. Most of the simpler suggestions how the model might be modified to increase its domain of application use the concept of a *mean free path* in a room.

Mean Free Path

The calculation leading to Eq. (6-1.9) indicates that the average rate at which acoustic energy is incident on the walls of the room per unit surface area is

† L. L. Beranek, "Audience and Seat Absorption in Large Halls," *J. Acoust. Soc. Am.*, **32**:661–670 (1960); *Music, Acoustics, and Architecture*, Wiley, New York, 1962, pp. 541–554.

Table 6-2 Absorbing-power increments due to persons and seats

Description	Absorbing-power increment, metric sabins					
	125 Hz	250 Hz	500 Hz	1000 Hz	2000 Hz	4000 Hz
Man standing, in heavy coat	0.17	0.41	0.91	1.30	1.43	1.47
Without coat	0.12	0.24	0.59	0.98	1.13	1.12
Musician, sitting, with instrument	0.60	0.95	1.06	1.08	1.08	1.08
Student, seated, including seat, high school	0.20	0.28	0.31	0.37	0.41	0.42
Elementary school	0.17	0.21	0.26	0.30	0.33	0.37
Person seated in church pew	0.23	0.25	0.31	0.35	0.37	0.35
Per m² of floor area, without audience, moderately upholstered chairs, 0.90 × 0.55 m	0.44	0.56	0.67	0.74	0.83	0.87
Cloth-covered seats with perforated bottoms	0.49	0.66	0.80	0.88	0.82	0.70
With audience, wooden chairs, 2/m²	0.24	0.40	0.78	0.98	0.96	0.87
1/m²	0.16	0.24	0.56	0.69	0.81	0.78
Moderately upholstered chairs	0.55	0.86	0.83	0.87	0.90	0.87

Source: H. Kuttruff, *Room Acoustics,* Applied Science, London, 1973, pp. 156–157; L. L. Beranek, *Acoustics,* McGraw-Hill, New York, 1954, pp. 300–301.

$(c/4)\bar{w}$, so $(c/4)\bar{w}S$ is the rate at which energy is incident on all walls, S being the total wall surface area. The ratio $cS/4V$ of this to the total energy $\bar{w}V$ in the room can be interpreted as an average rate (with a weighting described below) at which a "ray" of sound bouncing about the room undergoes reflections.

A simple derivation† supporting the above interpretation is as follows. Suppose the energy E in the room is divided into energies $E_1, E_2, E_3, \ldots$, each being associated with a distinct ray (see Fig. 6-4). If one ignores absorption, the energy associated with each ray stays constant. If the number of reflections ray r undergoes in time Δt is ΔN_r, the average energy-weighted number of reflections per ray in time Δt is

$$\langle \Delta N \rangle = \frac{\Sigma E_r \, \Delta N_r}{\Sigma E_r} \tag{6-2.1}$$

The numerator, however, is the total ray energy striking the walls in time Δt, or, from the discussion above, $(c/4)\bar{w}S \, \Delta t$, and the denominator is the total energy

† P. E. Sabine, *Acoustics and Architecture,* McGraw-Hill, New York, 1932, pp. 309–311. An earlier but dissimilar derivation leading to the same result was given by Jaeger, "Toward a Theory of Reverberation."

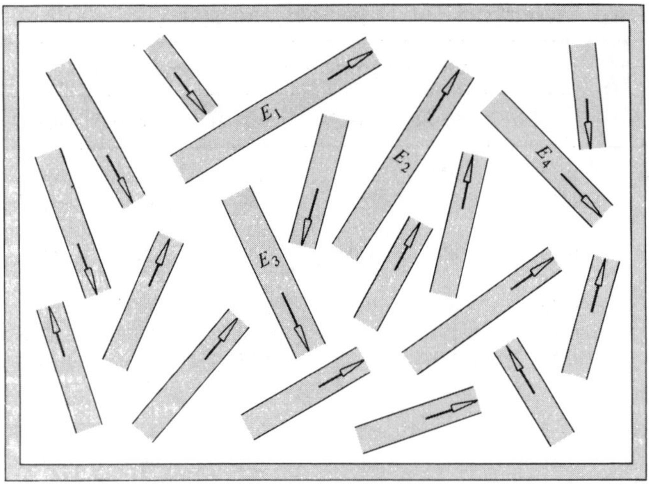

Figure 6-4 Partitioning of a room's acoustic energy into many rays, each of fixed energy; this idealization leads to $4V/S$ for the characteristic path length.

$\bar{w}V$ in the room; the right side is therefore $(cS/4V)\,\Delta t$. The relation $\langle dN/dt\rangle = cS/4V$ therefore results.

The distance a "ray" moving with the sound speed c travels in time $1/\langle dN/dt\rangle$ is

$$l_c = \frac{4V}{S} \tag{6-2.2}$$

and represents a *characteristic path length* for sound in a room. For a cubical room of length a on each side, one has $V = a^3$, $S = 6a^2$, so $l_c = \frac{2}{3}a$. For a spherical room, l_c is $\frac{4}{3}$ times the radius. For a rectangular room, l_c is between $\frac{2}{3}$ and 2 times the room's smallest dimension.

Various definitions† of a *mean free path* appear in the early literature on architectural acoustics, but the ones most meaningful within the context of the Sabine-Franklin-Jaeger model are those leading to the l_c above. The quantity l_c is not the average distance between reflections for any given ray, nor is it the average over rays of such an average distance; instead it is c times the reciprocal of an average collision frequency per ray of rays with walls. Consequently, l_c is the *reciprocal of the mean free reciprocal path length,* but to keep our terminology brief we refer to it as a mean free path or characteristic path length.

† A geometrical definition (not explicitly involving energy) leading also to $4V/S$ has been given by C. W. Kosten, "The Mean Free Path in Room Acoustics," *Acustica*, **10**:245–250 (1960). Various proposed definitions are reviewed by F. V. Hunt, "Remarks on the Mean Free Path Problem," *J. Acoust. Soc. Am.*, **36**:556–564 (1964).

Limitations of Sabine's Equation

A possible weak point in the derivation of the Sabine-Franklin reverberation time is the assumption that the energy-dissipation rate at time t depends on the simultaneous value of the energy density in the room. What is more nearly true is that it depends on the current values *near* each wall of the energy-density portion propagating toward the wall. But if the energy in the room is changing rapidly with time, the approximation of this local quantity by an average over room volume becomes suspect. One can argue, as in the previous section, that a time of the order of $3l_c/c$ or greater is required for the spatial distribution of energy to equilibrate whenever some change in the source output is made. Consequently, the model's predictions for reverberant decay may be invalid if the characteristic decay time τ is comparable to or less than $3l_c/c$ or, equivalently, if the average (surface-area-weighted) absorption coefficient is of the order of $\frac{1}{3}$ or greater.

Equations (6-1.4) often give a *higher* average energy-versus-time curve during reverberant decay than is measured and thus predict a longer time for $\bar{w}$ to decay by some fixed fraction. The energy incident on the walls is representative of the average energy density in the center of the room at a time of the order of $\frac{1}{2}l_c/c$ or more earlier. This average energy density at the earlier time is higher (during reverberant decay), so the energy incident on the walls is higher than was assumed in the derivation; the rate of energy dissipation is therefore also higher, and the energy in the room decreases faster than predicted by the Sabine-Franklin-Jaeger model.

Norris-Eyring Reverberation Time

A simple assumption† overcoming the limitations just described (but raising other objections) is that the energy incident per unit time on the walls decreases stepwise (see Fig. 6-5a) after the source has been turned off. For the first‡ l_c/c s, the directional energy density at the walls for propagation directions pointing into the walls is taken as $\bar{w}_{\text{init}}/4\pi$ and thus corresponds to energy not having suffered wall reflections since $t = 0$. During the next l_c/c s, all arriving energy is assumed to have suffered one and only one wall reflection, so the average energy density associated with it has decreased by a factor of $1 - \bar{\alpha}$, where $\bar{\alpha}$ is the area-averaged absorption coefficient. Thus, $D = (1 - \bar{\alpha})\bar{w}_{\text{init}}/4\pi$ for the second interval. Similarly, D is $(1 - \bar{\alpha})^2\bar{w}_{\text{init}}/4\pi$ for the next l_c/c s, etc.

The net energy absorbed in the first interval is $\bar{\alpha}V\bar{w}_{\text{init}}$; the energy remaining

† C. F. Eyring, "Reverberation Time in 'Dead' Rooms," *J. Acoust. Soc. Am.*, **1**:217–241 (1930). The first conception of Eq. (5) is attributed to R. F. Norris by C. A. Andree, ibid., **3**:549–550 (1932). Norris' version of the derivation is given as appendix II in V. O. Knudsen's *Architectural Acoustics*, Wiley, New York, 1932, pp. 603–605.

‡ The variant on the derivation of taking the first interval as $\frac{1}{2}l_c/c$, the rest as l_c/c, yields the same reverberation time.

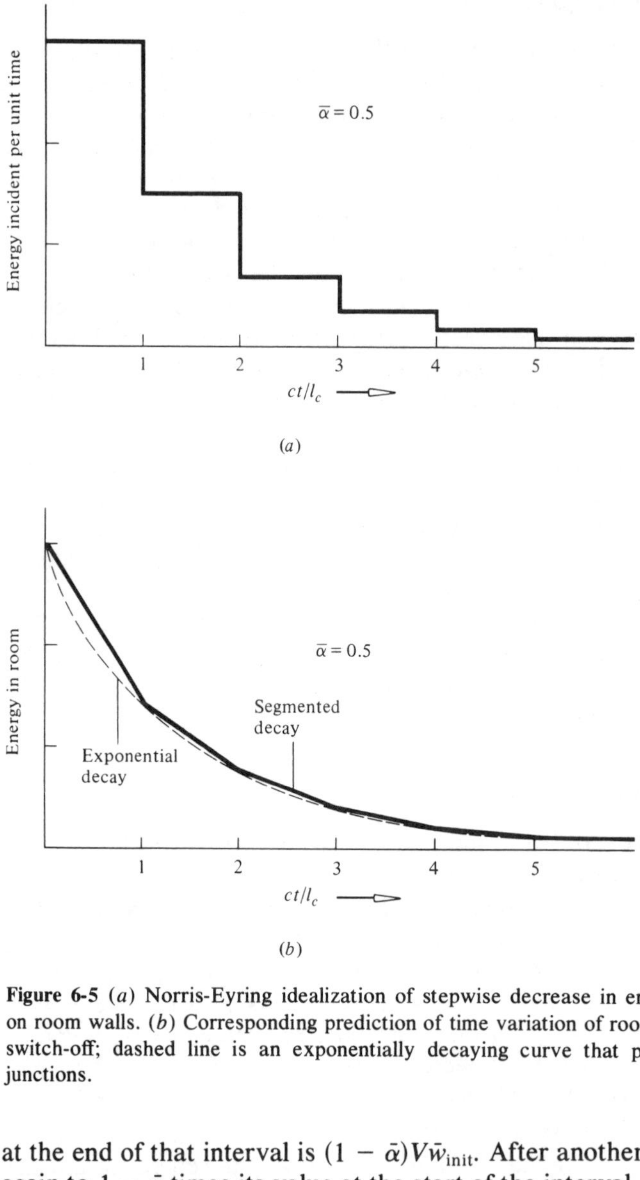

Figure 6-5 (*a*) Norris-Eyring idealization of stepwise decrease in energy incident per unit time on room walls. (*b*) Corresponding prediction of time variation of room's energy following source switch-off; dashed line is an exponentially decaying curve that passes through the segment junctions.

at the end of that interval is $(1 - \bar{\alpha})V\bar{w}_{\text{init}}$. After another interval, it is reduced again to $1 - \bar{\alpha}$ times its value at the start of the interval. Consequently, the net volume-averaged energy density remaining at time $t_N = Nl_c/c$ is

$$\bar{w}(t_N) = \bar{w}_{\text{init}}(1 - \bar{\alpha})^N \tag{6-2.3}$$

The stepwise variation in $\bar{\mathcal{P}}_d$ implies that $\bar{w}(t)$ decreases linearly with time between integer values of ct/l_c (Fig. 6-5b), the slope changing discontinuously at times nl_c/c. A good approximation to the overall decay curve results if one uses (3) even when N is not an integer, i.e.,

$$\bar{w}(t) = \bar{w}_{\text{init}}(1 - \bar{\alpha})^{ct/l_c} = \bar{w}_{\text{init}} e^{-t/\tau_{\text{NE}}} \tag{6-2.4}$$

$$\tau_{\text{NE}} = \frac{4V}{cS[-\ln(1 - \bar{\alpha})]} \tag{6-2.5}$$

The corresponding Norris-Eyring reverberation time T_{60} is $13.82\tau_{\text{NE}}$.

The Norris-Eyring reverberation time is the same as the Sabine-Franklin T_{60} except that $\bar{\alpha}$ has been replaced by $-\ln(1 - \bar{\alpha})$. The latter is approximately $\bar{\alpha} + \bar{\alpha}^2/2$ and differs from $\bar{\alpha}$ by less than 10 percent if $\bar{\alpha} < 0.2$. However, for $\bar{\alpha} = 0.3, 0.4, 0.5$, one has $-\ln(1 - \bar{\alpha})$ equal to 0.36, 0.51, 0.67, so the distinction becomes appreciable when $\bar{\alpha}$ is of the order of $\frac{1}{3}$ or greater. Since the Norris-Eyring T_{60} is less than the Sabine-Franklin T_{60}, it implies a more rapid decay of sound.

Rooms with Asymmetric Absorption†

The assumption that the energy incident per unit area and time is the same at any given time for all wall surfaces may be questioned if one surface (area S_1) has an absorption coefficient α_1 substantially different from the value α_0 for the remaining surfaces (area $S - S_1$).

If one idealizes the energy incident (per unit area and time) on any surface as decreasing stepwise in time (as in the derivation of the Norris-Eyring equation), the net energy absorbed during the second time interval is (see Fig. 6-6)

$$(-\Delta E)_2 = \Delta t \sum_{i,j} \alpha_j f_{ji}(1 - \alpha_i) S_i c \frac{\bar{w}_{\text{init}}}{4} \tag{6-2.6}$$

Here f_{ji} represents the fraction of the power $(1 - \alpha_i)S_i c\bar{w}_{\text{init}}/4$ reflected by the ith surface during the first time interval that is incident on the jth surface during the second time interval. These fractions are such that

$$\sum_j f_{ji} = 1 \qquad \sum_i f_{ji} S_i = S_j \qquad \text{where } f_{ji} = 0 \text{ if } i = j \tag{6-2.7}$$

The second relation ensures that the energy incident per unit time and area will be the same for all surfaces when α is the same for all surfaces; the third results because the reflected energy does not come directly back to the surface S_i. (Explicit expressions‡ for the f_{ji}, termed *radiation shape factors* in heat-transfer applications, in terms of quadruple integrals result from simple geometrical considerations; analytical formulas, tabulations, and curves exist in the litera-

† T. W. F. Embleton, "Absorption Coefficients of Surfaces Calculated from Decaying Sound Fields," *J. Acoust. Soc. Am.*, **50**:801–811 (1971).

‡ H. C. Hottel, "Radiant Heat Transmission," *Mech. Eng.*, **52**:699–704 (1930); D. C. Hamilton and W. R. Morgan, "Radiant-Interchange Configuration Factor," *Nat. Adv. Comm. Aeronaut. Rep.* NACA TN2836, Washington, 1952; Kreith, *Principles of Heat Transfer*, pp. 243–251.

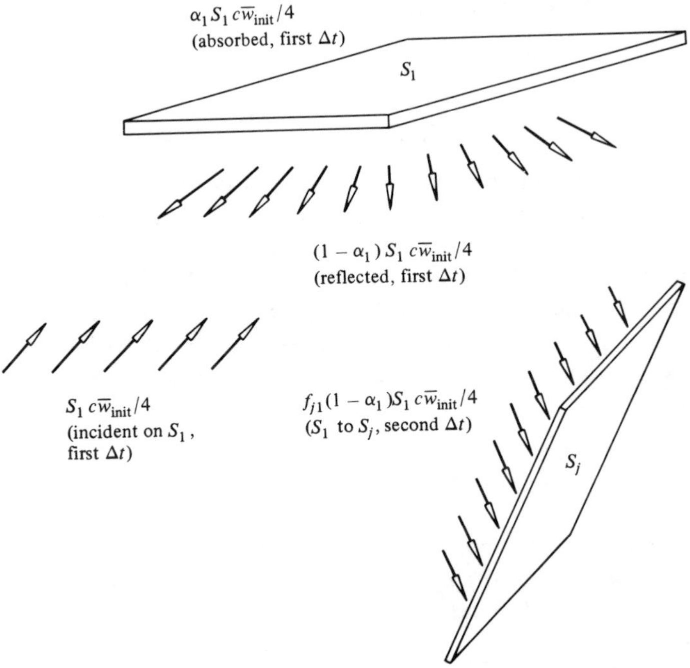

Figure 6-6 Partitioning of the energy reflected from surface S_1 during the first time interval. A fraction f_{j1} impinges on surface S_j during the second interval.

ture. However, the example discussed below, when only one surface has a dissimilar absorption coefficient, leads to results independent of the numerical values of the f_{ji}.)

The double sum in the expression (6) for $(-\Delta E)_2$, when all the α_i except α_1 have the same value α_0, reduces, after some algebra and with the help of Eqs. (7), to

$$(-\Delta E)_2 = [E_{\text{inc}}(2)][\bar{\alpha} + (\Delta\bar{\alpha})_E] \tag{6-2.8}$$

where $\bar{\alpha}$ is the area-averaged absorption coefficient, $E_{\text{inc}}(2)$ is the net energy incident on all surfaces during the second time interval, and

$$(\Delta\bar{\alpha})_E = \frac{(\alpha_1 - \alpha_0)^2}{1 - \bar{\alpha}} \left(\frac{S_1}{S}\right)^2 \tag{6-2.9}$$

Equation (8) allows the apparent absorption coefficient (net energy absorbed divided by net energy incident) during the second time interval to be identified as $\bar{\alpha} + (\Delta\bar{\alpha})_E$. A simple model results if this is assumed to be the fraction of energy absorbed during all later intervals; the rationale is that the asymmetry in the area distribution of incident energy is primarily caused by the most recent reflection; if the absorption coefficients were suddenly changed so that all the α_i became the same, the energy incident per unit area and time would be nearly the same for all wall surfaces after a time interval Δt.

With the assumption just described, the average energy per unit volume remaining in the room at time $t = N\,\Delta t$ for $N \gg 1$ is approximately $[1 - \bar{\alpha} - (\Delta\bar{\alpha})_E]^N$ times $\bar{w}_{\text{init}}$. Consequently, the train of reasoning leading to the Norris-Eyring reverberation time must be modified so that $\bar{\alpha}$ is replaced by $\bar{\alpha} + (\Delta\bar{\alpha})_E$. This modification, with $\Delta t = 4V/cS$, yields

$$T_{60} = \frac{(24 \ln 10)V/cS}{-\ln\,[1 - \bar{\alpha} - (\Delta\bar{\alpha})_E]} \tag{6-2.10}$$

The additional term $-(\Delta\bar{\alpha})_E$ in the argument of the logarithm is the only distinction between this and the Norris-Eyring reverberation time.

Example The floor (surface 1) of a cubical room has absorption coefficient α_1; the vertical walls and the ceiling each have absorption coefficient α_0. The quantity α_0 is known from previous measurements; one measures T_{60} and seeks to determine α_1. Estimate the error resulting from use of the Norris-Eyring model.

SOLUTION Let $\alpha_{1,\text{NE}}$ be the value of α_1 computed from Eq. (5) with $T_{60} = 13.82\tau_{\text{NE}}$ and with $\bar{\alpha} = \frac{1}{6}\alpha_1 + \frac{5}{6}\alpha_0$. Equation (10) would give the same numerical value for the argument of the logarithm as the Norris-Eyring model, so the corrected value of α_1 must be such that

$$\tfrac{1}{6}\alpha_{1,\text{NE}} = \tfrac{1}{6}\alpha_1 + (\Delta\bar{\alpha})_E \tag{6-2.11}$$

$$\frac{\alpha_{1,\text{NE}} - \alpha_1}{\alpha_1} = \frac{(\alpha_1 - \alpha_0)^2}{\alpha_1(6 - \alpha_1 - 5\alpha_0)} \tag{6-2.12}$$

Equation (12) follows from Eq. (11) with $(\Delta\bar{\alpha})_E$ taken from Eq. (9).

The fractional error in α_1 predicted by Eq. (12) vanishes when $\alpha_1 = \alpha_0$; if $\alpha_1 \gg \alpha_0$, it reduces to $\alpha_1/(6 - \alpha_1)$, which is still small if $\alpha_1 < 0.1$. If α_1 were of the order of 1, the predicted error would be close to 20 percent.

The Room Constant†

An extension of the Sabine-Franklin-Jaeger theory to take into account the field near the source begins with the premise that the reverberant field has no effect on direct wave or source power. At moderate distances from the source, the time-averaged radial component of intensity conforms to spherical spreading and is described by $\mathcal{P}_{\text{av}}Q_\theta/4\pi r^2$, where the *directivity factor* Q_θ is a function of direction whose integral over all solid angles pointing from the source *into* the room is 4π. For a spherically symmetric radiator some distance from any surface, Q_θ should be 1; for one resting on the floor, it should be 2. If r is large

† E. Dietze and W. D. Goodale, Jr., "The Computation of the Composite Noise Resulting from Random Variable Sources," *Bell Syst. Tech. J.*, **18**:605–623 (1939); A. London, "Methods for Determining Sound Transmission Loss in the Field," *J. Res. Natl. Bur. Stand.*, **26**:419–453 (1941); Beranek, *Acoustics*, McGraw-Hill, New York, 1954, pp. 313–324; R. W. Young, "Sabine Reverberation Equation and Sound Power Calculations," *J. Acoust. Soc. Am.*, **31**:912–921 (1959).

enough for this direct wave to be considered locally planar, the plane-wave relation $w_{\text{av}} = I_{r,\text{av}}/c$ applies, so the energy density associated with the direct wave is $\mathcal{P}_{\text{av}}Q_\theta/4\pi r^2 c$. The product of this with ρc^2 yields the corresponding mean squared pressure.

The time averages of the energy densities (or of the mean squared pressures) of the direct and reverberant fields are presumed to be additive. This is not exactly true, especially if the frequency band of interest is narrow or if the source is emitting a pure tone, but it may be regarded as approximately so if one thinks in terms of local spatial averages, for the reasons cited in the derivation of Eq. (6-1.7). The reverberant field consists of all energy reflected one or more times from the room's walls; it is assumed to be diffuse and to be such that local spatial averages are independent of position, and thus it is characterized by a uniform-reverberant-field energy density $\bar{w}_R$ (a spatial average). The energy density at any point in the room is then $\bar{w}_R$ plus the corresponding expression, $\mathcal{P}_{\text{av}}Q_\theta/4\pi r^2 c$, for the direct wave.

The power feeding and maintaining the reverberant field is the source power minus the energy lost per unit time on the first reflection. If we assume, in the absence of any evidence to the contrary, e.g., a highly directional source aimed at an open window, that the fraction of power lost on one reflection is the average wall-absorption coefficient $\bar{\alpha}$, then $(1 - \bar{\alpha})\bar{\mathcal{P}}$ is the rate at which energy is being added to the reverberant field. One may argue, as in Sec. 6-1, that the rate at which this reverberant energy is being dissipated is proportional to $\bar{w}_R$, the proportionality constant being $\bar{\alpha}Sc/4$. In the steady state, $d\bar{w}_R/dt = 0$; since the energy added per unit time equals the rate of dissipation, one obtains

$$\bar{w}_R = \frac{4\bar{\mathcal{P}}}{cR_{\text{rc}}} \qquad R_{\text{rc}} = \frac{\bar{\alpha}S}{1 - \bar{\alpha}} \tag{6-2.13}$$

Here the *room constant* R_{rc} (units of area) represents the room's absorbing power divided by $1 - \bar{\alpha}$.

With the local spatial average $\overline{p^2}$ of the mean squared pressure taken as the sum of the direct-field and reverberant-field contributions, each term being ρc^2 times the corresponding energy density, one finds, from the previously given expressions for the two energy densities, that

$$\overline{p^2} = \rho c\, \bar{\mathcal{P}} \left(\frac{Q_\theta}{4\pi r^2} + \frac{4}{R_{\text{rc}}} \right) \tag{6-2.14}$$

This formula gives an indication of how far from, or close to, the source one must be to be assured that the reverberant (or direct) field predominates. At the *radius of reverberation*, or *critical radius*,

$$r_0 = \left(\frac{R_{\text{rc}}Q_\theta}{16\pi} \right)^{1/2} \tag{6-2.15}$$

the two terms are of equal contribution, and the sound-pressure level is 3 dB higher than expected from either alone. At $2r_0$ the direct-field contribution is only one-fourth that of the near field, and the level is only $10 \log (1 + \frac{1}{4}) \approx 1$ dB

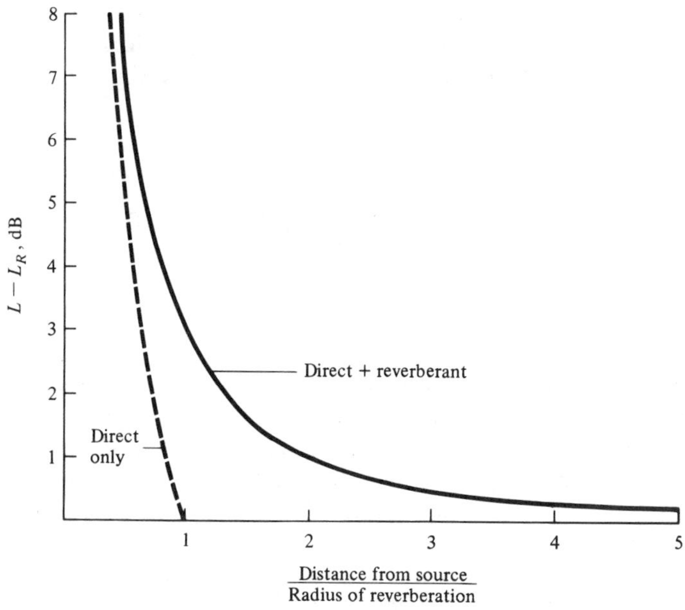

Figure 6-7 Sound-pressure level (relative to that of reverberant field) versus ratio of distance r from source to radius of reverberation r_0. Function plotted is $10 \log [(r_0/r)^2 + 1]$; dashed line, corresponding to direct field alone, is $10 \log [(r_0/r)^2]$.

higher than that of the reverberant field alone; at $3r_0$ the discrepancy is 0.5 dB; at $4r_0$ it is 0.3 dB; at $5r_0$ it is 0.2 dB. At $r_0/2$, $r_0/4$, and $r_0/8$, the levels are 7, 12, and 18 dB higher than that of the reverberant field alone and 1, 0.3, and 0.1 dB higher than that of the direct field alone (see Fig. 6-7). To determine the direct field of a source in a reverberant room to within 1 dB, one should pick a point at which the sound-pressure level is at least 7 dB greater than that typically measured at a distant point in the room or sufficiently close to the source for the sound-pressure level to increase by at least 5 dB when the distance from the source is halved. Alternatively, one can estimate r_0 in advance by taking $Q_\theta = 1$ (suspended source) or $Q_\theta = 2$ (source on floor) and by calculating the room constant from a reverberation-time measurement, using Eq. (6-1.12) and $R_{rc} = A_s/(1 - A_s/S)$.

If the room constant R_{rc} is to be derived from a reverberation-time measurement via the Sabine-Franklin equation, however, it is inconsistent to retain the factor $1 - \bar{\alpha}$ in the denominator in the definition (13) of R_{rc}. The model implicitly assumes $\bar{\alpha} \ll 1$, and since the factor $(1 - \bar{\alpha})^{-1}$ gives a correction of second order in $\bar{\alpha}$, that is, $\bar{\alpha}/(1 - \bar{\alpha}) \approx \bar{\alpha} + \bar{\alpha}^2$, one should disregard it unless the reverberation-time formula is itself accurate to second order. If $\bar{\alpha}_S$ is the value derived from the Sabine-Franklin formula, and if $\bar{\alpha}_S$ is greater than the actual $\bar{\alpha}$ for the room by some amount $\Delta\bar{\alpha}$, then $S\bar{\alpha}_S$ would be a valid second-order approximation to the room constant if $\Delta\bar{\alpha}/\bar{\alpha} = \bar{\alpha}$. According to the Norris-Eyring formula, $\Delta\bar{\alpha} \approx \frac{1}{2}\bar{\alpha}^2$, so $\Delta\bar{\alpha}/\bar{\alpha} \approx \frac{1}{2}\bar{\alpha}$. Furthermore, the Norris-

Eyring equation often tends to overestimate $\bar{\alpha}$, partly for the reasons cited in the derivation of Eq. (10), so $\Delta\bar{\alpha}/\bar{\alpha}$ is typically somewhat larger than $\frac{1}{2}\bar{\alpha}$. For such reasons and in the absence of any better model of comparable simplicity, it is usual practice to take $R_{\text{rc}} = S\bar{\alpha}_S$.

6-3 APPLICATIONS OF THE SABINE-FRANKLIN-JAEGER THEORY

Design and Correction of Rooms

Criteria for what constitutes good acoustics for rooms intended for specified purposes have been extensively developed since the time of Sabine and are discussed in various books and articles.† An extensive discussion of them is beyond the scope of the present text, but it should be noted that the reverberation time T_{60} plays a central role in the quantitative formulation of some of the simpler criteria (see Fig. 6-8).

An indication of why the reverberation time should be significant results from the transient solution of (6-1.3). That equation, with $4V/c\tau$ replacing A_s, can be rewritten as an ordinary differential equation for $\bar{w}e^{t/\tau}$ and subsequently integrates to

$$\bar{w}(t) = V^{-1} \int_{-\infty}^{t} e^{-(t-t')/\tau} \bar{\mathcal{P}}(t')\, dt' \qquad (6\text{-}3.1)$$

If $\bar{\mathcal{P}}$ has been constant for an indefinite time, one has the steady-state case and (1) reduces to

$$\bar{w}_{\text{tot}} = \frac{\bar{\mathcal{P}}\tau}{V} \qquad (6\text{-}3.2)$$

which can alternately be obtained from (6-1.3) by setting $d\bar{w}/dt = 0$ at the outset. (The subscript tot here implies that this is the energy density resulting from the total history of the source.)

The portion of this steady-state energy density generated by the source in the most recent interval of duration Δt results from a replacement of the lower integration limit in Eq. (1) by $t - \Delta t$, such that

$$\bar{w}_{\text{last}\,\Delta t} = \bar{w}_{\text{tot}}(1 - e^{-\Delta t/\tau}) \qquad (6\text{-}3.3)$$

† See, for example, Beranek, *Music, Acoustics, and Architecture;* W. Furrer, *Room and Building Acoustics and Noise Abatement,* Butterworths, Washington, 1964; A. Lawrence, *Architectural Acoustics,* Elsevier, Amsterdam, 1970; Knudsen, *Architectural Acoustics;* A. F. B. Nickson and R. W. Muncey, "Criteria for Room Acoustics," *J. Sound Vib.,* **1**:292–297 (1964); P. H. Parkin, W. E. Scholes, and A. C. Derbyshire, "The Reverberation Times of Ten British Concert Halls," *Acustica,* **2**:97–100 (1952); H. Bagenal and A. Wood, *Planning for Good Acoustics,* Methuen, London, 1931.

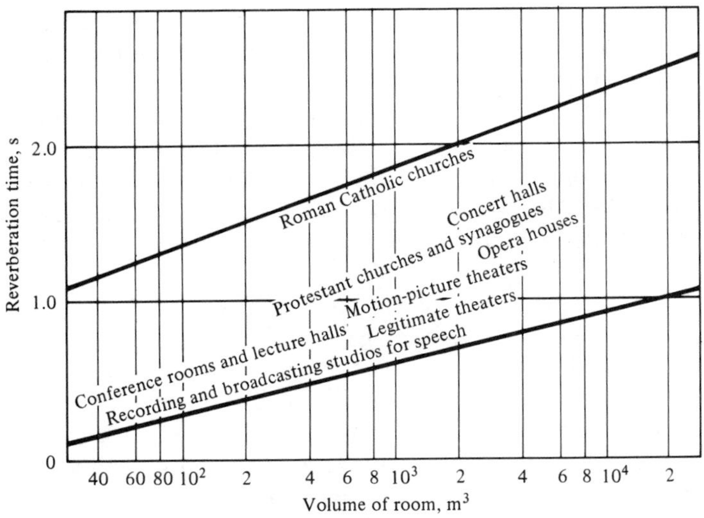

Figure 6-8 Optimum midfrequency (500 to 1000 Hz) reverberation times for fully occupied rooms versus volume. *(From L. L. Doelle, Environmental Acoustics, McGraw-Hill, New York, 1972, p. 56.)*

If the sound from the source is transmitting information, e.g., speech or music, "early" echoes reinforce the information and "late" echoes interfere. Consequently, one can conceive† of a value of Δt that splits the sound currently received into "useful" sound and interfering sound. The ratio of the energy densities associated with these two categories is identified from (3) as

$$\frac{\bar{w}_{\text{useful}}}{\bar{w}_{\text{interfering}}} = e^{\Delta t/\tau} - 1 \tag{6-3.4}$$

Since $\tau = T_{60}/(6 \ln 10)$, this indicates that, for specified Δt, the ratio of the useful to interfering energy is determined by the reverberation time; the larger the T_{60} the lower the ratio.

The auditory sensation adheres to no semblance of simple mathematical rules, but it is sometimes helpful‡ to view it as a system that responds to a running time average of some function (not necessarily the square) of the acous-

† This originated with C. Zwikker, "Partitioning of Loudspeaker Intensities," *Ingenieur* (*The Hague*), **44:**39–45 (1929), and has subsequently been applied by a number of investigators, e.g., R. Thiele, "Directional Distribution and Chronological Order of Sound Echoes in Rooms," *Acustica*, **3:**291–302 (1953); F. Santon, "Numerical Prediction of Echograms and the Intelligibility of Speech in Rooms," *J. Acoust. Soc. Am.*, **59:**1399–1405 (1976).

‡ W. A. Munson, "The Growth of Auditory Sensation," *J. Acoust. Soc. Am.*, **19:**584–591 (1947); J. J. Zwislocki, "Temporal Summation of Loudness: An Analysis," ibid., **46:**431–441 (1969); M. J. Penner, "A Power Law Transformation Resulting in a Class of Short-Term Integrators That Produce Time-Intensity Trades for Noise Bursts," ibid., **63:**195–201 (1978).

tic pressure outside the ear. For processing ordinary speech, existing data† suggest an integration time of the order of 50 ms. This integration time represents a plausible choice for the Δt in Eq. (4).

If the useful energy density masks the interfering energy density whenever the former is greater than or equal to, say, 5 times the latter, little additional improvement in the perception of information results when τ decreases below the value $\Delta t/(\ln 6)$ resulting when the right side of Eq. (4) is set equal to 5. This transitional value of τ, with $\Delta t = 50$ ms, corresponds to a reverberation time $T_{60} \approx 0.4$ s.

On the other hand, increasing T_{60}, given fixed room volume V and fixed source power output $\bar{\mathcal{P}}$, increases the average energy density in the room. Because the auditory system tends to perceive the information associated with louder sound better, the perception may increase somewhat if the reverberation time is increased beyond the lower value described above. If the reverberation time becomes too long, the information becomes garbled and perception decreases, even though the sound continues to become louder. Thus, for given V and $\bar{\mathcal{P}}$, there is an *optimum reverberation time*‡ for the room, which, according to the reasoning just described, should increase with increasing room volume. For small rooms, in situations where maximum perception of information is desired, e.g., speech, the optimum reverberation time is substantially less than 1 s.

For music, it is desirable that the information be partially smeared out to smooth over attack transients intrinsically associated with common types of musical instruments. Substantially less smearing is desired for chamber music than for orchestral music. The optimum reverberation time in any event should be higher for a given room volume for music reception than for speech reception and experiments have been performed to determine what this optimum should be.

There are other design considerations§ in architectural acoustics, but within the context of the Sabine-Franklin-Jaeger model (which assumes the sound to be perfectly diffuse and uniformly distributed) the only parameter to be considered for a room of fixed volume is the reverberation time. If the reverberation time differs from optimum, one seeks to change the absorbing power A_s by altering the wall covering; rooms are designed to achieve the optimum reverberation time.

Another category of application in this context is *noise reduction*. Factory rooms are typically constructed so that they have high reverberation times; a

† H. Haas, "On the Influence of a Simple Echo on the Comprehension of Speech," *Acustica*, **1**:49–58 (1951). The value of 50 ms is what was chosen (with reference to speech) as the break point in the partitioning of acoustic energy density into a useful and a disturbing part in Thiele, "Directional Distribution. . . ."

‡ S. Lifshitz, "Mean Intensity of Sound in an Auditorium and Optimum Reverberation," *Phys. Rev.*, **27**:618–621 (1926); W. A. MacNair, "Optimum Reverberation Time for Auditoriums," *J. Acoust. Soc. Am.*, **1**:242–248 (1930); J. P. Maxfield, "The Time Integral Basic to Optimum Reverberation Time," ibid., **20**:483–486 (1948).

§ See, for example, E. Meyer and H. Kuttruff, "Progress in Architectural Acoustics," in E. G. Richardson and E. Meyer (eds.), *Technical Aspects of Sound*, vol. 3, Elsevier, Amsterdam, 1962, pp. 221–337.

noise source in such a room produces sound levels at distant points substantially higher than would be received in an open space. The mean squared pressure at distances somewhat larger than the radius of reverberation, according to Eqs. (6-1.7) and (2), conforms on the average to the relation

$$\overline{p^2} = \frac{\rho c^2 \tau \bar{\mathcal{P}}}{V} = \frac{\rho c^2 T_{60} \bar{\mathcal{P}}}{(6 \ln 10) V} \tag{6-3.5}$$

so decreasing the reverberation time by a factor of K decreases the sound-pressure level by 10 log K decibels. If $\bar{\alpha}$ is much less than 1, an appreciable reduction is feasible. The decrease of T_{60} will have little effect on the noise in the immediate vicinity of the source, but if no one spends a considerable fraction of time at such points, this need not be taken into consideration. Otherwise, one would seek to reduce $\bar{\mathcal{P}}$ by altering or enclosing the source.

Measurement of Absorption Coefficients and Reverberation Times

The use of reverberation-time measurements to deduce absorption coefficients of wall coverings [see Eq. (6-1.14)] is a standard application of the Sabine-Franklin-Jaeger model. Typically, such measurements are carried out in reverberation chambers especially constructed for the purpose (see Fig. 6-9), and

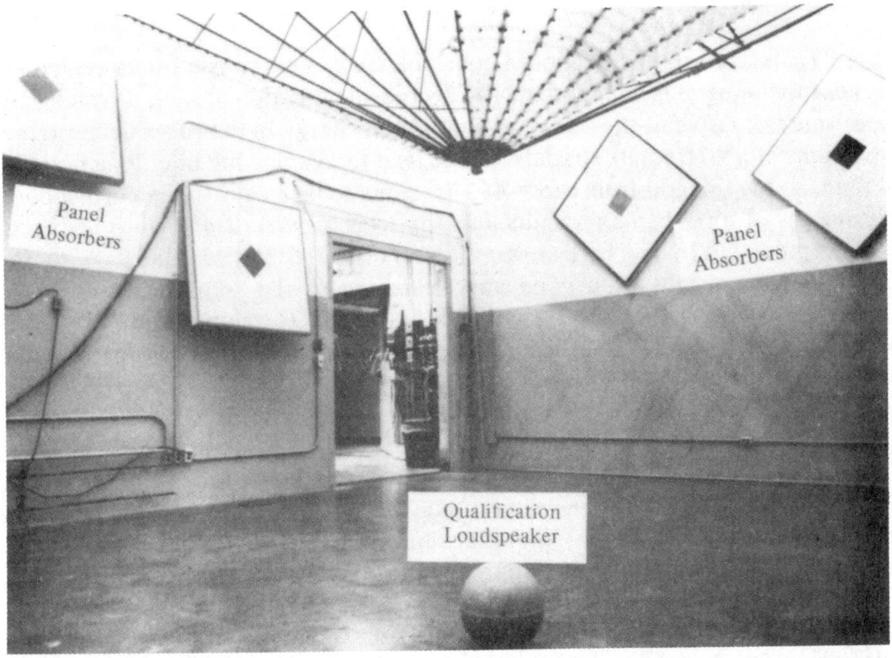

Figure 6-9 Reverberation room at Carrier Corporation, Syracuse, N.Y. The indicated qualification loudspeaker is for assessing conformance with standard criteria for reverberation rooms. Overhead is the rotating diffuser. [*J. T. Rainey, C. E. Ebbing, and R. A. Ryan, Noise Control Eng.*, 7:82 *(1976).*]

efforts are made to ensure that the assumptions inherent in the model are satisfied. To determine the reverberation time, one ideally wants a decay curve giving the average acoustic energy density or the volume average of p^2 versus time following source switch-off. This volume average can be approximated by the average (over microphones) of the running time averages of the squared acoustic pressure taken from several microphones judiciously spaced throughout the room or by the long-time average resulting when a microphone traverses a long path within the room. The latter technique is applicable if a steady-state source of known power output $\bar{\mathcal{P}}$ is used, the reverberation time being subsequently derived from Eq. (2).

How best to estimate the reverberation time, given one and only one source location and one and only one receiver location, is of practical interest for field applications; what is often done is to fire a pistol and to record A-weighted or octave-band sound-pressure levels versus time. The pistol shot injects acoustic energy E_{init} into the room, and, for times somewhat larger than $3l_c/c$, this can be presumed to fill the room uniformly. The instrumentation used to obtain the sound-pressure level versus time is invariably such that the resulting level corresponds to a short-term (characteristic averaging time of the order of 0.1 s) running time average of p^2. Typically, the decay curve is somewhat erratic, but a smoother curve results if one plots instead†

$$10 \log \left[\frac{1}{T_{\text{ref}}} \int_t^\infty \frac{p^2(t')}{p_{\text{ref}}^2}\, dt'\right] = 10 \log \left(\frac{1}{T_{\text{ref}}} \int_t^\infty 10^{L(t')/10}\, dt'\right) \qquad (6\text{-}3.6)$$

where T_{ref} is any arbitrarily chosen constant. Such a curve is a priori smoother because the integral is a monotonically decreasing function of time. It should be more representative of the decay of total sound energy in the room because the deviations of $p^2(t)$ from its spatial average tend to average out over long periods of time, so the integral from t to ∞ of $p^2(t)$ tends to be closer to the corresponding integral of $\overline{p^2(t)}$ than a typical value of $p^2(t)$ is to $\overline{p^2(t)}$. If $\overline{p^2(t)}$ does decay as $e^{-t/\tau}$, as predicted by Eq. (6-1.4), the integral of $\overline{p^2(t)}$ from t to ∞ is $\tau\overline{p^2(t)}$, so the integral above would be a good approximation to the sound-pressure level corresponding to $\overline{p^2(t)}$, plus a constant, $10 \log (\tau/T_{\text{ref}})$. The slope (negative) of the curve described by Eq. (6) therefore gives the decay rate in decibels per second and is equal to $60/T_{60}$.

Measurement of Source Power

The acoustic power $\bar{\mathcal{P}}$ of the source can be evaluated from Eq. (5), given measurements of T_{60} and $\overline{p^2}$. The latter, and therefore also $\bar{\mathcal{P}}$, depends on the location and orientation of the source, but one ideally‡ wants the free-field

† M. R. Schroeder, "New Method of Measuring Reverberation Time," *J. Acoust. Soc. Am.*, **37**:409–412 (1965); W. T. Chu, "Comparison of Reverberation Measurements Using Schroeder's Impulse Method and Decay-Curve Averaging Method," ibid., **63**:1444–1450 (1978).

‡ T. J. Schultz, "Sound Power Measurements in a Reverberant Room," *J. Sound Vib.*, **16**:119–129 (1971).

power output $\bar{\mathcal{P}}_{ff}$ that would result if the source were suspended in an open space or (a different $\bar{\mathcal{P}}_{ff}$) if the source were resting on a rigid infinite plane.

Some insight into whether $\bar{\mathcal{P}}$ is a good approximation to $\bar{\mathcal{P}}_{ff}$ results if one considers the source to be a vibrating solid whose surface motion is unaffected by the external pressure. The acoustic pressure on the surface of the solid can be taken as $p_{dir} + p_{rvrt}$ (dir for direct, rvrt for reverberant). Then the deviation $\Delta\bar{\mathcal{P}}$ of the acoustic power from $\bar{\mathcal{P}}_{ff}$ is the integral of $(p_{rvrt}v_n)_{av}$ over the surface area S_0 of the source.

To estimate the magnitude of $\Delta\bar{\mathcal{P}}$, we take the rms value of p_{rvrt}, from Eq. (5), to be $(\rho c^2\bar{\mathcal{P}}\tau/V)^{1/2}$. The source is taken to be a radially oscillating sphere of radius a, where $ka \ll 1$, so the rms value of v_n, from Eq. (4-1.5), equals $(4\pi\bar{\mathcal{P}}_{ff}/\rho c)^{1/2}(kS_0)^{-1}$. All phase differences between p_{rvrt} and v_n are considered equally likely, so the expected value of $(\Delta\bar{\mathcal{P}})^2$ is $\frac{1}{2}$ of what results of p_{rvrt} and v_n are in phase. Thus, the rms value of $\Delta\bar{\mathcal{P}}$ is

$$(\Delta\bar{\mathcal{P}})_{rms} = \frac{1}{\sqrt{2}}\left(\frac{\rho c^2\bar{\mathcal{P}}\tau}{V}\right)^{1/2}\left(\frac{4\pi\bar{\mathcal{P}}_{ff}}{\rho c}\right)^{1/2}\frac{S_0}{kS_0} = (\bar{\mathcal{P}}\bar{\mathcal{P}}_{ff})^{1/2}\left(\frac{2\pi c\tau}{k^2V}\right)^{1/2} \tag{6-3.7}$$

The criterion ıɔr $|\Delta\bar{\mathcal{P}}| \ll \bar{\mathcal{P}}_{ff}$ is therefore that $2\pi c\tau/k^2V \ll 1$ or, since $A_s = 4V/c\tau$, that

$$k^2A_s \gg 8\pi \tag{6-3.8}$$

Consequently, a measured $\bar{\mathcal{P}}$ will be close to $\bar{\mathcal{P}}_{ff}$ if the frequency generated is substantially larger than $c/(A_s)^{1/2}$.

The foregoing analysis presumes that the Sabine-Franklin-Jaeger model is applicable and that the source is some distance (relative to a wavelength) from any wall surface. A similar reasoning applied to dipole and quadrupole sources yields the same criterion. However, for larger sources, one finds the additional criterion $S_0 \ll A_s$.

If the criteria just stated are marginally met, the value of $\Delta\bar{\mathcal{P}}$ may be expected to fluctuate somewhat with source-position displacements over distances comparable to a wavelength and also to fluctuate with frequency; closer determination of $\bar{\mathcal{P}}_{ff}$ results from averaging over source positions and over finite frequency bands.

One refinement† is the use of (slowly) rotating vanes (see Fig. 6-9) in the reverberation chamber which cause the pressure patterns in the room to fluctuate without changing room volume or its reverberation time. Ideally, the rotation causes a long-time average to become representative of what would result from an average over both source position and microphone position, so the acoustic power computed from Eq. (5) would be closer to $\bar{\mathcal{P}}_{ff}$.

† J. Tichy, "Effects of Source Position, Wall Absorption, and Rotating Diffuser on the Qualifications of Reverberation Rooms," *Noise Control Eng.*, **7**:57–70 (1976); J. Tichy and P. K. Baade, "Effect of Rotating Diffusers and Sampling Techniques on Sound-Pressure Averaging in Reverberation Rooms," *J. Acoust. Soc. Am.*, **56**:137–143 (1974); C. E. Ebbing, "Experimental Evaluation of Moving Sound Diffusers for Reverberant Rooms," *J. Sound Vib.*, **16**:99–118 (1971).

Simultaneous Conversations in a Reverberant Room†

The theory of room acoustics gives quantitative insight into acoustical phenomena (*cocktail party effect*) occurring when many people are in one room and many conversations are simultaneously in progress. As the number of people increases, the overall sound level in the room increases, the interference from other conversations makes listening more difficult, talkers raise their voices, and people cluster closer together.

Suppose (see Fig. 6-10) there are N persons in the room, N/K persons per group, and K conversations simultaneously in progress; the acoustic power of each talker is $\bar{\mathcal{P}}$. A listener receives the direct sound from the nearest talker plus the reverberant sound from all the talkers. It is assumed that the radius of reverberation is substantially less than the spacing between clusters, so the sound from other talkers may be regarded as reverberant sound. The reverberant-sound energy density should be K times that due to any one talker, so the sound energy density in the vicinity of one such talker at a distance r should be the expression in Eq. (6-2.14) divided by ρc^2 with the second term multiplied by K, that is,

$$\bar{w} = \frac{\bar{\mathcal{P}}}{c}\left(\frac{Q_\theta}{4\pi r^2} + \frac{4K}{R_{\mathrm{rc}}}\right) \tag{6-3.9}$$

For simplicity, we take $Q_\theta = 1$ (spherical spreading).

The neglect of the direct field from neighboring clusters is justified if $1/4\pi d_{\mathrm{cl}}^2$ is less than $(\frac{1}{3})(4/R_{\mathrm{rc}})$ (so the reverberant field of any one cluster dominates its own direct field beyond a cluster spacing distance d_{cl}), that is, if

$$d_{\mathrm{cl}} > \left[\frac{9(\ln 10)V}{2\pi c T_{60}}\right]^{1/2} \tag{6-3.10}$$

For example, for a room 10 by 10 by 5 m with $V = 500\ \mathrm{m}^3$, $c = 342$ m/s, and $T_{60} = 1$ s, one would require $d_{\mathrm{cl}} > 2.2$ m for (9) to be valid.

An approximate criterion for one to comprehend a conversation is that the *signal-to-noise ratio* S/N exceed 1. This ratio is that of the energy density associated with the nearest talker to that of the other talkers. The appropriate expression deduced from Eq. (9) is

$$\mathrm{S/N} = \frac{(r_0/r)^2 + 1}{K - 1} \tag{6-3.11}$$

where $r_0 = (R_{\mathrm{rc}}/16\pi)^{1/2}$ is the radius of reverberation of the room.

The effect of the people in the room on the room constant can be taken into account by setting $R_{\mathrm{cr}} = R_{\mathrm{cr}}^0 + N\,\Delta A_s$, where ΔA_s is the incremental additional absorbing power per person. For a party that is not too crowded, this occu-

† I. Pollack and J. M. Pickett, "Cocktail Party Effect," *J. Acoust. Soc. Am.*, **29**:1262(A) (1957); W. R. MacLean, "On the Acoustics of Cocktail Parties," ibid., **31**:79–80 (1959); L. A. Crum, "Cocktail Party Acoustics," ibid., **57**:S20 (1975).

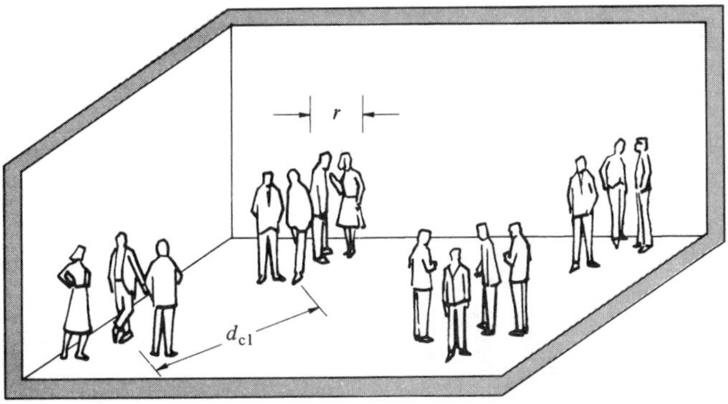

Figure 6-10 Parameters for discussion of cocktail party effect; N people are distributed among K clusters; d_{cl} denotes distance between clusters, and r denotes distance between people in the same cluster.

pancy correction is negligible. For example, for a room 10 by 10 by 5 m and with a reverberation time of 1 s, the room constant is 81 metric sabins, so if one takes $\Delta A_s \approx 0.5$ metric sabin (the value measured by Sabine), the number N of guests would have to be 160 in order that $N\,\Delta A_s \approx R_{cr}^0$ and this would correspond to 0.6 m^2 of floor area per person. Long before the party became so crowded, however, the signal-to-noise ratio of Eq. (11) would drop below 1 for any reasonable choices of listener-talker separation distance r and of N/K.

Disregarding the possible dependence of r_0 on N, one sees from the form of Eq. (11) that for any given choice of r the signal-to-noise ratio decreases as the number K of clusters increases. If one takes $r_0 = 1.3$ m (corresponding to the example above, with $R_{cr} = 81$ metric sabins) and takes $r = 0.6$ m, the signal-to-noise ratio is below 1 when K exceeds 6. With four persons per cluster, this would give $N = 24$ for the number of guests at this threshold of conversational frustration. If the number of guests exceeds this threshold, r must be decreased for intelligible conversation to be maintained, but eventually r must be so small that only one listener can stand sufficiently close to a talker.

An acoustically overcrowded party can be avoided by choosing a room with a sufficiently large room constant (as opposed to floor area) to accommodate the anticipated number of simultaneous conversations.

6-4 COUPLED ROOMS AND LARGE ENCLOSURES

Transmission of Reverberant Sound through a Panel

In noise-control applications, the sound that *escapes* from a room is often of major interest. To introduce the relevant principles, let us consider a room in which the reverberant energy density is $\bar{w}_{in}$. The energy incident per unit time on a panel of area ΔA, in accordance with the discussion leading to Eq. (6-1.9),

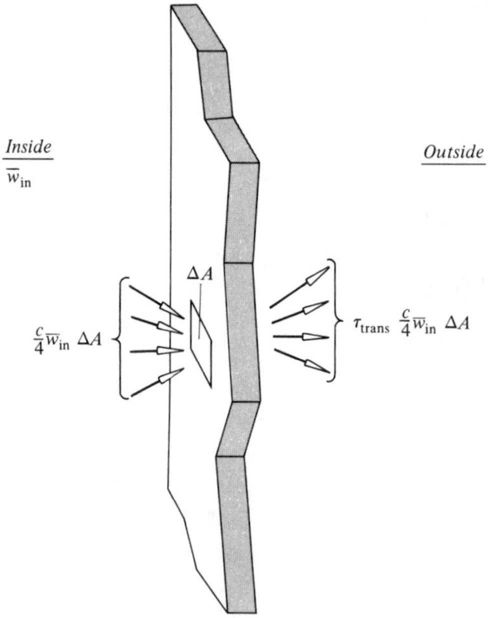

Figure 6-11 Reverberant-sound transmission through a wall. Interior field is assumed perfectly diffuse, so that energy incident per unit time on area ΔA is $(c/4)\bar{w}_{in}\ \Delta A$; fractions α_d, τ_{trans}, and r are dissipated, transmitted, and reflected.

should be $(c/4)\bar{w}_{in}\ \Delta A$; a fraction r is reflected, a fraction α_d is dissipated within the wall proper, and a fraction τ_{trans} is transmitted (see Fig. 6-11). In accord with the examples of plane-wave transmission discussed in Secs. 3-6 and 3-8, one expects an analog of the acoustic-energy-conservation principle to apply, so that these three fractions sum to 1.

The *transmission loss*† of the wall segment under consideration is defined as

$$R_{TL} = 10 \log \frac{1}{\tau_{trans}} \tag{6-4.1}$$

Ideally, this is an intrinsic frequency-dependent property of the material, but it can also depend on the panel's area, shape, and installation. It does, however, invariably satisfy a reciprocity relation

$$R_{TL}(\text{left} \rightarrow \text{right}) = R_{TL}(\text{right} \rightarrow \text{left}) \tag{6-4.2}$$

This is in accord with the results on plane-wave transmission described in Chap. 3 and can be inferred along more general lines in a manner similar to that of Sec. 4-9. Its intrinsic validity becomes plausible if one considers two rooms with no absorption separated by a panel; within each room there is initially the same acoustic energy density. The energy going from room 1 to room 2 must equal that going from room 2 to room 1 [hence Eq. (2)], or otherwise the energy

† E. Buckingham, "Theory and Interpretation of Experiments on the Transmission of Sound through Partition Walls," *Sci. Pap. Bur. Stand.* (*U.S.*), **20:**193–219 (1924–1926).

densities would become unequal; i.e., the panel would be performing similarly to a *Maxwell's demon*.†

If the panel dimensions are sufficiently large compared with a representative wavelength, the energy transmitted per unit time and wall area should be the integral over solid angle (direction **e** pointing obliquely into the wall and out of the room) of $\tau_{\text{trans}}(\mathbf{e})cD(\mathbf{e})\mathbf{e}\cdot\mathbf{n}_{\text{out}}$, where $\tau_{\text{trans}}(\mathbf{e})$ is the plane-wave acoustic-power transmission coefficient corresponding to incidence direction **e**. Thus, in a manner similar to that of the derivation of Eq. (6-1.10), one identifies the ratio of total energy transmitted to total energy incident as

$$\tau_{\text{trans,ri}} = \frac{\iint' \tau_{\text{trans}}(\mathbf{e})\mathbf{e}\cdot\mathbf{n}_{\text{out}}\, d\Omega}{\iint' \mathbf{e}\cdot\mathbf{n}_{\text{out}}\, d\Omega} \tag{6-4.3a}$$

$$= 2\int_0^{\pi/2} \tau_{\text{trans}}(\theta)\sin\theta\cos\theta\, d\theta \tag{6-4.3b}$$

If the other side of the wall bounding a room filled with diffuse sound is an open space without sources, the local volume average of the acoustic energy density just outside the wall is

$$\bar{w}_{\text{out}} = \frac{\bar{w}_{\text{in}}}{4\pi}\iint' \tau_{\text{trans}}(\mathbf{e})\, d\Omega \tag{6-4.4}$$

where the integral extends over directions pointing obliquely toward the open space. [The incident field is assumed to be made up of a large number of plane waves uniformly distributed in propagation direction, each of which generates a plane transmitted wave, with amplitude decreased by $[\tau_{\text{trans}}(\mathbf{e})]^{1/2}$, propagating in the same direction.] Consequently, the corresponding ratio of local volume averages of mean squared pressures is

$$\frac{\overline{(p^2)}_{\text{out}}}{\overline{(p^2)}_{\text{in}}} = \tfrac{1}{2}K\tau_{\text{trans,ri}} \tag{6-4.5}$$

$$K = \frac{\int_0^{\pi/2} \tau_{\text{trans}}(\theta)\sin\theta\, d\theta}{2\int_0^{\pi/2} \tau_{\text{trans}}(\theta)\sin\theta\cos\theta\, d\theta} \tag{6-4.6}$$

A rough approximation often used in the absence of a knowledge of the angular variation of τ_{trans} is to take $K = 1$ (resulting exactly when τ_{trans} is independent of θ), such that Eq. (5) yields

$$\bar{L}_{\text{out}} = \bar{L}_{\text{in}} - R_{\text{TL}} - 3\text{ dB} \tag{6-4.7}$$

† J. C. Maxwell, *Theory of Heat,* Longmans Green, London, 1871, p. 308. The demon is "a being whose faculties are so sharpened that he can follow every molecule in its course . . . who opens and closes [a] hole [connecting two portions of a vessel], so as to allow only the swifter molecules to pass from [side] A to [side] B, and only the slower ones to pass from [side] B to [side] A."

with $\bar{L}_{in}$ and $\bar{L}_{out}$ representing sound-pressure levels corresponding to $(\overline{p^2})_{in}$ and $(\overline{p^2})_{out}$.

Transmission Out through an Open Window

An extension of the above analysis applies to the field at larger distances from an open window (area ΔA). The energy passing through the window per unit time and propagating within solid angle $d\Omega$ is the same as that incident, or $(c\bar{w}_{in}/4\pi)\mathbf{e}\cdot\mathbf{n}_{out}\,d\Omega\,\Delta A$. At a large radial distance r from the window (where $r^2 \gg \Delta A$), this incremental power passes through a portion (area $r^2\,d\Omega$) of the sphere of radius r centered at the window. Hence, the intensity at large r should be

$$I_{r,\text{av}} = \frac{c\bar{w}_{in}}{4\pi}\frac{\mathbf{e}\cdot\mathbf{n}_{out}\,d\Omega\,\Delta A}{r^2\,d\Omega} = c\bar{w}_{in}\cos\theta\,\frac{\Delta A}{4\pi r^2} \tag{6-4.8}$$

where θ is the angle with the line normal to the window. Since the field locally resembles an outgoing spherical wave at large distances, and since $\bar{w}_{in}$ is $(\overline{p^2})_{in}/\rho c^2$, Eq. (8) implies

$$[p^2(r)]_{\text{av}} = (\overline{p^2})_{in}\cos\theta\,\frac{\Delta A}{4\pi r^2} \tag{6-4.9}$$

As an example, suppose a room with a sound level inside of 90 dB has an open window of area $\Delta A = 1\ \text{m}^2$. The sound level outside is not less than 50 dB unless r exceeds $10^2/(4\pi)^{1/2} = 28$ m.

Theory of Large Enclosures

A common procedure (see Fig. 6-12) for reducing the acoustic power radiating into the environment is to build an enclosure around the source. The simplest theory† of such enclosures assumes that the sound field within the enclosure is reverberant and that the actual acoustic power output of the source is unaltered by the presence of the enclosure.

The net energy per unit time escaping out of the enclosure, according to the discussion in the earlier part of this section, should be

$$\bar{\mathcal{P}}_{out} = \frac{c}{4}\bar{w}_{in}\iint \tau_{trans}\,dS \tag{6-4.10}$$

while that dissipated $\bar{\mathcal{P}}_d$ within the enclosure and not transmitted out is given by a similar expression involving the surface integral of α_d. The requirement that the actual sound power output $\bar{\mathcal{P}}_{actual}$ of the source equal $\bar{\mathcal{P}}_{out} + \bar{\mathcal{P}}_d$

† For analyses of enclosures that are *not* large compared to source dimensions, see R. S. Jackson, "The Performance of Acoustic Hoods at Low Frequencies," *Acustica,* **12:**139–152 (1962), "Some Aspects of the Performance of Acoustic Hoods," *J. Sound Vib.,* **3:**82–94 (1966); M. C. Junger, "Sound Transmission through an Elastic Enclosure Acoustically Closely Coupled to a Noise Source," *ASME Pap.* 70-WA/DE-12, American Society of Mechanical Engineers, New York, 1970.

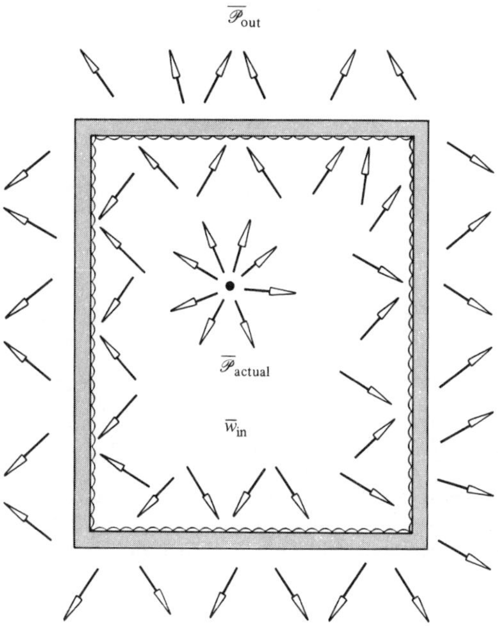

Figure 6-12 Idealized model of "large" enclosure; source power output $\bar{\mathcal{P}}_{actual}$ causes reverberant field of energy density $\bar{w}_{in}$ inside enclosure, while power $\bar{\mathcal{P}}_{out}$ escapes to external environment.

consequently yields the power ratio

$$\frac{\bar{\mathcal{P}}_{out}}{\bar{\mathcal{P}}_{actual}} = \frac{\iint \tau_{trans}\, dS}{\iint \tau_{trans}\, dS + \iint \alpha_d\, dS} \tag{6-4.11}$$

An implication of Eq. (11) is that no sound-power reduction is achieved unless there is some absorption. Thus, enclosure walls are typically lined with absorbing material. If the quotient $\bar{\tau}_{trans}/\bar{\alpha}_d$ of the area averages of τ_{trans} and α_d is small compared with 1, then the ratio $\bar{\mathcal{P}}_{out}/\bar{\mathcal{P}}_{actual}$ approaches $\bar{\tau}_{trans}/\bar{\alpha}_d$; for fixed $\bar{\alpha}_d$, an increase in the transmission loss of the walls results in more power reduction. If one thinks in terms of sound rays bouncing about inside the enclosure, an increased noise reduction caused by increased R_{TL} (decreased $\bar{\tau}_{trans}$) is associated with rays undergoing more reflections and thus losing more energy through dissipation at the walls before a significant fraction of their original energy can be transmitted out.

Coupled Rooms

If a source (acoustic power $\bar{\mathcal{P}}$) is in a room (see Fig. 6-13) separated by a panel of area ΔA from a second room, the difference of the sound-pressure levels in the two rooms can be predicted from considerations of acoustic-energy conser-

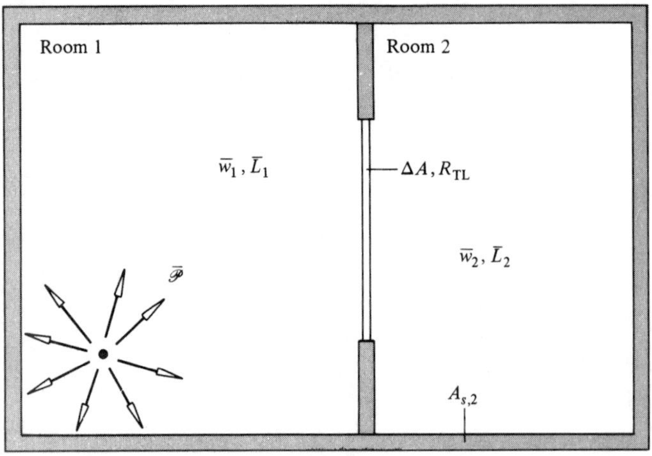

Figure 6-13 Adjacent rooms coupled by a panel of area ΔA with transmission loss R_{TL}. Source with power output $\bar{\mathcal{P}}$ causes energy densities $\bar{w}_1$ and $\bar{w}_2$ and sound-pressure levels $\bar{L}_1$ and $\bar{L}_2$. Noise reduction L_{NR}, equal to $\bar{L}_1 - \bar{L}_2$, is determined by ΔA, R_{TL}, and absorbing power $A_{s,2}$ of room 2.

vation. The appropriate generalization† of Eq. (6-1.3) for room 1 is

$$V_1 \frac{d\bar{w}_1}{dt} = -\frac{c}{4} A_{s,1}\bar{w}_1 - \frac{c}{4} \tau_{\text{trans}}\bar{w}_1 \Delta A + \frac{c}{4} \tau_{\text{trans}}\bar{w}_2 \Delta A + \bar{\mathcal{P}} \tag{6-4.12}$$

The first term on the right is the negative of the energy dissipated per unit time within room 1; the second is the negative of the rate at which energy is being transmitted from room 1 to room 2; the third is the rate at which energy is being transmitted from room 2 to room 1. Similarly, for room 2, one has

$$V_2 \frac{d\bar{w}_2}{dt} = -\frac{c}{4} A_{s,2}\bar{w}_2 + \frac{c}{4} \tau_{\text{trans}} \Delta A(\bar{w}_1 - \bar{w}_2) \tag{6-4.13}$$

In the steady-state situation, the second of the two equations above leads to

$$\frac{\bar{w}_2}{\bar{w}_1} = \frac{(\overline{p^2})_2}{(\overline{p^2})_1} = \frac{\tau_{\text{trans}} \Delta A}{\tau_{\text{trans}} \Delta A + A_{s,2}} \tag{6-4.14}$$

which is independent of $\bar{\mathcal{P}}$ and of the properties of room 1. The corresponding difference of the two sound levels, termed the *noise reduction* L_{NR}, is consequently

$$L_{NR} = \bar{L}_1 - \bar{L}_2 = R_{TL} + 10 \log \left(10^{-R_{TL}/10} + \frac{A_{s,2}}{\Delta A}\right) \tag{6-4.15}$$

† A. H. Davis, "Reverberation Equations for Two Adjacent Rooms Connected by an Incompletely Soundproof Partition," *Phil. Mag.*, (6)**50:**75–80 (1925).

The inverse relation, with R_{TL} expressed in terms of L_{NR} and $A_{s,2}/\Delta A$, is the basis for the common method of experimentally measuring the transmission loss of panels. [One measures $\bar{L}_1$ and $\bar{L}_2$ in the two coupled reverberant rooms of a specially designed TL facility with a sample panel forming part of the common wall (the rest of the wall being virtually nontransmissive); $A_{s,2}$ is found from measurement of the reverberation time of room 2.] Note that the noise reduction increases when $A_{s,2}$ increases. In the ideal case when $A_{s,2}$ is 0, the noise reduction is 0, regardless of the R_{TL} of the panel.

Reverberant Decay in Coupled Rooms†

If the source of sound in room 1 is suddenly turned off, the subsequent decay of $\bar{w}_1$ and $\bar{w}_2$ is governed by the two coupled differential equations (12) and (13) with $\bar{\mathcal{P}}$ set to zero. Their solution can be worked out by standard techniques‡ for systems of homogeneous ordinary differential equations with constant coefficients; one sets

$$(\bar{w}_1, \bar{w}_2) = (A_1, A_2)e^{-at} + (B_1, B_2)e^{-bt} \tag{6-4.16}$$

where a characteristic decay rate a and the corresponding eigenvector (A_1, A_2) are related such that

$$\begin{bmatrix} -V_1 a + \frac{c}{4}(A_{s,1} + \tau_{\mathrm{trans}}\,\Delta A) & -\frac{c}{4}\tau_{\mathrm{trans}}\,\Delta A \\ -\frac{c}{4}\tau_{\mathrm{trans}}\,\Delta A & -V_2 a + \frac{c}{4}(A_{s,2} + \tau_{\mathrm{trans}}\,\Delta A) \end{bmatrix} \begin{bmatrix} A_1 \\ A_2 \end{bmatrix} = \begin{bmatrix} 0 \\ 0 \end{bmatrix} \tag{6-4.17}$$

The same relation holds between b, B_1, and B_2. The quantities a and b are the two roots of the equation that results when the determinant of coefficients is set to zero; A_1/A_2 is determined subsequently from either of Eqs. (17). Initial values of $\bar{w}_1$ and $\bar{w}_2$ supply the remaining information necessary for determination of the four coefficients, A_1, A_2, B_1, B_2.

As long as $\tau_{\mathrm{trans}}\,\Delta A$ is somewhat less than $(V_1V_2)^{1/2}|A_{s,2}/V_2 - A_{s,1}/V_1|$ and is less than either $A_{s,2}$ or $A_{s,1}$, the decay constants a and b are approximately the reverberation times for the two rooms considered separately, but the coupling between the rooms implies that the decay of $\bar{w}_1$ or $\bar{w}_2$ can no longer be strictly considered as a single exponential decay. If, for example, $a \gg b$, $A_1 \gg B_1$, the energy density $\bar{w}_1$ at first decays nearly as e^{-at} but eventually as e^{-bt}.

† Davis, "Reverberation Equations . . . ,"; H. Kuttruff, *Room Acoustics*, Applied Science, London, 1973, pp. 119–123.

‡ See, for example, I. S. Sokolnikoff and R. M. Redheffer, *Mathematics of Physics and Modern Engineering*, 2d ed., McGraw-Hill, New York, 1966, pp. 148–151.

6-5 THE MODAL THEORY OF ROOM ACOUSTICS

The concept of a *room mode*† leads to a theory of room acoustics‡ intrinsically less approximate than the Sabine-Franklin-Jaeger model. Here we confine ourselves to a simple version of the modal theory that uses modes for a room with rigid walls. Below, we show that the use of such modes does not preclude the development of an approximate theory applicable to rooms with walls of finite impedance.

The Eigenvalue Problem

For a room with rigid walls, there are a multitude of particular solutions (labeled by $n = 1, 2, 3, \ldots$) of the homogeneous wave equation of the form

$$p = \Psi(\mathbf{x}, n)e^{-i\omega(n)t} \tag{6-5.1}$$

The *eigenfunction* $\Psi(\mathbf{x}, n)$ satisfies the Helmholtz equation and the rigid-wall boundary condition

$$[\nabla^2 + k^2(n)]\Psi(\mathbf{x}, n) = 0 \text{ in } V \qquad \nabla\Psi(\mathbf{x}, n) \cdot \mathbf{n}_{\text{out}} = 0 \text{ on } S \tag{6-5.2}$$

The eigenvalue $k^2(n)$, equal to $\omega^2(n)/c^2$, is one of a discrete series of real positive numbers for which a nontrivial solution of the boundary-value problem (2) exists. The determination of values of $k^2(n)$ and of the associated eigenfunctions is an *eigenvalue problem;* the field associated with a given $\Psi(\mathbf{x}, n)$ is a room mode.

Modes for a Rectangular Room

To exemplify the above remarks, we consider a rectangular room (Fig. 6-14) bounded by rigid walls lying along the planes $x = 0$, $x = L_x$, $y = 0$, $y = L_y$, $z = 0$, $z = L_z$. A possible $\Psi(\mathbf{x}, n)$ of the factored form $X(x)Y(y)Z(z)$ is substituted into the Helmholtz equation, such that subsequent division by Ψ yields

$$X^{-1}X''(x) + Y^{-1}Y''(y) + Z^{-1}Z''(z) + k^2 = 0 \tag{6-5.3}$$

Because the second, third, and fourth terms are independent of x, the x derivative of the first term is zero, so that term is a constant. Anticipating that this

† J. W. S. Rayleigh, *The Theory of Sound,* vol. 2, 2d ed., reprinted by Dover, New York, 1945, sec. 267. Earlier work by J. M. C. Duhamel gave eigenfunctions and natural frequencies for finite segments of rectangular and circular tubes with rigid walls but ends that were pressure-release surfaces ["On the Vibrations of a Gas in Cylindrical, Conical, etc., Tubes," *J. Math. Pures Appl.,* **14:**49–110 (1849), especially pp. 84–86]. The basic concept per se of a vibration mode as a building block in the description of a vibrating system with more than 1 degree of freedom dates back to Daniel Bernoulli's modal description of the vibrating string in 1753.

‡ K. Schuster and E. Waetzmann, "On Reverberation in Closed Spaces," *Ann. Phys.,* (5)**1:**671–695 (1929); M. J. O. Strutt, "On the Acoustics of Large Rooms," *Phil. Mag.,* (7)**8:**236–250 (1929); P. M. Morse, "Some Aspects of the Theory of Room Acoustics," *J. Acoust. Soc. Am.,* **11:**56–66 (1939).

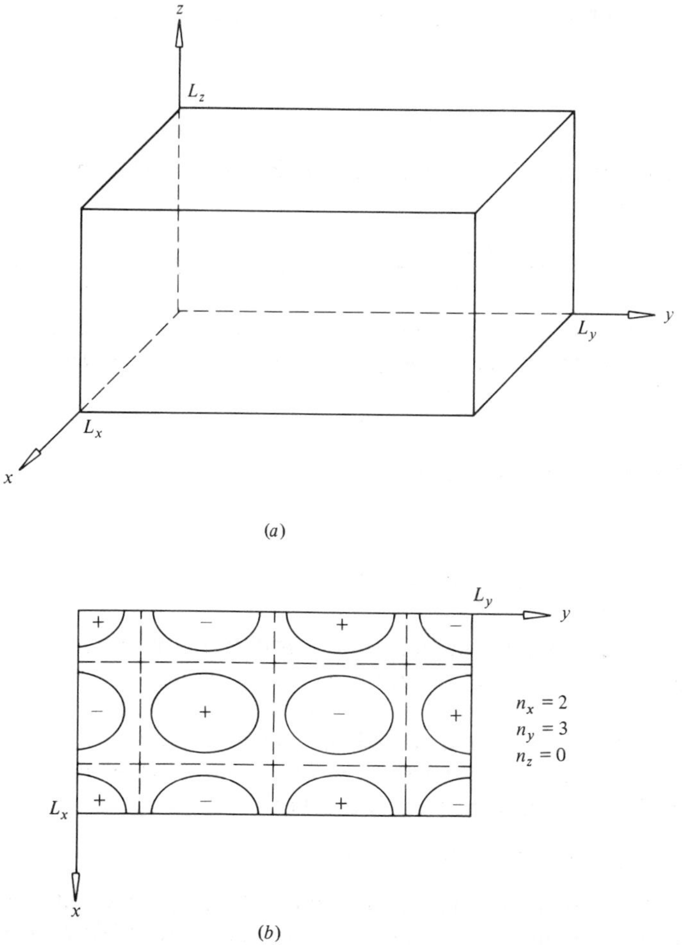

Figure 6-14 (*a*) Coordinate system and parameters for description of modes in a rectangular room L_x by L_y by L_z. (*b*) Sketch of $n_x = 2$, $n_y = 3$, $n_z = 0$ mode (independent of z coordinate). Dashed lines indicate acoustic-pressure nodes; indicated signs of eigenfunction result if p is taken as positive at the origin.

constant is negative, we write it as $-k_x^2$ and have

$$X''(x) + k_x^2 X(x) = 0 \tag{6-5.4}$$

Similar ordinary differential equations hold for $Y(y)$ and $Z(z)$, and from Eq. (3) we conclude that the three *separation constants* are related such that $k_x^2 + k_y^2 + k_z^2 = k^2$.

The solution of Eq. (4) ensuring that the boundary condition $\partial\Psi/\partial x = 0$ at $x = 0$ will be satisfied is of the form of a constant times $\cos k_x x$. The other boundary condition, $\partial\Psi/\partial x = 0$ at $x = L_x$, requires that $\sin k_x L_x = 0$. This gives $k_x = \pi n_x/L_x$, so $X(x)$ must be a constant times $\cos(n_x \pi x/L_x)$ for some integer n_x. Since similar considerations apply to $Y(y)$ and $Z(z)$, a possible eigenfunction $\Psi(\mathbf{x}, n)$ is

$$\Psi(\mathbf{x}, n_x, n_y, n_z) = A \cos \frac{n_x \pi x}{L_x} \cos \frac{n_y \pi y}{L_y} \cos \frac{n_z \pi z}{L_z} \qquad (6\text{-}5.5)$$

where A is an arbitrary constant. The corresponding eigenvalue, from the relation $k_x^2 + k_y^2 + k_z^2 = k^2$, is

$$k^2(n_x, n_y, n_z) = \pi^2 \left[\left(\frac{n_x}{L_x}\right)^2 + \left(\frac{n_y}{L_y}\right)^2 + \left(\frac{n_z}{L_z}\right)^2 \right] \qquad (6\text{-}5.6)$$

Any combination of integers n_x, n_y, n_z gives a mode. The index n in Eqs. (2) in this case is the set of these three integers (each assumed nonnegative to avoid redundancy).

Orthogonality of Modal Eigenfunctions

The identity

$$\int_0^{L_x} \cos \frac{n_x \pi x}{L_x} \cos \frac{n_x' \pi x}{L_x}\, dx = 0 \qquad n_x \neq n_x' \qquad (6\text{-}5.7)$$

(given $n_x \geq 0$, $n_x' \geq 0$) requires that the volume integral of the product of two eigenfunctions described by Eq. (6) be zero unless $n_x = n_x'$, $n_y = n_y'$, and $n_z = n_z'$.

To investigate the possibility of mutual orthogonality† of modal eigenfunctions for general shapes of rooms, we let $\Psi_1 = \Psi(\mathbf{x}, n_1)$ and $\Psi_2 = \Psi(\mathbf{x}, n_2)$ denote two eigenfunctions. Then from Eq. (2) it follows that

$$\Psi_2(\nabla^2 + k_1^2)\Psi_1 - \Psi_1(\nabla^2 + k_2^2)\Psi_2 = 0$$

But $\Psi_2 \nabla^2 \Psi_1 - \Psi_1 \nabla^2 \Psi_2$ is the divergence of $\Psi_2 \nabla \Psi_1 - \Psi_1 \nabla \Psi_2$, so an integration over room volume with subsequent application of Gauss' theorem and of the boundary condition yields

$$(k_1^2 - k_2^2) \iiint \Psi_1 \Psi_2 \, dV = 0 \qquad (6\text{-}5.8)$$

Thus, the integral must be zero if $k_1^2 \neq k_2^2$.

It is possible, e.g., for a cubic room, that two or more independent eigenfunctions correspond to the same eigenvalue. One can always select them, however, e.g., by the Schmidt orthogonalization process,‡ to be a linearly independent set and to be such that the volume integral of the product of any two different members of the set vanishes. Furthermore, since any $\Psi(\mathbf{x}, n)$ multiplied by a constant is still an eigenfunction, we assume that the multiplicative

† J. W. S. Rayleigh, "On the Fundamental Modes of a Vibrating System," *Phil. Mag.*, (5)**46:**434–439 (1873).

‡ See, for example, R. Courant and D. Hilbert, *Methods of Mathematical Physics*, vol. 1, Interscience, New York, 1953, p. 4.

constant has been chosen such that $\Psi(\mathbf{x}, n)$ is *normalized* to have a mean squared volume average of 1. With these choices, we have an *orthonormal set* satisfying

$$\iiint_V \Psi(\mathbf{x}, n)\Psi(\mathbf{x}, n')\, dV = \delta_{nn'} V \tag{6-5.9}$$

Another property of the set of eigenfunctions chosen in this manner is

$$\iiint_V \nabla\Psi(\mathbf{x}, n) \cdot \nabla\Psi(\mathbf{x}, n')\, dV = \delta_{nn'} k^2(n) V \tag{6-5.10}$$

The proof results from the consecutive replacements of $\nabla\Psi \cdot \nabla\Psi'$ by $\nabla \cdot (\Psi' \nabla\Psi) - \Psi' \nabla^2\Psi$ (a vector identity) and of $\nabla^2\Psi$ by $-k^2(n)\Psi$ (from the Helmholtz equation). The volume integral of the first term is transformed into a surface integral by Gauss' theorem and is recognized as being zero because of the boundary condition; the volume integral of the second term yields $\delta_{nn'} k^2(n) V$ because of Eq. (9), so Eq. (10) results.

Similarly, a multiplication of the Helmholtz equation by $\Psi^*(\mathbf{x}, n)$ and a subsequent integration over V yields

$$k^2(n) = \frac{\iiint \nabla\Psi(\mathbf{x}, n) \cdot \nabla\Psi^*(\mathbf{x}, n)\, dV}{\iiint |\Psi(\mathbf{x}, n)|^2\, dV} \tag{6-5.11}$$

so $k^2(n)$ must be real and positive. Since the Helmholtz equation then requires $\Psi^*(\mathbf{x}, n)$ to be an eigenfunction, we can always choose eigenfunctions to be real.

Modal Expansion of Functions

The modal eigenfunctions satisfying Eqs. (2) constitute a *complete set;* any well-behaved function $f(\mathbf{x})$ for points $\mathbf{x}$ within the room can be approximated† as a linear combination of the $\Psi(\mathbf{x}, n)$. An expansion coefficient a_n can be determined from the requirement

$$\iiint f(\mathbf{x})\Psi(\mathbf{x}, n)\, dV = \iiint \left[\sum_{n'} a_{n'} \Psi(\mathbf{x}, n')\right] \Psi(\mathbf{x}, n)\, dV$$

such that Eq. (9) yields

$$a_n = \frac{1}{V} \iiint f(\mathbf{x})\Psi(\mathbf{x}, n)\, dV \tag{6-5.12}$$

† The applicable theorem is that "the eigenfunctions of any self-adjoint differential system of the second order form a complete set." That Eqs. (2) describe a self-adjoint system follows from the equivalence of $\Psi \nabla^2\phi - \phi \nabla^2\Psi$ to the divergence of $\Psi \nabla\phi - \phi \nabla\Psi$ and from the vanishing of the normal component of the latter at the walls when both Ψ and ϕ satisfy the boundary condition. For a general proof, see I. Stakgold, *Boundary Value Problems of Mathematical Physics,* vol. 1, Macmillan, New York, 1967, pp. 212–220.

Field of a Point Source in a Room with Walls of Large Impedance

We now apply the mathematical apparatus of room modes to determine the field of a point source of angular frequency ω and monopole amplitude $\hat{S}$. The room walls are characterized by a specific impedance Z, possibly having different values on different surfaces, but being such that $|Z|/\rho c \gg 1$ so that the walls are nearly rigid. The complex pressure amplitude $\hat{p}$ satisfies the inhomogeneous Helmholtz equation with a source term $-4\pi\hat{S}\,\delta(\mathbf{x} - \mathbf{x}_0)$ on the right side. On the walls of the room, $\hat{p}$ satisfies the boundary condition (see Sec. 3-3) $\nabla\hat{p} \cdot \mathbf{n}_{\text{out}} = ik(\rho c/Z)\hat{p}$, where $\mathbf{n}_{\text{out}}$ is the unit normal pointing out of the room. The completeness property allows us to determine an expansion† for $\hat{p}(\mathbf{x})$ in terms of eigenfunctions $\Psi(\mathbf{x}, n)$ appropriate to the same room geometry but which satisfy the rigid-wall boundary condition of Eq. (2).

To develop expressions for the coefficients a_n, we follow a procedure similar to that for solving a boundary-value problem in terms of a Green's function, but we use an eigenfunction rather than a Green's function. Multiplying Eq. (4-3.4) by $\Psi(\mathbf{x}, n)$ and subsequently integrating over room volume, expressing $\Psi\,\nabla^2\hat{p}$ as $\hat{p}\,\nabla^2\Psi$ plus the divergence of $\Psi\,\nabla\hat{p} - \hat{p}\,\nabla\Psi$, then making use of Gauss' theorem and of the boundary condition of Eq. (2), we obtain

$$\iiint \hat{p}(\nabla^2 + k^2)\Psi(\mathbf{x}, n)\,dV + \iint \Psi(\mathbf{x}, n)\,\nabla\hat{p} \cdot \mathbf{n}_{\text{out}}\,dS$$
$$= -4\pi\hat{S} \iiint \Psi(\mathbf{x}, n)\delta(\mathbf{x} - \mathbf{x}_0)\,dV = -4\pi\hat{S}\,\Psi(\mathbf{x}_0, n) \qquad (6\text{-}5.13)$$

Further reduction results because $\nabla^2\Psi(\mathbf{x}, n)$ is $-k^2(n)\Psi(\mathbf{x}, n)$ and from the boundary condition $\nabla\hat{p} \cdot \mathbf{n}_{\text{out}} = (ik)(\rho c/Z)\hat{p}$.

Insertion of an eigenfunction expansion for $\hat{p}$ results in the coupled algebraic equations

$$[k^2 - k^2(n)]a_n + ik \sum_m B_{nm}a_m = \frac{-4\pi\hat{S}\,\Psi(\mathbf{x}_0, n)}{V} \qquad (6\text{-}5.14)$$

with the abbreviation

$$B_{nm} = \frac{1}{V} \iint \Psi(\mathbf{x}, n)\,\frac{\rho c}{Z}\,\Psi(\mathbf{x}, m)\,dS \qquad (6\text{-}5.15)$$

the integral extending over the surface area of the room.

For a room with nearly rigid walls, the coupling terms ($m \neq n$) in Eq. (14) are of minor importance; the possibility that some $k(n)$ may be close to k can be taken into account if we group the $m = n$ term, $ikB_{nn}a_n$, with the $[k^2 - k^2(n)]a_n$ term; we then solve the coupled equations by iteration, taking $B_{nm} = 0$ for $m \neq n$ in the first approximation. In such a manner, with the so-derived a_n's inserted into the expansion for $\hat{p}$, one obtains the approximate expression

$$\hat{p} = -4\pi\frac{\hat{S}}{V} \sum_n \frac{\Psi(\mathbf{x}, n)\Psi(\mathbf{x}_0, n)}{k^2 - k^2(n) + ikB_{nn}} \qquad (6\text{-}5.16)$$

† P. M. Morse and K. U. Ingard, "Linear Acoustic Theory" in S. Flügge (ed.), *Handbuch der Physik,* vol. 11, pt. 2 (*Akustik* I), Springer, Berlin, 1961, p. 60.

For a typical higher-order mode in a room, the local volume average of Ψ^2 is nearly independent of position. A similar statement holds for $|\Psi|^2$ at points on the walls, but the surface-area average is nearly twice the volume average. These remarks are supported by the rectangular-room eigenfunctions given by Eq. (5) when $n_x > 0$, $n_y > 0$, $n_z > 0$. (If one or more of the three indices is 0, the ratio is less than 2.) Such considerations suggest, for most of the modes of interest, that Eq. (15) (for $n = m$) can be approximated by

$$B_{nn} = \frac{2}{V} \iint \frac{\rho c}{Z}\, dS \tag{6-5.17}$$

Another approximate identification results from the assumption that the walls are locally reacting and from insertion† of the plane-wave absorption coefficient $\alpha(\theta)$, equal to $1 - |\mathcal{R}|^2$ and determined from Eq. (3-3.4), into Eq. (6-1.11), so that, with $\beta = \rho c/Z$ replacing $1/\zeta$,

$$\alpha_{\text{ri}} = 8\beta_R \int_0^{\pi/2} \frac{\cos^2\theta \sin\theta \, d\theta}{(\beta_R + \cos\theta)^2 + \beta_I^2} \approx 8\beta_R \tag{6-5.18}$$

The latter expression, applicable for the case of the nearly rigid wall, results when β_R and β_I are set to zero in the integrand. Our approximate expression (17) for B_{nn} therefore leads to

$$c \operatorname{Re} B_{nn} \approx \frac{c}{4V} \iint \alpha_{\text{ri}}\, dS \approx \frac{1}{\tau} \tag{6-5.19}$$

where τ is the characteristic time of the Sabine-Franklin-Jaeger model.

The imaginary part of B_{nn} is of minor consequence. The denominator factor in Eq. (16) can be written as $[k - k_{\text{sh}}(n)][k + k_{\text{sh}}(n) - Y_{nn}] + ikX_{nn}$, where the shifted eigenvalue $k_{\text{sh}}(n)$ is such that $k_{\text{sh}}^2 - k_{\text{sh}} Y_{nn} = k^2(n)$. Here X_{nn} and Y_{nn} are the real and imaginary parts of B_{nn}. One ordinarily is interested in values of k much greater than any $|Y_{nn}|$, so the term $-Y_{nn}$ in the factor $k + k_{\text{sh}}(n) - Y_{nn}$ can be discarded. For virtually all the terms contributing to the sum, $k_{\text{sh}}(n)$ can be approximated by $k(n) + Y_{nn}/2$. Since most of the Y_{nn} have nearly the same value, the resonant frequencies $ck_{\text{sh}}(n)$ have nearly the same spacing as the $\omega(n)$. Insofar as one is not concerned with a precise prediction of the resonance frequencies, the $k_{\text{sh}}(n)$ can be replaced by the $k(n)$ without changing the overall predictions of the modal formulation. Thus, the denominator factor is replaced by $k^2 - k^2(n) + ikX_{nn}$. With the additional approximation represented by Eq. (19), we accordingly obtain

$$\hat{p} \approx -4\pi \frac{\hat{S}}{V} \sum_n \frac{\Psi(\mathbf{x}, n)\Psi(\mathbf{x}_0, n)}{k^2 - k^2(n) + ik/c\tau} \tag{6-5.20}$$

† E. T. Paris, "On the Coefficient of Sound-Absorption Measured by the Reverberation Method," *Phil. Mag.*, (7)5:489–497 (1928).

Acoustic Energy in a Room

To express the time average of the acoustic energy in the room in terms of modes, one begins with the volume integral

$$E = \frac{1}{4\rho c^2} \iiint \left[|\hat{p}|^2 + \left(\frac{c}{\omega}\right)^2 \nabla\hat{p} \cdot \nabla\hat{p}^* \right] dV \tag{6-5.21}$$

Insertion of the appropriate expansions for $\hat{p}$ and $\hat{p}^*$ [sums over n and m of $a_n\Psi(\mathbf{x}, n)$ and $a_m^*\Psi(\mathbf{x}, m)$] yields a double sum over n and m, the cross terms of which vanish because of Eqs. (9) and (10), so we obtain

$$E = \frac{V}{4\rho c^2} \sum_n |a_n|^2 \left\{ 1 + \left[\frac{\omega(n)}{\omega} \right]^2 \right\} \tag{6-5.22}$$

The sums over n resulting from the 1 and the $[\omega(n)/\omega]^2$ terms in the coefficient of $|a_n|^2$ correspond to the potential energy E_P and the kinetic energy E_K.

If the field is that of a point source, appropriate values for the a_n are the coefficients of the $\Psi(\mathbf{x}, n)$ in Eq. (20). This replacement yields, for the potential energy E_P,

$$E_P = \frac{\overline{p^2} V}{2\rho c^2} = \frac{2\pi\bar{\mathscr{P}}_{\text{ff}}}{cV} \sum_n \frac{\Psi^2(\mathbf{x}_0, n)}{[k^2 - k^2(n)]^2 + k^2/c^2\tau^2} \tag{6-5.23}$$

where $\bar{\mathscr{P}}_{\text{ff}} = 2\pi|\hat{S}|^2/\rho c$ is the power the source radiates in a free-field environment.

The analogous sum for the kinetic energy E_K diverges because the fluid velocity in the vicinity of a point source varies as $1/r^2$. For large rooms and higher-frequency sources, however, a meaningful value† is obtained if one sums over only those $k(n)$ which are less than, say, $1/5r_0$, where r_0 is the radius of reverberation; the resulting truncated sum corresponds to the kinetic energy E_K' in the reverberant part of the field. The analogous truncation in Eq. (23) has negligible influence on E_P; the sum, $E' = E_P' + E_K'$, corresponds to the product of the energy density $\bar{w}$ introduced in Sec. 6-1 with the room-volume portion V' that excludes the source's immediate neighborhood.

Modal Description of Power Injection

The near field of a single-frequency point source has the characteristic form (discussed in Sec. 4-3)

$$\hat{p} = \frac{\hat{S}}{R} + \hat{S}f \qquad \hat{\mathbf{v}} = \frac{1}{\rho\omega}\left(\frac{i\hat{S}\mathbf{e}_R}{R^2} - i\hat{S}\,\nabla f \right) \tag{6-5.24}$$

where, as before, $\hat{S}$ is monopole amplitude, $R = |\mathbf{x} - \mathbf{x}_0|$ is distance from the source, and f is a function whose value and gradient are bounded at $\mathbf{x} = \mathbf{x}_0$. Starting from these general expressions and with consideration of the surface

† This was pointed out to the author by Preston W. Smith, Jr.

integral of $\frac{1}{2}$ Re $(\hat{p}^*\hat{\mathbf{v}} \cdot \mathbf{e}_R)$ over a sphere centered at the source, one can subsequently conclude, after taking the limit $R \to 0$, that the time-averaged power output of the source must be†

$$\bar{\mathcal{P}} = \bar{\mathcal{P}}_{\text{ff}} \left(\operatorname{Im} \frac{\hat{p}}{k\hat{S}} \right)_{\mathbf{x} \to \mathbf{x}_0} \tag{6-5.25}$$

where $\bar{\mathcal{P}}_{\text{ff}}$ is the power the source would radiate if it were in a free field environment. [If the source is in an unbounded region, $\hat{p} = \hat{S}R^{-1}e^{ikR}$ and Eq. (25) reduces to $\bar{\mathcal{P}}_{\text{ff}}$. Although $\hat{p}/\hat{S}$ diverges as $\mathbf{x} \to \mathbf{x}_0$, its imaginary part does not.]

In terms of room modes and in the approximation of the nearly rigid wall represented by Eq. (20), the above expression reduces to‡

$$\bar{\mathcal{P}} = \frac{4\pi\bar{\mathcal{P}}_{\text{ff}}}{V} \sum_n \frac{(1/c\tau)\Psi^2(\mathbf{x}_0, n)}{[k^2 - k^2(n)]^2 + k^2/c^2\tau^2} \tag{6-5.26}$$

6-6 HIGH-FREQUENCY APPROXIMATIONS

The principal formulas of the Sabine-Franklin-Jaeger model result when modal sums are approximated by integrals. The demonstration of this begins with the derivation of an expression for the number of room modes per unit frequency bandwidth.

The Modal Density

Let $N(\omega)$ denote the number of room modes whose natural frequencies are less than a given value of ω. For a rectangular room, Eq. (6-5.6) indicates that $N(\omega)$ is the total number of points in k_x, k_y, k_z space with coordinates $(n_x\pi/L_x, n_y\pi/L_y, n_z\pi/L_z)$ that lie in or on the boundaries of the first octant $(k_x \geq 0, k_y \geq 0, k_z \geq 0)$ at a radial distance less than ω/c (see Fig. 6-15). Each point lies in a rectangular box of dimensions $(\pi/L_x, \pi/L_y, \pi/L_z)$ with volume π^3/V, each box having only one such point, the set of all boxes filling the space. The box corresponding to the index triplet n_x, n_y, n_z confines k_x to between $(n_x - \frac{1}{2})\pi/L_x$ and $(n_x + \frac{1}{2})\pi/L_x$; analogous limits confine k_y and k_z.

The total volume in the k_x, k_y, k_z space occupied by all boxes whose center

† R. H. Lyon, "Statistical Analysis of Power Injection and Response in Structures and Rooms," *J. Acoust. Soc. Am.*, **45**:545–565 (1969).

‡ G. C. Maling, Jr., "Calculation of the Acoustic Power Radiated by a Monopole in a Reverberation Chamber," *J. Acoust. Soc. Am.*, **42**:859–865 (1967). The analogous result for a point dipole is given by S. N. Yousri and F. J. Fahy, "An Analysis of the Acoustic Power Radiated by a Point Dipole Source into a Rectangular Reverberation Chamber," *J. Sound Vib.*, **25**:39–50 (1972).

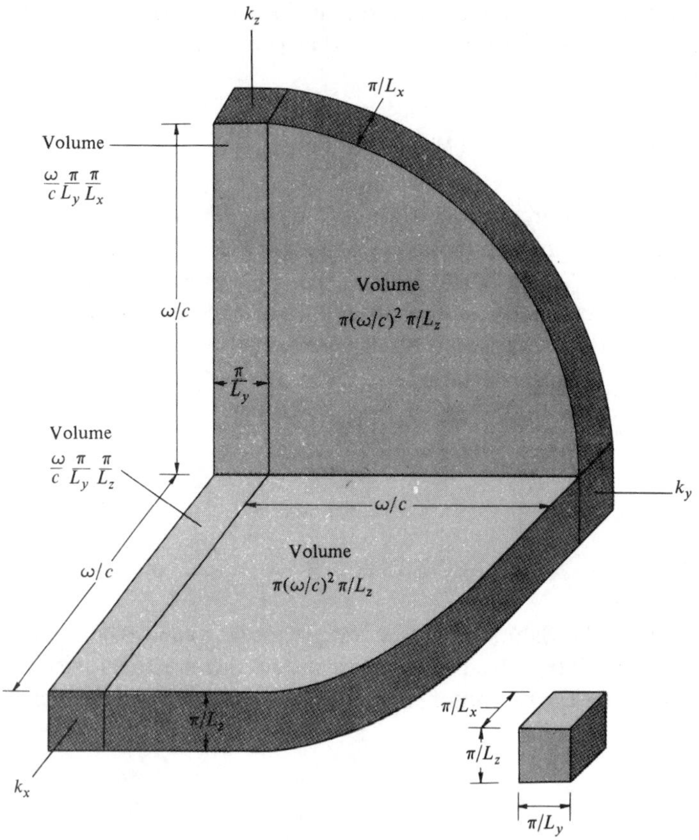

Figure 6-15 Sketch depicting some of the volume contributions in k_x, k_y, k_z space to the estimation of $N\pi^3/V$, where $N(\omega)$ is the number of room modes whose natural frequencies are less than ω; the room volume V is $L_xL_yL_z$. There is one mode associated with each rectangular block of dimensions π/L_x by π/L_y by π/L_z whose center lies within or on the boundary of the portion of the sphere of radius ω/c lying within the first octant of k_x, k_y, k_z space.

points satisfy the inequality consists approximately† of the sum of the following:

1. The volume $(\pi/6)(\omega/c)^3$ in an octant with radius ω/c
2. The sum of the volumes of three quarter-circle slabs of radius ω/c having thicknesses $\pi/2L_x$, $\pi/2L_y$, and $\pi/2L_z$, respectively
3. The sum of three volumes of rectangular columns each having length ω/c, the three cross-sectional areas being $\frac{1}{4}\pi^2/L_xL_y$, $\frac{1}{4}\pi^2/L_yL_z$, and $\frac{1}{4}\pi^2/L_xL_z$, respectively
4. A volume $\frac{1}{8}\pi^3/L_xL_yL_z$

† D.-Y. Maa, "Distribution of Eigentones in a Rectangular Chamber at Low Frequency Range," *J. Acoust. Soc. Am.*, **10**:235–238 (1939).

The estimated total number of modes $N(\omega)$, taken as the sum of these volumes divided by the volume π^3/V per point, is consequently

$$N(\omega) \approx \frac{1}{6}\frac{V}{\pi^2}\left(\frac{\omega}{c}\right)^3 + \frac{1}{16}\frac{S}{\pi}\left(\frac{\omega}{c}\right)^2 + \frac{1}{16}\frac{L}{\pi}\frac{\omega}{c} + \frac{1}{8} \qquad (6\text{-}6.1)$$

where $S = 2(L_x L_y + L_y L_z + L_z L_x)$ is the total surface area of the room and $L = 4(L_x + L_y + L_z)$ is the total length of all the edges in the room.

In the limit $V \gg 6S/(16\omega/c)$ (room dimensions large compared with a wavelength), the first term predominates. Although the above was derived for a rectangular room, the same leading term holds† for a room of any shape; that is, $(c/\omega)^3 N(\omega)/V$ approaches $1/6\pi^2$ in the limit of large ω.

The number of modes in a frequency band of width $\Delta\omega$ and centered at angular frequency ω can be estimated as $[dN(\omega)/d\omega]\,\Delta\omega$, with $N(\omega)$ taken as the leading term in the above. Thus, the average number of modes per unit angular frequency bandwidth (*modal density*) is

$$\frac{dN}{d\omega} = \frac{1}{2}\frac{V}{\pi^2}\frac{\omega^2}{c^3} = \frac{1}{(\Delta\omega)_{\text{mode}}} = \frac{1}{2\pi(\Delta f)_{\text{mode}}} \qquad (6\text{-}6.2)$$

Here $(\Delta f)_{\text{mode}}$ is the average spacing in hertz between successive room resonance frequencies. For example, in a room of volume 500 m^3 and near frequencies of 500 Hz, with c = 340 m/s, one has $(\Delta f)_{\text{mode}}$ = 0.025 Hz.

The Schroeder Cutoff Frequency

If the quantity $1/c\tau$ [see Eq. (6-5.20)] is sufficiently small compared with $k(n) - k(n-1)$ or $k(n+1) - k(n)$, a resonance is apparent whenever the source driving frequency ω is sufficiently close to the natural frequency $\omega(n)$. The nth term in sums such as those in Eqs. (6-5.23) and (6-5.26) becomes overwhelmingly larger than any other term as $\omega \to \omega(n)$ and the frequency dependence of p^2, of E', or of $\bar{\mathcal{P}}$ is approximately described by the factor $\{[k^2 - k^2(n)]^2 + k^2/c^2\tau^2\}^{-1}$. Near such a resonance this in turn is approximately $[c^2/2\omega(n)]^2\{[\omega - \omega(n)]^2 + (1/2\tau)^2\}^{-1}$. This factor is down to one-half its maximum value when $|\omega - \omega(n)| = 1/2\tau$, so the Q of the resonance is $\omega(n)\tau$ or $k(n)c\tau$; the bandwidth of the resonance peak is therefore

$$(\Delta\omega)_{\text{res}} = \frac{1}{\tau} \qquad (\Delta f)_{\text{res}} = \frac{6 \ln 10}{2\pi T_{60}} = \frac{2.20}{T_{60}} \qquad (6\text{-}6.3)$$

The latter, representing the bandwidth in hertz, is $(\Delta\omega)_{\text{res}}/2\pi$.

When the resonance peaks are closer together than the bandwidth associated with any one peak, the resonances are less evident. If the average spac-

† H. Weyl, "The Asymptotic Distribution Law for the Eigenvalues of Linear Partial Differential Equations (with Application to the Theory of Black Body Radiation)," *Math. Ann.*, **71**:441–479 (1912). A general proof is given by Courant and Hilbert, *Methods of Mathematical Physics*, vol. 1, pp. 429–445.

ing $(\Delta f)_{\mathrm{mode}}$ between peaks is of the order of or less than, say, $\frac{1}{3}(\Delta f)_{\mathrm{res}}$, the resonance peaks may be regarded† as a smoothed-out continuum. Since the average spacing $(\Delta f)_{\mathrm{mode}}$ decreases with increasing frequency, there is a frequency f_{Sch} (*Schroeder cutoff frequency*) below which $(\Delta f)_{\mathrm{res}} > 3(\Delta f)_{\mathrm{mode}}$ is not satisfied and above which it is. This frequency is identified, from Eqs. (2) and (3), as

$$f_{\mathrm{Sch}} = \left(\frac{c^3}{4 \ln 10}\right)^{1/2} \left(\frac{T_{60}}{V}\right)^{1/2} = c\left(\frac{6}{A_s}\right)^{1/2} \tag{6-6.4}$$

This, in SI units and with $c = 340$ m/s, becomes (in round numbers) $2000(T_{60}/V)^{1/2}$. Thus, for a room with $V = 500\ \mathrm{m}^3$ and with $T_{60} = 1$ s, the Schroeder cutoff frequency is 90 Hz. Note that the criterion $f \gg f_{\mathrm{Sch}}$ is equivalent to that previously derived in Sec. 6-3 for the deviation $\Delta\bar{\mathcal{P}}$ of the source power output to be small compared with $\bar{\mathcal{P}}_{\mathrm{ff}}$.

Approximation of Modal Sums by Integrals

What can be termed *Schroeder's rule* says that above the Schroeder cutoff frequency a sum over mode indices can be approximated by an integral. Suppose one has a sum of the generic form [see Eqs. (6-5.23) and (6-5.26)]

$$\mathrm{Sum} = \sum_n F(k(n), k)\Psi^2(\mathbf{x}_0, n) \tag{6-6.5}$$

and suppose also that there are a large number of terms of comparable magnitude for which $k(n)$ is between $k' - \Delta k'/2$ and $k' + \Delta k'/2$ for a $\Delta k'$ considerably less than k'. The number of terms corresponding to this wave-number interval is $c(dN/d\omega)_{\omega=ck'}\,\Delta k'$, where $dN/d\omega$ is the modal density of Eq. (2). If the average $\langle F\Psi^2\rangle_{k'}$ over the terms corresponding to such a wave-number interval varies slowly from interval to interval, the sum is approximately the integral

$$\mathrm{Sum} \to c\int_0^\infty \langle F\Psi^2\rangle_{k'} \left(\frac{dN}{d\omega}\right)_{\omega=ck'} dk' \tag{6-6.6}$$

The various assumptions just stated increase in validity the larger $k(n)$ is compared with $2\pi f_{\mathrm{Sch}}/c$. Insofar as the dominant contribution comes from terms where $k(n)$ is comparable to or larger than k, the integral (6) approximates the sum (5) with increasing success the larger the source frequency is compared with f_{Sch}. In the computation of the energies associated with the reverberant field, the upper limit should be replaced by a fraction (whose exact value should be of no consequence) of the reciprocal of the radius of reverberation.

† M. Schroeder, "The Statistical Parameters of Frequency Curves of Large Rooms," *Acustica*, **4**:594–600 (1954); M. R. Schroeder and K. H. Kuttruff, "On Frequency Response Curves in Rooms: Comparison of Experimental, Theoretical, and Monte Carlo Results for the Average Frequency Spacing between Maxima," *J. Acoust. Soc. Am.*, **34**:76–80 (1962). The first reference placed the transitional peak spacing at $\frac{1}{10}(\Delta f)_{\mathrm{res}}$, but this was changed to $\frac{1}{3}(\Delta f)_{\mathrm{res}}$ in the 1962 paper.

Because there is no systematic relation between the F's and Ψ^2's, the local average $\langle F\Psi^2\rangle_{k'}$ can be factored as $\langle F\rangle_{k'}\langle\Psi^2\rangle_{k'}$ to a good approximation if a great number of terms are involved. Thus, with $dN/d\omega$ taken from Eq. (2), one has

$$\text{Sum} \rightarrow \frac{V}{2\pi^2}\int_0^\infty F(k', k)R_P(k', \mathbf{x}_0)(k')^2\, dk' \tag{6-6.7}$$

where $R_P(k', \mathbf{x}_0)$ replaces $\langle\Psi^2\rangle_{k'}$ and is the average over n of those $\Psi^2(\mathbf{x}_0, n)$ for which $k(n)$ is in a small interval centered at k'.

Modal Averages of Squares of Eigenfunctions

The quantity $R_P(k, \mathbf{x}_0)$ can be alternately expressed as the ratio of the acoustic power output $\bar{\mathcal{P}}$ (time-averaged) of a monopole source at $\mathbf{x}_0$, with account taken of the proximity of the source to the nearest walls only, to the free-field acoustic power $\bar{\mathcal{P}}_{\text{ff}}$

$$R_P(k, \mathbf{x}_0) = \frac{\bar{\mathcal{P}}(k, \mathbf{x}_0)}{\bar{\mathcal{P}}_{\text{ff}}} \tag{6-6.8}$$

Here $k = \omega/c$, where ω is the frequency of the source generating power $\bar{\mathcal{P}}$.

The above assertion follows from the observation that the average of a large number N of $\Psi^2(\mathbf{x}, n)$ corresponding to nearly the same eigenvalue is approximately

$$\frac{1}{N}\sum_n \Psi^2(\mathbf{x}, n) \approx \frac{1}{N}\left|\sum_n \Psi(\mathbf{x}, n)e^{i\phi_n}\right|^2 = \left|\hat{q}(\mathbf{x})\right|^2$$

where the ϕ_n are randomly selected phase angles. The cross terms such as $2\Psi(\mathbf{x}, n)\Psi(\mathbf{x}, m)\cos(\phi_n - \phi_m)$, $n \neq m$, have a large variety of magnitudes and may have either sign, so they average out. The quantity $\hat{q}$ identified from the latter relation is an approximate solution of the Helmholtz equation whose normal derivative at the walls is zero. Within any localized region large compared with a wavelength, one can approximate $\hat{q}$ by a large number of plane waves uniformly distributed among propagation directions. Near the walls of the room, the relationships between the phases of these plane waves must be such that the boundary condition $\nabla\hat{q}\cdot\mathbf{n}_{\text{out}} = 0$ is satisfied. The overall volume average of $|\hat{q}|^2$ is 1, and for the most part $|\hat{q}|^2$ should be everywhere equal to its volume average, except near the walls of the room, where there are systematic relations between the phases of its constituent plane waves. Thus, $|\hat{q}(\mathbf{x})|^2 \rightarrow 1$ at distances far from a room boundary.

If $\mathbf{x}$ is near a particular wall, then [as in the derivation of Eq. (6-1.8)] the above reasoning suggests that $|\hat{q}(\mathbf{x})|^2$ is a constant times the average over incidence directions $\mathbf{n}$ of the mean squared pressure at $\mathbf{x}$ resulting when a plane wave of unit amplitude is incident obliquely on the wall with direction $\mathbf{n}$ and the wall is idealized as rigid. The multiplicative constant is chosen so that $|\hat{q}(\mathbf{x})|^2$ approaches 1 at large distances from the wall. Alternately, a unit-amplitude incident plane wave can be regarded as being generated by a point source of monopole amplitude $\hat{S} = d$ located at $\mathbf{x} - \mathbf{n}d$, where d is large. The principle of reciprocity requires the corresponding $|\hat{p}^2(\mathbf{x})|$ be the same as the $|\hat{p}^2(\mathbf{x} - \mathbf{n}d)|$ resulting when the point-source location is changed to $\mathbf{x}$. Consequently, the mean squared pressure at $\mathbf{x}$ due to a unit-amplitude incident plane wave is proportional to the far-field radiation pattern from a point source at $\mathbf{x}$, the proportionality factor being independent of direction. This implies that averaging over incidence directions is equivalent,† apart from a multiplicative constant, to determination of the power $\bar{\mathcal{P}}$ radiated from a source at $\mathbf{x}$. Since $|\hat{q}(\mathbf{x})|^2$ must approach 1 at large distances from the wall, and since $\bar{\mathcal{P}} \to \bar{\mathcal{P}}_{\mathrm{ff}}$ at such distances, one arrives at Eq. (8).

The correspondence described above requires $R_P(k, \mathbf{x})$ to be nearly 1 within the interior of the room, to be 2 on most wall surfaces, to be 4 along an intersection of two walls, and to be 8 at a corner where three walls meet. These values can be derived by the method of images (see Sec. 5-1) and are supported by calculations‡ of modal sums.

Evaluation of Modal Integrals

The integral in Eq. (7) approximates the sums, represented by Eqs. (6-5.23) and (6-5.26), that give $\overline{p^2}$ and $\bar{\mathcal{P}}$ for a point source in a room. For both cases, the function $F(k', k)$ is of the form

$$F(k', k) = \frac{K}{(k^2 - k'^2)^2 + k^2/c^2\tau^2} \tag{6-6.9}$$

Because $F(k', k)$ peaks strongly near $k' = k$ when $1/c\tau \ll k$, a good approximation results if we set $k' = k$ in the integrand except in the denominator factor, where we replace $k^2 - k'^2$ by $2k(k - k')$. Thus Eq. (7) becomes

$$\text{Sum} \to \frac{KV}{8\pi^2}\frac{\bar{\mathcal{P}}(k, \mathbf{x}_0)}{\bar{\mathcal{P}}_{\mathrm{ff}}}\int_0^\infty \frac{dk'}{(k - k')^2 + 1/(2c\tau)^2} \tag{6-6.10}$$

Given $k \gg 1/c\tau$, one may in addition make the further approximation of extending the lower limit to $-\infty$, so that the indicated integral [change integration variable to θ where $k' - k = (1/2c\tau)\tan\theta$] becomes $2\pi c\tau$.

In the application of the above analysis to the expressions, derivable from Eqs. (6-5.23) and (6-5.26) for the volume average of mean squared pressure and the acoustic-power output of a monopole source, the appropriate identifications for K are $4\pi\,\rho c\bar{\mathcal{P}}_{\mathrm{ff}}/V^2$ and $4\pi\bar{\mathcal{P}}_{\mathrm{ff}}/(c\tau V)$. Thus, the two quantities just mentioned become

$$\overline{p^2} = \rho c^2\tau\bar{\mathcal{P}}(k, \mathbf{x}_0)/V \qquad \bar{\mathcal{P}} = \bar{\mathcal{P}}(k, \mathbf{x}_0) \tag{6-6.11}$$

† R. V. Waterhouse, "Output of a Sound Source in a Reverberation Chamber and other Reflecting Environments," *J. Acoust. Soc. Am.*, **30**:4–13 (1958).

‡ See, for example, Maling, "Calculation of the Acoustic Power."

The potential energy E_P in the room is consequently $\frac{1}{2}\tau\bar{\mathscr{P}}$. An analogous derivation for the reverberant part of the kinetic energy leads with the summation truncation described in the previous section to

$$E'_K \approx \frac{\bar{\mathscr{P}}}{\pi c}\int_0^{k_m}\frac{(k'/k)^4\,dk'/k^2}{(1-k'/k)^2(1+k'/k)^2+1/(c\tau k)^2} \tag{6-6.12}$$

where the upper limit k_m should be much less than $\frac{1}{2}\pi c\tau k^2$ but much larger than $2/\pi c\tau$. The dominant contribution to the integration comes from k' near k, so an appropriate approximation sequence is to first set $k'/k = 1$ except in the factor $1 - k'/k$ and to *then* change the integration limits to $-\infty$ and ∞. Doing this yields $E'_K \approx \frac{1}{2}\tau\bar{\mathscr{P}}$, so $E' = E'_P + E'_K \approx \tau\bar{\mathscr{P}}$.

The similarity of the approximate relations derived above between $\bar{\mathscr{P}}$, $\overline{p^2}$, $E' = V\bar{w}$, and τ with what holds in steady-state circumstances for the Sabine-Franklin-Jaeger model demonstrates that the latter has a substantial basis in the wave theory of sound but holds only in the high-frequency limit, i.e., for f somewhat larger than f_{Sch}. While the analysis given here is for a constant-frequency point source, one can expect the same conclusions to apply to any type of source if the radiated frequencies are sufficiently high and the dimensions of the room sufficiently large. [However, the value of $\bar{\mathscr{P}}(k, \mathbf{x}_0)/\bar{\mathscr{P}}_{\text{ff}}$ will not necessarily be the same as what is derived for a monopole source. For a point dipole, for example, with its dipole-moment vector normal to the nearest wall, one would use Eq. (5-1.8*b*).]

6-7 STATISTICAL ASPECTS OF ROOM ACOUSTICS

Deviations of acoustic field quantities from the averages predicted by the Sabine-Franklin-Jaeger model are frequently given a statistical interpretation. Suppose, for example, that a source at $\mathbf{x}_0$ causes the contribution to the pressure from a given frequency band to be $p(\mathbf{x}, t|\mathbf{x}_0)$. The average over time and over listener position $\mathbf{x}$ of p^2 is predicted to be $\rho c^2\tau\bar{\mathscr{P}}/V$ by the reverberant-field model, but the model per se gives no information about how much a given average over time of $p^2(\mathbf{x}, t|\mathbf{x}_0)$ for fixed $\mathbf{x}$ and $\mathbf{x}_0$ may deviate from this double average.

A *probability density function* $w(q)$ for any field variable $q(\mathbf{x})$ can be constructed by measuring $q(\mathbf{x})$ at a large number of randomly selected points. The fraction of the total number of measured values between q_a and q_b is interpreted as the probability $P(q_b > q > q_a)$ that q falls within this range. The average probability per unit range of q is $P(q_b > q > q_a)/(q_b - q_a)$, and this ratio's value in the quasi limit of small $q_b - q_a$ is the probability density function $w(q)$ evaluated at the center of the interval. Thus, $w(q)\,dq$ is the probability that a random measurement is between $q - dq/2$ and $q + dq/2$.

The *expected value* of a function $f(q)$ can be written in two ways:

$$\langle f(q)\rangle = \int_{-\infty}^{\infty} f(q)w(q)\,dq = \frac{1}{V}\iiint f(q(\mathbf{x}))\,dV \tag{6-7.1}$$

The latter defines the "randomly selected points" to be such that the numbers of samples drawn from two subvolumes of equal size are the same.

One also defines a *joint-probability-density function* $w(q_1, q_2)$ for any two field variables $q_1(\mathbf{x})$ and $q_2(\mathbf{x})$ such that $w(q_1, q_2)\, dq_1\, dq_2$ is the probability that q_1 and q_2 simultaneously lie within the ranges $(q_1 - dq_1/2,\ q_1 + dq_1/2)$ and $(q_2 - dq_2/2,\ q_2 + dq_2/2)$. This function should be such that the expected value of any function $f(q_1, q_2)$ is the average over volume of $f(q_1(\mathbf{x}), q_2(\mathbf{x}))$. The integral of $w(q_1, q_2)$ over all values of q_2 yields the probability density function $w(q_1)$ for q_1.

Frequency Correlation

A starting point for the development of the principal hypotheses of statistical room acoustics may be taken as the expression (6-5.20) for the complex acoustic-pressure amplitude caused by a constant-frequency point source in the nearly rigid wall approximation. This we rewrite as

$$\hat{p}(\mathbf{x}, \omega|\mathbf{x}_0) = \frac{4\pi\hat{S}}{V}(a + ib) \tag{6-7.2}$$

where
$$\{a, b\} \approx \sum_n \{A_n, B_n\}\, \Psi(\mathbf{x}, n)\, \Psi(\mathbf{x}_0, n) \tag{6-7.3}$$

$$\{A_n, B_n\} = \frac{\{[k^2(n) - k^2],\ k/c\tau\}}{[k^2(n) - k^2]^2 + k^2/c^2\tau^2} \tag{6-7.4}$$

For given $\mathbf{x}$ and $\mathbf{x}_0$, the plots of $a(\omega)$ and $b(\omega)$ versus ω are calculable, but since the curves vary with $\mathbf{x}$, one may consider† $a(\omega)$ and $b(\omega)$ as stochastic processes. At frequencies somewhat above the Schroeder cutoff frequency, these are quasi-stationary processes because their statistical properties are insensitive to shifts in the frequency origin. Each process has zero mean since the spatial average is zero for each $\Psi(\mathbf{x}, n)$ (we assume that the zero-frequency mode is negligibly excited). Also, since a large number of terms contribute to their values, each of which could as well be negative as positive, one expects, with reference to various proofs under restricted conditions of the *central-limit theorem*,‡ that the pair $a(\omega)$, $b(\omega)$ forms a *joint gaussian process*. This implies, in particular, that if one lets each $q_1, q_2, \ldots, q_N$ denote either $a(\omega_i)$ or $b(\omega_i)$ for various selected frequencies ω_i, the joint-probability-density function for the set of q's is

$$w(q_1, q_2, \ldots, q_N) = (2\pi)^{-N/2} \det [M]^{-1/2} \exp\left(-\frac{1}{2}\sum_{i,j} [M^{-1}]_{ij} q_i q_j\right) \tag{6-7.5}$$

† M. Schroeder, "The Statistical Parameters of Frequency Curves of Large Rooms," *Acustica*, **4**:594–600 (1954).

‡ J. L. Doob, *Stochastic Processes*, Wiley, New York, 1953, pp. 71–72, 141.

where det $[M]$ and $[M^{-1}]$ denote the determinant and inverse, respectively, of a *correlation matrix* $[M]$ having elements $M_{ij} = \langle q_i q_j \rangle$. This, with the assumption that the processes are quasi-stationary, leads to the conclusion that the only statistical averages needed for a specification of all such probability density functions are the *frequency autocorrelation functions* $\langle a(\omega)a(\omega + \Delta\omega)\rangle$ and $\langle b(\omega)b(\omega + \Delta\omega)\rangle$ and the *frequency cross-correlation function* $\langle a(\omega)b(\omega + \Delta\omega)\rangle$.

Expressions for these functions follow from Eq. (1) and from the orthogonality and normalization of the $\Psi(\mathbf{x}, n)$. One has, for example [with $A_n(\omega)$ rewritten as $A(k(n), k)$],

$$\langle a(\omega)b(\omega + \Delta\omega)\rangle = \sum_n A(k(n), k)B(k(n), k + \Delta k)\Psi^2(\mathbf{x}_0, n) \qquad (6\text{-}7.6)$$

This sum is approximated by an integral in the manner described in the derivation of Eqs. (6-6.7), with $k(n) \to k'$, $\langle \Psi^2 \rangle \to R_P(k', \mathbf{x}_0)$, $\Delta n \to (V/2\pi^2)(k')^2\, dk'$. Since the overall integrand is for most intents zero unless k' is moderately close to k (given $|\Delta k|$ and $1/c\tau$ both substantially less than k), one sets $k' = k$ in the factors $R_P(k', \mathbf{x}_0)$ and $(k')^2$ at the outset and approximates

$$A(k', k) \approx \frac{2k(k' - k)}{4k^2(k' - k)^2 + k^2/c^2\tau^2} \qquad B(k', k) \approx \frac{k/c\tau}{4k^2(k' - k)^2 + k^2/c^2\tau^2} \qquad (6\text{-}7.7)$$

Also, since $\Delta k \ll k$, the only tangible effects of shifting k to $k + \Delta k$ arise in the factor $k' - k$; everywhere else in the expression for $B(k', k + \Delta k)$, one sets $k + \Delta k$ to k. A further approximation replaces the lower limit of integration by $-\infty$. Then, with a change of variable to β, where $\beta/2c\tau$ is $k' - k$, one obtains

$$\langle a(\omega)b(\omega + \Delta\omega)\rangle \approx \frac{V}{4\pi^2} R_P(k, \mathbf{x}_0)c\tau \int_{-\infty}^{\infty} \frac{\beta\, d\beta}{(\beta^2 + 1)[(\beta - 2\tau\,\Delta\omega)^2 + 1]} \qquad (6\text{-}7.8)$$

The indicated integral is performed by adding a semicircular arc ($\beta = Re^{i\phi}$, $0 < \phi < \pi$, $R \to \infty$) to the integration path such that the resulting contour encloses the poles (at $\beta = i$ and $\beta = 2\tau\,\Delta\omega + i$) in the upper half plane. The result, by the residue theorem, is $(\pi/2)\tau\,\Delta\omega/[1 + (\tau\,\Delta\omega)^2]$.

The evaluation of $\langle a(\omega)a(\omega + \Delta\omega)\rangle$ and $\langle b(\omega)b(\omega + \Delta\omega)\rangle$ is performed similarly, a distinction being that the β's of the numerator in Eq. (8) are replaced by $\beta(\beta - 2\tau\,\Delta\omega)$ and 1, respectively. The integral factor in both cases is $(\pi/2)/[1 + (\tau\,\Delta\omega)^2]$. The three correlation functions consequently vary with $\Delta\omega$ in the following manner:

$$\langle a(\omega)a(\omega + \Delta\omega)\rangle \approx \langle b(\omega)b(\omega + \Delta\omega)\rangle \approx \frac{\langle a^2(\omega)\rangle}{1 + (\tau\,\Delta\omega)^2} \qquad (6\text{-}7.9a)$$

$$\langle a(\omega)b(\omega + \Delta\omega)\rangle \approx \frac{\langle a^2(\omega)\rangle\tau\,\Delta\omega}{1 + (\tau\,\Delta\omega)^2} \qquad (6\text{-}7.9b)$$

The above expressions are applicable for estimation of the frequency autocorrelation function $\langle \overline{p^2}(\omega)\overline{p^2}(\omega + \Delta\omega)\rangle$ for the ensemble of frequency-response curves $\overline{p^2}(\omega, \mathbf{x})$. Here $p^2(\omega, \mathbf{x})$ is the time average of the squared acoustic pressure when the source's frequency is ω. If the source characteristics vary slowly with ω, and if they change negligibly over an interval $\Delta\omega$, then (for $\Delta\omega \ll \omega$)

$$\frac{\langle \overline{p^2}(\omega)\overline{p^2}(\omega + \Delta\omega)\rangle}{\langle \overline{p^2}(\omega)\rangle} = \frac{\langle [a^2(\omega) + b^2(\omega)][a^2(\omega + \Delta\omega) + b^2(\omega + \Delta\omega)]\rangle}{\langle a^2 + b^2\rangle^2} \tag{6-7.10}$$

To evaluate this, we use the relation,† applicable if x and y are any two random variables, for example, $a(\omega)$ and $a(\omega + \Delta\omega)$ or $b(\omega)$ and $a(\omega + \Delta\omega)$, with a joint gaussian probability distribution and zero mean, that

$$\langle x^2y^2\rangle = \langle x^2\rangle\langle y^2\rangle + 2\langle xy\rangle^2 \tag{6-7.11}$$

This, in conjunction with Eqs. (9), leads to $4\langle a^2\rangle^2\{1 + [1 + (\tau\,\Delta\omega)^2]^{-1}\}$ for the numerator of Eq. (10), so we obtain

$$\langle \overline{p^2}(\omega)\overline{p^2}(\omega + \Delta\omega)\rangle = \langle \overline{p^2}\rangle^2\{1 + [1 + (\tau\,\Delta\omega)^2]^{-1}\} \tag{6-7.12}$$

The Poisson Distribution

For pure-tone excitation above the Schroeder cutoff frequency, the mean squared acoustic pressure conforms to a Poisson distribution. The demonstration proceeds from the observation that

$$w(s) = \frac{d}{ds}\int_{-\infty}^{\infty}\!\!\int w(a, b)H(s - a^2 - b^2)\,da\,db \tag{6-7.13}$$

is the probability density function for $a^2 + b^2$. Here H is the Heaviside unit step function; the double integral is the probability that $a^2 + b^2 < s$; its derivative is thus the probability density function. Since the random variables a and b are uncorrelated for $\Delta\omega = 0$, since both individually correspond to a gaussian distribution with zero mean, and since both have the same mean squared value, the exponent in Eq. (5) in this particular case becomes $-(a^2 + b^2)/2\langle a^2\rangle$. One converts the integration variables in Eq. (13) to polar coordinates u, ϕ such that $a = u\cos\phi$, $b = u\sin\phi$, $da\,db = u\,du\,d\phi$, $a^2 + b^2 = u^2$, and then lets $u^2 = v$ such that $u\,du = \frac{1}{2}dv$; the ϕ integration gives a factor 2π; the v integration limits are 0 and s. The s differentiation then gives $\pi w(a, b)$ with $a^2 + b^2 = s$, so $w(s) = (1/\langle s\rangle)\exp(-s/\langle s\rangle)$, which is the probability density function for a Poisson distribution. Here $\langle s\rangle = 2\langle a^2\rangle$ is the average value $\langle a^2\rangle + \langle b^2\rangle$ of s.

† From (5) one has, for a bivariate gaussian distribution with $q_1 = x$, $q_2 = y$, $r = \langle xy\rangle/\langle y^2\rangle$,

$$\sum_{i,j}[M^{-1}]_{ij}q_iq_j = \frac{\langle y^2\rangle x^2 - 2\langle xy\rangle xy + \langle x^2\rangle y^2}{\langle x^2\rangle\langle y^2\rangle - \langle xy\rangle^2} = \frac{(x - ry)^2}{\langle (x - ry)^2\rangle} + \frac{y^2}{\langle y^2\rangle}$$

so $w(x, y)$ factors into a product of probability density functions for the statistically independent quantities $x - ry$ and y. Also, Eq. (5) yields $\langle y^4\rangle = 3\langle y^2\rangle^2$. Consequently, algebraic manipulation of the expression $\langle[(x - ry) + ry]^2y^2\rangle$ leads to Eq. (11).

Since the time average $\overline{p^2}$ of p^2 is a product of a nonrandom (i.e., independent of **x**) quantity with $a^2 + b^2$ and since, for any random variable x with probability density function $w_x(x)$, the probability density function $w_y(y)$ for $y = Kx$ is $w_x(y/K)/K$, such that $w_x(x)\,dx = w_y(y)\,dy$, the quantity $\overline{p^2}$ also conforms to a Poisson distribution, i.e.,

$$w(\overline{p^2}) = \frac{1}{\langle\overline{p^2}\rangle} \exp \frac{-\overline{p^2}}{\langle\overline{p^2}\rangle} \tag{6-7.14}$$

where $\langle\overline{p^2}\rangle$ is the spatial average of $\overline{p^2}$. (The overbar here implies a time average.)

The most probable value of $\overline{p^2}$ is 0, but since $\overline{p^2}$ is always nonnegative, the expected value is finite. The variance is

$$\langle(\overline{p^2} - \langle\overline{p^2}\rangle)^2\rangle = \langle(\overline{p^2})^2\rangle - \langle\overline{p^2}\rangle^2 = \langle\overline{p^2}\rangle^2 \tag{6-7.15}$$

since the integrals of xe^{-x} and x^2e^{-x} from 0 to ∞ are 1 and 2. Thus, the rms deviation of a measurement of $\overline{p^2}$ from $\langle\overline{p^2}\rangle$ is the same as $\langle\overline{p^2}\rangle$. [This is consistent with Eq. (12) in the limit $\tau\,\Delta\omega = 0$.] The probability that $\overline{p^2}$ exceeds $\langle\overline{p^2}\rangle$ is e^{-1} or 0.368, and the probability that it is less than the average is $1 - e^{-1} = 0.632$, so at a randomly selected point, it is nearly twice as probable that $\overline{p^2}$ will be less than the average rather than higher than the average.

The Poisson distribution requires also that the average sound-pressure level be 2.5 dB lower than that corresponding to $\langle\overline{p^2}\rangle$. To demonstrate this, let $z = \frac{1}{10}(\ln 10)(L - L_0)$, where L_0 is the sound level corresponding to the average $\langle\overline{p^2}\rangle$. Then, since $\overline{p^2}/\langle\overline{p^2}\rangle = 10^{(L-L_0)/10}$ is e^z, the probability density function for z is (see Fig. 6-16)

$$w(z) = \exp\left(\frac{-\overline{p^2}}{\langle\overline{p^2}\rangle}\right)\frac{d}{dz}\left(\frac{\overline{p^2}}{\langle\overline{p^2}\rangle}\right) = e^{z-e^z} \qquad -\infty < z < \infty \tag{6-7.16}$$

The expected value $\langle z\rangle$ for z (with a change of integration variable to $y = e^z$) is

$$\int_{-\infty}^{\infty} ze^{z-e^z}\,dz = \int_0^1 (\ln y)\frac{d}{dy}(1 - e^{-y})\,dy - \int_1^{\infty} (\ln y)\frac{d}{dy}e^{-y}\,dy = -\gamma \tag{6-7.17}$$

where $\gamma = 0.5772157 \cdots$ is the *Euler-Mascheroni* constant.† Since $10\gamma/(\ln 10)$ is 2.5, the average level $\langle L\rangle$ is $L_0 - 2.5$ dB. The probable deviation of a single measurement from L_0 is $\langle(L - L_0)^2\rangle^{1/2}$, which is the same as $[\langle(L - \langle L\rangle)^2\rangle + (\langle L\rangle - L_0)^2]^{1/2}$ or $[10/(\ln 10)][\langle(z - \langle z\rangle)^2\rangle + \gamma^2]^{1/2}$. The value $\pi^2/6$ for the quantity $\langle(z - \langle z\rangle)^2\rangle$ results from a lengthy computation,‡ so the net result is $\langle(L - L_0)^2\rangle^{1/2} = 6.1$ dB.

† E. T. Whittaker and G. N. Watson, *A Course of Modern Analysis,* 4th ed., Cambridge University Press, London, 1973, pp. 235–236, 243.

‡ H. Cramer, *Mathematical Methods of Statistics,* Princeton University Press, Princeton, N.J., 1946, p. 376. [Our $w(z)$ is Cramer's $j_1(z)$ with $\nu = 1$, such that his S_1 and S_2 are both zero.]

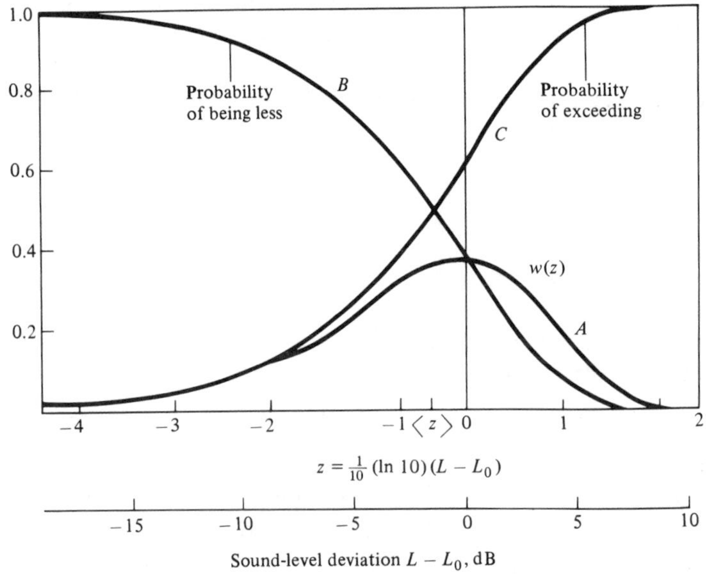

Figure 6-16 Implications of the Poisson distribution. Curve A: Probability density function $w(z)$ for $\frac{1}{10}$ (ln 10)$(L - L_0)$. Curve B: Probability $P(L)$ that measured sound-pressure level is less than L. Curve C: Probability $1 - P(L)$ that it is greater than L. The level L_0 corresponds to spatial average over entire room of mean squared acoustic pressure.

The rms deviation of L from $\langle L \rangle$ becomes† $[10/(\ln 10)]\pi/6^{1/2} = 5.6$ dB. The expected value of $(L - L_0)^2$, given $L > L_0$, is $(3.2 \text{ dB})^2$; given $L < L_0$, it is $(7.6 \text{ dB})^2$. Thus, if error brackets are to be placed on a data point, the upper bracket should be 7.6 dB above and the lower bracket 3.2 dB below, with a net spread of 10.8 dB.

Effect of Finite-Frequency Bandwidth

If the source is broadband, the variations in the mean squared pressure $\overline{p^2}$ corresponding to any finite-frequency band of bandwidth $\Delta\omega$ are considerably less than those for the constant-frequency case if $\tau\,\Delta\omega \gg 1$. To demonstrate this,‡ we consider a band extending from ω_1 to ω_2 over which the power output per unit frequency bandwidth is constant, such that the mean squared pressure for the band, according to Eqs. (2-7.7) and (2-9.6), is

$$\overline{p^2} = K \int_{\omega_1}^{\omega_2} [a^2(\omega, \mathbf{x}) + b^2(\omega, \mathbf{x})]\, d\omega \tag{6-7.18}$$

† This is in accord with measurements reported by P. Doak, "Fluctuations of the Sound Pressure Level in Rooms when the Receiver Position Is Varied," *Acustica,* **9:**1–9 (1959).

‡ M. R. Schroeder, "Effect of Frequency and Space Averaging on the Transmission Responses of Multimode Media," *J. Acoust. Soc. Am.*, **46:**277–283 (1969).

where K is independent of $\mathbf{x}$. The variance in $\overline{p^2}$ is then

$$\langle(\overline{p^2} - \langle\overline{p^2}\rangle)^2\rangle = K^2 \int_{\omega_1}^{\omega_2}\int \langle[f(\omega) - \langle f\rangle][f(\omega') - \langle f\rangle]\rangle \, d\omega \, d\omega' \quad (6\text{-}7.19)$$

where we abbreviate f for $a^2 + b^2$. A substitution from Eq. (12) then yields

$$\langle(\overline{p^2} - \langle\overline{p^2}\rangle)^2\rangle = \frac{\langle\overline{p^2}\rangle^2}{(\omega_2 - \omega_1)^2}\int_{\omega_1}^{\omega_2}\int [1 + \tau^2(\omega - \omega')^2]^{-1} \, d\omega \, d\omega' \quad (6\text{-}7.20)$$

The double integration can be performed by letting $x = (\omega - \omega_1)/(\omega_2 - \omega_1)$, $y = (\omega' - \omega_1)/(\omega_2 - \omega_1)$ be new integration variables (limits 0 and 1) such that $\omega - \omega' = (x - y)\,\Delta\omega$, where we write $\Delta\omega$ for $\omega_2 - \omega_1$. A further substitution of α for $x - y$ replaces the x integration by one on α from $-y$ to $1 - y$, so one has $0 < y < 1 - \alpha$ for α between 0 and 1 and $-\alpha < y < 1$ for α between -1 and 0. With this recognition, one can do the y integration first, keeping α fixed, the result being $1 - |\alpha|$, so

$$\langle(\overline{p^2} - \langle\overline{p^2}\rangle)^2\rangle = \langle\overline{p^2}\rangle^2 \int_{-1}^{1} (1 - |\alpha|)[1 + (\tau\,\Delta\omega)^2\alpha^2]^{-1} \, d\alpha \quad (6\text{-}7.21)$$

and here it is sufficient to integrate only from 0 to 1 and subsequently multiply the result by 2. A further change of integration variable to θ, where $\tan\theta = (d/d\theta)\ln(\sec\theta)$ replaces $(\tau\,\Delta\omega)\alpha$, yields (see Fig. 6-17)

$$\langle(\overline{p^2} - \langle\overline{p^2}\rangle)^2\rangle = \langle\overline{p^2}\rangle^2 \, V(\tau\,\Delta\omega) \quad (6\text{-}7.22)$$

$$V(\tau\,\Delta\omega) = \frac{2}{\tau\,\Delta\omega}\{\tan^{-1}(\tau\,\Delta\omega) - (\tau\,\Delta\omega)^{-1}\ln[1 + (\tau\,\Delta\omega)^2]^{1/2}\}$$
$$\approx \begin{cases} 1 - \frac{1}{6}(\tau\,\Delta\omega)^2 & \tau\,\Delta\omega \ll 1 \\ \dfrac{\pi}{\tau\,\Delta\omega} - \dfrac{2\ln(e\tau\,\Delta\omega)}{(\tau\,\Delta\omega)^2} & \tau\,\Delta\omega \gg 1 \end{cases} \quad (6\text{-}7.23)$$

The behavior when $\tau\Delta\omega \to 0$ is consistent with Eq. (15) for the single-frequency case. Also, the leading term $\pi/(\tau\,\Delta\omega)$ in the asymptotic expansion for $V(\tau\,\Delta\omega)$ is the same as would be obtained if there were $N = (\tau\,\Delta\omega)/\pi$ discrete widely spaced frequencies, each equally strongly excited. The leading term can also be written in terms of T_{60} and the bandwidth Δf in hertz as $3\ln 10/(T_{60}\,\Delta f)$ or as $6.9/(T_{60}\,\Delta f)$.

With the last recognition, one can conjecture that, in the limit of large $\tau\,\Delta\omega$, the probability density function $w(\overline{p^2})$ is the same as that of the sum of N independent random variables each having a Poisson distribution and the same mean, $\langle\overline{p^2}\rangle/N$. After a brief calculation similar to that in the derivation of Eq. (14), this conjecture leads to

$$w(\overline{p^2}) = \frac{1}{\Gamma(N)}\frac{N}{\langle\overline{p^2}\rangle}\left(\frac{N\overline{p^2}}{\langle\overline{p^2}\rangle}\right)^{N-1}\exp\left(-\frac{N\overline{p^2}}{\langle\overline{p^2}\rangle}\right) \quad (6\text{-}7.24)$$

where $\Gamma(N)$ [equal to $(N-1)!$ for integer N] is the gamma function. This reduces to Eq. (14) for $N = 1$ and has a mean of $\langle\overline{p^2}\rangle$ (as it should) and a variance of $\langle\overline{p^2}\rangle^2/N$. A comparison of the latter with Eq. (22) suggests that the

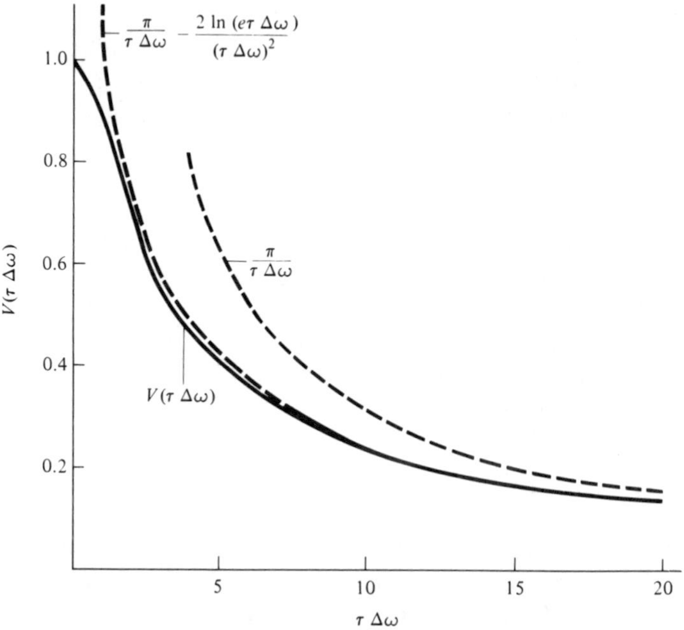

Figure 6-17 Function $V(\tau\,\Delta\omega)$ describing variance in $\overline{p^2}$ for sound of angular frequency bandwidth $\Delta\omega$ in a room with characteristic energy decay time τ. Also plotted are two approximate asymptotic expressions for the function.

above would be a fairly good approximate probability density function for arbitrary bandwidth if we set $N = 1/V(\tau\,\Delta\omega)$.

With $z = (\frac{1}{10}\ln 10)(L - L_0)$, as before, and with L_0 representing the sound-pressure level associated with $\langle\overline{p^2}\rangle$, the corresponding probability density function $N\exp(Nz - e^{Nz})$ has a mean of $-\gamma/N$ and a variance of $(\pi^2/6)/N^2$. Thus, the average sound-pressure level $\bar{L}$ is $L_0 - 2.5/N$ dB, and the rms deviation from $\bar{L}$ is $5.6/N$ dB.

Example For the third octave band with $f_0 = 250$ Hz in a room with a reverberation time of 1 s, what is the probability that L lies within ± 0.5 dB of L_0?

SOLUTION From the relations $T_{60} = (6\ln 10)\tau$ and $\Delta\omega = 2\pi(2^{1/6} - 2^{-1/6})f_0$ (third octave band) one determines $\tau\,\Delta\omega = 26.33$, and from $N = 1/V(\tau\,\Delta\omega)$ one finds $N = 9.35$. Since $\pm\frac{1}{2}$ dB corresponds to a z of $\pm(\ln 10)/20 = \pm 0.115$, the desired probability is the integral of $N\exp(Nz - e^{Nz})$ from -0.115 to 0.115; this integral is the difference of the values of $-\exp(-e^{Nz})$ (the indefinite integral) at $Nz = 1.1$ and $Nz = -1.1$, so the probability is 0.66. The probability of its lying within ± 1 dB of L_0 is similarly found to be 0.89. The corresponding probabilities for a pure tone ($N = 1$) would be 0.08 and 0.17.

6-8 SPATIAL CORRELATIONS IN DIFFUSE SOUND FIELDS

Our discussion of statistical room acoustics continues with an examination of the spatial variation of sound fields in reverberant rooms.

The Spatial Autocorrelation Function for Acoustic Pressure

The requisite statistical averages for the description of the spatial fluctuations result from the idealization of the sound field as a superposition of a large number of propagating plane waves, such that the acoustic pressure in the constant-frequency case has a complex amplitude given by Eq. (6-1.6). The autocorrelation function of the constant-frequency pressure field is the average over volume of the product of $p(\mathbf{x}, t)$ and $p(\mathbf{x} + \Delta\mathbf{x}, t + \Delta t)$ for fixed $\Delta\mathbf{x}$ and Δt; a derivation analogous to that of Eq. (6-1.7) yields

$$\langle p(\mathbf{x}, t)p(\mathbf{x} + \Delta\mathbf{x}, t + \Delta t)\rangle = \frac{1}{2}\sum_q |\hat{p}_q|^2 \cos \omega\left(\Delta t - \mathbf{n}_q \cdot \frac{\Delta\mathbf{x}}{c}\right) \qquad (6\text{-}8.1)$$

With the diffuse-field idealization, the cosine here is replaced by its average over propagation direction, and the sum of the $|\hat{p}_q|^2$ is replaced by $2\langle p^2\rangle$. The average over solid angle of $\cos [\omega(\Delta t - \mathbf{e} \cdot \Delta\mathbf{x}/c)]$ can be performed in spherical coordinates taking $\Delta\mathbf{x}$ in the z direction, so Eq. (1) reduces† to

$$\begin{aligned}\langle p(\mathbf{x}, t)p(\mathbf{x} + \Delta\mathbf{x}, t + \Delta t)\rangle &= \langle \overline{p^2}\rangle \frac{1}{2}\int_0^\pi \cos \omega\left(\Delta t - \frac{|\Delta\mathbf{x}|}{c}\cos\theta\right)\sin\theta\, d\theta \\ &= \langle \overline{p^2}\rangle \cos(\omega\, \Delta t)\,\frac{\sin k|\Delta\mathbf{x}|}{k|\Delta\mathbf{x}|} \qquad (6\text{-}8.2)\end{aligned}$$

The time periodicity with a period of $2\pi/\omega$ exhibited by the above autocorrelation function follows from the periodicity of the pressure. The spatially dependent factor is 1 when $|\Delta\mathbf{x}| = 0$ but equals 0 when $k|\Delta\mathbf{x}| = \pi, 2\pi, 3\pi, \ldots$ or when $|\Delta\mathbf{x}| = \lambda/2, \lambda, 3\lambda/2, \ldots$. Since the amplitude decreases to zero as $1/k|\Delta\mathbf{x}|$ when $k|\Delta\mathbf{x}| \to \infty$ (Fig. 6-18*a*), there is a basis for assuming that pressure measurements spaced more than several wavelengths apart are statistically independent.

An expression for the spatial autocorrelation function‡ $\langle \overline{p^2}(\mathbf{x})\overline{p^2}(\mathbf{x} + \Delta\mathbf{x})\rangle$ of the mean squared acoustic pressure results analogously from the superimposed-plane-waves hypothesis. With the recognition that the spatial average of the coupling factor $\exp [ik(\mathbf{n}_q - \mathbf{n}_{q'} + \mathbf{n}_r - \mathbf{n}_{r'}) \cdot \mathbf{x}]$ is negligibly small unless $q' = q$, $r' = r$ or $r' = q$, $r = q'$, one obtains as an intermediate

† R. K. Cook, R. V. Waterhouse, R. D. Berendt, S. Edelman, and M. C. Thompson, Jr., "Measurement of Correlation Coefficients in Reverberant Sound Fields," *J. Acoust. Soc. Am.*, **27**:1072–1077 (1955).

‡ D. Lubman, "Spatial Averaging in a Diffuse Sound Field," *J. Acoust. Soc. Am.*, **46**:532–534 (1969).

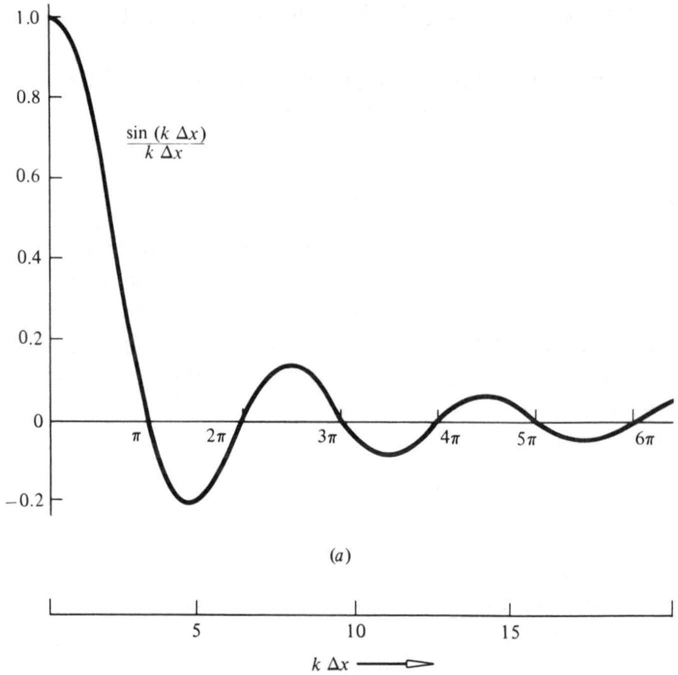

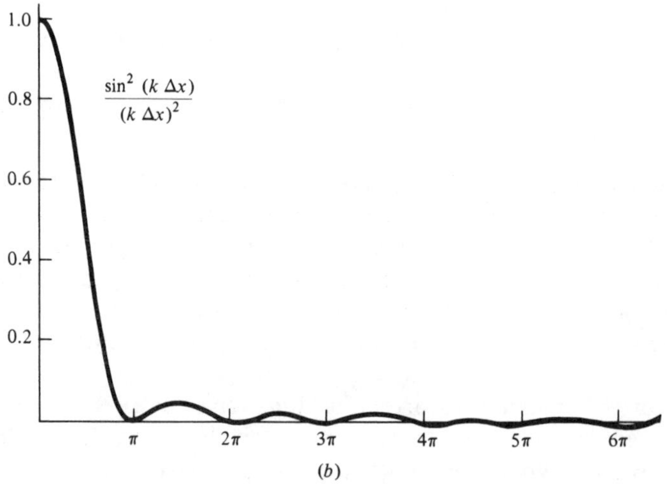

Figure 6-18 Spatial dependence of the autocorrelation functions of (*a*) acoustic pressure ($\Delta t = 0$) and (*b*) mean squared acoustic pressure in a constant-frequency sound field.

result

$$\langle \overline{p^2}(\mathbf{x})\overline{p^2}(\mathbf{x} + \Delta\mathbf{x})\rangle = \frac{1}{4}\sum_{q,r} |\hat{p}_q|^2|\hat{p}_r|^2 + \frac{1}{4}\sum_{q,r} |\hat{p}_q|^2|\hat{p}_r|^2 e^{ik(\mathbf{n}_r-\mathbf{n}_q)\cdot\Delta\mathbf{x}}$$

This in turn leads with the diffuse-field hypothesis to the expression (see Fig. 6-18*b*)

$$\langle \overline{p^2}(\mathbf{x})\overline{p^2}(\mathbf{x} + \Delta\mathbf{x})\rangle = \langle \overline{p^2}(\mathbf{x})\rangle^2 \left(1 + \left|\frac{1}{4\pi}\iint e^{ik\,\Delta\mathbf{x}\cdot\mathbf{e}}\,d\Omega\right|^2\right)$$

$$= \langle \overline{p^2}(\mathbf{x})\rangle^2 \left\{1 + \frac{\sin^2 k|\Delta\mathbf{x}|}{(k|\Delta\mathbf{x}|)^2}\right\} \tag{6-8.3}$$

Note that this function's limiting value of $2\langle\overline{p^2}\rangle$ when $|\Delta\mathbf{x}| = 0$ is consistent with Eq. (6-7.12).

The extension of the above result to when the field is composed of a band of frequencies proceeds from the notion of a spectral density, which implies

$$\left\langle\overline{p^2}(\mathbf{x})\overline{p^2}(\mathbf{x} + \Delta\mathbf{x})\right\rangle = \int_{\omega_1}^{\omega_2}\int \left\langle S_p(\omega, \mathbf{x})S_p(\omega', \mathbf{x} + \Delta\mathbf{x})\right\rangle d\omega\, d\omega' \tag{6-8.4}$$

Here $S_p(\omega, \mathbf{x})$ is such that its integral over ω gives $\overline{p^2}(\mathbf{x})$.

The average appearing in the above integrand can be written as $\langle S_p^2\rangle[1 + G(\omega, \omega', \Delta\mathbf{x})]$ with some choice of the function G. We assume that the spatial average of S_p^2 is independent of ω, so it is identified as $\langle\overline{p^2}\rangle^2/(\Delta\omega)^2$. Equations (6-7.12) and (3) require that G be $[1 + (\omega - \omega')^2\tau^2]^{-1}$ or $(\sin^2 k|\Delta\mathbf{x}|)/(k|\Delta\mathbf{x}|)^2$, when $\Delta\mathbf{x}$ is 0 or when $\omega = \omega'$. It must be 1 when both $\Delta\mathbf{x}$ and $\omega - \omega'$ are zero, and it must go to zero when $|\omega - \omega'|\tau$, $(\omega/c)|\Delta\mathbf{x}|$, or $(\omega'/c)|\Delta\mathbf{x}|$ becomes large. A simple approximate choice for G with these properties is the product of the two limiting functions corresponding to $\Delta\mathbf{x} = 0$ and $\omega - \omega' = 0$, with the replacement of k by $k_{\text{av}} = (\omega + \omega')/c$ in the latter. This synthesis yields

$$\left\langle S_p(\omega, \mathbf{x})S_p(\omega', \mathbf{x} + \Delta\mathbf{x})\right\rangle \approx \frac{\langle\overline{p^2}\rangle^2}{(\Delta\omega)^2}\left\{1 + [1 + \tau^2(\omega - \omega')^2]^{-1}\frac{\sin^2 k_{\text{av}}|\Delta\mathbf{x}|}{(k_{\text{av}}|\Delta\mathbf{x}|)^2}\right\} \tag{6-8.5}$$

For typical rooms, τ is invariably much larger than $|\Delta\mathbf{x}|/c$ for any $|\Delta\mathbf{x}|$ of interest. The factor $(\sin^2 k_{\text{av}}|\Delta\mathbf{x}|)/(k_{\text{av}}|\Delta\mathbf{x}|)^2$ may be considered as constant over the integration domain unless $(\Delta\omega/c)|\Delta\mathbf{x}|$ is comparable to 1 or (since $c\tau \gg |\Delta\mathbf{x}|$) unless $\tau\,\Delta\omega \gg 1$. In the latter case, the sharp peak in the factor $[1 + \tau^2(\omega - \omega')^2]^{1/2}$ at $\omega = \omega'$ allows one to consider the spatially dependent factor as being the same as if ω' were set equal to ω at the outset when one is doing, say, the ω' integration first. On this basis, we conclude that the value of the integral is unchanged for all practical purposes if the spatially dependent factor is replaced by its average over the frequency interval. Thus, with reference to the analysis leading to Eq. (6-7.22), we find that

Eq. (5) reduces to

$$\frac{\langle \overline{p^2}(\mathbf{x})\overline{p^2}(\mathbf{x} + \Delta\mathbf{x})\rangle}{\langle \overline{p^2}\rangle^2} \approx 1 + V(\tau\,\Delta\omega)F(k_1|\Delta\mathbf{x}|,\ k_2|\Delta\mathbf{x}|) \tag{6-8.6}$$

where $V(\tau\,\Delta\omega)$ is the function defined in Eq. (6-7.23) and we abbreviate

$$\begin{aligned} F(a, b) &= \frac{1}{b - a}\int_a^b \frac{\sin^2 x}{x^2}\,dx \\ &= \frac{1}{b - a}[\mathrm{Si}\,(2b) - \mathrm{Si}\,(2a) - b^{-1}\sin^2 b + a^{-1}\sin^2 a] \end{aligned} \tag{6-8.7}$$

where

$$\mathrm{Si}\,(y) = \int_0^y t^{-1}\sin t\,dt \tag{6-8.8}$$

is the *sine integral function*.

Spatial Averaging

If one measures $\overline{p^2}(\mathbf{x})$ at points $\mathbf{x}_1, \mathbf{x}_2, \ldots, \mathbf{x}_K$ and then averages them, the average being taken as an estimate of $\langle \overline{p^2}\rangle$, the variance associated with the estimate is

$$\left\langle \left(\frac{1}{K}\sum_i f_i - \langle f\rangle\right)^2\right\rangle = \frac{\langle f^2\rangle}{K^2}\sum_{ij}\left(\frac{\langle f_i f_j\rangle}{\langle f\rangle^2} - 1\right) \tag{6-8.9}$$

where we write f_i for $\overline{p^2}(\mathbf{x}_i)$. The rms relative error Δ_{rms} in the estimate is the square root of the above divided by $\langle f\rangle$. Thus, from Eq. (3), one obtains (for a pure tone)

$$\Delta_{\mathrm{rms}} = \frac{1}{K}\left[K + \sum_{i\neq j}\frac{\sin^2(k|\mathbf{x}_i - \mathbf{x}_j|)}{k^2|\mathbf{x}_i - \mathbf{x}_j|^2}\right]^{1/2} \tag{6-8.10}$$

A minimum value of $1/K^{1/2}$ for Δ_{rms} can be approximately achieved if one chooses the $\mathbf{x}_i$ and $\mathbf{x}_j$ such that each of the terms in the above sum ($i \neq j$) is much less than $1/K$. This would be so, for example, if $|\mathbf{x}_i - \mathbf{x}_j| \gg \lambda K^{1/2}/2\pi$.

A common method for spatial averaging is to move the microphone along a path at slow speed and to take the long-term time average of the received p^2. If the path of length L is straight, and if the signal is a pure tone, the expected rms relative error from this method is given by Eq. (10) with the sum expressed as a double integral and with the prescriptions $\Delta i/K \rightarrow dx/L$, $\Delta j/K \rightarrow dx'/L$, $|\mathbf{x}_i - \mathbf{x}_j| \rightarrow |x - x'|$, such that

$$\begin{aligned} (\Delta_{\mathrm{rms}})^2 &= \frac{1}{L^2}\int_0^L\!\!\int \frac{\sin^2[k(x - x')]}{k^2|x - x'|^2}\,dx\,dx' \\ &= 2\int_0^1 (1 - u)\,\frac{\sin^2 kLu}{(kLu)^2}\,du \end{aligned} \tag{6-8.11}$$

where the derivation of the second version is similar to that of Eq. (6-7.21). The integral over u can be expressed in terms of tabulated functions, but we confine ourselves here to limiting cases. For small kL, a power-series expansion and subsequent term-by-term integration yield

$$\Delta_{\text{rms}} \approx 1 - \tfrac{1}{36}(kL)^2 \qquad (6\text{-}8.12)$$

In the limit of large kL, the u in the factor $1 - u$ is of minor consequence. After its discard, the upper integration limit can be taken as ∞, so that Eq. (11) takes the form of $2/kL$ times the definite integral of $\xi^{-2}\sin^2\xi$, with ξ replacing kLu. The integral is a standard definite integral whose value is $\pi/2$, so the large-kL limit yields

$$\Delta_{\text{rms}} = \left(\frac{\pi}{kL}\right)^{1/2} = \left(\frac{\lambda}{2L}\right)^{1/2} \qquad (6\text{-}8.13)$$

If one wants the expected relative error to be less than 0.3, for example, one should choose L to be greater than $(\lambda/2)/(0.3)^2 = 5.5\lambda$.

Frequency Averaging versus Spatial Averaging

Since the variance in measurements of $\overline{p^2}$ decreases as the frequency bandwidth increases [see Eq. (6-7.23)], an average over frequency is roughly equivalent to an average over position. From a comparison of Eqs. (6-7.22) and (13), one arrives at the correspondence

$$k\,\Delta L \approx \tau\,\Delta\omega \qquad (6\text{-}8.14)$$

such that an average over a line of length ΔL leads to a prediction with the same probable error as an average over a frequency band of width $\Delta\omega$ if ΔL and $\Delta\omega$ are so related. Alternately, an insertion of τ from Eq. (6-1.4) transforms the above correspondence into

$$\frac{A_s\,\Delta L}{4V} \approx \frac{\Delta\omega}{\omega} \qquad (6\text{-}8.15)$$

This implies, for a cubic room with average absorption coefficient 0.1, that averaging along a line extending the length of the room is equivalent to averaging over a bandwidth of slightly less than $\frac{1}{4}$ octave. For broadband sources with power output per unit bandwidth slowly varying over $\frac{1}{4}$-octave intervals, the frequency average, i.e., a broadband measurement, with a single microphone position would normally be a simpler method of estimating the acoustic energy per unit frequency bandwidth accurately than a spatial average of contributions from a narrow band. However, if the sound is a pure tone, and if all the surfaces are motionless, e.g., no rotating vanes, some spatial averaging is necessary.

One consequence of the correspondence just described is that long-period time averages can replace spatial averages for any narrow-bandwidth sound field whose bandwidth in hertz is nevertheless substantially larger than $1/2\pi\tau$. Given that the nominal frequency of the sound is itself much greater than this bandwidth, the sound field may yet behave for other intents as a pure tone. Thus, for example, suppose one measured $p(\mathbf{x}_1, t)$ and $p(\mathbf{x}_2, t)$ at two typical points $\mathbf{x}_1$ and $\mathbf{x}_2$ for such a narrow-band sound field. Then one would expect, from Eq. (2), that†

$$\lim_{T\to\infty}\frac{1}{T}\int_0^T p(\mathbf{x}_1, t)p(\mathbf{x}_2, t)\,dt \approx \langle \overline{p^2}\rangle \frac{\sin k|\mathbf{x}_1 - \mathbf{x}_2|}{k|\mathbf{x}_1 - \mathbf{x}_2|}$$

$$\approx [\overline{p_1^2}(\mathbf{x}_1)\overline{p_2^2}(\mathbf{x}_2)]^{1/2} \frac{\sin k|\mathbf{x}_1 - \mathbf{x}_2|}{k|\mathbf{x}_1 - \mathbf{x}_2|} \tag{6-8.16}$$

provided $|\mathbf{x}_1 - \mathbf{x}_2|$ is somewhat less than $c/\Delta\omega$, where $\Delta\omega$ is the bandwidth of the sound.

PROBLEMS

6-1 An untreated room 6 m long, 5 m wide, and 3 m high has surfaces of average absorption coefficient $\alpha_0 = 0.01$. When all the sources of sound are on, the sound level is 90 dB. To reduce this level, the floor is covered with a carpet with absorption coefficient α_c. What should α_c be if the sound level is to be reduced to 80 dB?

6-2 The sound-pressure level in a factory room 10 by 10 by 4 m is typically 90 dB. The reverberation time for the room is 4 s. Estimate the sound power output of the sources in the room.

6-3 A reverberation time of 5 s is measured when four people are present in a room 5 by 5 by 4 m. What is the reverberation time when no one is present?

6-4 The total absorbing power of the surfaces of a room is 5 metric sabins. When a carpet of area 2 m² is hung on one wall of the room, the original reverberation time of 5 s drops to 4 s. What is the random-incidence absorption coefficient of the carpet?

6-5 In his original experiments, Sabine had no direct method of measuring sound level or source power output but nevertheless accurately measured reverberation times using two identical but widely spaced sound sources. Suppose when one source is excited and suddenly turned off, 3 s lapses before the sound in the room decreases to the threshold of audibility. If both sources are excited and suddenly turned off, the corresponding time is 4 s. What is the reverberation time of the room?

6-6 The sound-pressure level in a room is 90 dB. How much energy per unit time passes out through an open window of 1 m² area? What would the sound-pressure level be in the open space outside the room at a point 20 m from the window along a line making 45° with the unit normal to the window?

6-7 The reverberation time of a room is 4 s when the walls, floor, and ceiling all have absorption coefficient α_0. If half of the total surface area of the room is covered with an acoustic tile with absorption coefficient $4\,\alpha_0$, what will the reverberation time be?

6-8 Suppose that a sound source in a room excites plane waves that propagate only in the $+x$ and $-x$ directions. The two walls perpendicular to the x axis are a net distance L apart, and each has normal incidence absorption coefficient α. Determine an expression for the reverberation time T_{60} of the room for the described circumstances in the limit $\alpha \ll 1$.

6-9 A two-dimensional reverberant sound field is in a low-ceilinged room with parallel floor and ceiling. The field may be considered in any local region as being a superposition of a large number of plane waves, all of the same frequency and with propagation directions parallel to the floor.

(*a*) If the energy density in the room is $\bar{w}$, how much energy is incident per unit time and area on the average on the vertical walls of the room?

(*b*) If $\alpha(\theta)$ is the absorption coefficient for a plane wave at angle of incidence θ, what would be the fraction of incident energy absorbed for the two-dimensional random-incidence situation described above?

† Cook et al., "Measurement of Correlation Coefficient . . . ."

(c) Determine an expression for the reverberation time T_{60} for such a sound field in terms of the floor area of the room, the perimeter length of the floor, sound speed c, and the apparent absorption coefficient.

6-10 What would be the counterpart of the Norris-Eyring reverberation time for the one-dimensional field described in Prob. 6-8. What would be the appropriate modification if the two walls had different absorption coefficients?

6-11 Derive Eq. (6-6.12) and state whatever assumptions are required. Show that the integral expression leads to the approximate result $E'_K = \frac{1}{2}\tau\bar{\mathcal{P}}$.

6-12 Two rooms are connected by a panel of area 12 m². Each room has dimensions 4 by 4 by 3 m and an absorbing power of 1.2 metric sabins. What should the transmission loss of the panel be if the sound pressure level in room 2 is to be 60 dB when a source in room 1 causes a sound-pressure level of 90 dB within that room?

6-13 A sound source rests on the floor of a room with dimensions 5 by 6.28 by 4 m whose reverberation time is 3.22 s. If the sound level at a distance of 3 m from the source is 95 dB, what would you estimate for the sound level at a distance of 0.5 m from the source?

6-14 A limp panel, i.e., one that satisfies criteria for the mass law, has a transmission loss for normal incidence of R_0. Derive a simple expression for its random-incidence transmission loss.

6-15 The sound level in a factory room is 95 dB, but if all the windows are open simultaneously, the sound level drops to 90 dB. The dimensions of the room are 10 by 10 by 4 m, and the total area of the open windows is 10 m². Give an estimate for the reverberation time of the room when all windows are closed. What is the corresponding value of the average absorption coefficient of the room's surfaces?

6-16 A panel separating two rooms has an area of 5 m² and a transmission loss of 20 dB. Room 1 has a sound source in it and has a sound level at a representative point of 90 dB. Room 2 has no sound sources and has negligible absorption. What would you estimate for the sound level in room 2?

6-17 A small intense source of sound is in a room with a room constant of 25 metric sabins. A worker standing about 1 m from the source experiences a sound level of 95 dB. Assuming that the source rests on a nearly rigid floor, what reduction in sound level can be expected for this worker when the room constant is increased by a factor of 10?

6-18 A cocktail party for serious conversationalists is planned for a room 10 by 10 by 4 m with a reverberation time of 1.2 s. Previous parties have been such that attendees clustered in groups of four; typical listeners stand 0.5 m from the person they are trying to hear. What is the maximum number of guests that should be invited?

6-19 Two adjacent apartment living rooms have a common wall of area 20 m² with a transmission loss of 40 dB. Both rooms have absorbing power of 30 metric sabins. If a loud stereo in one room causes a sound level of 70 dB in the second room, what would you expect for the sound level in the room in which the stereo is being played?

6-20 The absorption coefficient of a particular surface is $0.1 \cos \theta$ when radiated by a plane wave at angle of incidence θ. What would be the corresponding random-incidence absorption coefficient?

6-21 The sound level in a room is 85 dB. What is the sound level just outside an exterior wall whose transmission loss is 30 dB?

6-22 The given wall of area A is of checkerboard construction such that a portion A_1 has a transmission loss R_1 while the remaining portion A_2 has a transmission loss R_2. What value would you assign for the transmission loss R_{TL} for the wall as a whole?

6-23 A cubic enclosure 2 m on each side is placed over a small sound source resting on a rigid floor. The transmission loss of the walls of the enclosure is 20 dB for each wall. What would the absorption coefficient of the inner lining of the enclosure have to be if its insertion loss (10 log of ratio of power transmitted out without enclosure to that with enclosure present) is to be 15 dB?

6-24 Determine the lowest 10 nonzero natural frequencies for a rectangular room of dimensions 4 by 5 by 7 m with rigid walls and give a plot of the number of modes having resonance frequency less than f versus frequency f. On the same graph plot both the asymptotic expression (6-6.1) and its leading term. Discuss whether the other terms represent an improvement to the fit. Are 10 points sufficient to test the derivation of the asymptotic expression?

6-25 Determine the natural frequencies and modal eigenfunctions for a rectangular swimming pool of dimensions L_x by L_y by L_z. The upper surface, $z = L_z$, is a pressure-release surface, while the remaining boundary surfaces are rigid.

6-26 For a cubical room with dimensions L on a side, determine a complete set of orthonormal eigenfunctions that correspond to the natural frequency $\omega = 5\pi c/L$.

6-27 The surfaces of a room, dimensions L_x by L_y by L_z, have specific acoustic impedance $z = 1000\rho c$. A point source of monopole amplitude $\hat{S}$ is placed close to the corner (0, 0, 0) and is driven at angular frequency $\omega = \pi c/L_x$. Estimate the resulting acoustic-pressure amplitude at the opposite corner (L_x, L_y, L_z). (Assume that only one mode is appreciably excited.)

6-28 A vertical line source in a rectangular room (floor dimensions L_x and L_y) excites only those modes for which the eigenfunction is independent of z. Derive an expression appropriate in the limit of large ω for the number $N(\omega)$ of such modes that have natural frequency less than ω rad/s.

6-29 A room with dimensions 20 by 30 by 10 m has a reverberation time of 3 s.

(*a*) What is the corresponding Schroeder cutoff frequency?

(*b*) If a pure tone of 250 Hz is played in the room and causes an average sound-pressure level of 80 dB, what is the probability that a given person will hear 70 dB or less.

(*c*) If a person at a distance of 1 m from you hears 85 dB, what is the probability that you will hear more than 90 dB?

CHAPTER

SEVEN

LOW-FREQUENCY MODELS OF SOUND TRANSMISSION

Acoustic phenomena are often interpreted in terms of concepts based on the assumption that the acoustic wavelength is large compared with a characteristic length. The radiation of sound from small vibrating bodies, discussed in Chap. 4, is an instance of this; other examples emerge in the present chapter. To establish a theoretical basis for examples involving low frequencies in pipes and ducts, we begin with a discussion of guided waves.

7-1 GUIDED WAVES

Sound waves in a duct can be described in terms of *guided wave modes*.† We here consider a duct (waveguide) of constant cross-sectional shape and area (see Fig. 7-1), aligned so that its walls (idealized as rigid) are parallel to the x axis.

† The concept originated in major part with J. W. S. Rayleigh, *The Theory of Sound,* vol. 2, 1878, 2d ed., 1896, reprinted by Dover, New York, 1945, secs. 268, 340. Existence of higher-order modes was demonstrated experimentally by H. E. Hartig and C. E. Swanson, " 'Transverse' Acoustic Waves in Rigid Tubes," *Phys. Rev.,* **54:**618–626 (1938). Such modes are of interest in regard to noise generated by turbomachinery, fans, compressors, and jet engines. See, for example, J. M. Tyler and T. G. Sofrin, "Axial Flow Compressor Noise Studies," *Soc. Automot. Eng. Trans.,* **70:**309–332 (1962).

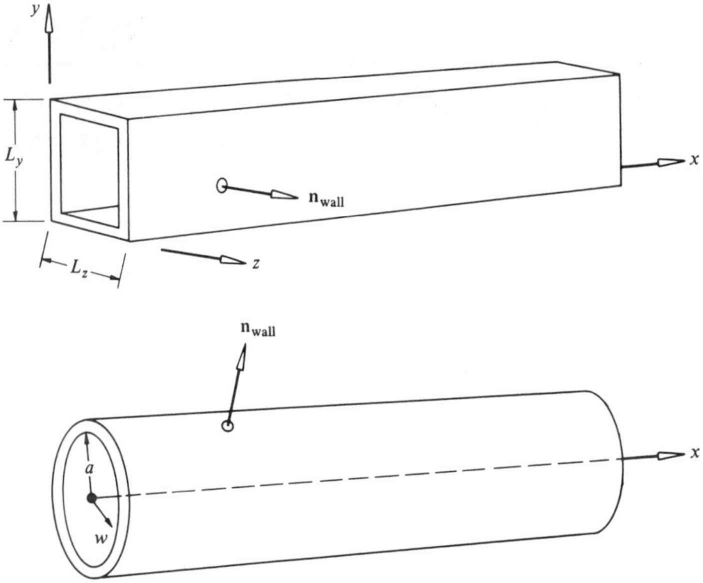

Figure 7-1 Duct of constant cross section: (*a*) rectangular duct, (*b*) circular duct.

Duct Cross-Sectional Eigenfunctions

Regardless of whether the cross-section is circular, rectangular, or less regularly shaped, one can construct appropriate separable solutions of the Helmholtz equation of the form

$$\hat{p}(x, y, z) = X_n(x)\Psi_n(y, z) \tag{7-1.1}$$

because the separation-of-variables technique described in Sec. 6-5 leads, for some separation constant α_n^2, to the differential equations

$$\left(\frac{\partial^2}{\partial y^2} + \frac{\partial^2}{\partial z^2}\right)\Psi_n + \alpha_n^2\Psi_n = 0 \tag{7-1.2a}$$

$$\frac{d^2X_n}{dx^2} + (k^2 - \alpha_n^2)X_n = 0 \tag{7-1.2b}$$

Furthermore, Eq. (1) will conform to the rigid-wall boundary condition if $\nabla\Psi_n \cdot \mathbf{n}_{\text{wall}} = 0$ at the duct walls.

The Ψ_n and α_n^2 are eigenfunctions and eigenvalues for a "two-dimensional room" with rigid walls, so in accordance with the remarks in Sec. 6-5, the α_n^2 are real and nonnegative and take on discrete values. The set of Ψ_n can be chosen as orthonormal, such that

$$\frac{1}{A}\iint \Psi_n(y, z)\Psi_{n'}(y, z)\, dA = \delta_{nn'} \tag{7-1.3}$$

where the integral extends over the cross-sectional area A of the duct. Furthermore, the $\Psi_n(y, z)$ form a complete set, so for any function $f(y, z)$, one has, when (y, z) lies in the duct,

$$f(y, z) = \sum_n a_n \Psi_n(y, z) \qquad a_n = \frac{1}{A} \iint f(y, z) \Psi_n(y, z)\, dA \tag{7-1.4}$$

Duct with Rectangular Cross Section

For a duct whose interior occupies the region $0 < y < L_y$, $0 < z < L_z$, the eigenfunctions and eigenvalues are identified from Eqs. (6-5.5) and (6-5.6) as

$$\Psi_n = K(n_y, n_z) \cos \frac{n_y \pi y}{L_y} \cos \frac{n_z \pi z}{L_z} \tag{7-1.5a}$$

$$\alpha_n^2 = \pi^2 \left[\left(\frac{n_y}{L_y} \right)^2 + \left(\frac{n_z}{L_z} \right)^2 \right] \tag{7-1.5b}$$

where the constant $K(n_y, n_z)$ is determined from Eq. (3). (If both n_y and n_z are zero, K is 1; if only one is zero, K is $2^{1/2}$; if both are nonzero, K is 2.)

Duct with Circular Cross Section

If the duct has a circular cross section† of radius a, Eq. (2a) is appropriately written in polar coordinates (r, ϕ) where $y = r\cos\phi$, $z = r\sin\phi$, for which the laplacian‡ in two dimensions is $\partial^2/\partial r^2 + r^{-1}\,\partial/\partial r + r^{-2}\,\partial^2/\partial\phi^2$. The resulting version of (2a) is further separable, so that a function $R(r)$ times either $\cos m\phi$ or $\sin m\phi$ is a possible solution. For the function Ψ_n to be single-valued and continuous in ϕ, the separation constant m must be an integer. The radial factor $R(r)$ satisfies the differential equation that results when $\partial^2/\partial\phi^2$ is replaced by $-m^2$:

$$\left[\frac{d^2}{dr^2} + \frac{1}{r}\frac{d}{dr} + \left(\alpha_n^2 - \frac{m^2}{r^2} \right) \right] R(r) = 0 \tag{7-1.6}$$

This is *Bessel's equation*;§ its only solution finite at $r = 0$ is $KJ_m(\alpha_n r)$, where K is a constant and J_m is the Bessel function of order m. The boundary condition requires $dR/dr = 0$ at $r = a$, so α_n must be such that $\alpha_n J_m'(\alpha_n a) = 0$. If η_{qm}

† J. W. S. Rayleigh, "Oscillations in Cylindrical Vessels," *Phil. Mag.*, (5)**1**:272–279 (1876); "On the Passage of Electric Waves through Tubes, or the Vibrations of Dielectric Cylinders," ibid., **43**:125–132 (1897). A related analysis for elastic waves in a solid cylinder was given by L. Pochhammer, "Concerning the Velocities of Small Vibrations in an Unlimited Isotropic Circular Cylinder," *J. Reine Angew. Math.*, **81**:324–336 (1876).

‡ This follows from p. 173n. with $\xi_1, \xi_2, \xi_3 = r, \phi, x$ and with $h_r = 1$, $h_\phi = r$, $h_x = 1$.

§ Derived by L. Euler in 1764 in an analysis of vibrations of a stretched membrane. That $J_m(\alpha_n r)$ is a solution follows from an explicit substitution of its power-series expansion into the differential equation. G. N. Watson, *A Treatise on the Theory of Bessel Functions*, 2d ed., Cambridge University Press, Cambridge, 1944, pp. 5, 6, 15–19.

denotes the qth root ($q = 1, 2, \ldots$) of $\eta_{qm} J'_m(\eta_{qm}) = 0$, the corresponding α_n is η_{qm}/a and the corresponding eigenfunction is

$$\Psi_n(r, \phi) = K_{qm} J_m\left(\frac{\eta_{qm} r}{a}\right) \begin{Bmatrix} \cos m\phi \\ \sin m\phi \end{Bmatrix} \tag{7-1.7}$$

For $m > 0$, the $\eta = 0$ root of $\eta J'_m(\eta) = 0$ leads to the trivial solution $\Psi_n = 0$, but setting η to 0 reduces $J_0(\eta r/a)$ to 1, so that Ψ_n in the $m = 0$, $\eta = 0$ case is a constant. The other roots ($\eta \neq 0$) are solutions of $J'_m(\eta) = 0$. Taking $q = 1$ as labeling the lowest root, one has in particular $\eta_{q0} = 0.0, 3.832, 7.016$; $\eta_{q1} = 1.841, 5.331, 8.536$; $\eta_{q2} = 3.054, 6.706, 9.969$ for $q = 1, 2, 3$. In the limit of large q (fixed m), roots can be determined from the asymptotic-series expression for the Bessel function and approach† $(q + m/2 - \frac{3}{4})\pi$.

Cutoff Frequencies and Evanescent Modes

Possible solutions of Eq. (2*b*) for the axial factor $X_n(x)$ are $\exp(\pm i\beta_n x)$, where $\beta_n = (k^2 - \alpha_n^2)^{1/2}$ for $k^2 > \alpha_n^2$ and $\beta_n = i(\alpha_n^2 - k^2)^{1/2}$ for $\alpha_n^2 > k^2$. A propagating guided wave is therefore described by the expression

$$p(x, y, z, t) = \operatorname{Re} B e^{-i\omega t} e^{i\beta_n x} \Psi_n(y, z) \tag{7-1.8}$$

providing $k^2 > \alpha_n^2$; the corresponding disturbance has a trace velocity (phase velocity) of $v_{\text{ph}} = \omega/(k^2 - \alpha_n^2)^{1/2}$ along the x axis. However, if $k^2 < \alpha_n^2$, the factor $\exp i\beta_n x$ becomes $\exp(-|\beta_n|x)$ and Eq. (8) then corresponds to a disturbance dying out exponentially with increasing x.

For a given frequency, there are a limited number of modes for which $\alpha_n^2 < k^2$. There is at least one, this being the *plane-wave*, or *fundamental, mode,* for which α_n is 0 and Ψ_n is constant. Modes for which $\alpha_n^2 > k^2$ are *evanescent,* while those for which $\alpha_n^2 < k^2$ are *propagating* modes. If ω is greater than the *cutoff frequency* $\omega_{c,n}$ given by $c\alpha_n$, the mode is propagating, but below that frequency it is evanescent. For all modes other than the plane-wave mode, propagation above the cutoff frequency is *dispersive*. Different frequencies correspond to different phase velocities and to different repetition lengths along the x axis. If $\alpha_n \neq 0$, a wave packet composed of a sum of waves of the form of Eq. (8), with n fixed but with various frequencies, would have a time-dependent signature that distorts with increasing propagation distance.

An evanescent mode transports no net acoustic energy. If p is given by Eq. (8), then v_x (derived from $\rho\, \partial v_x/\partial t = -\partial p/\partial x$) is given by an analogous expression but with B replaced by $\beta_n B/\omega\rho$. If β_n is imaginary, as for an evanescent mode, the time average $I_{x,\text{av}}$ of the x component of the acoustic intensity vanishes because p and v_x are 90° out of phase; the power transported through the duct, represented by the integral of $I_{x,\text{av}}$ over the cross-sectional area, is also zero.

† J. McMahon, "On the Roots of the Bessel and Certain Related Functions," *Ann. Math. (Charlottesville, Va.)*, **9**:23–30 (1894–1895).

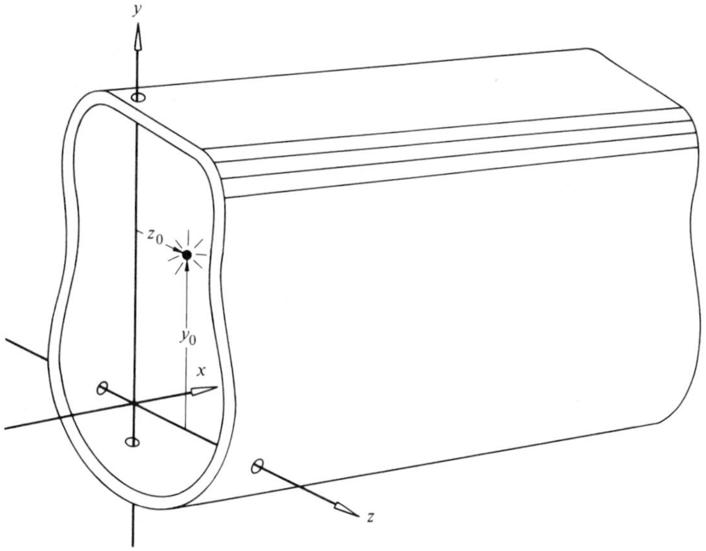

Figure 7-2 Point source in a duct.

In many situations of practical interest, the frequency is so low that the only propagating mode is the plane-wave mode. For a rectangular duct, this is so, according to Eq. (5*b*), if $\omega < c\pi/L_{max}$, where L_{max} is the maximum of L_y or L_z. For a circular duct of radius a, the dispersive modes are all evanescent if $\omega < 1.841c/a$. The latter criterion requires, for example, that the frequency be less than 1000 Hz for a 0.1-m-radius duct containing air at 20°C.

Point Source in a Duct

At large distances from a source within a duct, only the propagating modes need be considered. We illustrate this with an analysis† of the field (within a duct of infinite length) of a point source with angular frequency ω, monopole amplitude $\hat{S}$, located at y_0, z_0, with $x_0 = 0$ (see Fig. 7-2). The complex pressure amplitude $\hat{p}(x, y, z)$ can be expanded in the $\Psi_n(y, z)$ as in Eq. (4), with the coefficients a_n taken as functions $X_n(x)$ [not necessarily the same as those in Eq. (2*b*)].

If such a modal expansion is substituted into the Helmholtz equation with a point-source term $-4\pi\hat{S}\delta(\mathbf{x} - \mathbf{x}_0)$ on the right side, and if the result is multiplied by a particular $\Psi_n(y, z)$ and subsequently integrated over the cross-sectional area of the duct, one obtains, with use of Eqs. (2*a*) and (3), the inhomogeneous differential equation

$$\left[\frac{d^2}{dx^2} + (k^2 - \alpha_n^2)\right]X_n = -\frac{4\pi\hat{S}}{A}\Psi_n(y_0, z_0)\delta(x) \qquad \text{(7-1.9)}$$

† First discussed by M. Taylor, "On the Emission of Sound by a Source on the Axis of a Cylindrical Tube," *Phil. Mag.*, (6)**24**:655–664 (1912).

The solution for $x \neq 0$ satisfies the homogeneous equation (2*b*) and may be taken as a constant times exp $i\beta_n|x|$, such that it corresponds to a wave that either propagates away ($k^2 > \alpha_n^2$) or dies out exponentially ($k^2 < \alpha_n^2$) from the source. The multiplicative constant must be the same for $x > 0$ as for $x < 0$ to ensure X_n continuous at $x = 0$. The delta function requires, however, that dX_n/dx be discontinuous. Integration of both sides from $x = -\varepsilon$ to $x = \varepsilon$ yields (in the limit $\varepsilon \to 0$)

$$\left(\frac{dX_n}{dx}\right)_{+\varepsilon} - \left(\frac{dX_n}{dx}\right)_{-\varepsilon} \to -\frac{4\pi\hat{S}}{A}\Psi_n(y_0, z_0)$$

so the solution of (9) is

$$X_n = \frac{-2\pi\hat{S}\Psi_n(y_0, z_0)}{i\beta_n A} e^{i\beta_n|x|} \tag{7-1.10}$$

The resulting $\hat{p}$ is the sum over n of $X_n\Psi_n$.

The analogous expression for $\hat{v}_x$ derives from the x component of Euler's equation and from the solution for $\hat{p}$, the result being

$$\hat{v}_x = \pm\sum_n \frac{\beta_n}{\omega\rho} X_n(x)\Psi_n(y, z) \tag{7-1.11}$$

where the signs apply for $x > 0$ and $x < 0$, respectively. The quantity $\omega\rho/\beta_n$ is the characteristic *modal specific impedance* associated with the nth mode.

The power transmitted in the $+x$ direction is the area integral of $\frac{1}{2}\text{Re}\,\hat{p}\hat{v}_x^*$. Because of the orthogonality (3) of the modal eigenfunctions $\Psi_n(y, z)$, all the cross terms in the resulting double sum integrate to zero, so the power is the sum of the powers associated with the individual modes. Those associated with the evanescent modes vanish, however, since their modal specific impedances are imaginary. Consequently, one is left with

$$\mathscr{P}_{\text{right}} = \frac{2\pi^2|\hat{S}|^2}{A\omega\rho}\sum_n{}' \frac{\Psi_n^2(y_0, z_0)}{(k^2 - \alpha_n^2)^{1/2}} \tag{7-1.12}$$

for the power transmitted to the right of the source. The total power output, to the left and to the right, is twice this. (Here the prime on the sum implies that one include only terms for which $\alpha_n^2 < k^2$.) One implication is that the power output of the source suddenly jumps to a very large value whenever the driving frequency is increased from just below to just above any mode's cutoff frequency.

When the driving frequency is below the cutoff frequency for the first dispersive mode, such that only the plane-wave mode ($\alpha_n = 0$, $\Psi_n = 1$) is excited, the net power output $\mathcal{P}$, equal to $2\mathcal{P}_{\text{right}}$, reduces to†

$$\mathcal{P} = \frac{4\pi^2|\hat{S}|^2 c}{A\,\omega^2 \rho} = \frac{2\pi c^2}{\omega^2 A}\,\mathcal{P}_{\text{ff}} \tag{7-1.13}$$

where $\mathcal{P}_{\text{ff}} = 2\pi|\hat{S}|^2/\rho c$ is the power radiated by the source in a free-field environment. For the same circumstances, at distances sufficiently large for evanescent modes to be neglected, the complex pressure amplitude reduces to

$$\hat{p} = \frac{i(2\pi c\hat{S})}{\omega A}\,e^{i(\omega/c)|x|} \tag{7-1.14}$$

Because Re $[(i4\pi\hat{S}/\omega\rho)e^{-i\omega t}]$ is the time rate of change of the volume excluded by the source, the latter leads to the identification for the time-dependent acoustic pressure (at large $|x|$) as

$$p = \frac{\rho c}{2A}\left(\frac{dV_S}{dt}\right)_{t\to t-|x|/c} \tag{7-1.15}$$

This applies to sources that excite any combination of frequencies, providing each is below the cutoff frequency for the first dispersive mode. It can be compared with the corresponding expression $(\rho/4\pi R)\, d^2V_S/dt^2$ (with $t \to t - R/c$) for the acoustic pressure resulting from a monopole source in an unbounded medium [see Eq. (4-1.6)].

7-2 LUMPED-PARAMETER MODELS

A lumped-parameter model‡ uses a limited number of time-dependent aggregate variables rather than field quantities varying with both position and time. The partial-differential equations and boundary conditions interrelating the field quantities are replaced by ordinary differential equations interrelating the aggregate variables. The coefficients (*lumped-parameter elements*) in the latter description usually have a viable physical interpretation, either in terms of an analogous mechanical system or an analogous electrical system. Typically, lumped-parameter models are used when the frequency is such that $ka \ll 1$, where a is a characteristic dimension appropriate to the physical system.

An example of a lumped-parameter model would be a *spring*, whereby one idealizes an elastic solid of possibly complicated shape as a massless entity whose sole property, as regards the analysis of the behavior of the physical system of which it is a part, is its spring constant, i.e., incremental force

† Taylor, "On the Emission of Sound," derives this when the source is on the axis of a circular tube. The generalization to a duct of arbitrary cross-sectional shape is given (although without details of derivation) by H. Lamb, "The Propagation of Waves of Expansion in a Tube," *Proc. Lond. Math. Soc.*, (2)**37**:547–555 (1934).

‡ An extensive exposition of the concept is given by H. H. Woodson and J. R. Melcher, *Electromechanical Dynamics*, pt I, *Discrete Systems*, Wiley, New York, 1968, pp. 15–59.

required per incremental change in elongation; force and elongation replace stress and strain fields.

Volume Velocity and Average Pressure

In acoustics, the commonly used lumped-parameter variables are volume velocity and average pressure. For a surface S_1 terminated at its edges by a rigid surface (see Fig. 7-3), the volume velocity U_1 flowing across S_1 is defined as the integral

$$U_1 = \iint \mathbf{v} \cdot \mathbf{n} \, dS_1 \qquad (7\text{-}2.1)$$

The side of S_1 toward which the unit normal $\mathbf{n}$ points determines the positive sense of U_1. Since the surface integral of $\rho\mathbf{v} \cdot \mathbf{n}$ is the mass flowing across S_1 per unit time in the linear acoustics approximation (without ambient flow), U_1 would be the volume flowing across S_1 per unit time if the fluid were of ambient density.

The second variable one associates with the aggregate acoustic field over the surface S_1 is the average acoustic pressure p_1. This is the surface integral of $p\mathbf{v} \cdot \mathbf{n}$ divided by U_1, so it is a weighted (by $\mathbf{v} \cdot \mathbf{n}$) area average of p. The definition of p_1 is such that $p_1 U_1$ is the power transmitted across S_1 in the positive sense. In typical applications, S_1 is selected so that the pressure along it does not vary significantly and no distinction between pressure and average pressure is made.

Acoustic Impedance

In the description of lumped-parameter models that use volume velocity and pressure (we omit the qualifying adjective "average") as variables, a convenient concept is that of acoustic impedance Z_A. For the surface S_1, this

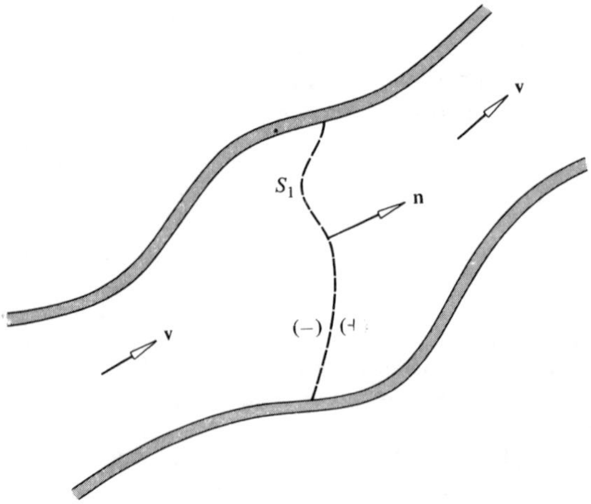

Figure 7-3 The volume velocity across S_1 is the area integral of $\mathbf{v} \cdot \mathbf{n}$, where $\mathbf{n}$ points normal to S_1 toward the + side.

frequency-dependent quantity is defined† as the ratio

$$Z_{A,1} = \frac{\hat{p}_1}{\hat{U}_1} \tag{7-2.2}$$

where $\hat{p}_1$ and $\hat{U}_1$ are either the complex amplitudes (constant-frequency disturbance) or the Fourier transforms (transient disturbance) of $p_1(t)$ and $U_1(t)$. The unit of $Z_{A,1}$ is 1 kg/($m^4 \cdot$ s). The reciprocal $\hat{U}_1/\hat{p}_1$ is called the *acoustic mobility* (rather than acoustic admittance). If the identifications of plus and minus sides of S_1 are interchanged, $Z_{A,1}$ changes sign.

Acoustical Two-Ports

Suppose one takes two surfaces S_1 and S_2 in an acoustical system (see Fig. 7-4*a*) and defines the plus and minus sides of each such that if U_1 is positive, volume will flow through S_1 toward S_2; positive U_2 corresponds to volume flowing from S_1 through S_2. The region between S_1 and S_2 is here regarded as a passive black box, which we call a *two-port*‡ and which will serve as our prototype of a lumped-parameter model.

The acoustic boundary-value problem for the black-box region, given pressures $p_1(t)$ and $p_2(t)$ on surfaces S_1 and S_2, should, according to the theorems developed in Sec. 4-5, have a unique solution, and from this solution one can determine U_1 and U_2. The linear nature of the governing partial differential equations and the boundary conditions requires that U_1 and U_2 be linear functions of p_1 and p_2. Thus, for the constant-frequency case, one should have§

$$\begin{bmatrix} \hat{U}_1 \\ \hat{U}_2 \end{bmatrix} = \begin{bmatrix} D_{11} & D_{12} \\ D_{21} & D_{22} \end{bmatrix} \begin{bmatrix} \hat{p}_1 \\ \hat{p}_2 \end{bmatrix} \tag{7-2.3}$$

where the *acoustic-mobility matrix* $[D]$ is a frequency-dependent property of the two-port. Considerations of reciprocity require, moreover, that $D_{12} = -D_{21}$.

Given the reciprocity requirement, Eqs. (3) can be written alternatively as

$$\hat{U}_1 = (Z_{\text{left}}^{-1} + Z_{\text{mid}}^{-1})\hat{p}_1 - Z_{\text{mid}}^{-1}\hat{p}_2 \tag{7-2.4a}$$

$$\hat{U}_2 = Z_{\text{mid}}^{-1}\hat{p}_1 - (Z_{\text{right}}^{-1} + Z_{\text{mid}}^{-1})\hat{p}_2 \tag{7-2.4b}$$

with a suitable definition of parameters Z_{left}, Z_{right}, and Z_{mid} in terms of D_{11}, D_{22}, and $D_{12} = -D_{21}$. These equations have a circuit analog‖ (see Fig. 7-4*b*) in

† G. W. Stewart, "Acoustic Wave Filters," *Phys. Rev.*, **20:**528–551 (1922).

‡ In electric-circuit theory, the term denotes any two-terminal-pair network. See, for example, H. H. Skilling, *Electrical Engineering Circuits,* Wiley, New York, 1957, pp. 537–572.

§ W. P. Mason, "A Study of the Regular Combination of Acoustic Elements, with Application to Recurrent Acoustic Filters, Tapered Acoustic Filters, and Horns," *Bell Syst. Tech. J.*, **6:**258–294 (1927).

‖ This is the conventional acoustic analogy. An *acoustic-mobility analogy* in which pressure → current, volume velocity → voltage is also occasionally used. The latter was introduced by F. A. Firestone, "A New Analogy between Mechanical and Electrical Systems," *J. Acoust. Soc. Am.*, **4:**249–267 (1932–1933).

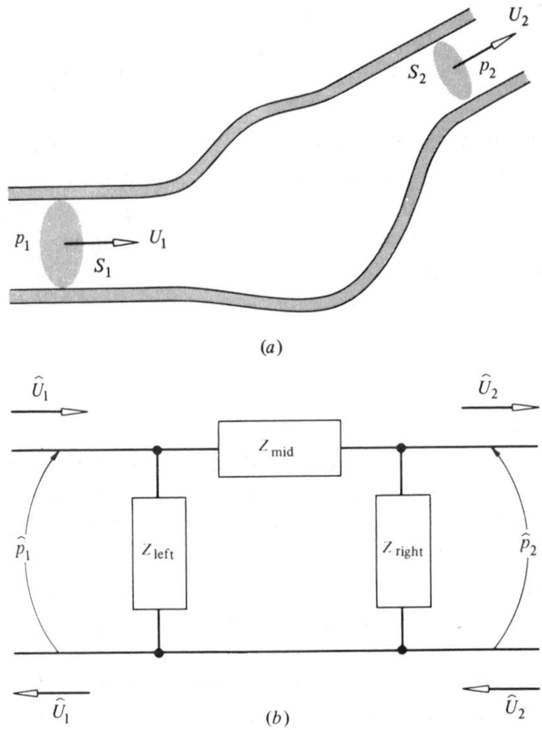

Figure 7-4 (*a*) Acoustical two-port in which position-independent pressures p_1 and p_2 are applied at surfaces S_1 and S_2; sense of positive volume flow is from S_1 toward S_2. (*b*) Corresponding electrical analog for constant-frequency case represented by a π network.

which $\hat{p}_1$ and $\hat{p}_2$ are voltages applied at the ends of a circuit two-port consisting of a π network with lumped impedances Z_{left}, Z_{mid}, and Z_{right}; $\hat{U}_1$ and $\hat{U}_2$ are currents flowing into and out of the two-port at its two ends. The analogy holds because circuit-theory principles (voltage at a node is univalued, and sum of currents flowing into a node is zero) applied to the circuit yield the same equations.

Once the impedances for our two-port are identified, the relation between the acoustic impedances $Z_{A,1}$ and $Z_{A,2}$ on surfaces S_1 and S_2 can be interpreted in terms of circuits. If the two-port in Fig. 7-4*b* has a load $Z_{A,2}$ on its right, $Z_{A,1}$ will be the equivalent impedance of a *one-port* in which $Z_{A,2}$ and Z_{right} are in parallel, the combination being in series with Z_{mid}, and that combination being in parallel with Z_{left}, such that

$$Z_{A,1} = \left\{ \frac{1}{Z_{\text{left}}} + \left[Z_{\text{mid}} + \left(\frac{1}{Z_{\text{right}}} + \frac{1}{Z_{A,2}} \right)^{-1} \right]^{-1} \right\}^{-1} \tag{7-2.5}$$

This is equivalent to what results from Eqs. (4) if one sets $\hat{p}_2 = Z_{A,2}\hat{U}_2$, then eliminates $\hat{U}_2$, and solves for $Z_{A,1} = \hat{p}_1/\hat{U}_1$.

Continuous-Volume-Velocity Two-Port

Of the two limiting cases of principal interest, one is that for which Z_{left} and Z_{right} are so large that they can be idealized as infinite and replaced by open

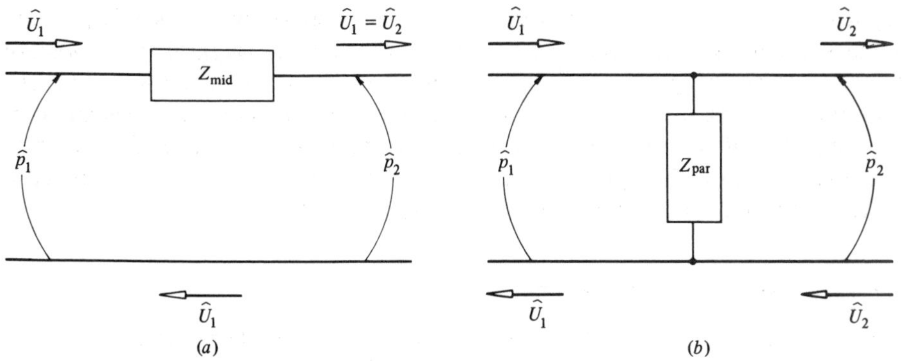

Figure 7-5 Circuit analogs for (*a*) a continuous-volume-velocity two-port and (*b*) a continuous-pressure two-port.

circuits in the circuit diagram, such that (see Fig. 7-5*a*)

$$\hat{U}_1 = \hat{U}_2 \qquad \hat{p}_1 - \hat{p}_2 = Z_{\text{mid}}\hat{U}_1 \qquad Z_{A,1} - Z_{A,2} = Z_{\text{mid}} \tag{7-2.6}$$

The latter idealization generally implies the assumption of incompressible flow. Suppose one has, for example, a volume V with openings of areas A_1 and A_2 on opposite sides, all other portions of the surface being rigid. Then the incompressible idealization would require, when one integrates $\nabla \cdot \mathbf{v}$ over the volume and uses Gauss' theorem, that $U_1 = U_2$.

If the volume is hollow, and if Euler's equation $\rho\, \partial\mathbf{v}/\partial t = -\nabla p$ applies throughout, $\nabla \times \mathbf{v} = 0$ for all time since it must have been zero in the remote past; so one can describe $\mathbf{v}$ in terms of a potential function $\Phi(\mathbf{x}, t)$ such that $\mathbf{v} = \nabla\Phi$, $p = -\rho\, \partial\Phi/\partial t$. (Here, as in previous sections of the text, ρ is understood to be the ambient density ρ_0.) Since $\nabla \cdot \mathbf{v} = 0$, one has $\nabla^2\Phi = 0$. Given that p is uniform over A_1 and A_2, these surfaces must have uniform potentials, which we denote by $\Phi_1(t)$ and $\Phi_2(t)$. The solution for $\Phi(\mathbf{x}, t)$, given Φ_1 and Φ_2, can be written as

$$\Phi(\mathbf{x}, t) = \Phi_1(t) + [\Phi_2(t) - \Phi_1(t)]f(\mathbf{x}) \tag{7-2.7}$$

where f($\mathbf{x}$) is independent of t, satisfies Laplace's equation, and equals 0 on A_1 and 1 on A_2; its normal derivative vanishes on all other boundary surfaces. Taking the gradient and time derivative of this and multiplying by ρ gives

$$\rho\frac{\partial\mathbf{v}}{\partial t} = [p_1(t) - p_2(t)]\,\nabla f \tag{7-2.8}$$

If one chooses any cross-sectional surface S_{mid} of V such that A_1 is on one side and A_2 is on the other and $\mathbf{n}$ points from the A_1 side to the A_2 side normal to the surface, then an area integral of the above leads to

$$p_1(t) - p_2(t) = M_A\frac{dU}{dt} \tag{7-2.9}$$

where $U = U_1 = U_2$ is the volume velocity flowing through the volume V from A_1 toward A_2 and M_A is ρ divided by the integral over S_{mid} of $\nabla f \cdot \mathbf{n}$. One can argue that the surface integral of $\nabla f \cdot \mathbf{n}$ is independent of the surface S_{mid} (in the same manner as one concludes that $U_1 = U_2$); so the integral is a constant appropriate to the geometry of the volume and to the choices for A_1 and A_2; consequently, the quantity M_A (*acoustic inertance*) is a constant independent of S_{mid} and of U. Rewriting Eq. (9) in terms of complex amplitudes and comparing the result with (6) then yields $Z_{\text{mid}} = -i\omega M_A$ as the acoustic impedance associated with this continuous-volume-velocity two port.

Continuous-Pressure Two-Port

The other limiting case corresponds to $Z_{\text{mid}} \to 0$. The short circuit allows a replacement of the parallel combination of Z_{left} and Z_{right} by a single impedance $Z_{\text{par}} = (Z_{\text{left}}^{-1} + Z_{\text{right}}^{-1})^{-1}$, so one has (Fig. 7-5*b*)

$$\hat{p}_1 = \hat{p}_2 = Z_{\text{par}}(\hat{U}_1 - \hat{U}_2) \qquad \frac{1}{Z_{A,1}} = \frac{1}{Z_{\text{par}}} + \frac{1}{Z_{A,2}} \tag{7-2.10}$$

A nontrivial situation ($Z_{\text{par}} \neq \infty$) to which such a model applies is when the inertial term in Euler's equation is negligible so $\nabla p = 0$ but the compressibility is not neglected; then the integral version of the conservation of mass equation (with ρ' replaced by p/c^2) would give

$$U_1 - U_2 = \frac{\partial p}{\partial t}\frac{V}{\rho c^2} = C_A\frac{\partial p}{\partial t} \tag{7-2.11}$$

with p uniform throughout the volume V of the two-port. Then Eq. (10) leads to the identification $Z_{\text{par}} = 1/(-i\omega C_A)$ with $C_A = V/\rho c^2$. The quantity C_A (*acoustic compliance*) corresponds to capacitance in the electric-circuit analog.

7-3 GUIDELINES FOR SELECTING LUMPED-PARAMETER MODELS

There are two principal idealizations made in the construction of lumped-parameter models: (1) the pressure changes very little over distances small compared with a wavelength, and (2) the sum of the volume velocities flowing out of a small volume is zero. The continuous-pressure two-port is based on the first idealization, the continuous-volume-velocity two-port on the second. In each case, one of the two idealizations is not made but is replaced by a coupling relation involving a complex impedance (a lumped-parameter element).

Continuity of Pressure

The premise that acoustic pressure does not "ordinarily" vary appreciably over distances much less than a wavelength can be examined by taking two points $\mathbf{x}_1$ and $\mathbf{x}_2$ at which the pressures are p_1 and p_2, respectively (see Fig. 7-6*a*). If one selects a path connecting $\mathbf{x}_1$ and $\mathbf{x}_2$ along which Euler's equation is

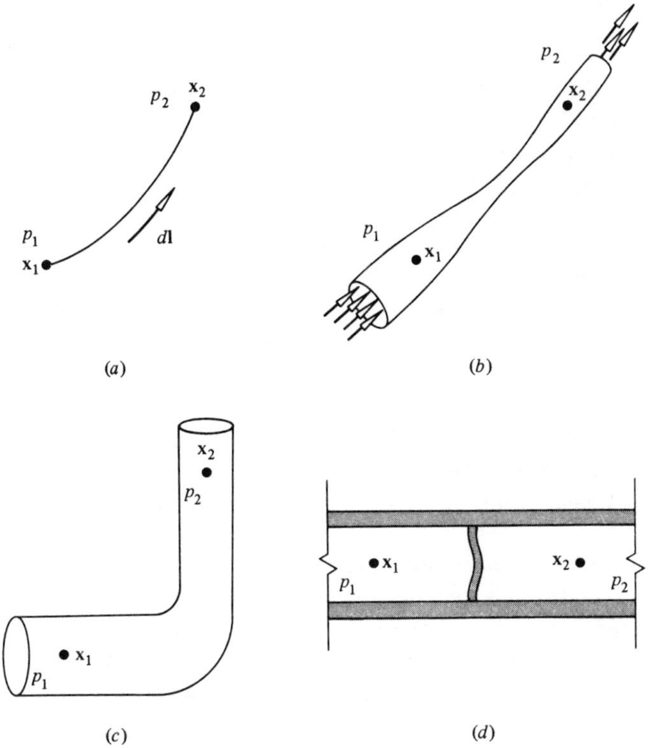

Figure 7-6 (*a*) Path connecting points $\mathbf{x}_1$ and $\mathbf{x}_2$ used in investigation of the magnitude of the difference of the acoustic pressures at the two points. (*b*) Streamtube of flow from $\mathbf{x}_1$ to $\mathbf{x}_2$. (*c*) Two ducts joined by an elbow. (*d*) Flexible plate extending across the cross section of a duct. The question considered is whether the pressures are nearly equal at $\mathbf{x}_1$ and $\mathbf{x}_2$.

a good approximation, then it should be so that (acoustic version of *Bernoulli's equation*)

$$\rho \frac{\partial}{\partial t} \int_{\mathbf{x}_1}^{\mathbf{x}_2} \mathbf{v} \cdot d\mathbf{l} = - \int_{\mathbf{x}_1}^{\mathbf{x}_2} \nabla p \cdot d\mathbf{l} = -(p_2 - p_1) \tag{7-3.1}$$

where $d\mathbf{l}$ represents the differential displacement along the path. Consequently, if the disturbance is of constant frequency, the magnitude $|\hat{p}_2 - \hat{p}_1|$ is bounded by $\rho c|\hat{\mathbf{v}}|_{\max} k \, \Delta s$, where Δs is net distance along the path and $|\hat{\mathbf{v}}|_{\max}$ is the maximum value of $|\hat{\mathbf{v}}|$ along the path.

Much closer than a wavelength in the present context means $k \, \Delta s \ll 1$. Granted this, one can regard the statement $\hat{p}_2 \approx \hat{p}_1$ as a good approximation if $\rho c|\hat{\mathbf{v}}|_{\max}$ is not substantially larger than either $|\hat{p}_2|$ or $|\hat{p}_1|$. Recall that, for a traveling plane wave, $|\hat{p}| = \rho c|\hat{\mathbf{v}}|$; the same holds for a traveling fundamental-mode wave in a duct. Thus, if $|\hat{\mathbf{v}}|$ is along the path of the same order of magnitude in relation to $|\hat{p}|$ as for a plane wave, the requirement $k \, \Delta s \ll 1$ leads to $p_1 \approx p_2$.

In other circumstances, $|\hat{\mathbf{v}}|_{max}$ can be estimated by assuming that the flow between the points is incompressible and taking the path to be a streamline. If one knows from other considerations that the velocities $\hat{\mathbf{v}}_1$ and $\hat{\mathbf{v}}_2$ are of the order of magnitude of $|\hat{p}_1|/\rho c$ and $|\hat{p}_2|/\rho c$ (as they will be if $\mathbf{x}_1$ and $\mathbf{x}_2$ are located in duct segments where the plane-wave mode dominates), the question reduces to whether a streamtube (Fig. 7-6*b*) surrounding the streamline narrows appreciably along the path. Conservation of mass implies that $|\hat{\mathbf{v}}|$ varies inversely as streamtube area, so a streamtube with a narrow constriction allows the possibility of a large pressure change between $\mathbf{x}_1$ and $\mathbf{x}_2$.

The foregoing analysis applies to two ducts joined by an elbow† (see Fig. 7-6*c*). Because the evanescent modes die out with distance, p will be uniform across either duct at a moderate distance (comparable to a cross-sectional dimension) from the elbow. The pressures at such points on opposite sides of the elbow are nearly the same if a streamtube connecting them or their neighbors is not constricted. However, if the elbow has a sizable constriction, the streamtube may narrow considerably in going through the elbow and one will not assume $\hat{p}_1 \approx \hat{p}_2$.

An extreme case where $\hat{p}_1 \approx \hat{p}_2$ is not indicated is when the geometry is such that the flow must pass through a small orifice. For example, if a duct has a rigid plate (Fig. 7-6*d*) extending across a cross section, the plate having a small hole in its center, then any streamtube passing through the orifice must be constricted. Other circumstances for which a substantial change in pressure might occur over a short distance are when there is *no* path connecting $\mathbf{x}_1$ and $\mathbf{x}_2$ along which Euler's equation is everywhere valid. Examples would be a flexible plate, membrane, or porous blanket extending across a duct.

Continuity of Volume Velocity

The idealization "ordinarily" made is that the net volume velocity flowing out of a volume (with dimensions much less than a wavelength) is zero. Situations for which this is a reasonable premise can be identified by integrating the conservation-of-mass relation over the volume (see Fig. 7-7). Starting from $\partial p/\partial t + \rho c^2 \nabla \cdot \mathbf{v} = 0$, one obtains (with an application of Gauss' theorem)

$$\sum U_n^{\text{out}} = -\frac{\partial}{\partial t}\iiint \frac{p}{\rho c^2}\, dV \tag{7-3.2}$$

where U_n^{out} is the volume velocity flowing out through the portion S_n of the surface bounding V.

Suppose there is only one opening of area A into the volume V, the remaining surface being rigid. We would normally regard the volume velocity flowing

† W. Lippert, "The Measurement of Sound Reflection and Transmission at Right-Angled Bends in Rectangular Tubes," *Acustica*, **4**:313–319 (1954); J. W. Miles, "The Diffraction of Sound due to Right-Angled Joints in Rectangular Tubes," *J. Acoust. Soc. Am.*, **19**:572–579 (1947). Lippert's fig. 7 (based on his data) and Miles' theory suggest that the continuity of pressure is a good approximation up to $ka \approx 1$, where a is the width of the duct.

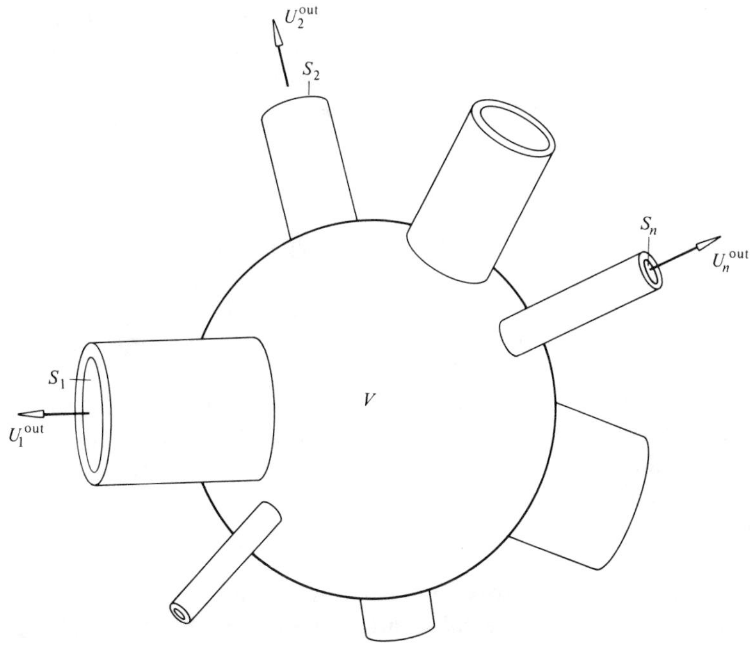

Figure 7-7 A volume V bounded partly by rigid boundaries and by surfaces S_1, S_2, ... The volume velocity flowing out of V through S_n is U_n^{out}.

out through this area as negligibly small if $|\hat{U}| \ll |\hat{p}|A/\rho c$, that is, $|Z_A|$ much larger than the value $\rho c/A$ expected for a plane wave in a duct of cross-sectional area A. Equation (2) shows this criterion is satisfied if kV/A is much less than unity; the lower the frequency the more likely this is to be so. However, even though the volume's dimensions may be much less than a wavelength, it is still possible (see Fig. 7-8) to have $kV/A \approx 1$ if the opening area A is a small fraction of the surface area of V. For such a situation, the assumption that the net volume velocity coming out of a small volume is zero should be reconsidered.

Returning to the general case where there is more than one opening, let us assume that the source of the disturbance transmits energy into the volume

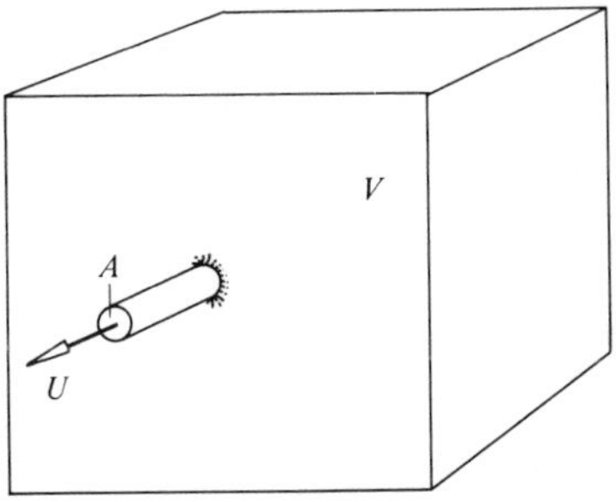

Figure 7-8 A volume with a single small opening for which the approximation that net volume velocity flowing out of a volume should be zero may not be valid.

through area A_1 and that a subsidiary analysis (taking into account the system's terminations) has determined what the acoustic impedances at all the other openings should be. Also, let us assume that the pressure is uniform throughout the junction region. The complex-amplitude version of Eq. (2) then gives

$$\frac{\hat{U}_1^{\text{out}}}{\hat{p}} = -\sum_{n=2}^{N} \frac{1}{Z_{A,n}^{\text{out}}} + \frac{ikV}{\rho c} \tag{7-3.3}$$

We do not expect the terms on the right to cancel each other, so insofar as we seek to determine the number on the left side, the $ikV/\rho c$ term can be neglected if *at least one* of the $Z_{A,n}^{\text{out}}$ is substantially less in magnitude than $\rho c/kV$. Even if this is not satisfied, a "satisfactory" estimate of $\hat{U}_1^{\text{out}}/\hat{p}$ to the order of the traveling-wave magnitude $A_1/\rho c$ is obtained with the $ikV/\rho c$ term neglected as long as $kV/A_1 \ll 1$. The approximation $\Sigma U_n^{\text{out}} = 0$ will therefore lead to the same implications as Eq. (2) if the terminal impedance on any opening is substantially smaller in magnitude than $\rho c/kV$ or if our concern is with the impedance the junction and appendages present to a subsystem coupled to the junction through an area large compared to kV.

Example: Duct with change in cross-sectional area In the duct sketched in Fig. 7-9, all indicated dimensions are substantially less than a wavelength, so evanescent modes are significant only between $x = -\delta_1$ and $x = \delta_2$. The plane-wave-mode disturbance in, say, the $x > \delta_2$ region is a superposition of plane waves traveling in the $+x$ and $-x$ directions, so an extrapolation of these waves back to $x = 0$ determines what the pressure and volume velocity (positive sense corresponding to flow in the $+x$ direction) corresponding to the plane-wave mode would be at $x = 0$. Furthermore, the orthogonality relation (7-1.3) leads to the conclusion that the other modes never contribute to the volume velocity, so the $x \to 0$ extrapolated volume velocity associated with the plane-wave mode should be the same in the limit $k\delta_2 \ll 1$ as the actual volume velocity at $x = 0$. The extrapolated pressure should be the area averaged pressure at $x = 0$. Such considerations in conjunction with Eq. (2) lead to the

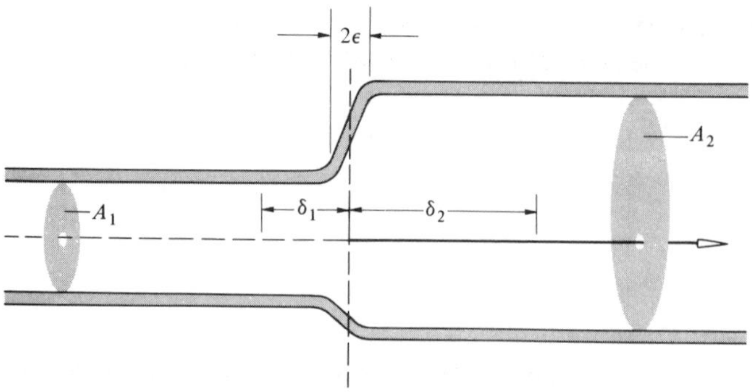

Figure 7-9 Duct with change in cross-sectional area.

conclusion that the volume velocity is continuous across the junction. Since the volume intrinsically associated with the junction is in effect zero, the right side of Eq. (2) gives no contribution. This reasoning still applies when the opening at the junction is obstructed, by a plate with an orifice, by a porous membrane, or by a flexible plate extending across the junction.

We cannot necessarily conclude, however, that the two plane-wave-mode pressures extrapolated to $x = 0$ should be the same; nevertheless, from Eq. (7-2.6) one can set

$$\hat{p}(0^-) - \hat{p}(0^+) = Z_J\hat{U}(0) \tag{7-3.4}$$

where Z_J is an acoustic impedance associated with the junction;† $\hat{p}(0^-)$ represents the plane-wave-mode pressure in the $x < 0$ duct segment extrapolated to $x = 0$, and $\hat{p}(0^+)$ is the corresponding extrapolated pressure for the $x > 0$ duct segment.

For an unobstructed junction, the simple rule that emerges from Eq. (1) is that an upper limit to $|Z_J|$ is $\rho ck(\delta_1 + \delta_2)/A_{\min}$ where $A_{\min}$ is the minimum duct area. This can be compared with the traveling-plane-wave impedances $\rho c/A_1$ and $\rho c/A_2$ for the two duct segments. If the estimated upper limit is substantially less than either of these, we replace (4) by $\hat{p}(0^-) = \hat{p}(0^+)$. A circumstance where this might not be valid would be where both ducts are circular and of radii a_1 and a_2, with $a_1 \gg a_2$. Then we can take $\delta_1 + \delta_2 \approx a_1$, so we would be concerned about the finite value of Z_J when ka_1 is comparable to a_2^2/a_1^2 or when the frequency $\omega/2\pi$ is comparable to or larger than a critical value of $(ca_2^2/a_1^3)/2\pi$.

For example, if a duct of 3 cm radius were joined to one of 10 cm radius, we would consider taking the junction's impedance into account at frequencies of the order of $(340)(\frac{3}{10})^2/[(2\pi)(0.1)] \approx 50$ Hz. In contrast, the lowest cutoff fre-

† For results applicable to cylindrical ducts, see F. Karal, "The Analogous Acoustical Impedance for Discontinuities and Constrictions of Circular Cross-Section," *J. Acoust. Soc. Am.*, **25**:327–334 (1953). Karal's approximate low-frequency result in the present notation is that the acoustic inertance [equal to $Z_J/(-i\omega)$] associated with a junction between joined circular cylinders of radii b and a (with $b < a$) with a common axis is of the form

$$M_A = \frac{8\rho}{3\pi^2 b} H\left(\frac{b}{a}\right)$$

where $H(b/a)$ is 1 when $b/a \to 0$ and decreases monotonically to zero as $b/a \to 1.0$. The general theory for arbitrary ka is developed by J. W. Miles; "The Reflection of Sound Due to a Change in Cross Section of a Circular Tube," ibid., **16**:14–19 (1944). A derivation based on the Schwarz-Christoffel transformation applied to a rectangular duct, occupying the region $0 < y < a$, $0 > z > d$, with a rigid partition at $x = 0$ having a slit of width b in its middle extending from $z = 0$ to $z = d$, $y = (a - b)/2$ to $y = (a + b)/2$, yields an acoustic inertance

$$M_A = \frac{2\rho}{\pi d} \ln\left[\csc\left(\frac{b}{a}\frac{\pi}{2}\right)\right]$$

which diverges logarithmically to ∞ as $b \to 0$. J. W. Miles, "The Analysis of Plane Discontinuities in Cylindrical Tubes, II," ibid., **17**:272–284 (1946); P. M. Morse and K. U. Ingard, *Theoretical Acoustics*, McGraw-Hill, New York, 1968, pp. 483–487.

quencies for dispersive modes in the two ducts are 3300 and 1000 Hz, respectively.

Reflection and Transmission at a Junction

The estimation of the amplitude of waves, transmitted and reflected at a junction, within the context of the model described by Eq. (4) proceeds along lines similar to those discussed in Secs. 3-3 and 3-6. If the incident wave comes from the $-x$ side, the resulting traveling wave on the other side of the junction causes the acoustic impedance for $x > 0$ to be $\rho c/A_2$. The impedance for the plane-wave mode in the $x < 0$ portion will therefore be $Z_J + \rho c/A_2$ at $x = 0^-$.

The pressure-amplitude reflection coefficient for the incident (plane-wave mode) wave can be written, with a suitable interpretation of symbols in Eq. (3-3.4), as

$$\mathcal{R} = \frac{Z_A(0^-) - \rho c/A_1}{Z_A(0^-) + \rho c/A_1} = \frac{Z_J + \rho c/A_2 - \rho c/A_1}{Z_J + \rho c/A_2 + \rho c/A_1} \tag{7-3.5}$$

The requirement that the volume velocity at $x = 0$ be $(1 - \mathcal{R})(A_1/\rho c)$ times the incident pressure amplitude and that the transmitted pressure amplitude at $x = 0^+$ be $\rho c/A_2$ times the volume velocity at $x = 0^+$ causes the ratio of transmitted pressure to incident pressure to be

$$\mathcal{T} = \frac{A_1}{A_2}(1 - \mathcal{R}) = \frac{2\rho c/A_2}{Z_J + \rho c/A_2 + \rho c/A_1} \tag{7-3.6}$$

In the usual case, when Z_J is neglected, $\mathcal{R}$ and $\mathcal{T}$ reduce to $(A_1 - A_2)/(A_1 + A_2)$ and $2A_1/(A_1 + A_2)$. The fraction of the incident power that is transmitted is $4A_1A_2/(A_1 + A_2)^2$.

7-4 HELMHOLTZ RESONATORS AND OTHER EXAMPLES

The Helmholtz Resonator

The classic model (see Fig. 7-10*a*) of a Helmholtz resonator† (a wine bottle being a ubiquitous example) consists of a rigid-walled volume connected to the external environment by a small opening, which may or may not have a neck. The overall dimensions are all much less than an acoustic wavelength. Within the vol-

† H. Helmholtz, "Theory of Air Oscillations in Tubes with Open Ends," *J Reine Angew. Math.*, **57**:1–72 (1860); *On the Sensations of Tone*, 4th ed., 1877, trans. A. J. Ellis, Dover, New York, pp. 42–44, 55, 372–374; M. S. Howe, "On the Helmholtz Resonator," *J. Sound Vib.*, **45**:427–440 (1976); U. Ingard, "On the Theory and Design of Acoustical Resonators," *J. Acoust. Soc. Am.*, **25**:1037–1062 (1953); A. S. Hersh and B. Walker, "Fluid Mechanical Model of the Helmholtz Resonator," *NASA* CR-2904 (1977). Applications to noise control are discussed by M. C. Junger, "Helmholtz Resonators in Load-Bearing Walls," *Noise Control Eng.*, **4**:17–25 (1975).

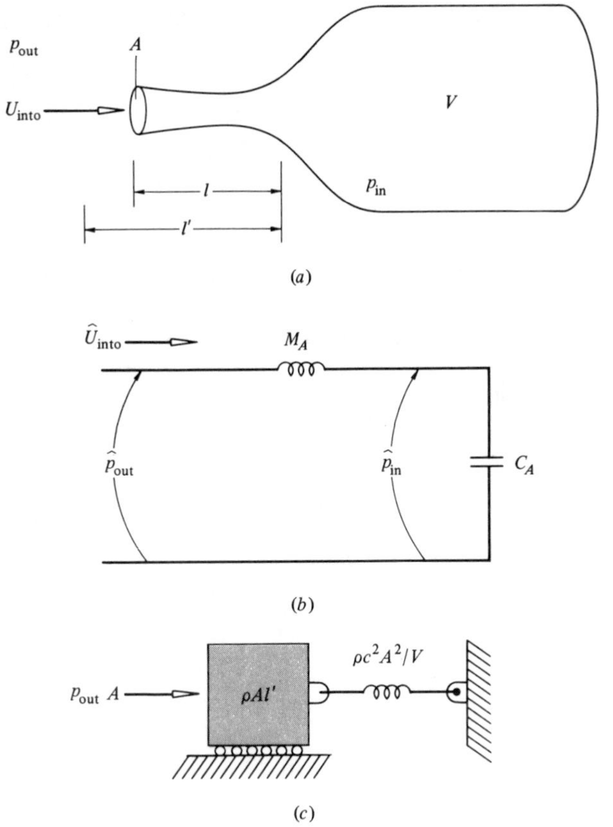

Figure 7-10 (*a*) Sketch of a Helmholtz resonator within which the pressure is p_{in} and through whose neck flows volume velocity U_{into}. (*b*) Electric-circuit analog. (*c*) Mechanical analog.

ume proper at points not near the opening, Eq. (7-3.1) suggests that the pressure should be spatially uniform; the analysis leading to Eq. (7-2.11) consequently requires the volume velocity U_{into} flowing into the volume to be $(V/\rho c^2)\partial p/\partial t$. The generalization of this relation that takes dissipation into account is $\hat{U}_{into} = \hat{p}_{in}/Z_{vol}$, where Z_{vol} is the acoustic impedance (with a positive real part) associated with the volume. Here, however, we restrict our attention to the ideal case, such that $Z_{vol} = 1/(-i\omega C_A)$, where the acoustic compliance C_A is $V/\rho c^2$.

Near the opening, possibly also within the neck, and just outside the opening in the external environment, the pressure may vary markedly with position. However, since the volume in that region is small ($k\,\Delta V/A \ll 1$), we model the region near the opening as a continuous-volume-velocity two-port. The complex pressure amplitude $\hat{p}_{out}$ somewhat outside the opening† is therefore related

† As is explained in the next section, the pressure amplitude at moderate distances r from the opening is of the form $\hat{A}(\mathbf{x}) + \hat{B}/r$, where $\hat{A}(\mathbf{x})$ is slowly varying with position $\mathbf{x}$ relative to the center of the opening, $\hat{B}$ is independent of position; the identification for $\hat{p}_{out}$ is $\hat{A}(\mathbf{0})$.

to $\hat{p}_{\text{in}}$ by the relation, $\hat{p}_{\text{out}} - \hat{p}_{\text{in}} = Z_{\text{op}}\hat{U}_{\text{into}}$, where Z_{op} is the opening's acoustic impedance.

If one neglects dissipation, Eq. (7-2.9) applies and Z_{op} is $-i\omega M_A$. If the opening has a long neck of length l, the inertance is nearly that of a duct segment of length l and area A within which the disturbance is in the plane-wave mode. For such a circumstance, but for $kl \ll 1$, the fluid in the neck behaves like a lumped mass ρAl caused to accelerate by the force $(p_1 - p_2)A$, where A is the neck cross-sectional area. The resulting acceleration of this lumped mass is $A^{-1}dU_{\text{into}}/dt$, so $\rho l \; dU_{\text{into}}/dt$ should be $(p_1 - p_2)A$ (mass times acceleration equals force). A comparison of such a relation with Eq. (7-2.9) leads to $\rho l/A$ for the neck's acoustic inertance M_A. If the neck is not long or is even nonexistent, one can still write $M_A = \rho l'/A$, where l' is an "effective neck length."

The definitions of Z_{op} and Z_{vol} taken together lead to

$$\hat{p}_{\text{out}} = Z_{\text{HR}}\hat{U}_{\text{into}} \qquad \hat{p}_{\text{in}} = \frac{Z_{\text{vol}}}{Z_{\text{HR}}}\hat{p}_{\text{out}} \qquad Z_{\text{HR}} = Z_{\text{vol}} + Z_{\text{op}} \tag{7-4.1}$$

Here Z_{HR} is the acoustic impedance just outside the opening of the resonator and HR stands for Helmholtz resonator. These relations correspond to a circuit diagram (Fig. 7-10*b*) of a continuous-volume-velocity two-port terminated by an impedance Z_{vol}. If Z_{vol} is taken as $1/(-i\omega C_A)$ and Z_{op} as $-i\omega M_A$, the analog is an *LC* circuit (inductor and capacitor in series). In the latter idealized case, the substitutions $-i\omega \to d/dt$ and $\hat{U}_{\text{into}} \to dX_{\text{into}}/dt$ yield

$$M_A \frac{d^2X_{\text{into}}}{dt^2} + \frac{1}{C_A} X_{\text{into}} = p_{\text{out}} \qquad p_{\text{in}} = C_A^{-1}X_{\text{into}} \tag{7-4.2}$$

where X_{into} denotes the *volume displacement*.

Alternatively, if $\xi_{\text{into}} = X_{\text{into}}/A$ denotes the average particle displacement in the opening, the first of these can be written

$$M_{\text{mech}}\frac{d^2\,\xi_{\text{into}}}{dt^2} + k_{\text{sp}}\xi_{\text{into}} = F_{\text{mech}} \tag{7-4.3}$$

where $M_{\text{mech}} = \rho Al'$ = apparent mass of fluid moving in vicinity of opening
$k_{\text{sp}} = \rho c^2A^2/V$ = apparent spring constant associated with compressible fluid in volume
$F_{\text{mech}} = p_{\text{out}}A$ = apparent force exerted on opening by pressure field outside opening

Thus the Helmholtz resonator can be interpreted (see Fig. 7-10*c*) as a forced harmonic oscillator, i.e., a mass and a spring moving under the influence of an external force.

The pressure p_{out} outside the opening is affected by the dynamic state of the resonator, but for simplicity we here regard p_{out} as being externally controlled. Consequently, if it is made to oscillate with angular frequency ω, Eq. (2) yields

$$X_{\text{into}} = C_Ap_{\text{in}} = \frac{p_{\text{out}}}{-\omega^2M_A + C_A^{-1}} \tag{7-4.4}$$

Resonance occurs when the denominator vanishes; this is at the resonance frequency ω_r, where

$$\omega_r = \frac{1}{(M_A C_A)^{1/2}} = \left(\frac{k_{\text{sp}}}{M_{\text{mech}}}\right)^{1/2} = c\left(\frac{A}{l'V}\right)^{1/2} \tag{7-4.5}$$

If ω is close to ω_r, the pressure oscillations inside the volume are considerably larger than just outside the opening. In addition, because the resonator's impedance Z_{HR} is $(-i\omega C_A)^{-1} - i\omega M_A$, Eq. (5) implies that Z_{HR} is 0 at the resonance frequency.

Helmholtz Resonator as a Side Branch

We next consider the example† of a long straight duct (of cross-sectional area A_D and extending along the x axis) that has a Helmholtz resonator attached to one of its walls in the vicinity of $x = 0$ (see Fig. 7-11a). Let $\hat{U}_{\text{HR}}$ denote the complex amplitude of the volume velocity flowing into the Helmholtz resonator; let $\hat{U}_D(0^-)$ and $\hat{U}_D(0^+)$ denote volume-velocity amplitudes in the duct just before and just after the junction with the resonator, positive sense corresponding to flow in the $+x$ direction. The discussion in Sec. 7-3 concerning volume velocities flowing out of a small volume suggests that volume velocity is locally conserved, so we set

$$\hat{U}_D(0^-) = \hat{U}_{\text{HR}} + \hat{U}_D(0^+) \tag{7-4.6}$$

Also, the pressure $p_D(x, t)$ in the duct is expected to be continuous at $x = 0$, and $\hat{p}_D(0)$ should be the pressure amplitude just outside the resonator opening; $\hat{p}_D(0^-)$, $\hat{p}_D(0^+)$, and $\hat{p}_{\text{out,HR}}$ are therefore all equal. Dividing both sides of (6) by the common pressure amplitude then gives (see Fig. 7-11b)

$$Z_A^{-1}(0^-) = Z_{\text{HR}}^{-1} + Z_A^{-1}(0^+) \tag{7-4.7}$$

where $Z_A(x)$ is the acoustic impedance in the duct.

Reflection and transmission of waves past the resonator is analyzed as described previously in the discussion of the effects of a change in duct cross-sectional area. The pressure-amplitude reflection coefficient is given by the first version of Eq. (7-3.5), which, from Eq. (7), leads to

$$\mathscr{R} = \frac{(Z_{\text{HR}}^{-1} + A_D/\rho c)^{-1} - \rho c/A_D}{(Z_{\text{HR}}^{-1} + A_D/\rho c)^{-1} + \rho c/A_D} = \frac{-\rho c/A_D}{2Z_{\text{HR}} + \rho c/A_D} \tag{7-4.8}$$

The pressure-amplitude transmission coefficient $\mathscr{T}$ is $1 + \mathscr{R}$ because the pressure amplitude at $x = 0^+$ is $(1 + \mathscr{R})\hat{p}_i(0^-)$. The fractions of incident power reflected and transmitted are $|\mathscr{R}|^2$ and $|\mathscr{T}|^2$; the fraction absorbed by the resonator is $1 - |\mathscr{R}|^2 - |\mathscr{T}|^2$.

† G. W. Stewart, "Acoustic Transmission with a Helmholtz Resonator or an Orifice as a Branch Line," *Phys. Rev.*, **27**:487–493 (1926).

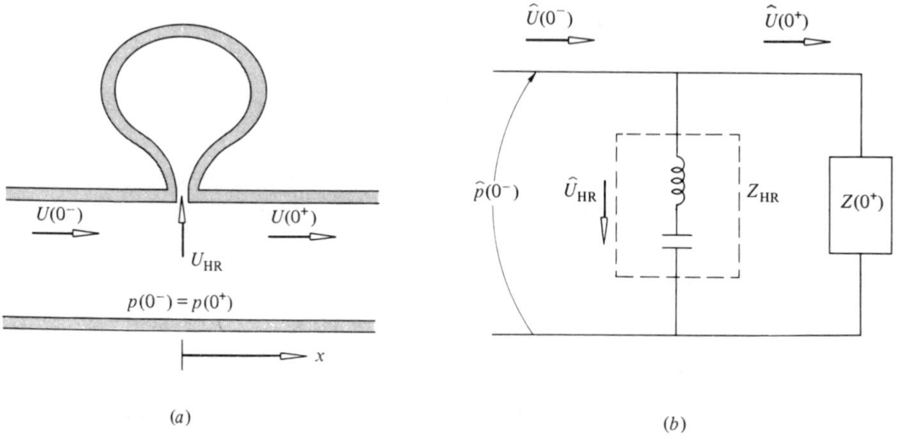

Figure 7-11 Helmholtz resonator as a sidebranch: (*a*) geometrical configuration; (*b*) equivalent circuit.

Near the resonance frequency of the resonator, $Z_{HR} \to 0$ (or becomes very small when energy dissipation is taken into account), so $\mathscr{R} \to -1$ (as for reflection by a pressure-release surface) and $\mathscr{T} \to 0$. Thus the resonator has the potentially useful property of causing nearly total reflection of acoustic waves at frequencies near its resonance frequency.

Composite Example†

Various seemingly complicated acoustical systems can be satisfactorily and simply analyzed by lumped-parameter techniques; an example is shown in Fig. 7-12*a*. A force of complex amplitude $\hat{F}$ and angular frequency ω drives a piston of mechanical mass M_P at one end of a short duct segment of cross-sectional area A. The other end is terminated by a closed cavity, while the middle of the duct has two side branches. The upper branch leads successively through two cavities connected by a narrow constriction. The duct (area A_L) in the lower branch has a porous membrane of flow resistance $\Delta p/v = R_f$ stretched across it. Beyond the membrane, the lower duct leads in an unspecified manner to the external environment, so that the (terminal) acoustic impedance just below the membrane appears to be Z_{term}.

The modeling of the system proceeds with the replacement of the driving force by a driving pressure of $\hat{F}/A$. The piston becomes an acoustic inertance of M_P/A^2. With each duct subsection or constriction one associates an acoustic inertance, denoted by M_{A1}, M_{A2}, etc. With the cavities one associates acoustic compliances C_{A1}, C_{A2}, C_{A3}. The porous membrane becomes an acoustic resistance $R_A = R_f/A_L$.

† For a number of similar examples, see L. L. Beranek, *Acoustics*, McGraw-Hill, New York, 1954, pp. 67–69, 437–442.

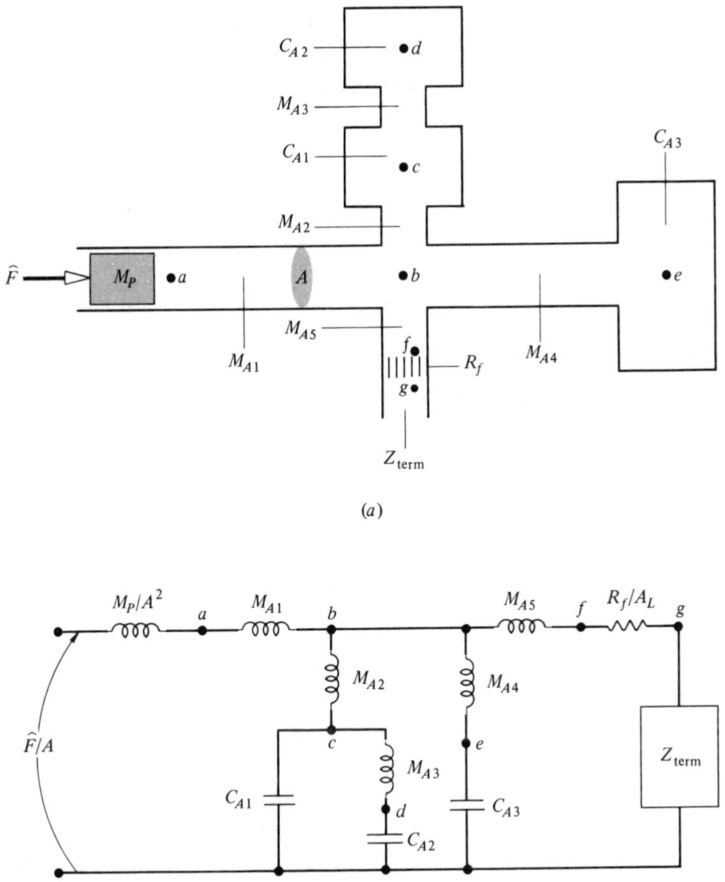

Figure 7-12 (*a*) Composite acoustical system discussed in the text. (*b*) Circuit representation of the lumped-parameter model. The voltages at the points *a*, *b*, *c*, . . . in the latter correspond to the acoustic pressures at the corresponding points in the acoustical system.

The circuit analog in Fig. 7-12*b* is a compact representation of all the equations constituting the model. The correspondences depicted between voltages in the circuit diagram and pressures at points in the acoustical system are in accord with the relations $\hat{p}_1 - \hat{p}_2 = Z_A\hat{U}_{12}$ and $U_1 - U_2 = \hat{p}/Z_A$ that hold for continuous-volume-velocity and continuous-pressure two-ports, respectively. Thus, for example, the current from *b* to *c* corresponds to the volume velocity U_{bc} flowing into the cavity with compliance C_{A1}. Part of this volume velocity accounts for the time rate of change of pressure in the cavity and corresponds to current flowing through C_{A1} in the circuit diagram; the other part of U_{bc} is U_{cd} and corresponds to the current flowing through M_{A3} and C_{A2} in the circuit diagram. From an analysis of the circuit equations, one can determine the mechanical impedance presented by the system to the force $\hat{F}$ and the net power generated by the force, as well as the volume velocity flowing through any portion of the system and the pressure at each designated point in the sketch.

7-5 ORIFICES

Another example for which a lumped-parameter model is applicable is the transmission of sound through an orifice† (hole) in an otherwise rigid thin plate (see Fig. 7-13); the orifice's cross-sectional dimensions (a denoting a representative value) are much less than $\lambda/2\pi$. Let the z axis be normal to the plate, the coordinate origin being centered at the orifice. The analysis here adopts the conceptual framework of matched asymptotic expansions (discussed previously in Sec. 4-7). We eventually concentrate on the case when the orifice is circular, but for the present we proceed without any special assumption concerning its shape.

Matched-Asymptotic-Expansion Solution for Orifice Transmission

On the $-z$ side of the plate, a wave with pressure $p_i(x, y, z, t)$ is incident and in the absence of the orifice creates a reflected wave with pressure $p_i(x, y, -z, t)$. We group these two (external) pressures together and call the sum $p_{\text{ext}}^{(-)}(\mathbf{x}, t)$. Given that the orifice is small, the resulting field at large distances $r \gg a$ from the orifice consists, in the region $z < 0$, approximately of the incident wave, the reflected wave, and an outgoing spherical wave. On the $z > 0$ side, the field in the same limit is a spherical wave. These two spherical waves are caused by the motion of fluid at the opening, so the result (5-3.3) based on the low-ka approximation to the Rayleigh integral is applicable. The surface integral appearing there over $\dot{v}_n$ is identified from the definition (7-2.1) as $-\dot{U}_{12}$ or $\dot{U}_{12}$ for the spherical waves propagating on the $-z$ and $+z$ sides of the plate, where U_{12} is the volume velocity flowing through the orifice from the $-z$ side to the $+z$ side. Thus, our expressions for the outer solutions at large r become

$$p \to [p_{\text{ext}}^{(-)}(\mathbf{x}, t), 0] \mp \frac{\rho}{2\pi r} \dot{U}_{12}\left(t - \frac{r}{c}\right) \qquad \begin{cases} z < 0 \\ z > 0 \end{cases} \tag{7-5.1}$$

These automatically satisfy the wave equation and, moreover, satisfy the boundary condition $\nabla p \cdot \mathbf{n} = 0$ on the plate boundary.

The inner solution for small ka is described by a velocity potential $\Phi(\mathbf{x}, t)$ that has an asymptotic expansion in powers of $1/r$, each term of which satisfies Laplace's equation. If we keep just the first two terms, we have

$$\Phi \to \Phi_{\infty}^{(-,+)} \pm \frac{U_{12}}{2\pi r} \qquad \begin{cases} z < 0 \\ z > 0 \end{cases} \tag{7-5.2}$$

† J. W. S. Rayleigh, "On the Passage of Waves through Apertures in Plane Screens, and Allied Problems," *Phil. Mag.*, (5)**43**:259–272 (1897).

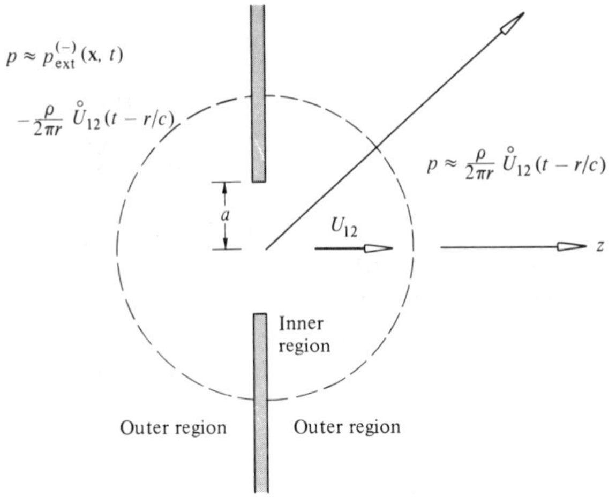

Figure 7-13 Geometry used in the discussion of sound transmission through an orifice.

where $\Phi_\infty^{(-)}$ and $\Phi_\infty^{(+)}$ are the asymptotic values of Φ on the $-z$ and $+z$ sides of the orifice. That the coefficients of $1/r$ are equal but opposite is in accord with the conservation of mass; the U_{12} appearing here must also be the volume velocity from $-z$ side to $+z$ side through the orifice, so it is the same as the U_{12} in Eq. (1). Because U_{12} must be linearly dependent on $\Phi_\infty^{(-)}$ and $\Phi_\infty^{(+)}$, and because it must be zero when the two asymptotic potentials are equal, one can set

$$\Phi_\infty^{(+)} - \Phi_\infty^{(-)} = \frac{M_{A,\text{or}}}{\rho} U_{12} \tag{7-5.3}$$

where the proportionality factor $M_{A,\text{or}}$ is the acoustic inertance intrinsically associated with the orifice.

Matching Eqs. (1) to Eqs. (2) consists of expanding Eqs. (1) in a power series in r (the leading term of which goes as $1/r$) and equating the coefficients of the r^{-1} and r^0 terms with those in the expansion of $-\rho\, \partial\Phi/\partial t$. Matching of the r^{-1} terms substantiates our use of the function $U_{12}(t - r/c)$ in Eqs. (1); the matching of the r^0 terms yields

$$[p_{\text{ext}}^{(-)}(\mathbf{0}, t), 0] \pm \frac{\rho}{2\pi c} \ddot{U}_{12} = -\rho \dot{\Phi}_\infty^{(-,+)} \tag{7-5.4}$$

When inserted into Eq. (3), these give for the constant-frequency case ($\partial/\partial t \to -i\omega$)

$$(-i\omega M_{A,\text{or}})\hat{U}_{12} = [\hat{p}_{\text{ext}}(0) - R_A^{(-)}\hat{U}_{12}] - (R_A^{(+)}\hat{U}_{12}) \tag{7-5.5}$$

$$R_A^{(-)} = R_A^{(+)} = \frac{\omega^2\rho}{2\pi c} = \frac{k^2\rho c}{2\pi} \tag{7-5.6}$$

The transient version of (5) is an ordinary differential equation for $U_{12}(t)$.

Acoustic-Radiation Resistance

The second term of Eq. (5-3.1) indicates that the real part of the acoustic radiation impedance associated with sound generation by fluid motion in the orifice must always be $\rho ck^2/2\pi$ to lowest nonvanishing order in ka, which is in accord with the values of $R_A^{(+)}$ and $R_A^{(-)}$ in Eq. (5). Also, these values yield $(\rho/2\pi c)(\dot{U}_{12}^2)_{av}$ for the averaged acoustic power radiated to each side of the orifice by the fluid motion. Because this power is the same as is carried away by each of the spherical waves in Eq. (1), the consistency of the solution represented by Eqs. (1) and (3) is further substantiated. Although $R_A^{(-)}$ and $R_A^{(+)}$ are identical, we make a distinction between the two corresponding terms in Eq. (5) because, in other instances, one or both of the acoustic resistance terms do not appear in the formulation.

Helmholtz Resonator with Baffled Opening

One such instance is when the orifice connects a Helmholtz resonator to an external environment (see Fig. 7-14). The "outer solution" for the interior of the resonator would be taken as that where p is spatially uniform and $\mathbf{v}$ is such that $\nabla \cdot \mathbf{v} = -(\rho c^2)^{-1}\, \partial p/\partial t$. Matching this with the inner solution, Eq. (2), gives $p_{in} = -\rho\dot{\Phi}_\infty^{(+)}$ and $U_{12} = \dot{p}_{in}V/\rho c^2$, so that the transient version of Eq. (5) becomes instead

$$p_{ext} - p_{in} = M_A\dot{U}_{12} - \frac{\rho}{2\pi c}\ddot{U}_{12} \tag{7-5.7}$$

with $\dot{p}_{in} = \rho c^2 U_{12}/V$. In this instance, the $R_A^{(+)}$ term in (5) is replaced by one involving the acoustic compliance of the resonator. To the external pressure field, the acoustic impedance $\hat{p}_{ext}/\hat{U}_{12}$ of the Helmholtz resonator appears to be $-i\omega M_A + 1/(-i\omega C_A) + R_A^{(-)}$, where C_A is the acoustic compliance asso-

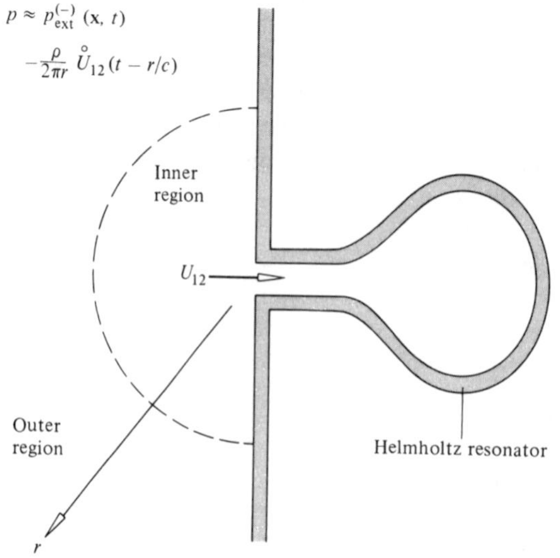

Figure 7-14 Orifice terminated by a Helmholtz resonator; U_{12} is volume velocity from external medium into the resonator.

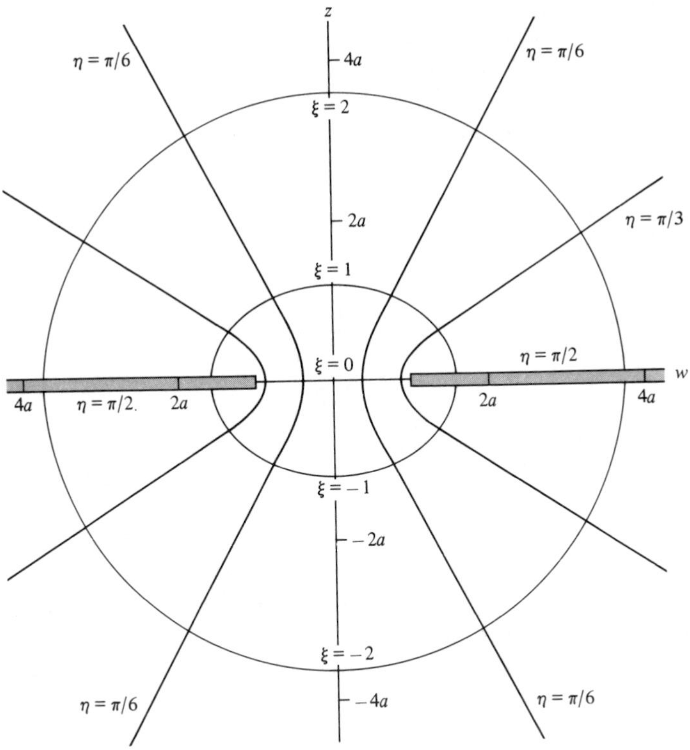

Figure 7-15 Oblate-spheroidal coordinate system used in the derivation of the acoustic inertance of a circular orifice. Note that ξ ranges from $-\infty$ to ∞, η from 0 to $\pi/2$. The plate is the surface $\eta = \pi/2$; the orifice corresponds to $\xi = 0$.

ciated with the volume and $R_A^{(-)}$ is the acoustic resistance given by Eq. (6). From this point of view, the resonator, even in the absence of fluid friction, is intrinsically a damped oscillator, the damping being associated with the radiation of sound from the mouth of the resonator.

Acoustic Inertance of a Circular Orifice in a Thin Plate

The incompressible-flow inner-region solution can be found in closed form when the orifice is circular (radius a), the plate thickness being idealized as infinitesimal. The appropriate coordinate system for a determination of Φ is oblate-spheroidal coordinates, such that $w = a \cosh \xi \sin \eta$, $x = w \cos \phi$, $y = w \sin \phi$, and $z = a \sinh \xi \cos \eta$, where† ξ ranges from $-\infty$ to ∞, η ranges from 0 to $\pi/2$, and ϕ ranges from 0 to 2π (see Fig. 7-15). The boundary condi-

† In the analysis (Sec. 4-8) of radiation from a vibrating circular plate, the range of ξ was taken to be from 0 to ∞ and the range of η to be between 0 and π. The distinction arises because we wish the coordinates to be continuous at all points not adjacent to solid boundaries. Here we wish ξ to be continuous at the orifice and accept the discontinuity of η at neighboring points on opposite sides of the plate.

tion on Φ corresponding to the presence of the rigid plate is $\partial\Phi/\partial\eta = 0$ at $\eta = \pi/2$ for all ξ. The requirement that the potential Φ approach asymptotic expressions of the form of Eqs. (2) at large r implies that Φ is independent of η and ϕ at large $|\xi|$. All this will be so if Φ is a function only of ξ (other than of time t). In this case, Laplace's equation reduces to

$$\frac{1}{\cosh \xi}\frac{d}{d\xi}\left(\cosh \xi \frac{d\Phi}{d\xi}\right) = 0 \tag{7-5.8}$$

which successively integrates† to

$$\frac{d\Phi}{d\xi} = \frac{B}{\cosh \xi} = B\frac{d(\sinh \xi)/d\xi}{1 + \sinh^2 \xi}$$

$$\Phi = D + B \tan^{-1} (\sinh \xi) \tag{7-5.9}$$

where D and B are constants, the arc tangent being understood to be between $-\pi/2$ and $\pi/2$. Note that the orifice ($\xi = 0$) is a surface of constant potential, as required by symmetry.

At large $|\xi|$, $w \to (a/2)e^{|\xi|} \sin \eta$, $|z| \to (a/2)e^{|\xi|} \cos \eta$, so $r \to (a/2)e^{|\xi|}$ and $\sinh \xi \to \pm r/a$, where the two signs correspond to $\xi > 0$ and $\xi < 0$ (or $z > 0$ and $z < 0$). Since $\tan^{-1} f \to \pi/2 - 1/f$ as $f \to +\infty$ and $\tan^{-1} f \to -\pi/2 + 1/|f|$ as $f \to -\infty$, one accordingly has, at large r, that $\Phi \to (D \mp B\pi/2) \pm Ba/r$ for $z < 0$ and $z > 0$. Comparison of these with Eq. (2) then gives $Ba = U_{12}/2\pi$, $\Phi_\infty^{(+)} - \Phi_\infty^{(-)} = B\pi$; Eq. (3) therefore yields‡

$$M_{A,\text{or}} = \frac{\rho}{2a} \tag{7-5.10}$$

as the acoustic inertance associated with the orifice. Since acoustic inertance is the pressure per unit volume acceleration, the quantity $(\pi a^2)^2 M_{A,\text{or}}$ or $(\rho)(\pi a^2)(\pi a/2)$ is the apparent mass of air oscillating back and forth through the

† H. Lamb, *Hydrodynamics*, 1879, 5th ed., 1932, sec. 108, pp. 144–145. Lamb's expression in the present notation is $\Phi = -B \cot^{-1} (\sinh \xi)$, which is $-B [\pi/2 - \tan^{-1} (\sinh \xi)]$, so our result differs from his by a constant whose value is immaterial insofar as $\mathbf{v} = \nabla\Phi$ is concerned. The solution dates back to E. Heine (1843).

‡ For the more general case of an elliptical orifice of area A and eccentricity e [defined such that $(1 - e^2)^{1/2}$ is ratio of minor axis to major axis] the result is

$$\frac{M_{A,\text{or}}}{\rho} = \frac{1}{2}\left(\frac{\pi}{A}\right)^{1/2} \frac{2}{\pi} K(e^2)(1 - e^2)^{1/4}$$

$$\approx \frac{1}{2}\left(\frac{\pi}{A}\right)^{1/2}\left(1 - \frac{e^4}{64} - \frac{e^6}{64} - \cdots\right)$$

where $K(e^2)$ is the complete elliptical integral of the first kind defined by Eq. (5-3.8). This is derived by Rayleigh, *Theory of Sound*, vol. 2, sec. 306. Rayleigh's discussion is in terms of a conductivity, which is the same as ρ divided by the acoustic inertance. His conclusion based on the above result is that it is a good approximation to take the conductivity as $2(A/\pi)^{1/2}$ [or to take $M_{A,\text{or}}$ as $(\rho/2)(\pi/A)^{1/2}$]. For a general review, see C. L. Morfey, "Acoustic Properties of Openings at Low Frequencies," *J. Sound Vib.*, 9:357–366 (1969).

orifice. This is the mass of fluid in a column of cross-sectional area πa^2 and length $\pi a/2$.

Diffraction of Plane Wave by a Circular Orifice

The foregoing results lead to the conclusion that if a plane wave $\hat{p}_i = Ae^{i\mathbf{k}\cdot\mathbf{x}}$ [such that $\hat{p}_{\text{ext}}^{(-)}(\mathbf{0})$ is $2A$] is incident on a plate with a circular orifice of radius a where $ka \ll 1$, the diffracted wave on the $z > 0$ side of the orifice is given by [see Eqs. (1), (5), (6), and (10)]

$$\hat{p} = \frac{-i\omega\rho}{2\pi} \frac{2A}{-i\omega(\rho/2a) + k^2\rho c/\pi} \frac{e^{ikr}}{r} \approx \frac{2aA}{\pi} \frac{e^{ikr}}{r} \tag{7-5.11}$$

The time-averaged transmitted power is $2\pi r^2|\hat{p}|^2/2\rho c$, or

$$\mathcal{P}_{\text{av}} = \frac{4a^2}{\pi} \frac{A^2}{\rho c} = \frac{8a^2}{\pi} I_{i,\text{av}} \tag{7-5.12}$$

This is $8/\pi^2 = 0.81$ times the acoustic power $\pi a^2 I_{i,\text{av}}$ incident on the aperture when $\mathbf{k}$ is parallel to $\mathbf{e}_z$. By contrast, the Kirchhoff approximation (see Sec. 5-2) would predict the volume velocity through the orifice to have an amplitude $(A/\rho c)\pi a^2$ and the transmitted power to be $(ka)^2(\pi a^2/4)A^2/\rho c$, or $(ka)^2/2$ times the incident power when the incoming wave is at normal incidence. Given $ka \ll 1$, the latter would be considerably smaller than is actually the case.

7-6 ESTIMATION OF ACOUSTIC INERTANCES AND END CORRECTIONS

In the absence of dissipative mechanisms, the only lumped-parameter element needed to describe a continuous-volume-velocity two-port is its acoustic inertance M_A. This is often difficult to calculate exactly (the circular-orifice example in the previous section being an exception), but there are applicable fluid-dynamic principles regarding incompressible flow for estimating and putting bounds on its value.

Principle of Minimum Kinetic Energy

Euler's equation leads to the conclusion $\nabla \times \mathbf{v} = 0$, so one can conceive of a velocity potential Φ such that $\mathbf{v} = \nabla\Phi$, $p = -\rho\, \partial\Phi/\partial t$. This conclusion is not changed if the flow is incompressible, so that $\nabla \cdot \mathbf{v} = 0$ replaces the mass-conservation equation, but there are also other conceivable incompressible flows satisfying the appropriate boundary conditions that are not potential flows. Of all such flows, however, the potential flow gives the minimum kinetic energy.†

† The theorem is due to W. Thomson (Lord Kelvin), "On the Vis-Viva [Kinetic Energy] of a Liquid in Motion," *Camb. Dublin Math. J.*, 1849; reprinted in *Mathematical and Physical Papers*, vol. 1, Cambridge University Press, Cambridge, 1882, pp. 107–112.

To demonstrate this, let $\mathbf{v}(\mathbf{x}, t)$ be a potential-flow field and imagine that a variation $\delta\mathbf{v}$ dependent on $\mathbf{x}$ is added to it. Both $\mathbf{v}$ and $\delta\mathbf{v}$ are incompressible flow fields, but $\nabla \times \delta\mathbf{v}$ is not necessarily zero.

The total kinetic energy $(\mathrm{KE})_{\mathrm{var}}$ associated with the varied field in a fixed volume V is

$$(\mathrm{KE})_{\mathrm{var}} = \iiint \tfrac{1}{2}\rho(\mathbf{v} + \delta\mathbf{v})^2\, dV \tag{7-6.1}$$

The cross term $\rho\mathbf{v} \cdot \delta\mathbf{v}$ in the integrand can be written as $\nabla \cdot (\rho\Phi\, \delta\mathbf{v})$ because $\mathbf{v} = \nabla\Phi$, $\nabla \cdot \delta\mathbf{v} = 0$; so its volume integral becomes a surface integral. Thus, since $\delta\mathbf{v} \cdot \delta\mathbf{v} \geq 0$, Eq. (1) yields the inequality

$$(\mathrm{KE})_{\mathrm{var}} \geq (\mathrm{KE})_{\mathrm{true}} + \iint \rho\Phi\, \delta\mathbf{v} \cdot \mathbf{n}\, dS \tag{7-6.2}$$

where the integral is over the surface S bounding V; the "true" kinetic energy corresponds to $\delta\mathbf{v} = 0$.

Suppose that the boundary conditions on some portions of S are those appropriate to a rigid boundary, so that $\mathbf{v} \cdot \mathbf{n} = 0$, while on all other portions $\mathbf{v} \cdot \mathbf{n}$ is known. Then, for *any* incompressible flow field (not necessarily irrotational) that satisfies the boundary conditions, the deviation $\delta\mathbf{v}$ of this $\mathbf{v}$ from the actual $\mathbf{v}$ must be such that $\delta\mathbf{v} \cdot \mathbf{n} = 0$ everywhere on S. Since the second term in Eq. (2) vanishes, the actual flow gives the *minimum* kinetic energy of all conceivable incompressible flows that satisfy the same boundary conditions.

Other applicable circumstances are when the boundary conditions are specified so that Φ has value Φ_1 on one portion S_1 of S and has value Φ_2 on another portion S_2 of S while $\mathbf{v} \cdot \mathbf{n} = 0$ on the remainder of S (see Fig. 7-16). Thus $\mathbf{n} \times \mathbf{v} = 0$ on S_1 and S_2. The solution of the boundary-value problem can be characterized by a volume velocity U_{12} flowing from S_1 to S_2. Any incompressible flow through V satisfying $\mathbf{v} \cdot \mathbf{n} = 0$ on all portions of S other than S_1 and S_2 also corresponds to some U_{12}. If the resulting U_{12} is the actual U_{12}, the surface integral of $\delta\mathbf{v} \cdot \mathbf{n}$ vanishes on both S_1 and S_2. Since Φ is constant on either S_1 and S_2, and since $\delta\mathbf{v} \cdot \mathbf{n} = 0$ on all other portions, the second term of

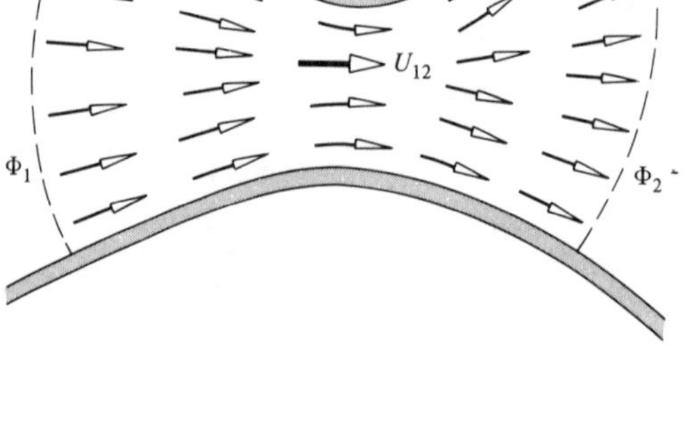

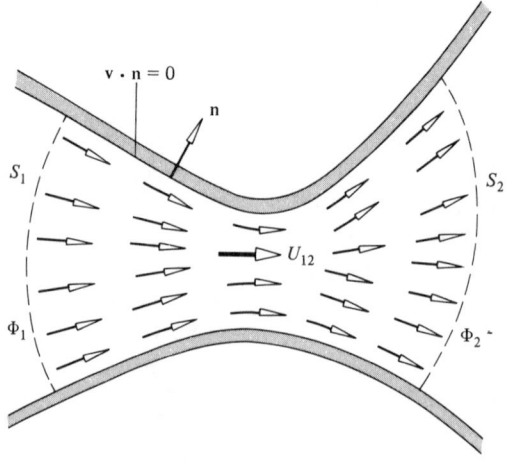

Figure 7-16 Circumstances for which the principle of minimum kinetic energy yields the principle of minimum acoustic inertance.

(2) must vanish. Therefore, regardless of the values of Φ_1 and Φ_2, the potential-flow field corresponding to a given U_{12} is the one of all such flow fields for which the kinetic energy is a minimum.

Principle of Minimum Acoustic Inertance

For the circumstances described above where Φ is constant on portions S_1 and S_2 and $\mathbf{v} \cdot \mathbf{n} = 0$ on the remainder of S, the acoustic inertance M_A, defined such that M_A/ρ is $(\Phi_2 - \Phi_1)/U_{12}$, can also be written† as $2\mathrm{KE}/U_{12}^2$. Consequently, the minimum-kinetic-energy principle yields

$$M_A \leq \frac{2(\mathrm{KE})_{\mathrm{var}}}{U_{12}^2} \tag{7-6.3}$$

Any incompressible (but not necessarily irrotational) flow field passing through V with $\mathbf{v} \cdot \mathbf{n} = 0$ on all portions of S other than S_1 and S_2 will give particular values of U_{12} and KE; with them one can calculate an estimate of M_A from Eq. (3). Since the true kinetic energy corresponding to the same U_{12} will be smaller, the estimated M_A will be an upper bound.

Effect of Relaxing of Constraints

A consequence of Eq. (3) is that any relaxing of constraints must decrease the acoustic inertance. Thus, for example, the geometry in Fig. 7-17*b* results in a lower acoustic inertance than that in Fig. 7-17*a*. To demonstrate this, let V_b be a control volume that corresponds to the less constrained flow; let V_a correspond to the constrained volume with the same choices for S_1 and S_2, so V_a is entirely confined within V_b. A possible flow through V_b corresponds to a potential flow through V_a but with nonmoving fluid in the regions of V_b not lying in volume V_a. Such a flow field when inserted into the right side of Eq. (3) would give the true acoustic inertance $M_{A,a}$ for V_a but must overestimate $M_{A,b}$ since it is not the true potential flow for V_b. Thus $M_{A,b} < M_{A,a}$. The proof also implies that an imposition of a constraint must increase the acoustic inertance.

Lower Bound for Acoustic Inertance

The principle of minimum kinetic energy gives a powerful method for obtaining an upper bound to M_A when the potential-flow boundary-value problem is not

† The proof begins with the requirement $\Phi \nabla^2 \Phi = 0$. With a vector identity and with $\mathbf{v} = \nabla\Phi$, this leads to

$$\tfrac{1}{2}\rho\nabla \cdot (\Phi \mathbf{v}) = \tfrac{1}{2}\rho v^2$$

Integration over the volume and subsequent application of Gauss' theorem yields

$$\tfrac{1}{2}\rho\Phi_2 U_{12} - \tfrac{1}{2}\rho\Phi_1 U_{12} = \mathrm{KE}$$

so the definition, $M_A/\rho = (\Phi_2 - \Phi_1)/U_{12}$, requires that $2\mathrm{KE}/U_{12}^2$ also be M_A.

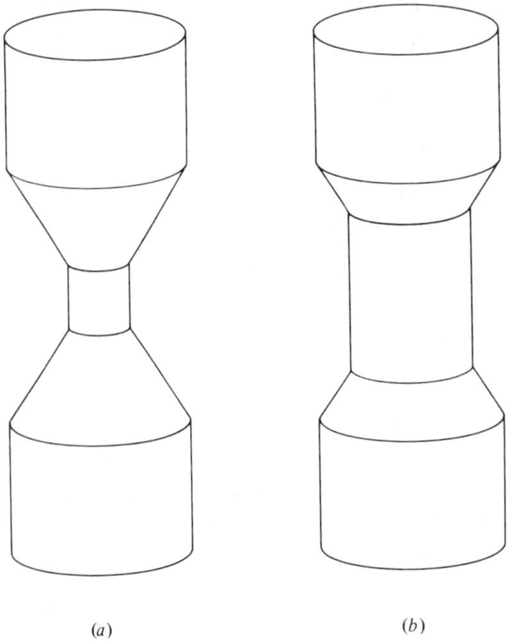

Figure 7-17 The geometry in (*a*) is such that the flow is constrained relative to that for the geometry in (*b*). The assertion is made that the less constrained geometry has the lower acoustic inertance.

easily solvable. Here we describe a theorem due to Rayleigh† that can yield a lower bound.

Let us suppose the volume V is divided (see Fig. 7-18) into two volumes V_{I} and V_{II} by a surface S_{mid} extending across its middle, so that fluid flowing from S_1 to S_2 must flow through S_{mid}. Although S_1 and S_2 are specified to be equipotential surfaces ($\mathbf{n} \times \mathbf{v} = 0$), one does not necessarily expect S_{mid} to be an equipotential also. However, if V_{I} were considered by itself, one might formally regard S_{mid} as being an equipotential and one could thereby associate an acoustic inertance $M_{A,\mathrm{I}}$ with volume V_{I}. Similarly, acoustic inertance $M_{A,\mathrm{II}}$ can be associated with V_{II}. The statement that can be made concerning the acoustic inertance M_A for the volume V as a whole is

$$M_A \geq M_{A,\mathrm{I}} + M_{A,\mathrm{II}} \tag{7-6.4}$$

so that the sum $M_{A,\mathrm{I}} + M_{A,\mathrm{II}}$ gives a lower bound for M_A.

To prove the assertion, let Φ_{I}, Φ_{II} be solutions for the boundary-value

† J. W. S. Rayleigh, "On the Theory of Resonance," *Phil. Trans. R. Soc. Lond.*, **161**:77–118 (1870); *Theory of Sound*, vol. 2, sec. 305. Rayleigh's statement of the theorem, paraphrased in the terminology of the present text, was that if the ambient density is diminished in any region, the acoustic inertance should also be decreased. The inertance would be the $M_{A,\mathrm{I}} + M_{A,\mathrm{II}}$ in Eq. (4) if ρ were formally considered to go to zero in a thin layer about the surface S_{mid}. Consequently, the actual inertance should be greater than or equal to $M_{A,\mathrm{I}} + M_{A,\mathrm{II}}$. In terms of the electrical analog, Rayleigh's assertion seems obvious, but the physical realization of such a limiting case in a fluid-dynamic context presents conceptual difficulties, so the theorem is here demonstrated without consideration of cases where the ambient density is nonuniform.

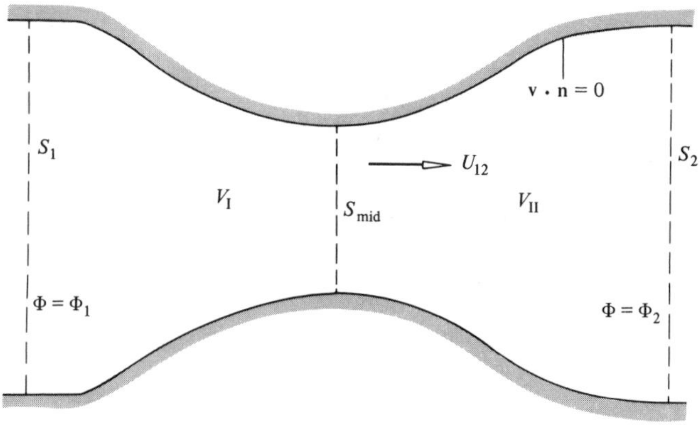

Figure 7-18 Geometry used in proof of Rayleigh's lower-bound theorem for acoustic inertances.

problem corresponding to volumes V_I and V_{II} and let Φ be the solution corresponding to volume V as a whole. It is assumed that each such solution corresponds to the same volume velocity. We denote the corresponding velocity fields by $\mathbf{v}_I$, $\mathbf{v}_{II}$, and $\mathbf{v}$. The kinetic energy KE for the boundary-value problem corresponding to volume V can be expressed in terms of those values $(\mathrm{KE})_I$ and $(\mathrm{KE})_{II}$ corresponding to the velocity fields $\mathbf{v}_I$ and $\mathbf{v}_{II}$ in volumes V_I and V_{II} as

$$\mathrm{KE} = \iiint \tfrac{1}{2}\rho(\mathbf{v}_I + \mathbf{v} - \mathbf{v}_I)^2\, dV_I + \iiint \tfrac{1}{2}\rho(\mathbf{v}_{II} + \mathbf{v} - \mathbf{v}_{II})^2\, dV_{II}$$

$$\geq (\mathrm{KE})_I + (\mathrm{KE})_{II} + \iiint \rho\mathbf{v}_I \cdot (\mathbf{v} - \mathbf{v}_I)\, dV_I + \iiint \rho\mathbf{v}_{II} \cdot (\mathbf{v} - \mathbf{v}_{II})\, dV_{II} \quad (7\text{-}6.5)$$

where the inequality follows from $(\mathbf{v} - \mathbf{v}_I)^2 \geq 0$, $(\mathbf{v} - \mathbf{v}_{II})^2 \geq 0$. In the third term, we use $\mathbf{v}_I = \nabla\Phi_I$, $\nabla \cdot (\mathbf{v} - \mathbf{v}_I) = 0$ to replace $\mathbf{v}_I \cdot (\mathbf{v} - \mathbf{v}_I)$ by its equivalent $\nabla \cdot [\Phi_I(\mathbf{v} - \mathbf{v}_I)]$, such that, with Gauss' theorem, we obtain

$$\iiint \rho\mathbf{v}_I \cdot (\mathbf{v} - \mathbf{v}_I)\, dV_I = \iint \rho\Phi_I(\mathbf{v} - \mathbf{v}_I) \cdot \mathbf{n}_I\, dS_1 + \iint \rho\Phi_I(\mathbf{v} - \mathbf{v}_I) \cdot \mathbf{n}_I\, dS_{\mathrm{mid}} \quad (7\text{-}6.6)$$

(Recall that $\mathbf{v} \cdot \mathbf{n}_I = \mathbf{v}_I \cdot \mathbf{n}_I = 0$ on the portions of the surface S_I of V_I other than S_1 and S_{mid}.) Here $\mathbf{n}_I$ denotes the unit outward normal on S_I. Since Φ_I is constant on either S_1 or S_{mid}, the definition of the volume velocity U_{12} requires that each of the integrals in Eq. (6) vanish. Since the same is true for the analogous integral over V_{II} in Eq. (5), that equation requires KE to be greater or equal to $(\mathrm{KE})_I + (\mathrm{KE})_{II}$. The identification of M_A as $2\mathrm{KE}/U_{12}^2$ and the hypothesis that the three kinetic energies each correspond to the same volume velocity then leads to Eq. (4).

Flanged Opening in a Duct

A circular duct (radius a) with a flanged opening (Fig. 7-19a) furnishes a simple example to which the above principles apply. The potential-flow problem in the vicinity of the opening is such that $\Phi \to \Phi_D + (U/\pi a^2)z$ for large negative z within the duct and $\Phi \to \Phi_\infty - U/2\pi r$ at large r in the half space outside the opening. The acoustic inertance M_A is defined for this example such that M_A/ρ

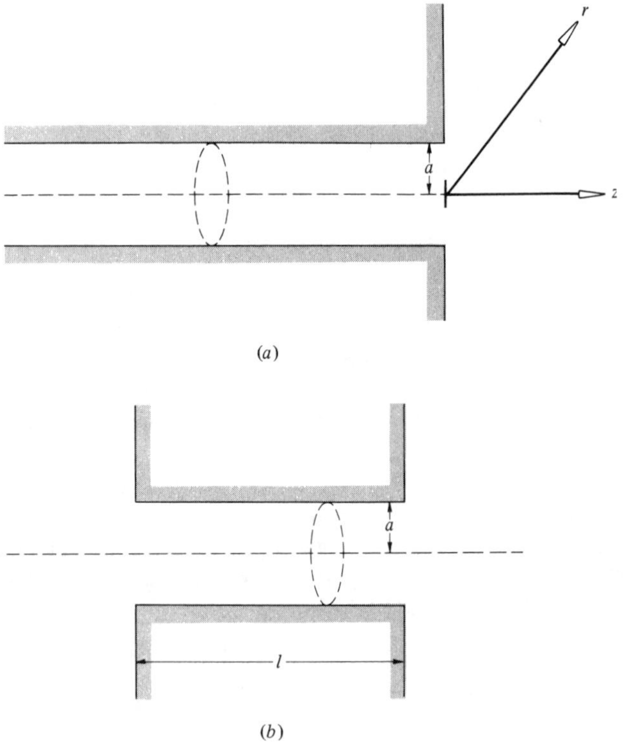

Figure 7-19 (*a*) A semi-infinite duct with a flanged opening. (*b*) A duct segment of finite length with flanges at both ends.

is $(\Phi_\infty - \Phi_D)/U$. To estimate its value by the principle of minimum acoustic inertance, we postulate an incompressible flow such that within the duct $v_z = U/\pi a^2$ is uniform over the cross section; the flow outside the opening is taken to be a potential flow. Conservation of mass across each differential area of the opening imposes $v_z = U/\pi a^2$ at $z = 0$ for $w < a$ as a boundary condition on the $z > 0$ solution. The existence of the flange requires $v_z = 0$ on the remainder of the $z = 0$ plane.

The potential flow outside the flange has a kinetic energy equal to the volume integral of $\frac{1}{2}\rho(\nabla\Phi)^2 = \frac{1}{2}\rho\,\nabla\cdot(\Phi\,\nabla\Phi)$. Gauss' theorem (with the choice of 0 for Φ_∞) converts this to an area integral over the opening of $-\frac{1}{2}\rho v_z \Phi$. At the opening, v_z is assumed equal to $U/\pi a^2$. The area integral of $-\rho\Phi$ is the time integral of the area integral of p; the latter is identified from the result for the vibrating circular piston in a rigid wall. Equation (5-3.10) yields in the low-frequency limit a value of $(8/3\pi)aU$ for the area integral of $-\Phi$, and the kinetic energy therefore becomes

$$\mathrm{KE} = \frac{1}{2}\frac{8}{3\pi^2}\frac{\rho}{a}U^2 \tag{7-6.7}$$

For the postulated flow field, this is the excess kinetic energy associated with the presence of the opening; Eq. (3) consequently yields

$$M_A \le \frac{8}{3\pi^2}\frac{\rho}{a} \tag{7-6.8}$$

To apply Rayleigh's lower-bound theorem, we take $S_{\rm mid}$ to be the opening. Our definition of acoustic inertance is such that there is no inertance associated with the duct ($M_{A,{\rm I}} = 0$), so the lower bound $M_{A,{\rm II}}$ is $(\Phi_\infty - \Phi_{\rm op})\rho/U$, where $\Phi_{\rm op}$ is the assumed uniform potential across the opening. This quantity $M_{A,{\rm II}}$, however, can be taken from the solution given in Sec. 7-5 for potential flow through a circular orifice in a thin rigid plate. That solution is such that the orifice is of uniform potential and $\Phi_\infty^{(+)} - \Phi_{\rm or} = \Phi_{\rm or} - \Phi_\infty^{(-)}$. Consequently, the inertance associated with the region $z > 0$ is one-half that given by Eq. (7-5.10). Thus, we obtain, from Eq. (4),

$$M_A \ge \frac{1}{4}\frac{\rho}{a} \tag{7-6.9}$$

This, in conjunction with Eq. (8), brackets M_A between $0.250\rho/a$ and $0.270\rho/a$. The actual value† is $0.261\rho/a$.

Circular Orifice in a Plate of Finite Thickness

This example (Fig. 7-19b) can be regarded as a short circular duct of length l ($kl \ll 1$) with flanges on both openings. If an incompressible flow is postulated that has uniform flow within the duct, the principle of minimum acoustic inertance applies and an analysis similar to that leading to Eq. (8) yields

$$M_A \le \frac{\rho l}{\pi a^2} + 2\frac{8}{3\pi^2}\frac{\rho}{a} \tag{7-6.10}$$

Rayleigh's lower-bound theorem similarly yields

$$M_A \ge \frac{\rho l}{\pi a^2} + \frac{1}{2}\frac{\rho}{a} \tag{7-6.11}$$

In the limit $l \to 0$, M_A is given by the expression (7-5.10) for an orifice in a thin plate, so Eq. (11) is exact in this limit. If l/a is large, the cross section in the middle of the duct should be of nearly uniform potential, so M_A should be twice the inertance of a duct segment of length $l/2$ with a flanged opening; the inertance due to each half is nearly $\rho(l/2)/\pi a^2$ plus the inertance intrinsically associated with a flanged opening. Taking King's result of $0.261\rho/a$ for the latter, we have

$$M_A \approx \frac{\rho l}{\pi a^2} + \frac{2(0.261)\rho}{a} \qquad l \gg a \tag{7-6.12}$$

† L. V. King, "On the Electrical and Acoustic Conductivities of Cylindrical Tubes Bounded by Infinite Flanges," *Phil. Mag.*, (7)**21:**128–144 (1936).

It cannot necessarily be assumed that this is either a lower bound or an upper bound for arbitrary l/a, but it is an overestimate in the limit $l/a \to 0$.

End Corrections

The acoustic inertance for the example above can be written in the form

$$M_A = \frac{\rho}{A}(l + \Delta l) \tag{7-6.13}$$

where A is the cross-sectional area of the duct and Δl is an end correction associated with the terminations of the duct at the two ends. If $l \gg (A)^{1/2}$, the remarks preceding Eq. (12) indicate that Δl is independent of l and furthermore can be decomposed into contributions $(\Delta l)_1$ and $(\Delta l)_2$ that are associated with each of the two ends. Thus, if one end of the duct opens with a flange into an unlimited space, the correction $(\Delta l)_1$ for this end is AM_{A1}/ρ, where M_{A1} is the acoustic inertance associated with the opening. For a circular duct of radius a with a flanged opening, Eqs. (8) and (9) yield an end correction $(\Delta l)_1$ with the limits $(8/3\pi)a$ and $(\pi/4)a$ or, equivalently, $0.85a$ and $0.79a$. King's exact result for $(\Delta l)_1$ is $0.82a$.

Another model of duct termination is that of a thin-walled hollow circular tube protruding into an open space. The absence of the constraining flange causes the acoustic inertance associated with the opening to decrease, so the end correction must be less than $0.82a$. There are no simple calculations that place more stringent bounds on the end correction, but an intricate exact solution† for the radiation of waves from an unflanged hollow tube yields, in the low-frequency limit,

$$\Delta l = (0.61 \cdots)a \qquad \text{unflanged opening} \tag{7-6.14}$$

The corresponding acoustic inertance is $\rho(\pi a^2)^{-1}\,\Delta l$ or $0.20\rho/a$.

Effective Neck Lengths of Helmholtz Resonators

In the discussion preceding Eq. (7-4.1), the acoustic inertance of a Helmholtz resonator is taken as $\rho l'/A$, where l' is an effective neck length. In the estimation of l' we distinguish cases where the actual neck length l is much less and much greater than the radius a of the opening. In both cases, a is assumed to be much less than the dimensions of the vessel. If $l \ll a$, the opening is similar to that of an orifice in a thin plate, so the appropriate estimate of the acoustic inertance is that of Eq. (11), which leads to $l + (\pi/2)a$ for l'.

If $l \gg a$, then $l' \approx l + (\Delta l)_1 + (\Delta l)_2$, where $(\Delta l)_1$ and $(\Delta l)_2$ are the end corrections associated with the inner and outer openings. The inner opening re-

† H. Levine and J. Schwinger, "On the Radiation of Sound from an Unflanged Circular Pipe," *Phys. Rev.*, **73**:383–406 (1948). The case when the tube walls are of finite thickness is analyzed by Y. Ando, "On the Sound Radiation from Semi-Infinite Pipe of Certain Wall Thickness," *Acustica*, **22**:219–225 (1970).

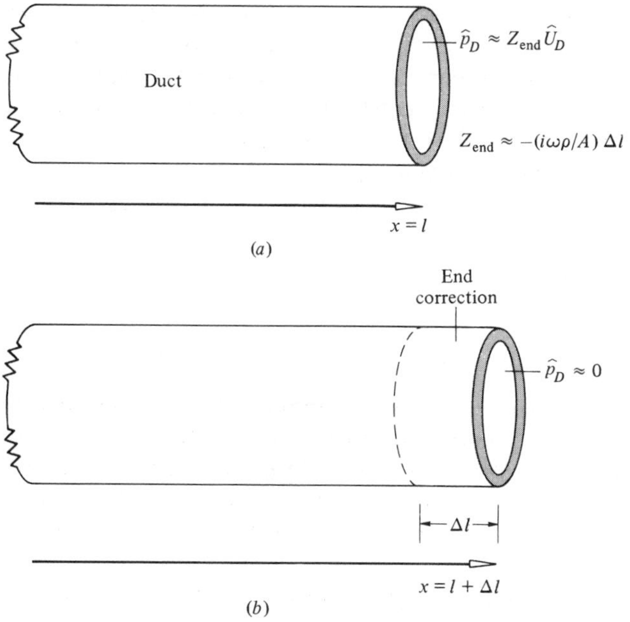

Figure 7-20 (*a*) Open-ended duct extending into open space. (*b*) Duct with end correction Δl that has equivalent acoustical properties if the end is taken to be a pressure-release surface.

sembles a flanged termination, so we set $(\Delta l)_1 = 0.82a$. This value would also apply for $(\Delta l)_2$ if the outer end of the neck terminates in a flange (a Helmholtz resonator with a baffled opening). If the neck is long and its walls are thin, the model of an unflanged opening is more appropriate so one would set $(\Delta l)_2 = 0.61a$. Thus, in the latter case, for example, one would have†

$$l' = l + 0.82a + 0.61a \qquad l \gg a \tag{7-6.15}$$

If the neck is not circular, the usual approximation is to replace a by $(A/\pi)^{1/2}$.

Boundary Conditions at Open Ends of Ducts

A classic example of the application of an end correction is at the open end of a duct (see Fig. 7-20). We begin with the observation that the end presents an acoustic impedance Z_{end} to any plane-wave-mode disturbance within the duct $(x < l)$, where, in the low-frequency limit,

$$Z_{end} = -i\omega M_A + R_A \qquad M_A = \frac{\rho}{A}\,\Delta l \tag{7-6.16a}$$

$$R_A = \frac{K\rho c k^2}{4\pi} \tag{7-6.16b}$$

† W. P. Mason, "The Approximate Networks of Acoustic Filters," *Bell Syst. Tech. J.*, **9**:332–340 (1930).

The acoustic radiation resistance R_A, according to Eq. (7-5.6), should be $\rho ck^2/2\pi$ if the opening has an infinite flange, so the parameter K is identified as 2 for that case. If the opening resembles a thin-walled tube protruding into space, the acoustic pressure at large distances from the opening is only half as large given the same volume velocity at the end, so K would then be 1. [The derivation is analogous to that ensuing from Eq. (7-5.1).]

The simplest end-correction approximation consists of the replacement† of the boundary condition

$$\frac{\hat{p}_D}{\hat{U}_D} = Z_{\text{end}} \qquad x = l \tag{7-6.17a}$$

by

$$\hat{p}_D = 0 \qquad x = l + \Delta l \tag{7-6.17b}$$

Here $\hat{p}_D(x)$ and $\hat{U}_D(x)$ are the plane-wave-mode pressure and volume-velocity amplitudes within the duct ($x < l$). Their values for $x > l$ are regarded as what would be extrapolated using the one-dimensional linear acoustic equations. Adopting the boundary condition (17b) is equivalent to regarding the end as being at $x = l + \Delta l$ and to assuming that this virtual end is a pressure-release surface.

Approximate justification of Eq. (17b) proceeds with the neglect of the radiation resistance, so that Eqs. (16a) and (17a) imply a zero value for $\hat{p}_D(l) + i\omega\rho\, \Delta l\, \hat{U}_D(l)/A$. But Euler's equation equates $i\omega\rho\hat{U}_D/A$ to $d\hat{p}_D/dx$, and $\hat{p}_D(l) + \Delta l(d\hat{p}_D/dx)_l$ is approximately $\hat{p}_D(l + \Delta l)$, so Eq. (17b) results.

Since the radiation resistance is proportional to k^2, its effects on the field within the duct are ordinarily minor at low frequencies. The exception is when the system is at resonance. Nevertheless, for the determination of the resonance frequencies, Eq. (17b) remains a good approximation at low frequencies and is preferable to taking the actual end at $x = l$ as a pressure-release surface.

7-7 MUFFLERS AND ACOUSTIC FILTERS

A muffler‡ is a device that reduces the sound emanating from the end of a pipe but which continues to allow the flow of gas through the pipe. In an idealized conceptual model of a muffler (see Fig. 7-21), the source is characterized by the volume velocity $U(t)$ injected into the exhaust system; each frequency compo-

† Helmholtz, "Theory of Air Oscillations"; Rayleigh, *The Theory of Sound,* vol. 2, sec. 314. The necessity for an end correction emerged with the experimental discovery by Felix Savart (1823) that the first velocity node is less than $\frac{1}{4}$ wavelength from the open end. The boundary condition of $p = 0$ at the open end (without end correction) was adopted by Daniel Bernoulli, Euler, and Lagrange in the eighteenth century.

‡ P. O. A. L. Davies, "The Design of Silencers for Internal Combustion Engines," *J. Sound Vib.,* **1:**185–201 (1964); T. F. W. Embleton, "Mufflers," in L. L. Beranek (ed.), *Noise and Vibration Control,* McGraw-Hill, New York, 1971, pp. 362–405; E. K. Bender and A. J. Bremmer, "Internal-Combustion Engine Intake and Exhaust System Noise," *J. Acoust. Soc. Am.,* **58:**22–30 (1975).

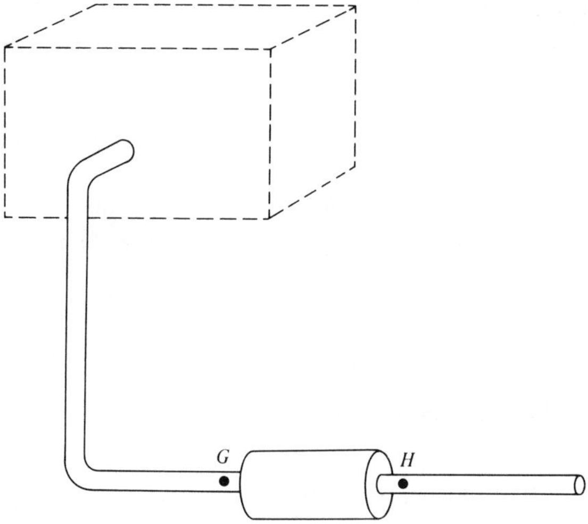

Figure 7-21 Simplified model of an exhaust system. The muffler is inserted between points G and H.

nent is assumed to propagate independently, and it is assumed that the muffler and the configuration of the pipe do not alter the spectral density of the volume velocity actually injected by the source. The interaction of the acoustic portion of the flow with the mean flow is also neglected.

The assumptions just stated imply, for any given muffler design, that there should be a direct proportionality between the same frequency components of volume velocities existing at two given points. Thus, we can characterize the source for our present purposes by what the spectral density would be at a given point if the pipe extended indefinitely without interruptions or changes in cross-sectional area. We choose this point G to be just upstream of where the muffler is to be inserted. The external sound radiation is determined by the spectral density of the volume velocity leaving the tail of the pipe, which in turn is determined by the ratio of the spectral density at the exit plane to that nominally expected at G. This ratio, however, can be derived from an analysis of constant-frequency sound propagation.

The Transmission Matrix and Its Consequences

The segment of the pipe that includes the muffler, extending between points G and H in Fig. 7-21, can be regarded as an acoustical two-port, so the matrix equation (7-2.3) applies. In an equivalent manner, we can write

$$\begin{bmatrix} \hat{p}_G \\ \hat{U}_G \end{bmatrix} = \begin{bmatrix} K_{11} & K_{12} \\ K_{21} & K_{22} \end{bmatrix} \begin{bmatrix} \hat{p}_H \\ \hat{U}_H \end{bmatrix} \tag{7-7.1}$$

where the quantities K_{ij} are frequency-dependent quantities embodying the acoustical properties of the muffler. Reciprocity requires that the matrix determinant be 1. Also, for a symmetric muffler, which looks the same from both ends, K_{11} and K_{22} must be identical.

The ratio $\hat{p}_H/\hat{U}_H$ is the acoustic impedance Z_H just downstream of the muffler presented by the tailpipe and the environment. The ratio $\hat{p}_G/\hat{U}_G$, derived from Eq. (1), is accordingly

$$\frac{\hat{p}_G}{\hat{U}_G} = \frac{K_{11}Z_H + K_{12}}{K_{21}Z_H + K_{22}} \tag{7-7.2}$$

If a wave is incident on the muffler at G from the upstream direction, then $\hat{U}_G = (1 - \mathcal{R}_G)\hat{U}_i$, where $\hat{U}_i$ is the portion of the volume velocity at G associated with this incident wave and where

$$\mathcal{R}_G = \frac{\hat{p}_G/\hat{U}_G - \rho c/A}{\hat{p}_G/\hat{U}_G + \rho c/A} \tag{7-7.3}$$

is the pressure-amplitude reflection coefficient for a wave incident on the muffler. Since Eq. (1) leads to

$$(1 - \mathcal{R}_G)\hat{U}_i = (K_{21}Z_H + K_{22})\hat{U}_H \tag{7-7.4}$$

we accordingly find

$$\frac{2\rho c}{A}\frac{\hat{U}_i}{\hat{U}_H} = K_{11}Z_H + K_{12} + \frac{\rho c}{A}K_{21}Z_H + \frac{\rho c}{A}K_{22} \tag{7-7.5}$$

Insertion Loss

The acoustic-pressure amplitude in the far field is directly proportional to the volume velocity $|\hat{U}_H|$ just downstream of the muffler. Consequently, the performance of the muffler is characterized by the ratio of $|\hat{U}_H|^2$ to what its value would be without the muffler. With the assumption (discussed below) that $|\hat{U}_i|$ is unaffected by the muffler and with the recognition that $[K]$ is the unit matrix when the muffler is not present, one finds from (5) that the reciprocal of this ratio is

$$\frac{|K_{11}Z_H + K_{12} + (\rho c/A)K_{21}Z_H + (\rho c/A)K_{22}|^2}{|Z_H + \rho c/A|^2} \tag{7-7.6}$$

The assumption that $\hat{U}_i$ is unaffected by the muffler's presence is equivalent to the expectation that waves reflected back to the source by the muffler have negligible amplitude when they eventually return to the muffler. Circumstances for which the assumption is valid are when the pipe upstream of the muffler is such that a traveling wave experiences, say, 5 dB attenuation or more on one round trip. A similar assumption that further simplifies the analysis is that there is sufficient attenuation along the tailpipe to ensure that whatever is transmitted beyond the muffler at H does not return to the muffler (anechoic termination). This allows us to assume a traveling plane wave at H such that $Z_H = \rho c/A$. Both assumptions are traditional† in muffler design but warrant reconsideration

† D. D. Davis, G. M. Stokes, D. Moore, and G. L. Stevens, "Theoretical and Experimental Investigation of Mufflers with Comments on Engine-Exhaust Muffler Design," *Nat. Advis. Comm. Aeronaut. Rep.* 1192, Washington, 1954; G. W. Stewart, "Acoustic Wave Filters," *Phys. Rev.*, **20:**528–551 (1922).

in particular cases. They are adopted here to obtain an unencumbered perspective on muffler performance.

With the assumptions just described, the insertion loss of the muffler, defined as the sound-pressure-level drop caused by its insertion, is 10 times the logarithm of the expression (7-7.6) with Z_H replaced by $\rho c/A$, that is,

$$\mathrm{IL} = 10 \log \left(\tfrac{1}{4} \left| K_{11} + K_{22} + \frac{\rho c}{A} K_{21} + \frac{A}{\rho c} K_{12} \right|^2 \right) \tag{7-7.7}$$

Since we are assuming anechoic termination of the muffler, insertion loss is the same as transmission loss. The objective of a good muffler design is that IL be very low for low frequencies, so the steady flow is not inhibited, but IL be high at those acoustic frequencies which convey the dominant portion of the noise. Thus the muffler should perform like a low-pass filter.

Reactive and Dissipative Mufflers

A reactive muffler is one for which the dissipation in the muffler can be neglected. In this event, the parameters Z_{left}, Z_{right}, and Z_{mid} in Eqs. (7-2.4) are all imaginary numbers, and consequently one finds K_{11} and K_{22} to be real and K_{12} and K_{21} to be imaginary. A reactive muffler reduces the sound power entering the muffler by altering the acoustic impedance Z_G at the entrance of the muffler. For example, if Z_G were zero, no power would pass into the muffler. Any plane wave incident on the muffler would undergo perfect reflection. Even if the attenuation in the upstream pipe were insignificant, this would still reduce the power radiated out of the tailpipe, because the created standing wave would have a pressure at the source nearly 90° out of phase with the source's volume velocity.

A dissipative muffler, on the other hand, does not appreciably alter the power entering the muffler but instead dissipates it before it leaves the muffler. The simplest idealization of a dissipative muffler is a lined segment of pipe of length L that attenuates the amplitude of a traveling plane wave by a factor of $e^{-\alpha L}$ without appreciably reflecting the sound or altering the ratio of pressure to volume velocity. The transmission loss in this case is easily seen to be 10 times $\log e^{2\alpha L}$, but it is instructive to see how the result follows from the formulation developed above.

Letting $\hat{p}_a$ and $\hat{p}_b$ be the amplitudes at G and H, respectively, of two coexisting plane waves traveling downstream and upstream, respectively, we find

$$\left(\hat{p}_G, \frac{\rho c \hat{U}_G}{A}\right) = \hat{p}_a \pm \hat{p}_b e^{ikL} e^{-\alpha L} \qquad \left(\hat{p}_H, \frac{\rho c \hat{U}_H}{A}\right) = \hat{p}_a e^{ikL} e^{-\alpha L} \pm \hat{p}_b$$

Elimination of $\hat{p}_a$ and $\hat{p}_b$ from these and a comparison with Eq. (1) yields

$$[K] = \begin{bmatrix} \cos(kL + i\alpha L) & -i\dfrac{\rho c}{A}\sin(kL + i\alpha L) \\ -i\dfrac{A}{\rho c}\sin(kL + i\alpha L) & \cos(kL + i\alpha L) \end{bmatrix} \tag{7-7.8}$$

so the insertion loss of Eq. (7) reduces to $(10 \log e)(2\alpha L)$. Thus, the larger αL the larger the insertion loss. The power entering the muffler is larger by a factor of $10^{IL/10}$ than that leaving the muffler.

Helmholtz Resonators as Filters

The theory of a Helmholtz resonator as a side branch, developed in Sec. 7-4, leads to $\hat{p}_G = \hat{p}_H$, $\hat{U}_G = \hat{p}_G/Z_{HR} + \hat{U}_H$, where Z_{HR} is the acoustic impedance of the resonator. Consequently, we identify $K_{11} = 1$, $K_{12} = 0$, $K_{21} = 1/Z_{HR}$, $K_{22} = 1$, and Eq. (7) yields

$$10^{IL/10} = \frac{|Z_{HR} + \frac{1}{2}\rho c/A|^2}{|Z_{HR}|^2} \tag{7-7.9}$$

$$= 1 + \frac{1}{4\beta^2(f/f_r - f_r/f)^2} \tag{7-7.9a}$$

where $\beta^2 = (M_A/C_A)(A/\rho c)^2$ and $2\pi f_r = (M_A C_A)^{-1/2}$. In the second version, we have explicitly inserted the expression $(-i\omega C_A)^{-1} - i\omega M_A$ for the acoustic impedance Z_{HR} of the Helmholtz resonator.

The Helmholtz resonator primarily filters out frequencies close to the resonance frequency f_r. The infinite insertion loss predicted at the resonance frequency is consistent with the prediction that the resonator acts as a perfect reflector at such a frequency. However, if β is large compared with 1, the bandwidth over which appreciable insertion loss occurs is small compared with f_r.

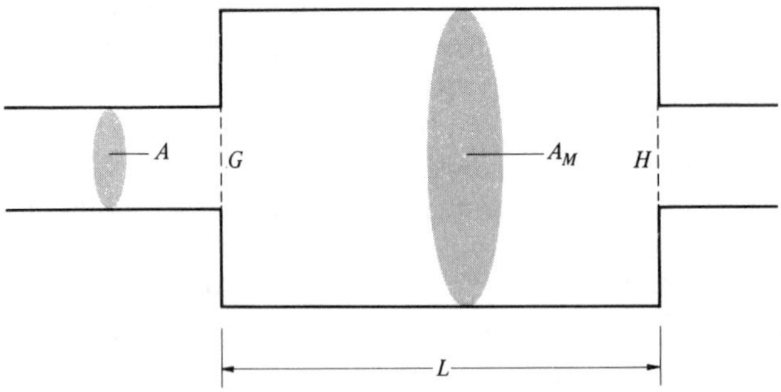

Figure 7-22 Geometry of an expansion-chamber muffler.

Expansion-Chamber Muffler

Another simple prototype (see Fig. 7-22) of a muffler consists of a duct of length L and of larger area A_M inserted between pipes of area A. With the neglect of the acoustic inertances at the duct junctions, the matrix $[K]$ for such a muffler can be identified from Eq. (8) with α set to zero and with A replaced by A_M. Subsequent insertion of these expressions into (7) yields

$$10^{\mathrm{IL}/10} = \cos^2 kL + \tfrac{1}{4}(m + m^{-1})^2 \sin^2 kL = 1 + \tfrac{1}{4}(m - m^{-1})^2 \sin^2 kL \qquad (7\text{-}7.10)$$

where we use m for the area expansion ratio A_M/A. This gives zero insertion loss when kL is a multiple of π; the insertion loss is periodic in f with a period of $c/2L$. A maximum occurs when f is an odd multiple of $c/4L$, such that L is an odd multiple of quarter wavelengths. The maximum predicted insertion loss is $10 \log [(m + m^{-1})^2/4$ and is accordingly determined by the area expansion ratio. Values of $m = 4, 9, 16, 25$, and 36 correspond to peak insertion losses of 6.5, 13.2, 18, 22, and 25 dB.

Commercial Muffler Designs

The analysis of actual commercial mufflers (see Fig. 7-23) is often complicated by multiple chambers and perforated pipes. The muffler insertion loss, moreover, is often significantly affected by the ambient flow and by nonlinear effects. However, some insight if not accurate predictions can still be obtained with the

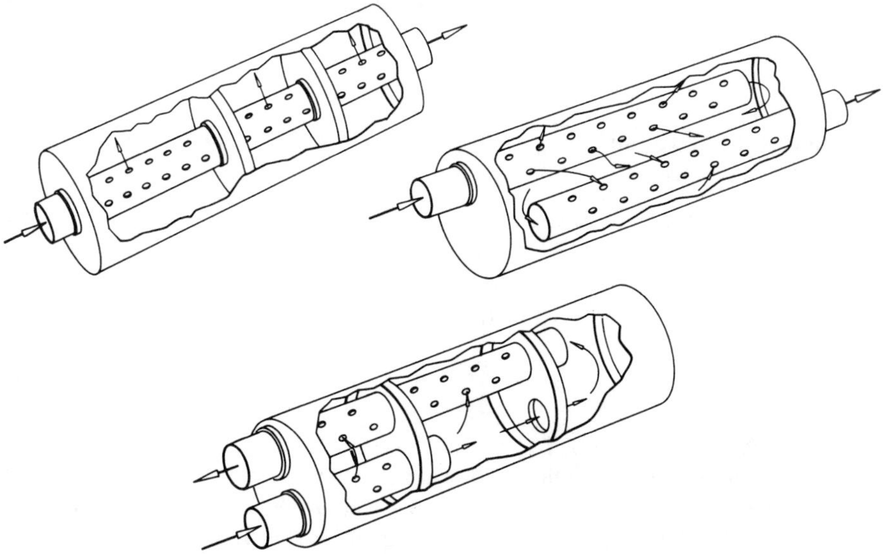

Figure 7-23 Sketches of commercial mufflers. [*From T. F. W. Embleton, "Mufflers" in L. L. Beranek (ed.), Noise and Vibration Control, McGraw-Hill, New York, 1971, p. 379.*]

classical lumped-parameter techniques. To determine† the $[K]$ matrix, one assumes that, within each segment, the pressure p is uniform over the cross section but not the same inside and outside a perforated pipe. Within such a pipe, the volume velocity parallel to the axis suffers a discontinuity at each orifice, the discontinuity equaling the volume velocity through the orifice. The latter's complex amplitude is in turn given by $(\hat{p}_{\text{in}} - \hat{p}_{\text{out}})/(-i\omega M_A)$, where the orifice's acoustic inertance M_A is of the order of $\rho/2a$. When a pipe extends only partway into a concentric chamber, the volume velocities up axis for pipe and for surrounding chamber must sum to that down axis for the chamber, as if three ducts of areas A_{pipe}, A_{out}, and $A_{\text{pipe}} + A_{\text{out}}$ met at a common junction. The three corresponding pressures are assumed to be the same at the junction.

Example The straight-through muffler in Fig. 7-24 is analyzed by associating volume velocities $\hat{U}_{\text{ch}}(x)$ and $\hat{U}_{\text{pipe}}(x)$ with the chamber (area A_{out}) and pipe (area A_{pipe}). The large number of perforations is taken into account in a smeared-out manner by replacing the mass-conservation equations with

$$\frac{A_{\text{out}}}{\rho c^2}(-i\omega\hat{p}_{\text{ch}}) + \frac{d\hat{U}_{\text{ch}}}{dx} = \frac{n(\hat{p}_{\text{pipe}} - \hat{p}_{\text{ch}})}{-i\omega M_A} \qquad (7\text{-}7.11a)$$

$$\frac{A_{\text{pipe}}}{\rho c^2}(-i\omega\hat{p}_{\text{pipe}}) + \frac{d\hat{U}_{\text{pipe}}}{dx} = \frac{n(\hat{p}_{\text{ch}} - \hat{p}_{\text{pipe}})}{-i\omega M_A} \qquad (7\text{-}7.11b)$$

where n is the number of perforations per unit length of pipe axis. Since Euler's equation still holds for the interior and exterior regions, one has

$$-i\omega\rho\hat{U}_{\text{ch}} = -A_{\text{out}}\frac{d\hat{p}_{\text{ch}}}{dx} \qquad (7\text{-}7.12a)$$

$$-i\omega\rho\hat{U}_{\text{pipe}} = -A_{\text{pipe}}\frac{d\hat{p}_{\text{pipe}}}{dx} \qquad (7\text{-}7.12b)$$

Elimination of $\hat{U}_{\text{ch}}$ and $\hat{U}_{\text{pipe}}$ from Eqs. (11) and (12) yields two coupled wave equations, general solutions of which are

$$\hat{p}_{\text{pipe}} = A\cos kx + B\sin kx + A_{\text{out}}C\cos\beta x + A_{\text{out}}D\sin\beta x \qquad (7\text{-}7.13a)$$

$$\hat{p}_{\text{ch}} = A\cos kx + B\sin kx - A_{\text{pipe}}C\cos\beta x - A_{\text{pipe}}D\sin\beta x \qquad (7\text{-}7.13b)$$

where A, B, C, D are arbitrary constants, k is ω/c, and

$$\beta^2 = k^2 - \frac{n\rho}{M_A}(A_{\text{pipe}}^{-1} + A_{\text{out}}^{-1}) \qquad (7\text{-}7.14)$$

The boundary conditions, $\hat{U}_{\text{ch}} = 0$ at $x = 0$ and at $x = L$, give two relations between the four constants, while two other relations result from $\hat{U}_{\text{pipe}} = \hat{U}_H$ at $x = L$ and from $\hat{p}_{\text{pipe}} = \hat{p}_H$ at $x = L$. Consequently, the constants, A, B, C, D become linear combinations of $\hat{U}_H$ and $\hat{p}_H$. Equations (12b) and (13b) with such

† A detailed discussion along similar lines but with nonlinear orifice impedance and ambient flow taken into account is given by J. W. Sullivan, "A Method of Modeling Perforated Tube Muffler Components," *J. Acoust. Soc. Am.*, **66**:772–788 (1979).

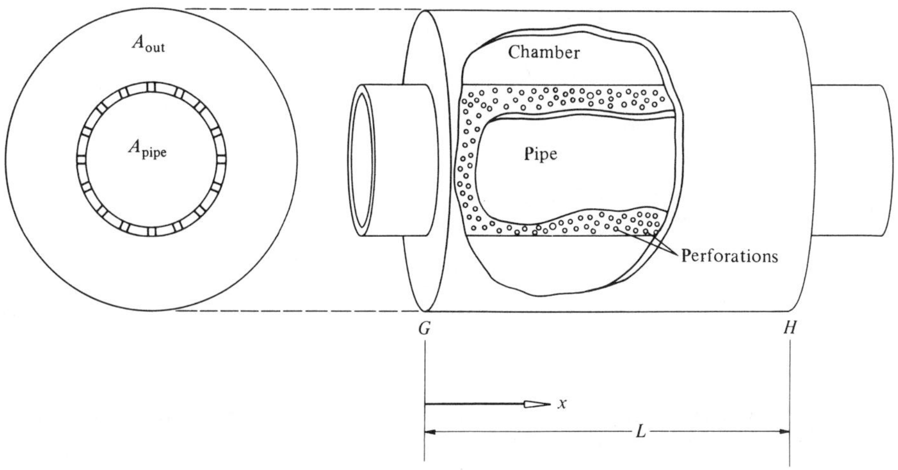

Figure 7-24 Parameters characterizing a simplified model of a straight-through muffler.

substitutions and with x set to zero therefore yield equations of the form (1). The matrix $[K]$ can subsequently be identified and the insertion loss can be determined from Eq. (7). (Since the intent here is only to describe the analytical method, the algebra is not carried through.)

7-8 HORNS

A horn† (see Fig. 7-25) is an impedance-matching device that increases the acoustic power output of a source and gives a directional preference to the radiated power. To understand the rationale underlying the first of these properties, consider a small acoustic source of fixed volume-velocity amplitude $\hat{U}$ whose power output is

$$\mathcal{P} = \tfrac{1}{2}|\hat{U}|^2 \operatorname{Re} Z \tag{7-8.1}$$

where Z is the acoustic impedance presented to the source by its external environment. For a radially oscillating sphere of radius a, Eq. (4-1.4) implies that

$$Z = \frac{\rho c}{4\pi a^2}\frac{(ka)^2 - ika}{(ka)^2 + 1} \approx \frac{\rho c k^2}{4\pi}\left(1 - \frac{i}{ka}\right) \tag{7-8.2}$$

when the source is in a free environment; the second version results when $ka \ll 1$. The time-averaged power radiated is therefore $(\rho c/8\pi)k^2|\hat{U}|^2$ in the low-frequency limit, which is characteristic of any monopole source. When mounted on a rigid wall, the source produces twice this power. In contrast, the

† For a historical overview, see J. K. Hilliard, "Historical Review of Horns used for Audience-Type Sound Reproduction," *J. Acoust. Soc. Am.*, **59**:1–8 (1976).

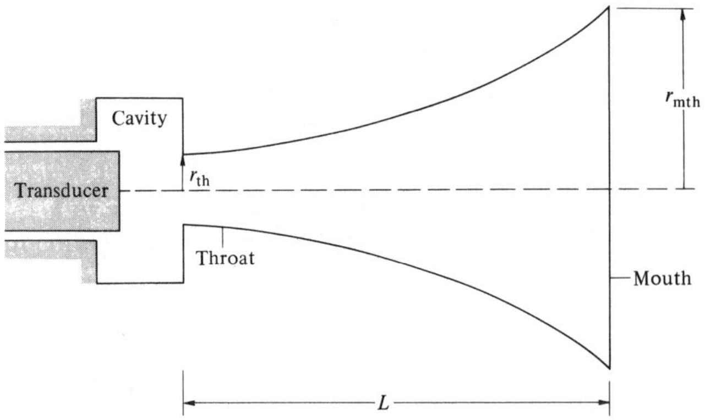

Figure 7-25 Schematic description of a horn and of its coupling to a transducer. [*After C. T. Molloy, J. Acoust. Soc. Am.*, **22:**551 *(1950)*.]

power output when the source is at the rigid end of a tube of cross-sectional area A and of unbounded length is $\rho c|\hat{U}|^2/2A$ [twice that given by Eq. (7-1.13)], providing the frequency is lower than the cutoff frequency for the first dispersive mode. If $k^2A \ll 2\pi$, a source in a duct is a much more powerful generator of acoustic energy than when it is in an open environment.

Such an enhancement in power output does not necessarily result when the source is connected to the external environment by a duct segment of *finite* length the far end of which is open. Reflections of sound from the open end alter the impedance at the source position so that Re Z is not in general $\rho c/A$. For a duct of constant cross-sectional area and of length L, the impedance at the source is given by†

$$Z = \frac{\rho c}{A} \frac{Z_{\text{end}} \cos kL - i(\rho c/A) \sin kL}{(\rho c/A) \cos kL - iZ_{\text{end}} \sin kL} \tag{7-8.3}$$

For a narrow tube, the end impedance Z_{end}, given by Eq. (7-6.16a), is small in magnitude compared with $\rho c/A$, so Re Z is typically (Re Z_{end})/($\cos^2 kL$), which is much less than $\rho c/A$ except near the resonance frequencies. However, the resonance peaks are narrow, so the tube is unsatisfactory as a coupling device if one wants a substantial power amplification with minor frequency distortion over a broad frequency band.

A duct of variable cross section can circumvent this difficulty if (ideally) the cross-sectional area varies slowly enough to prevent internal reflections and if the mouth at the far end is wide enough to ensure negligible reflection at the abrupt termination. The graphs‡ (see Fig. 7-26) of the real and imaginary parts

† This follows from Eqs. (7-7.1) and (7-7.8) with α set to 0, with $\hat{p}_H/\hat{U}_H$ set to Z_{end}, and with $\hat{p}_G/\hat{U}_G$ set to Z.

‡ C. T. Molloy, "Response Peaks in Finite Horns," *J. Acoust. Soc. Am.*, **22:**551–557 (1950); H. Levine and J. Schwinger, "On the Radiation of Sound from an Unflanged Circular Pipe," *Phys. Rev.*, **73:**383–406 (1948).

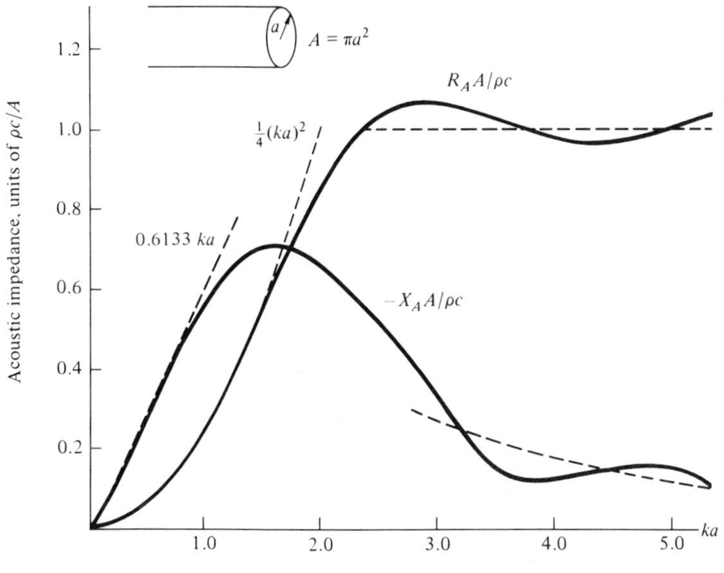

Figure 7-26 Real and imaginary parts of the acoustic impedance Z in units of $\rho c/A$ at the mouth of an open-ended unflanged thin-walled circular tube (radius a). [*After C. T. Molloy, J. Acoust. Soc. Am.*, **22:***552 (1950); low-frequency limits based on results of H. Levine and J. Schwinger, Phys. Rev.*, **73:***383 (1948).*]

of $AZ_{\text{mth}}/\rho c$ versus ka for the acoustic impedance at the mouth (mth) of an open-ended unflanged thin-walled circular tube when a plane wave is incident from along its axis suggest that the squared magnitude $|\mathcal{R}|^2$ of the reflection coefficient will be less than 0.25 if $ka > 2$. If $ka = 1$, $|\mathcal{R}|^2$ is of the order of $\frac{1}{2}$. Below $ka = 1$, there may still be some overall amplification of the radiated acoustic power if the area at the mouth is still large compared with the area at the source, but the plot of $\mathcal{P}$ for fixed $|\hat{U}|$ versus frequency will exhibit distinct resonances. Consequently, the usually stated design criterion† is that ka should be greater than 1 at the mouth for the lowest frequency radiated. For a frequency of 100 Hz in air with a sound speed of 340 m/s, this implies that the mouth diameter should be of the order of 1 m. In practice, however, smaller diameters are often used, it being asserted‡ that the resonance peaks are not noticeable to the human ear if

† Beranek, *Acoustics*, p. 268.

‡ C. R. Hanna and J. Slepian, "The Function and Design of Horns for Loud Speakers," *Trans. Am. Inst. Elec. Eng.*, **43:**393–411 (1924): "Variations in acoustic power of the order of ten to one between 200 and 4000 cycles are not noticed by the ear, however, and the departure from a uniform response can be kept within this range by a proper design of the horn." The figure 10 : 1 is at variance with the original conception of the decibel as the minimum increment of sound level detectable by the human ear but may be appropriate for broadband sound. Beranek (*Acoustics*, p. 280) chooses a design in one of his examples for which the variation is 2 : 1 and refers to such as "fairly well damped" resonances.

the power variation with frequency is substantially less than 10 : 1. Also, the coupling of the transducer to the horn through the throat and the circuitry associated with the transducer can be designed (so that the complex ratio of $\hat{U}$ to the signal amplitude is frequency-dependent) to minimize the variations caused by the resonances.

The Webster Horn Equation

Most analyses of horns are based on a quasi-one-dimensional model of sound propagation in a rigid-walled duct (see Fig. 7-27) of variable cross-sectional area $A(x)$. To derive the governing equation,† one integrates the wave equation for the acoustic pressure over the volume of a duct segment between x and $x + \Delta x$. Gauss' theorem is then used to change the volume integral of $\nabla^2 p$ to a surface integral of $\nabla p \cdot \mathbf{n}$. But since $\nabla p \cdot \mathbf{n} = 0$ on the walls of the duct, one is left with the differences of the integrals of $\partial p/\partial x$ over the cross section at $x + \Delta x$ and x. Dividing by Δx and taking the limit as $\Delta x \to 0$ then yields

$$\frac{\partial}{\partial x} \iint \frac{\partial p}{\partial x}\, dA - \frac{1}{c^2}\frac{\partial^2}{\partial t^2} \iint p\, dA = 0 \qquad (7\text{-}8.4)$$

The approximation is made that p is uniform over the cross section, and the above reduces to the Webster horn equation

$$\frac{1}{A}\frac{\partial}{\partial x}\left(A \frac{\partial p}{\partial x}\right) - \frac{1}{c^2}\frac{\partial^2 p}{\partial t^2} = 0 \qquad (7\text{-}8.5)$$

$$\left\{\frac{\partial^2}{\partial x^2} + \frac{1}{4A^2}[(A')^2 - 2AA''] - \frac{1}{c^2}\frac{\partial^2}{\partial t^2}\right\} A^{1/2} p = 0 \qquad (7\text{-}8.5a)$$

where in the second version (derived from the first), the primes denote differentiation with respect to x. This is supplemented by Euler's equation

$$\rho \frac{\partial v_x}{\partial t} = -\frac{\partial p}{\partial x} \qquad \rho \frac{\partial U}{\partial t} = -A \frac{\partial p}{\partial x} \qquad (7\text{-}8.6)$$

when a determination of the volume velocity is desired.

The criterion for the applicability of Eq. (5), that the fractional change of p over a cross section be small, leads (after a brief analysis of the linear acoustic equations) for cylindrically symmetric disturbances in a duct of radius $r(x)$ to

$$\frac{\frac{1}{2} r r' (\partial p/\partial x)_{\text{rep}}}{p_{\text{rep}}} \ll 1 \qquad \frac{krr'}{2} \ll 1 \qquad \frac{(r')^2}{2} \ll 1 \qquad (7\text{-}8.7)$$

where the quantities $(\partial p/\partial x)_{\text{rep}}$ and p_{rep} denote representative magnitudes of $\partial p/\partial x$ and p; the second version results if one assumes $(\partial p/\partial x)_{\text{rep}}/p_{\text{rep}} \approx k$, as for a plane wave. The third version results in the low-frequency limit if one takes $r = r'x$ with r' constant and uses the outgoing spherical-wave expression

† A. G. Webster, "Acoustical Impedance, and the Theory of Horns and of the Phonograph," *Proc. Natl. Acad. Sci. (U.S.)*, 5:275–282 (1919).

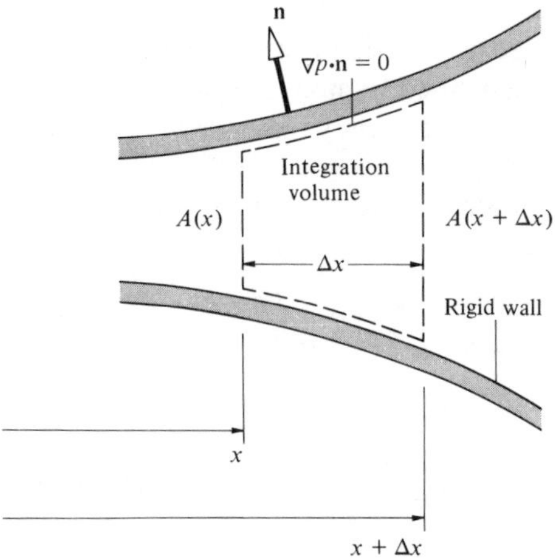

Figure 7-27 Conceptual model used for derivation of the Webster horn equation.

$x^{-1}e^{ikx}$ for p_{rep}. [While Eq. (5) formally applies to propagation in a conical tube of solid angle $\Delta\Omega$ with $A \to x^2\,\Delta\Omega$ when x is radial distance from the apex, the interpretation adhered to here for A is area of a planar cross section transverse to a fixed cartesian axis. A wide-angled cone of slowly varying solid angle is therefore precluded from consideration.]

Salmon's Family of Horns†

Circumstances for which the Webster horn equation is most easily solved are those for which the coefficient $(1/4A^2)[(A')^2 - 2AA'']$ in Eq. (5*a*) is constant. If we set this‡ to $-m^2$ and replace A by πr^2, we obtain the ordinary differential equation

$$\frac{d^2 r}{dx^2} = m^2 r \tag{7-8.8}$$

The solution of this for $r(x)$ is

$$r = r_{\text{th}}(\cosh mx + T \sinh mx) \tag{7-8.9}$$

where r_{th} is the radius at the throat ($x = 0$) and $r_{\text{th}}Tm$ is dr/dx at $x = 0$. The case $m = 0$ yields the solution

$$r = r_{\text{th}} + \left(\frac{dr}{dx}\right)_{\text{th}} x \tag{7-8.10}$$

which describes a *conical horn*. For $m > 0$, the special cases of $T = 1$ and

† V. Salmon, "Generalized Plane Wave Horn Theory" and "A New Family of Horns," *J. Acoust. Soc. Am.*, **17**:199–211, 212–218 (1946).

‡ One can also set it to $+m^2$, in which case $r(x)$ is $r_{\text{th}}(\cos mx + T \sin mx)$. This is discussed by B. N. Nagarkar and R. D. Finch, "Sinusoidal Horns," *J. Acoust. Soc. Am.*, **50**:23–31 (1971), who point out that the bell of an English horn is a sinusoidal horn.

$T = 0$ yield the *exponential horn,* where $r = r_{\text{th}}e^{mx}$, and the *catenoidal horn,* where $r = r_{\text{th}} \cosh mx$. In the former case, m is called the *flare constant.*

For any member of Salmon's family described by nonzero m, the solutions $\hat{p}$ of the Webster horn equation in the constant-frequency case are linear combinations of $A^{-1/2}e^{i\gamma x}$ and $A^{-1/2}e^{-i\gamma x}$, where $\gamma = (k^2 - m^2)^{1/2}$ for $k > m$ and $\gamma = i(m^2 - k^2)^{1/2}$ for $k < m$. Thus, one obtains the transmission relation [by a derivation similar to that of Eq. (3-4.14)]

$$\begin{bmatrix} A^{1/2}\hat{p} \\ (A^{1/2}\hat{p})' \end{bmatrix}_{x=0} = \begin{bmatrix} \cos \gamma L & -\gamma^{-1} \sin \gamma L \\ \gamma \sin \gamma L & \cos \gamma L \end{bmatrix} \begin{bmatrix} A^{1/2}\hat{p} \\ (A^{1/2}\hat{p})' \end{bmatrix}_{x=L} \tag{7-8.11}$$

This equation leads in turn to the impedance translation relation

$$\left(\frac{i\omega\rho}{ZA} + \frac{r'}{r}\right)_{\text{th}} = i\gamma \frac{1 + \varepsilon}{1 - \varepsilon} \tag{7-8.11a}$$

$$\varepsilon = e^{2i\gamma L} \left(\frac{i\omega\rho/ZA + r'/r - i\gamma}{i\omega\rho/ZA + r'/r + i\gamma}\right)_{\text{mth}} \tag{7-8.11b}$$

where the subscripts th and mth refer to the throat and mouth. This suffices to determine the throat impedance for any member of Salmon's family of horns.

Concept of a Semi-Infinite Horn

The quantity ε may be small in magnitude compared with 1 in either of two limiting circumstances. In the high-frequency limit, where $k^2 \gg m^2$, γ is approximately k and r'/r is small compared with k. If the mouth is sufficiently wide, the quantity $i\omega\rho/ZA$ at the mouth is also nearly ik, so the terms $i\omega\rho/ZA$ and $-i\gamma$ tend to cancel in the numerator. Since $|e^{2i\gamma L}|$ is equal to 1, the result is that $|\varepsilon|$ is small.

The other limiting case is that where $k < m$, so $\gamma \to i|\gamma|$, but L is large enough to ensure that $e^{-|\gamma|L} \ll 1$. In either case, one can expand $(1 + \varepsilon)/(1 - \varepsilon)$ in a power series such that, to first order in ε,

$$\left(\frac{i\omega\rho}{ZA}\right)_{\text{th}} = i\gamma - \left(\frac{r'}{r}\right)_{\text{th}} + 2i\gamma\varepsilon \tag{7-8.12}$$

With the "small" first-order term in ε discarded, the above is what would have resulted if one had ignored the impedance boundary condition at the outset and had required instead that $A^{1/2}\hat{p}$ be of the form of a constant times $e^{i\gamma x}$ within the horn, i.e., either an outgoing dispersive wave or an evanescent wave that decreases exponentially with increasing x. If one disregards the inapplicability of the Webster horn equation at large L and overlooks the fact that $e^{i\gamma x}$ satisfies the Sommerfeld radiation condition only in the limit $k \gg m$, the solution $e^{i\gamma x}$ for $A^{1/2}\hat{p}$ can be loosely interpreted as that appropriate for a horn of infinite length, i.e., a semi-infinite horn. The concept is useful because it leads to a simple expression for the throat impedance that has some validity in limiting cases, as explained above.

The Cutoff Frequency

The semi-infinite-horn model predicts Re $Z_{th} = 0$ and therefore no power output [in accordance with Eq. (1)] when $k < m$ or, equivalently, when $f < f_c$, where the cutoff frequency is given by

$$f_c = \frac{cm}{2\pi} \tag{7-8.13}$$

This prediction results because Eq. (12) with ε set to 0 has a right side that is purely real when $k < m$ $[i\gamma = -(m^2 - k^2)^{1/2}]$. Zero power transmission below the cutoff frequency is not absolutely correct, but the prediction indicates that relatively small power output results for a long horn unless k is of the order of m or larger. Equation (12) yields, to first order in ε, an expression for Re Z_{th} that varies with $|\gamma|$ primarily as $e^{-2|\gamma|L}$ when $e^{-2|\gamma|L}$ is small and $k < m$. Thus, for a long horn, where mL is somewhat larger than 1, there is a rapid decrease of power transmission as the frequency decreases below the cutoff frequency.

In the other limit, when k is large, the semi-infinite-horn model predicts

$$Z_{th} = \frac{\rho c}{A_{th}} \frac{k}{\gamma + i(r'/r)_{th}} \to \rho c/A_{th} \tag{7-8.14}$$

The limiting expression is the same as for radiation into an infinitely long duct of constant cross-sectional area A_{th}. Note that, for a catenoidal horn, $(r'/r)_{th}$ is zero, so Z_{th} is formally infinite according to this model when $k = m$ and is purely real (resistive) above the cutoff frequency. For the exponential horn, $(r'/r)_{th} = m$, so Z_{th} reduces to

$$Z_{th} = \frac{\rho c}{A_{th}} \frac{\gamma - im}{k} \tag{7-8.15}$$

and Re Z_{th} is 0 at $k = m$ and increases with k.

To illustrate the transition between the model represented by Eq. (11*a*) and the semi-infinite-horn model, some numerical examples† are given in Fig. 7-28 for an exponential horn where $mr_{th} = \frac{1}{30}$; the mouth impedances are taken from Fig. 7-26.

Other Considerations in Horn Design

Electroacoustic transducers are typically coupled to horns through a small cavity. The coupling can be modeled by an acoustic compliance C_A in parallel with the impedance $-i\omega M_A + Z_{th}$. The compliance can be taken as $V/\rho c^2$ from Eq. (7-2.11); an estimate of the acoustic inertance M_A would be $0.261\rho/r_{th}$ in accord with the model of a circular duct with a flanged opening discussed in Sec. 7-6. The overall acoustic impedance seen by the transducer diaphragm

† Similar examples are exhibited by H. F. Olson, "Horn Loud Speakers," *RCA Rev.*, **1**(4):68–83, April, 1937. Examples for the catenoidal horn are given by G. J. Thiessen, "Resonance Characteristics of a Finite Catenoidal Horn," *J. Acoust. Soc. Am.*, **22**:558–562 (1950).

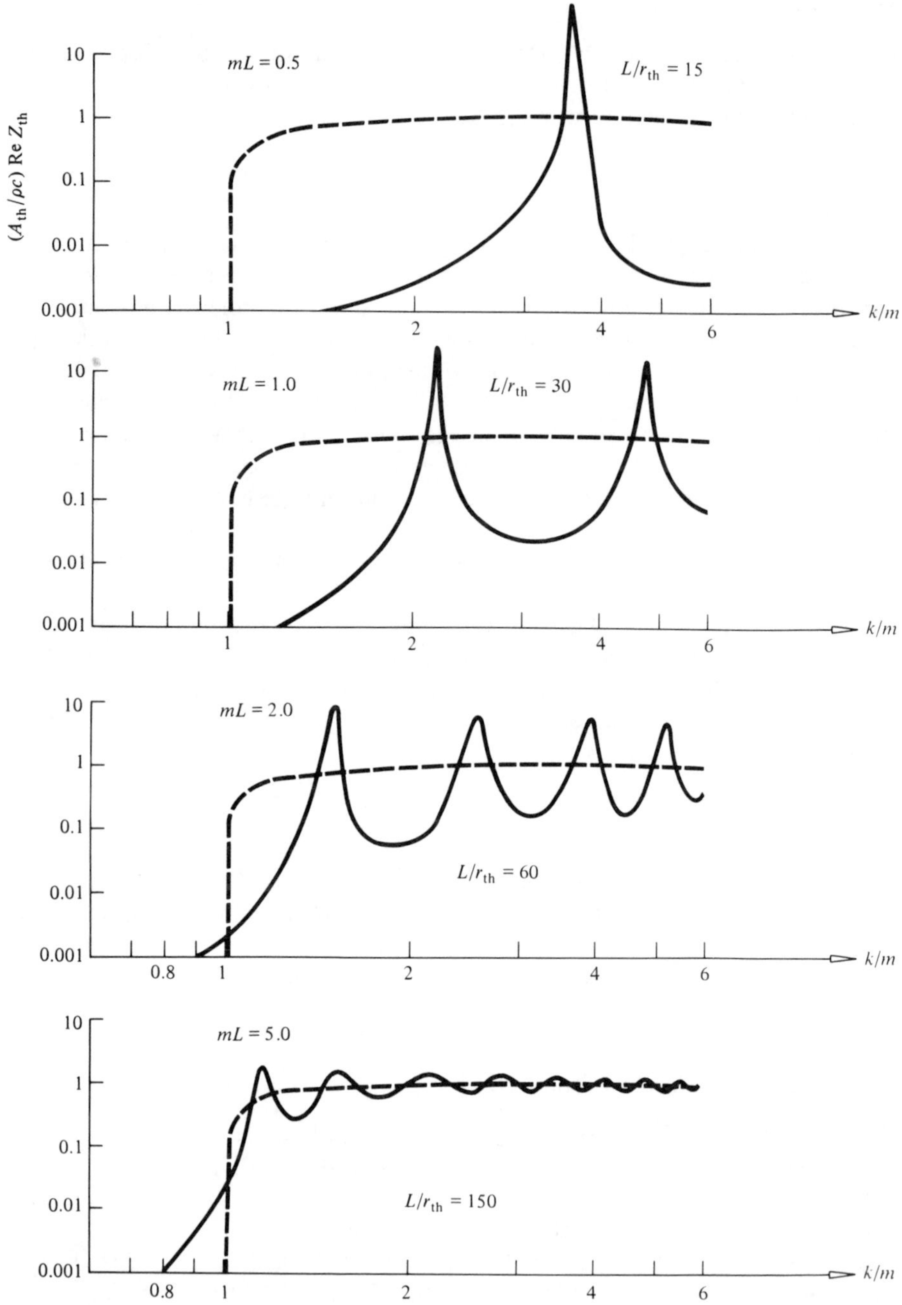

Figure 7-28 Real part of throat impedance, units of $\rho c/A_{th}$, of an exponential horn with flare constant $m = (30\, r_{th})^{-1}$ versus k/m (frequency in units of nominal cutoff frequency, $cm/2\pi$) for various choices of horn length L. (*a*) $Lm = 0.5$; (*b*) $Lm = 1.0$; (*c*) $Lm = 2.0$; (*d*) $Lm = 5.0$. Dashed line corresponds to the semi-infinite horn limit.

would be

$$Z_{\text{dia}} = [-i\omega C_A + (Z_{\text{th}} - i\omega M_A)^{-1}]^{-1} \tag{7-8.16}$$

The selection of the throat radius, which governs the throat impedance in the high-frequency limit, is constrained by the choice of the cutoff frequency, the length of the horn, and the mouth radius. The cutoff frequency f_c determines m; for fixed type of radius profile and for given m and L, the mouth radius is directly proportional to r_{th}. Consequently, a smaller throat radius leads to a mouth impedance departing more from the ideal value of $\rho c/A_{\text{mth}}$ that would give no plane-wave reflection. To circumvent this difficulty, acoustical radiation systems frequently use two horns, one designed for low frequencies and the other for high frequencies, with cross-over circuitry to channel each frequency within the overall signal to the appropriate horn.

Because horn lengths required for the achievement of good impedance matching at low frequencies are often unwieldy, many commercially marketed horns are of a folded design,† so that the propagation direction reverses once or twice before the wave leaves the mouth, although the wave continually passes through regions with gradually increasing cross-sectional area.

Another consideration affecting the choice of throat radius is that of nonlinear distortion.‡ One cause of such distortion is the amplitude dependence of the compliance of the cavity, that is, $V/\rho_0 c^2 \to V/[\gamma(p_0 + p')]$ if the horn is operating in air of specific-heat ratio γ. Another nonlinear effect is that the speed of the wave propagating down the horn depends on amplitude, such that $c \to c + \beta p'/\rho_0 c^2$, where β is a positive constant intrinsic to the fluid. (This is explained in Chap. 11.) The pressure peaks therefore tend to overtake the troughs with increasing propagation distance, a tendency partially offset by the amplitude decrease with propagation distance through a horn of expanding area. The primary result of both effects is the generation of the second harmonic (twice the frequency) of the original signal. The distortion increases with the transducer driving amplitude, so the design must take into account the peak power required.

PROBLEMS

7-1 A source that nominally generates 1 mW acoustic power in open air at a frequency of 100 Hz is placed in the center of a very long rectangular duct with cross-sectional dimensions of 0.1 by 0.2 m. (Take $c = 340$ m/s and $\rho = 1.2$ kg/m^3.)

(*a*) What propagating modes are excited?

(*b*) How much acoustic power is generated?

† R. W. Carlisle, "Method of Improving Acoustic Transmission in Folded Horns," *J. Acoust. Soc. Am.*, **31**:1135–1137 (1959).

‡ A. L. Thuras, R. T. Jenkins, and H. T. O'Neil, "Extraneous Frequencies Generated in Air Carrying Intense Sound Waves," *J. Acoust. Soc. Am.*, **6**:173–180 (1935); S. Goldstein and N. W. McLachlan, "Sound Waves of Finite Amplitude in an Exponential Horn," ibid., 275–278 (1935).

7-2 A high-frequency source emitting sound of 8000 Hz frequency is at a randomly selected point in the duct of Prob. 7-1. Estimate the number of propagating duct modes that are excited.

7-3 A model for fan noise in a circular duct (radius a and aligned parallel to the z axis) due to Tyler and Sofrin, "Axial Flow Compressor Noise Studies," is based on the concept of *spinning modes*. A simplified version of the theory takes the z component v_z of fluid velocity at the fan end $(z = 0)$ of the duct to be

$$v_z = V_0 \cos[n(\phi - \Omega t)]$$

where Ω is fan angular speed and n is number of blades.

(*a*) What frequencies are generated according to this model?

(*b*) Give a general expression (involving Bessel functions) for acoustic pressure at an arbitrary point in the duct (assumed to be of infinite length).

(*c*) Under what circumstances will only one propagating spinning mode be excited?

7-4 An acoustic dipole of nominal power output $\mathcal{P}_{\text{ff}}$ in a free-field environment is placed in the center of a long circular duct (radius a) and is aligned with its dipole-moment vector parallel to the duct axis. The dipole generates angular frequency ω, where ω is less than the lowest cutoff frequency for any dispersive mode.

(*a*) What is the power output of the dipole?

(*b*) How would this answer be affected if the dipole were aligned transverse to the axis?

7-5 A semi-infinite rectangular duct (dimensions a by $2a$) is capped at the $x = 0$ end by a flat rigid wall.

(*a*) If a harmonic point source is located on the duct centerline at $x_0 = \lambda/3$, what will the resulting pressure amplitude at large x be? Let $\mathcal{P}_{\text{ff}}$ be the free-field acoustic power output; assume that the source angular frequency $\omega = 2\pi c/\lambda$ is low enough for only the plane-wave mode to propagate.

(*b*) How does the answer change if x_0 becomes $\lambda/2$?

7-6 Verify (with as much generality as you wish) that the acoustic-mobility matrix $[D]$ for an acoustical two-port satisfies the reciprocity requirement $D_{12} = -D_{21}$.

7-7 The mechanical analog of an acoustical two-port is sketched in the accompanying figure.

(*a*) Sketch a possible acoustical system to which the analog applies.

(*b*) Is this a continuous-pressure two-port or a continuous-volume-velocity two-port?

(*c*) Sketch the circuit analog for the system.

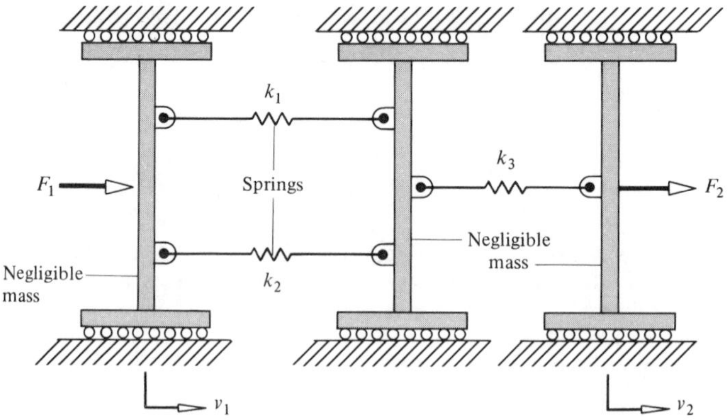

Problem 7-7

7-8 (*a*) If a duct segment of length L and cross-sectional area A with a plane-wave-mode disturbance within it is modeled as an acoustical two-port, what are the appropriate identifications for the elements Z_{left}, Z_{mid}, Z_{right} in Fig. 7-4 for arbitrary kL?

(*b*) Show that the circuit analog in the low-frequency limit consists of two capacitors and an inductor.

(*c*) What is the corresponding mechanical analog?

(*d*) How do your results in (*b*) and (*c*) compare with results when the flow is considered incompressible? When the internal pressure gradients are neglected?

7-9 Three pipes of cross-sectional areas A_1, A_2, and A_3 are joined in a Y configuration and contain fluid of ambient density ρ and sound speed c. Consider the dimensions of the junction and the diameters of the three pipes to be all substantially less than a characteristic wavelength. Sound is incident from the far end of the first pipe; the conditions are such that there are no reflected waves from the far ends of pipes 2 and 3. What fraction of the incident acoustic power is transmitted into pipe 2?

7-10 A long circular duct of radius a is filled with air of ambient density ρ and sound speed c. At $x = 0$ the duct has stretched across it a thin membrane with negligible mass under tension T N/m. The nature of the membrane is such that it deflects an average distance $\bar{y}$ given by

$$\bar{y} = \frac{\Delta p}{8T} a^2$$

when there is a net pressure drop Δp across it. If a plane wave of angular frequency ω is incident from the far left, what fraction of the incident power will be transmitted to the air on the right side of the membrane? (Consider $ka \ll 1$.)

7-11 The side branch to an infinitely long pipe of cross-sectional area A is another pipe of cross-sectional area A_b. If this side branch is regarded as a muffler, what is the corresponding insertion loss?

7-12 The influence of a side branch on acoustic waves in a duct system is such that it causes the acoustic impedance in the duct just to the left of the branch to be Z_L when that just to the branch's right is Z_R and when the source is also on the left side. In terms of ρ, c, Z_L, Z_R, and A (duct cross-sectional area), what fraction of the incident acoustic power is transmitted out of the duct into the side branch?

7-13 The incompressible potential flow through a slit of width b in a thin rigid partition extending across a rectangular duct of dimensions a by d is described in parametric form $(0 < y < a/2, \eta \geq 0)$ by the equations (see accompanying figure)

$$\Phi = B \ln (\xi^2 + \eta^2)^{1/2}$$

$$x + iy = \frac{a}{\pi} \ln \frac{[(\zeta - \alpha^2)^{1/2} + (\zeta - \alpha^{-2})^{1/2}]\zeta^{1/2}}{\alpha^{-1}(\zeta - \alpha^2)^{1/2} + \alpha(\zeta - \alpha^{-2})^{1/2}}$$

$$\alpha = \tan\left(\frac{b}{a}\frac{\pi}{4}\right) \qquad \zeta = \xi + i\eta$$

where the mapping (Schwarz-Christoffel transformation) described by the second equation is such that the center of the duct ($y = a/2$, all x) corresponds to the negative ξ axis in the complex ζ plane. Show that this solution leads to the acoustic inertance given on page 329*n*. For what ranges of frequency could one ignore the presence of the constriction?

7-14 A long rectangular tube, cross-sectional area A, has a circular patch of area $A_p = 0.1A$ on one of its walls replaced by an attenuating device. The principal mechanical property of the device, which resembles a very lightweight piston mounted flush with the duct wall, is that excess pressure in the duct causes it to move outward with velocity $v = pA_p/b$, where b is a dashpot constant (force per velocity). If a plane wave of angular frequency ω is incident from the left, what fractions of the incident power are reflected, absorbed, and transmitted beyond the device? Give your answer in terms of ω, A, c, ρ, and b and consider all applicable dimensions to be much smaller than c/ω.

7-15 A Helmholtz resonator (volume V) has two circular mouths, each of radius a and with negligible neck length. The separation distance between the two orifices is large compared with a. If a turbulent pressure field $P_{ext} \cos \omega t$ is simultaneously at the two mouths, near what value of ω would you expect resonance to occur?

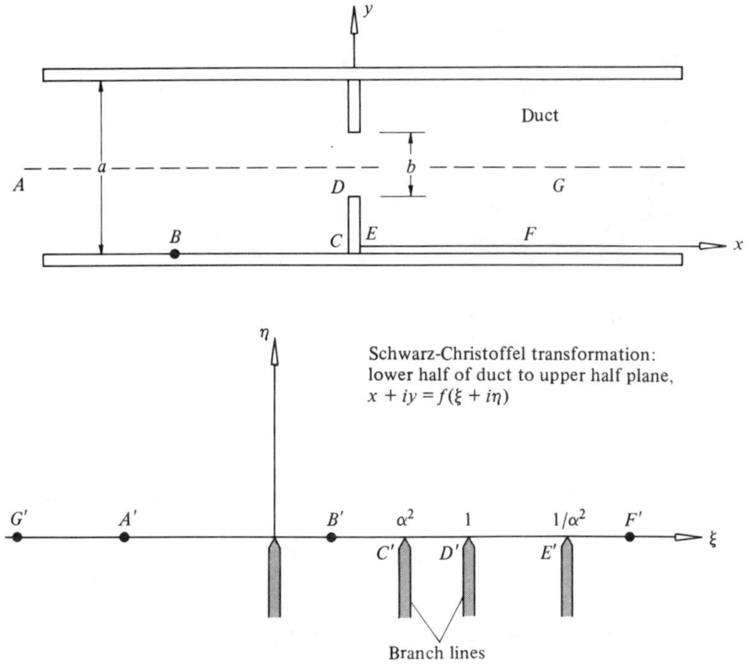

Problem 7-13

7-16 A generalization of a Helmholtz resonator that takes into account the elasticity of its walls assumes that the volume inside the bottle increases by $\Delta V = G\,\Delta p$ when the pressure inside increases by Δp, where G is a constant. If the resonator has volume V, mouth cross-sectional area A, and effective neck length l', what are (*a*) its acoustical impedance and (*b*) its resonance frequency with the wall elasticity taken into account? (*c*) Relative to what combination of ρ, c, A, l', and V should G be small if wall elasticity is to be neglected?

7-17 A Helmholtz resonator has volume V, neck cross-sectional area A, and resonance frequency f_r. In terms of these quantities and of c and ρ, determine (*a*) resonator neck inertance M_A, (*b*) effective neck length l', and (*c*) ratio of acoustic pressure inside to fluctuating pressure outside (just above the neck) when the neck is oscillating at the resonance frequency. In part (*c*) assume that the mouth has a wide flange and that the principal cause of energy loss is acoustic radiation from the mouth.

7-18 The internal friction of a Helmholtz resonator with a resonance frequency of 250 Hz and a volume of 5×10^{-4} m³ is such that, at resonance, the pressure amplitude inside is 15 times that outside.

(*a*) If the acoustic impedance of the resonator is of the form

$$Z_A = R_A - i\left(\omega M_A - \frac{1}{\omega C_A}\right)$$

what are R_A, M_A, and C_A?

(*b*) What is the Q of the resonator? (Take $\rho = 1.2$ kg/m³ and $c = 340$ m/s.)

7-19 Two Helmholtz resonators (see accompanying figure), each of volume V, are connected by a neck with acoustic inertance M_A. The first resonator also has a mouth (inertance M_A) that opens into the external environment.

(*a*) Sketch the circuit analog for this system.

(*b*) Determine the acoustic impedance at the open mouth and sketch its magnitude versus frequency.

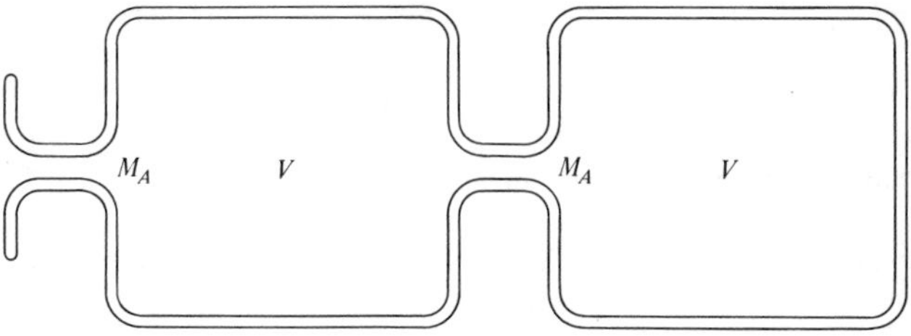

Problem 7-19

(*c*) At what frequencies, if any, does the impedance vanish?

(*d*) What are the relative phases of the pressures in the two volumes when the system is oscillating at each such frequency?

7-20 For a given fixed frequency, the acoustic impedance Z_{HR} of a Helmholtz resonator attached as a side branch to a duct of cross-sectional area A is purely imaginary (reactive). Plane waves incident within the duct from the left are partially reflected, such that only a fraction α_T of the incident power is transmitted beyond the resonator. In terms of α_T, A, and ρc, what are the possible values of Z_{HR}?

7-21 To reduce the low-frequency noise transmitted by a square duct of cross-sectional dimensions 0.4 by 0.4 m, a resonance chamber of volume V is fitted over a 2-cm-radius hole on the side of the duct.

(*a*) If the chamber performs as a Helmholtz resonator without a neck, what should V be for nearly total reflection of 60-Hz noise?

(*b*) If the chamber is designed in this manner, what fraction of incident power is transmitted past the resonator when the frequency is 120 Hz?

(*c*) Suppose one uses three such resonators instead of one, spaced at intervals that correspond to $\frac{1}{4}$ wavelength at 120 Hz. What fraction of incident power will be transmitted at 120 Hz?

7-22 Discuss the example of sound transmission past a junction between two ducts using the framework and terminology of matched asymptotic expansions. In particular, explain how one would define and derive an acoustic inertance associated with the junction from the incompressible-potential-flow solution for the junction. Your definition should lead (and you should demonstrate that this is so) to

$$M_{A,J} = \frac{2(\mathrm{KE})_{\mathrm{excess}}}{U_{12}^2}$$

where $(\mathrm{KE})_{\mathrm{excess}}$ is the excess kinetic energy caused by the presence of the junction and U_{12} is the volume velocity through the junction.

7-23 A reverberant room contains sound of predominantly 500 Hz at a sound-pressure level of 80 dB. One of the walls (concrete, 15 cm thick) has a 1-cm-radius hole leading to the outside.

(*a*) How much acoustic power leaks through the hole?

(*b*) If the wall dimensions are 4 by 3 m, what is its apparent transmission loss due to the presence of the hole?

7-24 A plane wave impinges at angle of incidence θ on a flat rigid surface that has a circular patch of radius a at its center. At the frequency ω of interest, the patch behaves like a pressure-release surface. Given that $ka \ll 1$, determine the effect of the patch on the reflected (or scattered) wave field. If the incident wave has intensity I_{av}, how much power is scattered by the patch?

7-25 Two long square ducts (each of cross-sectional dimensions w by w) are side by side and share a common wall. An orifice of radius a through this wall couples the two ducts so that a wave traveling through one causes waves to propagate away from the orifice in the other. Derive an expression applicable to low frequencies for the sound-pressure-level difference between the two ducts when the sound source is in one of the ducts.

7-26 Suppose that the orifice considered in Sec. 7-5 has a porous blanket of flow resistance R_f extending across it. For the circumstances adopted in the derivation of Eq. (7-5.11), determine expressions for the rate of energy dissipation by the blanket and for the power transmitted to the other side of the plate.

7-27 A circular duct of radius b has a rigid partition extending across its cross section, within which is a circular orifice, centered at the duct axis, of radius a. Determine an upper bound for the acoustic inertance of the orifice. What nontrivial limiting expression should describe the inertance in the limit of small a/b?

7-28 Karal's low-frequency result cited on page 329*n*. for the acoustic inertance associated with the junction between two cylindrical ducts is slightly in error in the limit $b/a \ll 1$. What should the result in this limit be?

7-29 A long circular duct of radius a opens with a wide flange into an unbounded space ($z > 0$). A plane wave of angular frequency $\omega = ck$ is incident from the $-z$ end of the duct toward the opening. Derive an approximate formula valid for $ka \ll 1$ for the fraction of the incident power radiated out of the end of the pipe.

7-30 A piston oscillates with displacement amplitude of 0.0001 m at one end of a thin-walled rigid circular tube of radius 0.05 m. The end of the tube extends without a flange into open air of ambient density 1.2 kg/m^3 and sound speed 340 m/s.

(*a*) What should the length of the tube be if its lowest resonance frequency is to be 250 Hz?

(*b*) What acoustic power is generated by the piston when it is oscillating at 250 Hz in such a tube?

(*c*) What is the Q of the resonance?

(*d*) What is the next highest resonance frequency for the tube?

7-31 A single-expansion-chamber reaction muffler is to be designed to provide at least 10 dB transmission loss for all frequencies between 500 and 1500 Hz. The smallest possible expansion-area ratio $m = A_M/A$, given $A_M > A$, compatible with this design objective is most desirable. What values of L (expansion chamber length) and m would you select? Take the speed of sound of the air in the muffler to be 340 m/s.

7-32 A segmented duct has cross-sectional area A_1 for $x < 0$, area A_2 for $0 < x < \lambda/2$, area A_3 for $\lambda/2 < x < 3\lambda/4$, and area A_4 for $x > 3\lambda/4$, where λ denotes an acoustic wavelength. If a plane wave is incident from the left ($x < 0$) through the segment of area A_1, what fraction of the incident power is transmitted to the segment of area A_4?

7-33 Derive an energy-conservation corollary for the Webster horn model represented by Eqs. (7-8.5) and (7-8.6). What does the model imply concerning the time average of pU for constant-frequency disturbances?

7-34 A horn's cross-sectional area $A(x)$ is described by αx, where α is a constant. Show that the solution of Webster's horn equation for the constant-frequency case can be expressed in terms of Bessel functions and Neumann functions (Bessel functions of the second kind).

7-35 The diaphragm of a transducer has area A_{dia} and is coupled to a horn of throat area A_{th} via a cavity of volume V. Driving frequencies of interest are such that neither kV/A_{dia} or kV/A_{th} is necessarily small, although k^3V, k^2A_{dia}, and k^2A_{th} are each much less than 1. Analysis of the system gives an acoustic inertance M_A for the flow from the cavity into the horn. The acoustic impedance in the horn just beyond the throat is that appropriate to a semi-infinite exponential horn of flare constant m. Discuss how the system's performance varies with the cavity volume V when the driving frequency is $\frac{1}{5}$, equal to, and 5 times the nominal cutoff frequency of the horn. (Make whatever assumptions seem reasonable concerning the other parameters of the system.)

7-36 A perforated pipe of radius b has n holes per unit length, each of radius a. If the pipe is in an open space, and if planar waves of constant frequency are made to propagate down the pipe, what relation should hold between wave number k and angular frequency ω? Derive a suitable wave equation using approximations analogous to those that yield Eqs. (7-7.11). Is there a cutoff frequency for plane-wave propagation down the pipe? If so, adopt some plausible values for the system's parameters and estimate the cutoff frequency's order of magnitude.

7-37 Determine an expression for the insertion loss for the model of a straight-through muffler (Fig. 7-24) represented by Eqs. (7-7.11*a*) to (7-7.14). Sketch IL versus kL for $A_{\text{out}}/A_{\text{pipe}} = 3$, $n\rho/M_A = 100A_{\text{pipe}}/L^2$.

CHAPTER

EIGHT

RAY ACOUSTICS

8-1 WAVEFRONTS, RAYS, AND FERMAT'S PRINCIPLE

The concept of a *wavefront* plays a central role in that branch of acoustical theory known as *geometrical acoustics* or *ray acoustics*. A wavefront is any moving surface along which a waveform feature is being simultaneously received (see Fig. 8-1). For example, if the time history of acoustic pressure has a single pronounced peak that arrives at $\mathbf{x}$ at time $\tau(\mathbf{x})$, the set of all points satisfying $t = \tau(\mathbf{x})$ describes the corresponding wavefront at time t. For a constant-frequency disturbance, the wavefronts are surfaces along which the phase of the oscillating acoustic pressure everywhere has the same value. It is not necessarily assumed that the amplitude along a wavefront is constant or that the wavefront is planar; however, the theory described below tacitly assumes that the amplitude varies only slightly over distances comparable to a wavelength and that the radii of curvature of the wavefront are substantially larger than a wavelength.

Ray Paths in Moving Media

The theory of plane-wave propagation described in Sec. 1-7 predicts that wavefronts move with speed c when viewed in a coordinate system in which the ambient medium appears at rest. If the ambient medium is moving with velocity $\mathbf{v}$, the wave velocity $c\mathbf{n}$ seen by someone moving with the fluid becomes†

† G. G. Stokes, "On the Effect of Wind on the Intensity of Sound," *Rep. Br. Assoc. Adv. Sci., 27th Meet., Dublin, 1857,* pt II, *Misc. Commun.*, pp. 22–23; G. Jaeger, "On the Propagation of Sound in Moving Fluid," *Sitzungsber. Kais. Akad. Wiss. (Vienna), Math-Naturwiss. Kl.*, sec. IIa, **105:**1040–1046 (1896); E. H. Barton, "On the Refraction of Sound by Wind," *Phil. Mag.*, (6)**1:**159–165 (1901).

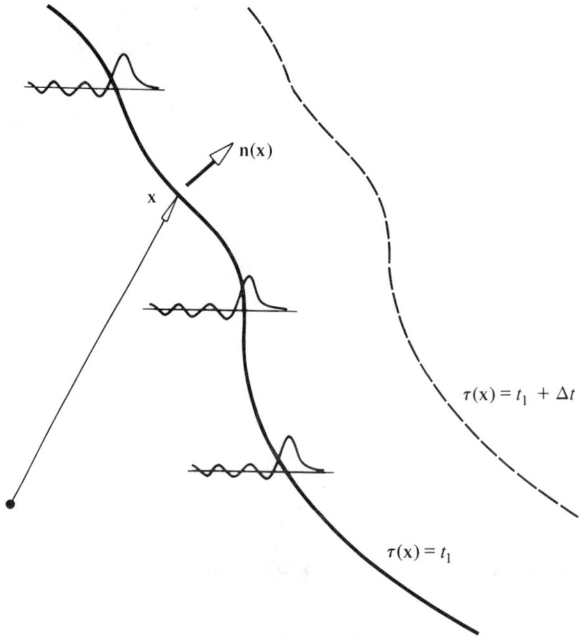

Figure 8-1 Concept of a wavefront. Points over which the wavefront simultaneously passes receive the same waveform feature at the same time.

$\mathbf{v} + c\mathbf{n}$ in a coordinate system at rest. Here $\mathbf{n}$ is the unit vector normal to the wavefront; it coincides with the direction of propagation if the coordinate system is moving with the local ambient fluid velocity $\mathbf{v}$. However, the direction of propagation perceived by a stationary observer is not necessarily the same as that of $\mathbf{n}$. The latter is independent of the velocity of the frame of reference, but the direction of propagation is not. (Throughout the following four sections, the subscript on $\mathbf{v}_0$ is omitted.)

Let $\mathbf{x}_P(t)$ be a moving point (Fig. 8-2) that lies on the wavefront $t = \tau(\mathbf{x})$ at an initial time. Then, according to the reasoning outlined above, $\mathbf{x}_P(t)$ will always lie on the moving wavefront if its velocity is

$$\frac{d\mathbf{x}_P}{dt} = \mathbf{v}(\mathbf{x}_P, t) + \mathbf{n}(\mathbf{x}_P, t)c(\mathbf{x}_P, t) = \mathbf{v}_{\text{ray}} \tag{8-1.1}$$

Here we allow for the possibility that $\mathbf{v}$ and c may vary with both position and time. The line described in space by $\mathbf{x}_P(t)$ versus t is a *ray path;* the function $\mathbf{x}_P(t)$ is a *ray trajectory*. The speed of the wavefront normal to itself is the dot product of the right side of (1) with $\mathbf{n}$; this product equals $c + \mathbf{v} \cdot \mathbf{n}$, which is less than the magnitude $|c\mathbf{n} + \mathbf{v}|$ of the ray velocity $\mathbf{v}_{\text{ray}}$.

Equation (1) suffices to determine the wavefront at successive times and represents an extension of Huygens' principle. For inhomogeneous media, however, it is awkward to use by itself because it requires a knowledge of $\mathbf{n}$ at

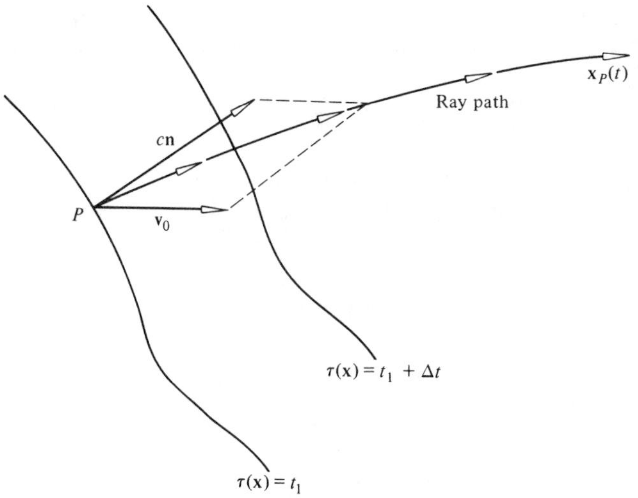

Figure 8-2 Concept of a ray path. The point $\mathbf{x}_P(t)$ moves with velocity $c\mathbf{n} + \mathbf{v}$ such that it is always on wavefront $\tau(\mathbf{x}) = t$ and in so doing traces out a ray path.

each instant along the path (which would require the construction of the wavefront surface in the vicinity of the ray at closely spaced time intervals). To circumvent this, we derive an additional differential equation that allows the prediction of the time rate of change of $\mathbf{n}$. Instead of dealing with $\mathbf{n}$ directly, we use a *wave-slowness vector*† $\mathbf{s}(\mathbf{x}) = \nabla\tau(\mathbf{x})$, which is parallel to $\mathbf{n}$ because $\nabla\tau$ is perpendicular to the surface $t = \tau(\mathbf{x})$.

The label "wave-slowness" applies because the reciprocal of $|\mathbf{s}|$ is the speed $c + \mathbf{n} \cdot \mathbf{v}$ with which the wavefront moves normal to itself. The demonstration of this proceeds from a consideration of the wavefront at closely spaced times t and $t + \Delta t$. For a given ray trajectory $\mathbf{x}_P(t)$, the position at $t + \Delta t$ is approximately $\mathbf{x}_P(t) + \dot{\mathbf{x}}_P(t)\Delta t$, so $t + \Delta t \approx \tau(\mathbf{x}_P + \dot{\mathbf{x}}_P\Delta t)$, which in turn is approximately $\tau(\mathbf{x}_P) + \Delta t\, \dot{\mathbf{x}}_P \cdot \nabla\tau$. However, $t = \tau(\mathbf{x}_P)$ and $\nabla\tau = \mathbf{s}$, so this requires that $\nabla\tau \cdot \dot{\mathbf{x}}_P = 1$ or, from (1), that

$$\mathbf{s} \cdot (c\mathbf{n} + \mathbf{v}) = 1 \qquad c\mathbf{s} \cdot \mathbf{n} = 1 - \mathbf{v} \cdot \mathbf{s} \tag{8-1.2}$$

for any given point on the waveform at any given time. Since $\mathbf{s}$ is parallel to $\mathbf{n}$, one has $\mathbf{s} = (\mathbf{s} \cdot \mathbf{n})\mathbf{n}$ and $\mathbf{n} = \mathbf{s}/(\mathbf{s} \cdot \mathbf{n})$, and the above therefore yields

$$\mathbf{s} = \frac{\mathbf{n}}{c + \mathbf{v} \cdot \mathbf{n}} \qquad \mathbf{n} = \frac{c\mathbf{s}}{\Omega} \tag{8-1.3}$$

where

$$\Omega = 1 - \mathbf{v} \cdot \mathbf{s} = 1 - \mathbf{v} \cdot \nabla\tau = \frac{c}{c + \mathbf{v} \cdot \mathbf{n}} \tag{8-1.4}$$

† For a plane wave of constant frequency, $\mathbf{s}$ is $\mathbf{k}/\omega$, so it is parallel to the phase velocity and equal in magnitude to the reciprocal of the phase speed. The terminology dates back to L. Cagniard, *Réflexion et réfraction des ondes séismiques progressive,* Gauthier-Villars, Paris, 1939, trans. E. A. Flinn and C. H. Dix, McGraw-Hill, New York, 1962.

Equation (3) substantiates the assertion that $|\mathbf{s}|^{-1} = c + \mathbf{n} \cdot \mathbf{v}$. Also, because $\mathbf{n} \cdot \mathbf{n} = 1$ and $\mathbf{s} = \nabla\tau$, the above relations give

$$s^2 = \frac{\Omega^2}{c^2} \qquad (\nabla\tau)^2 = \frac{\Omega^2}{c^2} \tag{8-1.5}$$

This partial-differential equation is the *eikonal equation*, $\tau(\mathbf{x})$ being the *eikonal*.†

A differential equation for the time rate of change of s along a ray trajectory can be derived‡ starting from

$$\frac{d\mathbf{s}(\mathbf{x}_P)}{dt} = (\dot{\mathbf{x}}_P \cdot \nabla)\mathbf{s} = c(\mathbf{n} \cdot \nabla)\mathbf{s} + (\mathbf{v} \cdot \nabla)\mathbf{s} \tag{8-1.6}$$

where all the indicated quantities are understood to be evaluated at $\mathbf{x}_P(t)$. Because $\mathbf{n}$ is in the direction of s, the first term has a factor $(\mathbf{s} \cdot \nabla)\mathbf{s}$, which can be expressed

$$(\mathbf{s} \cdot \nabla)\mathbf{s} = -\mathbf{s} \times (\nabla \times \mathbf{s}) + \tfrac{1}{2}\nabla s^2 = 0 + \tfrac{1}{2}\nabla \frac{\Omega^2}{c^2} = -\frac{\Omega}{c^2}\nabla(\mathbf{v} \cdot \mathbf{s}) - \frac{\Omega^2}{c^3}\nabla c \tag{8-1.7}$$

where we recognize that $\nabla \times (\nabla\tau) = 0$ and we substitute for s^2 from Eq. (5). Subsequent insertion of Eq. (7) and of $\mathbf{n} = c\mathbf{s}/\Omega$ into Eq. (6) yields

$$\frac{d\mathbf{s}}{dt} = -\frac{\Omega}{c}\nabla c - \nabla(\mathbf{v} \cdot \mathbf{s}) + (\mathbf{v} \cdot \nabla)\mathbf{s} \tag{8-1.8}$$

A further reduction follows from the vector identity [of which that in Eq. (7) is a special case]

$$\nabla(\mathbf{v} \cdot \mathbf{s}) = \mathbf{v} \times (\nabla \times \mathbf{s}) + \mathbf{s} \times (\nabla \times \mathbf{v}) + (\mathbf{v} \cdot \nabla)\mathbf{s} + (\mathbf{s} \cdot \nabla)\mathbf{v} \tag{8-1.9}$$

where the first term is zero because s is a gradient.

† In optical literature, the eikonal $W(\mathbf{x})$ is defined to be $c_0\tau(\mathbf{x})$, where c_0 is a reference (constant) wave speed, e.g., the speed of light in vacuo. Equation (5) then, with v set to 0, would yield $(\nabla W)^2 = (c_0/c)^2$, where c_0/c is the *index of refraction*. The introduction of a reference sound speed, however, seems superfluous in the present context, so $\tau(\mathbf{x})$ is here referred to as the eikonal. See M. Born and E. Wolf, *Principles of Optics*, 4th ed., Pergamon, Oxford, 1970, pp. 110–112. The term was introduced into optics by H. Bruns in 1895; the concept, however, is due to W. R. Hamilton (1832). The version given here of the eikonal equation was derived for motion of weak discontinuities in a fluid by G. S. Heller, "Propagation of Acoustic Discontinuities in an Inhomogeneous Moving Liquid Medium," *J. Acoust. Soc. Am.*, **25**:950–951 (1953), and by J. B. Keller, "Geometrical Acoustics, I: The Theory of Weak Shock Waves," *J. Appl. Phys.*, **25**:938–947 (1954).

‡ The earliest of the many different published derivations is E. A. Milne, "Sound Waves in the Atmosphere," *Phil. Mag.*, (6)**42**:96–114 (1921). The analysis of ray paths in a moving stratified fluid dates back to Jaeger, "On the Propagation of Sound," and Barton, "On the Refraction of Sound by Wind," and to S. Fujiwhara, "On the Abnormal Propagation of Sound Waves in the Atmosphere," *Bull. Cent. Meteorol. Obs. Jap.*, vol. 1, no. 2 (1912); vol. 4, no. 2 (1916), and R. Emden, "Contributions to the Thermodynamics of the Atmosphere, II: On the Propagation of Sound in a Wind-Moving Polytropic Atmosphere," *Meterorol. Z.*, **53**:13–29, 74–81, 114–123 (1918). For a medium without ambient flow, the ray equations date back to Snell, Huygens, and W. R. Hamilton, although they were rarely applied to the propagation of sound in inhomogeneous media until the twentieth century.

The *ray-tracing equations* are Eqs. (1) and (8), which we here write, with the substitution $\mathbf{n} = c\mathbf{s}/\Omega$ and with the identity (9), as†

$$\frac{d\mathbf{x}}{dt} = \frac{c^2\mathbf{s}}{\Omega} + \mathbf{v} \tag{8-1.10a}$$

$$\frac{d\mathbf{s}}{dt} = -\frac{\Omega}{c}\nabla c - \mathbf{s} \times (\nabla \times \mathbf{v}) - (\mathbf{s}\cdot\nabla)\mathbf{v} \tag{8-1.10b}$$

or (in cartesian coordinates)

$$\frac{ds_i}{dt} = -\frac{\Omega}{c}\frac{\partial c}{\partial x_i} - \sum_{j=1}^{3} s_j \frac{\partial}{\partial x_i} v_j \tag{8-1.10b'}$$

(Here and in what follows the subscript P is omitted.) These equations do not depend on the spatial derivatives of $\mathbf{s}$; so if $c(\mathbf{x}, t)$ and $\mathbf{v}(\mathbf{x}, t)$ are specified, and if a ray position $\mathbf{x}$ and wave-slowness vector $\mathbf{s}$ are specified at time t_0, Eqs. (10) can be integrated in time to determine $\mathbf{x}$ and $\mathbf{s}$ at any subsequent instant; no information concerning neighboring rays is required. These are nonlinear, but they are ordinary differential equations of first order, so they are amenable to standard numerical techniques of integration.‡

Fermat's Principle

If l denotes distance along a ray path, then $d\mathbf{x}/dl$ (abbreviated here as $\mathbf{x}'$) denotes ray direction. The ray-speed magnitude v_{ray} satisfying Eq. (1) is therefore such that

$$c\mathbf{n} = v_{\text{ray}}\mathbf{x}' - \mathbf{v} \tag{8-1.11}$$

However, $\mathbf{n}\cdot\mathbf{n}$ is 1 and $\mathbf{x}'\cdot\mathbf{x}'$ is also 1, so v_{ray} satisfies the quadratic equation

$$v_{\text{ray}}^2 - 2v_{\text{ray}}\,\mathbf{v}\cdot\mathbf{x}' - (c^2 - v^2) = 0$$

† These are a special case of the general ray equations for propagation of a wave packet of slowly varying frequency $\omega(\mathbf{x}, t)$ and wave number $\mathbf{k}(\mathbf{x}, t)$ in a time-dependent inhomogeneous anisotropic medium. If $F(\omega, \mathbf{k}, \mathbf{x}, t) = 0$ describes the dispersion relation at time t near point $\mathbf{x}$, rays are given by the equations (in cartesian coordinates)

$$\frac{d\omega}{dt} = -\frac{\partial F/\partial t}{\partial F/\partial\omega} \qquad \frac{dx_i}{dt} = -\frac{\partial F/\partial k_i}{\partial F/\partial\omega} \qquad \frac{dk_i}{dt} = \frac{\partial F/\partial x_i}{\partial F/\partial\omega}$$

In our particular case, $F = (\omega - \mathbf{v}\cdot\mathbf{k})^2 - c^2k^2 = 0$ comes from the eikonal equation. For a derivation, see G. B. Whitham, "Group Velocity and Energy Propagation for Three-Dimensional Waves," *Common. Pure Appl. Math.*, **14**:675–691 (1961); "A Note on Group Velocity," *J. Fluid Mech.*, **9**:347–352 (1960). Various versions of the second ray-tracing equation (10*b*) are reviewed and shown to be equivalent by R. Engelke, who gives a derivation of his own in "Ray Trace Acoustics in Unsteady Inhomogeneous Flow," *J. Acoust. Soc. Am.*, **56**:1291–1292 (1974).

‡ See, for example, R. W. Hamming, "Numerical Solution of Ordinary Differential Equations," in M. Klerer and G. A. Korn (eds.), *Digital Computer User's Handbook*, McGraw-Hill, New York, 1967, chap. 2.6; C. B. Moler and L. P. Solomon, "Use of Splines and Numerical Integration in Geometrical Acoustics," *J. Acoust. Soc. Am.*, **48**:739–744 (1970).

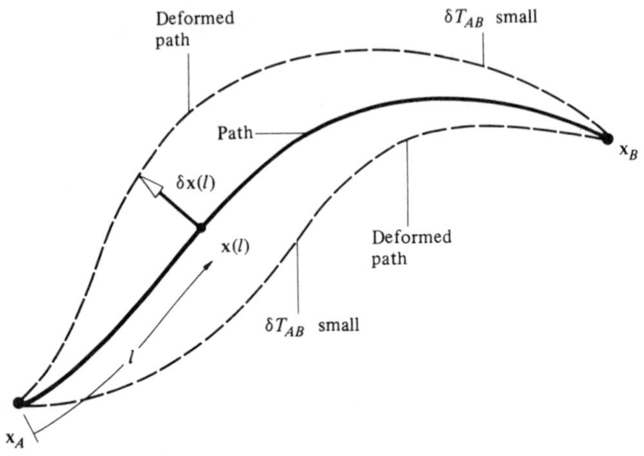

Figure 8-3 *Fermat's principle:* the travel time of the actual ray path connecting two points is stationary with respect to small virtual changes.

whose positive solution, given $c^2 > v^2$, is

$$v_{\text{ray}} = \mathbf{v} \cdot \mathbf{x}' + [c^2 - v^2 + (\mathbf{v} \cdot \mathbf{x}')^2]^{1/2} \tag{8-1.12}$$

The time that a ray takes to go from $\mathbf{x}_A$ to $\mathbf{x}_B$ is consequently

$$T_{AB} = \int_{l_A}^{l_B} \frac{dl}{\mathbf{v} \cdot \mathbf{x}' + [c^2 - v^2 + (\mathbf{v} \cdot \mathbf{x}')^2]^{1/2}} \tag{8-1.13}$$

Here we assume that c and $\mathbf{v}$ are functions only of position, such that for a given ray path they can be regarded as functions of distance l along the path.

Fermat's principle† is that the actual ray path connecting $\mathbf{x}_A$ and $\mathbf{x}_B$ is such that it renders the travel-time integral T_{AB} stationary with respect to small virtual changes in the path. If a small variation $\mathbf{x}(l) \to \mathbf{x}(l) + \delta\mathbf{x}(l)$ is imposed on the actual path (see Fig. 8-3), the resulting variation δT_{AB} should be zero to first order in the $\delta\mathbf{x}$.

A proof for when the path has no intermediate reflections proceeds with change of integration variable to the projection q of the ray path on the straight line connecting $\mathbf{x}_A$ and $\mathbf{x}_B$, such that dl becomes $(\mathbf{x}_q \cdot \mathbf{x}_q)^{1/2}\, dq$ and $\mathbf{x}'$ becomes $\mathbf{x}_q/(\mathbf{x}_q \cdot \mathbf{x}_q)^{1/2}$, where $\mathbf{x}_q$ is the derivative of $\mathbf{x}$ with respect to q. The travel time T_{AB} then becomes the integral from 0 to $|\mathbf{x}_B - \mathbf{x}_A|$ over q of $L(\mathbf{x}_q, \mathbf{x})$, where

$$L(\mathbf{x}_q, \mathbf{x}) = \frac{x_q^2}{\mathbf{v} \cdot \mathbf{x}_q + [(c^2 - v^2)x_q^2 + (\mathbf{v} \cdot \mathbf{x}_q)^2]^{1/2}} \tag{8-1.14}$$

The requirement that the travel time be stationary then leads to the Euler-

† Pierre de Fermat (1657) originally conjectured that the optical travel time is a minimum (*principle of least time*), but it was later recognized by W. R. Hamilton (1833) that there are exceptions to this and that the correct statement is that the actual path is stationary with respect to other adjacent paths. The proof that the principle also applies to acoustic waves in moving media is due to P. Uginčius, "Ray Acoustics and Fermat's Principle in a Moving Inhomogeneous Medium," *J. Acoust. Soc. Am.*, **51**:1759–1763 (1972).

Lagrange equation†

$$\frac{d}{dq}\frac{\partial L}{\partial \mathbf{x}_q} - \frac{\partial L}{\partial \mathbf{x}} = 0 \tag{8-1.15}$$

(Here $\partial L/\partial \mathbf{x}$ denotes the vector with components $\partial L/\partial x$, $\partial L/\partial y$, $\partial L/\partial z$.) Algebraic manipulations with the relations and definitions derived earlier in this section reduce the partial derivatives of the function $L(\mathbf{x}_q, \mathbf{x})$ to

$$\frac{\partial L}{\partial \mathbf{x}_q} = \frac{\mathbf{n}}{\mathbf{n}\cdot\mathbf{v}_{\text{ray}}} = \mathbf{s} \tag{8-1.16a}$$

$$\frac{\partial L}{\partial \mathbf{x}} = -\frac{dl/dq}{v_{\text{ray}}}\left[\frac{\Omega}{c}\nabla c + \mathbf{s}\times(\nabla\times\mathbf{v}) + (\mathbf{s}\cdot\nabla)\mathbf{v}\right] \tag{8-1.16b}$$

so Eq. (15) is equivalent to the ray-tracing equation (10*b*). Fermat's principle is therefore a consequence of the ray equations.

In a wider sense, Fermat's principle also applies to ray paths whose directions change abruptly. It leads to the predictions, inferred earlier (Chap. 3) from the trace-velocity matching principle, that angle of reflection equals angle of incidence (law of mirrors) upon reflection at a flat surface and that angle of refraction is related to angle of incidence by Snell's law (in the absence of ambient flow) on transmission through a planar interface. The principle also correctly predicts paths by which diffracted waves can reach a listener.

Example A source and listener (see Fig. 8-4) are at heights h and z above a plane interface separating two fluids with sound speeds c_{I} and c_{II}, where $c_{\mathrm{II}} > c_{\mathrm{I}}$. Two of the stationary paths are the direct path and the reflected path. Another possibility is a path that goes from source to interface along a line that makes an angle θ with the vertical, then proceeds just below the surface along a horizontal line, and then emerges into medium I along a path that proceeds from surface to listener at an angle ϕ with the vertical. The travel time along such a path is

$$T_{AB} = \frac{h}{c_{\mathrm{I}}\cos\theta} + \frac{r - h\tan\theta - z\tan\phi}{c_{\mathrm{II}}} + \frac{z}{c_{\mathrm{I}}\cos\phi} \tag{8-1.17}$$

where r is the total horizontal distance. The requirement that T_{AB} be stationary with respect to variations in θ leads to the equation $\partial T_{AB}/\partial\theta = 0$ or, after some algebra, to $\sin\theta = c_{\mathrm{I}}/c_{\mathrm{II}}$. Consequently, θ is the critical angle $\theta_c = \sin^{-1}(c_{\mathrm{I}}/c_{\mathrm{II}})$, that is, the angle at which the reflection-coefficient magnitude first becomes 1. The requirement $\partial T_{AB}/\partial\phi = 0$ similarly leads to $\phi = \sin^{-1}(c_{\mathrm{I}}/c_{\mathrm{II}})$. The only constraint on the solution is that the travel time along the middle segment must be positive, so r must exceed $(h + z)\tan\theta_c$.

† For introductory discussions of the calculus of variations, see J. Mathews and R. L. Walker, *Mathematical Methods of Physics*, Benjamin, New York, 1965, pp. 304–326; S. H. Crandall, D. C. Karnopp, E. F. Kurtz, Jr., and D. C. Pridmore-Brown, *Dynamics of Mechanical and Electromechanical Systems*, McGraw-Hill, New York, 1968, pp. 1–35, 417–424. There is an analogy between Eq. (15) and Lagrange's equations of classical mechanics, between $L(\mathbf{x}_q, \mathbf{x})$ and a lagrangian, and between Fermat's principle and Hamilton's principle.

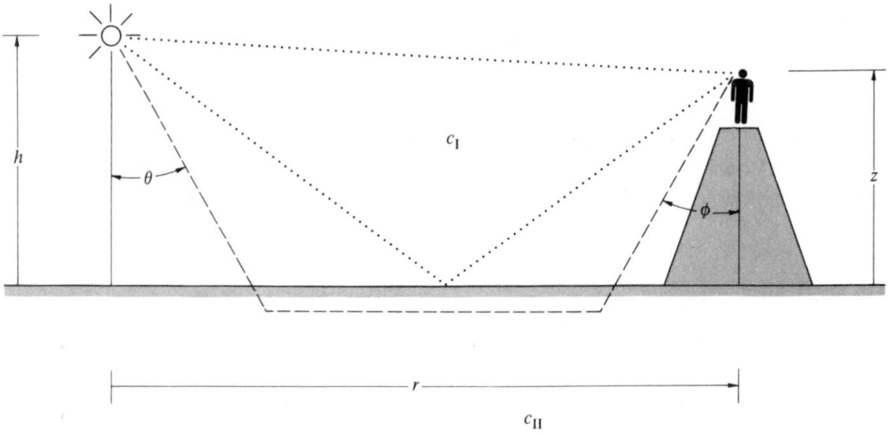

Figure 8-4 Possible ray paths connecting source and listener above a plane interface separating two dissimilar fluids.

This *refraction arrival path,*† which we here infer from Fermat's principle, lies outside the domain of what is normally referred to as geometrical acoustics. The existence of such a path, however, is confirmed by the solution of the boundary-value problem for a transient point source above a plane interface above two fluids. If r is sufficiently large, the first arrival comes with a travel time given by Eq. (17), with θ and ϕ set to θ_c, and arrives from a direction that is proceeding obliquely upward at an angle of θ_c with the vertical.

The applicability of Fermat's principle to the prediction of paths like that of the refraction arrival is a principal tenet of the *geometrical theory of diffraction.*‡ A *diffracted ray* is a ray which originates at an interface, a surface, or an edge and which propagates with all the attributes of a ray generated by a real source but which is created by a process inexplicable (and therefore labeled as diffraction) within the confines of the ordinary geometrical acoustics theory. The portion of the refraction arrival path from the interface to the listener is an example of a diffracted ray.

† C. B. Officer, *Introduction to the Theory of Sound Transmission,* McGraw-Hill, New York, 1958, pp. 195–201; W. M. Ewing, W. S. Jardetzky, and F. Press, *Elastic Waves in Layered Media,* McGraw-Hill, New York, 1957, pp. 93–102; K. O. Friedrichs and J. B. Keller, "Geometrical Acoustics, II: Diffraction, Reflection, and Refraction of a Weak Spherical or Cylindrical Shock at a Plane Interface," *J. Appl. Phys.,* **26:**961–966 (1955). Applications of the refraction arrival to geophysical exploration date back to A. Mohorovičić (1910).

‡ J. B. Keller, "Geometrical Theory of Diffraction," *J. Opt. Soc. Am.,* **52:**116–130 (1962); "A Geometrical Theory of Diffraction" in L. M. Graves (ed.), *Calculus of Variations and Its Applications, Proc. Symp. Appl. Math.,* vol. 8, McGraw-Hill, New York, 1958, pp. 57–52; G. L. James, *Geometrical Theory of Diffraction for Electromagnetic Waves,* Peregrinus, Stevenage, England, 1976, pp. 97–98, 130–131, 169–171.

8-2 RECTILINEAR SOUND PROPAGATION

For a homogeneous medium in which c and $\mathbf{v}$ are constant, a consequence of the second ray-tracing equation (8-1.10b) is that $\mathbf{s}$ and $\mathbf{n}$ are constant. The ray velocity $d\mathbf{x}/dt$ is also constant, and the ray paths are straight lines. This deduction, for the circumstances just described, is the law of *rectilinear propagation of sound.*

Parametric Description of Wavefronts

Suppose a wavefront (moving toward larger values of z) is given by $z = f(x, y)$ at $t = 0$. The ambient velocity $\mathbf{v}$ is zero, and c is constant. It is desired to describe the wavefront at some later time t (see Fig. 8-5).

The ray passing through a point $\mathbf{x}_P$ on the initial wavefront is moving in the

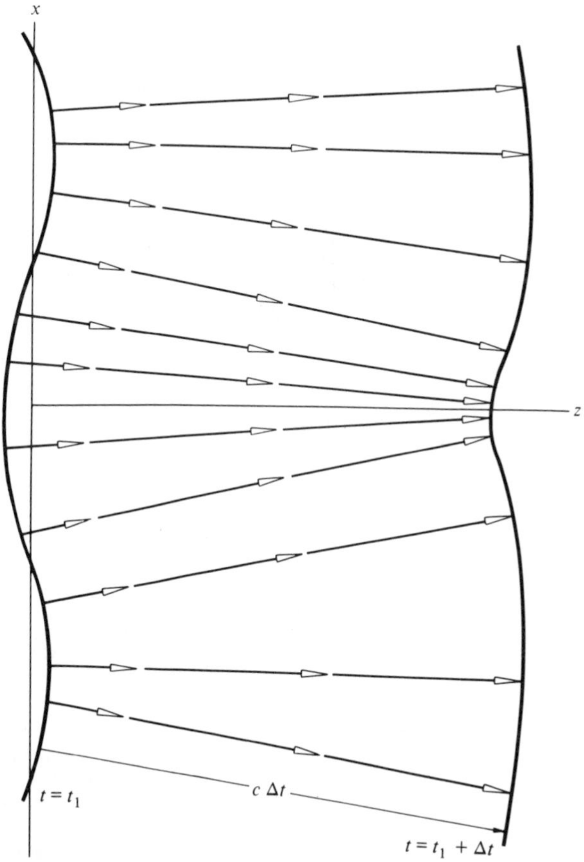

Figure 8-5 Construction of a wavefront at time t when the wavefront at time $t = 0$ is given. The ambient fluid velocity is zero and the ambient sound speed is constant.

direction $\mathbf{n}$, where (with $f_x = \partial f/\partial x$)

$$\mathbf{n} = \left\{ \frac{\nabla[z - f(x, y)]}{|\nabla[z - f(x, y)]|} \right\}_{\mathbf{x}=\mathbf{x}_P} = \frac{\mathbf{e}_z - f_x\mathbf{e}_x - f_y\mathbf{e}_y}{(1 + f_x^2 + f_y^2)^{1/2}} \tag{8-2.1}$$

At time t, the ray is at $\mathbf{x} = \mathbf{x}_P + ct\mathbf{n}$. If we let α and β represent x_P and y_P, this position can be written

$$\mathbf{x}(\alpha, \beta, t) = \alpha\mathbf{e}_x + \beta\mathbf{e}_y + f(\alpha, \beta)\mathbf{e}_z + \frac{ct(\mathbf{e}_z - f_\alpha\mathbf{e}_x - f_\beta\mathbf{e}_y)}{(1 + f_\alpha^2 + f_\beta^2)^{1/2}} \tag{8-2.2}$$

This gives a parametric description of the wavefront at time t through the parameters α and β; any choice of α and β generates a point on the wavefront. Thus, an analytical expression replaces Huygens' graphical construction.

Variation of Principal Radii of Curvature along a Ray

Any surface locally resembles an elliptical bowl (concave or convex) or a saddle and has two principal radii of curvature. If one picks any point (Fig. 8-6) on the surface, chooses it to be the origin, and lets the z direction be perpendicular to the surface at that point, the x and y axes can always be selected in

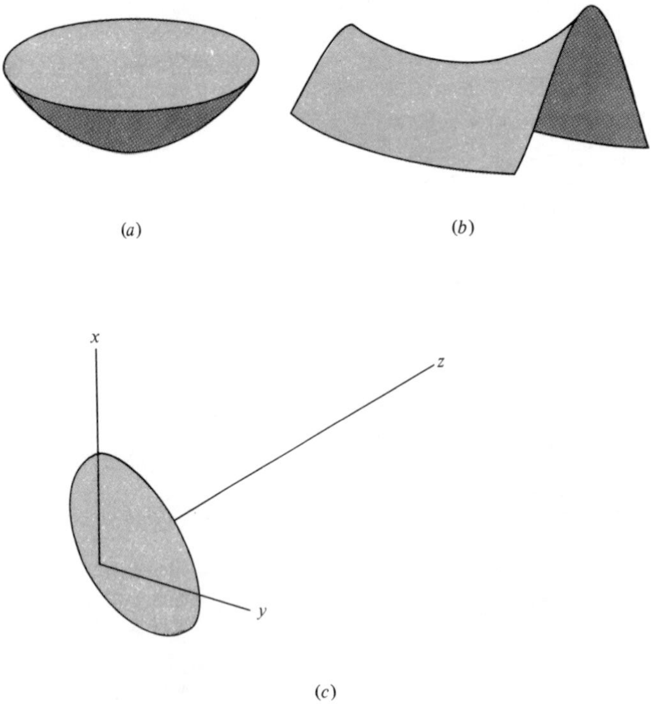

Figure 8-6 Characteristic local shapes of surfaces: (a) elliptical bowl; (b) saddle shape; (c) geometry used in the discussion of the variation of wavefront radii of curvature along a ray.

such a way that the surface near the selected point can be described to second order in x and y by

$$z = \frac{x^2}{2r_1} + \frac{y^2}{2r_2} \tag{8-2.3}$$

where r_1 and r_2 (possibly negative) are the two principal radii of curvature. [The identification follows since a circle in the xz plane of radius r_1 that is tangential to the $z = 0$ plane is given by $(z - r_1)^2 + x^2 = r_1^2$ or by $z = x^2/2r_1$ for $z \ll r_1$, $|x| \ll r_1$.]

The variation of r_1 and r_2 along a ray moving through a homogeneous quiescent medium can be deduced from Eq. (2). One chooses the coordinate system so that the ray passes through the origin at $t = 0$ in the $+z$ direction and $f(\alpha, \beta)$ equals $\alpha^2/2r_1^0 + \beta^2/2r_2^0$ (to second order in α and β). Then, to second order in α and β, the z component of (2) yields

$$z = ct + \frac{\alpha^2}{2r_1^0}\left(1 - \frac{ct}{r_1^0}\right) + \left(\frac{\beta^2}{2r_2^0}\right)\left(1 - \frac{ct}{r_2^0}\right) \tag{8-2.4}$$

However, to first order (which is all that is required) in α and β, the x and y components of Eq. (2) yield $\alpha(1 - ct/r_1^0)$ and $\beta(1 - ct/r_2^0)$ for x and y; thus to second order in x and y one has

$$z = ct + \frac{\frac{1}{2}x^2}{r_1^0 - ct} + \frac{\frac{1}{2}y^2}{r_2^0 - ct} \tag{8-2.5}$$

Since this is of the same form as Eq. (3), the directions associated with the principal radii of curvature remain constant along any given ray. The radii themselves decrease by ct during time t; or, equivalently, after the ray has traveled distance Δz, they are each decreased by Δz. This assumes that the wavefront is concave along the ray of interest. If it is convex or saddle-shaped such that, say, $r_1^0 < 0$, $|r_1|$ increases with the distance of propagation, the incremental increase equaling the incremental change of distance along the ray. A decrease of wavefront curvature radius is associated with a focusing of rays and an increase with a defocusing.

Caustics

Equation (5) indicates that if, say, $r_1^0 > 0$ and $r_2^0 > r_1^0$, the wavefront will develop a cusp ($r_1 = 0$) at time $t = r_1^0/c$. Points at which this occurs are points at which adjacent rays intersect. The locus of all such points, each of which corresponds to a given ray proceeding out from the original wavefront, is a *caustic surface* (see Fig. 8-7). Since the wavefront has a cusp at the point where it touches a caustic, the assumption that the wavefront everywhere locally resembles a propagating plane wave is no longer approximately valid and the basic tenets of geometrical acoustics are inapplicable. The extension of the theory to cover such contingencies is deferred to Sec. 9-4.

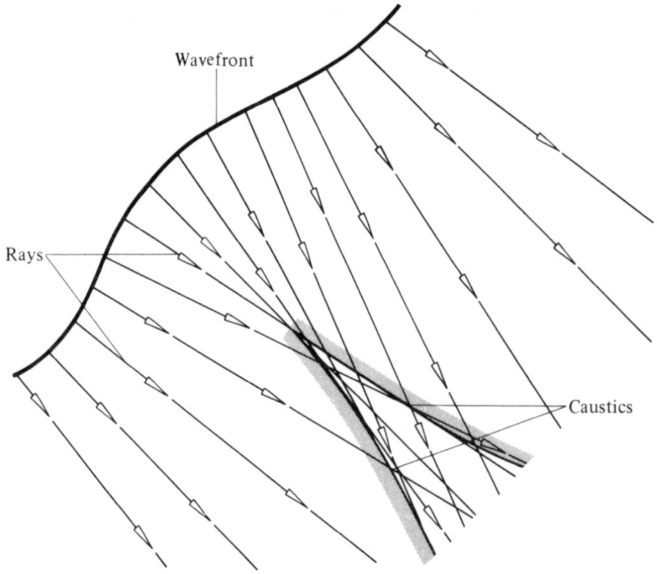

Figure 8-7 Formation of a caustic. [*From A. D. Pierce, J. Acoust. Soc. Am.*, **44:***1055 (1968).*]

The geometrical-acoustics prediction, however, of where the caustics occur is of intrinsic interest because it indicates where abnormally high amplitudes can be expected. Since the concept of a caustic applies also to rays in inhomogeneous media, the location and meteorological circumstances of intrinsically noisy activities,† e.g., static tests of large rocket engines, are often carefully selected so that distant populated areas are not touched by caustics.

Example A wavefront $z = f(x)$ has a concave radius of curvature $R(x)$ with a minimum value R_0 at $x = 0$. The z axis is perpendicular to the wavefront at $x = 0$; also, the origin is selected so that $f(0) = 0$. We seek to describe the caustic in the vicinity of the point $x = 0$, $z = R_0$ (see Fig. 8-8).

SOLUTION Since the ray passing through the wavefront at $x = \alpha$ touches the caustic when $ct = R(\alpha)$, Eqs. (2) yield

$$x = \alpha - R(\alpha)f'(\alpha)[1 + (f')^2]^{-1/2} \tag{8-2.6a}$$

$$z = f(\alpha) + R(\alpha)[1 + (f')^2]^{-1/2} \tag{8-2.6b}$$

(primes denoting derivatives with respect to α) as the parametric description of

† R. N. Tedrick, "Meteorological Focusing of Acoustic Energy," *Sound: Uses Control,* **2**(6):24–27 (1963); J. Reed, "Climatology of Airblast Propagations from Nevada Test Site Nuclear Airbursts," *Rep.* SC-RR-69-572, Sandia Laboratories, Albuquerque, 1969, available from National Technical Information Services, Washington, Accession No. N70-29525.

the caustic. If these are expanded in a power series in α, we find

$$
\begin{aligned}
x &\approx \alpha - (R_0 + \tfrac{1}{2}R_0'' \alpha^2)(\alpha f_0'' + \tfrac{1}{6}f_0^{\text{iv}} \alpha^3)[1 - \tfrac{1}{2}(f_0'' \alpha)^2] \\
&\approx (1 - R_0 f_0'')\alpha - [\tfrac{1}{2}R_0'' f_0'' + \tfrac{1}{6}R_0 f_0^{\text{iv}} - \tfrac{1}{2}(f_0'')^3 R_0]\alpha^3
\end{aligned}
$$

$$
\begin{aligned}
z &\approx \tfrac{1}{2}f_0'' \alpha^2 + (R_0 + \tfrac{1}{2}R_0''\alpha^2)[1 - \tfrac{1}{2}(f_0'' \alpha)^2] \\
&\approx R_0 + [\tfrac{1}{2}f_0'' - \tfrac{1}{2}(f_0'')^2 R_0 + \tfrac{1}{2}R_0'']\alpha^2
\end{aligned}
$$

with the zero subscript implying evaluation at $\alpha = 0$. Note that the geometry requires f_0, f_0', f_0''', and R_0' each to be zero.

Since the radius of curvature of a line is given by

$$R(\alpha) = \frac{[1 + (f')^2]^{3/2}}{f''(\alpha)} \tag{8-2.7}$$

one finds $f_0'' = 1/R_0$ and $R_0^3 f_0^{\text{iv}} = 3 - R_0 R_0''$, so the above approximate description of the caustic reduces to

$$x \approx -\frac{1}{3}\frac{R_0''}{R_0}\alpha^3 \qquad z - R_0 \approx \tfrac{1}{2}R_0'' \alpha^2 \tag{8-2.8}$$

The caustic is consequently given by

$$x = \mp\left(\frac{8}{9R_0^2 R_0''}\right)^{1/2}(z - R_0)^{3/2} \tag{8-2.9}$$

in the vicinity of $z = R_0$, $x = 0$.

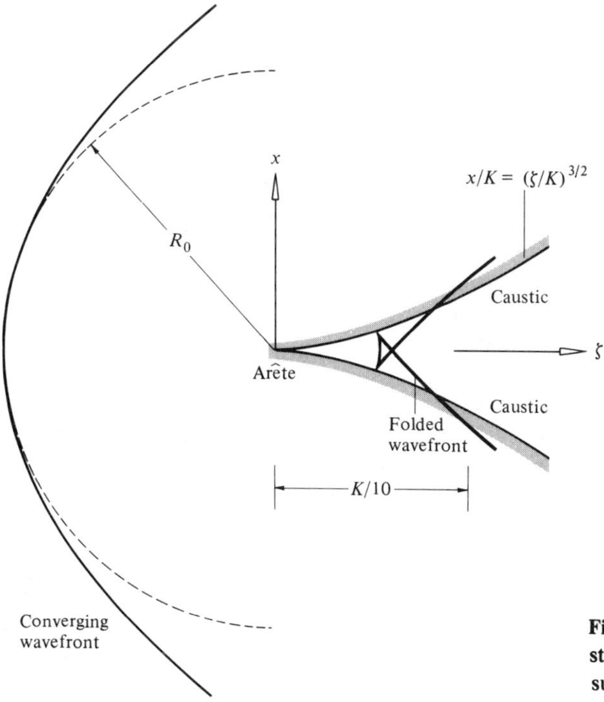

Figure 8-8 Geometry adopted for study of the shape of a caustic surface near its vertex.

The characteristic cusp with which the two branches of the caustic meet is sometimes called an *arête*.† Beyond the arête and between the two branches, three rays, rather than one, pass through each point, and the wavefront has a folded form.‡

8-3 REFRACTION IN INHOMOGENEOUS MEDIA

That sound waves *refract* (change their propagation direction) on passing through an interface separating two fluids with different sound speeds is discussed in Sec. 3-6. In continuous media, refraction is characterized by a gradual bending of ray paths rather than by an abrupt change of direction. Here we explore the implications of the ray-tracing equations as regards such ray bending.

Refraction by Sound-Speed Gradients

When the ambient fluid velocity is zero, and when the sound speed is independent of time, the wave slowness $\mathbf{s}$ becomes $\mathbf{n}/c$ and Eqs. (8-1.10) reduce to

$$\frac{d\mathbf{x}}{dt} = c^2\mathbf{s} \qquad \frac{d\mathbf{s}}{dt} = -\frac{1}{c}\nabla c \tag{8-3.1}$$

To determine the influence of the sound-speed gradient on the bending of rays, we consider the ray that initially passes through the origin in the $+x$ direction, such that $\mathbf{s} = \mathbf{e}_x/c(0)$ at $t = 0$. Then, to first order in t, the second of Eqs. (1) yields

$$\mathbf{s} = \frac{1}{c}\mathbf{e}_x - \frac{1}{c}\nabla c\, t \tag{8-3.2}$$

where c and its derivatives (c_x, c_y, c_z) are understood to be evaluated at (0, 0, 0). It accordingly follows from the equations for dy/dt and dz/dt that y and z are proportional to t^2 for small t. Then, because $x = ct$ to lowest order, the first of Eqs. (1) yields, to lowest nonvanishing order in x,

$$y = -\frac{1}{2}\frac{c_y}{c}x^2 \qquad z = -\frac{1}{2}\frac{c_z}{c}x^2 \tag{8-3.3}$$

which are the equations of parabolas.

Suppose, moreover, that one has selected the coordinate axes in such a way that, at $x = 0$, $c_z = 0$ and ∇c is parallel to $\mathbf{e}_y$. Then the ray path is locally

† W. D. Hayes, in "Round Table Discussion on Sonic Boom Problems," *Aircraft Engine Noise and Sonic Boom, AGARD Conf. Proc.*, **42:**36–38 (1969).

‡ See, for example, the shadowgraph by W. J. Pierson, Jr. of water waves focused by passage over a bottom protuberance, given by J. J. Stoker, *Water Waves*, Interscience, New York, 1957, p. 135. Analogous features appear in schlieren photographs of shock waves after passage through jets; see, for example, S.-L. V. Hall, "Distortion of the Sonic Boom Pressure Signature by High-Speed Jets," *J. Acoust. Soc. Am.*, **63:**1749–1752 (1978).

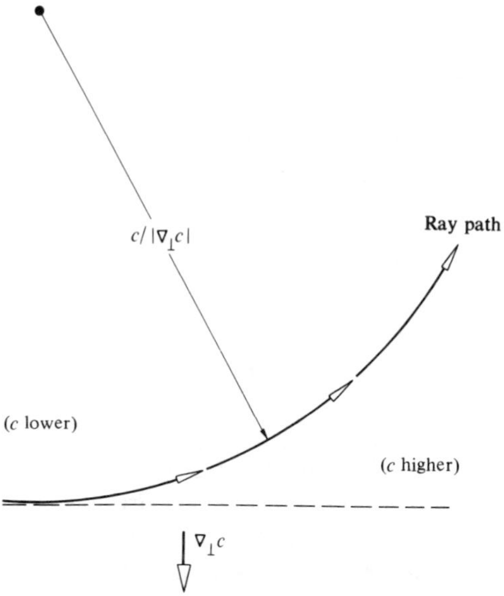

Figure 8-9 Ray-path curvature in a medium with spatially varying sound speed. Ray bends in plane of transverse gradient $\nabla_\perp c$ and of ray path, away from direction of $\nabla_\perp c$ with a radius of curvature equal to $c/|\nabla_\perp c|$.

curved toward negative y if $c_y > 0$ and curved toward positive y if $c_y < 0$. In either case the radius of curvature of the ray path is $c/|c_y|$ (see Fig. 8-9).

The above discussion leads to the conclusion that if a sound ray is moving through a medium with variable sound speed, the ray curves away from its direction of propagation if the component $\nabla_\perp c$ of ∇c transverse to the direction of propagation is nonzero. The ray bends in the plane of $\nabla_\perp c$ and the local ray path but away from the direction of $\nabla_\perp c$, toward the lower-sound-speed side. The radius of curvature of the ray path is $c/|\nabla_\perp c|$, or $c/(|\nabla c| \sin \theta_0)$, where θ_0 is the angle between the ray direction and the direction of ∇c.

The bending of rays toward regions of lower sound speed is explicable in terms of wavefronts. Since the portion of the wavefront on the low-sound-speed side of a ray is moving slower, the wavefront must tilt toward that side. Since the ray (given $\mathbf{v} = 0$) remains normal to the wavefront, it bends in that direction.

Rays in a Medium with Constant-Sound-Speed Gradient†

When ∇c is everywhere the same, the ray path is always a perfect arc of a circle. To demonstrate this, it is sufficient to assume that c varies only with z and that the ray is moving in the xz plane, so $s_y = 0$. Equation (8-1.5) with $\mathbf{v} = 0$ therefore gives $s_z^2 = c^{-2} - s_x^2$, and so the relation $s_z/s_x = dz/dx$ [from Eq. (1)]

† A tabulation of sound-speed profiles for which the ray-tracing equations can be integrated in closed form is given by A. Barnes and L. P. Solomon, "Some Curious Analytical Ray Paths for Some Interesting Velocity Profiles in Geometrical Acoustics," *J. Acoust. Soc. Am.*, **53**:147–155 (1973).

yields

$$\left(\frac{dz}{dx}\right)^2 - \frac{1}{c^2 s_x^2} = -1 \tag{8-3.4}$$

Furthermore, the second of Eqs. (1) predicts that s_x is constant when $c = c(z)$.

That Eq. (4) describes a circle when dc/dz is constant results because the algebraic equation

$$(x - a)^2 + (z - b)^2 = r_c^2$$

has the property

$$\left(\frac{dz}{dx}\right)^2 = \left(\frac{x-a}{z-b}\right)^2 = \frac{r_c^2}{(z-b)^2} - 1 \tag{8-3.5}$$

Consequently, a comparison of Eqs. (4) and (5) indicates that if $c = c_0 - \alpha z$ (such that $\nabla c = -\alpha \mathbf{e}_z$ is constant), the integral of Eq. (4) is a circle of radius $r_c = 1/\alpha s_x$ centered at a point on the line (see Fig. 8-10) at the virtual height $z = c_0/\alpha$ where the sound speed extrapolates to zero. Of the possible rays passing through the point, those moving perpendicular to the sound-speed gradient bend the most.

Refraction by Wind Gradients

Let us next consider a ray that passes through the origin at $t = 0$ with wavefront normal direction $\mathbf{n}_0$. The corresponding initial value of the wave-slowness

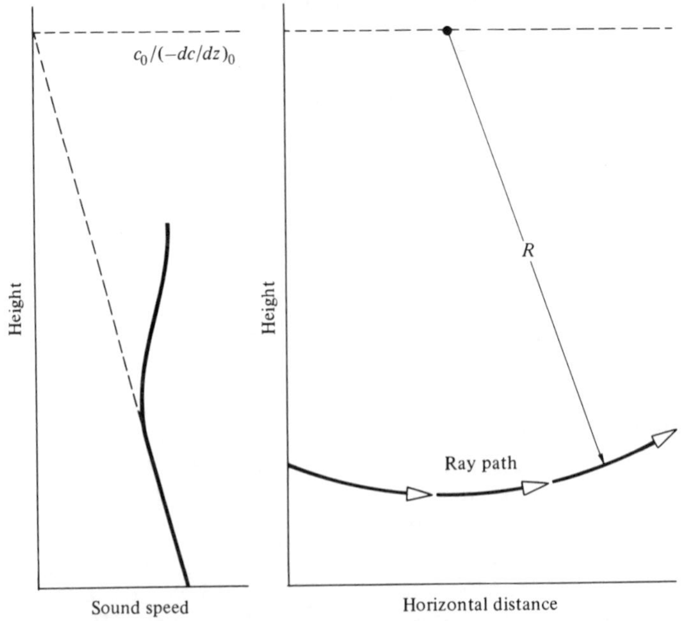

Figure 8-10 For a medium in which sound speed varies linearly with height, ray path is arc of circle centered at height where extrapolated sound speed goes to zero.

vector is determined from Eq. (8-1.3); Eq. (8-1.10b') therefore integrates to first order in t to

$$\mathbf{s} \approx (c + \mathbf{v} \cdot \mathbf{n}_0)^{-1}[\mathbf{n}_0 - t\,\nabla(c + \mathbf{v} \cdot \mathbf{n}_0)] \tag{8-3.6}$$

Equation (8-1.10a) consequently yields the power-series expansion

$$\mathbf{x} \approx (c\mathbf{n}_0 + \mathbf{v})t + \tfrac{1}{2}t^2\,[(\mathbf{v}_{\text{ray}} \cdot \nabla)(c\mathbf{n}_0 + \mathbf{v}) - c\,\nabla_\perp(c + \mathbf{v} \cdot \mathbf{n}_0)] \tag{8-3.7}$$

where $\nabla_\perp = \nabla - \mathbf{n}_0(\mathbf{n}_0 \cdot \nabla)$ is the gradient transverse to $\mathbf{n}_0$, and $\mathbf{v}_{\text{ray}}$ is $c\mathbf{n}_0 + \mathbf{v}$. All coefficients and derivatives are understood to be evaluated at the origin.

The plane of bending of the ray is that containing the two vectors $\dot{\mathbf{x}}$ and $\ddot{\mathbf{x}}$ that appear as coefficients of t and $\frac{1}{2}t^2$ in Eq. (7). The ray bends toward the direction of the component $\ddot{\mathbf{x}}_\perp$ of $\ddot{\mathbf{x}}$ that is transverse to $\dot{\mathbf{x}}$; the radius of curvature r_c is $\dot{\mathbf{x}} \cdot \dot{\mathbf{x}}/|\ddot{\mathbf{x}}_\perp|$.

Many ambient velocity fields of interest are approximately such that $(\mathbf{v} \cdot \nabla)\mathbf{v} = 0$, so $\mathbf{v}$ varies negligibly with translation along the direction of flow. With this assumption and with the neglect of the slight difference between $\mathbf{n}_0$ and the direction of $\dot{\mathbf{x}}$, Eq. (7) leads to

$$\ddot{\mathbf{x}}_\perp \approx c[(\mathbf{n}_0 \cdot \nabla)\mathbf{v} - \nabla(c + \mathbf{v} \cdot \mathbf{n}_0)_\perp] \approx -c\,\nabla_\perp c - c\mathbf{n}_0 \times (\nabla \times \mathbf{v}) \tag{8-3.8}$$

This applies, in particular, if $|\mathbf{v}| \ll c$ or if $\mathbf{n}_0$ is parallel to $\mathbf{v}$. From this relation one concludes that the ray curves in a direction which is opposite to that of $\nabla_\perp c + \mathbf{n}_0 \times (\nabla \times \mathbf{v})$, with a radius of curvature approximately equal to c divided by the magnitude of this vector.

As an example, suppose $\mathbf{n}_0 = \mathbf{e}_z \cos\theta + \mathbf{e}_x \sin\theta$ and that c, v_x, and v_y depend only on vertical distance z, while $v_z = 0$. Then Eq. (8) reduces to

$$\ddot{\mathbf{x}}_\perp = -c\left(\frac{dc}{dz}\sin\theta + \frac{dv_x}{dz}\right)\mathbf{e}_2 + c\left(\frac{dv_y}{dz}\cos\theta\right)\mathbf{e}_y \tag{8-3.9}$$

where $\mathbf{e}_2$, equal to $\mathbf{e}_z \sin\theta - \mathbf{e}_x \cos\theta$, is the unit vector in the xz plane that is perpendicular to $\mathbf{n}_0$.

The y component of $\ddot{\mathbf{x}}_\perp$ is associated with the ray's sideways drift caused by crosswinds; it is often of minor consequence, either because rays of interest are nearly horizontal ($\cos\theta$ is small) or because the net shift in ray direction due to this component averages out to nearly zero. Its neglect leads to a radius of curvature† equal to

$$r_c = \frac{c}{(dc/dz)\sin\theta + dv_x/dz} \tag{8-3.10}$$

A positive value implies downward bending; a negative value implies upward bending.

A further approximation, valid for rays proceeding in nearly horizontal directions, is to replace $\sin\theta$ by 1, so that $c\sin\theta + v_x$ is replaced by $c + v_x$ in

† B. Gutenberg, "Propagation of Sound Waves in the Atmosphere," *J. Acoust. Soc. Am.*, **14**:151–155 (1942).

the above. This leads to the simple rule that the ray undergoes refraction as if it were moving in a medium with no winds but with an effective sound speed $c_{\text{eff}} = c + v_x$, where v_x is the component of the wind velocity in the vertical plane containing the ray. From this viewpoint, wind-speed gradients and sound-speed gradients have the same influence on sound rays. However, if θ is less than, say, 30°, the influence of a wind-speed gradient is substantially greater than that of a sound-speed gradient of the same magnitude.

8-4 RAYS IN STRATIFIED MEDIA

The ambient properties of the atmosphere and of the oceans (see Fig. 8-11) vary primarily with height or depth, and the ambient fluid velocity is primarily horizontal. Consequently, the stratified-fluid model discussed above [with $c = c(z)$, $\mathbf{v} = \mathbf{v}(z)$, and $v_z = 0$] is commonly used in approximate analyses of sound propagation.

The Ray Integrals

For a stratified fluid, the ray-tracing equation (8.1.10b') requires that s_x and s_y both be constant along any given ray. This can be viewed either as a consequence of the trace-velocity matching principle discussed in Sec. 3-5 or as a generalization of Snell's law. Furthermore, once s_x and s_y are specified, s_z can be determined as a function of height z from Eq. (8-1.5), i.e.,

$$s_z = \pm \left[\left(\frac{\Omega}{c}\right)^2 - s_x^2 - s_y^2\right]^{1/2} \tag{8-4.1}$$

(Note that $1 - \mathbf{v} \cdot \mathbf{s}$ is independent of s_z since $\mathbf{v}$ does not have a z component.) Thus, Eqs. (8-1.10b) can be regarded as solved, and from Eqs. (8-1.10a) one obtains† [with $dx/dz = (dx/dt)/(dz/dt)$ and $dt/dz = 1/(dz/dt)$]

$$\frac{dx}{dz} = \frac{c^2 s_x + \Omega v_x}{c^2 s_z} \qquad \frac{dt}{dz} = \frac{\Omega}{c^2 s_z} \tag{8-4.2}$$

with an analogous equation for dy/dz.

Since the right sides of Eqs. (2) are functions only of z, one can determine x, y, and t as functions of z (and of s_x and s_y) by direct integration, e.g.,

$$x = x_0 + \int_{z_0}^{z} \frac{c^2 s_x + (1 - \mathbf{v} \cdot \mathbf{s}) v_x}{c^2 s_z}\, dz \tag{8-4.3}$$

where x_0 is the value of x at height z_0.

† These were first derived by Fujiwhara, "On the Abnormal Propagation of Sound."

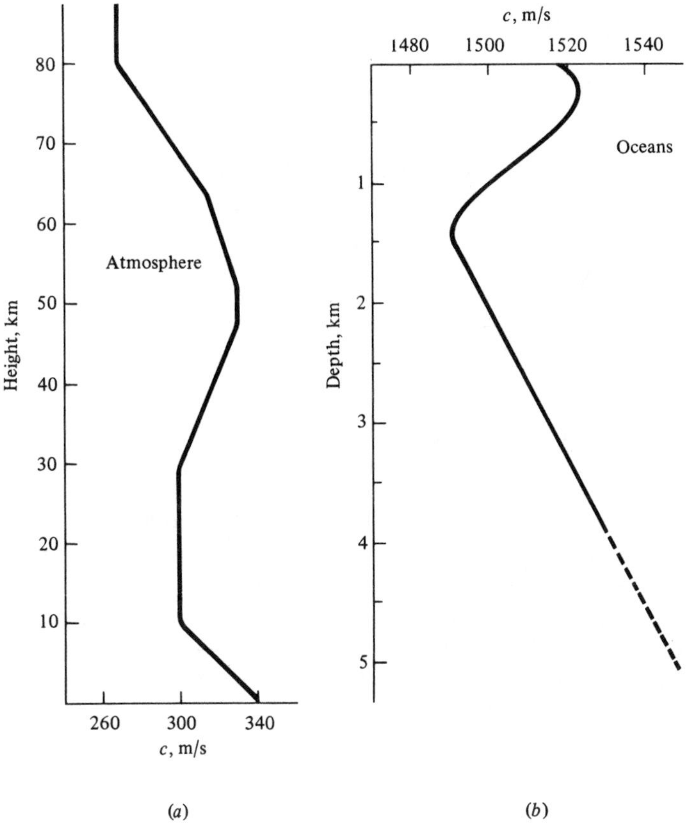

Figure 8-11 Representative sound-speed-versus-height profiles for (*a*) the atmosphere and (*b*) the oceans. These profiles are typical, but there is considerable variability with seasons, geographical location, and meterological conditions, especially near the ground or the sea surface. The sound speed in the atmosphere increases again with increasing height above 90 km. [*Based on tables and figures in A. E. Cole, A. Court, and A. J. Kantor, "Model Atmospheres," chap. 2 in S. L. Valley (ed.), Handbook of Geophysics and Space Environments, Air Force Cambridge Research Laboratories, 1965, and by M. Ewing and J. L. Worzel, "Long Range Sound Transmission," in Propagation of Sound in the Ocean, Geological Society of America, Memoir 27, 1948.*]

Channeling of Ray Paths

In the application of Eqs. (3) and its counterparts, one must take into account the fact that a ray is confined to a height region for which $s_z^2 \geq 0$. For an actual ray, x, y, and t may not be single-valued functions of z since s_z changes sign whenever the ray reaches a height (*turning point*) at which s_z^2 goes to zero (see Fig. 8-12). The initial position and direction of the ray determine s_x and s_y and the initial sign for s_z. Providing that $1 - \mathbf{v} \cdot \mathbf{s} > 0$, the sign of s_z will be the same as that of dt/dz [see Eq. (2)] and will therefore be positive for a ray proceeding obliquely up and negative for one proceeding obliquely down.

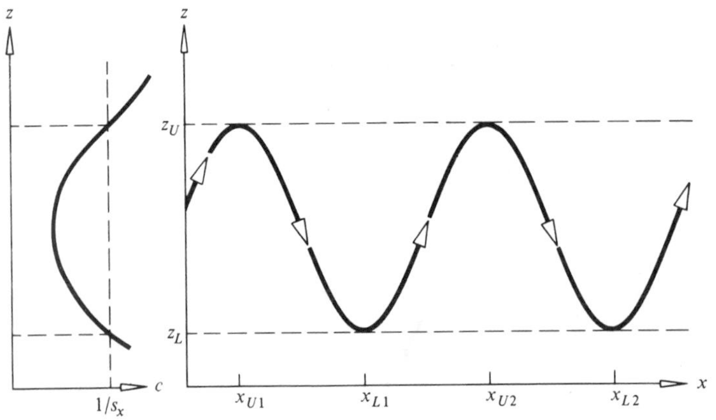

Figure 8-12 Ray channeled between turning points.

Suppose the initial sign is positive. Then Eq. (3) and its counterparts describe the ray trajectory up until it reaches that height (providing one exists) at which s_z^2 first becomes zero. At that point the ray trajectory is horizontal and curving down, so it must thereafter return to lower heights. Let z_U be the height of this upper turning point, and let x_{U1}, y_{U1}, t_{U1} be the values of x, y, and t at which it is first reached. Thereafter, s_z is negative, and subsequent values of x for the next segment of the ray trajectory are given by

$$x = x_{U1} + \int_z^{z_U} \frac{c^2 s_x + \Omega v_x}{c^2 |s_z|} \, dz \tag{8-4.4}$$

Analogous formulas hold for the corresponding values of y and t. Such relations hold up until the ray reaches that lower turning point z_L (if one exists) at which s_z^2 again becomes zero and at which s_z again changes sign.

Note that although s_z vanishes at z_U, integrals like that in Eq. (4) are nevertheless finite. Near z_U, the denominator factor $|s_z|$ goes to zero as $(z_U - z)^{1/2}$, so the integrand remains integrable.

If a ray trajectory has both upper and lower turning points, it is *channeled*. Such a trajectory will be periodic both in time and in horizontal displacement. The net x displacement in going from the lower turning point to the upper turning point is the same as that from the upper turning point to the lower turning point and is the same for every such segment of the ray path. The same statement holds for y displacements and travel-time segments. The *average horizontal velocity* of the ray is

$$\mathbf{v}_H = \frac{(\Delta x)_{L\to U}\mathbf{e}_x + (\Delta y)_{L\to U}\mathbf{e}_y}{(\Delta t)_{L\to U}} \tag{8-4.5}$$

where $(\Delta t)_{L\to U}$ is the net time required to go from the lower turning point to the upper turning point.

If the ray reaches a horizontal interface, such as the upper surface of the ocean for underwater sound propagation, the wave associated with it will be partially reflected and partially transmitted. However, so far as the reflected wave is concerned, its wavefronts will also be locally planar and can also be described in terms of rays. Thus, the incident ray gives rise to a reflected ray that represents a continuation of the incident path back into the fluid. The trace-velocity matching principle requires s_x and s_y to be the same for the reflected ray as for the incident ray, so the only ray parameter that changes on ray reflection is s_z, which simply changes sign. However, at such a surface, s_z does not go to zero, as is the case for internal reflection.

Given the presence of interfaces, one has the possibility† of a ray being channeled between an upper interface and a lower turning point, an upper interface and a lower interface, etc.

Rays in Fluids without Ambient Flow

When the medium has no ambient fluid velocity, the ray path is always in the same vertical plane and one can orient the coordinate system so that $s_y = 0$. Then s_x can be identified [see Eq. (8-1.3)] as $\pm(\sin\theta)/c$, where θ is the angle between the ray direction and the vertical; s_x is positive for a ray proceeding obliquely in the $+x$ direction. The constancy of s_x along a ray is thus identical to the elementary version of Snell's law for refraction at an interface between two fluids. Equation (1) also reduces to

$$s_z = \pm(c^{-2} - c_0^{-2}\sin^2\theta_0)^{1/2} \tag{8-4.6}$$

where $\sin\theta_0$ is the value of $\sin\theta$ at the height where the sound speed is c_0. The ray is confined to a height regime for which

$$c^2(z) \le \frac{c_0^2}{\sin^2\theta_0} \tag{8-4.7}$$

and turning points occur at heights where the equality holds. Consequently, any region of height in which the profile of c versus z has a minimum is a potential sound-speed channel, e.g., the SOFAR channel in the ocean. Also, if the sound speed at some depth below an interface has a higher value than that just below the interface, a ray can be channeled between the interface and the higher-sound-speed region. The region in which the ray is channeled can in each case be determined from Eq. (7) without an explicit determination of the path.

† Terminology in underwater sound classifies rays by their upper and lower turning points. A ray that goes from source to a lower internal turning point, then to the surface, where it is reflected, is an RSR ray (refracted-surface-reflected). A ray that traverses between upper and lower internal turning points is a SOFAR ray. A channeled ray is an SLR (surface-limited ray) or a BLR (bottom-limited ray) if its upper turning point is the ocean surface or if its lower turning point is the ocean bottom, respectively. See, for example, Officer, *Introduction to the Theory of Sound Transmission*, pp. 98–101, 155–161; W. H. Munk, "Sound Channel in an Exponentially Stratified Ocean, with Application to SOFAR," *J. Acoust. Soc. Am.*, **55**:220–226 (1974).

Example: Axial rays Suppose $c(z)$ has a minimum value of c_0 at $z = 0$ and that near the minimum $c = c_0 + \alpha^2 z^2$. A model profile† which exhibits such properties and which is amenable to analytic investigation is that where $1/c^2$ equals $(1/c_0)^2(1 - z^2/L^2)$, with $L^2 = c_0/2\alpha^2$. For such a model, Eqs. (2), with $dt/dx = (dt/dz)/(dx/dz)$, can be rewritten with the help of Eq. (6) as

$$\frac{dx}{dz} = \frac{\pm\sin\theta_0}{(\cos^2\theta_0 - z^2/L^2)^{1/2}} \qquad \frac{dt}{dx} = \frac{1 - z^2/L^2}{c_0 \sin\theta_0} \tag{8-4.8}$$

The first leads to the differential equation

$$\sin^2\theta_0 \left(\frac{dz}{dx}\right)^2 + \frac{z^2}{L^2} = \cos^2\theta_0$$

which has the solution

$$z = L\cos\theta_0 \sin\frac{x - x_0}{L\sin\theta_0} \tag{8-4.9}$$

Thus the ray path crosses $z = 0$ at intervals of $(\Delta x)_{U\to L}$ of $\pi L \sin\theta_0$; the path-repetition distance is twice this. The time required for the ray to travel the horizontal distance $(\Delta x)_{U\to L}$ is just this distance times the average, over x, of dt/dx [see Eq. (8)]. Since the average of z^2, from Eq. (9), is $\frac{1}{2}L^2\cos^2\theta_0$, one accordingly finds the average horizontal velocity to be

$$v_H = \frac{(\Delta x)_{U\to L}}{(\Delta t)_{U\to L}} = \frac{2c_0\sin\theta_0}{1 + \sin^2\theta_0} \tag{8-4.10}$$

The above results strictly apply only if $c(z)$ is given by $c_0/[1 - (z/L)^2]^{1/2}$, but the conclusion, that $(\Delta x)_{U\to L}$ approaches $\pi(c/c'')_0^{1/2}$ as $\theta_0 \to \pi/2$, applies to rays channeled in any region‡ where $c/(d^2c/dz^2)$ at the sound-speed minimum has the same value. (This presumes that c, c', and c'' are continuous.) If $c(z)$ is even about the altitude of its minimum, and if the source is on the channel's axis (where c is smallest), the skip distance $\pi(c/c'')_0^{1/2}$ for the axial ray, $\theta_0 = 90°$, is an extremal; adjacent rays intersect on the axis at this distance and at its multiples, so a sequence of caustics must appear at horizontal distances of $n\pi(c/c'')_0^{1/2}$, where $n = 1, 2, \ldots$. If the profile is not symmetric, however, this is not necessarily the case (see Fig. 8-13). Nevertheless, each channeled ray must graze a caustic somewhere between its first and second turning points.

† R. R. Goodman and L. R. B. Duykers, "Calculation of Convergent Zones in a Sound Channel," *J. Acoust. Soc. Am.*, **34**:960–962 (1962).

‡ How this limit is approached is explored in detail by M. A. Pederson, "Ray Theory Applied to a Wide Class of Velocity Functions," *J. Acoust. Soc. Am.*, **43**:619–634 (1968); "Theory of the Axial Ray," ibid., **45**:157–176 (1969); (with D. White) "Ray Theory for Sources and Receivers on an Axis of Minimum Velocity," ibid., **48**:1219–1248 (1970).

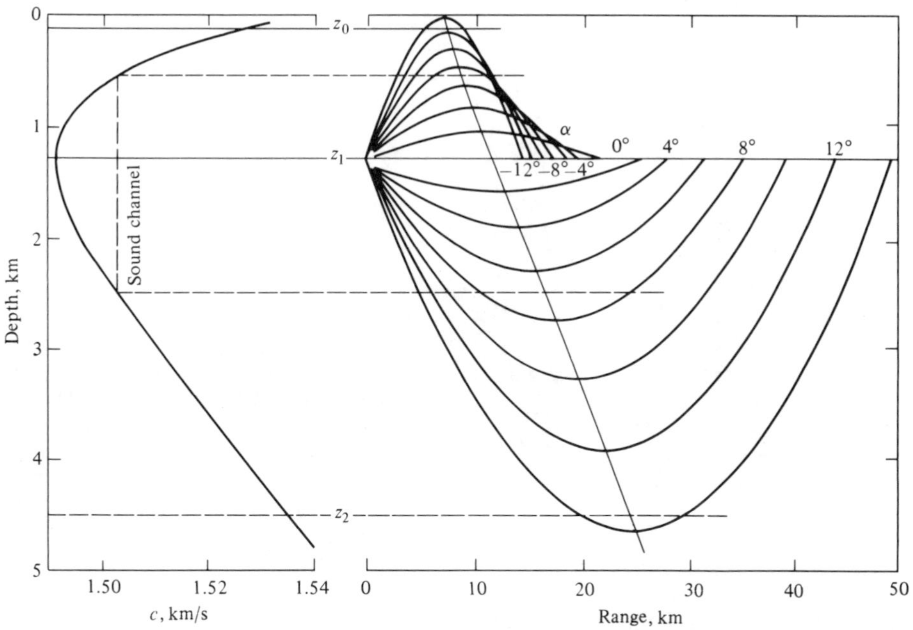

Figure 8-13 Model underwater SOFAR channel and corresponding ray paths from source at depth z_1 of minimum sound speed. (Note compressed horizontal scale.) Each ray is labeled by the angle α it initially makes with the horizontal; positive α means that the ray is initially propagating obliquely downward; $\alpha = 0°$ ray is horizontal and remains at depth z_1. Sound speed $c(z)$ is $c_1[1 + \varepsilon(\eta + e^{-\eta} - 1)]$ with $c_1 = 1.492$ km/s, $\varepsilon = 0.0074$, $\eta = (z - z_1)/(z_1/2)$, $z_1 = 1.3$ km. Selected profile is such that the caustic surface lies above $z = z_1$ and the point where the $\alpha = 0°$ ray grazes the caustic is not a vertex (arête) of the caustic. The focusing on the channel axis is therefore considerably weaker than for a channel symmetric about z_1. [*From W. H. Munk, J. Acoust. Soc. Am.,* **55**:*222* (*1974*).]

Abnormal Sound

Audible sound is often received at distances of 200 to 300 km from large explosions, even though the sound may be inaudible at closer distances (see Fig. 8-14). The analysis† (*air seismology*) of the arrival times, angles of incidence, and locations of reception of this abnormal sound is a principal tool for studying the meteorology of the upper atmosphere.

To explain the phenomenon, let us for simplicity ignore crosswinds, so that rays from the source stay within a vertical plane. A ray proceeding in the xz plane from a source on the ground will be such that the angle θ, between unit

† F. J. Whipple, "The Propagation of Sound to Great Distances," *Q. J. R. Meteorol. Soc.*, **61**:285–308 (1935); E. F. Cox, "Abnormal Audibility Zones in Long Distance Propagation through the Atmosphere," *J. Acoust. Soc. Am.*, **21**:6–16, 501 (1949); A. P. Crary and V. C. Bushnell, "Determination of High-Altitude Winds and Temperature in the Rocky Mountain Area by Acoustic Soundings," *J. Meteorol.*, **12**:463–471 (1955); W. L. Donn and D. Rind, "Natural Infrasound as an Atmospheric Probe," *Geophys. J. R. Astron. Soc.*, **26**:111–133 (1971).

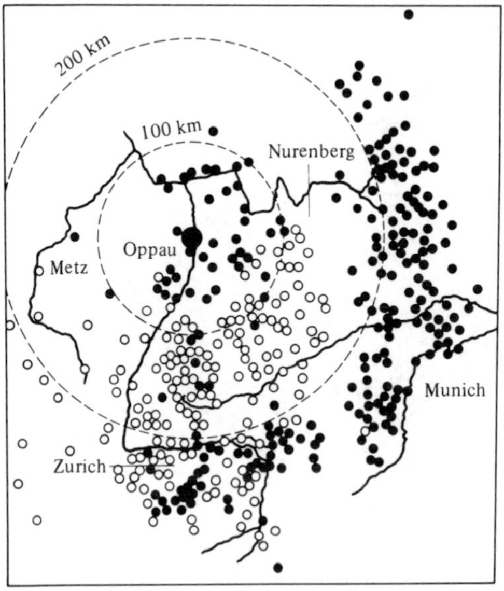

Figure 8-14 Locations where sound was heard (black dots) and not heard (open circles) following an explosion at Oppau, Germany, on Sept. 21, 1921. The anomalous zone of audibility, to the east and south, beyond 200 km is explained by a model atmosphere in which stratospheric winds are blowing toward the east. [*From R. K. Cook, Sound,* **1:***13 (1962).*]

wavefront normal **n** and the vertical, satisfies

$$s_x = \frac{\sin \theta}{c + v_x \sin \theta} = \text{const} \tag{8-4.11}$$

Although the ray direction is in general slightly different from that of **n**, it is horizontal when **n** is horizontal. Thus, the ray with initial angle θ_0 turns back to the ground when it reaches turning-point height z_{tp} that satisfies

$$c(z_{\text{tp}}) + v_x(z_{\text{tp}}) = \frac{c_g}{\sin \theta_0} \tag{8-4.12}$$

where c_g is the sound speed at the ground. (The wind speed near the ground is here considered negligible.)

For the atmosphere at middle latitudes, the effective sound-speed profile $c(z) + v_x(z)$ typically has a shape like those sketched in Fig. 8-15. Whether the peak value that occurs between 30 and 60 km altitude exceeds the value at the ground depends on the direction associated with increasing x, with the season of year, and with latitude. Since $c + v_x$ typically decreases with height in the lower portion of the atmosphere (the *troposphere*), a zone of silence is formed on the ground at intermediate distances from the source (see Fig. 8-16). This is sometimes offset† by local meteorological conditions close to the ground; the profiles in the first 3 km fluctuate in a less systematic fashion. However, those rays leaving the source with elevation angles of 10° or greater are generally not refracted back to the ground until they have reached altitudes of 30 km or higher.

† See, for example, T. F. W. Embleton, G. J. Thiessen, and J. E. Piercy, "Propagation in an Inversion and Reflections at the Ground," *J. Acoust. Soc. Am.*, **59:**278–282 (1976).

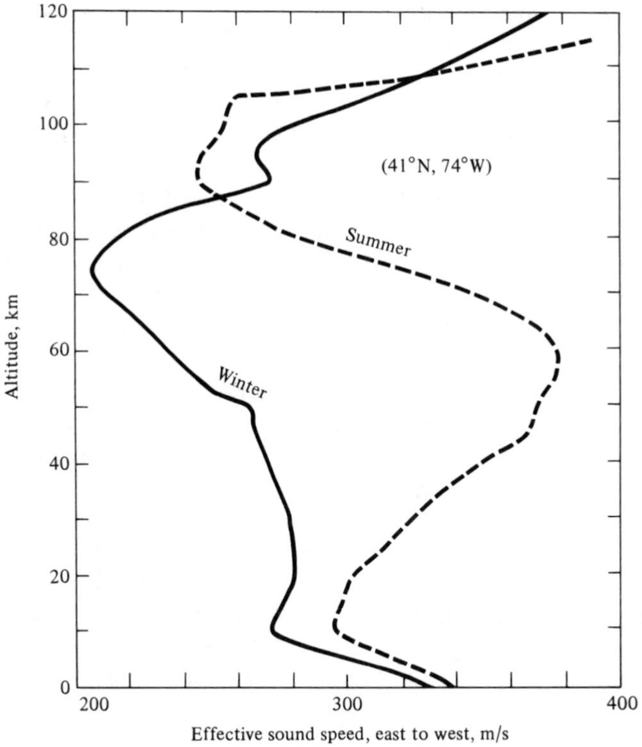

Figure 8-15 Model atmospheric profiles of effective sound speed versus height for propagation east to west in northeastern United States. [*From D. Rind and W. L. Donn, J. Atmos. Sci.*, **32:** *1695 (1975).*]

The existence of ray paths that proceed from the ground to the stratosphere then to ground requires, from Eq. (12), that $c + v_x$ at some altitude exceed the ground-level sound speed c_g. The apparent angle of incidence θ_0 of the arriving sound (determined from measurement of wavefront horizontal transit speed $1/s_x$ across an array of microphones) yields $c(z_{\text{tp}}) + v_x(z_{\text{tp}})$. The arrival time is

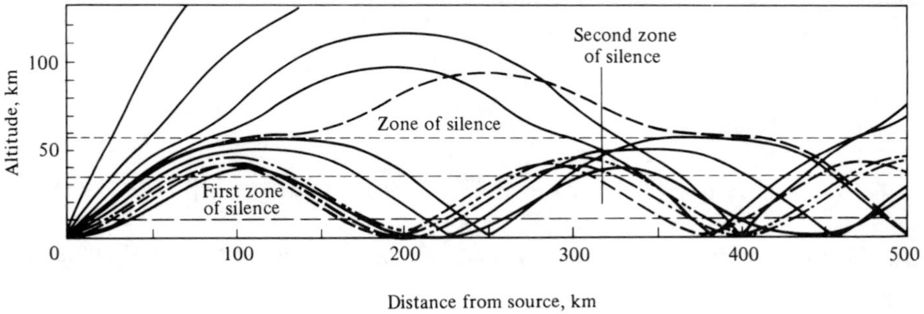

Figure 8-16 Representative ray paths east to west in Northern Hemisphere in summer. [*From B. Gutenberg, in T. F. Malone (ed.), Compendium of Meteorology, American Meteorological Society, Boston, 1951, p. 374.*]

invariably substantially later (typically about 1 min) than would be expected for a wave traveling (creeping) directly along the ground with the sound speed. Such *creeping waves* (see Sec. 9-5) are frequently detected with sensitive instrumentation when geometrical-acoustics considerations would preclude their existence, but their amplitudes are very weak. The geometrical-acoustics model retains its validity insofar as dominant arrivals are concerned.

The striking feature of a zone within which abnormal sound is received is its abrupt onset at a distance of the order of 200 km (see Fig. 8-16). The existence of such a critical range follows from ray-theory computations of the horizontal range $R(\theta_0)$ (*skip distance*) a ray must travel before it returns to the ground. For a profile in which $c + v_x$ decreases monotonically to a minimum value and then increases with further altitude increase until it reaches a maximum value greater than c_g at altitude z_m, a range $R(\pi/2)$ corresponding to grazing incidence $\theta_0 = \pi/2$ will exist and be of the order of 200 km or more. As θ_0 decreases, R will at first decrease until it reaches some minimum value $R_{\min}$; thereafter it increases up to the range $R(\theta_{0,m})$, where $\theta_{0,m}$ is the value of θ_0 for which Eq. (12) predicts that z_{tp} equals z_m. As θ_0 decreases below $\theta_{0,m}$, the range takes a sudden large jump [turning point at a much higher altitude, where $c + v_x$ once again reaches $c(z_m) + v_x(z_m)$], so that the zone of abnormal audibility is limited by the ranges $R_{\min}$ and $R(\theta_{0,m})$. Since $R(\theta_0)$ has a minimum, a caustic must touch the ground at range $R_{\min}$. The abnormal sound is consequently loudest just beyond the inner boundary of the abnormal-audibility zone.

8-5 AMPLITUDE VARIATION ALONG RAYS

Wave Amplitudes in Homogeneous Media

To gain insight into how wave amplitudes vary along ray paths, we consider a constant-frequency wave moving in a fluid with constant sound speed and ambient density and for which the ambient fluid velocity is zero. The acoustic pressure therefore satisfies the wave equation (1-6.1) and has a complex spatially dependent amplitude $\hat{p}(\mathbf{x})$ that satisfies the Helmholtz equation (1-8.13). The insertion† of

$$\hat{p}(\mathbf{x}) = P(\mathbf{x}, \omega)e^{i\omega\tau(\mathbf{x})}$$

into the latter yields

$$\nabla^2 P + i\omega(2\nabla P \cdot \nabla\tau + P\,\nabla^2\tau) - \omega^2 P\left[(\nabla\tau)^2 - \frac{1}{c^2}\right] = 0 \qquad (8\text{-}5.1)$$

To solve this in the high-frequency limit, we assume the existence of an asymp-

† A. Sommerfeld and J. Runge, "Application of Vector Calculus to the Fundamentals of Geometrical Optics," *Ann. Phys.*, (4)**35**:277–298 (1911).

totic expansion for P:

$$P(\mathbf{x}, \omega) = P_0(\mathbf{x}) + \frac{1}{\omega}P_1(\mathbf{x}) + \frac{1}{\omega^2} P_2(\mathbf{x}) + \cdots \tag{8-5.2}$$

This is then substituted into (1), and it is required that the resulting coefficient of each power of ω vanish identically. The first two in the infinite sequence of equations so derived involve only τ and P_0; we assume that P_0 is an adequate approximation for P, so we keep only the first two equations and therein replace P_0 by P; the resulting equations are

$$(\nabla\tau)^2 = \frac{1}{c^2} \tag{8-5.3a}$$

$$2\nabla P \cdot \nabla\tau + P\,\nabla^2\tau = 0 \qquad \text{or} \qquad \nabla\cdot(P^2\nabla\tau) = 0 \tag{8-5.3b}$$

Note that these equations also result from equating the coefficients of ω^2 and ω in Eq. (1) to zero. The second version of Eq. (3b) follows from a multiplication of the first version by P.

Equation (3a) is the *eikonal equation* (8-1.5) with the ambient velocity set to zero, so its solution can be given in terms of rays. Once any wavefront surface is specified and a value of τ is associated with it, the value of $\tau(\mathbf{x})$ for any position $\mathbf{x}$ can be determined by finding that ray connecting the originally specified wavefront with the point $\mathbf{x}$. If the ray passes through point $\mathbf{x}_0$ on the originally specified wavefront, and if $\tau(\mathbf{x}_0) = \tau_0$, $\tau(\mathbf{x})$ is τ_0 plus the travel time at speed c along the ray from $\mathbf{x}_0$ to $\mathbf{x}$.

The solution of Eq. (3b) can be developed in terms of *ray-tube areas*. With the ray passing from $\mathbf{x}_0$ to $\mathbf{x}$ one associates a ray tube (Fig. 8-17) consisting of all rays passing through a tiny area $A(\mathbf{x}_0)$ centered at $\mathbf{x}_0$ transverse to the ray path. When the ray tube reaches $\mathbf{x}$, its cross-sectional area will be $A(\mathbf{x})$. One integrates Eq. (3b) over the volume of the ray-tube segment connecting $\mathbf{x}_0$ and $\mathbf{x}$ and applies Gauss' theorem to convert it into a surface integral. Then, since the ray path is everywhere in the direction of $\nabla\tau = \mathbf{s}$, the surface integral over the

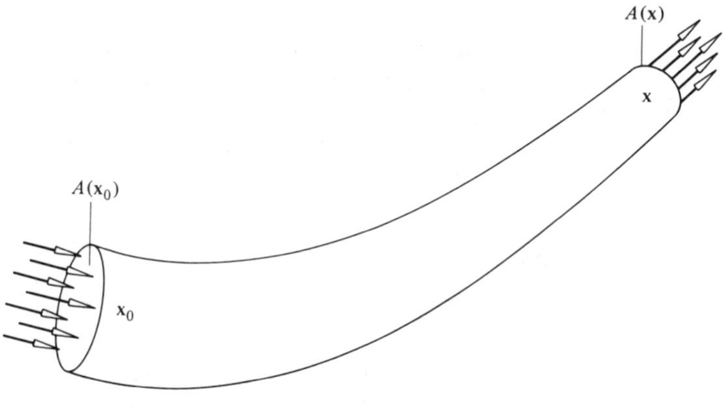

Figure 8-17 Sketch of a ray tube.

sides of the ray-tube segment vanishes identically and one is left with contributions from just the two ends. Thus, one has

$$P^2(\mathbf{x}_0)A(\mathbf{x}_0)(\nabla\tau \cdot \mathbf{n})_{\mathbf{x}_0} = P^2(\mathbf{x})A(\mathbf{x})(\nabla\tau \cdot \mathbf{n})_{\mathbf{x}}$$

where **n** is the unit vector in the direction of the ray or, equivalently (because there is no ambient flow), the unit vector normal to the wavefront. However, $\nabla\tau \cdot \mathbf{n} = \mathbf{s} \cdot \mathbf{n}$ is here $1/c$ [from Eq. (8-1.3)], and since c is constant, the above reduces to

$$P(\mathbf{x}) = P(\mathbf{x}_0)\left[\frac{A(\mathbf{x}_0)}{A(\mathbf{x})}\right]^{1/2} \tag{8-5.4}$$

Thus, wave amplitude varies along a ray in inverse proportion to the square root of the ray-tube area. If the ray-tube area grows smaller (*focuses*), the amplitude increases.

The volume integral of $\nabla^2\tau$ over the ray-tube segment is similarly found to be $(1/c)[A(\mathbf{x}) - A(\mathbf{x}_0)]$. Thus, for any short tube segment of length dl and therefore of (approximate) volume $A(\mathbf{x})\,dl$, one has

$$\nabla^2\tau = \frac{1}{cA}\frac{dA}{dl} \tag{8-5.5}$$

where $A(l)$ is ray-tube area at distance l along the ray. Moreover, if the coordinate system is chosen so that the z axis points in the ray direction and x and y axes in the principal curvature directions, the point of interest being taken as the origin, then near that point, Eq. (8-2.3) yields

$$\tau \approx \text{const} + \frac{1}{c}\left(z - \frac{x^2}{2r_1} - \frac{y^2}{2r_2}\right) \tag{8-5.6}$$

where r_1 and r_2 are the two principal radii of curvature (positive if concave) of the wavefront at (0, 0, 0). Consequently, we conclude that

$$c\,\nabla^2\tau = -\left(\frac{1}{r_1} + \frac{1}{r_2}\right) \tag{8-5.7}$$

In addition, since $r_1 = r_1^0 - l$ and $r_2 = r_2^0 - l$ [from Eq. (8-2.5)], one can replace $-1/r_1$ by $(d/dl)(\ln r_1)$. With a similar replacement for $-1/r_2$ and with $c\nabla^2\tau$ replaced [from Eq. (5)] by $(d/dl)(\ln A)$, integration of Eq. (7) leads to the conclusion that (A/r_1r_2) is independent of l, so the ratio of ray-tube areas in Eq. (4) is the same as $r_1^0r_2^0/r_1r_2$ (see Fig. 8-18). Therefore the amplitude along the ray is

$$P(\mathbf{x}) = \left[\frac{r_1^0r_2^0}{(r_1^0 - l)(r_2^0 - l)}\right]^{1/2} P(\mathbf{x}_0) \tag{8-5.8}$$

and varies inversely as the geometric mean of the two principal radii of wavefront curvature.

The above can be generalized to a superposition of different frequencies or to a transient waveform. Since ray paths and travel times are independent of frequency, and since amplitude ratios at different points on the ray path are also

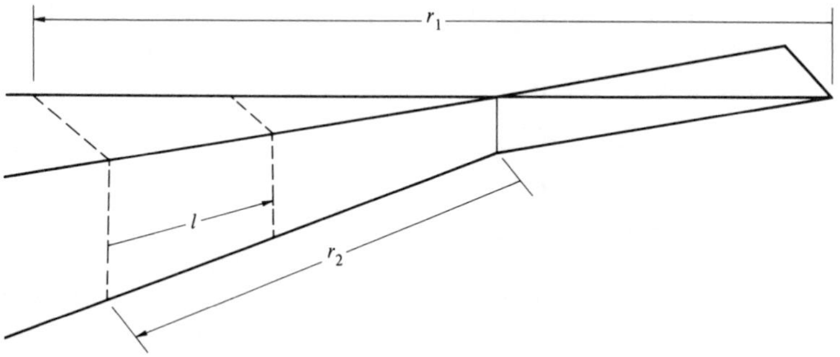

Figure 8-18 Geometrical proof that ray tube area is proportional to the product of the wavefront's principal radii of curvature.

independent of frequency, the solution of the wave equation in the geometrical-acoustics approximation is

$$p = B(l, \xi)f(t - \tau, \xi) \tag{8-5.9}$$

where the parameter ξ (or, strictly speaking, pair of parameters, ξ_1, ξ_2) distinguishes different rays. The waveform shape $f(t - \tau, \xi)$ is the same along any given ray, but the amplitude factor $B(l, \xi)$ varies with distance l along the ray. If $f(t - \tau, \xi)$ is chosen so that it equals $p(\mathbf{x}, t)$ at the initial point ($l = 0$) on the ray, then B is the coefficient of $P(\mathbf{x}_0)$ in Eq. (8) and τ is $\tau_0 + l/c$.

Energy Conservation along Rays

Although the analog of the above derivation can be carried through for propagation in a medium in which $c(\mathbf{x})$ and $\rho(\mathbf{x})$ are slowly varying functions of position (we continue to assume no ambient flow), the following heuristic derivation based on the conservation of acoustic energy may be more enlightening. Let us assume at the outset that

$$p(\mathbf{x}, t) = B(\mathbf{x})f(t - \tau, \xi) \tag{8-5.10}$$

where τ is a solution of the eikonal equation and ξ is a constant along any given ray. The requirement that this describe a propagating plane wave in any local region [via Eq. (1-7.8)] means that the acoustically induced fluid velocity must be identified as $(\mathbf{n}/\rho c)p$ or $(B/\rho)\nabla\tau f$, since $\mathbf{n}$ is $c\nabla\tau$. The energy density and intensity associated with this wave disturbance can consequently be identified from Eqs. (1-11.3) [using $(\nabla\tau)^2 = 1/c^2$] as

$$w = \frac{B^2}{\rho c^2} f^2(t - \tau, \xi) \qquad \mathbf{I} = \mathbf{n}cw \tag{8-5.11}$$

The acoustic-energy-conservation theorem $\partial w/\partial t + \nabla \cdot \mathbf{I} = 0$ then gives

$$2\frac{B^2}{\rho c^2} f \frac{\partial f}{\partial t} + f^2 \nabla \cdot \left(\frac{B^2}{\rho} \nabla\tau\right) + 2\frac{B^2}{\rho}(\nabla\tau \cdot \nabla f)f = 0 \tag{8-5.12}$$

If one ignores the weak dependence of f on position through $\xi(\mathbf{x})$, then $\nabla f = -\partial f/\partial t \ \nabla\tau$; the first and third terms in the above cancel [since $(\nabla\tau)^2 = 1/c^2$], and one is left with

$$\nabla \cdot \left(\frac{B^2}{\rho}\nabla\tau\right) = 0 \tag{8-5.13}$$

which is analogous to the relation $\nabla \cdot \mathbf{I}_{av}$ derived in Chap. 1.

Integration of Eq. (13) over a ray-tube segment leads, in a manner similar to that yielding Eq. (4), to the conclusion that $(B^2/\rho c)A$ is constant along any ray tube, where A is ray-tube cross-sectional area. Thus, if $\mathbf{x}_0$ and $\mathbf{x}$ are any two points along the same ray,

$$B(\mathbf{x}) = \left[\frac{(A/\rho c)_{\mathbf{x}_0}}{(A/\rho c)_{\mathbf{x}}}\right]^{1/2} B(\mathbf{x}_0) \tag{8-5.14}$$

gives the general law of variation of pressure amplitude along a ray in an inhomogeneous quiescent medium. For a constant-frequency wave, this relation can be interpreted as the requirement that the time-averaged energy per unit time flowing along a ray tube be independent of distance along the ray.†

8-6 WAVE AMPLITUDES IN MOVING MEDIA

Linear Acoustics Equations for Moving Media

To determine the effects of steady but inhomogeneous ambient flows on wave amplitudes in the geometrical-acoustics approximation, we begin with the nonlinear fluid-dynamic equations introduced in Chap. 1. With the various idealizations described there, they can be written

$$\frac{D\mathbf{v}}{Dt} + \frac{1}{\rho}\nabla p + g\mathbf{e}_z = 0 \tag{8-6.1a}$$

$$\frac{D\rho}{Dt} + \rho\nabla \cdot \mathbf{v} = 0 \tag{8-6.1b}$$

$$\frac{Ds}{Dt} = 0 \tag{8-6.1c}$$

$$p = p(\rho, s) \tag{8-6.1d}$$

† Sometimes labeled as *Green's law* for acoustic waves because of George Green's analogous result for shallow-water waves: "On the Motion of Waves in a Canal of Variable Depth and Width," *Trans. Camb. Phil. Soc.* (1837), reprinted in N. M. Ferrers (ed.), *Mathematical Papers of the Late George Green*, Macmillan, London, 1871, pp. 225–230. Green's laws in physical systems are reviewed by H. M. Paynter and F. D. Ezekiel, "Water Hammer in Nonuniform Pipes as an Example of Wave Propagation in Gradually Varying Media," *Trans. Am. Soc. Mech. Eng.*, **80:**1585–1595 (1958).

Here, to demonstrate that gravity has no explicit influence on propagation in the high-frequency limit, the gravitational force per unit mass ($-g\mathbf{e}_z$, g being acceleration due to gravity and $\mathbf{e}_z$ being the unit vector in the vertical direction) is included with Euler's equation. If one follows the general procedure outlined in Sec. 1-5, sets $\mathbf{v} = \mathbf{v}_0(\mathbf{x}) + \mathbf{v}'(\mathbf{x}, t)$, $p = p_0(\mathbf{x}) + p'(\mathbf{x}, t)$, etc., and requires Eqs. (1) to be satisfied identically by the ambient state, then, to first order in the acoustic perturbation, one has

$$D_t\mathbf{v}' + \mathbf{v}' \cdot \nabla \mathbf{v}_0 + \frac{1}{\rho_0} \nabla p' - \frac{\rho'}{\rho_0^2} \nabla p_0 = 0 \tag{8-6.2a}$$

$$D_t\rho' + \mathbf{v}' \cdot \nabla \rho_0 + \rho' \nabla \cdot \mathbf{v}_0 + \rho_0 \nabla \cdot \mathbf{v}' = 0 \tag{8-6.2b}$$

$$D_t s' + \mathbf{v}' \cdot \nabla s_0 = 0 \tag{8-6.2c}$$

$$p' = c^2\rho' + \left(\frac{\partial p}{\partial s}\right)_0 s' \tag{8-6.2d}$$

Here the sound speed c and the thermodynamic coefficient $(\partial p/\partial s)_0$ are functions of position; $D_t = \partial/\partial t + \mathbf{v}_0 \cdot \nabla$ represents the time derivative following the ambient flow.

Equation (2*d*) allows the elimination of ρ' from Eqs. (2*a*) and (2*b*). In regard to the first and third terms in Eq. (2*b*), the substitution yields

$$c^2(D_t\rho' + \rho' \nabla \cdot \mathbf{v}_0) = D_t p' - \left(\frac{\partial p}{\partial s}\right)_0 D_t s' + c^2 p' \nabla \cdot \frac{\mathbf{v}_0}{c^2} - c^2 s' \nabla \cdot \left[\frac{\mathbf{v}_0}{c^2}\left(\frac{\partial p}{\partial s}\right)_0\right]$$

Also, because (1*d*) is satisfied in the ambient state, the ambient gradients ∇p_0, $\nabla \rho_0$, and ∇s_0 satisfy the same relation as p', ρ', and s' do in Eq. (2*d*). Consequently, Eq. (2*c*) yields

$$-\left(\frac{\partial p}{\partial s}\right)_0 D_t s' = \mathbf{v}' \cdot \nabla p_0 - c^2 \mathbf{v}' \cdot \nabla \rho_0$$

and Eq. (2*b*) reduces to

$$D_t p' + \mathbf{v}' \cdot \nabla p_0 + c^2 p' \nabla \cdot \frac{\mathbf{v}_0}{c^2} + \rho_0 c^2 \nabla \cdot \mathbf{v}' - s' c^2 \nabla \cdot \left[\frac{1}{c^2}\left(\frac{\partial p}{\partial s}\right)_0 \mathbf{v}_0\right] = 0 \tag{8-6.3}$$

With such substitutions, the equations resulting from Eqs. (2*a*) and (2*b*) can be approximated consistent with the notion of a slowly varying medium if terms of second order in spatial derivatives of ambient variables are discarded. Here any spatial derivative of any ambient variable is first order. Since s' would be zero for an acoustic wave in a homogeneous medium, its departures from a zero value are due to spatial variations of the ambient variables; consequently, s' is also first order. A term like the last term in Eq. (3) is then second order and is therefore discarded. The resulting equations are

$$D_t\mathbf{v}' + (\mathbf{v}' \cdot \nabla)\mathbf{v}_0 + \frac{1}{\rho_0}\nabla p' - \frac{p'}{(\rho_0 c)^2}\nabla p_0 = 0 \tag{8-6.4a}$$

$$D_t p' + \mathbf{v}' \cdot \nabla p_0 + c^2 p' \nabla \cdot \frac{\mathbf{v}_0}{c^2} + \rho_0 c^2 \nabla \cdot \mathbf{v}' = 0 \tag{8-6.4b}$$

Conservation of Wave Action

The above equations, with some further approximations, lead to a conservation law† similar to that in Sec. 1-11. Taking the dot product of (4*a*) with $\rho_0\mathbf{v}'$, multiplying (4*b*) by $p'/\rho_0 c^2$, and adding the two equations yields

$$\left(\frac{\partial}{\partial t} + \mathbf{v}_0 \cdot \nabla\right)w - (v')^2(\mathbf{v}_0 \cdot \nabla)\frac{\rho_0}{2} - (p')^2(\mathbf{v}_0 \cdot \nabla)(2\rho_0 c^2)^{-1} + \nabla \cdot \mathbf{I} + \rho_0\mathbf{v}' \cdot [(\mathbf{v}' \cdot \nabla)\mathbf{v}_0] + \rho_0^{-1}(p')^2 \nabla \cdot \frac{\mathbf{v}_0}{c^2} = 0 \tag{8-6.5}$$

where $w = \frac{1}{2}\rho_0(v')^2 + (p')^2/2\rho_0 c^2$ and $\mathbf{I} = p'\mathbf{v}'$ represent what the energy density and intensity would be when viewed by someone moving with the ambient flow. If we limit our attention to a field that everywhere locally resembles a traveling plane wave, then in all the smaller terms involving spatial derivatives of ambient variables it is a consistent approximation to set $\mathbf{v}' = \mathbf{n}p'/\rho_0 c$ and $(p')^2 = \rho_0 c^2 w$. (Both relations hold for a homogeneous medium, even when $\mathbf{v}_0$ is not zero.) This substitution then yields

$$\frac{\partial w}{\partial t} + \mathbf{v} \cdot \nabla w - w\left[\frac{1}{\rho}(\mathbf{v} \cdot \nabla)\frac{\rho}{2} + \rho c^2(\mathbf{v} \cdot \nabla)(2\rho c^2)^{-1}\right] + \nabla \cdot \mathbf{I} + w\mathbf{n} \cdot [(\mathbf{n} \cdot \nabla)\mathbf{v}] + c^2 w \nabla \cdot \frac{\mathbf{v}}{c^2} = 0 \tag{8-6.6}$$

where we resume the custom of omitting the subscripts on ρ_0 and $\mathbf{v}_0$ whenever the possibility of confusing them with other quantities is negligible.

In regard to the next to the last term in Eq. (6), the unit vector $\mathbf{n}$ can alternately be written as $(c/\Omega)\mathbf{s}$, where $\mathbf{s} = \nabla\tau$. Also $\mathbf{s} \cdot [(\mathbf{s} \cdot \nabla)\mathbf{v}]$ is $(\mathbf{s} \cdot \nabla)(\mathbf{s} \cdot \mathbf{v}) - \mathbf{v} \cdot [(\mathbf{s} \cdot \nabla)\mathbf{s}]$ from a vector identity. In the first of these two terms, $\mathbf{s} \cdot \mathbf{v}$ can be replaced by $1 - \Omega$; in the second term, $(\mathbf{s} \cdot \nabla)\mathbf{s}$ can be replaced by $\frac{1}{2}\nabla(\Omega^2/c^2)$ from Eq. (8-1.7). Consequently, one obtains

$$\mathbf{n} \cdot [(\mathbf{n} \cdot \nabla)\mathbf{v}] = \Omega c\mathbf{n} \cdot \nabla \frac{1}{\Omega} + \frac{\Omega}{c}\mathbf{v} \cdot \nabla \frac{c}{\Omega} \tag{8-6.7}$$

To the same order of approximation as to which Eq. (6) was derived, one

† C. J. R. Garrett, "Discussion: The Adiabatic Invariant for Wave Propagation in a Nonuniform Moving Medium," *Proc. R. Soc. Lond.*, **A299:**26–27 (1967); F. P. Bretherton and C. J. R. Garrett, "Wavetrains in Inhomogeneous Moving Media," ibid., **A302:**529–554 (1969). The derivation here is similar to that of W. D. Hayes, "Energy Invariant for Geometric Acoustics in a Moving Medium," *Phys. Fluids,* **11:**1654–1656 (1968).

can also set $\mathbf{n}w = c^{-1}\mathbf{I}$ in a term like $w\mathbf{n} \cdot \nabla(1/\Omega)$ that vanishes when the medium is homogeneous. Then, with the substitutions just described, Eq. (6) reduces to

$$\frac{\partial w}{\partial t} + \mathbf{v} \cdot \nabla w + w\mathbf{v} \cdot \nabla \left[\ln \rho^{-1/2} + \ln (\rho c^2)^{1/2} + \ln \frac{c}{\Omega} \right] + \nabla \cdot \mathbf{I} + \Omega \mathbf{I} \cdot \nabla \frac{1}{\Omega} + c^2 w \, \nabla \cdot \frac{\mathbf{v}}{c^2} = 0$$

which, with further manipulation, yields

$$\frac{\partial}{\partial t}\left(\frac{w}{\Omega}\right) + \nabla \cdot \left(\frac{\mathbf{I} + \mathbf{v}w}{\Omega}\right) = 0 \qquad (8\text{-}6.8)$$

If $\mathbf{v} = 0$, the above reduces to the law of conservation of acoustic energy,† Eq. (1-11.2). Although we here have an added factor of $1/\Omega$ in each term, the equation is still a conservation law because it is a sum of a time derivative and a divergence. An interpretation of what physical quantity is being conserved follows from consideration of the constant-frequency case and multiplication of both sides by $1/\omega$, so that the resulting equation resembles (8) with Ω replaced by $\omega\Omega$. The quantity $\omega\Omega$ or $\omega - \omega\mathbf{v} \cdot \nabla\tau$ (abbreviated here as ω^*) can be regarded as the frequency one would measure if one were moving with the ambient flow since the operation of $\partial/\partial t + \mathbf{v} \cdot \nabla$ on $\exp[-i\omega(t - \tau)]$ is equivalent to a multiplication by $-i\omega^*$. [The exponential factor with $t \to t - \tau(\mathbf{x})$ describes the predominant spatial dependence in the geometrical-acoustics approximation for a disturbance of constant frequency; $p(\mathbf{x}, t)$ should be of the form, Re $\{B(\mathbf{x}) \exp[-i\omega(t - \tau)]\}$, where $B(\mathbf{x})$ is slowly varying.]

There exists in mechanics a theory of adiabatic invariance which originated with Boltzmann in a thermodynamic context and which was subsequently further developed by Ehrenfest and Burgers‡ for application to the old quasi-

† While Eq. (8) is approximate and holds only in the geometrical-acoustics approximation, an exact acoustic-energy corollary of the linear acoustic equations for an inhomogeneous steady ambient flow does exist in the form of a sum of a time derivative and a divergence, although the resulting expression involves Clebsch potentials that are not local properties of the acoustic field: W. Möhring, "Toward an Energy Statement for Sound Propagation in Stationary Flowing Media," *Z. Angew. Math. Mech.*, **50:**T196–T198 (1960); "Energy Flux in Duct Flow," *J. Sound Vib.*, **18:**101–109 (1971); "On Energy, Group Velocity, and Small Damping of Sound Waves in Ducts with Shear Flow," ibid., **20:**93–101 (1973). A simpler corollary holds for potential isentropic flows: L. A. Chernov, "The Flux and Density of Acoustic Energy in Moving Media," *Zh. Tech. Fiz.*, **16:**733–736 (1946); R. W. Cantrell and R. W. Hart, "Interaction between Sound and Flow in Acoustic Cavities: Mass, Momentum and Energy Considerations," *J. Acoust. Soc. Am.*, **36:**697–706 (1964). Other energy statements for moving fluids are given by O. S. Ryshov and G. M. Shefter, "On the Energy of Acoustic Waves Propagating in Moving Media," *J. Appl. Math. Mech. (USSR)*, **26:**1293–1309 (1962), and by C. L. Morfey, "Acoustic Energy in Non-uniform Flows," *J. Sound Vib.*, **14:**159–170 (1971).

‡ L. Boltzmann, "On the Mechanical Significance of the Second Law of Heat Theory," *Sitzungsber. Kais. Akad. Wiss., Math. Naturwiss. Kl.*, pt. 2, **53:**195–220 (1866); "On the Priority of the Discovery of the Relation between the Second Law of the Mechanical Heat Theory and the Principle of Least Action," *Ann. Physik. Chem.*, **143:**211–230 (1871); P. Ehrenfest. "Boltzmann

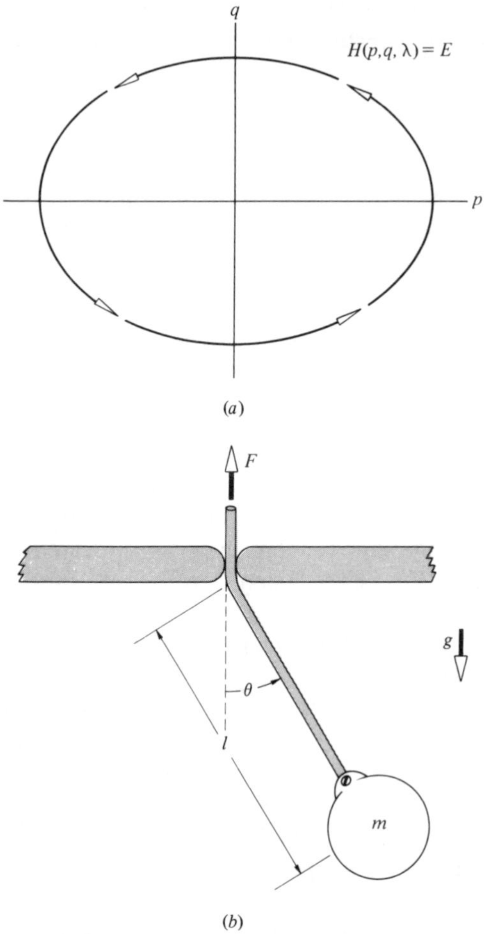

Figure 8-19 (*a*) Curve in phase plane described by $H(p,\ q,\ \lambda) = E$. (*b*) Example for which the action variable is an adiabatic invariant.

classical quantum mechanics, i.e., that before the epochal work (1923–1926) of de Broglie, Heisenberg, Schrödinger, Born, Jordan. The simplest version of this theory† applies to a 1-degree-of-freedom system (Fig. 8-19) described by a *hamiltonian* $H(q, p, \lambda)$ depending on a generalized coordinate q, on its conjugate momentum p, and on some parameter λ that varies slowly with time. For fixed

Theorem and Energy Quanta," *K. Akad. Wet. Amsterdam, Proc. Sec. Sci.*, **16**:591–597 (1914); "Adiabatic Invariants and Quantum Theory," *Ann. Phys.*, (4)**51**:327–352 (1916); J. M. Burgers, "The Adiabatic Invariants of Conditionally Periodic Systems," ibid., **52**:195–202 (1917). Some special cases are discussed by J. W. S. Rayleigh, "On the Pressure of Vibrations," *Phil. Mag.*, (6)**3**:338–346 (1902). The acoustical version of the theorem is given by W. E. Smith, "Generalization of the Boltzmann-Ehrenfest Adiabatic Theorem in Acoustics," *J. Acoust. Soc. Am.*, **50**:386–388 (1971). Its principal application to acoustics before the development of the concept of wave action was in the theory of radiation pressure. See, for example, R. T. Beyer, "Radiation Pressure: The History of a Mislabeled Tensor," *J. Acoust. Soc. Am.*, **63**:1025–1030 (1978).

† L. D. Landau and E. M. Lifshitz, *Mechanics*, Pergamon, Oxford, 1960, pp. 154–156; E. J. Saletan and A. H. Cromer, *Theoretical Mechanics*, Wiley, New York, 1971, pp. 259–263.

λ, the equation $H(q, p, \lambda) = E$, where E (identified as energy) is constant, describes a curve in a *phase space* described by coordinates p and q. It is assumed that this curve is closed. The product of $1/2\pi$ with the area enclosed in phase space by a curve of given constant E and λ defines an action variable $I(\lambda, E)$. The theory predicts that if λ varies slowly enough with t, then E varies in such a manner that I remains nearly constant in time, so one would say that action is conserved. For the harmonic oscillator, the hamiltonian is $p^2/2m + \frac{1}{2}kq^2$, where m is mass and k is spring constant. The curve $H = E$ in phase space then describes an ellipse of area $\pi(2mE)^{1/2}(2E/k)^{1/2}$. The action variable is therefore $I = E/\omega$, where $\omega = (k/m)^{1/2}$ is the natural frequency of the oscillator. Thus, for example, if k is a slowly varying function of t, one expects E/ω to remain constant throughout the motion.

The theory applies in particular to a pendulum mass m suspended by a string whose length $l(t)$ is varied slowly by pulling the string through a small hole in the ceiling. If the amplitudes of oscillation are small, the hamiltonian is $\frac{1}{2}p_\theta^2/ml^2 + \frac{1}{2}mgl\,\theta^2$, where $p_\theta = ml^2\dot{\theta}$, θ is the angular deviation of the string from the vertical and g is the acceleration due to gravity. For harmonic oscillations of frequency $\omega = (mgl/ml^2)^{1/2} = (g/l)^{1/2}$, the energy E is $\frac{1}{2}mgl\theta_{\text{max}}^2$. The adiabatic invariance of $I = E/\omega$ requires that θ_{max} change with t so that $l^{3/2}\theta_{\text{max}}$ remains constant.

Because w/ω^* resembles an action variable per unit volume, the conservation relation of Eq. (8), with $\Omega \rightarrow \omega^*$, is regarded as a law of conservation of wave action; w/ω^* is the wave action per unit volume, or wave-action density, while $(\mathbf{I} + \mathbf{v}w)/\omega^*$ is the wave-action flux.

Equation (8), with $\Omega \rightarrow \omega^*$, although here derived for circumstances of steady flow, applies† also to a wave packet of nearly constant frequency traveling in a medium whose properties are slowly varying functions of both position and time. As the packet moves, the frequency viewed by an observer at rest changes because of the time dependence of the sound speed and ambient velocity. However, if w and $\mathbf{I}$ are defined as above, and if ω^* is taken as the frequency (also time-dependent) measured by someone moving with the ambient flow, the conservation of wave action still holds. The plausibility of this assertion should be evident since what appears to be an inhomogeneous time-independent flow to someone at rest appears to be changing with time when viewed in a moving reference frame. Since w, $\mathbf{I}$, and ω^* are invariant under changes of reference frame, Eq. (8), with $\Omega \rightarrow \omega^*$, should be also. If one considers the various quantities in that equation to be functions of $\mathbf{x}'$, t' where $\mathbf{x}' = \mathbf{x} - \mathbf{v}_f t$, $t' = t$, and the frame velocity $\mathbf{v}_f$ is constant, then $\nabla = \nabla'$, $\partial/\partial t = \partial/\partial t' - \mathbf{v}_f \cdot \nabla'$, and the wave-action-conservation equation is transformed into

$$\frac{\partial}{\partial t'}\left(\frac{w}{\omega^*}\right) + \nabla' \cdot \left(\frac{\mathbf{I} + \mathbf{v}'w}{\omega^*}\right) = 0 \qquad \text{(8-6.9)}$$

† Hayes, "Energy Invariant for Geometric Acoustics in a Moving Medium." (See also on p. 375*n*.)

where $\mathbf{v}' = \mathbf{v} - \mathbf{v}_f$ represents the ambient velocity viewed in a reference frame moving with velocity $\mathbf{v}_f$ relative to the original reference frame.

The Blokhintzev Invariant

Given that one has selected a reference frame in which the ambient medium appears to be time-independent, an advantage of the law of conservation of wave action in the form of Eq. (8) is that it also applies to transient disturbances. Thus, if one sets

$$p' = P(\mathbf{x})f(t - \tau(\mathbf{x}), \xi) \qquad \mathbf{v}' = \frac{\mathbf{n}p'}{\rho c} \tag{8-6.10}$$

where f is an arbitrary function (composed, however, primarily of high frequencies) and ξ is constant along any given ray, an equation for $P(\mathbf{x})$ results from a substitution of these expressions into Eq. (8). Following this procedure and neglecting terms involving $\nabla\xi$, we obtain

$$w = \left(\frac{P^2}{\rho c^2}\right)f^2 \qquad \mathbf{I} + \mathbf{v}w = \mathbf{v}_{\text{ray}}w$$

$$\nabla \cdot \left(\frac{\mathbf{I} + \mathbf{v}w}{\Omega}\right) = f^2\nabla \cdot \left(\frac{P^2\mathbf{v}_{\text{ray}}}{\rho c^2\Omega}\right) - 2\left(\frac{P^2\mathbf{v}_{\text{ray}}}{\rho c^2\Omega}\right) \cdot \nabla\tau f \frac{\partial f}{\partial t} \tag{8-6.11}$$

However, since $\mathbf{v}_{\text{ray}} \cdot \nabla\tau = 1$ [see Eq. (8-1.3)], the second term on the right side of (11) is $-\partial(w/\Omega)/\partial t$, so Eq. (8) yields†

$$\nabla \cdot \left(\frac{P^2\mathbf{v}_{\text{ray}}}{\rho c^2\Omega}\right) = 0 \tag{8-6.12}$$

which is one of the fundamental equations of geometrical acoustics. If the ambient velocity is set to zero, this reduces to the previously derived Eq. (8-5.13).

If one integrates Eq. (12) over the volume of a ray-tube segment and follows the procedure described in the previous section, the conclusion is reached that the *Blokhintzev invariant*‡

$$\frac{P^2|\mathbf{v}_{\text{ray}}|A}{(1 - \mathbf{v} \cdot \nabla\tau)\rho c^2} = \text{const} \tag{8-6.13}$$

is constant along any given infinitesimal ray tube of variable cross-sectional area A. [Alternate versions of this conclusion result with the replacement of $\mathbf{v}_{\text{ray}}$ by $\mathbf{v} + c\mathbf{n}$ or of Ω by $c/(c + \mathbf{v} \cdot \mathbf{n})$.]

† A rigorous derivation leading to the same result follows the general procedure outlined by S. Weinberg, "Eikonal Method in Magnetohydrodynamics," *Phys. Rev.*, **126**:1899–1909 (1962).

‡ D. I. Blokhintzev, *Acoustics of a Nonhomogeneous Moving Medium*, Leningrad, 1946; trans. NACA TM 1399, National Advisory Committee for Aeronautics, Washington, especially pp. 35–40; "The Propagation of Sound in an Inhomogeneous and Moving Medium, I," *J. Acoust. Soc. Am.*, **18**:322–328 (1946).

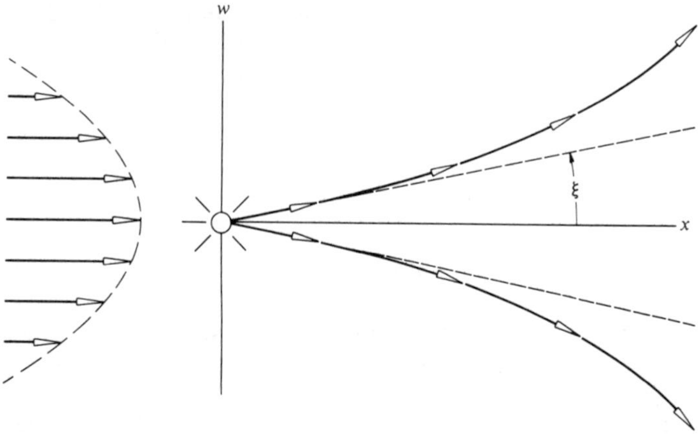

Figure 8-20 Ray paths from a point source on the axis of a symmetric jet. Here ξ is the angle the ray initially makes with the direction of flow.

Example: Point source in a jet† Sound is emanating from a small source at the origin in a medium of constant sound speed and ambient density (see Fig. 8-20). The ambient fluid velocity is in the $+x$ direction and varies with the radial coordinate $r = (y^2 + z^2)^{1/2}$ such that $v_x(r)$ has a maximum along the x axis. Describe the variation of the mean squared pressure along the x axis.

SOLUTION Because of the cylindrical symmetry, each ray leaving the source stays within a plane passing through the x axis. The refraction is therefore the same as if the ray were moving in a stratified medium; thus Eqs. (8-4.1) and (8-4.2) apply but with $z \to r$, $s_z \to s_r$. With the abbreviations M and L for v_x/c and $1/cs_x$, these equations yield

$$\frac{dr}{dx} = \frac{(L - M + 1)^{1/2}(L - M - 1)^{1/2}}{1 - M^2 + ML} \tag{8-6.14}$$

The flow Mach number $M(r)$ has a maximum at $r = 0$, so we write

$$M \approx M_0 - \tfrac{1}{2}(M_0 + 1)^2\alpha^2 r^2 \tag{8-6.15}$$

where M_0 is $M(0)$ and α is a constant. The additional factor $(M_0 + 1)^2$ is for analytical convenience. The other quantity L, in Eq. (14), is a constant for any given ray; for the ray lying on the $+x$ axis, dr/dx is 0, so the axial ray's L is $M_0 + 1$. Since we are only interested in rays within a small ray tube centered at the $+x$ axis, we accordingly set

$$L = (M_0 + 1) + \tfrac{1}{2}(M_0 + 1)^2\xi^2 \tag{8-6.16}$$

where the ray parameter ξ is considered small compared with 1.

† J. Atvars, L. K. Schubert, and H. S. Ribner, "Refraction of Sound from a Point Source Placed in an Air Jet," *J. Acoust. Soc. Am.*, **37**:168–170 (1965); L. K. Schubert, "Numerical Study of Sound Refraction by a Jet Flow, I: Ray Acoustics," ibid. **51**:434–446 (1972).

The substitution of (15) and (16) into (14) and the subsequent discard, in factors of the order of 1, of small terms proportional to $\alpha^2 r^2$ and ξ^2 yields the approximate ray-path equation

$$\frac{dr}{dx} = (\alpha^2 r^2 + \xi^2)^{1/2} \tag{8-6.17}$$

which in turn integrates to

$$r = \frac{\xi}{\alpha} \sinh \alpha x \tag{8-6.18}$$

with the condition that the ray pass through the source position. The initial slope dr/dx of the ray is ξ, but refraction causes the ray to bend away from the x axis; $\sinh \alpha x$ is larger than αx.

The cross-sectional area of the tube containing all rays with $\xi < \xi_0$ is πr^2, where r is as given by Eq. (18) with $\xi \to \xi_0$. All the other factors, except P^2, in the Blokhintzev invariant are independent of x for the ray proceeding along the x axis. Consequently, the mean squared pressure varies with x as

$$(p^2)_{\text{av},r=0} = \frac{\text{const}}{\alpha^{-2} \sinh^2 \alpha x} \tag{8-6.19}$$

For small x, this corresponds to spherical spreading (const$/x^2$), but at larger x the decrease is exponential.

The model just discussed gives a partial explanation for why the noise from a jet leaving a nozzle has an anomalous zone of relative quiet at large distances downstream and at small angles with respect to the jet's axis.

Why sound from a source near the ground is louder downwind than upwind is explained in a similar manner.† The wind velocity increases with height, so rays initially proceeding downwind in directions that are nearly horizontal are refracted down; the drop-off with distance is less than that of spherical spreading. Upwind, the opposite effect occurs.

8-7 SOURCE ABOVE AN INTERFACE

Another example illustrating some of the geometrical-acoustics concepts introduced in previous sections is that of an isotropic point source located at height h above a plane interface (Fig. 8-21). The nominal location of the interface is the $z = 0$ plane, and the source location is $(0, 0, h)$. If the interface separates two fluids,‡ both are assumed to have zero ambient fluid velocity; fluid I above the

† Stokes, "On the Effect of Wind . . . ," 1857; H. Bateman, "The Influence of Meteorological Conditions on the Propagation of Sound," *Mon. Weather Rev.*, **42**:258–265 (1914).

‡ The full-wave solution dates back to A. Sommerfeld's analysis of the analogous electromagnetic-wave problem: "On the Spreading of Waves in the Wireless Telegraphy," *Ann. Phys.*, (4)**28**:665–736 (1909). A detailed description is given by L. M. Brekhovskikh, *Waves in Layered Media*, Academic, New York, 1960, pp. 234–302. For numerical results, see M. S. Wein-

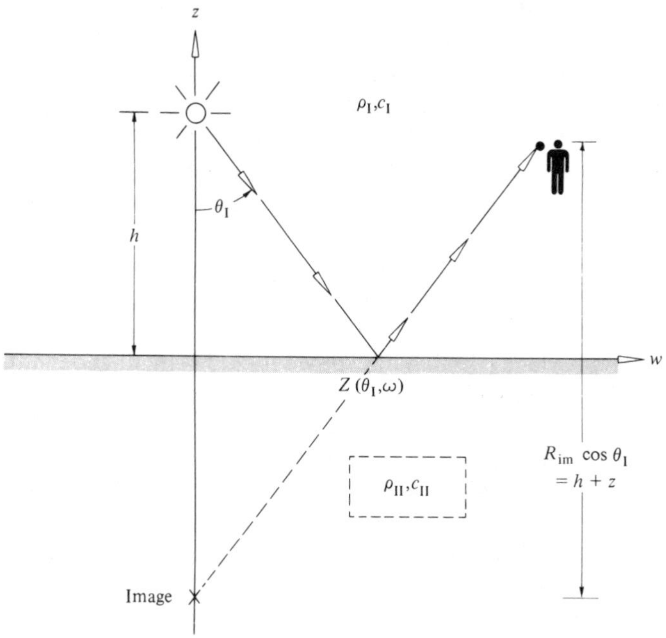

Figure 8-21 Point source above a plane interface.

interface has sound speed c_I and ambient density ρ_I; c_{II} and ρ_{II} denote the corresponding quantities below the interface. The example applies in particular to the problem of predicting the sound underwater caused by a source in air above the water's surface. Near the souce, where the direct wave predominates, the acoustic pressure p is $f(t - R/c_I)/R$, where $f(t)$ is a function characteristic of the source.

Sound Field above the Interface

In the upper medium, the received sound arrives via a direct ray and via a ray that goes from the source to the interface and back to the observation point. Because angle of incidence equals angle of reflection, this reflected ray appears to emanate from an image source at $(0, 0, -h)$. We neglect any ray displacement tangential† to the surface during reflection. Another assumption is that the

stein and A. G. Henney, "Wave Solution for Air-to-Water Sound Transmission," *J. Acoust. Soc. Am.*, **37:**899–901 (1965); J. V. McNicholas, "Lateral Wave Contribution to the Underwater Signature of an Aircraft," ibid., **53:**1755 (1973).

† A narrow beam of sound incident obliquely on a surface does undergo a tangential displacement; the cross-sectional distribution of the energy in the beam is also altered: A. Schoch, "Sideways Displacement of a Totally Reflected Ray of Ultrasound Waves," *Acustica*, **2:**18–22 (1952); M. A. Breazeale, J. Adler, and L. Flax, "Reflection of a Gaussian Utrasonic Beam from a Liquid-Solid Interface," *J. Acoust. Soc. Am.*, **56:**866–872 (1974). The effect is of minor consequence, however, for a very wide beam of sound or for a spherical wave incident on the interface.

change in wave amplitude and phase on reflection is the same as for a plane wave at the same angle of incidence. Ray-tube areas along the two rays are proportional to R^2 and R_{im}^2, respectively, where R and R_{im} are distances from the source and image source. The only modification caused by an interface that is not perfectly reflecting is that the complex amplitude of each frequency component of the reflected wave is multiplied by $\mathcal{R}(\theta_{\text{I}}, \omega)$, where $\mathcal{R}(\theta_{\text{I}}, \omega)$ is the pressure-amplitude reflection coefficient when the angle of incidence (medium I) is θ_{I}. This, according to Eq. (3-3.4) is given by

$$\mathcal{R}(\theta_{\text{I}}, \omega) = \frac{Z(\theta_{\text{I}}, \omega) - \rho_{\text{I}} c_{\text{I}}/(\cos \theta_{\text{I}})}{Z(\theta_{\text{I}}, \omega) + \rho_{\text{I}} c_{\text{I}}/(\cos \theta_{\text{I}})} \tag{8-7.1}$$

where, for a locally reacting surface, the specific impedance Z is independent of θ_{I}, while for an interface between two fluids $\rho_{\text{II}} c_{\text{II}}/Z$ is a function of $(c_{\text{II}}/c_{\text{I}}) \sin \theta_{\text{I}}$ (see Sec. 3-6). The identification for θ_{I} is such that $\cos \theta_{\text{I}}$ and $\sin \theta_{\text{I}}$ are $(h + z)/R_{\text{im}}$ and w/R_{im}. Here $w = (x^2 + y^2)^{1/2}$ is cylindrical distance from the vertical line passing through the source, and $R_{\text{im}} = [(h + z)^2 + w^2]^{1/2}$ is distance from the image source.

For waves of constant frequency, where $f(t) = \text{Re} \hat{f} e^{-i\omega t}$, the solution in the geometrical-acoustics approximation for the complex-pressure amplitude $\hat{p}$ is given, according to the discussion above, by

$$\hat{p} = \hat{f} R^{-1} e^{i(\omega/c_{\text{I}})R} + \hat{f} \mathcal{R}(\theta_{\text{I}}, \omega) R_{\text{im}}^{-1} e^{i(\omega/c_{\text{I}})R_{\text{im}}} \tag{8-7.2}$$

The validity of this is suspect whenever it predicts an unusually small value of $\hat{p}$ since any corrections based on a full-wave analysis† could then be an appreciable fraction of the total acoustic-pressure amplitude. An instance of this would be the field near $z = 0$ (such that $R \approx R_{\text{im}}$) when $\mathcal{R}(\theta_{\text{I}}, \omega)$ is close to -1. This occurs, for example, for reflection from a locally reacting surface when $\cos \theta_{\text{I}} \ll \rho_{\text{I}} c_{\text{I}}/|Z|$ (or $h + z \ll w \rho_{\text{I}} c_{\text{I}}/|Z|$). Here we exclude such cases from our consideration.

The transient solution corresponding to the above results if one takes $\hat{f}$ and $\hat{p}$ to be the Fourier transforms of $f(t)$ and $p(\mathbf{x}, t)$. After application of the Fourier integral theorem, Eq. (2-8.4), one finds

$$p = \frac{1}{R} f\left(t - \frac{R}{c_{\text{I}}}\right) + \frac{1}{R_{\text{im}}} g\left(t - \frac{R_{\text{im}}}{c_{\text{I}}}, \theta_{\text{I}}\right) \tag{8-7.3}$$

† K. U. Ingard, "On the Reflection of a Spherical Wave from an Infinite Plane," *J. Acoust. Soc. Am.*, **23**:329–335 (1951); A. Wenzel, "Propagation of Waves along an Impedance Boundary," ibid., **55**:956–963 (1974); S.-I. Thomasson, "Reflection of Waves from a Point Source by an Impedance Boundary," ibid., **59**:780–785 (1976). A principal feature of the latter formulations is a surface wave that propagates along the boundary. A detailed discussion of the limitations of the geometrical-acoustics solution is given by M. E. Delany and E. N. Bazley, "Monopole Radiation in the Presence of an Absorbing Plane," *J. Sound Vib.*, **13**:269–279 (1970). How the geometrical-acoustics model can be extended to incorporate multiple reflections is discussed by Delany and Bazley in "A Note on the Sound Field Due to a Point Source inside an Absorbent-Lined Enclosure," ibid., **14**:151–157 (1971).

where the waveform $g(t, \theta_I)$ corresponding to the reflected wave is the inverse Fourier transform of the product of $\mathcal{R}(\theta_I, \omega)$ and the Fourier transform of $f(t)$. For reflection from an interface between two fluids, when $\sin\theta_I < c_I/c_{II}$, $\mathcal{R}(\theta_I)$ is real and independent of frequency, so $g(t, \theta_I) = \mathcal{R}(\theta_I)f(t)$. If, however, $c_{II}/c_I > 1$ and $\sin\theta_I > c_I/c_{II}$, the function $g(t, \theta_I)$ is given by Eq. (3-6.12) in terms of the Hilbert transform of $f(t)$.

Field below the Interface†

If the interface separates two different fluids, the wave arrives at a point $(x, y, -d)$ at depth d below the interface along a refracted path that crosses the interface at intermediate radial distance w_i making angles θ_I and θ_{II} with the vertical above and below the interface, respectively (see Sec. 3-6). These two angles are related by Snell's law and are such that $h\tan\theta_I$ and $d\tan\theta_{II}$ are w_i and $w - w_i$. Given w, h, d, c_I, and c_{II}, these relations and Snell's law suffice to determine θ_I, θ_{II}, and w_i uniquely, regardless of whether $c_{II} > c_I$ or $c_I > c_{II}$. There is one and only one ray passing through any given point below the surface.

To determine ray-tube-area variation along such a ray, consider two rays leaving the source at angles θ_I and $\theta_I + \delta\theta_I$, both rays having the same azimuth angle ϕ (see Fig. 8-22). They cross the interface at cylindrical distances $h\tan\theta_I$ and $h\tan\theta_I + h(\sec^2\theta_I)\,\delta\theta_I$ [recall that $(d/d\theta)\tan\theta$ is $\sec^2\theta$] and subsequently propagate in the refracted directions θ_{II} and $\theta_{II} + \delta\theta_{II}$, where (take differentials of Snell's law)

$$c_I^{-1}\cos\theta_I\,\delta\theta_I = c_{II}^{-1}\cos\theta_{II}\,\delta\theta_{II} \tag{8-7.4}$$

Also, the two rays cross the plane $z = -d$ at radial distances of w and $w + \delta w$, where

$$w = h\tan\theta_I + d\tan\theta_{II} \tag{8-7.5a}$$

$$\begin{aligned}\delta w &= h\sec^2\theta_I\,\delta\theta_I + d\sec^2\theta_{II}\,\delta\theta_{II} \\ &= \left(h\sec^2\theta_I + d\frac{c_{II}}{c_I}\cos\theta_I\sec^3\theta_{II}\right)\delta\theta_I\end{aligned} \tag{8-7.5b}$$

The corresponding values of w_i and δw_i result from setting d equal to 0 in these expressions. The perpendicular separation of the two rays is $(\cos\theta_{II})\,\delta w$ at depth d.

A ray tube can be taken as all rays leaving the source with azimuth angles between ϕ and $\phi + \delta\phi$, angles with the vertical between θ_I and $\theta_I + \delta\theta_I$. Because of the cylindrical symmetry, each ray stays in the same vertical plane. Since the azimuthal width of the tube at cylindrical distance w is $w\,\delta\phi$, the

† A. A. Hudimac, "Ray Theory Solution for the Sound Intensity in Water Due to a Point Source above It," *J. Acoust. Soc. Am.*, **29**:916–917 (1957); R. J. Urick, "Noise Signature of an Aircraft in Level Flight over a Hydrophone in the Sea," ibid., **52**:993–999 (1972); R. W. Young, "Sound Pressure in Water from a Source in Air and Vice Versa," ibid., **53**:1708–1716 (1973).

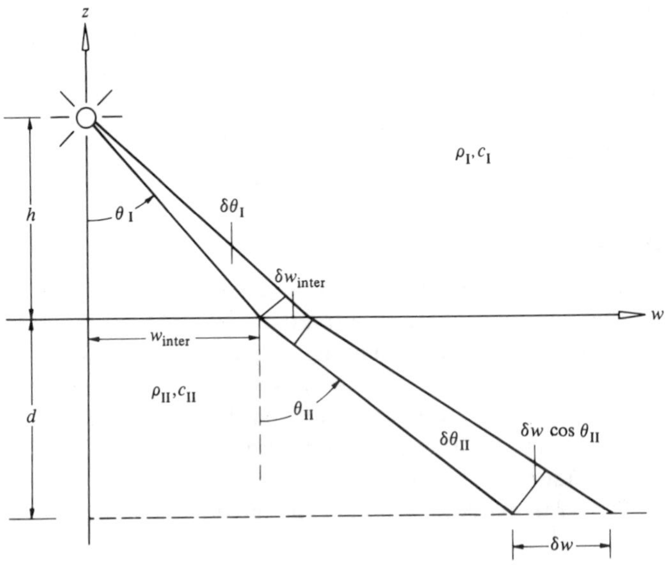

Figure 8-22 Ray geometry for two adjacent rays that propagate from a source at height h through an interface ($z = 0$) to a depth d.

ray-tube area just before the ray crosses the interface is $(w_i\,\delta\phi)(\delta w_i \cos\theta_I)$. Just after it crosses the interface it is $(w_i\,\delta\phi)(\delta w_i \cos\theta_{II})$. When it reaches depth d, the ray-tube area is $(w\,\delta\phi)(\delta w \cos\theta_{II})$. Thus, in going from just below the interface to depth d, the ray-tube area increases by a factor of $w\,\delta w/(w_i\,\delta w_i)$ and, in accord with Eq. (8-5.4), the pressure amplitude must decrease by a factor of $(w_i\,\delta w_i)^{1/2}/(w\,\delta w)^{1/2}$.

The acoustic pressure just when the ray reaches the interface is that of the direct wave alone, $R^{-1}f(t - R/c_I)$, where $R = h \sec\theta_I$, multiplied by the pressure-amplitude transmission coefficient $\mathcal{T}(\theta_I)$ appropriate to angle of incidence θ_I

$$\mathcal{T}(\theta_I) = \frac{2\rho_{II}c_{II}/(\cos\theta_{II})}{\rho_I c_I/(\cos\theta_I) + \rho_{II}c_{II}/(\cos\theta_{II})} \qquad (8\text{-}7.6)$$

Thereafter, the ray moves with speed c_{II} in direction θ_{II}; at depth d the net travel time from the source to depth d is $(h/c_I)\sec\theta_I + (d/c_{II})\sec\theta_{II}$. The time dependence of the signature must be that of $f(t)$ with t replaced by t minus this travel time.

The geometrical-acoustics solution to the problem can now be taken as pressure at the interface $(h \sec\theta_I)^{-1}f(t - (h/c_I)\sec\theta_I)$ [but with the additional shift in argument of $f(t)$ just described] times the transmission coefficient (6) times the amplitude-diminution factor $(w_i\,\delta w_i)^{1/2}/(w\,\delta w)^{1/2}$ for additional ray-tube spreading in the propagation from the interface to depth d. In this manner, one obtains

$$p = \frac{\mathcal{T}(\theta_{\rm I})f\left(t - \dfrac{h}{c_{\rm I}}\sec\theta_{\rm I} - \dfrac{d}{c_{\rm II}}\sec\theta_{\rm II}\right)}{\left(h + d\dfrac{\tan\theta_{\rm II}}{\tan\theta_{\rm I}}\right)^{1/2}\left(h\sec^2\theta_{\rm I} + d\dfrac{c_{\rm II}}{c_{\rm I}}\cos\theta_{\rm I}\sec^3\theta_{\rm II}\right)^{1/2}}$$

$$= \frac{\mathcal{T}(\theta_{\rm I})\cos\theta_{\rm I} f\left(t - \dfrac{h}{c_{\rm I}}\sec\theta_{\rm I} - \dfrac{d}{c_{\rm II}}\sec\theta_{\rm II}\right)}{\left(h + d\dfrac{c_{\rm II}}{c_{\rm I}}\cos\theta_{\rm I}\sec\theta_{\rm II}\right)^{1/2}\left(h + d\dfrac{c_{\rm II}}{c_{\rm I}}\cos^3\theta_{\rm I}\sec^3\theta_{\rm II}\right)^{1/2}} \qquad (8\text{-}7.7)$$

where in the second version use has been made of Snell's law.

In order to apply Eq. (7) to the prediction of the sound field at a given point, one must first determine $\theta_{\rm I}$ and $\theta_{\rm II}$ in terms of w, h, and d from Snell's law and from Eq. (5a). In general, this requires a numerical solution, but limiting cases are amenable to analytical approximation. In particular, if the point of observation is directly below the source ($w = 0$), one has $\theta_{\rm I} = \theta_{\rm II} = 0$ and Eq. (7) reduces to

$$p = \frac{2\rho_{\rm II}c_{\rm II}}{\rho_{\rm I}c_{\rm I} + \rho_{\rm II}c_{\rm II}} \frac{f(t - h/c_{\rm I} - d/c_{\rm II})}{(c_{\rm II}/c_{\rm I})[d + (c_{\rm I}/c_{\rm II})h]} \qquad (8\text{-}7.8)$$

This varies with depth d as a spherically symmetric wave radiating from a source at virtual height $(c_{\rm I}/c_{\rm II})h$.

8-8 REFLECTION FROM CURVED SURFACES

The major features of reflection from a curved surface are amenable to geometrical-acoustics techniques when the surface's radii of curvature are large compared with a wavelength. The chief assumption is that the reflection on any limited portion of the surface is locally the same as for plane-wave reflection from a flat surface with the same unit outward-normal vector. Here we consider the curved surface to be rigid, and we assume the ambient fluid medium to be homogeneous and without ambient flow.

General Geometrical Considerations

Let $\mathbf{x}_S$ be a point on the curved surface, let $\mathbf{n}_S(\mathbf{x}_S)$ be the unit outward normal (into the fluid) of the surface at $\mathbf{x}_S$, and let $\mathbf{n}_i(\mathbf{x}_S)$ be the direction of the incident sound ray that hits the surface at $\mathbf{x}_S$ (see Fig. 8-23). According to the law of mirrors, the unit vector $\mathbf{n}_r(\mathbf{x}_S)$ in the direction of the reflected ray must have the same tangential component as $\mathbf{n}_i(\mathbf{x}_S)$ but the opposite normal component. If one changes $\mathbf{x}_S$ to $\mathbf{x}_S + \delta\mathbf{x}_S$, the three unit vectors $\mathbf{n}_i$, $\mathbf{n}_r$, and $\mathbf{n}_S$ undergo incremental variations $\delta\mathbf{n}_i$, $\delta\mathbf{n}_r$, and $\delta\mathbf{n}_S$. For sufficiently small $\delta\mathbf{x}_S$, these are related by the differential versions of the equations requiring $\mathbf{n}_i + \mathbf{n}_r$ to be tangential to the

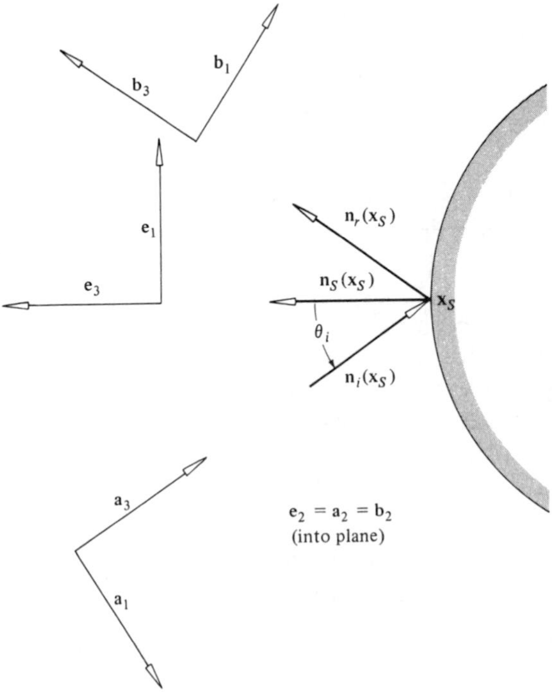

Figure 8-23 Geometry of incident and reflected rays in the vicinity of a curved surface.

surface, $\mathbf{n}_r - \mathbf{n}_i$ to be normal to the surface, and the unit vector to have unit length, i.e.,

$$(\delta\mathbf{n}_i + \delta\mathbf{n}_r) \cdot \mathbf{n}_S + (\mathbf{n}_i + \mathbf{n}_r) \cdot \delta\mathbf{n}_S = 0 \tag{8-8.1a}$$

$$(\delta\mathbf{n}_r - \delta\mathbf{n}_i) \times \mathbf{n}_S + (\mathbf{n}_r - \mathbf{n}_i) \times \delta\mathbf{n}_S = 0 \tag{8-8.1b}$$

$$\mathbf{n}_i \cdot \delta\mathbf{n}_i = \mathbf{n}_r \cdot \delta\mathbf{n}_r = \mathbf{n}_S \cdot \delta\mathbf{n}_S = 0 \tag{8-8.1c}$$

To solve the above equations for $\delta\mathbf{n}_r$, we introduce unit vectors $\mathbf{e}_1$, $\mathbf{e}_2$, $\mathbf{e}_3$, $\mathbf{a}_1$, $\mathbf{a}_2$, $\mathbf{a}_3$, $\mathbf{b}_1$, $\mathbf{b}_2$, and $\mathbf{b}_3$, where $\mathbf{e}_3$ is vector $\mathbf{n}_S$ normal to the surface at $\mathbf{x}_S$, $\mathbf{e}_1$ is the unit vector tangential to the surface in the direction of $\mathbf{n}_i + \mathbf{n}_r$ at $\mathbf{x}_S$, and $\mathbf{a}_3$ and $\mathbf{b}_3$ are unit vectors in the directions of $\mathbf{n}_i$ and $\mathbf{n}_r$, respectively, at $\mathbf{x}_S$. The unit vector $\mathbf{e}_2$ equals $\mathbf{a}_2$ and $\mathbf{b}_2$ and is such that $\mathbf{e}_1 \times \mathbf{e}_2 = \mathbf{e}_3$; the vector $\mathbf{a}_1$ is such that $\mathbf{a}_2 \times \mathbf{a}_3 = \mathbf{a}_1$. An analogous definition holds for $\mathbf{b}_1$. If θ_i denotes the angle of incidence of the wave at $\mathbf{x}_S$, the definitions are such that

$$\begin{bmatrix} \mathbf{a}_1 \\ \mathbf{a}_3 \end{bmatrix} = \begin{bmatrix} \mp\cos\theta_i & -\sin\theta_i \\ \sin\theta_i & \mp\cos\theta_i \end{bmatrix} \begin{bmatrix} \mathbf{e}_1 \\ \mathbf{e}_3 \end{bmatrix} \tag{8-8.2}$$

where the upper signs in the matrix product yield $\mathbf{a}_1$ and $\mathbf{a}_3$; the lower signs yield $\mathbf{b}_1$ and $\mathbf{b}_3$.

These unit vectors allow the substitution of $\mathbf{e}_3$, $(2\sin\theta_i)\mathbf{e}_1$, and $(2\cos\theta_i)\mathbf{e}_3$

for $\mathbf{n}_S$, $\mathbf{n}_i + \mathbf{n}_r$, and $\mathbf{n}_r - \mathbf{n}_i$ in Eqs. (1*a*) and (1*b*). Equations (1*c*) require that $\delta\mathbf{n}_i$ have only $\mathbf{a}_1$ and $\mathbf{a}_2$ components, that $\delta\mathbf{n}_r$ have only $\mathbf{b}_1$ and $\mathbf{b}_2$ components, and that $\delta\mathbf{n}_S$ have only $\mathbf{e}_1$ and $\mathbf{e}_2$ components. Insertion of these identifications into Eqs. (1*a*) and (1*b*) yields the two scalar equations

$$\mathbf{b}_1 \cdot \delta\mathbf{n}_r = -\mathbf{a}_1 \cdot \delta\mathbf{n}_i + 2\mathbf{e}_1 \cdot \delta\mathbf{n}_S \tag{8-8.3a}$$

$$\mathbf{b}_2 \cdot \delta\mathbf{n}_r = \mathbf{a}_2 \cdot \delta\mathbf{n}_i + (2\cos\theta_i)\mathbf{e}_2 \cdot \delta\mathbf{n}_S \tag{8-8.3b}$$

Next note that, near the point $\mathbf{x}_S$, any incident wavefront reaching $\mathbf{x}_S$ at time $\delta t = 0$ can be described by

$$c\,\delta t = \delta\mathbf{x} \cdot \mathbf{a}_3 + \frac{1}{2}\sum_{\mu,\nu=1}^{2} g^i_{\mu\nu}(\delta\mathbf{x} \cdot \mathbf{a}_\mu)(\delta\mathbf{x} \cdot \mathbf{a}_\nu) \tag{8-8.4}$$

where $g^i_{\mu\nu} = g^i_{\nu\mu}$ are the components of the curvature tensor of the incident wavefront and $\delta\mathbf{x} = \mathbf{x} - \mathbf{x}_S$ is here not restricted to be tangential to the reflecting surface. (The two eigenvalues of the 2×2 curvature matrix† are the reciprocals of the surface's principal radii of curvature, that is, $g_{11}g_{22} - g^2_{12} = 1/r_a r_b$ and $g_{11} + g_{22} = 1/r_a + 1/r_b$, where r_a and r_b are both positive if the surface is convex.)

The gradient of the right side of Eq. (4) is $\mathbf{n}_i(\mathbf{x}_S + \delta\mathbf{x})$, or $\mathbf{a}_3 + \delta\mathbf{n}_i$, when $\delta\mathbf{x} = \delta\mathbf{x}_S$ is tangential to the surface. Consequently, the components of $\delta\mathbf{n}_i$ are

$$\delta\mathbf{n}_i \cdot \mathbf{a}_1 = g^i_{11}\,\delta\mathbf{x}_S \cdot \mathbf{a}_1 + g^i_{12}\,\delta\mathbf{x}_S \cdot \mathbf{a}_2 \tag{8-8.5a}$$

$$\delta\mathbf{n}_i \cdot \mathbf{a}_2 = g^i_{21}\,\delta\mathbf{x}_S \cdot \mathbf{a}_1 + g^i_{22}\,\delta\mathbf{x}_S \cdot \mathbf{a}_2 \tag{8-8.5b}$$

These, along with the analogous relations for the appropriate components of $\delta\mathbf{n}_r$ and $\delta\mathbf{n}_S$, recast Eqs. (3) into the matrix relation

$$\begin{bmatrix} g^r_{11} & g^r_{12} \\ g^r_{21} & g^r_{22} \end{bmatrix} \begin{bmatrix} \delta\mathbf{x}_S \cdot \mathbf{b}_1 \\ \delta\mathbf{x}_S \cdot \mathbf{b}_2 \end{bmatrix} = \begin{bmatrix} -g^i_{11} & -g^i_{12} \\ g^i_{21} & g^i_{22} \end{bmatrix} \begin{bmatrix} \delta\mathbf{x}_S \cdot \mathbf{a}_1 \\ \delta\mathbf{x}_S \cdot \mathbf{a}_2 \end{bmatrix} + 2\begin{bmatrix} g^S_{11} & g^S_{12} \\ g^S_{21}\cos\theta_i & g^S_{22}\cos\theta_i \end{bmatrix} \begin{bmatrix} \delta\mathbf{x}_S \cdot \mathbf{e}_1 \\ \delta\mathbf{x}_S \cdot \mathbf{e}_2 \end{bmatrix} \tag{8-8.6}$$

From this equation, two equations result for each of the cases: $\delta\mathbf{x}_S$ in the $\mathbf{e}_1$ direction and $\delta\mathbf{x}_S$ in the $\mathbf{e}_2$ direction. Solution of these four equations for g^r_{11}, g^r_{12},

† If the lines on the surface corresponding to principal radii r_a and r_b coincide with the $\mathbf{a}_1$ and $\mathbf{a}_2$ directions, respectively, then $g_{11} = 1/r_a$, $g_{22} = 1/r_b$, and $g_{12} = g_{21} = 0$. If one must rotate the tangential coordinate axes counterclockwise through an angle ϕ about the surface normal for them to coincide with the principal directions, then

$$g_{11} = r_a^{-1}\cos^2\phi + r_b^{-1}\sin^2\phi \qquad g_{12} = (r_a^{-1} - r_b^{-1})\cos\phi\sin\phi$$

$$\begin{bmatrix} g_{11} & g_{12} \\ g_{21} & g_{22} \end{bmatrix} = \begin{bmatrix} \cos\phi & -\sin\phi \\ \sin\phi & \cos\phi \end{bmatrix} \begin{bmatrix} r_a^{-1} & 0 \\ 0 & r_b^{-1} \end{bmatrix} \begin{bmatrix} \cos\phi & \sin\phi \\ -\sin\phi & \cos\phi \end{bmatrix}$$

Regardless of the value of ϕ, the determinant (*gaussian curvature*) is $1/r_a r_b$, and the trace is $r_a^{-1} + r_b^{-1}$.

g^r_{21}, and g^r_{22}, yields†

$$\begin{bmatrix} g^r_{11} & g^r_{12} \\ g^r_{21} & g^r_{22} \end{bmatrix} = \begin{bmatrix} g^i_{11} & -g^i_{12} \\ -g^i_{21} & g^i_{22} \end{bmatrix} + 2\begin{bmatrix} g^S_{11} \sec \theta_i & g^S_{12} \\ g^S_{21} & g^S_{22} \cos \theta_i \end{bmatrix} \tag{8-8.7}$$

This gives us a general law for how the wavefront curvature changes on reflection from a curved surface.

When the reflecting surface is perfectly flat (zero curvature tensor), the second matrix term on the right is zero and the curvature of the reflected wavefront is the same as that of the incident wavefront. The change of sign of the off-diagonal components is because left appears right and vice versa when viewed in a mirror.

If the incident wave is a plane wave, $[g^i]$ is zero. If it is a diverging spherical wave, then $g^i_{11} = g^i_{22} = 1/R_i$ and $g^i_{12} = g^i_{21} = 0$, where R_i is the incident wave's radius of curvature at the point $\mathbf{x}_S$. Similarly, if the reflecting surface is spherical and convex, one has $g^S_{11} = g^S_{22} = 1/R_S$, $g^S_{12} = g^S_{21} = 0$. Thus for a spherical wave incident on a sphere, Eq. (7) predicts that the reflected wave is concave with its principal radii of curvature equal to $[1/R_i + 2\ (\sec\theta_i)/R_S]^{-1}$ and $[1/R_i + 2\ (\cos\ \theta_i)/R_S]^{-1}$. If $\theta_i = 0$ (normal incidence), the reflected wave is locally spherical with both radii of curvature equal to $(1/R_i + 2/R_S)^{-1}$. In particular, if the incident wave is planar ($R_i = \infty$), the two radii for the reflected wavefront are both $R_S/2$.

Ray-Tube Area after Reflection

To determine the reflected wave amplitude after subsequent propagation through a distance l, one needs the ratio $A(l)/A(0)$ of ray-tube area at distance l to that at the point of reflection, which, from Eq. (8-5.8), is

$$\frac{A(l)}{A(0)} = \frac{(K_1^{-1} + l)(K_2^{-1} + l)}{K_1^{-1}K_2^{-1}} = 1 + l(K_1 + K_2) + l^2K_1K_2 \tag{8-8.8}$$

where K_1 and K_2 are the reciprocals of the two principal radii of curvature of the wavefront just after reflection. However, since $K_1 + K_2$ is $g^r_{11} + g^r_{22}$ and K_1K_2 is the determinant of $[g^r]$, this can be rewritten

$$\frac{A(l)}{A(0)} = \det \begin{bmatrix} 1 + lg^r_{11} & lg^r_{12} \\ lg^r_{21} & 1 + lg^r_{22} \end{bmatrix}$$

$$= \det \begin{bmatrix} 1 + (g^i_{11} + 2g^S_{11} \sec \theta_i)l & (-g^i_{12} + 2g^S_{12})l \\ (-g^i_{21} + 2g^S_{21})l & 1 + (g^i_{22} + 2g^S_{22} \cos \theta_i)l \end{bmatrix} \tag{8-8.9}$$

where the second version follows from Eq. (7).

For reflection of a spherical wave from a spherical surface, the off-diagonal

† G. A. Deschamps, "Ray Techniques in Electromagnetics," *Proc.* IEEE, **60**:1022–1035 (1972). The original derivation is due to A. Gullstrand, "The General Optical Imaging System," *K. Sven. Vetenskapakad. Hangl.*, (4)**55**:1–139 (1915).

elements of $[g^i]$ and $[g^S]$ are zero, while their diagonal elements are $1/R_i$ and $1/R_S$; therefore the above reduces to

$$\frac{A(l)}{A(0)} = [1 + (R_i^{-1} + 2R_S^{-1} \sec \theta_i)l][1 + (R_i^{-1} + 2R_S^{-1} \cos \theta_i)l] \quad (8\text{-}8.10)$$

The corresponding result for when the incident wave is planar is obtained by setting $R_i^{-1} = 0$.

If the reflecting surface is a cylinder, not necessarily of circular cross section, the two principal radii of curvature at the surface are R_C and ∞. If we let ϕ denote the angle between the plane of incidence and the line passing through the reflection point parallel to the cylinder axis, such that $g_{11}^S = 0$ when $\phi = 0$ and $g_{11}^S = 1/R_C$ when $\phi = \pi/2$, then $g_{11}^S = R_C^{-1} \sin^2 \phi$, $g_{22}^S = R_C^{-1} \cos^2 \phi$, and $g_{12}^S = g_{21}^S = \pm R_C^{-1} \sin \phi \cos \phi$. Consequently, when a spherical wave is incident, Eq. (9) reduces to

$$\frac{A(l)}{A(0)} = (1 + R_i^{-1}l)[1 + R_i^{-1}l + 2lR_C^{-1}N(\phi, \theta_i)] \quad (8\text{-}8.11)$$

$$N(\phi, \theta_i) = \sin^2 \phi \sec \theta_i + \cos^2 \phi \cos \theta_i = \frac{1 - (\mathbf{n}_i \cdot \mathbf{e}_C)^2}{-\mathbf{n}_S \cdot \mathbf{n}_i} \quad (8\text{-}8.12)$$

where $\mathbf{e}_C$ is the unit vector parallel to the cylinder axis. Again, the expression appropriate to when a plane wave is incident results with $R_i^{-1} \to 0$.

With $A(l)/A(0)$ determined, the pressure signature associated with the reflected wave is

$$p_r(\mathbf{x}_S + \mathbf{n}_r l, t) = \left[\frac{A(0)}{A(l)}\right]^{1/2} p_i\left(\mathbf{x}_S, t - \frac{l}{c}\right) \quad (8\text{-}8.13)$$

This corresponds to what would be received at a point $\mathbf{x} = \mathbf{x}_S + \mathbf{n}_r l$, where $\mathbf{n}_r = \mathbf{n}_i - 2(\mathbf{n}_S \cdot \mathbf{n}_i)\,\mathbf{n}_S$ is related to $\mathbf{n}_i$ by the law of mirrors.

Echoes from Curved Surfaces

As an application of the above formulation, we consider a small source at a distance R from the nearest point on a curved surface. At that point, the surface has principal radii of curvature $R_{S,\mathrm{I}}$ and $R_{S,\mathrm{II}}$. If $f(t - r/c)/r$ denotes the incident wave, the echo returned back to the source will be

$$p_r = \left[\frac{A(0)}{A(R)}\right]^{1/2} \frac{f(t - 2R/c)}{R} \quad (8\text{-}8.14)$$

In this example, it is possible to orient the coordinate system so that $[g^S]$ is diagonal. The angle θ_i is 0; l and R_i are both R, so Eq. (9) yields

$$p_r = \frac{f(t - 2R/c)}{2R(1 + R/R_{S,\mathrm{I}})^{1/2}(1 + R/R_{S,\mathrm{II}})^{1/2}} \quad (8\text{-}8.15)$$

Thus, the echo will be smaller by $10 \log [(1 + R/R_{S,\mathrm{I}})(1 + R/R_{S,\mathrm{II}})]$ dB relative

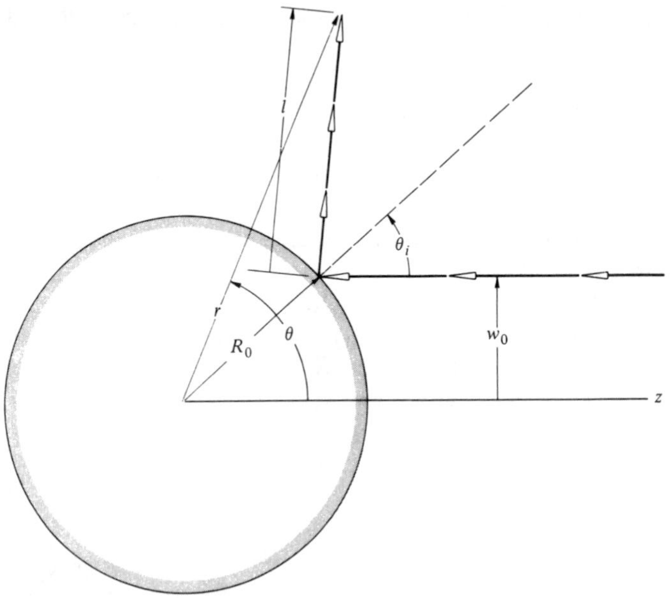

Figure 8-24 Parameters used in the geometrical-acoustics theory of reflection from a rigid sphere.

to what would be expected for reflection from a flat surface. If R is much less than either $R_{S,\text{I}}$ or $R_{S,\text{II}}$, the surface may be idealized as flat.

Sound Beam Incident on a Sphere

A collimated beam of sound is incident from the $+z$ direction on a sphere of radius R_0 (see Fig. 8-24), the beam's diameter being larger than $2R_0$. The time-averaged intensity of the incident wave in the vicinity of the sphere is I_i, and the intensity I_r of the reflected wave is to be estimated at radial distances r much larger than R_0. We are here interested in the short-wavelength limit† and accordingly use geometrical acoustics.

The ray of sound incident at a distance w_0 (less than R_0) from the z axis will strike the surface at an angle of incidence θ_i where $\theta_i = \sin^{-1}(w_0/R_0)$ and will reflect such that it makes an angle of $2\theta_i$ with the z axis. After a subsequent propagation distance l, it will pass through a point at $z = R_0 \cos\theta_i + l\cos 2\theta_i$, $w = R_0 \sin\theta_i + l\sin 2\theta_i$, or, in spherical coordinates, where $r^2 = R_0^2 + l^2 + 2R_0 l \cos\theta_i$ and $\theta = \tan^{-1}(w/z)$. If $l \gg R_0$, then $r \approx l + R_0 \cos\theta_i$ and $\theta \approx 2\theta_i$. Thus, we can set $\theta_i \approx \theta/2$, $l \approx r - R_0 \cos(\theta/2)$, so with $R_i^{-1} = 0$ and $R_S = R_0$,

† Full-wave results for intermediate values of kR_0 are tabulated by H. Stenzel, "On the Disturbance Caused by a Sound Field Incident on a Rigid Sphere," *Elektr. Nachrichtentech.*, **15**:71–78 (1938); *Leitfaden zur Berechnung von Schallvorgängen*, Springer, Berlin, 1939, pp. 104–114. Some of Stenzel's results are given in Sec. 9-1 of the present text.

Eq. (10) becomes

$$\frac{A(l)}{A(0)} \approx \left(2\frac{r}{R_0}\sec\frac{\theta}{2} - 1\right)\left(1 - 2\cos^2\frac{\theta}{2} + \frac{2r}{R_0}\cos\frac{\theta}{2}\right)$$

The quantity $[A(0)/A(l)]^{1/2}$ is thus approximately $R_0/2r$, providing θ is such that $2r/R_0 \gg \sec(\theta/2)$. (This excludes angles close to π). The net travel time along the path from where the incident ray crosses the plane $z = R_0$ to the point (r, θ) is $[l + R_0(1 - \cos\theta_i)]/c \approx r/c + (R_0/c)[1 - 2\cos(\theta/2)]$. Consequently, Eq. (13) yields

$$p_r(r, \theta, t) \approx \frac{R_0}{2r} p_i\left(0, t - \frac{r}{c} + 2\frac{R_0}{c}\cos\frac{\theta}{2}\right) \tag{8-8.16}$$

where $p_i(0, t)$ is what the incident pressure would be at the origin without the sphere. The cos $(\theta/2)$ factor in the retarded time implies that surfaces of constant r are not surfaces of constant phase, but the phase variation should be negligible for transverse displacements of the order of a wavelength.

If the incident plane wave is of constant frequency or is a superposition of constant-frequency waveforms, we can identify I_i as $(p_i^2/\rho c)_{\text{av}}$ and I_r as $(p_r^2/\rho c)_{\text{av}}$, giving

$$I_r \approx \left(\frac{R_0}{2r}\right)^2 I_i \tag{8-8.17}$$

for values of θ somewhat less than π. The net acoustic power reflected by the sphere is therefore

$$\mathcal{P}_r = 4\pi r^2 I_r = (\pi R_0^2) I_i \tag{8-8.18}$$

which is the net acoustic power incident on the front (projected area πR_0^2) of the sphere.

PROBLEMS

8-1 Show that the unit normal $\mathbf{n}$ to a wavefront varies with time along a ray according to the differential equation (in cartesian coordinates)

$$\frac{d\mathbf{n}}{dt} = -[\nabla - \mathbf{n}(\mathbf{n}\cdot\nabla)]c - \sum_k n_k[\nabla - \mathbf{n}(\mathbf{n}\cdot\nabla)]v_k$$

$$\frac{dn_x}{dt} = \left[-(n_y^2 + n_z^2)\frac{\partial}{\partial x} + n_x n_y \frac{\partial}{\partial y} + n_x n_z \frac{\partial}{\partial z}\right](c + \mathbf{n}\cdot\mathbf{v})$$

where n_x, n_y, and n_z are formally treated as constant in carrying out the differentiation. [R. Engelke, *J. Acoust. Soc. Am.*, **56**:1291–1292 (1974).]

8-2 Show that when there is no ambient flow a ray path satisfies the differential equation

$$\frac{d}{dl}\left(c^{-1}\frac{d\mathbf{x}}{dl}\right) = \nabla c^{-1}$$

where l is distance along the path. (P. G. Frank, P. G. Bergmann, and A. Yaspan, "Ray Acoustics," reprinted in R. B. Lindsay, *Physical Acoustics,* Dowden, Hutchinson and Ross, Stroudsburg, Penn., 1974).

8-3 Show that the differential equation in Prob. 8-2 results from Fermat's principle. Carry through the derivation in detail starting with Eq. (8-1.13) with $\mathbf{v}$ set to zero.

8-4 Show that the ray-tracing equations (8-1.10) follow from the relations on p. 375*n*.

8-5 In an isentropic ideal gas with steady irrotational ($\nabla \times \mathbf{v} = 0$) ambient flow, the sound speed c and ambient velocity $\mathbf{v}$ are related by

$$\frac{2c^2}{\gamma - 1} + v^2 = K$$

where K is a constant. Verify this relation and show that the ray-tracing equations lead to

$$\frac{d\mathbf{n}}{dt} = \mathbf{n}\times\left(\mathbf{n}\times\left\{\left[\mathbf{n} - \frac{(\gamma-1)}{2}\frac{\mathbf{v}}{c}\right]\cdot\nabla\right\}\mathbf{v}\right)$$

8-6 A ray is moving in a cylindrically symmetric medium for which c depends only on the radial distance w and for which $\mathbf{v} = 0$. For a ray path lying in the $z = 0$ plane, verify that $w\mathbf{e}_\phi \cdot \mathbf{n}/c$ is constant along the ray.

8-7 For a quiescent medium in which sound speed varies only with radial distance r (spherical coordinates), determine whether or not a given ray path always lies within a single plane.

8-8 For a medium whose ambient properties are described in cylindrical coordinates by $c = c(r)$, $v_\phi = u(r)$, $v_r = v_z = 0$, determine what ray properties are constant along a given ray. (What replaces Snell's law?) [R. B. Lindsay, *J. Acoust. Soc. Am.*, **20:**89–94 (1948); R. F. Salant, ibid., **46:**1153–1157 (1969).]

8-9 Supply all necessary algebraic details for the proof that the ray-tracing equation (8-1.10*b*) follows from the Euler-Lagrange equations (8-1.15) and from Eq. (8-1.14).

8-10 Use Fermat's principle to prove that angle of incidence equals angle of reflection.

8-11 Use Fermat's principle to prove that when source and listener lie on opposite sides of a plane interface separating two dissimilar homogeneous quiescent fluids, angle of incidence and angle of refraction of the connecting ray path are related by Snell's law.

8-12 Two points are at equal distances L from the center of a solid sphere of radius R. They are on opposite sides of the sphere and lie on a common axis ($L > R$). Given that the ambient medium has constant sound speed c and no flow, determine the minimum travel time between the two points. What is the corresponding ray path?

8-13 A wavefront moving in the $+z$ direction in a homogeneous nonmoving medium is described by

$$z = \frac{x^2/2R}{1 + 10x^2/R^2}$$

at $t = 0$. Sketch the wavefront at times $0.9R/c$, $1.0R/c$, and $1.1R/c$ and discuss possible physical interpretations of the results.

8-14 A wavefront moving in the $+z$ direction in a homogeneous nonmoving medium is described at time $t = 0$ by $z = f(x)$.

(*a*) Show that the ray passing through the point $x = \alpha$, $z = f(\alpha)$ at time $t = 0$ will graze a caustic at time

$$t = \frac{(1 + f_\alpha^2)^{3/2}}{cf_{\alpha\alpha}}$$

(given $f_{\alpha\alpha} > 0$).

(*b*) Show also that the caustic surface is described by the parametric equations

$$x = \alpha - \frac{f_\alpha(1 + f_\alpha^2)}{f_{\alpha\alpha}} \qquad z = f + \frac{1 + f_\alpha^2}{f_{\alpha\alpha}}$$

(*c*) Determine and plot the caustic surface for the example described in Prob. 8-13.

8-15 Given a model atmosphere without winds for which $c(z)/c_0$ is 1 for $0 < z < H$ and is $0.9 + 0.1z/H$ for $z > H$, determine the horizontal skip distance $R(\theta_0)$ versus initial angle of incidence θ_0. Is there a minimum range for the reception of abnormal sound on the ground? Assume that the source is on the ground. [L. M. Brekhovskikh, *Sov. Phys. Usp.*, **70**:159–166 (1960).]

8-16 A sound source is at $x = 0$, $y = 0$, $z = h$ above a rigid ground in a medium for which $c(z)$ is described up to any height of interest by $(1 - z/H)c_0$, where $H > h$.

(*a*) Show that points on the ground at horizontal distances greater than $(2hH - h^2)^{1/2}$ do not receive any direct waves.

(*b*) What broken ray path conforming to Fermat's principle would connect the source with a point on the ground at a range greater than $(2hH - h^2)^{1/2}$?

(*c*) Determine an expression for the travel time along such a ray path.

8-17 A stratified medium without ambient flow has a sound speed $c(z)$ given by $c_0 \cosh(z/H)$. Determine the ray path in the xz plane that passes through the origin making an angle of θ_0 with respect to the vertical.

8-18 A source and receiver are separated by a distance d and are at equal heights h above the ground. The sound speed $c(z)$ increases linearly with height as $c_0 + \alpha z$. Let a particular ray be reflected at the surface once and only once between source and receiver and let the reflection point be at a horizontal distance x from the source.

(*a*) Show that x satisfies the cubic equation

$$2x^3 - 3dx^2 + (2b^2 + d^2)x - b^2 d = 0$$

where $b^2 = h^2 + 2h/\gamma$ and $\gamma = \alpha/c_0$.

(*b*) Determine the possible ray paths corresponding to the roots of this equation. Under what circumstances are three different paths possible? (Embleton, Thiessen, and Piercy, "Propagation in an Inversion.")

8-19 A model for an underwater surface channel takes sound speed c as increasing linearly with depth z, such that $c = c_0 + \alpha z$.

(*a*) Show that if the sound source is at the surface, a ray making initial angle θ_0 with the vertical has a path given in parametric form through a parameter θ by

$$x = x_n(\theta, \theta_0) = nR(\theta_0) + c_0 \frac{\cos\theta_0 - \cos\theta}{\alpha \sin\theta_0}$$

$$z = z_n(\theta, \theta_0) = c_0 \frac{\sin\theta - \sin\theta_0}{\alpha \sin\theta_0} \qquad R(\theta_0) = \frac{2c_0 \cot\theta_0}{\alpha}$$

for $nR(\theta_0) < x < (n + 1)R(\theta_0)$ and where θ ranges from θ_0 to $\pi - \theta_0$. Here $n = 0, 1, 2, \ldots$ defines the nth branch of the ray; $R(\theta_0)$ is the ray's skip distance.

(*b*) Show that caustics correspond to the lines

$$x^2 = 4n(n + 1)\left(\frac{2c_0 z}{\alpha} + z^2\right)$$

for $n = 1, 2, \ldots$. [D. Raphael, *J. Acoust. Soc. Am.*, **48**:1249–1256 (1970).]

8-20 A sound source at the origin is surrounded by a medium for which $c(z)$ is $c_0(1 - z/H)$ and $\rho(z)$ is constant for a wide range of altitudes both above and below the source. If $\mathcal{P}$ is the power radiated by the source, what would one expect for the mean squared acoustic pressure at a horizontal distance x from the source?

8-21 For the circumstances described in Prob. 8-20 determine whether any of the rays leaving the source encounter a caustic.

8-22 (*a*) Show that the wavefronts for the circumstances described in Prob. 8-20 are given by

$$\tau(w, z) = \frac{2H}{c_0} \tanh^{-1}\left[\frac{w^2 + z^2}{(2H - z)^2 + w^2}\right]^{1/2}$$

where w corresponds to horizontal distance.

(*b*) Verify that each wavefront is a sphere whose center lies on the z axis. [D. H. Wood, *J. Acoust. Soc. Am.*, **47**:1448–1452 (1970).]

8-23 A source of sound lies a distance d below the water surface. In the absence of reflections from the air-water interface the acoustic pressure would be $f(t - R/c_w)/R$, where R is the distance from the source and c_w is the water's speed of sound. The sound speed c_a in the atmosphere is constant, but the ambient density ρ_a varies with height z as $\rho_{a,0}e^{-z/H}$, where H is a constant.

(*a*) Using geometrical-acoustics techniques, determine the acoustically induced fluid velocity at height $10H$ directly above the source.

(*b*) Suppose a source of the same power output is placed just above the surface. Would it cause a greater or a smaller disturbance at the considered altitude than the subsurface source does? (Take d to be much less than H.)

8-24 An intrinsically omnidirectional point source lies at the origin in an unbounded medium for which sound speed $c(z)$ and ambient density $\rho(z)$ vary only with height z. Show that the mean squared acoustic pressure along the z axis is

$$(p^2)_{\text{av}} = \frac{\mathcal{P}_{\text{av}}\rho(z)c(z)c^2(0)}{4\pi\left(\int_0^z c\,dz\right)^2}$$

where $\mathcal{P}_{\text{av}}$ is the time-averaged acoustic power output of the source.

8-25 A plane interface $z = 0$ separates a medium with no ambient flow (c_{I}, ρ_{I}, $z < 0$) from one with constant ambient horizontal flow velocity (c_{II}, ρ_{II}, $\mathbf{v}_{\text{II}}$, $z > 0$). Prove that if a plane wave is incident from the first medium, the time-average rate at which wave action arrives per unit interface area with the incident wave equals the sum of the corresponding quantities carried away by the reflected and transmitted waves.

8-26 (*a*) Show that with the neglect of gravity and if the ambient state is isentropic (s_0 constant) and irrotational ($\nabla \times \mathbf{v}_0 = 0$), Eqs. (8-6.2) lead to

$$\frac{\partial \mathbf{v}'}{\partial t} + \nabla\left(\mathbf{v}_0 \cdot \mathbf{v}' + \frac{p'}{\rho_0}\right) = 0 \qquad \frac{\partial \rho'}{\partial t} + \nabla \cdot (\rho_0\mathbf{v}' + \mathbf{v}_0\rho') = 0 \qquad \nabla \times \mathbf{v}' = 0 \qquad p' = \rho' c^2$$

(*b*) Show that these equations have the corollary

$$\frac{\partial \mathcal{W}}{\partial t} + \nabla \cdot \mathbf{J} = 0$$

$$\mathcal{W} = \tfrac{1}{2}\rho_0(v')^2 + \frac{(p')^2}{2\rho_0 c^2} + \frac{p'\mathbf{v}' \cdot \mathbf{v}_0}{c^2}$$

$$\mathbf{J} = (p' + \mathbf{v}_0 \cdot \mathbf{v}'\,\rho_0)\left(\mathbf{v}' + \frac{p'\mathbf{v}_0}{\rho_0 c^2}\right)$$

(*c*) Is the energy statement in part (*b*) consistent with the wave-action-conservation law of Eq. (8-6.8)? (Chernov, "The Flux and Energy Density. . . .")

8-27 The generalization of the Webster horn model, when a duct of cross-sectional area $A(x)$ has an ambient flow $v_0(x)$, is

$$\frac{A}{c^2}\frac{\partial p'}{\partial t} + \frac{\partial}{\partial x}\left[A\left(\rho_0 v' + \frac{v_0 p'}{c^2}\right)\right] = 0$$

$$\frac{\partial v'}{\partial t} + \frac{\partial}{\partial x}\left(\frac{p'}{\rho_0} + v_0 v'\right) = 0$$

(*a*) Derive these equations from Eqs. (8-6.2), making whatever approximations are necessary.

(*b*) Determine an energy corollary from these equations.

(*c*) Verify that the energy corollary is consistent with the wave-action-conservation principle when waves are presumed to be propagating in the $+x$ direction without reflection.

(*d*) What is the Blokhintzev invariant for this model?

8-28 A plastic lens is to be placed on a transducer face to focus an ultrasound beam on a point 30 cm distant. The beam propagates through water, sound speed 1500 m/s; the plastic has sound speed 2600 m/s and density 1200 kg/m^3. Using geometrical-acoustics concepts (such as Fermat's principle), design a lens-thickness–versus–radius profile that should accomplish the focusing.

8-29 For the example discussed in Sec. 8-8 of sound reflection from a rigid sphere, determine a simple approximate expression for the geometrical-acoustics prediction of the field near the shadow-zone boundary ($w - R_0 \ll R_0$, $z \le 0$). Take the incident wave to be of constant frequency with a complex pressure amplitude $\hat{p}_i$ of Pe^{-ikz}, where P is a constant, and take into account the interference of the reflected and incident waves. Assume that kR_0 is large and use cylindrical coordinates.

8-30 A point source is at distance $d = 4\lambda$ from the axis of a rigid cylinder of radius R_C. Take R_C to be 3λ, the cylinder to be aligned along the x axis, and the source to be at $(0, 0, d)$.

(*a*) Determine and sketch the far-field radiation pattern of the source-cylinder combination in the plane $y = 0$.

(*b*) What is the corresponding pattern in the plane $x = 0$? Use the geometrical-acoustics approximation but take into account the interference of direct and reflected waves.

8-31 A source is at height $H/10$ above the ground in an atmosphere where the sound speed c is $c_0(1 - z/H)$. The ground is locally reacting and has specific impedance $5\rho_0 c_0$. The source is intrinsically omnidirectional and has a time-averaged power output $\mathcal{P}$. Determine the geometrical-acoustics prediction for the mean squared acoustic pressure on the ground as a function of horizontal distance w from the source.

8-32 *Spherical aberration.* A plane wave proceeding originally in the $-z$ direction reflects from a hemispherical bowl described by $z = -(R_0^2 - w^2)^{1/2}$, where R_0 is radius of the bowl and w is radial distance in cylindrical coordinates. Discuss the location and shape of whatever caustics are formed by the reflected wave.

CHAPTER

NINE

SCATTERING AND DIFFRACTION

An obstacle or inhomogeneity in the path of a sound wave causes *scattering* if secondary sound spreads out from it in a variety of directions. Such an inhomogeneity could be, for example, a fish in the ocean, a region of turbulence in the atmosphere, or a red corpuscle in a bloodstream. The smearing of propagation directions that results when a sound beam reflects from a rough surface is also recognized as scattering.

The present chapter begins with a discussion (Sec. 9-1) of scattering of sound by small isolated bodies and inhomogeneities. The basic experimental configurations for the study of scattering are then discussed in Sec. 9-2. The *Doppler effect* and, in particular, the frequency shift caused by a scatterer's motion occupy our attention in Sec. 9-3.

The remainder of the chapter is concerned with *diffraction* phenomena. The term as used here applies to contexts where major features of the propagation and of the overall acoustic field are well described by ray-acoustic concepts. Diffraction is then the label assigned to those features of the field which the ray model fails to explain. A common example is the field in the *shadow zone* of a large solid object obstructing direct rays radiating from the source.

Examples of diffraction previously discussed in the present text are transverse spreading (Sec. 5-8) of a beam of sound radiated by a baffled piston in a wall and transmission (Sec. 7-5) through an orifice. The analysis of diffraction phenomena resumes here with discussions of fields near caustics (Sec. 9-4) and of the penetration of sound into shadow zones bordered by limiting rays that tangentially graze smooth surfaces (Sec. 9-5).

Subsequent sections analyze the fundamental problem of diffraction by a wedge, which furnishes a building block for synthesis of models for diffraction by objects whose sides meet at edges. Limiting cases of high-frequency diffraction introduce the basic vocabulary associated with the subject and serve as benchmarks for the estimation of magnitudes and for the interpretation of experiments.

9-1 BASIC SCATTERING CONCEPTS

A dominant feature in many scattering phenomena is that (except when resonances are excited) low frequencies scatter much less than high frequencies. The understanding of this led Tyndall and Rayleigh† to an explanation for the color of the sky. Light from the sky is scattered light; higher-frequency blue light scatters more than lower-frequency red light; hence the sky is blue.

Low-frequency (small ka) scattering is often referred to as *Rayleigh scattering* because of Rayleigh's fundamental contributions to the basic theory, which he developed for acoustics‡ as well as for optics.

Scattering by a Rigid Object§

A prototype for Rayleigh scattering is a constant-frequency plane wave proceeding in direction $\mathbf{e}_k$ (wave-number vector $\mathbf{k} = k\mathbf{e}_k$) that impinges on a rigid immovable body centered at the origin (see Fig. 9-1). The overall acoustic pressure is written

$$\hat{p} = Be^{i\mathbf{k}\cdot\mathbf{x}} + \hat{p}_{sc}(\mathbf{x}) \qquad (9\text{-}1.1)$$

where B is the peak amplitude of the incident wave p_i and $\hat{p}_{sc}(\mathbf{x})$ is the scattered wave's complex amplitude.

The function $\hat{p}_{sc}(\mathbf{x})$ satisfies the Helmholtz equation and the Sommerfeld radiation condition. Also, the $\nabla\hat{p}\cdot\mathbf{n} = 0$ requirement for a rigid surface imposes

$$\nabla\hat{p}_{sc}\cdot\mathbf{n} = -iBe^{i\mathbf{k}\cdot\mathbf{x}}\mathbf{k}\cdot\mathbf{n} \qquad (9\text{-}1.2)$$

† J. W. Strutt, Lord Rayleigh, "On the Light from the Sky, Its Polarization and Colour," *Phil. Mag.*, (4)**41**:107–120 (1871); "On the Transmission of Light through an Atmosphere Containing Small Particles in Suspension, and on the Origin of the Blue of the Sky," ibid., (5)**47**:375–384 (1899); V. Twersky, "Rayleigh Scattering," *Appl. Opt.*, **3**:1150–1162 (1964).

‡ J. W. S. Rayleigh, "Investigation of the Disturbance Produced by a Spherical Obstacle on the Waves of Sound," *Proc. Lond. Math. Soc.*, **4**:253–283 (1872); "On the Passage of Waves through Apertures in Plane Screens and Allied Problems," *Phil. Mag.*, (5)**43**:259–272 (1897).

§ H. Lamb, *The Dynamical Theory of Sound*, 2d ed., 1925, reprinted by Dover, New York, 1960, pp. 244–248; J. Van Bladel, "On Low-Frequency Scattering by Hard and Soft Bodies," *J. Acoust. Soc. Am.*, **44**:1069–1073 (1968); D. A. Darling and T. B. A. Senior, "Low-Frequency Expansions for Scattering by Separable and Nonseparable Bodies," ibid., **37**:228–234 (1965); A. F. Stevenson, "Solution of Electromagnetic Scattering Problems as Power Series in the Ratio (Dimension of Scatterer)/Wavelength," *J. Appl. Phys.*, **24**:1134–1151 (1953). The discussion in the present text is indebted to F. Obermeier, "Determination of the Scattering of a Plane Sound Wave by a Hard Sphere with the Assistance of the Method of Matched Asymptotic Expansions," unpublished (c. 1975).

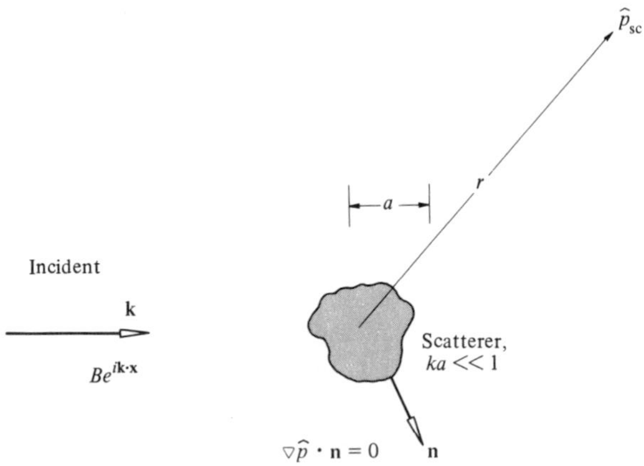

Figure 9-1 Scattering of a plane wave by a rigid immovable object small compared with a wavelength.

at the body's surface S (unit normal $\mathbf{n}$ pointing into fluid). Determination of $\hat{p}_{sc}$ is equivalent to determination of the field of a vibrating body of the same size and shape whose normal velocity is the negative of what is associated with the incident wave.

The expansion of the exponent in Eq. (2) to first order in k yields

$$\hat{\mathbf{v}}_{sc}\cdot\mathbf{n} = -\frac{B}{\rho c}\,\mathbf{e}_k\cdot\mathbf{n} - i\,\frac{B}{\rho c}\,(\mathbf{k}\cdot\mathbf{x})\mathbf{e}_k\cdot\mathbf{n} \tag{9-1.3}$$

The first term corresponds to rigid-body translation back and forth parallel to $\mathbf{e}_k$ with a velocity amplitude $-B/\rho c$ and, taken by itself, produces dipole radiation (to lowest nonvanishing order in ka, as explained in Sec. 4-7). Although the second term, which leads to monopole radiation, is smaller than the first by a factor of the order of ka, both have comparable influence on the far field because monopoles radiate more efficiently than dipoles. An approximation to the lowest order in ka results with the discard of terms of higher than the first order and with the neglect of higher-order multipoles for the two remaining terms.

The monopole portion, calculated with the complex-amplitude version of the leading term in Eq. (4-7.10), yields

$$\hat{p}_{sc,mono} = \frac{-k^2Be^{ikr}}{4\pi r}\iint(\mathbf{e}_k\cdot\mathbf{x})\mathbf{e}_k\cdot\mathbf{n}\,dS = \frac{-k^2BV}{4\pi r}\,e^{ikr} \tag{9-1.4}$$

with the aid of Gauss' theorem and the identity $\nabla\cdot[(\mathbf{e}_k\cdot\mathbf{x})\mathbf{e}_k] = 1$; here V denotes the total scattering body volume.

The dipole term results from Eq. (4-7.12), whose complex-amplitude version with the appropriate substitution from Eq. (3) yields

$$\hat{p}_{sc,dipole} = \frac{-ikB}{4\pi}\,\nabla\cdot[(\mathbf{M}\cdot\mathbf{e}_k)r^{-1}e^{ikr}] \tag{9-1.5a}$$

$$M_{\mu\nu} = V\delta_{\mu\nu} + W_{\mu\nu} \qquad \mathbf{M}\cdot\mathbf{e}_k = \sum_{\mu\nu}\mathbf{e}_\mu M_{\mu\nu}\mathbf{e}_\nu\cdot\mathbf{e}_k \tag{9-1.5b}$$

The matched asymptotic expansion procedure outlined in Sec. 4-7 guarantees that the tensor **W** is derivable from the solution for the incompressible potential flow caused by translational motion of the body. The *entrained-mass tensor*† $\rho\mathbf{W}$ is such that $\rho\mathbf{W}\cdot\dot{\mathbf{v}}_C$ is the force **F** exerted on the fluid by the body when it experiences acceleration $\dot{\mathbf{v}}_C$. The necessity for a tensor arises because **F** may have components transverse to $\dot{\mathbf{v}}_C$.

Since the components of **M** scale as a^3, the monopole and dipole terms are of comparable magnitude. The sum of these,

$$\hat{p}_{sc} = \frac{-k^2B}{4\pi}\left[V - \mathbf{e}_r\cdot\mathbf{M}\cdot\mathbf{e}_k\left(1+\frac{i}{kr}\right)\right]\frac{e^{ikr}}{r} \tag{9-1.6}$$

implies a far-field scattered-wave amplitude proportional to k^2a^3/r.

Particular matrix expressions for the tensor **M**,

$$\tfrac{3}{2}V\begin{bmatrix}1 & 0 & 0\\ 0 & 1 & 0\\ 0 & 0 & 1\end{bmatrix} \qquad \tfrac{8}{3}a^3\begin{bmatrix}0 & 0 & 0\\ 0 & 0 & 0\\ 0 & 0 & 1\end{bmatrix} \tag{9-1.7}$$

correspond, respectively, to a sphere [see Eq. (4-2.14)] and to a thin disk of radius a oriented transverse to the z axis [see Eq. (4-8.11)]. The reciprocity principle guarantees that such matrices are symmetric, so that selection of the coordinate system can be such that the matrix is diagonal. For a body of revolution centered at the z axis, the matrix is also such that $M_{xx} = M_{yy}$ (see Fig. 9-2).

The versions‡ of Eq. (6) that result for the sphere and disk examples (with $\mathbf{e}_k = \mathbf{e}_z$) just mentioned are, respectively,

$$\hat{p}_{sc} = \frac{-k^2B}{4\pi}\left(\tfrac{4}{3}\pi a^3\right)\left[1 - \tfrac{3}{2}\cos\theta\left(1+\frac{i}{kr}\right)\right]\frac{e^{ikr}}{r} \tag{9-1.8}$$

$$\hat{p}_{sc} = \frac{k^2B}{4\pi}\frac{8a^3}{3}\cos\theta\left(1+\frac{i}{kr}\right)\frac{e^{ikr}}{r} \tag{9-1.9}$$

† The symbols adopted here are those of T. B. A. Senior, "Low-Frequency Scattering," *J. Acoust. Soc. Am.*, **53**:742–747 (1973). Senior refers to **M** as the *magnetic-polarizability tensor* and to **W** as the *virtual-mass tensor*.

‡ Both results are due to Rayleigh (1872, 1897). A generalization of the sphere result to include viscosity is due to C. J. T. Sewell, "The Extinction of Sound in a Viscous Atmosphere by Small Obstacles of Cylindrical and Spherical Form," *Phil. Trans. R. Soc. Lond.*, **A210**:239–270 (1910); a concise account is given by H. Lamb, *Hydrodynamics, 6th ed., 1932,* reprinted by Dover, New York, 1945, pp. 657–659. The required modification of Eq. (8) for a freely suspended sphere that includes viscosity and also the acoustically induced motion is (with $ka \ll 1$)

$$\tfrac{3}{2}\cos\theta \to \frac{(m-m_d)\tfrac{3}{2}K_{vis}\cos\theta}{m - m_d + \tfrac{3}{2}m_dK_{vis}} \qquad K_{vis} = 1 + \frac{3i}{\beta a} - \frac{3}{\beta^2a^2} \qquad \beta = e^{i\pi/4}\left(\frac{\omega\rho}{\mu}\right)^{1/2}$$

where μ = viscosity
m = sphere's mass
m_d = mass of fluid displaced by sphere

The immovable-sphere result is obtained in the limit $m/m_d \to \infty$. The inviscid result is obtained in the limit $|\beta a| \to \infty$, so that $K_{vis} \to 1$.

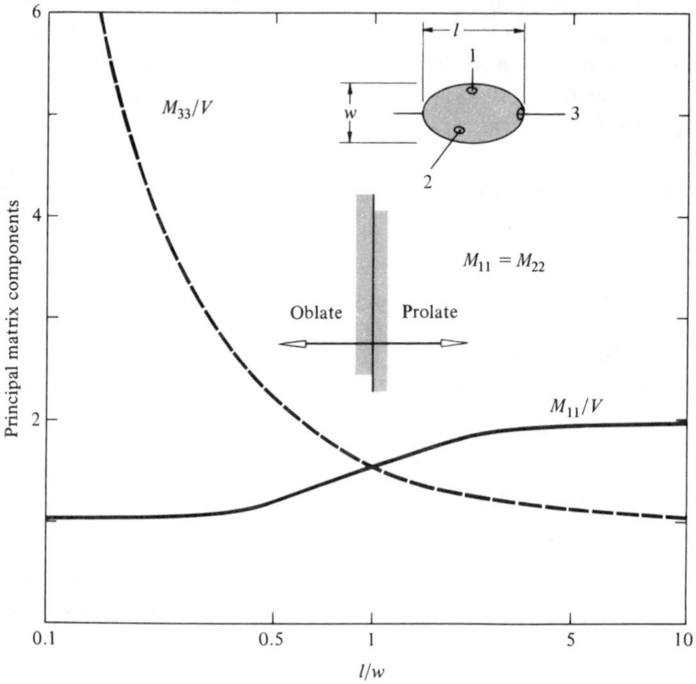

Figure 9-2 Principal components of the matrix **M** that appears in expression for dipole portion of field scattered by a body in the $ka \ll 1$ limit; $\rho\mathbf{M} - \rho V\mathbf{I}$ is the entrained-mass tensor. Plot is for spheroids (prolate if $l > w$; oblate if $l < w$) that are bodies of revolution (length l, maximum diameter w) about the x_3 axis. The volume V is $\frac{4}{3}\pi(w/2)^2(l/2)$. For the sphere $(l/w = 1)$, both M_{11}/V and M_{33}/V are 1; for the disk $(l/w \to 0)$, $M_{33} \to \frac{8}{3}(w/2)^3$, so $M_{33}/V \to (2/\pi)(w/l)$. [*From T. B. A. Senior, J. Acoust. Soc. Am.,* **53**:*745 (1973).*]

Here $\cos\theta$ is $\mathbf{e}_k \cdot \mathbf{e}_r$, such that θ is the angle the scattered direction makes with the incident direction. The monopole term is absent in the latter because the disk has no volume.

Scattering Cross Section

The time-averaged intensity I_{sc} of the scattered wave at large r, equal to the asymptotic value of $\frac{1}{2}|\hat{p}_{sc}|^2/\rho c$, is proportional to the time-averaged incident intensity I_i, decreases with r as $1/r^2$, and also depends in general on the direction from the scatterer to where the scattered pressure is measured. The quotient $r^2 I_{sc}/I_i$, representing the power scattered per unit solid angle and per unit incident intensity, is referred to as the *differential cross section* $d\sigma/d\Omega$, while the integral over solid angle of $d\sigma/d\Omega$ is referred to as the *scattering cross section* σ. The latter term† is also used in literature emphasizing analogies with radar

† Compare the definitions on pp. 818 and 509, respectively, of *International Dictionary of Applied Mathematics,* Van Nostrand, Princeton, N.J., 1960, and *IEEE Standard Dictionary of Electrical and Electronics Terms,* Wiley, New York, 1972.

applications for the directionally dependent quantity $4\pi\, d\sigma/d\Omega$; to avoid confusion, the alternative terms *backscattering cross section* σ_{back} and *bistatic cross section* σ_{bi} are here used for $4\pi\, d\sigma/d\Omega$ when the direction toward the receiver extends back toward the source and at an angle from the source, respectively. For an *isotropic scatterer,* for which $d\sigma/d\Omega$ is independent of direction and equal to $\sigma/4\pi$, the backscattering cross section and the bistatic cross section are the same as the scattering cross section σ.

Closely related to the backscattering cross section is the *target strength,* measured in decibels and defined so that

$$\text{TS} = 10 \log \frac{\sigma_{\text{back}}}{4\pi R_{\text{ref}}^2} \tag{9-1.10}$$

where the reference length R_{ref} is taken as 1 m in present-day literature.† The ratio in the argument of the logarithm can also be regarded as the differential cross section in the backscattering direction divided by a reference differential cross section of 1 m^2/sr. If L_i is the incident sound-pressure level at the scatterer, and if $L_{\text{back}}(R_0)$ is the sound-pressure level of the backscattered wave at distance R_0 from the scatterer, then the definition of target strength implies that

$$\text{TS} = L_{\text{back}}(R_0) + 10 \log \frac{R_0^2}{R_{\text{ref}}^2} - L_i \tag{9-1.11}$$

providing the scattered wave decreases with distance as in spherical spreading.

The differential cross section $d\sigma/d\Omega$ for the low-frequency scattering by a rigid immovable body evolves out of Eq. (6) to the expression

$$\frac{d\sigma}{d\Omega} = \frac{k^4}{16\pi^2} |V - \mathbf{e}_r \cdot \mathbf{M} \cdot \mathbf{e}_k|^2 \tag{9-1.12}$$

while the backscattering cross section results with $\mathbf{e}_r$ set to $-\mathbf{e}_k$ and with a subsequent multiplication by 4π, such that

$$\sigma_{\text{back}} = \frac{k^4}{4\pi} |V + \mathbf{e}_k \cdot \mathbf{M} \cdot \mathbf{e}_k|^2 \qquad \text{backscatter} \tag{9-1.13}$$

The predicted frequency dependence, as f^4, holds also for the scattering cross section σ. The required angular integration of $d\sigma/d\Omega$ becomes simpler with the z axis selected in the direction of $\mathbf{M} \cdot \mathbf{e}_k$, so that $\mathbf{e}_r \cdot \mathbf{M} \cdot \mathbf{e}_k$ is $|\mathbf{M} \cdot \mathbf{e}_k| \cos\theta$. The cross term integrates to zero (since $\cos\theta$ is odd about $\theta = \pi/2$), so the scattered acoustic powers associated with the monopole and dipole contributions are additive. These two remaining terms integrate to simple expressions because the average of $\cos^2\theta$ over the surface of a sphere is $\frac{1}{3}$ and because the

† C. S. Clay and H. Medwin, *Acoustical Oceanography: Principles and Applications,* Wiley, New York, 1977, pp. 180–183. The reference length of 1 yd (0.9144 m) is used in earlier literature. See, for example, J. W. Horton, *Fundamentals of SONAR,* 2d ed., United States Naval Institute, Annapolis, Md., 1959, pp. 41, 56–57, 329–330. Note that although Clay and Medwin's definition of backscattering cross section differs by a factor of 4π from that used here, the above definition of target strength is the same as theirs.

total solid angle about a point is 4π; the overall result is therefore

$$\sigma = \frac{k^4}{4\pi}[V^2 + \tfrac{1}{3}(\mathbf{M}\cdot\mathbf{e}_k)^2] \tag{9-1.14}$$

The scattering cross section σ, defined above as the scattered power per unit incident intensity, is the apparent area blocking the incident wave. The values resulting from Eqs. (7) for this parameter are

$$\sigma = \begin{cases} \tfrac{7}{9}(\pi a^2)(ka)^4 & \text{sphere} \quad (9\text{-}1.15a) \\ \tfrac{16}{27}(\pi a^2 \cos^2\theta_k)(ka)^4/\pi^2 & \text{disk} \quad (9\text{-}1.15b) \end{cases}$$

where θ_k is the angle between the disk's symmetry axis and the incident propagation direction. The scattering cross sections in these two cases are smaller, by factors of $\frac{7}{9}(ka)^4$ and $\frac{16}{27}(ka)^4\pi^{-2}\cos\theta_k$, than the projected areas πa^2 and $\pi a^2 \cos\theta_k$ the scattering body presents to the incident wave. The common factor $(ka)^4$ substantiates the conclusion that small obstacles appear even smaller to an incident wave.

Higher-Frequency Scattering

In the limit of large ka, geometrical-acoustics considerations require

$$\sigma \to 2A_{\text{proj}} \qquad \sigma_{\text{back}} \to \pi R_{S,\text{I}} R_{S,\text{II}} \tag{9-1.16}$$

The latter expression presumes that there is only one point on the near side of the scatterer where the unit normal points back toward the source [see Eq. (8-8.9)]; the principal radii of curvature at that point are $R_{S,\text{I}}$ and $R_{S,\text{II}}$; the surface is assumed to be convex. The factor of 2 multiplying the projected area A_{proj} in the expression for the scattering cross section arises because the definition in Eq. (1) of $\hat{p}_{\text{sc}}$ and the existence of the shadow require the scattered field to be nearly opposite to the incident field behind the body (on the side facing away from the source). The scattered power behind the body is therefore the projected area times the incident intensity, which is the same as the acoustic power reflected by the illuminated part of the body; hence the factor 2.

The transition between high- and low-frequency limits is not amenable to simple generalizations, but some insight results from an examination of numerical calculations for the rigid-sphere example. The solution† of the resulting boundary-value problem takes the form of a sum over products of spherical harmonics and spherical Hankel functions. For small ka, the first two terms, as further approximated by Eq. (8), suffice, but many terms must be summed when ka is of the order of 1 or larger. The computational results plotted in Fig. 9-3 are of $(d\sigma/d\Omega)^{1/2}/a$; also shown are the analogous limiting versions for the

† H. Stenzel, "On the Perturbation of the Sound Field Caused by a Rigid Sphere," *Elektr. Nachrichtentech.*, **15**:71–78 (1938); N. A. Logan, "Survey of Some Early Studies of the Scattering of Plane Waves by a Sphere," *Proc. IEEE*, **53**:773–785 (1965). The derivation is outlined by P. M. Morse and H. Feshbach, *Methods of Theoretical Physics*, vol. 2, McGraw-Hill, New York, 1953, pp. 1483–1484.

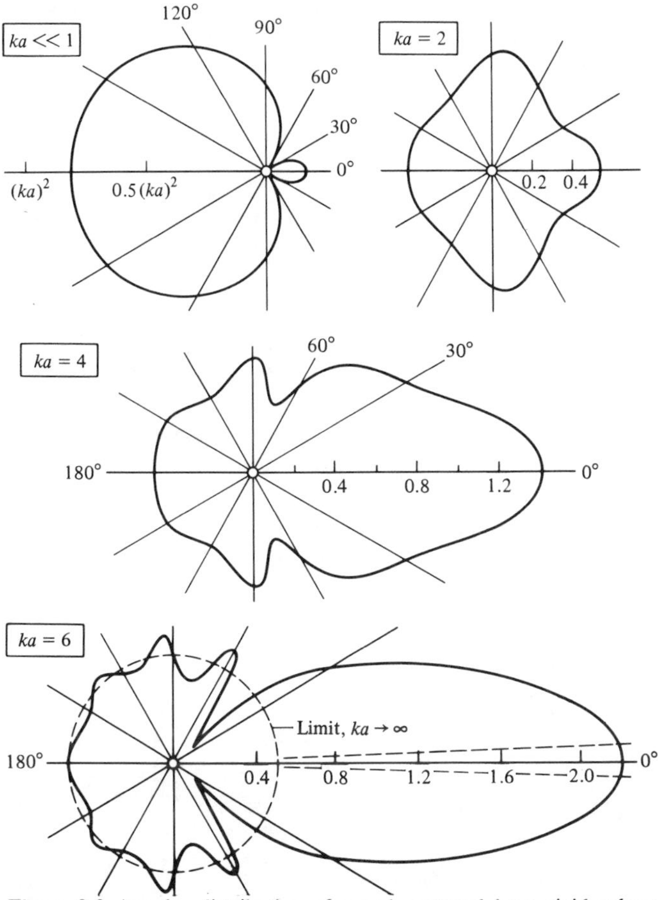

Figure 9-3 Angular distribution of sound scattered by a rigid sphere of radius a. The quantity $(d\sigma/d\Omega)^{1/2}/a$ is plotted versus the polar angle θ, where $d\sigma/d\Omega$ is the differential cross section: $\theta = 0$ corresponds to scattering in the forward direction, $\theta = 180°$ to backscatter. The plots for $ka = 2$, 4, and 6 are based on calculations of H. Stenzel (1938). [Michael Myers reports calculations (private communication, October 1988) demonstrating that Stenzel's calculations are appreciably inaccurate in the presentation of details.]

Rayleigh-scattering limit and the geometrical acoustics limit, these being

$$\frac{1}{a}\left(\frac{d\sigma}{d\Omega}\right)^{1/2} \rightarrow \begin{cases} \frac{1}{3}(ka)^2|1 - \frac{3}{2}\cos\theta| & ka \ll 1 \qquad (9\text{-}1.17a) \\ \frac{1}{2} + \pi^{1/2}\Delta(\theta) & ka \gg 1 \qquad (9\text{-}1.17b) \end{cases}$$

Here $\Delta(\theta)$ is a singular function concentrated at $\theta = 0$ and defined so that the integral of $\Delta^2(\theta)$ over solid angle is 1.

Scattering by Inhomogeneities

To study acoustic scattering by a departure of the medium from spatial homogeneity, we suppose that $\rho(\mathbf{x})$ and $c(\mathbf{x})$ differ near the origin from their prevalent uniform media values ρ_0 and c_0. The wave equation for an inhomogeneous quiescent medium (see Prob. 1-6)

$$\rho\nabla\cdot\left(\frac{1}{\rho}\nabla p\right) - \frac{1}{c^2}\frac{\partial^2 p}{\partial t^2} = 0 \tag{9-1.18}$$

leads for the constant-frequency case to

$$\nabla^2\hat{p} + k^2\hat{p} = k^2\Delta_1\hat{p} + \nabla\cdot(\Delta_2\nabla\hat{p}) \tag{9-1.19a}$$

$$k = \frac{\omega}{c_0} \qquad \Delta_2 = 1 - \frac{\rho_0}{\rho} \qquad \Delta_1 = 1 - \frac{\rho_0 c_0^2}{\rho c^2} \tag{9-1.19b}$$

where the right side of Eq. (19a) vanishes except near the origin. The two right-side terms are associated with monopole and dipole scattering, respectively. In what follows, the spatial dimension a characterizing the extent of the inhomogeneity is such that $kac_0/c \gg 1$ and $(ka)^2\rho_0/\rho \gg 1$ everywhere. As before, the incident acoustic pressure has complex amplitude $Be^{i\mathbf{k}\cdot\mathbf{x}}$, so $\hat{p} - Be^{i\mathbf{k}\cdot\mathbf{x}}$ satisfies the Sommerfeld radiation condition.

The formal recognition of the right side of the above as a source term allows the Green's function solution, Eq. (4-3.13), to transform Eq. (19a) into the integral equation

$$\hat{p} = Be^{i\mathbf{k}\cdot\mathbf{x}} - \frac{k^2}{4\pi}\iiint \Delta_1(\mathbf{x}_s)\hat{p}(\mathbf{x}_s)R^{-1}e^{ikR}\,dV_s$$
$$- \frac{1}{4\pi}\nabla\cdot\left[\iiint \Delta_2(\mathbf{x}_s)\nabla_s\hat{p}(\mathbf{x}_s)R^{-1}e^{ikR}\,dV_s\right] \tag{9-1.20}$$

This in turn yields the asymptotic (large r) expression for the scattered wave

$$\hat{p}_{sc} \approx \frac{-k^2B}{4\pi}\left[V_{\text{eff}} - \mathbf{e}_r\cdot\mathbf{M}_{\text{eff}}\cdot\mathbf{e}_k\left(1 + \frac{i}{kr}\right)\right]\frac{e^{ikr}}{r} \tag{9-1.21}$$

where

$$V_{\text{eff}} = \frac{1}{B}\iiint \Delta_1(\mathbf{x}_s)\hat{p}(\mathbf{x}_s)\,dV_s \tag{9-1.22a}$$

$$\mathbf{M}_{\text{eff}}\cdot\mathbf{e}_k = \frac{1}{ikB}\iiint \Delta_2(\mathbf{x}_s)\nabla_s\hat{p}(\mathbf{x}_s)\,dV_s \tag{9-1.22b}$$

Note that Eq. (21) is of the same form as Eq. (6). The coefficients are understood to be evaluated in the limit $ka \to 0$, so the scattering cross section here also is proportional to f^4.

In regard to the evaluation of the above coefficients, a solution technique applicable when Δ_1 and Δ_2 are not necessarily small follows the matched-asymptotic-expansion procedure outlined in Sec. 4-7. The differential equations for successive terms in the inner expansion result from insertion of a power series in k into Eq. (19a). Outer boundary conditions for this sequence of differential equations follow from the requirement that the inner solution for large r/a match the outer solution $Be^{i\mathbf{k}\cdot\mathbf{x}} + \hat{p}_{sc}$, with $\hat{p}_{sc}$ represented by Eq. (21), in the limit of small kr. In this manner, one finds the inner expansion to first order in k to be

$$\hat{p}_{\text{inner}} \approx B + iB\mathbf{k} \cdot \mathbf{\Phi}(\mathbf{x}) \tag{9-1.23}$$

where the μth component of the vector $\mathbf{\Phi}(\mathbf{x})$ satisfies

$$\nabla \cdot [(1 - \Delta_2)\, \nabla \Phi_\mu] = 0 \qquad \Phi_\mu(\mathbf{x}) - x_\mu \to 0 \qquad r \to \infty \tag{9-1.24}$$

The identification of the first term in Eq. (23) results from requiring the solution of the differential equation (19a) with $k^2 \to 0$ to match $Be^{i\mathbf{k}\cdot\mathbf{x}}$ in the limit of small r. Note that the first-order term in Eq. (23) must also satisfy the same $k \to 0$ partial-differential equation. The outer boundary condition in Eq. (24) results because $iB\mathbf{k} \cdot \mathbf{\Phi}(\mathbf{x})$ must asymptotically equal the first-order term $iB\mathbf{k} \cdot \mathbf{x}$ of the power-series expansion of $Be^{i\mathbf{k}\cdot\mathbf{x}}$.

Equation (23) allows the coefficients in Eqs. (22) to become†

$$V_{\text{eff}} = \iiint \Delta_1(\mathbf{x})\, dV \tag{9-1.25a}$$

$$M_{\text{eff},\,\mu\nu} = \iiint \Delta_2(\mathbf{x})\, \frac{\partial \Phi_\nu(\mathbf{x})}{\partial x_\mu}\, dV \tag{9-1.25b}$$

The symmetry of the tensor $\mathbf{M}_{\text{eff}}$ is a derivable consequence of Eqs. (24) and (25b).

The explicit expression (25a) for the effective volume of the scatterer can alternatively be interpreted as

$$V_{\text{eff}} = -\rho_0 c_0^2\, \Delta C_A \tag{9-1.26}$$

where ΔC_A is the increase of the acoustic compliance of a volume enclosing the inhomogeneity. Here acoustic compliance is defined (see Sec. 7-2) as volume decrease per unit increase in external pressure. If the scatterer is rigid, the compliance is reduced by $V/\rho_0 c_0^2$, so that V_{eff} is just the volume V of the scatterer, which is consistent with the result in Eq. (6). If the scatterer is more compliant than the ambient medium, $(\rho c^2)_{\text{sc}} < \rho_0 c_0^2$ and ΔC_A becomes positive, so V_{eff} is a negative number and its label as an effective volume becomes a misnomer. The symbol V_{eff} is retained here, however, as it makes identification from Eqs. (12) and (14) for the scattering cross section easy.

Spherical Inhomogeneity

Solution for the $\Phi_\nu(\mathbf{x})$ in general requires further approximation or numerical integration. An exception is that of the homogeneous sphere, such that $\Delta_2 = \epsilon$ for $r < a$ and $\Delta_2 = 0$ for $r > a$, where ϵ is constant. The symmetry permits the

† J. W. S. Rayleigh, "On the Incidence of Aerial and Electric Waves upon Small Obstacles in the Form of Ellipsoids or Elliptic Cylinders, and on the Passage of Electric Waves through a Circular Aperture in a Conducting Screen," *Phil. Mag.*, (5)**44**:28–52 (1897).

substitution $\Phi_\mu(\mathbf{x}) = x_\mu g(r)/r$, yielding the ordinary differential equation

$$\frac{d}{dr}\left[(1 - \Delta_2)r^2 \frac{dg}{dr}\right] - 2(1 - \Delta_2)g = 0 \tag{9-1.27}$$

with the derivable restrictions that $g(r)$ and $(1 - \Delta_2)r^2\, dg/dr$ be continuous at $r = a$. Solutions of the equation are $g = \alpha r$ for $r < a$ and $g = \beta r + \gamma/r^2$ for $r > a$. The outer boundary condition requires $\beta = 1$; the continuity requirements yield $\alpha = 3/(3 - \epsilon)$ and $\gamma = [\epsilon/(3 - \epsilon)]a^3$. The substitution of $\Phi_\nu = [3/(3 - \epsilon)]x_\nu$ for $r < a$ into Eq. (25*b*) then yields

$$M_{\mathrm{eff},\mu\nu} = \frac{3\epsilon}{3 - \epsilon} V\,\delta_{\mu\nu} = \frac{3(m - m_d)}{2m + m_d} V\,\delta_{\mu\nu} \tag{9-1.28}$$

where m = mass of foreign sphere
m_d = mass of ambient fluid it displaces
$V = \frac{4}{3}\pi a^3$

Inertia Effect for Freely Suspended Particle

The preceding result, Eq. (28), is the same as for a freely suspended rigid sphere, and its interpretation is facilitated by the derivation† that proceeds from such a viewpoint. Little additional complexity results if the body is nonspherical, but we do assume that its geometry is such that the incident acoustic wave causes no torque to be exerted about its center of mass and that the product of the tensor $\mathbf{W}$ with the unit vector $\mathbf{e}_k$ is also in direction $\mathbf{e}_k$; we therefore write $\mathbf{W} \cdot \mathbf{e}_k = W\mathbf{e}_k$ in what follows.

If $\boldsymbol{\xi}$ denotes the body's center-of-mass position, Newton's second law requires that

$$m\ddot{\boldsymbol{\xi}} = -V\,\nabla p_i - \mathbf{F}_{\mathrm{sc}} \tag{9-1.29}$$

The first term is the small-*ka* approximation to the force exerted on the body by the incident wave; $-\mathbf{F}_{\mathrm{sc}}$ is the force exerted on the body by the scattered wave's pressure at the surface. The definition of the entrained-mass tensor requires, however, that

$$\mathbf{F}_{\mathrm{sc}} = \rho\mathbf{W} \cdot (\ddot{\boldsymbol{\xi}} - \dot{\mathbf{v}}_i) \tag{9-1.30}$$

where $\mathbf{v}_i$ is the fluid velocity associated with the incident wave and ρ is the ambient density of the surrounding fluid. Elimination of $\mathbf{F}_{\mathrm{sc}}$ from the two above equations, replacement of ∇p_i by $-\rho\dot{\mathbf{v}}_i$, and a time integration yield

$$m(\dot{\boldsymbol{\xi}} - \mathbf{v}_i) + \rho\mathbf{W} \cdot (\dot{\boldsymbol{\xi}} - \mathbf{v}_i) = -(m - m_d)\mathbf{v}_i \tag{9-1.31}$$

The above equation and the assumed properties of $\mathbf{W}$ in turn require $\dot{\boldsymbol{\xi}}$ to be parallel to $\mathbf{v}_i$, with the result

$$m_d(\ddot{\boldsymbol{\xi}} - \dot{\mathbf{v}}_i) + \mathbf{F}_{\mathrm{sc}} = -\frac{m - m_d}{m + \rho W}\,\rho\mathbf{M} \cdot \dot{\mathbf{v}}_i \tag{9-1.32}$$

† Lamb, *Hydrodynamics*, 6th ed., p. 514.

This, however, is the relevant quantity as regards the dipole radiation by the scatterer since the boundary condition $\dot{\boldsymbol{\xi}} \cdot \mathbf{n} = \mathbf{v}_{sc} \cdot \mathbf{n} + \mathbf{v}_i \cdot \mathbf{n}$ enables us to regard such radiation as being generated by a rigid body translating with velocity $\dot{\boldsymbol{\xi}} - \mathbf{v}_i$. The resulting dipole field is given by Eq. (4-7.12) with $\dot{\mathbf{v}}_C$ replaced by $\ddot{\boldsymbol{\xi}} - \dot{\mathbf{v}}_i$. Since $\dot{\mathbf{v}}_i$ has complex amplitude $-i\omega(\mathrm{B}/\rho c)\mathbf{e}_k$, we conclude, after a comparison with Eq. (5a), that the only change required in Eq. (6) is that the immovable-body **M** tensor be multiplied by

$$K_{\text{inertia}} = \frac{m - m_d}{m + \rho W} \tag{9-1.33}$$

For the transversely oscillating rigid sphere, ρW is $\frac{1}{2} m_d$ and $M_{\mu\nu}$ is $\frac{3}{2} V \delta_{\mu\nu}$; so the above is consistent with Eq. (28).

Resonant Scattering

The foregoing derivation for V_{eff} leading from Eq. (22a) to Eq. (25a) requires the pressure near the scatterer to be not appreciably different from that of the incident wave. Since the magnitude of the monopole term at the edge of the scatterer is of the order of $k^2|B|V_{\text{eff}}/4\pi a$, this requires $|\Delta_1|k^2a^2$ to be small. A circumstance where this may be violated, with ka nevertheless small, is a *bubble* (see Fig. 9-4a), within which the ambient density is much less than that of the surrounding medium. (An example would be a gas bubble in water.) Then for a narrow range of frequencies, yet with $ka \ll 1$, it is possible to have a monopole term of inordinately large amplitude.

To isolate the monopole portion of the wave scattered by a bubble, we average the incident wave over the surface of a sphere so that $e^{i\mathbf{k}\cdot\mathbf{x}}$ is replaced by $(kr)^{-1} \sin kr$. Since the bubble is assumed spherically symmetric, the monopole portion of the incident and scattered fields becomes

$$(\hat{p}_i + \hat{p}_{sc})_{\text{mono}} = \begin{cases} B \dfrac{\sin kr}{kr} + \hat{S}\, \dfrac{e^{ikr}}{r} & r > a \qquad (9\text{-}1.34a) \\[2ex] D \dfrac{\sin k_b r}{k_b r} & r < a \qquad (9\text{-}1.34b) \end{cases}$$

where $\hat{S}$ and D are constants and $k_b = \omega/c_b$ is the wave number appropriate to the interior of the bubble. Both expressions are spherically symmetric solutions of the appropriate Helmholtz equation (see Sec. 1-12). The scattered part of the exterior-region solution conforms to the Sommerfeld radiation condition; the interior-region solution is required to be finite at the origin.

Determination of the coefficients $\hat{S}$ and D results from imposition of the requirements that $\hat{p}$ and $(1/\rho)\ \partial\hat{p}/\partial r$ be continuous at $r = a$. Limiting our consideration to frequencies such that ka and $k_b a$ are both small, we rewrite Eqs. (34) as

$$(\hat{p}_i + \hat{p}_{sc})_{\text{mono}} \approx \begin{cases} B - \frac{1}{6}B(kr)^2 + \dfrac{\hat{S}}{r} + ik\hat{S} & r > a,\ kr \ll 1 \qquad (9\text{-}1.35a) \\[2ex] D - \frac{1}{6}D(k_b r)^2 & r < a,\ k_b a \ll 1 \qquad (9\text{-}1.35b) \end{cases}$$

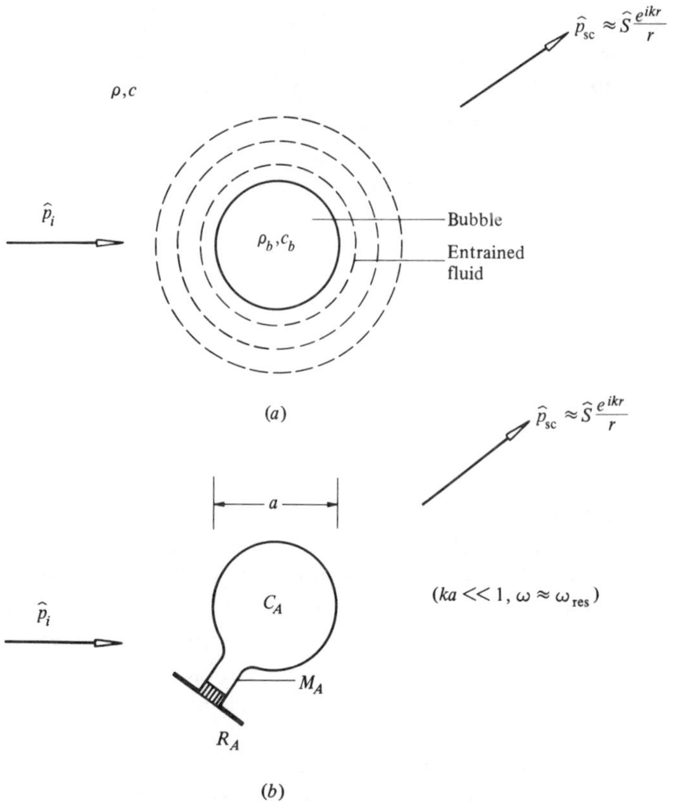

Figure 9-4 Parameters and concepts adopted in the discussion of resonant scattering by (*a*) a bubble; and (*b*) a Helmholtz resonator.

so that the continuity conditions yield

$$B + \frac{\hat{S}}{a} + ik\hat{S} \approx D \tag{9-1.36a}$$

$$\tfrac{1}{3}Bk^2a + \frac{\hat{S}}{a^2} \approx \frac{\rho}{3\rho_b} Dk_b^2 a \tag{9-1.36b}$$

The solution† for $\hat{S}$ in the same approximation is

† An appropriate idealization for the incorporation of thermal conductivity into the model is that the bubble-temperature fluctuation vanishes at the interface. Techniques similar to those described in Secs. 10-3 to 10-5 then yield for the replacement of Eq. (37)

$$\hat{S} = \frac{-(k^2/4\pi)V_b[1 - (\rho c^2/\rho_b c_b^2)\psi]B}{1 - \frac{1}{3}(k_b a)^2(\rho/\rho_b)(1 + ika)\psi}$$

$$\psi = 1 + (\gamma_b - 1)f(e^{i\pi/4}\phi_b) \qquad f(u) = 3(u^{-2} - u^{-1}\cot u)$$

where $\phi_b = (\omega\rho c_p/\kappa)_b^{1/2}a$, with c_p denoting the specific heat, γ_b denoting the specific-heat ratio, and

$$\hat{S} = \frac{-(k^2/4\pi)V_b[1 - (\rho c^2/\rho_b c_b^2)]B}{1 - \frac{1}{3}(k_b a)^2(\rho/\rho_b)(1 + ika)} \tag{9-1.37}$$

which yields

$$\hat{p}_{\text{sc,mono}} = \frac{(k^2/4\pi)\rho c^2 \, \Delta C_A \, B}{1 - \omega^2 M_A C_A - i\omega C_A R_A} \frac{e^{ikr}}{r} \tag{9-1.38}$$

with the identifications

$$M_A = \frac{3\rho V_b}{(4\pi a^2)^2} \qquad C_A = \frac{V_b}{\rho_b c_b^2} \qquad R_A = \frac{\rho c k^2}{4\pi} \tag{9-1.39}$$

for the acoustic inertance, acoustic compliance, and acoustic (radiation) resistance associated with the bubble. Here V_b is the bubble volume.

The above expression for M_A is consistent with the model of a bubble in which the fluid velocity in the external fluid varies with radius r as $1/r^2$, as in potential flow. The kinetic energy associated with an interface velocity v_S is then

$$\tfrac{1}{2}\rho v_S^2 4\pi \int_a^\infty \left(\frac{a^2}{r^2}\right)^2 r^2 \, dr \approx \tfrac{3}{2} m_d v_S^2$$

where $m_d = \rho V_b$ is the mass displaced by the bubble. Energy-conservation considerations therefore suggest that $3m_d\dot{v}_S$ is the difference $4\pi a^2 \, \Delta p$ of pressure forces inside and outside the bubble. The volume velocity is $4\pi a^2 v_S$, so the acoustic inertia, defined as Δp divided by the time derivative of volume velocity, is $3m_d$ divided by the surface area squared, as in Eq. (39).

The acoustic resistance in the above formulation is similarly explained as that associated with a monopole radiating into an unbounded space. (This follows from the result $\hat{p}_{\text{in},2} = ik\hat{S}$ derived in Sec. 4-7, with $\hat{S}$ identified as the complex amplitude of $\rho/4\pi$ times the time derivative of the volume velocity.)

A resonance in the scattering occurs when ω^2 is near $(M_A C_A)^{-1}$ or when the

---

κ denoting the thermal conductivity. For small bubbles such that $\phi_b \ll 1$, the bubble oscillates isothermally rather than adiabatically, so that $\psi \approx \gamma_b$. Equation (37) applies in the limit $\phi_b \gg 1$, so that $\psi \approx 1$. Values of the complex function are

ϕ_b	0	2	4	6	8	10
$f(e^{i\pi/4}\phi_b)$	(1, 0)	(0.91, 0.23)	(0.54, 0.34)	(0.35, 0.27)	(0.27, 0.22)	(0.21, 0.18)

The function $f(u)$ approximates to $1 + (u^2/15)$ at small u and to $-3/u$ at large ϕ_b. The imaginary part has a peak value of 0.36 at $\phi_b = 3.41$; the corresponding value for the real part is 0.63. Viscosity is ordinarily of minor influence for bubble scattering. The basic theory underlying the formula cited is due in major part to C. Devin, Jr., "Survey of Thermal, Radiation, and Viscous Damping of Pulsating Air Bubbles in Water," *J. Acoust. Soc. Am.*, **31**:1654–1667 (1959); additional clarification and numerical results are given by A. I. Eller, "Damping Constants of Pulsating Bubbles," ibid., **47**:1469–1470 (1970).

frequency f in hertz is near the bubble resonance frequency†

$$f_b = \frac{c_b}{2\pi a}\left(\frac{3\rho_b}{\rho}\right)^{1/2} \tag{9-1.40}$$

Equation (38) also applies to scattering at near-resonance frequencies by an isolated Helmholtz resonator (see Fig. 9-4*b*) provided the ΔC_A in the numerator is replaced by the acoustic compliance C_A of the resonator's cavity. A derivation based on the method of matched asymptotic expansions proceeds similarly to what is given in Sec. 7-5 for scattering by a Helmholtz resonator mounted on a wall. In the present case, the modified version of Eq. (38) yields

$$\hat{p}_{\text{sc,mono}} = \frac{ik\rho c}{4\pi r}\frac{Be^{ikr}}{Z_{\text{HR}} + \rho ck^2/4\pi} \tag{9-1.41}$$

where Z_{HR} is the acoustic impedance of the Helmholtz resonator.

Near the resonant frequency, the scattered wave is overwhelmingly monopole, so the scattered field is spherically symmetric and the scattering cross section is $4\pi r^2|\hat{p}_{\text{sc,mono}}|^2/B^2$. With radiation damping as the only damping mechanism taken into account, one finds from Eqs. (38) and (41) that σ is bounded‡ by $4\pi/k^2 = \lambda^2/\pi$.

Analogous considerations apply to scatterers that radiate as dipoles or quadrupoles when driven at a resonance frequency. A solid sphere suspended by a spring, for example, should radiate primarily as a dipole when the incident wave's frequency equals the system's resonance frequency. Similarly, apropos of the legendary story of the operatic tenor whose voice could shatter wine glasses, the scattered resonance sound in such a situation would most likely have been quadrupole radiation. The guiding principle for prediction of the scattered field's radiation pattern is that the scattering body is caused to vibrate

† M. Minnaert, "On Musical Air-Bubbles and the Sounds of Running Water," *Phil. Mag.*, (7)**16:**235–248 (1933). The generalization that correctly takes surface tension into account is

$$f_b = \frac{\rho^{-1/2}}{2\pi a}\left[3\left(\rho_b c_b^2 + \frac{n_b 2\sigma}{a}\right) - \frac{2\sigma}{a}\right]^{1/2}$$

where σ is surface tension in newtons per meter and n_b is the derivative $\partial(\rho_b c_b^2)/\partial p_b$, carried out at constant temperature and evaluated at the ambient pressure and temperature of the external fluid. It is understood also that $\rho_b c_b^2$ here denotes the value corresponding to the ambient external temperature T_0 and pressure p_0, so that $\rho_b c_b^2 = \gamma_b p_0$ and $n_b = \gamma_b$ for a gas bubble. An incorrect expression frequently seen in the literature forgets to account for the difference between the external and internal ambient pressures. The above result, attributed to J. M. Richardson (before 1947), is derived by R. W. Robinson and R. H. Buchanan, "Undamped Free Pulsations of an Ideal Bubble," *Proc. Phys. Soc. Lond.*, **B69:**893–900 (1956). Typical values of σ for an air-water interface are 0.076, 0.073, and 0.070 N/m at 0, 20, and 40°C, so surface tension becomes important for underwater bubbles only if $a < 10^{-5}$m.

‡ H. Lamb, "A Problem in Resonance, Illustrative of the Theory of Selective Absorption of Light," *Proc. Lond. Math. Soc.*, **32:**11–20 (1900); J. W. S. Rayleigh, "Some General Theorems concerning Forced Vibrations and Resonance," *Phil. Mag.*, (6) **3:**97–117 (1902); *Theory of Sound*, vol. 2, 2d ed., reissue of 1926, pp. 284A–284D.

as in its corresponding natural mode of free vibration when it is excited by a resonance frequency.

9-2 MONOSTATIC AND BISTATIC SCATTERING-MEASUREMENT CONFIGURATIONS

Instrumentation configurations for studies of scattering are broadly classified as *monostatic* and *bistatic*.† If the transmitter and receiver are at the same or at closely spaced points, the configuration is monostatic. If they are at widely spaced points, it is bistatic. In the discussion here, to emphasize the similarities in concept with other types of remote sensing systems, the sound-generation apparatus is referred to as the *transmitter* and the reception apparatus is referred to as the *receiver* unless special reference is being made to acoustical or electroacoustical properties.

Monostatic Pulse-Echo Sounding

In the prototype pulse-echo sounding experiment, a directional transmitter is aimed at a distant scattering object (see Fig. 9-5*a*). At time $t = 0$ the transmitter sends out a pulse of duration τ and of nearly constant angular frequency $\omega = 2\pi f$, where the ratio of τ to the period $1/f$ is much larger than 1. The distance r_s to the scatterer, moreover, is in turn somewhat larger than $c\tau/2$ and is such that the scatterer is in the transmitter's far field. The acoustic pressure incident near the scatterer is therefore describable (see Sec. 1-12) by

$$p_i = Dr^{-1}F\left(\theta, \phi, t - \frac{r}{c}\right) \tag{9-2.1}$$

where the function F is nonzero only if $0 < t - r/c < \tau$ and oscillates with angular frequency ω throughout the pulse interval; its normalization is such that the time average of F^2 is 1 for the time interval and for the direction toward the scatterer, taken here as $\theta = 0$. The constant D is then such that $(D^2/\rho c)/r_s^2$ is the incident wave's average intensity at the scatterer during the irradiation interval.

The scatterer's dimensions are regarded here as sufficiently small compared to r_s for the incident wave to appear locally planar, so that the definitions introduced in the preceding section apply. The scattered-wave intensity I_{sc} varies with direction and with radial distance r from the scatterer as $(d\sigma/d\Omega)I_i/r^2$, where the differential cross section $d\sigma/d\Omega$ can alternatively be expressed as $\sigma_{back}/4\pi$ for the backscattered direction. Thus, the intensity scat-

† The terminology comes from radar. See, for example, M. I. Skolnik, *Introduction to Radar Systems*, McGraw-Hill, New York, 1962, pp. 585–586. The "static" qualification in the adjectives monostatic and bistatic was originally intended to distinguish ground-based radar systems from airborne radar systems.

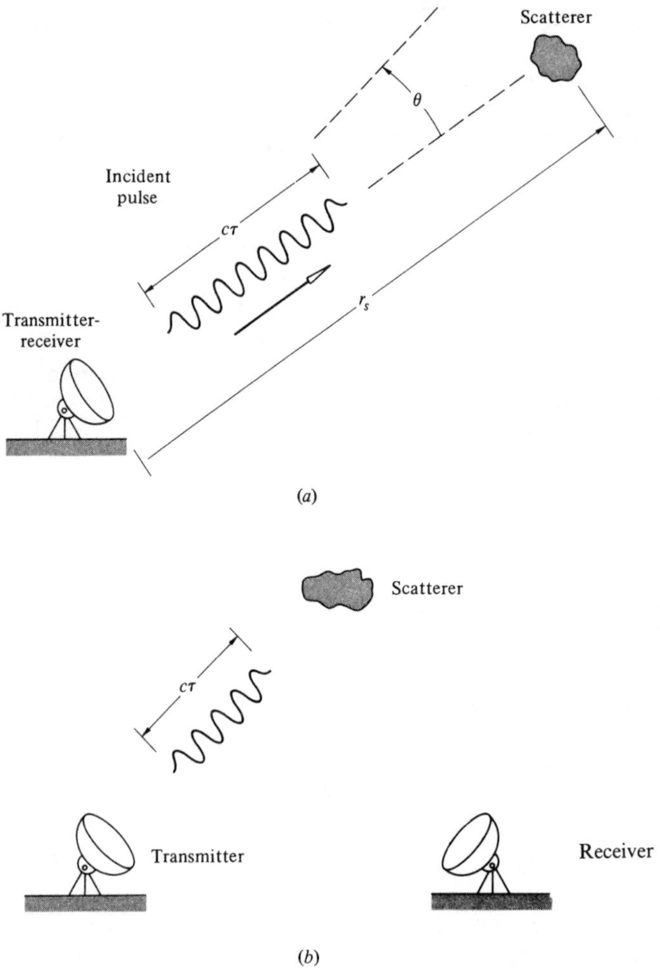

Figure 9-5 Instrumentation configurations for the study of scattering: (*a*) monostatic and (*b*) bistatic.

tered back to the transmitter becomes

$$I_{\text{back}} = \frac{D^2}{\rho c} r_s^{-2} \frac{\sigma_{\text{back}}}{4\pi} r_s^{-2} \tag{9-2.2}$$

during the interval when $0 < t - 2r_s/c < \tau$. Because r_s is larger than $c\tau/2$, the backscattered pulse does not overlap the incident pulse, and so the operation mode of the transducer can be switched to that for reception in the interval between termination of the transmission and first arrival of the echo.

The overall delay time, when multiplied by c, yields $2r_s$, so that the additional measurement of the echo's intensity, in conjunction with Eq. (2), suffices to determine the backscattering cross section σ_{back}.

Example The transducer in a SONAR (*so*und *na*vigation *r*anging) system when transmitting causes an rms acoustic pressure p_{rms} within the central beam at far-field distance r such that $p_{\text{rms}}r = 100$ N/m. If the peak backscattered signal arrives after net delay time of 3 s with an rms pressure of 10^{-5} Pa, what is the target strength, backscattering cross section, and distance to the scatterer?

SOLUTION The peak backscattered signal results when the beam points toward the scatterer, so the normalization of F requires the D in Eq. (1) to be 100 N/m. The time delay is understood to include the pulse duration, so that $2r_s$ should be 3×1500 m, yielding $r_s = 2250$ m, with $c = 1500$ m taken for the speed of sound in water. Then, since I_{back} is $(\rho c)^{-1}$ times the square of the backscattered rms pressure, Eq. (2) yields

$$10^{-5} = (100)\left(\frac{1}{2250}\right)^2\left(\frac{\sigma_{\text{back}}}{4\pi}\right)^{1/2}$$

which in turn yields $\sigma_{\text{back}}/4\pi = 0.256\ \text{m}^2$, $\sigma_{\text{back}} = 3.22\ \text{m}^2$, and TS $= -5.9$ dB, with 1 m taken as the reference length in Eq. (9-1.10).

Inhomogeneities and the Born Approximation

Scattering-measurement systems for inhomogeneous media are usually designed so that the signal received during any given time interval is virtually certain to be that which was scattered within a known spatial region (*scattering volume*) within the propagation medium. The size and dimensions of this scattering volume are controlled by the radiation pattern of the transmitter, by the directivity pattern of the receiving system, by the duration and signature characteristics of the incident pulse, and by the sampling interval and signal-processing system for the echo signal (see Fig. 9-6). For the present, we assume that such a design is achieved and, moreover, that the signal extracted from the echo has proceeded along a straight line from transmitter to scattering volume and from there along a straight line to the receiving system. Thus, we neglect *multiple scattering,* whereby the propagation direction changes more than once in the sound wave's progress from transmitter to receiver.

The *Born approximation* accompanies the assumption that the wave scattered by the volume is independent of what has been scattered elsewhere. The term, which originated with the analogous quantum-mechanical scattering problem,† in the present context implies what results when $\hat{p}(\mathbf{x}_s)$ under the integral sign in Eq. (9-1.20) is replaced by the complex amplitude of the incident wave alone. Doing such is the same as solving the integral equation by iteration, with the first iteration accepted as satisfactory. Although this requires in general that the scattered wave in the steady state be much weaker than the incident wave wherever the dominant inhomogeneities occur, no simple criteria involving magnitudes of Δ_1 and Δ_2 establish the upper limits of the approxima-

† M. Born, "Quantum Mechanics of Collision Processes," *Z. Phys.*, **38:**803–827 (1926); J. Mathews and R. L. Walker, *Mathematical Methods of Physics*, Benjamin, New York, 1965, p. 289.

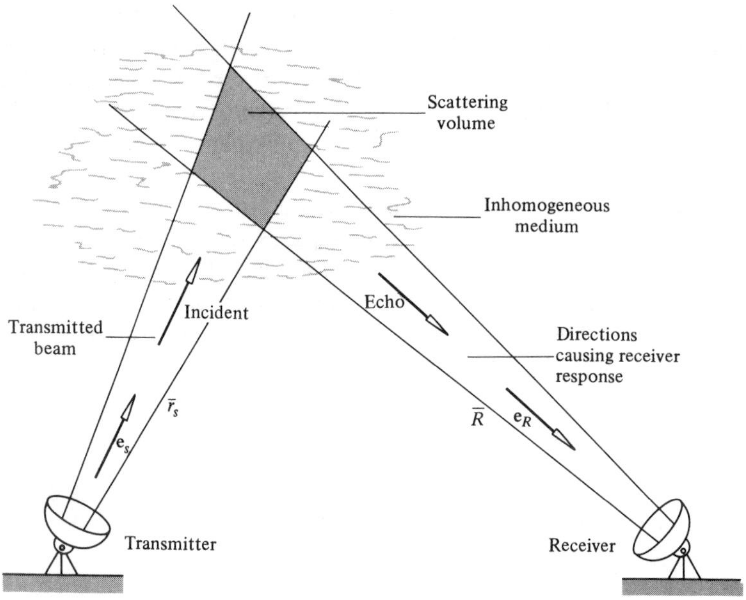

Figure 9-6 Scattering volume in a bistatic sounding experiment when the transmitter and receiver are both characterized by narrow beam widths.

tion's validity. It should, however, yield a good estimate of the scattered field if $|\Delta_1| \ll 1$ and $|\Delta_2| \ll 1$ and if the path integrals of both $k|\Delta_1|$ and $k|\Delta_2|$ are small compared with unity.

The modification of Eq. (9-1.20) to when the pulse in Eq. (1) is incident, with subsequent application of the Born approximation, yields the scattered wave in the form (applicable for bistatic as well as monostatic configurations)

$$p_{\mathrm{sc}} \approx \frac{-k^2 D}{4\pi} \iiint{}' \Delta_{\mathrm{eff}}(\mathbf{x}_s) \frac{F(\theta_s, \phi_s, t - r_s/c - R/c)}{r_s R}\, dV_s \tag{9-2.3}$$

$$\Delta_{\mathrm{eff}}(\mathbf{x}_s) = \Delta_1(\mathbf{x}_s) - \mathbf{e}_s \cdot \mathbf{e}_R \Delta_2(\mathbf{x}_s) \tag{9-2.4}$$

Here r_s is the distance $|\mathbf{x}_s|$ from the origin (center of transmitter) to the scattering point; $R = |\mathbf{x} - \mathbf{x}_s|$ is the distance from scattering point $\mathbf{x}_s$ to reception point $\mathbf{x}$; the unit vectors $\mathbf{e}_s$ and $\mathbf{e}_R$ point from the origin to $\mathbf{x}_s$ and from $\mathbf{x}_s$ to $\mathbf{x}$. The derivation here neglects the transverse gradients of F and assumes† that r_s and

† The stated assumption is what enables Δ_1 and Δ_2 to be combined into a single function $\Delta_{\mathrm{eff}}(\mathbf{x}_s)$. A more comprehensive model that takes into account ambient-flow deviations $\delta\mathbf{v}$ (as in turbulence, for example) from a state nominally at rest yields (approximately)

$$\Delta_{\mathrm{eff}}(\mathbf{x}_s) \approx \Delta_1(\mathbf{x}_s) - (\mathbf{e}_s \cdot \mathbf{e}_R) \left[\Delta_2(\mathbf{x}_s) - \frac{2}{c_0}\, \mathbf{e}_s \cdot \delta\mathbf{v} \right]$$

$$\approx \frac{\delta(\rho c^2)}{\rho c^2} - (\mathbf{e}_s \cdot \mathbf{e}_R) \left[\frac{\delta(\rho c^2)}{\rho c^2} - \frac{2}{c}\, \delta(c + \mathbf{e}_s \cdot \mathbf{v}) \right]$$

R are both much larger than $1/k$ for points within the scattering volume. (Note that the $\mathbf{e}_s$ and $\mathbf{e}_R$ appearing here are analogous to the $\mathbf{e}_k$ and $\mathbf{e}_r$ in the preceding section.)

For the monostatic configuration, $\mathbf{e}_s \cdot \mathbf{e}_R$ approximates to -1, so that Eqs. (9-1.19*b*) yield

$$\Delta_{\text{eff}}(\mathbf{x}) \approx \frac{2\delta(\rho c)}{\rho c} \qquad \text{backscatter} \tag{9-2.5}$$

where $\delta(\rho c)$ is the deviation of the characteristic impedance of the medium from its nominal value ρc (the subscript zero being now omitted). This concurs with the results in Sec. 3-6 for reflection at normal incidence from an interface separating two fluids, where reflection arises from discontinuities in impedance rather than in sound speed or density per se.

For a single clustered inhomogeneity of dimensions much smaller than a wavelength, the Born approximation leads to the replacement of the Φ_ν in Eq. (9-1.25*b*) by x_ν, so that Eq. (9-1.21) agrees with Eq. (3) above. The resulting Born approximation for the backscattering cross section, identified from Eqs. (9-1.13) and (9-1.25), is consequently

$$\sigma_{\text{back}} = \left[\frac{1}{\pi^{1/2}} \frac{k^2}{\rho c} \iiint \delta(\rho c)\, dV\right]^2 \tag{9-2.6}$$

Scattering Volumes Delimited by Electroacoustic Transducers

In order to refine the concept of a scattering volume further, it is convenient to regard the transmitter and receiver explicitly as electroacoustic transducers (see Sec. 4-10), so that a single function $i_{\text{tr}}(t)$, *loudspeaker excitation current,* characterizes the transmission, and a second function $e_{\text{rec}}(t)$, *microphone open-circuit voltage,* characterizes the reception. Analogous quantities can be defined for mechano-acoustic transducers: a rigid piston oscillating in a finite baffle is characterized by a normal velocity $v_n(t)$; one acting as a receiver is characterized by the force exerted on the piston face by the impinging sound wave, the piston being held virtually motionless. The physical design of the two transducers is immaterial for the discussion that follows provided the time-dependent functions we use are linearly related to the transmitted and incident acoustic fields, but the electroacoustic realizations of the model are most representative of typical applications.

When driven at constant frequency ω by a current of complex amplitude $\hat{i}_{\text{tr}}$,

---

so that the scattering can be regarded as being caused by fluctuations in bulk modulus ρc^2 and in the wave speed $c + \mathbf{e}_s \cdot \mathbf{v}$ in the direction of propagation. For derivations leading to this, see G. K. Batchelor, "Wave Scattering dúe to Turbulence," in F. S. Sherman (ed.), *Symposium on Naval Hydrodynamics,* National Academy of Sciences, Washington, 1957, pp. 409–423; E. H. Brown and F. F. Hall, Jr., "Advances in Atmospheric Acoustics," *Rev. Geophys. Space Phys.,* **16:**47–110 (1978).

the transmitting transducer produces a far-field radiated acoustic pressure

$$\hat{p} = \frac{-i\omega\rho}{4\pi} M_{tr}\hat{F}_{tr}(\theta, \phi)\,(r^{-1}e^{ikr})\hat{\imath}_{tr} \tag{9-2.7}$$

in the direction with angular coordinates θ, ϕ. Here $\hat{F}_{tr}(\theta, \phi)$, whose phase is of minor interest, is normalized so that the transmitter radiation pattern $|\hat{F}_{tr}|^2$ is 1 when $\theta = 0$. The remaining constant factor $\omega\rho M_{tr}/4\pi$ is determined by the ratio $r|\hat{p}|/|\hat{\imath}_{tr}|$ along that axis. The quantity M_{tr} so introduced is a convenient description of the transducer's ability to transform electric current into far-field pressure (as explained below).

The analogous description of a receiving transducer sets

$$\hat{e}_{rec} = M_{rec}\hat{F}_{rec}(\theta, \phi)\hat{p} \tag{9-2.8}$$

to describe the voltage caused by a plane wave nominally having amplitude $\hat{p}$ at the transducer face and arriving from direction θ, ϕ. Here the receiver directivity function $|\hat{F}_{rec}|^2$ is normalized to 1 at $\theta = 0$. The constant M_{rec} is the microphone response at normal incidence, with units of volts per pascal. Equivalently, if a point source of volume velocity amplitude (source strength) $\hat{U}$ is located a great distance away at a point with coordinates (r, θ, ϕ) so that the $\hat{p}$ in Eq. (8) is $-(i\omega\rho/4\pi)\hat{U}r^{-1}e^{ikr}$, that equation becomes

$$\hat{e}_{rec} = \frac{-i\omega\rho}{4\pi} M_{rec}\hat{F}_{rec}(\theta, \phi)(r^{-1}e^{ikr})\hat{U} \tag{9-2.8a}$$

Comparison of this equation with Eq. (7) and reference to the reciprocity theorems of Secs. 4-9 and 4-10 indicate that if a transducer is a reciprocal transducer, then

$$M_{tr} = M_{rec} \qquad \hat{F}_{tr}(\theta, \phi) = \pm\hat{F}_{rec}(\theta, \phi) \tag{9-2.9}$$

Although we do not necessarily assume that the transducers are reciprocal in the discussion below, the possibility provides the rationale for the use of the symbol M_{tr} in Eq. (7).

The incorporation of Eqs. (7) and (8) into the scattering model proceeds with the observation, from Eq. (3), that the scattered wave originates from a distributed source with source volume velocity per unit volume (source strength density)

$$\frac{dU_s}{dV_s} = \frac{\Delta_{eff}(\mathbf{x}_s)}{\rho c^2}\frac{\partial}{\partial t}p_i(\mathbf{x}_s, t) \tag{9-2.10}$$

The receiver voltage is the superposition of the incremental contributions (8a) from each elemental volume; the incident pressure is as given by Eq. (7). An appropriate relabeling and juxtaposition of coordinate systems consequently yields

$$\hat{e}_{\text{rec}} = \frac{i\omega\rho k^2}{(4\pi)^2} M_{\text{rec}} M_{\text{tr}} \hat{i}_{\text{tr}} \iiint \hat{F}_{\text{rec}} \hat{F}_{\text{tr}} \Delta_{\text{eff}} \frac{e^{ik(R+r_s)}}{r_s R} \, dV_s \tag{9-2.11}$$

$$e_{\text{rec}}(t) = \frac{-\rho k^2}{(4\pi)^2} M_{\text{rec}} M_{\text{tr}} \iiint |\hat{F}_{\text{rec}} \hat{F}_{\text{tr}}| \frac{\Delta_{\text{eff}}}{r_s R} \frac{d}{dt} i_{\text{tr}} \, dV_s \tag{9-2.11a}$$

The second version, in which di_{tr}/dt is evaluated at $t - R/c - r_s/c - \varepsilon/\omega$, is a restatement of the first with the time dependence explicitly inserted, ε representing the position-dependent phase of $\hat{F}_{\text{rec}}(\mathbf{e}_R)\hat{F}_{\text{tr}}(\mathbf{e}_s)$. The unit vectors $\mathbf{e}_R$ and $\mathbf{e}_s$ (denoting directions) and the distances R and r_s here have the same meanings as in Eq. (3).

Although both versions of Eq. (11) are derived for constant-frequency propagation, the latter version should also apply to pulse propagation, whereby $i_{\text{tr}}(t)$ is of nearly constant frequency in the interval $0 < t < \tau$ and is zero or nearly zero outside that interval. The voltage output recorded during any small interval centered at t depends primarily on the scattering within a volume (see Fig. 9-7) between the ellipsoids $t = (R + r_s)/c$ and $t = \tau + (R + r_s)/c$. The volume is further restricted if (as is typically the case and as is assumed in what follows) the transmitter and receiver patterns are narrow-beam and if, for the bistatic case, the beams are directed to intersect in a localized region centered at a point $\bar{\mathbf{x}}_s$ and at distances $\bar{r}_s$ and $\bar{R}$ from the transmitter and receiver. For the monostatic case, we consider the beams to be coaxial and rely on the finite pulse duration to delimit the scattering to a finite volume.

Since the scattering reaching a receiver in the bistatic configuration comes from a finite volume regardless of whether or not the pulse duration is short, for simplicity we here discuss first bistatic sounding assuming constant-frequency transmission. Since $|\hat{F}_{\text{tr}}|$ and $|\hat{F}_{\text{rec}}|$ are 1 for direction $\bar{\mathbf{e}}_s$ from origin to $\bar{\mathbf{x}}_s$ and for direction $\bar{\mathbf{e}}_R$ from $\bar{\mathbf{x}}_s$ to the receiver center, the scattering volume consists

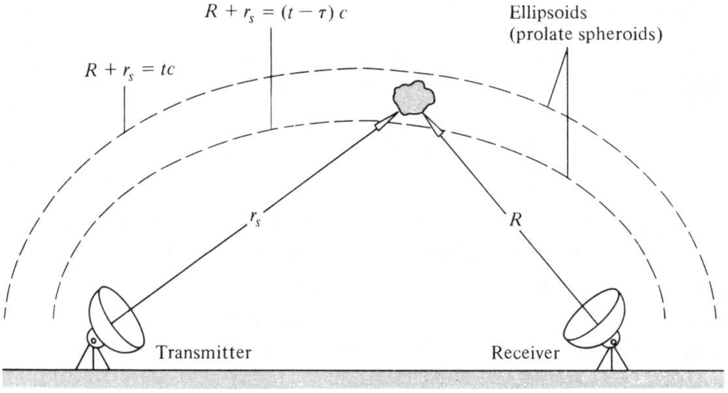

Figure 9-7 Concentric prolate spheroids delimiting region of possible scatterer locations for a bistatic pulse-sounding experiment. The given circumstances are such that the pulse transmission began at time 0 and ended at time τ; reception is taking place at time t.

primarily of all points where $|\hat{F}_{tr}| \cdot |\hat{F}_{rec}|$ is greater than, say, $\frac{1}{2}$. An estimate of its size is

$$\Delta V_s = \iiint |\hat{F}_{tr}|^2 |\hat{F}_{rec}|^2 \, dV_s \tag{9-2.12}$$

as explained below, in the derivation of Eq. (19).

The assumption that the scattering volume has dimensions much smaller than $\bar{r}_s$ and $\bar{R}$ allows the r_s and R in the denominator of the integrand in Eq. (11) to be replaced by $\bar{r}_s$ and $\bar{R}$. Additional substitutions from Eqs. (7) and (8) consequently yield

$$\hat{p}_{sc,ap} = \frac{\hat{p}_i}{(4\pi)^{1/2}} \frac{e^{ik\bar{R}}}{\bar{R}} \Psi \tag{9-2.13}$$

$$\Psi = \frac{-k^2}{(4\pi)^{1/2}} \iiint |\hat{F}_{tr}\hat{F}_{rec}| e^{i\epsilon} \Delta_{eff} e^{ik(R-\bar{R}+r_s-\bar{r}_s)} \, dV_s \tag{9-2.14}$$

where $\hat{p}_{sc,ap}$ is the apparent pressure impinging on the receiver from the direction of the scattering volume center and $\hat{p}_i$ is the incident wave's acoustic pressure at the volume's center $\bar{\mathbf{x}}_s$. The distinction between $\hat{p}_{sc,ap}$ and $\hat{p}_{sc}$ arises because the receiver weighs pressure contributions associated with different arrival directions differently.

Acoustic Radar Equation

The above formulation extends readily to monostatic sounding with a reciprocal transducer from a single localized scatterer at a point with coordinates $\bar{r}_s$, $\bar{\theta}_s$, $\bar{\phi}$. The quantity Ψ in Eq. (14) is replaced by one such that

$$|\Psi|^2 = |\hat{F}(\bar{\theta}_s, \bar{\phi}_s)|^4 \sigma_{back} \tag{9-2.15}$$

where σ_{back} is as given by Eq. (6) for a small weak inhomogeneity. Equation (13) then yields the *acoustic radar equation*†

$$\frac{I_{sc}}{(4\pi r^2 I_i)_0} \frac{(e^2_{rec})_{av}}{(e^2_{rec})_{av,0}} = \frac{1}{(4\pi)^2} \frac{\sigma_{back}}{\bar{r}_s^4} |\hat{F}(\bar{\theta}_s, \bar{\phi}_s)|^4 \tag{9-2.16}$$

where

$$\frac{(e^2_{rec})_{av}}{(e^2_{rec})_{av,0}} = \frac{I_{sc,ap}}{I_{sc}}$$

is the ratio of mean square voltage recorded to what would have been recorded if a signal had been of the same intensity incident from $\theta = 0$. Here I_{sc} is the actual acoustic intensity returning to the transducer, while $I_{sc,ap}$ is its apparent value when the returning wave is regarded as having come from the $\theta = 0$ direction. The quantity $(4\pi r^2 I_i)_0$, equal to $4\pi r^2$ times the transmitted intensity in

† So called because it is the acoustical counterpart of the *free-space radar transmission equation* (widely referred to as the *radar equation*) given, for example, by D. E. Kerr, *Propagation of Short Radio Waves,* McGraw-Hill, New York, 1951, reprinted by Dover, New York, 1965, p. 33. Kerr's equation, rewritten in the present text's notation and with application of his eqs. (13), (14), and (19), is the same as our Eq. (16).

the $\theta = 0$ direction at far-field distance r, can be regarded as acoustic power output times the *directive gain* associated with that direction; $(4\pi r^2 I_i)_0|\hat{F}(\bar{\theta}_s, \bar{\phi}_s)|^2$ is power output times directive gain associated with the direction $\bar{\theta}_s$, $\bar{\phi}_s$.

Incoherent Scattering

If the inhomogeneities causing scattering are dispersed throughout the scattering volume, the relative phases of contributions from different volume elements in Eq. (14) are approximately taken into account with the substitution

$$R - \bar{R} + r_s - \bar{r}_s \approx (\bar{\mathbf{e}}_s - \bar{\mathbf{e}}_R) \cdot \boldsymbol{\xi} \qquad (9\text{-}2.17)$$

which results from a truncated power-series expansion in the components of $\boldsymbol{\xi} = \mathbf{x}_s - \bar{\mathbf{x}}_s$. With the abbreviation $\Delta\mathbf{k}$ to represent the change $(\bar{\mathbf{e}}_R - \bar{\mathbf{e}}_s)k$ in wave-number vector undergone during the scattering, Eq. (14) yields

$$|\Psi|^2 = \frac{k^4}{4\pi}\int \cdots \int \Phi(\boldsymbol{\xi})\Phi^*(\boldsymbol{\xi}')\Delta_{\text{eff}}(\boldsymbol{\xi})\Delta_{\text{eff}}(\boldsymbol{\xi}')e^{i\Delta\mathbf{k}\cdot(\boldsymbol{\xi}'-\boldsymbol{\xi})}\, dV_{\xi}\, dV_{\xi'} \qquad (9\text{-}2.18)$$

where we also use the abbreviation $\Phi(\xi)$ for $|\hat{F}_{\text{tr}}\hat{F}_{\text{rec}}|e^{i\epsilon}$ as evaluated at the position $\bar{\mathbf{x}}_s + \boldsymbol{\xi}$.

If $\Delta_{\text{eff}}(\boldsymbol{\xi})$ in different regions appears to be statistically indistinguishable, the idealization of a random medium† is appropriate. The notion of a statistically homogeneous random process whose correlation disappears over a relatively short distance allows $\Delta_{\text{eff}}(\boldsymbol{\xi})\Delta_{\text{eff}}(\boldsymbol{\xi}')$ to be replaced by its ensemble average or, equivalently, by the local spatial average of $\Delta_{\text{eff}}(\boldsymbol{\xi})\Delta_{\text{eff}}(\boldsymbol{\xi} + \Delta\boldsymbol{\xi})$; this average is the spatial autocorrelation function $R(\Delta\boldsymbol{\xi};\ \Delta_{\text{eff}})$. The *incoherent-scattering model,* whereby the acoustic power scattered by moderately distant inhomogeneities are additive, results when the autocorrelation function is negligibly small for any $\Delta\boldsymbol{\xi}$ whose magnitude is comparable to or larger than a characteristic length over which $\Phi(\boldsymbol{\xi})$ changes appreciably. Such assumptions reduce Eq. (18) to

$$|\Psi|^2 = \eta(k, \Delta\mathbf{k})\, \Delta V_s \qquad (9\text{-}2.19)$$

where ΔV_s is as defined by Eq. (12) and where

$$\eta(k, \Delta\mathbf{k}) = \frac{k^4}{4\pi}\iiint R(\Delta\boldsymbol{\xi};\ \Delta_{\text{eff}})e^{i\Delta\mathbf{k}\cdot\Delta\boldsymbol{\xi}}\, d(\Delta\xi_x)\, d(\Delta\xi_y)\, d(\Delta\xi_z) \qquad (9\text{-}2.20)$$

$$= \tfrac{1}{4}\pi^2 k^4 S(\Delta\mathbf{k};\ \Delta_{\text{eff}}) \qquad (9\text{-}2.20a)$$

† Frequently cited references on acoustic waves in random media are L. A. Chernov, *Wave Propagation in a Random Medium,* 1958, 1960 trans. R. A. Silverman, reprinted by Dover, New York, 1967; V. I. Tatarski, *Wave Propagation in a Turbulent Medium,* 1961 trans. R. A. Silverman, reprinted by Dover, New York, 1967; V. I. Tatarski, *The Effects of the Turbulent Atmosphere on Wave Propagation,* 1967, 1970, 1971 trans. by Israel Program for Scientific Translations, available from U.S. Department of Commerce, National Technical Information Service, Springfield, VA 22151. Brown and Hall, "Advances in Atmospheric Acoustics," list and appraise much of the literature pertaining to the subject.

Here $S(\Delta\mathbf{k};\, \Delta_{\text{eff}})$, defined implicitly by the two versions of Eq. (20), is recognized with reference to the Wiener-Khintchine theorem (see Sec. 2-10) as the spectral density of $\Delta_{\text{eff}}(\boldsymbol{\xi})$ in wave-number space. The normalization adopted is such that

$$\langle \Delta_{\text{eff}}^2 \rangle = \int\!\!\int_0^\infty\!\!\int S(\Delta\mathbf{k};\, \Delta_{\text{eff}})\, d(\Delta k_x)\, d(\Delta k_y)\, d(\Delta k_y) \tag{9-2.21}$$

gives the mean squared value of $\Delta_{\text{eff}}(\boldsymbol{\xi})$. The nominal propagation-medium selection is here assumed to yield $\langle \Delta_{\text{eff}} \rangle = 0$.

Equation (19), in conjunction with Eq. (13), leads to the *bistatic acoustic sounding equation*

$$\frac{I_{\text{sc,ap}}}{(4\pi r^2 I_i)_0} = \frac{\eta\, \Delta V_s}{(4\pi)^2 \bar{r}_s^2 \bar{R}^2} \tag{9-2.22}$$

with η identified as the apparent bistatic cross section per unit volume. The implication here that the scattered intensity is proportional to scattering volume, which is the distinguishing feature of incoherent scattering, requires that inhomogeneities causing scattering be randomly dispersed and that any correlation length associated with the inhomogeneities be small compared with the scattering volume's dimensions. In contrast, if the scattering volume is small in terms of a correlation length, the far-field acoustic-pressure contributions scattered by different volume elements are in phase and reinforce each other; the scattering is then coherent, and the apparent bistatic cross section is proportional to the square of the scattering volume.

The Echosonde Equation

The incoherent-scattering idealization allows a lucid interpretation of pulse-echo measurements of scattering from inhomogeneous media. Equation (13), with such an idealization, implies that for the monostatic case

$$\delta\left(\frac{E}{A}\right)_{\text{sc,ap}} = \frac{(4\pi r^2 I_i)_0\, \delta t}{(4\pi)^2} \frac{\eta |\hat{F}|^4\, \delta V_s}{r_s^4} \tag{9-2.23}$$

is the apparent backscatter energy received per unit area due to scattering during time interval δt from volume element δV_s at a distant point r_s, θ_s, ϕ_s. The quantity $(4\pi r^2 I_i)_0$ is representative of the power the transmitter was radiating at time $t - 2r_s/c$. Hence the total apparent backscattered energy per unit area received up to time t is

$$\left(\frac{E}{A}\right)_{\text{sc,ap}} = \frac{1}{(4\pi)^2} \int\!\!\int\!\!\int \frac{\eta |\hat{F}|^4}{r_s^4} \left[\int_{-\infty}^{t-2r_s/c} (4\pi r^2 I_i)_0\, dt'\right] dV_s \tag{9-2.24}$$

where $(4\pi r^2 I_i)_0$ is zero up until $t' = 0$ and is zero for $t' > \tau$. Taking the time derivative and subsequently transforming the r_s integration into one over

$t' = t - 2r_s/c$ yields

$$I_{\mathrm{sc,ap}} = \frac{c/2}{(4\pi)^2} \int_0^{\tau} \left[\iint \frac{\eta |\hat{F}|^4}{r_s^2} \, d\Omega_s \right] (4\pi r^2 I_i)_0 \, dt' \tag{9-2.25}$$

for the apparent backscattered intensity. The quantity in brackets here is understood to be evaluated at $r_s = (t - t')c/2$. Consideration is limited to reception times t that are greater than the pulse duration τ.

If the time t is further taken to be much greater than τ, Eq. (25) approximates to†

$$I_{\mathrm{sc,ap}} \approx \frac{c\tau/2}{(4\pi)^2} \frac{\eta \, \Delta\Omega_s}{\bar{r}_s^2} (4\pi r^2 I_i)_0 \tag{9-2.26}$$

with

$$\Delta\Omega_s = \iint |\hat{F}|^4 \, d\Omega_s \tag{9-2.27}$$

interpreted as the solid angle being probed. The quantity $(4\pi r^2 I_i)_0$ now represents a time average, over the pulse duration τ, of transmitted power times directive gain of the transmitter. The radial distance $\bar{r}_s$ is approximately $ct/2$ and represents an average distance to the scattering volume [extending from $r_s = (t - \tau)c/2$ to $r_s = tc/2$]. The quantity η is the average backscattering cross section per unit volume [Eq. (20a) with $\Delta\mathbf{k} = 2k\mathbf{e}_z$, where $\mathbf{e}_z$ points along the beam's axis] for the spherical shell of solid angle $\Delta\Omega_s$. As before, $I_{\mathrm{sc,ap}}$ is the apparent acoustic intensity of the backscattered wave at the transducer, with account taken of the directional response characteristics during reception.

The applicability of the incoherent-scattering assumption to the derivation of Eq. (26) requires $c\tau$ and $\bar{r}_s(\Delta\Omega_s)^{1/2}$ be large compared with a correlation length of the inhomogeneities. If this is not so but the scattering medium is nevertheless random, the prediction (26) is an ensemble average of possible outcomes.

The generalization of the above considerations to pulse-echo sounding with the bistatic configuration yields

$$I_{\mathrm{sc,ap}} = \frac{\eta}{(4\pi)^2} (4\pi r^2 I_i)_0 \iiint'' \frac{|\hat{F}_{\mathrm{tr}}|^2 |\hat{F}_{\mathrm{rec}}|^2}{r_s^2 R^2} \, dV_s \tag{9-2.28}$$

where the double prime on the integral indicates that the region of integration is restricted to that lying between the prolate spheroids $r_s + R = tc$ and $r_s + R = (t - \tau)c$.

The similarities of Eq. (28) with Eq. (26) are emphasized with the introduction of a dimensionless *aspect factor* $\mathcal{A}$, equal to

$$\mathcal{A} = \frac{\bar{R}^2}{(c\tau/2)\, \Delta\Omega_{\mathrm{tr}}} \iiint'' \frac{|\hat{F}_{\mathrm{tr}}|^2 |\hat{F}_{\mathrm{rec}}|^2}{r_s^2 R^2} \, dV_s \tag{9-2.29}$$

where

$$\Delta\Omega_{\mathrm{tr}} = \iint |\hat{F}_{\mathrm{tr}}|^2 d\Omega_s \tag{9-2.30}$$

† This is analogous to the definition in R. E. Huschke (ed.), *Glossary of Meteorology*, American Meteorological Society, Boston, 1959, of the *radar storm-detection equation*, with η identified as the *radar reflectivity* of the echoing volume of the storm per unit volume. A derivation due to H. Goldstein appears in Kerr, *Propagation of Short Radio Waves*, pp. 588–591.

is the apparent beam width in sterradians of the transmitted beam and $\bar{R}$ is the distance from the receiver to the intersection of the transmitted beam's axis with the spheroid $R + r_s = ct$. Insertion of these definitions into Eq. (28) yields the *echosonde equation*†

$$I_{\text{sc,ap}} \approx \frac{c\tau/2}{(4\pi)^2} \frac{\eta\, \Delta\Omega_{\text{tr}} \mathscr{A}}{\bar{R}^2} (4\pi r^2 I_i)_0 \tag{9-2.31}$$

For the monostatic configuration, the aspect factor $\mathscr{A}$ becomes $\Delta\Omega_s/\Delta\Omega_{\text{tr}}$; when a reciprocal transducer is used to receive as well as transmit, the $\mathscr{A}$ must be less than 1 but approaches 1 for a sharp-edged beam. If $|\hat{F}|^2$ varies with angle as $\exp(-\alpha\theta^2)$, where α is somewhat larger than 1, then $\mathscr{A}$ is $\frac{1}{2}$; if it varies as $1/(1 + \alpha\theta^2)^2$, then $\mathscr{A}$ is $\frac{1}{3}$.

Example A transmitter in air sends out a 5-kHz pulse of 10 W acoustic power and pulse length $c\tau = 3.3$ m. The transmitter beam with width $\Delta\Omega_{\text{tr}}$ of the order of 0.1 sr is aimed at an angle of $\gamma_{\text{tr}} = 45°$ with the ground. An omnidirectional receiver at a distance d of 100 m (see Fig. 9-8) receives sound of intensity $I_{\text{sc}} = 10^{-14}$ W/m^2 after an interval of $l/c = 420$ ms. Make the idealizations that sound travels with constant speed of 340 m/s and that attenuation is negligible, to obtain a lower limit for the bistatic cross section per unit volume causing the scattering.

SOLUTION The problem statement and trigonometric principles require

$$(l - \bar{r}_s)^2 = \bar{r}_s^2 + d^2 - 2\bar{r}_s d \cos\gamma_{\text{tr}} \qquad \bar{R} = l - \bar{r}_s \tag{9-2.32}$$

$$\bar{r}_s = \frac{l^2 - d^2}{2l - 2d\cos\gamma_{\text{tr}}} \qquad h = \bar{r}_s \sin\gamma_{\text{tr}} \qquad \sin\gamma_{\text{rec}} = \frac{h}{\bar{R}} \tag{9-2.33}$$

The omnidirectional receiver assumption implies $|\hat{F}_{\text{rec}}|^2 = 1$. The aspect factor accordingly reduces to approximately $\mathscr{A} \approx \delta r_s/(c\tau/2)$, where δr_s is the incre-

† The label is attributed to W. D. Neff by Brown and Hall in their review, "Advances in Atmospheric Acoustics." The correspondence of our Eq. (31) with the Brown and Hall version emerges with the neglect of background winds and of attenuation along paths from transmitter to scattering volume and from scattering volume to receiver. The following identifications of Brown and Hall's symbols with those used here also apply:

$$\frac{\mathscr{P}_R}{g\varepsilon_R A_R} = I_{\text{sc,ap}} \qquad \varepsilon_T \mathscr{P}_T = (4\pi r^2 I_i)_0 \frac{\Delta\Omega_{\text{tr}}}{4\pi} \qquad R_s = \bar{R} \qquad \sigma_s = \frac{\eta}{4\pi} \qquad l_p = c\tau$$

where $\mathscr{P}_R$ is received electric power, ε_R is acoustical-to-electrical conversion efficiency of the receiver when a plane wave is incident at normal incidence, the acoustical power being taken as receiver area A_R times incident intensity. The g is a receiver directivity gain equal to ratio of apparent incident intensity at the receiver to actual incident intensity; $\varepsilon_T \mathscr{P}_T$ is acoustic power transmitted; $\mathscr{P}_T$ is electric power consumed by transmitter; l_p is pulse length; and σ_s is the differential scattering cross section per unit volume. Appropriate translations between terminology and symbols for analogous concepts that have arisen in other subfields (underwater acoustics, ultrasonic nondestructive testing, biomedical ultrasonics) are usually easily effected if the principles and approximations leading to Eq. (31) are understood.

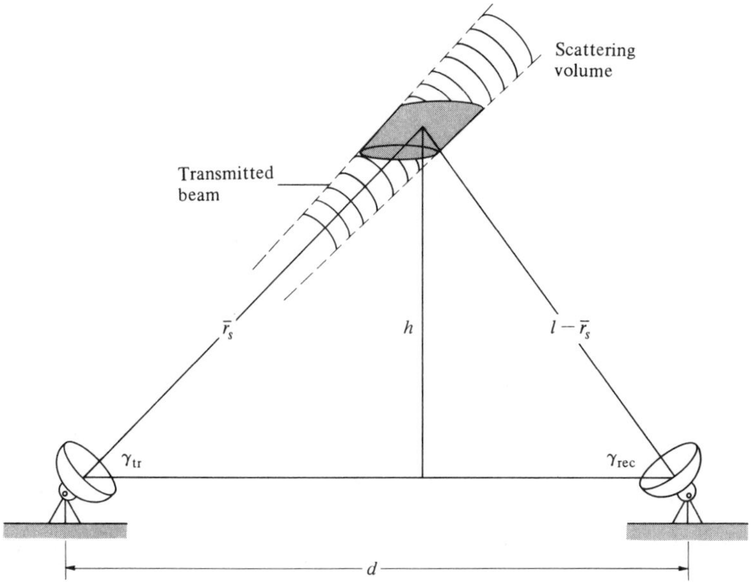

Figure 9-8 Parameters used in example discussed in text. Quantities d, γ_{tr}, and l are specified; the width of the transmitted beam is also given. The objective is to determine the bistatic cross section per unit volume.

mental radial distance the transmitter beam traverses in going through the scattering volume. Taking the differential of the expression for r_s in terms of l yields

$$\delta r_s = \frac{l^2 + d^2 - 2ld \cos \gamma_{tr}}{2(l - d \cos \gamma_{tr})^2} \delta l \qquad \mathscr{A} = \frac{l^2 + d^2 - 2ld \cos \gamma_{tr}}{(l - d \cos \gamma_{tr})^2} \tag{9-2.34}$$

where the latter results from the identification of δl as $c\tau$. Inserting the numbers cited above then gives $\bar{r}_s = 72$ m, $\bar{R} = 71$ m, $h = 51$ m, $\gamma_{rec} = 46°$, and $\mathscr{A} = 2.0$. Consequently, Eq. (31) states that

$$10^{-14} = \frac{(3.3/2)\eta}{(4\pi)^2} \frac{(2.0)(4\pi)(10)}{(71)^2}$$

which yields $\eta = 1.9 \times 10^{-11}$ m^2/m^3.

9-3 THE DOPPLER EFFECT

The classic prototype for the Doppler effect† (frequency shift associated with motion) is a constant-frequency sound source moving at constant subsonic

† Johann C. Doppler, who first propounded the principle in 1842 (although for a phenomenon that it is inadequate to explain fully), gives an account of it in "Remarks on My Theory of the Colored Light from Double Stars, with Regard to the Objections Raised by Dr. Ballot of Utrecht,"

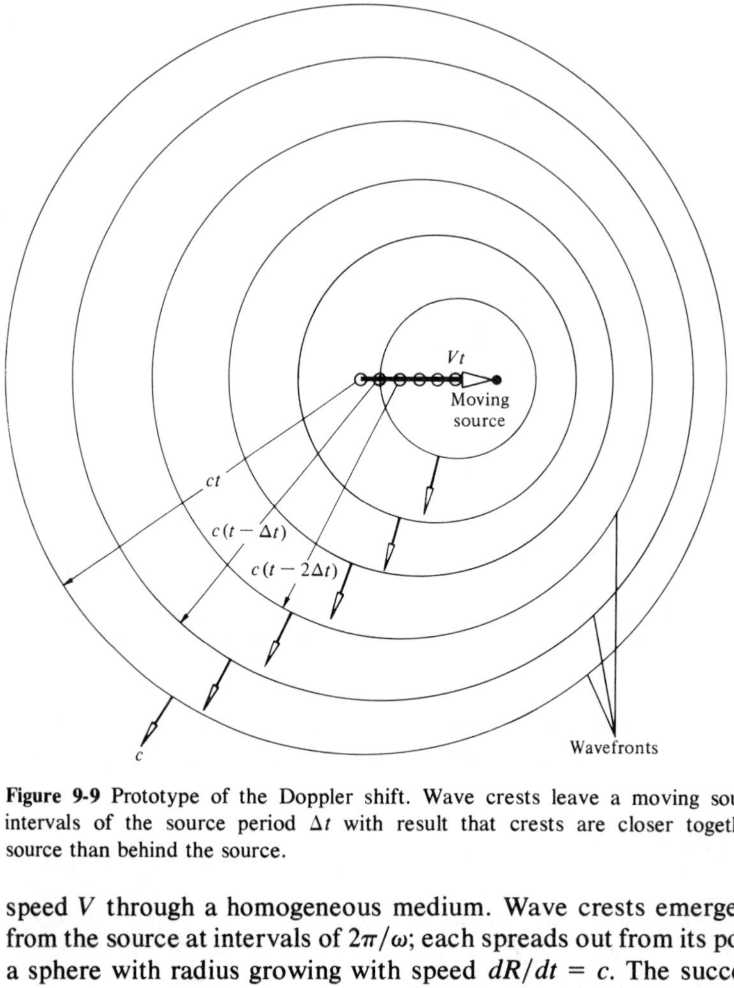

Figure 9-9 Prototype of the Doppler shift. Wave crests leave a moving source (speed V) at intervals of the source period Δt with result that crests are closer together ahead of the source than behind the source.

speed V through a homogeneous medium. Wave crests emerge (see Fig. 9-9) from the source at intervals of $2\pi/\omega$; each spreads out from its point of origin as a sphere with radius growing with speed $dR/dt = c$. The successively generated spheres are closer together ahead of the source but farther apart behind the source. Since the number of crests passing a stationary listener per unit time determines the frequency associated with the disturbance, the frequency received is higher ahead of the source but lower behind the source. A common instance of this *Doppler shift* is the drop in frequency of a train whistle as heard by someone when a locomotive speeds by.

Ann. Phys. Chem., **68:**1–35 (1846). A historical appraisal is given by J. Scheiner, "Johann Christian Doppler and the Principle Named after Him," *Himmel Erde*, **8:**260–271 (1896). Rayleigh, *Theory of Sound*, vol. 2, 2d ed., pp. 154–156, summarizes early experimental work on the acoustical Doppler effect by B. Ballot, S. Russell, E. Mach, R. König, and A. M. Mayer. See also the historical comments by A. Wood, *Acoustics*, 2d ed., 1960, Dover, New York, 1966, pp. 324–331. For discussions of the Doppler shift in electromagnetism from the viewpoint of the special theory of relativity, see J. D. Jackson, *Classical Electrodynamics*, Wiley, New York, 1962, pp. 360–364; and D. S. Jones, *The Theory of Electromagnetism*, Pergamon, Oxford, 1964, pp. 115–130.

Doppler Shift for a Moving Source

The magnitude of the frequency shift for the circumstances just described can be predicted by an extension of the geometric-acoustics model introduced in Sec. 8-1. Near the source trajectory, taken as the x axis, the phase $\phi(\mathbf{x}, t)$ of the disturbance is $\omega_0\tau$ at $x = V\tau$, where ω_0 is the intrinsic frequency at the source. Since surfaces of constant phase move at constant speed c, one accordingly has the parametric description

$$\phi(\mathbf{x}, t) = \omega_0 \tau \qquad (9\text{-}3.1a)$$

$$|\mathbf{x} - V\tau \mathbf{e}_x| = (t - \tau)c \qquad (9\text{-}3.1b)$$

Since the latter implies

$$(x - V\tau)^2 + r^2 = (t - \tau)^2 c^2 \qquad (9\text{-}3.2)$$

with $r^2 = y^2 + z^2$ and $\tau = \phi/\omega_0$, it in turn yields

$$\frac{\phi(\mathbf{x}, t)}{\omega_0} = \frac{c^2 t - Vx}{c^2 - V^2} - \left[\frac{x^2 + r^2 - t^2c^2}{c^2 - V^2} + \left(\frac{c^2 t - Vx}{c^2 - V^2}\right)^2\right]^{1/2} \qquad (9\text{-}3.3)$$

The sign of the radical here is selected to be such that $\phi/\omega_0 \to t - (x^2 + r^2)^{1/2}/c$ in the limit $V \to 0$, as required by Eq. (1b).

The frequency $\omega(\mathbf{x}, t)$ perceived by a stationary listener is $\partial\phi/\partial t$, with $\mathbf{x}$ held fixed in the differentiation. Although this can be derived directly from Eq. (3), it is more instructive to extract ω by implicit differentiation of Eq. (1b); doing so gives

$$-\frac{\omega}{\omega_0}\frac{V\mathbf{e}_x \cdot \mathbf{R}}{R} = \left(1 - \frac{\omega}{\omega_0}\right)c \qquad (9\text{-}3.4)$$

so that

$$\omega = \frac{\omega_0 c}{c - V\cos\theta} \qquad (9\text{-}3.5)$$

Here $\mathbf{R} = \mathbf{x} - V\tau\mathbf{e}_x$ is the vector-ray displacement to the reception point $\mathbf{x}$ from the point where the ray left the source's trajectory; the angle $\theta(\mathbf{x}, t)$ is that between the vector $\mathbf{R}$ and the velocity $V\mathbf{e}_x$. The frequency shift therefore depends on only the velocity component directed toward the listener. The result holds regardless of the detailed time history of the trajectory; the Doppler-shifted frequency at a given time and position is affected only by the source's velocity and frequency at the instant of generation of the wavelet currently being received. The source does not have to be traveling with constant velocity or in a straight line for Eq. (5) to apply;† however, determination of the point on the trajectory from which the wavelet originates requires additional labor to match the kinematics, possibly a graphical solution if the motion is not rectilinear.

† See, for example, M. V. Lowson, "The Sound Field for Singularities in Motion," *Proc. R. Soc. Lond.*, **A286**:559–572 (1965).

Galilean Transformations

A transformation from one coordinate system to a second moving at constant speed relative to the first, with the classical assumption that velocities add vectorially, is a galilean transformation. A Doppler shift accompanies any such change in coordinate system because the frequency is not a galilean invariant.

Let $\mathbf{x}_2(\mathbf{x}_1, t)$ describe the position in coordinate system 2 of a fixed point $\mathbf{x}_1$ in coordinate system 1 such that

$$\mathbf{x}_2(\mathbf{x}_1, t) = \mathbf{x}_1 - (t - t_0)\mathbf{V}_{2;1} \tag{9-3.6}$$

with $\mathbf{x}_2$ equaling $\mathbf{x}_1$ at time t_0 and with $\mathbf{V}_{2;1}$ denoting the velocity of the second system's axes with respect to the first. If $\phi_1(\mathbf{x}_1, t)$ and $\phi_2(\mathbf{x}_2, t)$ describe phases of the same acoustic disturbance, the fact that wave crests appear as wave crests regardless of the coordinate system's velocity requires

$$\phi_1(\mathbf{x}_2 + (t - t_0)\mathbf{V}_{2;1}, t) = \phi_2(\mathbf{x}_2, t) \tag{9-3.7}$$

In either coordinate system, the wave-number vector $\mathbf{k}$ is such that $\mathbf{k} = -\nabla\phi$, while the angular frequency ω is such that $\omega = \partial\phi/\partial t$. Consequently, differentiating Eq. (7) with respect to one of the components of $\mathbf{x}_2$ or with respect to t and then setting t to t_0 yield

$$\mathbf{k}_2 = \mathbf{k}_1 \qquad \omega_2 = \omega_1 - \mathbf{V}_{2;1} \cdot \mathbf{k}_1 \tag{9-3.8}$$

for the galilean transformations of wave-number vector and angular frequency.

A derivation of Eq. (5) from Eq. (8) proceeds with the selection of a system moving with the source as coordinate system 1 and with a system at rest as coordinate system 2. For coordinate system 1, the boundary conditions imposed by the vibrating source can be replaced by normal displacement oscillations on a motionless surface; a linear acoustic model therefore applies, and the disturbance appears to have angular frequency ω_0 everywhere, even though the ambient medium is moving. (This presumes low-amplitude oscillations and neglects any turbulence associated with the ambient flow past the source.) In coordinate system 2, on the other hand, the absence of an ambient flow allows one to use the plane-wave relation $\omega\mathbf{n} = c\mathbf{k}$ and to equate ray direction with that of $\mathbf{k}$. Thus we have

$$\mathbf{k}_2 = \mathbf{k}_1 = \frac{\omega_2}{c}\mathbf{e}_R \qquad \omega_1 = \omega_0 \tag{9-3.9}$$

where $\mathbf{e}_R$ is the unit vector $\mathbf{R}/R$ appearing in Eq. (4). Then, since $\mathbf{V}_{2;1} = -\mathbf{V}$, where $\mathbf{V}$ is the velocity of the source with respect to a motionless ambient medium, Eq. (8) yields

$$\omega_2 = \omega_0 + \mathbf{V} \cdot \mathbf{e}_R \frac{\omega_2}{c} \qquad \omega_2 = \frac{\omega_0}{1 - \mathbf{V} \cdot \mathbf{e}_R/c} \tag{9-3.10}$$

which is equivalent to Eq. (5).

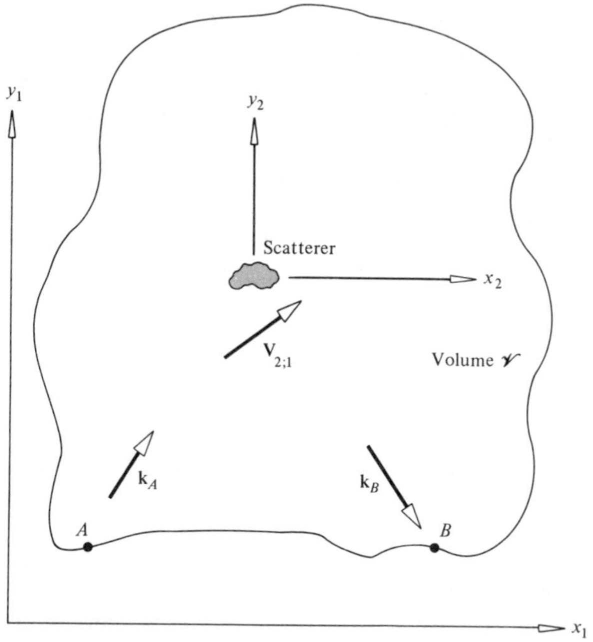

Figure 9-10 Construction used to derive Doppler shift of pulse scattered by an inhomogeneity drifting along with the ambient flow at velocity $\mathbf{V}_{2,1}$ relative to transmitter and receiver. Volume $\mathscr{V}$ and coordinate system x_2, y_2, z_2 move so that the scatterer appears at rest. Ray from source to scatterer enters $\mathscr{V}$ at A; ray from scatterer to receiver leaves $\mathscr{V}$ at B.

Echoes from Moving Targets

A scatterer's motion can cause a Doppler shift in the echo detected by a distant receiver.† This in turn allows a deduction from the echo's frequency of one of the velocity components. The relation between the two can be understood from the consideration of a coordinate system (labeled by 2) moving with the scatterer (see Fig. 9-10). A volume $\mathscr{V}$ surrounding the scatterer is presumed to be such that within it and in terms of coordinate system 2 the medium's properties and the scatterer's nominal location are time-independent.

The discussion here presumes a bistatic measurement configuration; the ray connecting transmitter and scatterer enters $\mathscr{V}$ at point A; that connecting scatterer and receiver leaves $\mathscr{V}$ at B. Since the scatterer is moving, A and B also

† Acoustical applications of the Doppler effect date back to World War II reports on underwater sound by C. H. Eckart and C. L. Pekeris; citations and a brief summary of wartime work are given by E. Gerjuoy and A. Yaspan, *Physics of Sound in the Sea,* 1946, vol. 8 of *Summary Technical Report of Division 6,* National Defense Research Committee (U.S.), reprinted 1969, U.S. Government Printing Office, Washington, pp. 329–331, 552. Underwater acoustic applications and related system problems are summarized by J. W. Horton, *Fundamentals of Sonar,* 2d ed., United States Naval Institute, Annapolis, Md., 1959, pp. 364–378, and by C. S. Clay and H. Medwin, *Acoustical Oceanography: Principles and Applications,* Wiley-Interscience, New York, 1977, pp. 334–337.

move with respect to the transmitter and receiver. We accordingly further refine the definition of their positions so that (1) A and B are not moving in terms of coordinate system 2 and (2) they occupy appropriate instantaneous positions in terms of coordinate system 1. The latter are determined by the choice of the time of echo reception and by the time history of the corresponding broken-ray trajectory connecting transmitter to scatterer to receiver. Here coordinate system 1 is that in which the transmitter and receiver appear motionless.

When examined in terms of coordinate system 2, the incident and scattered waves appear to have the same frequency, which we here denote as ω_2. Thus, with $\mathbf{k}_A$ and $\mathbf{k}_B$ denoting the incident and scattered signals' wave-number vectors at A and B, respectively, the galilean transformation relations (8) imply

$$\omega_{A,1} = \omega_2 + \mathbf{V}_{2;1} \cdot \mathbf{k}_A \qquad \omega_{B,1} = \omega_2 + \mathbf{V}_{2;1} \cdot \mathbf{k}_B \tag{9-3.11}$$

for the angular frequencies at A and B as measured in coordinate system 1. These, however, are the transmitted and received frequencies, ω_{tr} and ω_{rec}, while $\mathbf{V}_{2;1}$ is the velocity $\mathbf{V}_{\text{sc}}$ of the scatterer, so elimination of ω_2 yields

$$\omega_{\text{rec}} - \omega_{\text{tr}} = \mathbf{V}_{\text{sc}} \cdot (\mathbf{k}_B - \mathbf{k}_A) \tag{9-3.12}$$

The simplest idealization accompanying the application of the above relation is that, apart from the scatterer and its wake, the ambient medium is homogeneous and at rest relative to the transmitter and receiver. Then $\mathbf{k}_A$ and $\mathbf{k}_B$ become $(\omega_{\text{tr}}/c)\mathbf{n}_i$ and $(\omega_{\text{rec}}/c)\mathbf{n}_{\text{sc}}$, where $\mathbf{n}_i$ and $\mathbf{n}_{\text{sc}}$ are unit vectors in the directions of the incident and scattered waves. The Doppler shift to first order in $\mathbf{V}_{\text{sc}}/c$ accordingly satisfies

$$\frac{\omega_{\text{rec}} - \omega_{\text{tr}}}{\omega_{\text{tr}}} = \frac{\mathbf{V}_{\text{sc}}}{c} \cdot (\mathbf{n}_{\text{sc}} - \mathbf{n}_i) \tag{9-3.13}$$

$$= \frac{\mathbf{V}_{\text{sc}}}{c} \cdot \mathbf{e}_{\text{bi}}\, 2 \sin \tfrac{1}{2}\, \Delta\theta \tag{9-3.13a}$$

where, in the latter version, the deflection angle $\Delta\theta$ is that between $\mathbf{n}_{\text{sc}}$ and $\mathbf{n}_i$, while $\mathbf{e}_{\text{bi}}$ is the unit vector in the direction $\mathbf{n}_{\text{sc}} - \mathbf{n}_i$, which bisects the triangle with sides $\mathbf{n}_i$, $\mathbf{n}_{\text{sc}}$ and $\mathbf{n}_i + \mathbf{n}_{\text{sc}}$ and is perpendicular to $\mathbf{n}_i + \mathbf{n}_{\text{sc}}$.

For the monostatic echo-sounding configuration, $\mathbf{n}_{\text{sc}}$ is $-\mathbf{n}_i$, and so the right side of Eq. (13) becomes $-2\mathbf{V}_{\text{sc}} \cdot \mathbf{n}_i/c$, with the result that the Doppler shift is proportional to twice the component of the scatterer's velocity toward the transmitter.

Doppler-Shift Velocimeters

In typical applications† where the Doppler shift is used to measure ambient fluid velocity, the scatterer is presumed to be drifting along with the flow but

† The applications currently receiving principal attention in the archival literature are the measurement of flow velocities in blood vessels and the remote sensing of tropospheric winds; these date back to S. Satamura, "Study of the Flow Patterns in Peripheral Arteries by Ul-

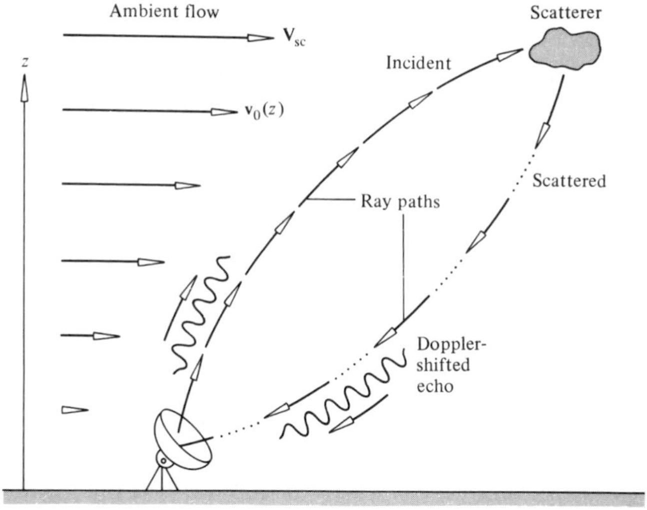

Figure 9-11 Sketch exemplifying how the Doppler shift evolves in a monostatic pulse-echo experiment when the scatterer is drifting along with the fluid. Because of z-dependent ambient flow, the ray path from transmitter to scatterer is not the same as the echo path from scatterer back to transmitter. If the difference between the two wave-number vectors has a nonzero component parallel to the scatterer velocity $\mathbf{V}_{sc}$, a Doppler shift results.

transmitter and receiver are outside the flow. The measurements ordinarily require the idealization (see Fig. 9-11) that the ambient velocity and acoustical properties appear unidirectional and stratified in the plane that contains transmitter and scatterer and is tangential to the scatterer's velocity vector. The same should apply for the plane containing receiver, scatterer, and the scatterer's velocity. Then the translational invariance parallel to $\mathbf{V}_{sc}$ between the transmitter and scatterer yields a version of Snell's law (in accordance with the trace-velocity matching principle discussed in Sec. 3-5) that the component of $\mathbf{k}$ parallel to $\mathbf{V}_{sc}$ should be constant all along the incident-wave path. This will hold even if the ray path is refracted† or if the propagation is not wholly describable in terms of concepts of geometrical acoustics. An analogous deduction concerns the scattered-wave path. The corresponding wave-number-

trasonics," *J. Acoust. Soc. Jap.*, **15**:151–158 (1959); D. L. Franklin, W. A. Schlegel, and R. F. Rushner, "Blood Flow Measured by Doppler Frequency Shift of Backscattered Ultrasound," *Science*, **132**:564–565 (1961); G. Kelton and P. Bricout, "Wind Velocity Measurements Using Sonic Techniques," *Bull. Am. Meteorol. Soc.*, **45**:571–580 (1964); and C. G. Little, "Acoustic Methods for the Remote Probing of the Lower Atmosphere," *Proc. IEEE*, **57**:571–578 (1969). Other applications discerned from recent patents listed in the Review of Acoustical Patents section, *J. Acoust. Soc. Am.*, **66**:615 (1979); **64**:711 (1978), are the measurement of subsurface-ocean-current velocities and the measurement of flow rates in ducts.

† T. M. Georges and S. F. Clifford, "Acoustic Sounding in a Refracting Atmosphere," *J. Acoust. Soc. Am.*, **52**:1397–1405 (1972); "Estimating Refractive Effects in Acoustic Sounding," ibid., **55**:934–936 (1974).

vector components, before and after scattering, however, are not necessarily the same.

Given that the transmitter and receiver are each in a region without ambient flow and given the idealizations just described, Eq. (12) yields (to first order in the Doppler shift)

$$V = \frac{(\omega_{\text{rec}} - \omega_{\text{tr}})c}{\omega_{\text{tr}}(\mathbf{e}_V \cdot \mathbf{n}_{\text{rec}} - \mathbf{e}_V \cdot \mathbf{n}_{\text{tr}})} \tag{9-3.14}$$

for the speed of the flow at the position of the scatterer. Here $\mathbf{e}_V$ is the unit vector in the direction of the flow; the unit vectors, $\mathbf{n}_{\text{tr}}$ and $\mathbf{n}_{\text{rec}}$, denote propagation directions of the transmitted and received waves at the transmitter and receiver, respectively. Ideally, the sounding experiment's design is such that $\mathbf{n}_{\text{tr}}$ and $\mathbf{n}_{\text{rec}}$ (or at least their components along the direction of interest) are well-defined quantities. Then if $\mathbf{e}_V$ is known, a measurement of the Doppler shift determines V.

If the direction of $\mathbf{V}$ is not known but the flow is stratified with the z coordinate, the x and y components of $\mathbf{V}$ are determined by two separate experiments: one with $\mathbf{n}_{\text{tr}}$ and $\mathbf{n}_{\text{rec}}$ lying in the xz plane, the other with them lying in the yz plane. Equation (14) applies to the first experiment's results with V replaced by V_x and with $\mathbf{e}_V$ replaced by $\mathbf{e}_x$.

Example: Volume-Blood-Flow-Computation An experimental procedure devised by D. W. Baker† for measuring volume of blood flowing per unit time in a blood vessel is as follows. The transducer used consists of two separate but closely spaced ceramic elements in a common housing; one element is used as a transmitter, the other as a receiver. For our present purposes, the system can be regarded as monostatic and as highly directional. Since the vessel is not visible, it is first necessary to locate it, to determine its orientation and radius. Moving the transducer over the surface of the skin and monitoring the intensity of Doppler-shifted echoes determines the vertical plane containing the vessel (with the skin's surface defining the horizontal plane). The transducer is then switched to a pulse-echo mode of operation, and its beam is kept confined to the previously determined vertical plane (see Fig. 9-12). The distance r of the transducer from the vessel centerline when the transducer is pointing at angle θ with the flow is determined from the time average of the echo delays from the near and far sides of the vessel. When the transducer is rotated through angle $\Delta\theta$, r decreases to r_1 while θ changes to $\theta_1 = \theta + \Delta\theta$. The quantities r, r_1, and $\Delta\theta$ are measured; how does one infer θ? Next the echo corresponding to the angle θ is monitored and the spectral density of the Doppler-shifted portion of the echo is used to derive an average frequency shift $\overline{\Delta f}$ for the backscatter from the flowing blood. How does one use this to determine the volume flow rate in the vessel?

† D. W. Baker, "Pulsed Ultrasonic Doppler Blood-Flow Sensing," *IEEE Trans. Sonics Ultrason.*, **SU-17**:170–185 (1970).

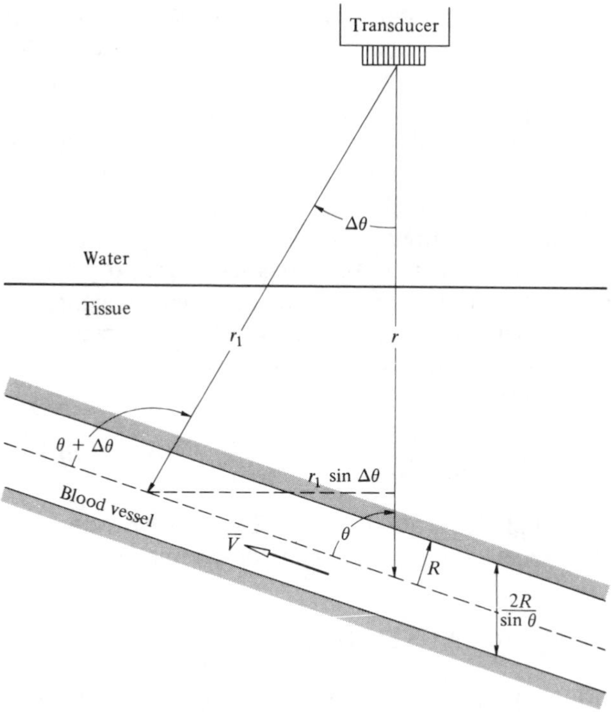

Figure 9-12 Ultrasonic determination of the volume of blood flowing per unit time through a blood vessel.

SOLUTION To determine the angle θ, we ignore the minor variations of the sound speed and density in tissue and blood from the values appropriate to water and assume the sound speed c to be 1500 m/s; refraction is negligible. Measurements of time delays are therefore equivalent to measurements of distance intervals (divided by c). A brief trigonometric analysis demonstrates that $r_1 \sin \Delta\theta$ and $r - r_1 \cos \Delta\theta$ are lengths of opposite and adjacent sides of a right-angle triangle with interior angle θ. Hence

$$\theta = \tan^{-1} \frac{r_1 \sin \Delta\theta}{r - r_1 \cos \Delta\theta} \tag{9-3.15}$$

The radius R of the vessel is then deduced from the time delay Δt of the echoes from the near and far sides of the vessel when the transducer beam makes angle θ with the vessel centerline; $c\ \Delta t$ should be $4R/(\sin\theta)$, so $R = \frac{1}{4}c\ \Delta t \sin\theta$. The extra factor of $\frac{1}{2}$ is because the second echo traverses an extra distance of one round trip across the vessel; the factor $\sin\theta$ is because the ray traverses the vessel obliquely.

Scattering of sound by blood† is caused by red cells (erythrocytes); normal

† K. K. Shung, R. A. Sigelmann, and J. M. Reid, "Angular Dependence of Scattering of Ultrasound from Blood," *IEEE Trans. Biomed. Eng.*, **BME-24:**325–331 (1977); E. L. Cartstensen,

human blood, although predominantly water, contains 5×10^{15} red cells per cubic meter; a typical cell has a volume of 87×10^{-18} m^3, a density of 1092 kg/m^3, and an apparent adiabatic bulk modulus ρc^2 exceeding that of water by a factor of 1.35. The surrounding fluid (blood plasma) has a density of 1021 kg/m^3 and a bulk modulus only 1.13 times that of water; the scattering is significantly affected by the fluid's viscosity, which is 1.8 times that of water. However, insofar as the present example is concerned, the only necessary assumptions are that the red cells are uniformly distributed across the stream and that the backscattered power from any area element of the cross section is proportional to the area. This allows the conclusion that $\overline{\Delta f}$ (the averaging being weighted by the spectral density) is proportional to $\bar{V}$, the cross-sectional area average of the flow velocity. If the flow is obliquely toward the transducer, $\mathbf{e}_V \cdot \mathbf{n}_{tr}$ is $-\cos\theta$ and $\mathbf{e}_V \cdot \mathbf{n}_{rec}$ is $+\cos\theta$; so Eq. (14) yields

$$\bar{V} = \frac{\overline{\Delta f}\, c}{2 f_{tr} \cos\theta} \tag{9-3.16}$$

for $\bar{V}$, where f_{tr} is the transmitted frequency in hertz. The corresponding volumetric flow rate Q is $\pi R^2 \bar{V}$, where the radius R is as determined above.

9-4 ACOUSTIC FIELDS NEAR CAUSTICS

The geometrical-acoustics model, described in the previous chapter, leads to the implausible prediction that amplitudes are infinite along surfaces (caustics) where adjacent rays intersect and where ray-tube areas vanish. Such hypothetical surfaces (which can emerge even in the middle of a homogeneous medium) do, however, describe the central structural forms to which characteristic wave patterns† are attached. Because such patterns develop where geometrical acoustics would at first glance be regarded as applicable but where it is actually *not* applicable, the patterns are diffraction phenomena.

We initially limit our considerations to a homogeneous nonmoving medium and to a constant-frequency field independent of the z coordinate, so that the rays are all straight lines parallel to the xy plane. The portion of the overall field of interest is that associated (see Fig. 9-13) with a family of rays each member of which is tangential to a curved caustic surface. On the convex side, two rays

---

K. Li, and H. P. Schwan, "Determination of the Acoustic Properties of Blood and Its Components," *J. Acoust. Soc. Am.*, **25**:286–299 (1953).

† The theory dates back to G. B. Airy, "On the Intensity of Light in the Neighborhood of a Caustic," *Trans. Camb. Phil. Soc.*, **6**:379–401 (1838); the exposition here is largely inspired by that of R. B. Buchal and J. B. Keller, "Boundary Layer Problems in Diffraction Theory," *Comm. Pure Appl. Math.*, **13**:85–144 (1960). The limitation that the choice for the caustic's radius of curvature is not precisely defined when one seeks to determine the field at some distance from the caustic and when R_c varies along the caustic is overcome in D. Ludwig, "Uniform Asymptotic Expansions at a Caustic," *Commun. Pure Appl. Math.*, **19**:215–250 (1966); and Yu. A. Kravtsov, "Two New Asymptotic Methods in the Theory cf "'ave Propagation in Inhomogeneous Media (Review)," *Sov. Phys. Acoust.*, **14**(1):1–17 (1968).

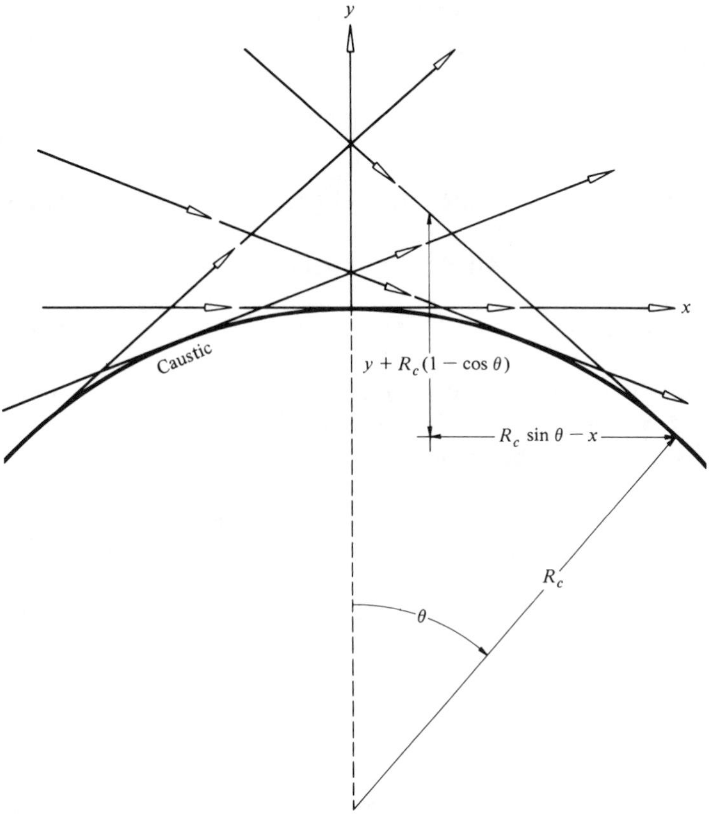

Figure 9-13 Ray geometry in the vicinity of a caustic when the ambient medium is homogeneous.

pass through each point. One ray has yet to touch the caustic; the other has already touched it. On the concave side, there are no rays of the considered family. Within any small region, the caustic surface is characterized by its radius of curvature R_c, which is assumed much larger than $1/k$.

We orient our coordinate system in such a way that the point of interest on the caustic is the origin and such that the caustic is tangential to the x axis and bends into the region $y < 0$. Let the eikonal $\tau(\mathbf{x})$ associated with the incident rays be 0 at the origin. The gradient $\nabla\tau$ is tangent to the caustic with a positive x component and has magnitude $1/c$, so

$$\nabla\tau \approx (\mathbf{e}_x \cos\theta - \mathbf{e}_y \sin\theta)/c \tag{9-4.1}$$

where $\theta(x, y)$ is such that

$$\frac{y + (R_c - R_c\cos\theta)}{R_c\sin\theta - x} = \tan\theta$$

When $x = 0$, this yields $\cos\theta = R_c/(y + R_c)$. The corresponding value of $\sin\theta$ is $(2yR_c + y^2)^{1/2}/(y + R_c)$, which is approximately $(2y/R_c)^{1/2}$. Consequently, along the line $x = 0, y > 0$, Eq. (1) integrates to $c\tau = -(8y^3/9R_c)^{1/2}$. Since the eikonal equation $(\nabla\tau)^2 = 1/c^2$ requires $\partial\tau/\partial x = (1/c)(1 - 2y/R_c)^{1/2}$ along the same line, near the origin one has

$$\tau \approx \frac{1}{c}\left[x - \frac{yx}{R_c} - \left(\frac{8y^3}{9R_c}\right)^{1/2}\right] \tag{9-4.2}$$

Ray-tube area near a caustic is proportional [see Eq. (8-5.7)] to $-1/\nabla^2\tau$, and so the above predicts that it varies with y as $y^{1/2}$. Consequently, the geometrical-acoustics prediction of the incident field near a caustic is

$$\hat{p}_i = P\left(\frac{R_c}{16y}\right)^{1/4} e^{ik(x-yx/R_c)}e^{-i(2/3)|\eta|^{3/2}} \tag{9-4.3}$$

where $|\eta| = (2k^2/R_c)^{1/3}y$ $(y > 0)$. The normalization takes P to be the amplitude at a distance $R_c/16$ from the caustic. Alternatively, since $y \approx l^2/2R_c$, where l is distance the ray has yet to travel before it touches the caustic, P corresponds to $l \approx R_c/8^{1/2}$.

The field associated with rays propagating away from the caustic also obeys the laws of geometrical acoustics at moderate values of y. The eikonal for these rays (apart from a possible additive constant) must be of the form of Eq. (2) but with the last term changed in sign. Thus, the overall field near the origin should asymptotically ($ky \gg 1$, $y \ll R_c$, $x \ll R_c$) be

$$\hat{p}_i + \hat{p}_{\text{away}} = P\left(\frac{R_c}{16y}\right)^{1/4} e^{ik(x-yx/R_c)}\left(e^{-i(2/3)|\eta|^{3/2}} + \mathcal{R}e^{i(2/3)|\eta|^{3/2}}\right) \tag{9-4.4}$$

where $\mathcal{R}$ is a constant. Our task is to find a solution of the Helmholtz equation valid near the origin that asymptotically approaches Eq. (4) at moderate positive values of y and approaches 0 at large negative values of y (on the nonilluminated side).

If one assumes that $\hat{p}$ is of the form $e^{ik(x-yx/R_c)}F(x, y)$ and inserts this into the Helmholtz equation, the result is a cumbersome partial-differential equation for F. However, with the neglect of fourth and higher-order terms in $(1/kR_c)^{1/3}$ and with the restriction to values of y and x of the order of $1/k$ or less, considerable simplification results. Since we anticipate that the magnitude of $\partial F/\partial y$ will be of the order of $k(kR_c)^{-1/3}$ times that of F, we discard terms like $-(2ikx/R_c)\,\partial F/\partial y$ and $(-k^2x^2/R_c^2)F$ in comparison with, say, $\partial^2F/\partial y^2$. This allows a solution not depending on x and yields the ordinary differential equation

$$\frac{d^2F}{dy^2} + \frac{2k^2y}{R_c}F = 0 \qquad \frac{d^2F}{d\eta^2} - \eta F = 0 \tag{9-4.5}$$

where the second version follows from the first with the abbreviation

$$\eta = -\left(\frac{2k^2}{R_c}\right)^{1/3} y \tag{9-4.6}$$

which is consistent with the use of $|\eta|$ in Eqs. (3) and (4).

The Airy Function

Apart from a multiplicative constant, the only solution of Eq. (5) having the desired property of going to 0 as $\eta \to \infty$ ($y \to -\infty$) is the Airy function, defined† for real η as

$$\mathrm{Ai}(\eta) = \frac{1}{\pi} \int_0^\infty \cos\left(\frac{s^3}{3} + \eta s\right) ds \tag{9-4.7a}$$

$$= \frac{1}{2\pi} \int_{C_{\mathrm{Ai}}} e^{i(s^3/3+\eta s)}\, ds \tag{9-4.7b}$$

In the latter version, which holds for arbitrary complex η, the contour C_{Ai} begins at $|s| = \infty$ on the line where the phase of s is $5\pi/6$ and terminates at $|s| = \infty$ on the line where the phase of s is $\pi/6$. A demonstration that the second version is equivalent to the first for real η results from a deformation of C_{Ai} to the real axis; that either satisfies Eq. (5) follows from

$$\frac{\partial^2}{\partial \eta^2} e^{i(s^3/3+\eta s)} = \left(i \frac{\partial}{\partial s} + \eta\right) e^{i(s^3/3+\eta s)}$$

An asymptotic expression for $\mathrm{Ai}(\eta)$ at large $|\eta|$ is derived from Eq. (7*b*) for when $-2\pi/3 < \phi < 2\pi/3$ (ϕ denoting phase of η) by deforming C_{Ai} to a steepest-descent path,‡ $s = s(l)$ with l real, passing through the saddle point at $s = e^{i\pi/2}\eta^{1/2}$, at which $ds/dl = e^{-i\phi/4}$. Since the integrand is sharply peaked at the saddle point, $s^3/3 + \eta s$ can be approximated by $i(\frac{2}{3})\eta^{3/2} + i|\eta|^{1/2}l^2$, where l is distance along the path from the saddle point. Thus, we find

$$\mathrm{Ai}(\eta) \to \frac{e^{-(2/3)\eta^{3/2}}}{2\pi^{1/2}\eta^{1/4}} \qquad -\frac{2\pi}{3} < \phi < \frac{2\pi}{3} \tag{9-4.8}$$

If $2\pi/3 < \phi < 4\pi/3$, contour C_{Ai} is stretched so that its midpoint extends to $-i\infty$ on the negative imaginary axis. The left segment is deformed to a steepest-descent path passing through the saddle point at $s = e^{i\pi/2}\eta^{1/2}$, at which $ds/dl = e^{-i\phi/4}$, while the right segment is deformed to one passing through a saddle point at $s = -e^{i\pi/2}\eta^{1/2}$, at which $ds/dl = e^{i\pi/2}e^{i\phi/4}$. Then, with approximations similar to those described above, we find

$$\mathrm{Ai}(\eta) \to \frac{1}{2\pi^{1/2}\eta^{1/4}} \left(e^{-(2/3)\eta^{3/2}} + ie^{(2/3)\eta^{3/2}}\right) \qquad \frac{2\pi}{3} < \phi < \frac{4\pi}{3} \tag{9-4.9}$$

The apparent discontinuity along the lines $\phi = 2\pi/3$ and $\phi = 4\pi/3$ suggested by a comparison of Eqs. (8) and (9) is nonexistent because $e^{(2/3)\eta^{3/2}}$ is negligibly small at large $|\eta|$ along the first line and because the second of the two terms

† Various definitions are in the literature. That adopted here is as given by H. A. Antosiewićz, in M. Abramowitz and I. Stegun (eds.), *Handbook of Mathematical Functions,* Dover, New York, 1965, pp. 446–452, 475–478.

‡ G. F. Carrier, M. Krook, and C. E. Pearson, *Functions of a Complex Variable,* McGraw-Hill, New York, 1966, pp. 263–266.

constituting (9), when evaluated at $\phi = 4\pi/3$, is the same as the first term evaluated at $\phi = -2\pi/3$.

If η is real and negative, Eq. (9), with $\eta = |\eta|e^{i\pi}$, yields

$$\mathrm{Ai}(\eta) \to \frac{e^{i\pi/4}}{2\pi^{1/2}|\eta|^{1/4}}\left(e^{-i(2/3)|\eta|^{3/2}} - ie^{i(2/3)|\eta|^{3/2}}\right)$$

$$= \frac{1}{\pi^{1/2}|\eta|^{1/4}}\cos\left(\tfrac{2}{3}|\eta|^{3/2} - \frac{\pi}{4}\right) \qquad \eta < 0 \qquad (9\text{-}4.10)$$

Thus, $\mathrm{Ai}(\eta)$, when considered as a function of real η, is oscillatory for $\eta < 0$ (see Fig. 9-14), Ai(0) is 0.355 · · · ; for subsequent negative values of η, $\mathrm{Ai}(\eta)$ rises to a peak value of 0.536 at $\eta = -1.019$, reaches its first zero at $\eta = -2.338$, reaches a minimum value of -0.419 at $\eta = -3.248$, reaches a second zero at $\eta = -4.088$, and reaches a second maximum value of 0.380 at $\eta = -4.820$. The nth zero occurs asymptotically at $\eta = -(\frac{3}{2})^{2/3}[n - \frac{1}{4})\pi]^{2/3}$. For the problem of interest here, an increment $\Delta\eta$ corresponds to $k\ \Delta y = (kR_c/2)^{1/3}(-\Delta\eta)$ where $kR_c \gg 1$, so that intervals between successive undula-

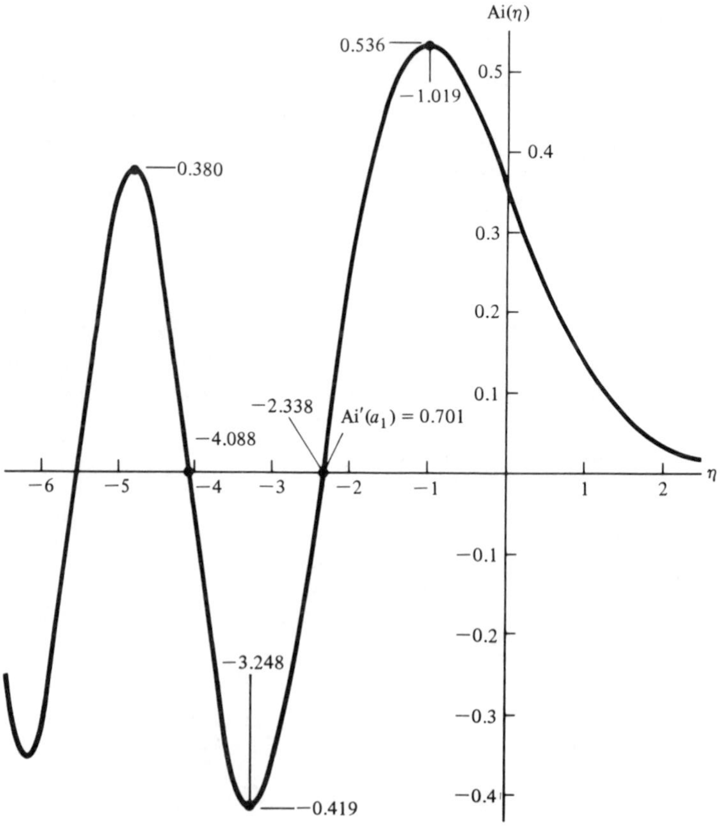

Figure 9-14 The Airy function for real values of its argument.

tions along a line transverse to the caustic are of the order of a wavelength or greater.

The first version of Eq. (10) is comparable to the geometrical-acoustics solution of Eq. (4) for the field on the illuminated side of the caustic. Consequently, we have

$$\hat{p} = P\pi^{1/2}2^{1/12}e^{-i\pi/4}(kR_c)^{1/6}e^{ik(x-yx/R_c)}\text{Ai}(\eta) \qquad (9\text{-}4.11)$$

as the solution of the Helmholtz equation that matches Eq. (4) in the limit $ky \gg 1$. Equation (10) also requires, in Eq. (4), the identification $\mathcal{R} = e^{-i\pi/2}$.

Since the maximum value of $\text{Ai}(\eta)$ is 0.536, the peak pressure magnification† at a caustic is $0.536\pi^{1/2}2^{1/12}(kR_c)^{1/6}$, or $1.01(kR_c)^{1/6}$, relative to what the geometric-acoustics model would predict for the incident wave at a transverse distance of $R_c/16$ from a caustic or at a propagation distance of $R_c/8^{1/2} = R_c/2.83$ from where the ray grazes the caustic. The indicated sixth-root dependence on frequency of this magnification is very weak; increasing the frequency by a factor of 10 increases the magnification by a factor of only 1.47.

Generalization to Inhomogeneous Media

If an inhomogeneous medium varies slowly over distances comparable to a wavelength, the acoustic pressure in any local region approximately satisfies the wave equation, providing the medium appears locally at rest in the selected (possibly moving) coordinate system. The rays are curved, but as indicated by the analysis in Sec. 8-3, the plane of curvature and the radius of curvature will be nearly the same for each ray in the vicinity of any given fixed point substantially removed from the source.

For most situations of interest, an appropriate idealization is that each line on the caustic surface traced out by successively intersecting adjacent rays lies locally in the same plane as the curved rays that graze the caustic. Another idealization is that the curvature of the caustic surface is such that the propagation direction of a grazing ray coincides with one of the principal directions of curvature. Then, with an appropriate coordinate system, the geometry of the ray system in the vicinity of a point on the caustic‡ is as sketched in Fig. 9-15; the ray proceeding locally in the $+x$ direction and grazing the caustic at the origin has a radius of curvature R_{ray}; the caustic has a principal radius of curvature R_c at the same point, the sign convention being such that positive R_{ray} corresponds to a bending in the $+y$ direction; positive R_c corresponds to a bending in the $-y$ direction.

† A comparable analysis based on the uniform asymptotic expression is given by D. Ludwig, "Strength of Caustics," *J. Acoust. Soc. Am.*, **43:**1179–1180 (1968).

‡ Caustics in inhomogeneous media are discussed by B. D. Seckler and J. B. Keller, "Geometrical Theory of Diffraction in Inhomogeneous Media," *J. Acoust. Soc. Am.*, **31:**192–205 (1959); "Asymptotic Theory of Diffraction in Inhomogeneous Media," ibid., **31:**206–216 (1959); D. A. Sachs and A. Silbiger, "Focusing and Refraction of Harmonic Sound and Transient Pulses in Stratified Media," ibid., **49:**824–840 (1971).

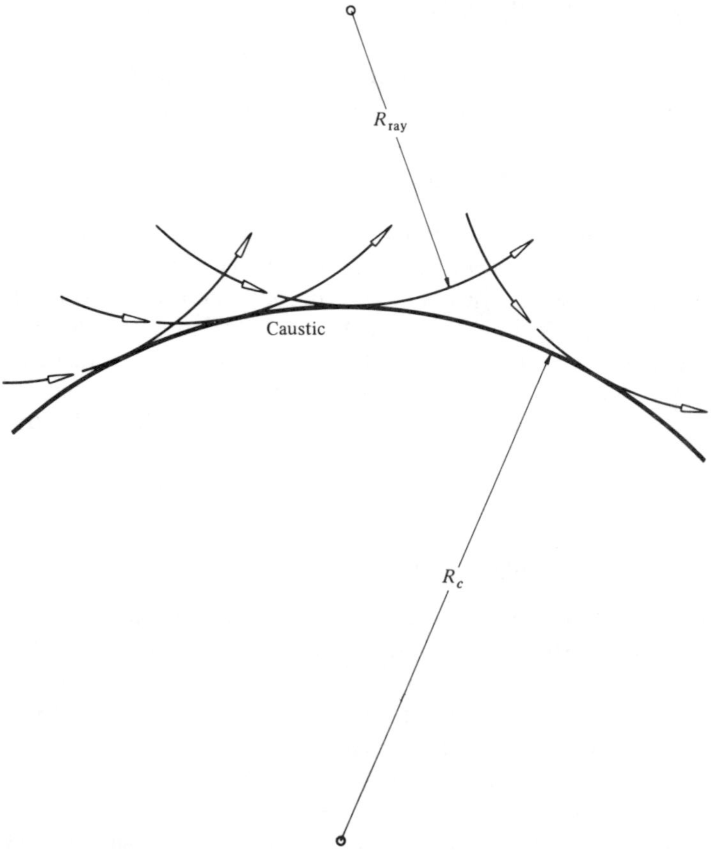

Figure 9-15 Curved rays near a curved caustic in an inhomogeneous medium. Circumstances assumed in the sketch are for when the sound speed decreases on the illuminated side with distance from the caustic.

For the ray grazing the caustic at the origin, its distance $|y'|$ from the nearest point on the caustic increases with x approximately as

$$|y'| = \frac{x^2}{2R'_c} \qquad \frac{1}{R'_c} = \frac{1}{R_{ray}} + \frac{1}{R_c} \tag{9-4.12}$$

Thus, if one were to choose a curvilinear orthogonal coordinate system ($x' \approx x + yx/R_{ray}$ and $y' \approx y - x^2/2R_{ray}$) such that $x' \approx x, y' \approx y$ in the vicinity of the origin and the ray passing through the origin appears to be straight, the apparent radius of curvature of the caustic would be R'_c. Since the form of the wave equation is only slightly altered by the switch in coordinate system, the analysis leading to Eq. (11) is still applicable, providing one substitutes R'_c for R_c. Since $x' - x'y'/R'_c$ is equivalent in this approximation to $x - xy/R_c$, the substitution need not be made in the exponential factor providing one interprets

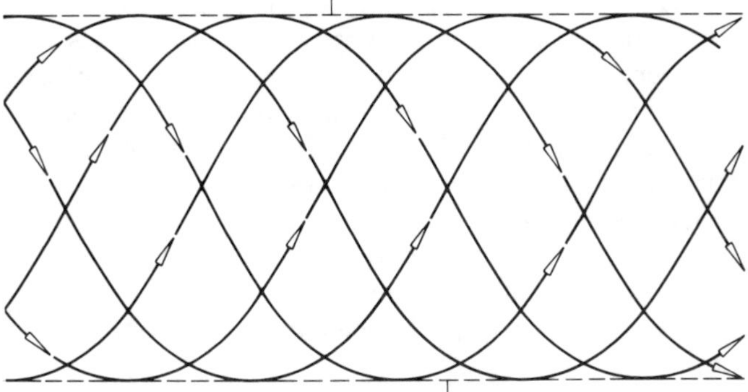

Figure 9-16 Caustics formed by a family of similar rays cycling between upper and lower turning points in a height region where the sound-speed profile has a minimum.

Eq. (11) in terms of the original coordinate system; y is still regarded as the transverse distance of the point of observation from the caustic.

Field near a Turning Point

An application of the analogy just described would be when the ray system consists (see Fig. 9-16) of a family of similar rays cycling† between upper (y_U) and lower (y_L) turning points in a region where the sound speed $c(y)$ has a minimum between y_L and y_U. Successive rays differ only by a displacement parallel to the x axis, so the planes $y = y_L$ and $y = y_U$ are caustics; $1/R_c = 0$ for both surfaces. Consequently, R_c' is R_{ray}, which, from Eq. (8-3.3), is $c/|dc/dy|$ evaluated at the caustic. Equation (11) therefore becomes

$$\hat{p}_{U,L} \approx P_{U,L}\pi^{1/2}2^{1/12}e^{-i\pi/4}\left(\frac{\omega}{|dc/dy|}\right)_{U,L}^{1/6} e^{ikx}\,\text{Ai}(\eta_{U,L}) \tag{9-4.13}$$

with
$$\eta_U = -\left(2\omega^2\left|\frac{dc}{dy}\right|\right)_U^{1/3}\frac{y_U - y}{c} \qquad \eta_L = -\left(2\omega^2\left|\frac{dc}{dy}\right|\right)_L^{1/3}\frac{y - y_L}{c} \tag{9-4.14}$$

where $k = \omega/c$ and $c = c(y_L) = c(y_U)$.

These expressions hold only near y_U and y_L, respectively. Also, within this context, the quantities P_L and P_U should be regarded as slowly varying functions of x. Given fixed y_U and y_L, they may be independent of x if the net phase shift along a complete ray cycle is an integer multiple of 2π, but this occurs only

† N. A. Haskell, "Asymptotic Approximation for the Normal Modes in Sound Channel Wave Propagation," *J. Appl. Phys.*, **22**:157–168 (1951); I. Tolstoy, "Phase Changes and Pulse Deformations in Acoustics," *J. Acoust. Soc. Am.*, **44**:675–683 (1968).

for certain discrete frequencies. Alternatively, for fixed ω, it occurs for certain discrete values of $k = \omega/c(y_L) = \omega/c(y_U)$; each such value $k_n(\omega)$ of k corresponds, however, to a different pair of turning points. Channeled waves with dependence on x as e^{ikx} [where $k = k_n(\omega)$] are *natural guided modes* analogous to the waveguide modes discussed in Sec. 7-1. Their existence does not depend on the validity of the geometrical-acoustics approximation or on the presence of two internal turning points. Such natural modes furnish a cogent explanation of acoustic fields at large horizontal distances from sources in the atmosphere and oceans. A discussion† of how they emerge in theoretical formulations is beyond the scope of this text, but the analysis in the following section (directed toward a different problem) bears some similarity to the guided-mode theory of long-range sound propagation.

Phase Shift at a Caustic

The identification of $\mathcal{R} = e^{-i\pi/2}$ in Eq. (4) implies that a ray undergoes a phase drop of $\pi/2$ every time it grazes a caustic. Thus, the net phase change over a long path is $\omega\,\Delta\tau - n\pi/2$, where $\Delta\tau$ is the travel time predicted by the ray-tracing equations and n is the number of caustics grazed along this path. The $\pi/2$ phase shift at a caustic is consistent with the purely geometrical-acoustics prediction that the amplitude varies inversely with the square root of ray-tube area. Beyond the caustic, the ray-tube area is formally negative, so predictions like that of Eq. (8-5.4) would still apply if we interpreted $(-|A|)^{-1/2}$ as $e^{-i\pi/2}|A|^{-1/2}$; the analysis leading to Eq. (11) tells us which of the two possible square roots of -1 should be used.

With this prescription, the geometrical-acoustic formulation‡ can be used even when caustics are present. For a given far-field point, one determines all the possible ray paths connecting source and receiver, computes amplitude and phases (using the geometrical-acoustics theory and setting $A = |A|$) for each ray's contribution, shifts the phases by integer multiples of $\pi/2$ to account for the caustics, and then superimposes the various individual ray contributions. This assumes that the receiver is not near a caustic; if it is, the contribution from two of the rays is replaced by an expression of the form of Eq. (11). The Blokhintzev invariant for each ray tube is determined from the wave field at moderately close distances to the source before refraction has an appreciable effect on wave amplitudes.

The $\pi/2$ phase shift at a caustic has a significant effect on waveforms from a transient source§ (a detonation, for example). Suppose a distant point receives

† See, for example, L. M. Brekhovskikh, *Waves in Layered Media*, Academic, New York, 1960, pp. 454–460.

‡ I. M. Blatstein, A. V. Newman, and H. Uberall, "A Comparison of Ray Theory, Modified Ray Theory, and Normal-Mode Theory for a Deep-Ocean Arbitrary Velocity Profile," *J. Acoust. Soc. Am.*, **55**:1336–1338 (1974).

§ R. M. Barash, "Evidence of Phase Shift at Caustics," *J. Acoust. Soc. Am.*, **43**:378–380 (1968); R. H. Mellen, "Impulse Propagation in Underwater Sound Channels," *ibid.*, **40**:500–501 (1966).

two distinct arrivals corresponding to two different ray paths; the first ray to arrive never grazed a caustic, but the second did so once. Nominally, one would expect the two waveforms to be similar, differing only in arrival times and peak amplitudes, but the second ray's remembrance of its $\pi/2$ phase shift at the caustic changes this expectation. If the first arrival p_1 is $f(t - \tau_1)$ and has a Fourier transform $\hat{f}(\omega)e^{i\omega\tau_1}$, the second arrival p_2 will have a Fourier transform $K\hat{f}(\omega)e^{i\omega\tau_2}e^{-i\pi/2}$, for $\omega > 0$, where K is a positive constant. However, p_2 is real, so its Fourier transform for $\omega < 0$ is the complex conjugate (causing $e^{-i\pi/2} \to e^{i\pi/2}$) of that for $\omega > 0$. Thus, we have $p_2 = Kf_H(t - \tau_2)$, where $f_H(t)$ is the Hilbert transform of $f(t)$ (see Sec. 3-6). Although $p_2(t)$ is dissimilar to $p_1(t)$, there is a definite mathematical relation between the two waveform shapes.

If the second ray had encountered two caustics instead of only one, the net experienced phase shift would be π, corresponding to a change in sign, so that the second arrival's waveform would resemble the negative of that of the first $(p \to -p)$.

9-5 SHADOW ZONES AND CREEPING WAVES

To obtain insight into how sound penetrates into shadow zones (regions without direct rays from the source), we begin with the particular example† of a source near the ground at height z_0 in a medium whose sound speed $c(z)$ decreases linearly with height at lower altitudes. (The analysis applies to an underwater source in water with sound speed increasing linearly with increasing depth, but for simplicity we refer to z as the upward direction and to the surface $z = 0$ as the ground.) This decrease causes the rays initially leaving the source in nearly horizontal directions to bend upward with a curvature radius of $R = c/|dc/dz|$. One ray, the limiting ray, barely grazes the ground, leaving a shadow zone (see Fig. 9-17) consisting of points where $w > (2Rz_0)^{1/2} + (2Rz)^{1/2}$, given that the horizontal distance w is substantially less than R. The analysis below is directed toward the prediction of the resulting field in such a shadow zone when w is substantially larger than a wavelength.

Point Source above a Locally Reacting Surface in a Stratified Medium

The source (monopole amplitude $\hat{S}$) is emitting sound of angular frequency ω, so with the neglect of density gradients, the complex amplitude of the acoustic pressure satisfies the inhomogeneous Helmholtz equation with $-4\pi\hat{S}\delta(x)\delta(y)\delta(z - z_0)$ on the right side and with k^2 replaced by $\omega^2/c^2(z)$.

The expression adopted as a starting point for development of a solution is

† C. L. Pekeris, "Theory of Propagation of Sound in a Half-Space of Variable Sound Velocity under Conditions of Formation of a Shadow Zone," *J. Acoust. Soc. Am.*, **18**:295–315 (1946); D. C. Pridmore-Brown and U. Ingard, "Sound Propagation into a Shadow Zone in a Temperature-Stratified Atmosphere above a Plane Boundary," ibid., **27**:36–42 (1955).

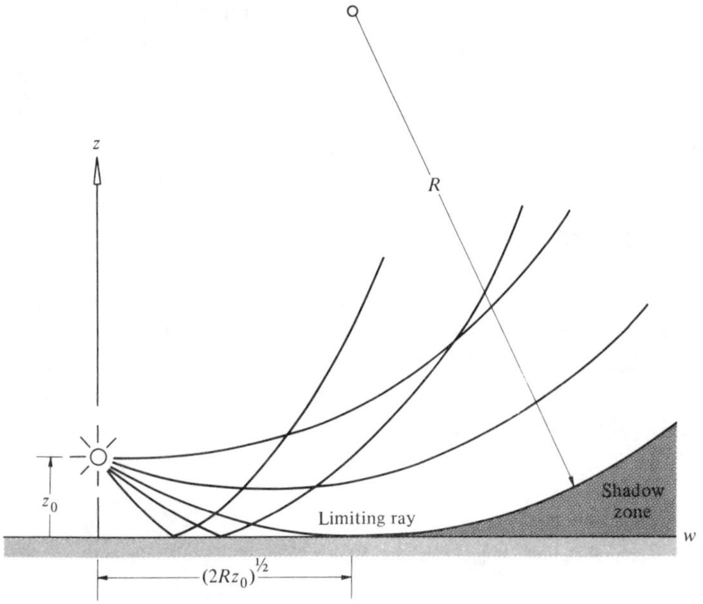

Figure 9-17 Shadow zone resulting from a source at height z_0 above a plane bounding a fluid in which the sound speed decreases linearly with height.

a double Fourier transform in x and y:

$$\hat{p} = -\frac{\hat{S}}{\pi} \lim_{\varepsilon \to 0} \int_{-\infty}^{\infty}\int e^{-\varepsilon^2(\alpha^2+\beta^2)} e^{i\alpha x} e^{i\beta y} Z(z, \alpha, \beta)\, d\alpha\, d\beta \tag{9-5.1}$$

This will satisfy the inhomogeneous Helmholtz equation if the function Z satisfies

$$\frac{d^2Z}{dz^2} + \left[\frac{\omega^2}{c^2(z)} - k^2\right] Z = \delta(z - z_0) \tag{9-5.2}$$

where k^2 is used as an abbreviation for $\alpha^2 + \beta^2$. The demonstration that such yields a solution rests on the identification for the Dirac delta function

$$\lim_{\varepsilon \to 0} \int_{-\infty}^{\infty} e^{-\varepsilon^2\alpha^2} e^{i\alpha x} d\alpha = 2\pi\delta(x) \tag{9-5.3}$$

developed in Sec. 2-8.

Since the field is cylindrically symmetric, Eq. (1) is unchanged if we replace y by 0 and set $x = w$. Changing the integration variables to k and θ, where $\alpha = k \cos \theta$ and $\beta = k \sin \theta$, allows one integration (that over θ) to be performed, since, from (2), Z may be presumed independent of θ. The integral over θ from 0 to 2π of exp $(ikw \cos \theta)$ is $2\pi J_0(kw)$ [see Eq. (5-4.6)], so we obtain

$$\hat{p} = -\hat{S} \lim_{\varepsilon \to 0} \int_0^{\infty} e^{-\varepsilon^2 k^2} 2\, J_0(kw) Z(z, k)\, k\, dk \tag{9-5.4}$$

The restriction of our interest to larger values of $\omega w/c(0)$ suggests a replacement of the Bessel function by its asymptotic limit,† which in turn decomposes into

$$J_0(\eta) \approx \left(\frac{1}{2\pi}\right)^{1/2} e^{-i\pi/4} \left[\frac{1}{\eta^{1/2}} e^{i\eta} - \frac{1}{(-\eta)^{1/2}} e^{-i\eta}\right] \tag{9-5.5}$$

where, for $\eta > 0$, $(-\eta)^{1/2}$ is understood to be $e^{i\pi/2}\eta^{1/2}$. Thus, with $Z(z, k)$ regarded as an even function of k, we can rewrite (4) as

$$\hat{p} \approx -\left(\frac{2}{\pi w}\right)^{1/2} \hat{S} e^{-i\pi/4} \int_{-\infty}^{\infty} k^{1/2} e^{ikw} Z(z, k)\, dk \tag{9-5.6}$$

where $k^{1/2}$ is $e^{i\pi/2}|k|^{1/2}$ when k is negative. This integral is now regarded as a contour integral with $k^{1/2} = |k|^{1/2} \exp(i\phi_k/2)$ and with the phase ϕ_k of k restricted to values between $-\pi/2$ and $3\pi/2$. The convergence factor $\exp(-\epsilon_k^2 k^2)$ in Eq. (4) is discarded because if the convergence is marginal, the contour can always be deformed away from the real axis so that e^{ikw} goes exponentially to zero when $|k| \to \infty$ on either end of the contour.

As regards the function $Z(z, k)$ that satisfies Eq. (2), we can conceive, when k is real and positive, of two solutions, $\psi(z, k)$ and $\Phi(z, k)$, of the homogeneous equation that satisfy an upper boundary condition conforming to the Sommerfeld radiation condition and a lower boundary condition at $z = 0$, respectively. The upper boundary condition is that (for real k) ψ either dies out exponentially or represents a wave propagating obliquely upward; the lower boundary condition corresponding to a locally reacting surface of specific impedance Z_S is that

$$\frac{d\Phi}{dz} + i\frac{k_0 \rho c}{Z_S}\Phi = 0 \qquad z = 0 \tag{9-5.7}$$

where $k_0 = \omega/c(0)$; for a rigid surface, $d\Phi/dz = 0$ ($Z_S \to \infty$), while for a pressure-release surface (as for the ocean's upper surface), $\Phi = 0$ at $z = 0$. The functions ψ and Φ for complex k are understood to be analytic except at branch lines, none of which are constructed so that they cross the real axis.

The solution $Z(z, k)$ of the inhomogeneous equation (2) is $A\psi(z, k)$ for $z > z_0$ and is $B\Phi(z, k)$ for $z < z_0$, where the constants A and B are such that Z is continuous at z_0 but has a discontinuity in slope there of 1. Thus, we have

$$Z(z, k) = \frac{\psi(z_>, k)\Phi(z_<, k)}{[(d\psi/dz)\Phi - (d\Phi/dz)\psi]_{z_0}} \tag{9-5.8}$$

with $z_<$ and $z_>$ representing the smaller and larger of z_0 and z. Since both ψ and Φ satisfy the homogeneous-differential-equation version of (2), the denominator expression (the *wronskian* of ψ and Φ) in (8) is independent of z_0; Eq. (7) therefore allows it to be reexpressed as

$$\left(\frac{d\psi}{dz}\Phi - \frac{d\Phi}{dz}\psi\right)_{z_0} = \left(\frac{d\psi}{dz} + \frac{ik_0\rho c}{Z_S}\psi\right)_0 \Phi(0, k) \tag{9-5.9}$$

† See p. 225*n*. and Eq. (5-7.8).

Insofar as we are interested only in the disturbance at lower altitudes, we suppose $c(z)$ to decrease indefinitely with increasing height. This idealization makes it possible to predict whether a given candidate for $\psi(z, k)$ will satisfy the upper boundary condition from its behavior at moderately small values of z; the differential equation is approximated by replacing $1/c^2(z)$ by $[1/c^2(0)](1 + 2z/R)$, where $R = c(0)/|dc/dz|_{z=0}$ is the radius of curvature of the ray initially propagating horizontally from the source. With this approximation, the homogeneous equation becomes

$$\frac{d^2\psi}{dz^2} + \left(k_0^2 - k^2 + \frac{2k_0^2 z}{R}\right)\psi = 0 \qquad (9\text{-}5.10)$$

where we abbreviate k_0 for $\omega/c(0)$. The differential equation is of the same form as in Eq. (9-4.5); one possible solution is the Airy function $\mathrm{Ai}(\tilde{\tau} - y)$, where

$$\tilde{\tau} = (k^2 - k_0^2)l^2 \qquad y = \frac{z}{l} \qquad l = \left(\frac{R}{2k_0^2}\right)^{1/3} \qquad (9\text{-}5.11)$$

are convenient abbreviations. Other solutions are $\mathrm{Ai}((\tilde{\tau} - y)e^{i2\pi/3})$ and $\mathrm{Ai}((\tilde{\tau} - y)e^{-i2\pi/3})$. There are only two linearly independent solutions; any constant times a solution is also a solution. Two recommended† after a study of various solutions of similar problems are

$$v(\eta) = \pi^{1/2}\,\mathrm{Ai}(\eta) \qquad w_1(\eta) = 2\pi^{1/2}e^{i\pi/6}\,\mathrm{Ai}(\eta e^{i2\pi/3}) \qquad (9\text{-}5.12)$$

with $\eta = \tilde{\tau} - y$.

Fock's $w_1(\eta)$ is chosen because it has the asymptotic behavior [derivable from Eqs. (9-4.8) and (9-4.9)]

$$w_1(\tilde{\tau} - y) \to \frac{e^{i\pi/4}}{y^{1/4}}e^{i(2/3)y^{3/2}}e^{-i\tilde{\tau}y^{1/2}} \qquad y \to \infty \qquad (9\text{-}5.13)$$

which is representative of a wave propagating obliquely upward. Consequently, $w_1(\tilde{\tau} - y)$ is an appropriate $\psi(z, k)$.

The function $\Phi(z, k)$ that satisfies Eq. (7) can be taken as

$$\Phi(z, k) = v(\tilde{\tau} - y) - \frac{v'(\tilde{\tau}) - qv(\tilde{\tau})}{w_1'(\tilde{\tau}) - qw_1(\tilde{\tau})}\,w_1(\tilde{\tau} - y) \qquad (9\text{-}5.14)$$

with the abbreviation $q = ik_0 l\rho c/Z_S$. Such substitutions reduce Eq. (8) to

$$Z(z, k) = \frac{w_1(\tilde{\tau} - y_>)\Phi(z_<, k)l}{v'(\tilde{\tau})w_1(\tilde{\tau}) - w_1'(\tilde{\tau})v(\tilde{\tau})} = -w_1(\tilde{\tau} - y_>)\Phi(z_<, k)l \qquad (9\text{-}5.15)$$

The second version results because the wronskian $v'w_1 - w_1'v$ of the two solutions of the Airy differential equation is a constant; its value of -1 can be

† V. A. Fock, *Electromagnetic Diffraction and Propagation Problems*, Pergamon, London, 1965, pp. 237, 379–381; N. A. Logan, General Research in Diffraction Theory, vol. 1, *Lockheed Missiles Space Div. Rep.* LMSD-288087, December 1959, pp. 5-1 to 5-13, available from National Technical Information Service, Springfield, VA 22161, accession number AD 241228.

derived after an insertion of the asymptotic formulas into the wronskian expression.

For lower-altitude reception sites, within and near the shadow zone, the dominant contribution to the integral (6) comes ·om values of k that are not substantially different from k_0. This is anticipated because the integral can be regarded as a superposition of plane and evanescent waves and because waves propagating with horizontal phase velocities of the order of $c(0) = \omega/k_0$ predominate near the ground at larger horizontal distances. Consequently, we make approximations consistent with such an anticipation at the outset; the results eventually derived will support the hypothesis. In particular, we replace the multiplicative factor $k^{1/2}$ in the integrand by $k_0^{1/2}$, and we approximate $k^2 - k_0^2 = (k + k_0)(k - k_0)$ in expression (11) for $\tilde{\tau}$ by $2k_0(k - k_0)$ so that $\tilde{\tau} \to \tau$, where $\tau = (2k_0 l^2)(k - k_0)$.

Changing the integration variable to τ in Eq. (6) consequently reduces the complex pressure amplitude to a standard expression

$$\hat{p} = \frac{\hat{S}}{w} e^{ik_0 w} V(\xi, y_0, y, q) \tag{9-5.16}$$

where

$$V(\xi, y_0, y, q) = e^{-i\pi/4}\left(\frac{\xi}{\pi}\right)^{1/2} \int_{-\infty}^{\infty} e^{i\xi\tau} w_1(\tau - y_>) \left[v(\tau - y_<) - \frac{v'(\tau) - qv(\tau)}{w_1'(\tau) - qw_1(\tau)} w_1(\tau - y_<) \right] d\tau \tag{9-5.17}$$

is Fock's form† of the *van der Pol–Bremmer diffraction formula*. Here ξ abbreviates $w/2k_0 l^2$ or, equivalently, $\xi = (k_0 R/2)^{1/3} w/R$; the quantity y_0 is z_0/l, so that $y_0 = (2k_0^2 R^2)^{1/3}(z_0/R)$.

Residues Series for the Shadow Zone

The definitions (12) and the asymptotic relations (9-4.8) and (9-4.9) lead to the conclusion that the integrand in Eq. (17) goes to zero as $\tau \to \infty$ in the upper half plane, Im $\tau > 0$, if $\xi - y_0^{1/2} - y^{1/2} > 0$. The latter is equivalent to the condition $w > (2Rz_0)^{1/2} + (2Rz)^{1/2}$ that the listener is in the shadow zone. The integral for such circumstances can be evaluated by a contour deformation and becomes $2\pi i$ times the sum of those residues corresponding to poles in the upper half plane. Such poles are the zeros τ_n (for $n = 1, 2, \ldots$) of the expression $w_1'(\tau) - qw_1(\tau)$ that appears in the integrand's denominator.

Near $\tau = \tau_n$, the denominator function $w_1' - qw_1$ approximates to

† Fock, *Electromagnetic Diffraction and Propagation Problems*, pp. 239–241; B. van der Pol and H. Bremmer, "Propagation of Radio Waves over a Finitely Conducting Spherical Earth," *Phil. Mag.*, (7) **25**:817–837 (1938). Other representations of analogous formulas are reviewed by Logan, *Lockheed Missles Space Div. Rep.* LMSD-288087. That latter's authoritative analysis of the interrelations between various published diffraction formulas compels acceptance of his nomenclature choices.

$[w_1''(\tau_n) - qw_1'(\tau_n)](\tau - \tau_n)$ or, because $w_1'(\tau_n) = qw_1(\tau_n)$ and because $w_1''(\tau)$ is $\tau w_1(\tau)$ from the differential equation (9-4.5), to $(\tau_n - q^2)w_1(\tau_n)(\tau - \tau_n)$. This makes an implicit identification possible for the residues. Also, the wronskian relation $v'w_1 - w_1'v = -1$ and the definition $w_1'(\tau_n) = qw_1(\tau_n)$ requires that $v'(\tau_n) - qv(\tau_n) = -1/w_1(\tau_n)$. Consequently, the residue series representation for V becomes

$$V(\xi, y_0, y, q) = (4\pi\xi)^{1/2}e^{i\pi/4} \sum_n \frac{e^{i\tau_n\xi}\, w_1(\tau_n - y_0)w_1(\tau_n - y)}{(\tau_n - q^2)[w_1(\tau_n)]^2} \tag{9-5.18}$$

where it is understood that $\xi > y_0^{1/2} + y^{1/2}$. Alternately, because $w_1(\tau_n)$ is $w_1'(\tau_n)/q$, we can replace the denominator in the above by $[(\tau_n/q^2) - 1][w_1'(\tau_n)]^2$. The first version is appropriate for the limiting case $q \to 0$, $Z_S \to \infty$, which corresponds to a rigid ground; the second version is appropriate for the limiting case $q \to \infty$, $Z_S \to 0$, which corresponds to a pressure-release surface.

Since one or the other of the two limiting cases† just mentioned approximate most circumstances of interest, and since the zeros of the Airy function $\mathrm{Ai}(\eta)$ or its derivative $\mathrm{Ai}'(\eta)$ are all real, we replace τ_n by $b_n e^{-i2\pi/3}$ in what follows. For the rigid surface, b_n is a_n', where $a_1', a_2', \ldots$ are the roots of $\mathrm{Ai}'(a_n') = 0$, while for the pressure-release surface, b_n is a_n, where $a_1, a_2, \ldots$ are the roots of $\mathrm{Ai}(a_n) = 0$. These identifications follow from Eq. (12) and the requirement that the τ_n satisfy $w_1'(\tau_n) - qw_1(\tau_n) = 0$. Since the a_n' and the a_n are all negative, each of the corresponding τ_n will lie in the first quadrant of the complex τ plane along the line where the phase of τ is $\pi/3$. The imaginary parts of successive τ_n's therefore increase with successive n, so if ξ is sufficiently large, given fixed y and y_0, the sum (18) approximates to just its leading term. In this manner, we obtain for the rigid boundary and pressure-release surfaces, respectively,

$$V(\xi, y_0, y, 0) \approx (4\pi\xi)^{1/2}e^{-i\pi/12} \exp(ia_1'\xi e^{-i2\pi/3}) \frac{f_1(y_0)f_1(y)}{(-a_1')} \tag{9-5.19a}$$

$$V(\xi, y_0, y, \infty) \approx (4\pi\xi)^{1/2}e^{-i\pi/12} \exp(ia_1\xi e^{-i2\pi/3})g_1(y_0)g_1(y) \tag{9-5.19b}$$

where we use the abbreviations

$$f_1(y) = \frac{\mathrm{Ai}(a_1' - ye^{i2\pi/3})}{\mathrm{Ai}(a_1')} = \frac{w_1(a_1'e^{-i2\pi/3} - y)}{2\pi^{1/2}e^{i\pi/6}\,\mathrm{Ai}(a_1')} \tag{9-5.20a}$$

$$g_1(y) = \frac{\mathrm{Ai}(a_1 - ye^{i2\pi/3})}{\mathrm{Ai}'(a_1)} = \frac{w_1(a_1e^{-i2\pi/3} - y)}{2\pi^{1/2}e^{i\pi/6}\,\mathrm{Ai}'(a_1)} \tag{9-5.20b}$$

† The b_n are the roots of $\mathrm{Ai}'(b) + ie^{i\pi/3}(\rho c/Z_S)k_0 l\,\mathrm{Ai}(b) = 0$, so for $|\rho c/Z_S|k_0 l \ll 1$ (nearly rigid surface), one has $b_n \approx a_n' + e^{-i\pi/6}(\rho c/Z_S)k_0 l/a_n'$, while for $|\rho_0 c/Z_S|k_0 l \gg 1$ (nearly soft surface), one has $b_n \approx a_n + e^{i\pi/6}Z_S/\rho c k_0 l$. Since $k_0 l = (k_0 R/2)^{1/3}$ increases with frequency, any surface of finite impedance will appear nearly soft within the context of the present theory if the frequency is sufficiently high. For a frequency of 1000 Hz, for $c = 340$ m/s, and for a ground impedance of $Z_S = 5\,\rho c(1 + i)$ (see Fig. 3-5), $|\rho c/Z_S|k_0 l$ is $0.30R^{1/3}$, where R is curvature radius in meters. Thus, for an atmospheric profile where $R > 10{,}000$ m, the boundary condition is more properly idealized as that of a pressure-release surface. This was pointed out by R. Onyeonwu, "Diffraction of Sonic Boom Past the Nominal Edge of the Corridor," *J. Acoust. Soc. Am.*, **58**:326–330 (1975).

with $e^{i2\pi/3} = (-1 + i\sqrt{3})/2$, $a_1' = -1.0188$, $\mathrm{Ai}(a_1') = 0.5357$, $a_1 = -2.3381$, and $\mathrm{Ai}'(a_1) = 0.7012$. In either case, the truncation is a justifiable approximation† if $\xi - y_0^{1/2} - y^{1/2}$ is somewhat larger than 1.

If both y and y_0 are moderately large, the functions $w_1(b_1e^{-i2\pi/3} - y)$ and $w_1(b_1e^{-i2\pi/3} - y_0)$ can be replaced by asymptotic expressions of the form of Eq. (13). Doing so reduces the leading term of Eq. (18) to

$$V \approx \frac{e^{i\pi/12}\xi^{1/2}e^{i(2/3)y^{3/2}}e^{i(2/3)y_0^{3/2}}}{K_1(q)y_0^{1/4}y^{1/4}} \exp\left[e^{-i\pi/6}b_1(\xi - y_0^{1/2} - y^{1/2})\right] \quad (9\text{-}5.21)$$

$$K_1(q) = (4\pi)^{1/2}(-b_1 + q^2e^{i2\pi/3})[\mathrm{Ai}(b_1)]^2 \quad (9\text{-}5.22)$$

$$= (4\pi)^{1/2}\left(1 - \frac{b_1}{q^2}e^{-i2\pi/3}\right)[\mathrm{Ai}'(b_1)]^2 \quad (9\text{-}5.22a)$$

where the two versions are appropriate to the limits $q \to 0$ (rigid surface) and $q \to \infty$ (pressure-release surface), respectively. In particular, $K_1(0) = 1.036$ and $K_1(\infty) = 1.743$.

Creeping Waves

An implication of Eqs. (16) and (18) is that within the shadow zone and on the surface, the amplitude of acoustic pressure or of any other acoustic field quantity must asymptotically decrease with distance w along the surface as $w^{-1/2}e^{-\alpha w}$, where the attenuation coefficient α (nepers per meter) is given by

$$\alpha = \mathrm{Re}(-e^{-i\pi/6}b_1)\left(\frac{k_0}{2R^2}\right)^{1/3} \quad (9\text{-}5.23)$$

$$= \frac{n}{2c}f^{1/3}\left(-\frac{dc}{dz}\right)_0^{2/3} \quad (9\text{-}5.23a)$$

with $n = 2\pi^{1/3}\,\mathrm{Re}(-e^{-i\pi/6}b_1)$ and with f denoting the frequency in hertz. For a rigid surface, n is 2.58, while for a pressure release surface, n is 5.93. The corresponding speed (phase velocity) at which lines of constant phase move along the surface is similarly deduced to be

$$v_{\mathrm{ph}} = \frac{c(0)}{1 + \mathrm{Im}(e^{-i\pi/6}b_1)/(2k_0^2R^2)^{1/3}} \quad (9\text{-}5.24)$$

and is always less than the sound speed $c(0)$.

The weak attenuation and slightly retarded phase velocity are two distinguishing characteristics of a creeping wave.‡ Such waves move along surfaces

† The criterion that emerges from a comparison of Eqs. (18) and (21) is that

$$|\exp[e^{-i\pi/6}(b_2 - b_1)(\xi - y_0^{1/2} - y^{1/2})]| \ll 1$$

which is approximately satisfied if $\xi - y_0^{1/2}$ is larger than $2/\{\mathrm{Re}\,[(-b_2 + b_1)e^{-i\pi/6}]\}$; this quantity equals 1.034 and 1.3198 for the rigid surface and for the pressure-release surface, respectively.

‡ The term *Kriechwelle* was introduced by W. Franz and K. Depperman, "Theory of Diffraction by a Cylinder with Consideration of the Creeping Wave," *Ann. Phys.*, (6)**10**:361–373 (1952). The prediction of such waves dates back to G. N. Watson, "The Diffraction of Electric Waves by

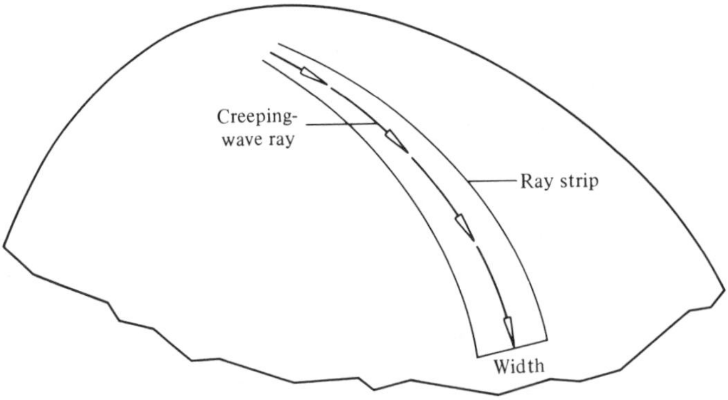

Figure 9-18 Concept of a creeping wave propagating along a surface. If the sound speed is constant, the creeping-wave ray is a geodesic. The amplitude on the surface decreases as the reciprocal of the square root of strip width and decreases exponentially with distance along path.

with ray paths (see Fig. 9-18) that are everywhere perpendicular to surfaces of constant phase (given an absence of ambient flow tangential to the surface). In the example considered here, the creeping-wave rays are straight horizontal lines extending radially from the source, but in other instances the rays curve along the surface. In addition to a weak exponential decay with propagation distance, the amplitude along the surface varies inversely with the square root of the perpendicular distance (ray-strip width) between adjacent rays propagating along the surface. In the above example, ray-strip width is proportional to w, so a factor of $w^{-1/2}$ emerges from the insertion of (21) into (16).

For propagation along a curved surface in a homogeneous medium,† the requirement that the creeping-wave rays be perpendicular to the surfaces of constant phase and that they move with a speed nearly equal to the sound speed leads to the recognition that the paths are *geodesics;* the path connecting two points on the surface is the shortest of all possible paths. (This property is analogous to *Fermat's principle of least time.*) For the two idealizations of principal interest, a sphere and a circular cylinder, the paths are great circles and helices, respectively.

If a creeping wave is propagating along a curved surface in a homogeneous medium, one can locally orient the coordinate system and origin so that the surface is given by $z = -x^2/2R_1 - y^2/2R_2$, where R_1 and R_2 denote the surface's two principal radii of curvature. The disturbance near the origin is taken of the form $e^{ik_x\xi}e^{ik_y\eta}F(\zeta)$, where $\xi \approx x - xz/R_1$, $\eta \approx y - yz/R_2$, and

---

the Earth," *Proc. R. Soc. Lond.*, **A95**:83–99 (1919). That the wave penetrating into the shadow zone above a plane boundary in a stratified medium can be regarded as a creeping wave has been pointed out by G. D. Malyuzhinets, "Development in Our Concepts of Diffraction Phenomena (On the 130th Anniversary of the Death of Thomas Young)," *Sov. Phys. Usp.*, **69**:749–758 (1959).

† R. M. Lewis, N. Bleistein, and D. Ludwig, "Uniform Asymptotic Theory of Creeping Waves," *Commun. Pure Appl. Math.*, **20**:295–328 (1967); J. B. Keller, "Diffraction by a Convex Cylinder," *IRE Trans. Antennas Prop.*, **4**:312–321 (1956); B. R. Levy and J. B. Keller, "Diffraction by a Smooth Object," *Commun. Pure Appl. Math.*, **12**:159–209 (1959).

$\zeta \approx z + x^2/2R_1 + y^2/2R_2$. Approximations† similar to those described in the derivations of Eqs. (10) and (9-4.5) then result in the differential equation

$$\frac{d^2F}{d\zeta^2} + \left(k_0^2 - k^2 + \frac{2k^2\zeta}{R_{\text{eff}}}\right) F = 0 \tag{9-5.25}$$

where

$$k_0^2 = \frac{\omega^2}{c^2} \qquad k^2 = k_x^2 + k_y^2 \qquad R_{\text{eff}}^{-1} = R_1^{-1}\cos^2\theta_k + R_2^{-1}\sin^2\theta_k \tag{9-5.26}$$

where θ_k is the direction of (k_x, k_y) relative to the x axis. Given $\zeta \ll R_{\text{eff}}$, the k^2 in the last term can be approximated by k_0^2, so one recovers Eq. (10) but with a new interpretation of R; ζ is interpreted as distance transverse to the surface. The boundary condition and the selection of the least attenuated wave then leads to an Airy function of the form $w_1(\tau_1 - \zeta/l_{\text{eff}})$ just as in the leading term of Eq. (18), only with z/l replaced by ζ/l_{eff}, where $l_{\text{eff}} = (R_{\text{eff}}/2k_0^2)^{1/3}$.

Ray Shedding by a Creeping Wave

The implication of Eq. (21) is that deep within the shadow zone but not near the surface (z somewhat larger than l) the disturbance propagates along ordinary geometrical-acoustics rays. The origin of these rays,‡ however, is not the source but the creeping wave (see Fig. 9-19). This identification emerges if we write the product (16) with the insertion of Eq. (21) for V as

$$\hat{p} = \frac{e^{i\pi/12}(R^2/4k_0)^{1/6}\hat{S}e^{-\alpha\Delta w}\exp\left[i\omega\tau_{\text{TR}}(z_0) + i(\omega/v_{\text{ph}})\Delta w + i\omega\tau_{\text{TR}}(z)\right]}{w^{1/2}[K_1(q)/2^{1/2}][(2Rz_0)(2Rz)]^{1/4}} \tag{9-5.27}$$

where

$$\Delta w = w - (2Rz_0)^{1/2} - (2Rz)^{1/2} \tag{9-5.28a}$$

$$c_0\tau_{\text{TR}}(z) = (2Rz)^{1/2} + \frac{2}{3}\left(\frac{2z^3}{R}\right)^{1/2} \tag{9-5.28b}$$

Here $(2Rz_0)^{1/2}$ is horizontal distance from the source to the edge of the shadow zone; $(2Rz)^{1/2}$ is horizontal distance from surface to listener along a ray that leaves the ground at the grazing angle and subsequently passes through the listener position. Such a ray would leave the ground at $(w_0, 0)$, where $w_0 = w - (2Rz)^{1/2}$. The quantity $\tau_{\text{TR}}(z)$ can be identified as the travel time along such a ray segment. The latter follows from Eqs. (8-4.2), which predict that

† In the analogous theory of radio-wave propagation along the surface of a spherical earth, this is known as the *earth-flattening approximation*: J. C. Schelleng, C. R. Burrows, and E. B. Ferrell, "Ultra-Short-Wave Propagation," *Proc. Inst. Radio Eng.*, **21**:427–463 (1933); C. L. Pekeris, "Accuracy of the Earth-Flattening Approximation in the Theory of Microwave Propagation," *Phys. Rev.*, **70**:518–522 (1946); "The Field of a Microwave Dipole Antenna in the Vicinity of the Horizon," *J. Appl. Phys.*, **18**:667–680 (1947).

‡ That shedded rays are present has been demonstrated by schlieren photographs of acoustic pulses incident on cylinders. See, for example, W. G. Neubauer, "Experimental Measurement of 'Creeping' Waves on Solid Aluminum Cylinders in Water Using Pulses," *J. Acoust. Soc. Am.*, **44**:298–299 (1968).

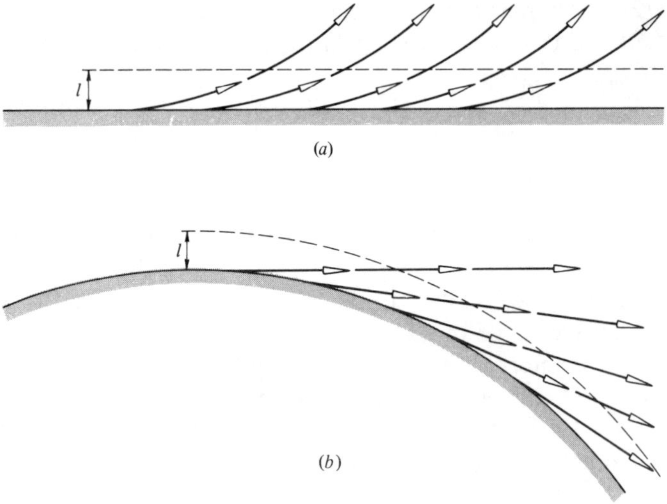

Figure 9-19 Shedding of rays by a creeping wave: (a) flat surface bounding a fluid where the sound speed increases linearly with height; (b) curved surface bounding a fluid of constant sound speed.

$d\tau_{TR}/dw$ will be c_0/c^2 since $s_w = 1/c_0$ for a ray initially tangential to the surface. The quantity c_0/c^2 is approximately $(1 + 2z/R)/c_0$, but z is $(w - w_0)^2/2R$ along the ray, so $d\tau_{TR}/dw$ integrates to $c_0\tau_{TR} = (w - w_0) + \frac{1}{3}(w - w_0)^2/R^2$. Then, replacing $w - w_0$ by $(2Rz)^{1/2}$, we obtain Eq. (28b).

Similarly $\tau_{TR}(z_0)$ corresponds to travel time along the ray that goes from source to edge of shadow zone at the surface, a horizontal distance of $(2Rz_0)^{1/2}$. The phase change $\omega\tau_{TR}(z_0) + (\omega/v_{ph})\,\Delta w + \omega\tau_{TR}(z)$ therefore corresponds to a broken ray path that travels from source to ground with the sound speed, then along the ground a distance Δw with the phase velocity v_{ph}, and then from ground to listener with the sound speed.

The above observation yields the interpretation that the sound reaching the listener at w, z is shed by the creeping wave at w_0, 0. This view is further supported by the attenuation factor $e^{-\alpha\Delta w}$. The disturbance at w, z is carried by the creeping wave over only the interval $[(2Rz_0)^{1/2}, 0]$ to $[w_0, 0]$, a net distance of Δw.

The factors of $w^{1/2}$ and $(2Rz)^{1/4}$ in the denominator of Eq. (27) are similarly interpreted in terms of geometrical acoustics; their product is proportional to the square root of the ray-tube area associated with the ray passing through the listener location. Two rays successively shed at w_0 and $w_0 + \delta w_0$ will have an approximate perpendicular separation $\delta z \approx -\delta[(w - w_0)^2/2R]$, or $(w - w_0)(\delta w_0/R)$, after traversing a distance $w - w_0$. Thus ray-tube area varies with z as $w - w_0$ or as $(2Rz)^{1/2}$. The cylindrical spreading (which began at the source) creates the other factor of w in the ray-tube-area expression.

9-6 SOURCE OR LISTENER ON THE EDGE OF A WEDGE

A prototype for theories of diffraction by edges is that of the field in the vicinity of a rigid wedge-shaped obstacle. The edge of the wedge coincides with the z axis; one face occupies the half plane $y = 0$, $x > 0$ in such a way that it is given by $\phi = 0$ in a cylindrical coordinate system, $x = r \cos \phi$, $y = r \sin \phi$. The other face is at $\phi = \beta$, so the wedge exterior consists of points for which ϕ is between 0 and β (see Fig. 9-20).

Exact solutions of the wave equation for the exterior region of such a wedge are somewhat intricate, but simple expressions emerge for various limiting cases. We here begin with the simplest, that where either the source or the listener is on the edge.

Source on Edge

Let a point source of time-dependent monopole amplitude $S(t)$ be at $z = z_S$ on the edge ($r_S = 0$), the source being such that $p(\mathbf{x}, t)$ would be $S(t - R/c)/R$ without the wedge present, with R denoting the radial distance $[r^2 + (z - z_S)^2]^{1/2}$ from the source.

The boundary condition at the wedge faces is satisfied by the free-space solution, because it predicts a radial flow. However, the free-space solution does not give the correct rate of mass flow out from the source (through a surface close to the source) into the region exterior to the wedge. The net rate $\dot{m}(t)$ that mass flows from the source must, according to Eq. (4-3.9) and Euler's equation, be such that

$$\frac{d\dot{m}}{dt} = 4\pi S(t) \qquad (9\text{-}6.1)$$

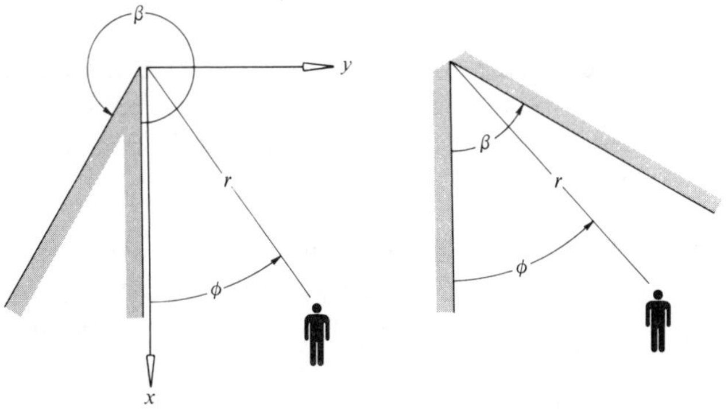

Figure 9-20 Parameters for description of propagation in wedge-limited regions: (a) propagation outside a wedge of exterior angle β; (b) propagation inside a wedge.

The definition of $S(t)$ is such that this holds regardless of the location of the source and in particular when the source is adjacent to a solid surface. This is consistent, for example, with what is obtained when a source is near a flat rigid plane and the field is determined by the method of images.

When the source is on the edge, the expelled mass flows into a solid angle of

$$\Delta\Omega = \int_0^{\beta}\int_0^{\pi} \sin\theta \, d\theta \, d\phi = 2\beta \tag{9-6.2}$$

rather than into 4π sr, as for free-space radiation. Thus the time rate of change of mass flow rate per solid angle must be $4\pi S(t)/2\beta$ when the wedge is present and is enhanced relative to the free-space case by a factor of $2\pi/\beta$. Such reasoning results in the solution†

$$p = \frac{2\pi}{\beta}\frac{S(t - R/c)}{R} \tag{9-6.3}$$

for the acoustic pressure resulting from a point source on a wedge of exterior angle β, where $0 < \beta < 2\pi$.

Since the time-averaged acoustic intensity is proportional to the mean squared acoustic pressure for a spherically spreading wave, Eq. (3) implies that the intensity is enhanced by a factor of $(2\pi/\beta)^2$ relative to when the source is in a free environment. The energy spreads into 2β sr, so the enhancement of the acoustic power output is

$$\frac{\mathcal{P}}{\mathcal{P}_{\text{ff}}} = \frac{2\beta R^2(2\pi/\beta)^2 I_{\text{ff}}}{4\pi R^2 I_{\text{ff}}} = \frac{2\pi}{\beta} \tag{9-6.4}$$

which is consistent with what would be derived by the method of images (see Sec. 5-1) for the special cases $\beta = \pi$ and $\beta = \pi/2$.

Listener on the Edge

When the listener (rather than the source) is on the edge, Eq. (3) also applies, because of the principle of reciprocity; R is interpreted, as before, as distance from source to listener. The field, however, will not be spherically symmetric unless the source is also on the edge.

Example Suppose a wave from a distant source impinges on a thin rigid screen. The acoustic-pressure amplitude at a given point on the edge would nominally be P_0 without the barrier present. What is its value at the same point when the screen is present?

† Rayleigh, *The Theory of Sound,* vol. 2, pp. 112–113. The applicability of Rayleigh's analysis for a point source at the vertex of a rigid cone of given solid angle to a source on a wedge's edge is pointed out by R. V. Waterhouse, "Diffraction Effects in a Random Sound Field," *J. Acoust. Soc. Am.*, **35:**1610–1620 (1963).

SOLUTION In this case, β is 2π, so reciprocity considerations and Eq. (3) imply that the amplitude will also be P_0 when the screen is present. There may be a marked change in the amplitude at points not on the edge of the screen, however.

9-7 CONTOUR-INTEGRAL SOLUTION FOR DIFFRACTION BY A WEDGE

To solve the more difficult boundary-value problem† with a harmonic point source at an arbitrary point (z_S set to zero for simplicity) near a rigid wedge, we seek a complex pressure amplitude $\hat{p}$ that satisfies the Helmholtz equation (1-8.13) everywhere outside the wedge except at $\mathbf{x}_S$; near $\mathbf{x}_S$ it should be of the form $\hat{S}/|\mathbf{x} - \mathbf{x}_S|$ plus a bounded function; it should also satisfy the rigid-wall boundary condition $\partial\hat{p}/\partial\phi = 0$ at the faces ($\phi = 0$, $\phi = \beta$) of the wedge. In addition, at large distances from the source and the edge, the solution must satisfy the Sommerfeld radiation condition.

To describe the solution, it is convenient to introduce a *wedge index* $\nu = \pi/\beta$ ($\geq\frac{1}{2}$) and a function $\mathcal{R}(\zeta)$, where

$$\mathcal{R}(\zeta) = (r^2 + r_S^2 - 2rr_S \cos\zeta + z^2)^{1/2} \tag{9-7.1}$$

is the distance in the free-space Green's function

$$\mathcal{G}(\zeta) = \frac{1}{\mathcal{R}(\zeta)} e^{ik\mathcal{R}(\zeta)} \tag{9-7.2}$$

Thus $\mathcal{R}(\phi - \phi_S)$ represents the direct distance between source and listener; $\hat{S}\mathcal{G}(\phi - \phi_S)$ would be the solution without the wedge present. We shall be interested in values of $\mathcal{R}(\zeta)$ when ζ is complex, and in order to specify uniquely which square root is implied by (1), we define $\mathcal{R}(\zeta)$ so that it is positive for real ζ and analytic except at branch cuts [at which the phase of $\mathcal{R}(\zeta)$ has a discontinuity of π] that extend vertically up and down from *branch points* above and below the real axis, respectively (see Fig. 9-21). These branch points, at which $\mathcal{R} = 0$, are found from (1) to be at $2\pi l \pm i\alpha$, where l is any integer and where

$$\alpha = \cosh^{-1} \frac{r^2 + r_S^2 + z^2}{2rr_S} \tag{9-7.3}$$

The function $\mathcal{G}(\zeta - \phi)$ satisfies the Helmholtz equation, so the superposition principle requires any contour integral of the form

$$\hat{p} = \hat{S} \int_C f(\zeta)\mathcal{G}(\zeta - \phi)\, d\zeta \tag{9-7.4}$$

† H. M. MacDonald, "A Class of Diffraction Problems," *Proc. Lond. Math. Soc.*, **14**:410–427 (1915); T. J. I'A. Bromwich, "Diffraction of Waves by a Wedge," ibid., **14**:450–463 (1915). A bibliography including references to earlier work by H. Poincaré (1892), A. Sommerfeld (1896), and MacDonald (1902) is given by H. G. Garnir, *Bull. Soc. R. Sci. Liege*, **21**:207–231 (1952).

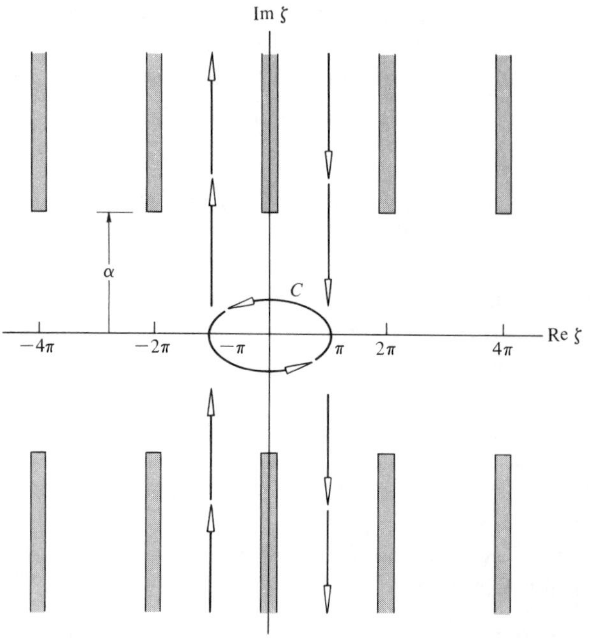

Figure 9-21 Branch cuts in the complex ζ plane for the function $\mathscr{G}(\zeta)$. Indicated closed contour is appropriate for integer wedge index ν. The contributions from the two vertical contours passing through $-\pi$ and π cancel each other for ν an integer.

to satisfy the Helmholtz equation, given position-independent contour C and function $f(\zeta)$. This expression, moreover, will satisfy the Sommerfeld radiation condition. Alternatively, we may change the variable of integration to $\zeta - \phi$, rename it as ζ, and have

$$\hat{p} = \hat{S} \int_{C_\phi} f(\zeta + \phi)\mathscr{G}(\zeta)\, d\zeta \tag{9-7.5}$$

Insofar as C_ϕ can be deformed without crossing any poles or branch cuts into a contour C independent of ϕ for any ϕ between 0 and β, we can take C_ϕ to be independent of ϕ and the same as the original contour C. The task is then to find appropriate $f(\zeta + \phi)$ and C in order that Eq. (5), with $C_\phi \to C$, will represent a solution of the boundary-value problem posed above.

Method of Images for Integer Wedge Index

If the wedge index ν is an integer, the problem can be solved by the method of images introduced in Sec. 5-1. Locations of the $2\nu - 1$ images (see Fig. 9-22*a*) required to ensure that the boundary conditions will be satisfied are found in a manner similar to that used to develop (Sec. 3-4) the solution for the transient disturbance caused by a vibrating piston in a tube with a rigid end. The solution for integer ν is consequently

$$\hat{p} = \hat{S} \sum_{m=0}^{\nu-1} \left[\mathscr{G}\left(\frac{2m\pi}{\nu} - \phi_S - \phi\right) + \mathscr{G}\left(\frac{2m\pi}{\nu} + \phi_S - \phi\right)\right] \tag{9-7.6}$$

Alternatively, we can express the sum by a contour integral

$$\hat{p} = \frac{\hat{S}}{2\pi i} \int_C \mathscr{G}(\zeta)[h(\zeta + \phi + \phi_S) + h(\zeta + \phi - \phi_S)]\, d\zeta \tag{9-7.7}$$

where $h(\zeta)$ has poles at $\zeta = 2m\beta = 2\pi m/\nu$ and the residue of $h(\zeta)$ at each such pole is unity. The contour C is understood to encircle one pole each for which $m = 0 \pmod{\nu}$, $m = 1 \pmod{\nu}$, . . . , $m = \nu - 1 \pmod{\nu}$ (see Fig. 9-22b). A

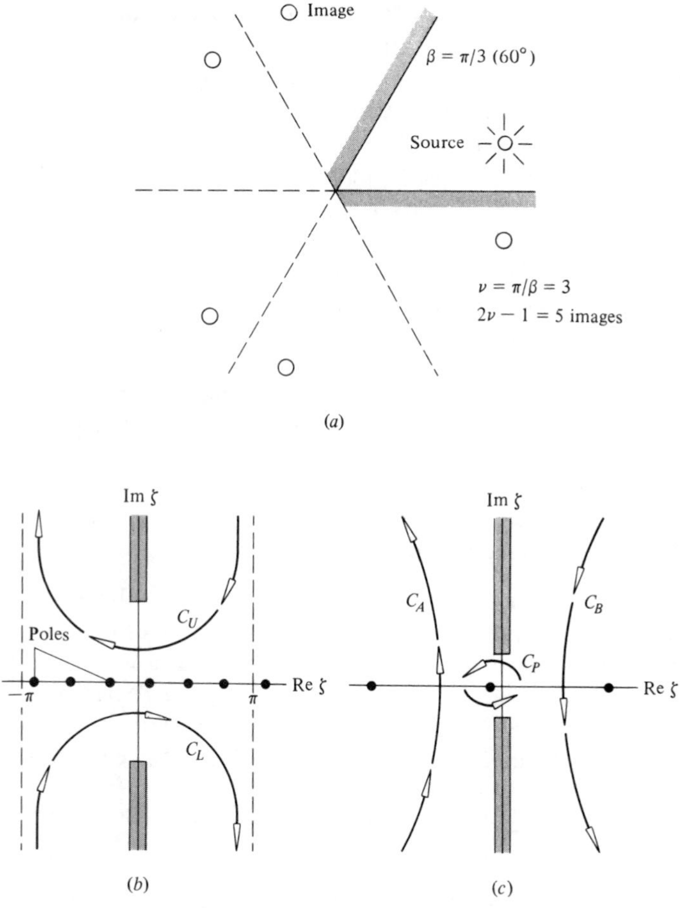

Figure 9-22 (a) Images for a source within a 60° wedge ($\beta = \pi/3$, $\nu = 3$). (b) Deformed contour for integration that yields the sum of free-space fields of the source and its images. (c) Deformed contour appropriate for when the listener is arbitrarily close to the source. Contour C_P gives field with $1/R$ singularity; contours C_A and C_B give finite contributions at the source.

choice for $h(\zeta)$ is

$$h(\zeta) = \frac{\nu}{2} \cot\left(\frac{\nu}{2}\zeta\right) \tag{9-7.8}$$

The residue at the pole, $\zeta = 2m\beta$, is 1 because $\cos m\pi = (-1)^m$ and because $\sin[(\nu/2)\zeta] \to (-1)^m[(\nu/2)\zeta - m\pi]$ as $\zeta \to 2m\pi/\nu$. The additional restriction that $h(\zeta)$ repeat itself at intervals of 2π assures that this choice for $h(\zeta)$ is unique except for an arbitrary additive constant, which is of no consequence. A possible choice for the contour C is one encircling all poles between $-\pi$ and π.

The closed-contour choice for C is satisfactory for integer ν, but when ν is a noninteger, the number of enclosed poles varies with ϕ and the integral therefore becomes a discontinuous function of ϕ. To circumvent this difficulty, we pick another integration contour that does not cross the real axis. For integer ν, we note that the integrand repeats itself at intervals of 2π, so integration along a downward path from $\pi + i\infty$ to $\pi - i\infty$ will exactly cancel one along an upward path from $-\pi - i\infty$ to $-\pi + i\infty$. Thus the value of (7) for integer ν is unchanged if we add additional contours that go parallel to the imaginary axis up and down the lines $\zeta_R = -\pi$ and $\zeta_R = \pi$, respectively. The overall contour can then be split into contours C_U and C_L, where C_U goes from $\pi + i\infty$ to π, then arcs above the real axis from π to $-\pi$, then goes from $-\pi$ to $-\pi + i\infty$; C_L is C_U's inversion ($\zeta \to -\zeta$) through the origin. Alternatively, since $\mathcal{G}(\zeta) \to 0$ as $\zeta_I \to \infty$ for $-2\pi < \zeta_R < -\pi$ and $0 < \zeta_R < \pi$, we can deform C_U to any contour (see Fig. 9-22*b*) that starts at $\zeta_I = \infty$ for some ζ_R between 0 and π, then goes down and passes below the branch point at $i\alpha$, and then goes back to $\zeta_I \to +\infty$ in the region where ζ_R is between -2π and $-\pi$. The corresponding deformed C_L can be taken as the inversion of C_U, starting at $\zeta_I = -\infty$ with $-\pi > \zeta_R < 0$, passing above $-i\alpha$, and ending at $\zeta_I \to -\infty$ with $\pi < \zeta_R < 2\pi$.

Generalization to Noninteger Wedge Indices

The claim is now made that Eq. (7) with $h(\zeta)$ given by Eq. (8) and with $C = C_U + C_L$ (where C_U and C_L are the contours described above) is also the solution of the boundary-value problem for arbitrary ν (including $\nu < 1$). (Recall that the preceding derivation presumed that ν is an integer.) To verify that our candidate solution has the requisite properties, we first note that $\mathcal{R}(\zeta) = \mathcal{R}(-\zeta)$ and that C_U is the inversion of C_L, so that (7) can be reexpressed

$$\hat{p} = \frac{\hat{S}}{2\pi i} \int_{C_L} \mathcal{G}(\zeta)\Sigma h \, d\zeta \tag{9-7.9}$$

$$\Sigma h = \sum_{n,m=1}^{2} \frac{\nu}{2} \cot\left(\frac{\nu}{2}\,[\zeta + (-1)^n\phi + (-1)^m\phi_S]\right) \tag{9-7.10}$$

In the latter expression, the sum extends over all sign combinations of $\pm\phi \pm \phi_S$. Note that the sum includes the terms $h(\zeta + \phi + \phi_S)$, $h(\zeta + \phi - \phi_S)$, $-h(-\zeta + \phi + \phi_S)$, and $-h(-\zeta + \phi - \phi_S)$, where the last two are the inversions $\zeta \to -\zeta$, with a sign change (since $d\zeta$ on C_U goes to $-d\zeta$ on C_L when $\zeta \to -\zeta$) of the first and second terms.

Expression (9) satisfies the Helmholtz equation because $\mathcal{G}(\zeta)$ satisfies

$$\left(\frac{\partial^2}{\partial r^2} + \frac{1}{r}\frac{\partial}{\partial r} + \frac{1}{r^2}\frac{\partial^2}{\partial \zeta^2} + \frac{\partial^2}{\partial z^2} + k^2\right)\mathcal{G}(\zeta) = 0$$

which in turn implies

$$(\nabla^2 + k^2)\hat{p} = \frac{\hat{S}}{2\pi i}\int_{C_L} \frac{1}{r^2}\left[\mathcal{G}(\zeta)\frac{\partial^2}{\partial \phi^2}\Sigma h - \Sigma h \frac{\partial^2 \mathcal{G}}{\partial \zeta^2}\right] d\zeta$$

Since $\mathcal{G}(\zeta)$ vanishes exponentially at the endpoints of C_L, the second term above can be integrated by parts twice, thereby transferring the operator $\partial^2/\partial\zeta^2$ from $\mathcal{G}$ to Σh. The integrand then contains the factor

$$\left(\frac{\partial^2}{\partial \phi^2} - \frac{\partial^2}{\partial \zeta^2}\right)\Sigma h = 0$$

which (as demonstrated in Sec. 1-7) is identically zero because Σh is a sum of terms that depend on ζ and ϕ only through one of the combinations $\zeta + \phi$ or $\zeta - \phi$.

Next we check that Eq. (9) exhibits the proper singular behavior near the source location. When $r \to r_S$, $z \to 0$, $\phi \to \phi_S$, one finds that $\alpha \to 0$ and that a pole of $h(\zeta + \phi - \phi_S)$ approaches the origin. To isolate the effect of the pole, we deform $C_L + C_U$ into $C_A + C_B + C_P$, where C_A, C_B, and C_P are as sketched in Fig. 9-22c. The contributions from C_A and C_B are bounded while that from C_P gives $\hat{S}\mathcal{G}(\phi_S - \phi)$, which is just the direct wave from the source.

The boundary condition $\partial\hat{p}/\partial\phi = 0$ at $\phi = 0$ is guaranteed by Eq. (10) because $h(\zeta)$ is an odd function of its argument, so Σh is even in ϕ for fixed ζ and ϕ_S. The other boundary condition, $\partial\hat{p}/\partial\phi = 0$ at $\phi = \beta$, follows because $h(\zeta)$ is periodic in ζ with period 2β; if one replaces ϕ by $2\beta - \phi$ in Σh, uses the periodicity property, and recognizes that each term is odd in its argument, one finds Σh unchanged, so it must be even about $\phi = \beta$.

The Sommerfeld radiation condition is satisfied by Eq. (9) because $\mathcal{G}(\zeta)$ for all finite ζ satisfies this condition. (The asymptotic expressions derived in the following section support this inference.)

The limit $r_S \to 0$ of Eq. (9) must, according to Eq. (9-6.3), yield $(2\pi/\beta)\hat{S}R^{-1}e^{ikR}$. That such is indeed the case is demonstrated beginning with the expansion† (Im $\zeta < 0$)

† A. A. Tuzhilin, "New Representations of Diffraction Fields in Wedge-Shaped Regions with Ideal Boundaries," *Sov. Phys. Acoust.*, **9**:168–172 (1963). Equation (11) is most easily derived from a power-series expansion of Σh in $u = e^{-i\nu\zeta}$, with the cotangents in Eq. (10) expressed in terms of exponentials. Tuzhilin's expression (given without a derivation) for what is here termed $I_{\nu n}$ is

$$I_{\nu n} = i\left(\frac{\pi k}{2R_1}\right)^{1/2} \sum_{s=0}^{\infty} \frac{H^{(1)}_{n\nu+1/2+2s}(kR_1)}{s!\,\Gamma(n\nu + 1 + s)}\left(\frac{krr_S}{2R_1}\right)^{\nu n+s}$$

where R_1 is $(r^2 + r_S^2 + z^2)^{1/2}$ and $H^{(1)}_\mu(kR_1)$ is the Hankel function of the first kind with (noninteger)

$$\Sigma h = 2i\nu \sum_{n=0}^{\infty} \varepsilon_n e^{-i\nu n\zeta} \cos \nu n\phi \cos \nu n\phi_S \tag{9-7.11}$$

such that

$$\hat{p} = \frac{2\pi}{\beta} \hat{S} \sum_{n=0}^{\infty} \varepsilon_n \cos \nu n\phi \cos \nu n\phi_S \, I_{\nu n} \tag{9-7.12}$$

with

$$I_{\nu n} = \frac{1}{2\pi} \int_{C_L} \mathscr{G}(\zeta) e^{-i\nu n\zeta} \, d\zeta \tag{9-7.13}$$

Here ε_n is 1 for $n = 0$ and is 2 for $n \geq 1$. The result derived in Sec. 9-6 emerges when $r_S \to 0$, because $I_0 \to R^{-1}e^{ikR}$ and $I_{\nu n} \to 0$ $(n \geq 1)$ in this limit.

9-8 GEOMETRICAL-ACOUSTIC AND DIFFRACTED-WAVE CONTRIBUTIONS FOR THE WEDGE PROBLEM

Here the contour solution for a point source in the vicinity of a wedge is applied to determine an asymptotic approximation for the field. The contour C_L in Eq. (9-7.9) can be deformed† into one crossing the real axis at $\zeta = 0$ and at $\zeta = \pi$, provided one adds an additional contour that encircles the poles between 0 and π in the counterclockwise sense. Since $\mathscr{G}(\zeta) \to 0$ as $\zeta_I \to \infty$ for ζ_R between 0 and π, the deformed C_L contour can be split into left and right segments that terminate and originate at $\zeta = \pi/2 + i\infty$ (see Fig. 9-23). The left segment can be taken as symmetric with respect to inversions through the origin and the integrand is odd in ζ; thus the integral along the left segment vanishes identically, and we are left with a contour C_π (the right segment) plus a counterclockwise contour encircling the poles between 0 and π.

The Geometrical Acoustics Portion of the Field

The poles of Σh occur when $\sin [(\pi/2\beta)(\zeta \pm \phi \pm \phi_S)]$ vanishes or when $\zeta \pm \phi \pm \phi_S$ is $2\beta l$, where l is any integer; the residue of Σh at each such pole is unity; however, we must include only poles at points ζ_P, where $0 < \zeta_P < \pi$.

---

index μ. The properties of the latter are such that

$$I_{\nu n} \approx \frac{e^{-i\nu n\pi/2}}{\Gamma(1 + \nu n)} \left(\frac{krr_S}{2R_1}\right)^{\nu n} R_1^{-1} e^{ikR_1}$$

when kR_1 is large compared with 1 and when $krr_S/2R_1$ is small. See, for example, G. N. Watson, *A Treatise on the Theory of Bessel Functions*, 2d ed., Cambridge University Press, Cambridge, 1922, p. 197.

† F. J. W. Whipple, "Diffraction by a Wedge and Kindred Problems," *Proc. Lond. Math. Soc.*, **16**:481–500 (1919).

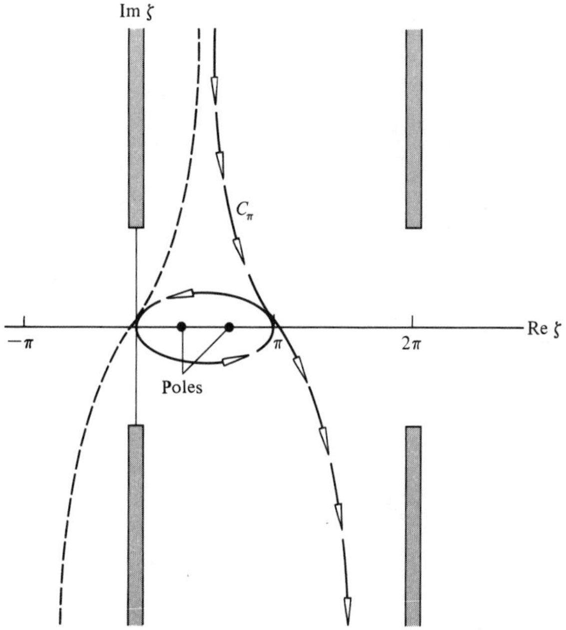

Figure 9-23 Deformed contour in the complex ζ plane, yielding asymptotic representation for sound diffraction by a rigid wedge. The integral along the contour segment passing through the origin vanishes because of symmetry.

The residue theorem accordingly yields, for the geometrical-acoustics field,

$$\hat{p}_{\text{GA}} = \hat{S} \sum_l{}' \mathscr{G}(2\beta l - \phi_S - \phi) + \hat{S} \sum_l{}' \mathscr{G}(2\beta l + \phi_S - \phi) \tag{9-8.1}$$

where both sums extend over all values of l for which the indicated argument is between $-\pi$ and π.

Each included term represents a spherical wave diverging from an image and corresponds to a possible ray path that connects source and listener. The direct-ray term $\hat{S}\mathscr{G}(\phi_S - \phi)$ corresponds to the $l = 0$ term in the second sum and is present only if $|\phi - \phi_S| < \pi$. The ray reflected once from the $\phi = 0$ face corresponds to the $l = 0$ term in the first sum and is present only if $\phi_S + \phi < \pi$. The ray reflected once from the $\phi = \beta$ face corresponds to the $l = 1$ term in the first sum and is present only if $2\beta - \phi_S - \phi < \pi$. (Recall that both ϕ and ϕ_S are between 0 and β.) A similar physical interpretation can be given for each of the other terms.

If $\beta > \pi/2$ ($\nu < 2$), the only possible terms are those where the arguments of the Green's functions are $\phi_S - \phi$ (direct), $\phi + \phi_S$ (0 face), $2\beta - \phi_S - \phi$ (β face), $2\beta + \phi_S - \phi$ (0 face then β face), and $-2\beta + \phi_S - \phi$ (β face then 0 face). The second and third possibilities both occur if either the fourth or fifth is realized, but the fourth and fifth are mutually exclusive. If β is between $\pi/2$ and π, there is always one singly reflected ray path, but a doubly reflected path is possible only if $|\phi_S - \phi| > 2\beta - \pi$. For this range of β, there are two or three paths if $|\phi_S - \phi|$ is less than $2\beta - \pi$ and four paths if $|\phi_S - \phi| > 2\beta - \pi$ (see Fig. 9-24a).

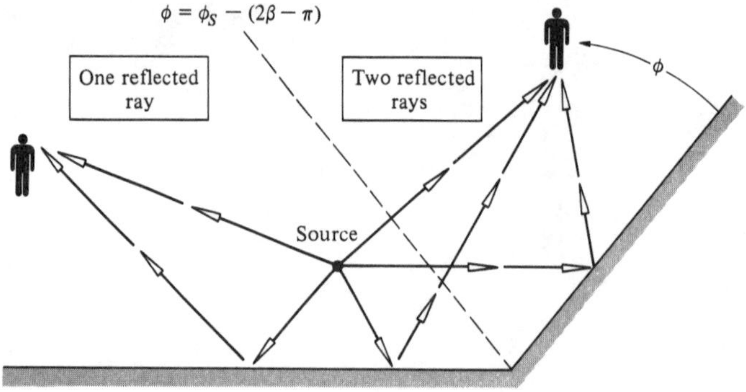

(a)

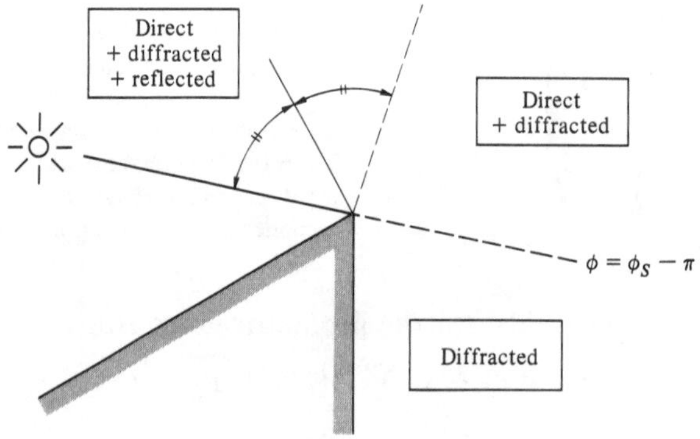

(b)

Figure 9-24 (a) Possible singly reflected ray paths connecting source and listener when the wedge angle β is between $\pi/2$ and π. (b) Ranges of ϕ and ϕ_S in which various wave contributions are expected for wedge with exterior angle β greater than π.

If $\beta > \pi$ ($\nu < 1$), so that source and listener are in the exterior region of a wedge, one has a direct ray if $|\phi - \phi_S| < \pi$, a ray reflected from the $\phi = 0$ face if $\phi + \phi_S < \pi$, and a ray reflected from the $\phi = \beta$ face if $\phi + \phi_S > 2\beta - \pi$; the last two possibilities are mutually exclusive. If $\phi + \phi_S$ is between $2\beta - \pi$ and π, moreover, there is no reflected path. If $\phi_S > \pi$ and $\phi < \phi_S - \pi$, or if $\phi_S < \beta - \pi$ and $\phi > \pi + \phi_S$, there is neither a direct path nor a reflected path (see Fig. 9-24b). In such circumstances, the listener is in a shadow zone, and any nontrivial estimation of the acoustic field requires an evaluation of the contribution to $\hat{p}$ from the contour C_π.

The Diffracted Wave

The contour C_π term is simply what is left over when one has constructed a solution according to geometrical-acoustic principles; consequently it is identified as the diffracted wave $\hat{p}_{\text{diffr}}$. A less abstract representation results from the deformation of C_π to coincide with the line $\zeta_R = \pi$. If one sets $\zeta = \pi - is$, then $d\zeta = -i\, ds$ and s range from $-\infty$ to ∞. Since $\cos(\pi - is)$ is $-\cosh s$, the quantity $\mathscr{G}(\pi - is)$ is even in s, so we need only keep terms even in s in the remainder of the integrand. Thus, after some manipulation of trigonometric identities, we can make the substitution

$$\cot\left[\frac{\nu}{2}(x - is)\right] \to \frac{\sin \nu x}{\cosh \nu s - \cos \nu x}$$

and the diffracted-wave contribution becomes†

$$\hat{p}_{\text{diffr}} = -\frac{\hat{S}}{4\beta}\int_{-\infty}^{\infty} \mathscr{G}(\pi - is) \sum_{q=1}^{4} \frac{\sin \nu x_q}{\cosh \nu s - \cos \nu x_q}\, ds \tag{9-8.2}$$

and we use the abbreviations x_1, x_2, x_3, and x_4 for $\pi + \phi + \phi_S$, $\pi - \phi - \phi_S$, $\pi + \phi - \phi_S$, and $\pi - \phi + \phi_S$.

Next we can combine the x_1 and x_2 terms together and the x_3 and x_4 terms together, with the result

$$\hat{p}_{\text{diffr}} = \frac{\hat{S}\sin \nu\pi}{2\beta}\int_{-\infty}^{\infty} \mathscr{G}(\pi - is)[F_\nu(s, \phi + \phi_S) + F_\nu(s, \phi - \phi_S)]\, ds \tag{9-8.3}$$

where we use the abbreviation

$$F_\nu(s, \phi) = \frac{(\cos \nu\pi - \cos\nu\phi) - (\cosh \nu s - 1)\cos \nu\phi}{(\cosh \nu s - 1)^2 + 2(\cosh \nu s - 1)(1 - \cos \nu\phi \cos \nu\pi) + (\cos \nu\pi - \cos \nu\phi)^2} \tag{9-8.4}$$

The presence of the factor $\sin \nu\pi$ here demonstrates explicitly that there is no diffracted-wave contribution if ν is an integer.

The transient solution‡ for the wedge-diffraction problem follows directly from Eqs. (1) and (2) if the source time variation $S(t)$ is regarded as the integral from $-\infty$ to ∞ of $\hat{S}(\omega)e^{-i\omega t}$.

† Numerical calculations (which agree remarkably with experimental results) of this integral have been carried out by P. Ambaud and A. Bergassoli, "The Problem of the Wedge in Acoustics," *Acustica*, **27**:291–298 (1972).

‡ To derive the transient expression for the diffracted wave, one uses the symmetry of the integrand in Eqs. (2) and (3) and replaces the integration range from 0 to ∞ with a simultaneous multiplication by 2. Then one changes the variable of integration to $\xi = \mathscr{R}(\pi - is)$ so that

$$\cosh s = 1 + \frac{\xi^2 - L^2}{2rr_S} \qquad s = 2\tanh^{-1}\left(\frac{\xi^2 - L^2}{\xi^2 - Q^2}\right)^{1/2}$$

$$L^2 = (r + r_S)^2 + z^2 \qquad Q^2 = (r - r_S)^2 + z^2$$

$$\mathscr{G}(\pi - is)\, ds = \frac{2e^{i\omega\xi/c}\, d\xi}{(\xi^2 - L^2)^{1/2}(\xi^2 - Q^2)^{1/2}}$$

Asymptotic Expression for the Diffracted Wave

To derive an approximate expression for the diffracted wave in the limit where both kr and kr_S are large compared with 1, we regard s as a complex variable and deform the integration contour to a steepest-descent path along which the real part of $\mathcal{R}$ is constant and equal to its value at $s = 0$ but on which the imaginary part increases without limit as one moves in either direction away from $s = 0$. Then $|e^{ik\mathcal{R}}|$ decreases in the most rapid manner achievable by a contour deformation. Since for small s

$$\mathcal{R}(\pi - is) = (r^2 + r_S^2 + 2rr_S \cosh s + z^2)^{1/2} \approx L + \frac{rr_S}{2L} s^2 \tag{9-8.5}$$

the path considered makes an angle of $\pi/4$ with the real axis at $s = 0$. [Here L^2 is used as an abbreviation for $(r + r_S)^2 + z^2$.]

If $krr_S/2L \gg 1$, the dominant contribution to the integral comes from very small values of s, so in the denominator of $\mathcal{R}^{-1}e^{ik\mathcal{R}}$ it is sufficient to set $s = 0$ so that $\mathcal{R}^{-1}$ becomes L^{-1}. Also it is sufficient to use Eq. (5) as an approximation for the $\mathcal{R}$ in the exponent. However, for the factors $F_\nu(s, \phi \pm \phi_S)$ the possibility exists that for certain values of $\phi \pm \phi_S$, where $\cos \nu\pi = \cos \nu(\phi \pm \phi_S)$, the integrand may be singular at $s = 0$, so we keep the s^2 term in the denominator. In the numerator, it is sufficient to set $s = 0$. Then, with the aid of the algebraic identity $(M + is)^{-1} + (M - is)^{-1} = 2M/(M^2 + s^2)$ we have

$$F_\nu(x, \phi) \approx \frac{1}{2\nu(1 - \cos \nu\pi \cos \nu\phi)^{1/2}} \left[\frac{1}{M_\nu(\phi) + is} + \frac{1}{M_\nu(\phi) - is}\right] \tag{9-8.6}$$

$$M_\nu(\phi) = \frac{\cos \nu\pi - \cos \nu\phi}{\nu(1 - \cos \nu\pi \cos \nu\phi)^{1/2}} \tag{9-8.7}$$

Consequently, the diffracted wave becomes

$$\hat{p}_{\text{diffr}} = \frac{\hat{S}}{2\pi} \frac{e^{ikL}}{L} \sum_{+,-} \frac{\sin \nu\pi}{[1 - \cos \nu\pi \cos \nu(\phi \pm \phi_S)]^{1/2}} \int_{-\infty}^{\infty} \frac{e^{i(\pi/2)\Gamma^2 s^2}\, ds}{M_\nu(\phi \pm \phi_S) + is} \tag{9-8.8}$$

---

where ξ ranges from L to ∞. Then the Fourier integral theorem (2-8.4) allows the identification

$$p_{\text{diffr}} = -\frac{1}{\beta}\int_L^\infty S\left(t - \frac{\xi}{c}\right) K_\nu(\xi, L, Q, \phi, \phi_S)\, d\xi$$

$$K_\nu = \frac{1}{(\xi^2 - L^2)^{1/2}(\xi^2 - Q^2)^{1/2}} \sum_{q=1}^{4} \frac{\sin \nu x_q}{\cosh \nu s - \cos \nu x_q}$$

where s is given in terms of ξ. The unit impulse response results with $S(t - \xi/c)$ set to $\delta(t - \xi/c)$, so that

$$p_{\text{diffr,ui}} = -\frac{c}{\beta} K_\nu(ct, L, Q, \phi, \phi_S) H(ct - L)$$

where H denotes the Heaviside unit step function. Equivalent expressions are derived using a different method by M. A. Biot and I. Tolstoy, "Formulation of Wave Propagation in Infinite Media by Normal Coordinates with an Application to Diffraction," *J. Acoust. Soc. Am.*, **29**:381–391 (1957).

where we again take advantage of the symmetry of the contour and where we abbreviate

$$\Gamma = \left(\frac{krr_S}{\pi L}\right)^{1/2} = \left(\frac{2rr_S}{\lambda L}\right)^{1/2} \tag{9-8.9}$$

A further change of integration variable to u such that $s = (2/\pi)^{1/2}\Gamma^{-1}e^{i\pi/4}u$ reduces $\hat{p}_{\text{diffr}}$ to the form†

$$\hat{p}_{\text{diffr}} = \hat{S}\,\frac{e^{ikL}}{L}\,\frac{e^{i\pi/4}}{\sqrt{2}}\sum_{+,-}\frac{\sin\nu\pi}{[1-\cos\nu\pi\cos\nu(\phi\pm\phi_S)]^{1/2}}\,A_D(\Gamma M_\nu(\phi\pm\phi_S)) \tag{9-8.10}$$

where $A_D(X)$ is the diffraction integral

$$A_D(X) = \frac{1}{\pi 2^{1/2}}\int_{-\infty}^{\infty}\frac{e^{-u^2}\,du}{(\pi/2)^{1/2}X - e^{-i\pi/4}u} = \text{sign}\,(X)\,[f(|X|) - ig(|X|)] \tag{9-8.11}$$

which previously appeared as Eq. (5-8.9) and which is discussed in some detail in Sec. 5-8.

Equation (10) gives us a uniform asymptotic expression for the diffracted field, valid for large values of Γ and for any wedge angle β between 0 and 2π. The total asymptotic solution is $\hat{p}_{\text{GA}} + \hat{p}_{\text{diffr}}$, where $\hat{p}_{\text{GA}}$ is given by Eq. (1).

Physical Interpretation of the Diffracted Wave

If the quantities $M_\nu(\phi \pm \phi_S)$ are not small in magnitude, the diffraction integral $A_D(X)$ is approximated by its asymptotic form $1/\pi X$ and $\hat{p}_{\text{diffr}}$ reduces to

$$\hat{p}_{\text{diffr}} = \frac{\hat{S}}{2\beta}\left(\frac{2\pi}{kLrr_S}\right)^{1/2} e^{i(kL+\pi/4)}D_\nu(\phi, \phi_S) \tag{9-8.12}$$

$$D_\nu(\phi, \phi_S) = \frac{\sin\nu\pi}{\cos\nu\pi - \cos\nu(\phi+\phi_S)} + \frac{\sin\nu\pi}{\cos\nu\pi - \cos\nu(\phi-\phi_S)} \tag{9-8.13}$$

The decrease in amplitude with increasing frequency here displayed is in accord with the notion that the geometrical-acoustics solution is a high-frequency approximation; the diffracted wave vanishes if $k \to \infty$.

† The result is due in essence to W. Pauli, "On Asymptotic Series for Functions in the Theory of Diffraction of Light, *Phys. Rev.*, **54**:924–931 (1938). Various different versions existing in the literature are equivalent in the limit of large Γ because if $F(\phi)$ is any nonzero function that equals 1 whenever $\psi(\phi)$ is 0, then

$$A_D(\Gamma F\psi) \approx F^{-1}A_D(\Gamma\psi)$$

and because, if $1/\psi(\phi) = 1/\psi_1(\phi) + 1/\psi_2(\phi)$, where ψ_1 and ψ_2 have different zeros, then

$$A_D(\Gamma\psi) \approx \frac{\psi_1+\psi_2}{\psi_1-\psi_2}\,[A_D(\Gamma\psi_2) - A_D(\Gamma\psi_1)]$$

The version in the text applies for any ν, ϕ, and ϕ_S.

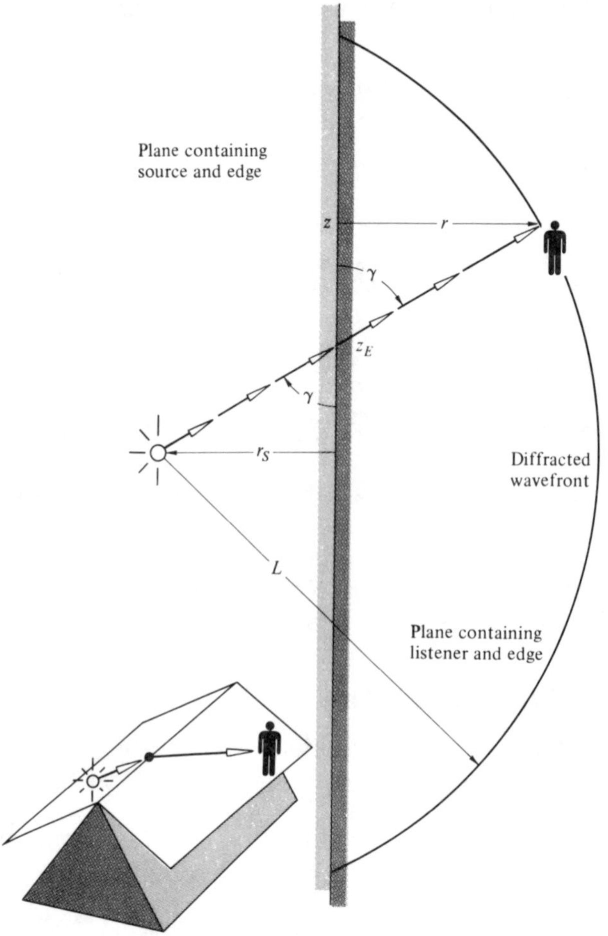

Figure 9-25 Broken ray path from source to edge to listener in shadow zone; angle γ is made by both segments with the edge. The listener lies on a diffracted wavefront at a point where the two principal radii of curvature are L and r.

The diffracted wave, however, can also be interpreted in terms of geometrical-acoustic concepts. The quantity $L = [(r + r_S)^2 + z^2]^{1/2}$ is the shortest distance of a broken line that goes from the source to the edge and thence to the listener (see Fig. 9-25). This diffracted path touches the edge at $z_E = [r_S/(r + r_S)]z$, and there both incident and diffracted rays make the same angle, $\gamma = \tan^{-1} [(r + r_S)/z]$, with the diffracting edge. (This is *Keller's law of edge diffraction* and follows from the extended interpretation of Fermat's principle discussed in Sec. 8-1.)

Since the phase variation of $\hat{p}_{\text{diffr}}$ is predominantly that of e^{ikL}, the diffracted wavefronts are surfaces of constant L. Thus, diffracted rays move in the direction of ∇L, or of

$$\mathbf{n} = \frac{r + r_S}{L}\mathbf{e}_r + \frac{z}{L}\mathbf{e}_z = \frac{r\mathbf{e}_r + (z - z_E)\mathbf{e}_z}{[r^2 + (z - z_E)^2]^{1/2}} \tag{9-8.14}$$

The latter version substantiates the assertion that the diffracted ray originates at the point z_E on the edge. (Recall that in a homogeneous medium the rays are straight lines.)

Since a surface of constant L (with r_S fixed) is circularly symmetric (see Fig. 9-25), one of the two principal radii of curvature at a given point on the wavefront is r. Since any cross section through the z axis is an arc of a circle centered at $(-r_S, 0)$, the other principal radius of curvature is the circle radius or L. Thus the quantity $(rL)^{1/2}$ is the geometric mean of the two principal radii of curvature. Since this is proportional to the square root of ray-tube area, the amplitude variation with r and z in the approximation represented by Eq. (12) is wholly consistent with the geometrical-acoustic prediction of Eq. (8-5.8). Thus one can conclude that, for the most part, the diffracted wave propagates according to the laws of geometrical acoustics.

With the interpretation just described, one can reconstruct the expression (12), starting from the premise that near the edge the diffracted field is

$$\hat{p}_{\text{diffr}} \approx \frac{\hat{p}_{\text{inc}} e^{i\pi/4}}{2\beta r^{1/2}} \left[\frac{2\pi}{k \sin\gamma}\right]^{1/2} D_\nu(\phi, \phi_S) \tag{9-8.15}$$

where $\hat{p}_{\text{inc}}$ is the incident wave's complex amplitude at the point where the diffracted ray leaves the edge and γ is the angle that the ray makes with the edge. The angle-dependent factor D_ν here implies that the edge acts as a directional source of acoustic energy.

In the same spirit, one concludes that Eq. (15) holds for a wave incident from any source, regardless of whether the source can be idealized as omnidirectional. In particular, it is applicable when the incident wave is regarded as either an obliquely incident plane wave or a cylindrical wave. In each such case one determines the diffracted wave path and the point z_E on the edge at which the received diffracted ray originates along with the incident acoustic pressure at this point. The apparent value of r_S, determined by the local variation of γ with distance along the edge, is $r_S = \mp(\sin^2\gamma)/(d\gamma/dz_E)$, where the two sign choices apply to when the incident ray is proceeding obliquely in the $+z$ or the $-z$ direction. With such a substitution, Eq. (15) leads, for larger r, to

$$\hat{p}_{\text{diffr}} = \frac{\hat{p}_{\text{inc}}}{2\beta} \frac{(2\pi)^{1/2} e^{i(ks+\pi/4)}}{(kr)^{1/2}(\sin\gamma \mp s\, d\gamma/dz_E)^{1/2}} D_\nu(\phi, \phi_S) \tag{9-8.16}$$

where $s = r/(\sin\gamma)$ is distance along the diffracted ray from the edge. The applicable result for an incident plane wave is obtained by setting $d\gamma/dz_E = 0$.

The factor $D_\nu(\phi, \phi_S)$ becomes singular if $\cos\nu\,(\phi \mp \phi_S) = \cos\nu\pi$ or, equivalently if $2\beta l \pm \phi \pm \phi_S = \pi$ for any integer l and any sign combination. This, however, is just the condition that a pole in the ζ plane be at $\zeta = \pi$ and thus lie on the contour C_π. Alternatively, any value of ϕ for which such a condition holds marks the transition between the presence or absence of some geometrical-acoustics ray path. Thus, if the region of absence of such a ray is regarded as a shadow zone for such a geometrical-acoustics wave, the transitional value of ϕ corresponds to the shadow-zone boundary. When there are no geometrical-acoustic paths on one side of the boundary, the region there is one of total shadow (from the standpoint of geometrical acoustics).

The use of the diffraction integral $A_D(X)$ rather than its asymptotic expression $1/(\pi X)$ in Eq. (12), on the other hand, leads to a finite prediction for the diffracted wave. Since $A_D(X)$ is discontinuous at $X = 0$ $[A_D(0^+) = (1 - i)/2$, $A_D(0^-) = -(1 - i)/2]$, the quantity $\hat{p}_{\text{diffr}}$ will be discontinuous at each shadow-zone boundary. The discontinuity at any such ϕ is

$$\Delta p_{\text{diffr}} = \hat{S}\,\frac{e^{ikL}}{L} = (\hat{p}_{\text{diffr}})_{M_\nu=0^+} - (\hat{p}_{\text{diffr}})_{M_\nu=0^-} \tag{9-8.17}$$

since $\cos \nu(\phi \pm \phi_S)$ is $\cos \nu\pi$ and $1 - \cos \nu(\phi \pm \phi_S) \cos \nu\pi$ is $\sin^2 \nu\pi$ if $M_\nu(\phi \pm \phi_S)$ is 0. (The shadow-zone boundaries predicted for integer ν are merged in pairs such that the discontinuity from illumination to shadow for any one geometrical-acoustics term is exactly canceled by a discontinuity from shadow to illumination for a second geometrical-acoustics term; the geometrical-acoustics sum for integer ν has no discontinuities.)

The overall solution is continuous, so that each discontinuity in $\hat{p}_{\text{diffr}}$ is compensated by an equal and opposite discontinuity in $\hat{p}_{\text{GA}}$. To demonstrate this, let a shadow-zone boundary be at, say, $\phi_{\text{sz}} = 2\beta l + \phi_S - \pi$. Then if ϕ is slightly less than ϕ_{sz}, one will be in the shadow zone for the geometrical-acoustics term $\hat{S}\mathscr{G}(2\beta l + \phi_S - \phi)$. The net discontinuity in $\hat{p}_{\text{GA}}$ at ϕ_{sz} is accordingly $\hat{S}L^{-1}e^{ikL}$. However, $M_\nu(\phi - \phi_S)$ for ϕ near ϕ_{sz} has a sign opposite to that of $\phi - \phi_{\text{sz}}$, so Eq. (17) predicts $\Delta\hat{p}_{\text{diffr}}$ to be opposite to $\Delta\hat{p}_{\text{GA}}$ such that the sum $\hat{p}_{\text{GA}} + \hat{p}_{\text{diffr}}$ is continuous at $\phi = \phi_{\text{sz}}$.

Although the diffracted field near shadow-zone boundaries cannot be wholly interpreted in terms of diffracted rays emanating from the edge, we can nevertheless reexpress Eq. (10) in terms of parameters characterizing such rays. In particular, one can write, in a manner similar to Eq. (16),

$$\hat{p}_{\text{diffr}} = \frac{\hat{p}_{\text{inc}}e^{i(ks+\pi/4)}}{\sqrt{2}}\,\frac{\sin\gamma}{\sin\gamma \mp s\,d\gamma/dz_E}\sum_{+,-}\frac{\sin\nu\pi\,A_D(\Gamma M_\nu(\phi \pm \phi_S))}{[1 - \cos\nu\pi\cos\nu(\phi \pm \phi_S)]^{1/2}} \tag{9-8.18}$$

with

$$\Gamma = \left[\frac{(kr/\pi)\sin^2\gamma}{\sin\gamma \mp s\,d\gamma/dz_E}\right]^{1/2} \tag{9-8.19}$$

where the various symbols appearing here have the same meaning as in Eq. (16). Again, the result for an incident plane wave† is obtained by setting $d\gamma/dz_E = 0$.

† For plane waves incident on a thin screen ($\beta = 2\pi$, $\nu = \frac{1}{2}$), Eq. (18) reduces to the *exact*

9-9 APPLICATIONS OF WEDGE-DIFFRACTION THEORY

The asympotic expression for diffraction by a wedge simplifies further when the listener is near the shadow-zone boundary, yielding a method for rapid estimation of a barrier's insertion loss. The present section discusses this simplification and gives examples of how geometrical acoustics can augment the basic model so that it applies to situations in which edge diffraction takes place under less idealized circumstances.

Insertion Loss of Single-Edged Barriers

We first consider the case when source and listener are at distant points on opposite sides of an acute wedge with exterior angle β. The source angle ϕ_S is between π and β but close to neither limit. Estimates are desired regarding the effectiveness of the wedge as a barrier to sound when the listener is near or only slightly within the shadow zone, such that ϕ is less than $\phi_S - \pi$ but yet not close to the nearer side ($\phi = 0$) of the wedge.

Because we are interested in the behavior near the edge of the shadow zone, we write $\Delta\phi = \phi - (\phi_S - \pi)$ and regard $|\Delta\phi|$ as small compared with 1. Then the $\phi - \phi_S$ term dominates in Eq. (9-8.10), so we discard the $\phi + \phi_S$ term. We set $\Delta\phi = 0$ in the coefficient of $A_D(\Gamma M_\nu)$, but since $M_\nu(\phi - \phi_S)$ vanishes when $\Delta\phi = 0$, we express $M_\nu(\phi - \phi_S)$ to first order in $\Delta\phi$. Such steps reduce the overall field near the shadow-zone boundary to

$$\hat{p} \approx \hat{S}\,\frac{e^{ikR}}{R}\,H(X) + \hat{S}\,\frac{e^{ikL}}{L}\,\frac{e^{i\pi/4}}{\sqrt{2}}\,A_D(-X) \tag{9-9.1}$$

where $X = \Gamma\,\Delta\phi$ and R is the direct path distance from the source; the Heaviside unit step function $H(X)$ is 0 in the shadow zone and 1 in the illuminated region.

Expansion of R in a power series in $\Delta\phi$ yields, to second order,

$$\begin{aligned} R &= [r^2 + r_s^2 + z^2 - 2rr_S\cos(\pi - \Delta\phi)]^{1/2} \\ &\approx [L^2 - rr_S(\Delta\phi)^2]^{1/2} \approx L - \frac{1}{2}\,\frac{rr_S}{L}\,(\Delta\phi)^2 \end{aligned} \tag{9-9.2}$$

so the $X = \Gamma\,\Delta\phi$ in Eq. (1), with Γ taken from Eq. (9-8.9), is such that

$$X^2 = \frac{2k}{\pi}\,(L - R) = 2N_F \qquad N_F = \frac{L - R}{\lambda/2} \tag{9-9.3}$$

---

result

$$\hat{p}_{\text{diffr}} = \frac{-\hat{p}_{\text{inc}}\,e^{i(ks+\pi/4)}}{\sqrt{2}} \sum_{+,-} A_D\left[\left(\frac{4kr}{\pi}\sin\gamma\right)^{1/2}\cos\tfrac{1}{2}(\phi \pm \phi_S)\right]$$

which, for normal incidence ($s = r$, $\sin\gamma = 1$), was first derived by A. Sommerfeld, "Mathematical Theory of Diffraction," *Math. Ann.*, **47**:317–374 (1896).

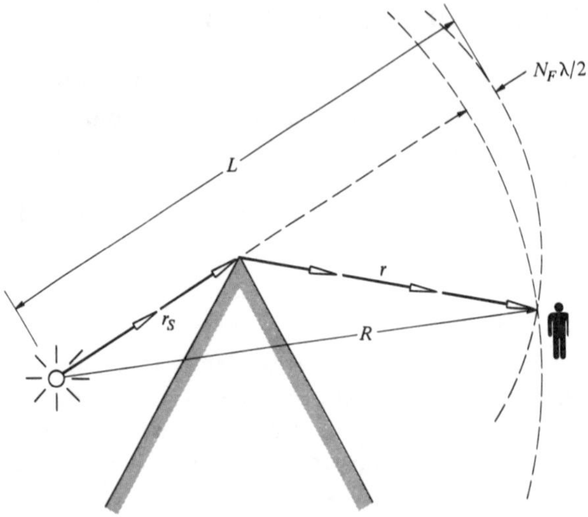

Figure 9-26 Geometrical definition of Fresnel number $N_F = (L - R)/(\lambda/2)$ for circumstances when the z coordinates of source and listener are the same. Indicated circular arcs have radii of r and R. The path length L is $r_S + r$.

The quantity N_F is identified as the *Fresnel number* (see Fig. 9-26), i.e., excess distance of shortest diffracted path from source to edge to listener in units of half wavelengths; this appears also in the discussion in Sec. 5-8 of radiation from a baffled piston source.

Since $A_D(X)$ is odd in X, since $k(L - R) \approx (\pi/2)X^2$, and since $L/R \approx 1$ for the listener locations of interest, Eq. (1) reduces to

$$\hat{p} \approx \hat{S}\,\frac{e^{ikR}}{R}\left[H(X) - \frac{e^{i\pi/4}}{2^{1/2}}\,A_D(X)e^{i(\pi/2)X^2}\right] \tag{9-9.4}$$

The quantity appearing here in brackets is the same as in Eq. (5-8.18), so the field near the shadow-zone boundary is similar to that at the edge of a "beam" of sound radiated by a baffled piston.

The *insertion loss* of the barrier, as predicted† by the approximation above, is

$$\text{IL} = -10\log\left|H(X) - \frac{e^{i\pi/4}}{2^{1/2}}\,A_D(X)e^{i(\pi/2)X^2}\right|^2 \tag{9-9.5}$$

or 10 times the logarithm of the reciprocal of the characteristic single-edge diffraction pattern plotted in Fig. 5-13.

Within the illuminated region the insertion loss oscillates between negative and positive values because of the interference between direct and diffracted waves. The peak negative insertion loss, occurring at $X = +1.2$ ($N_F = 0.7$), is $-10 \log 1.28 \approx -1$ dB. The insertion loss is 0 dB at $X = 0.8$ ($N_F = 0.3$) and is positive for all other X closer to, and into, the shadow zone. The approxima-

† Z. Maekawa, "Noise Reduction by Screens," pap. F-13, *Proc. 5th Int. Congr. Acoust.*, G. Thone, Liège, 1965. The discrepancies between Maekawa's Kirchoff-theory result, Eq. (5) above, and his empirical chart of thin-screen barrier attenuation versus Fresnel number are explained by U. J. Kurze, "Noise Reduction by Barriers," *J. Acoust. Soc. Am.*, **55**:504–518 (1974).

tions (5-8.13) lead to

$$\mathrm{IL} \approx 20 \log 2 - \frac{20}{\ln 10} X \approx 6 - 8.7X \tag{9-9.6}$$

for X near 0, such that $\mathrm{IL} \approx 6 + 12.28\,(N_F)^{1/2}$ on the shadow side and for small Fresnel number. This, however, is a fair approximation only up to $N_F \approx 0.1$. For larger values of N_F, the asymptotic formulas of Eq. (5-8.12) become increasingly valid, so that

$$\mathrm{IL} \approx 10 \log (4\pi^2 N_F) \approx 16 + 10 \log N_F \tag{9-9.7}$$

is a good approximation for $N_F > 2$ on the shadow side. (This presumes, however, that $|\Delta\phi|$ remains small.)

Equations (4) and (5) are remarkable in that they are independent of the wedge exterior angle β. In the small $\Delta\phi$ limit, all wedges diffract the same. A *diffraction boundary layer* that marks the transition from illumination to shadow can be regarded as a function of only one dimensionless parameter, which can be taken as the Fresnel number N_F. Such conclusions are the same as those yielded by the Fresnel-Kirchhoff approximation (see Sec. 5-2), so the claim that the latter can be valid for small deflections is substantiated.

Although the above analysis presumes that $|\Delta\phi|$ is small, it does not require $\Gamma|\Delta\phi|$ to be small; so the use of asymptotic expressions for $f(X)$ and $g(X)$ in the derivation of Eq. (7) is not inconsistent. It would not be unreasonable, given that, say, 2 dB accuracy is acceptable, to apply Eq. (7) for any point in the shadow zone where $|\Delta\phi|$ is less than, say, 20° provided $\Gamma|\Delta\phi|$ exceeds 2.

Example: Barrier on Rigid Ground An omnidirectional source resting on the ground and generating 500-Hz sound is 15 m from a barrier 5 m high. A point 20 m farther on the opposite side of the barrier at 2 m height would receive a sound-pressure level of 90 dB re 20 μPa without the barrier present. Estimate the level when the barrier is present.

SOLUTION There are two diffracted paths† connecting the source, edge, and listener (see Fig. 9-27), the second having an intermediate ground reflection between the edge and listener. The two path lengths, L_1 and L_2, are $[(15)^2 + (5)^2]^{1/2} + [(20)^2 + (3)^2]^{1/2} = 36.04$ m and $[(15)^2 + (5)^2]^{1/2} + [(20)^2 + (7)^2]^{1/2} = 37.00$ m. The two direct distances, R_1 and R_2, are both $[(35)^2 + (2)^2]^{1/2} = 35.06$ m. Consequently, with c taken as 340 m/s so that $\lambda = 0.68$ m, the two Fresnel numbers are $N_{F1} = 2.88$ and $N_{F2} = 5.71$. The two waves arrive with amplitudes corresponding to sound-pressure levels, from Eq. (7), of $90 - 10 \log [(4\pi^2)(2.88)] = 69.4$ dB and $90 - 10 \log [(4\pi^2)(5.71)] = 66.5$ dB. Their phase difference is $(kL_2 + \pi/4) - (kL_1 + \pi/4)$, according to Eq. (1)

† A general analysis when neither source or listener is on the ground and when the ground has finite impedance is given by H. G. Jonasson, "Sound Reduction by Barriers on the Ground," *J. Sound Vib.*, **22:**113–126 (1972).

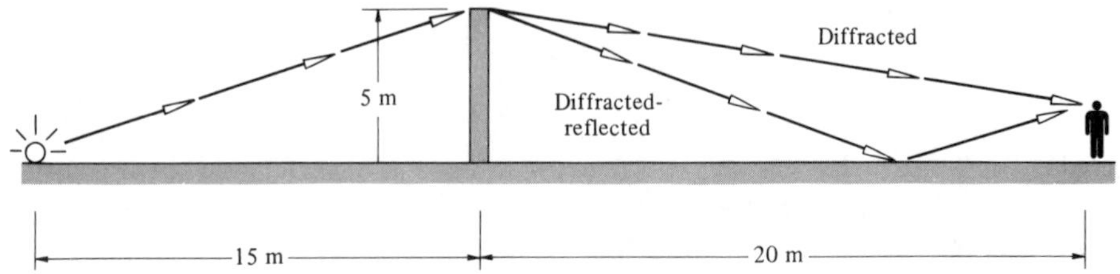

Figure 9-27 Possible paths connecting source and listener and passing over a barrier on the ground; the source is on the ground, and the listener is above the ground. Distances cited correspond to the example discussed in the text.

and to the asymptotic approximation $1/\pi X$ for $A_D(X)$, or $(2\pi/0.68)(0.96) = 8.87$ rad $(508 - 360 = 148°)$. The sound-pressure level corresponding to the algebraic sum of the two diffracted arrivals is therefore

$$\begin{aligned} L_p &= 10 \log |10^{69.4/20} + e^{i8.87} 10^{66.5/20}|^2 \\ &= 10 \log [10^{69.4/10} + 10^{66.5/10} + 2(10^{69.4/10}\, 10^{66.5/10})^{1/2} \cos 148°] \\ &= 64.1 \text{ dB} \end{aligned} \tag{9-9.8}$$

If the ground on the listener side of the barrier were perfectly absorbing instead of perfectly reflecting, L_p at the considered reception site would be 69.4 dB instead (5.3 dB higher).

Far Field of a Source on the Side of a Building

The sound reaching a distant listener in front of a building (see Fig. 9-28) from a source on the side is described by Eq. (9-8.10) with $\beta = 3\pi/2$, $\nu = 2/3$, and $\phi_S = \beta$, as for a source on one side of a 90° (interior-angle) wedge. If the listener is sufficiently distant and the ground is perfectly reflecting, the field is described by twice that of Eq. (9-8.10). In the limit $r \gg r_S$, the parameter Γ reduces to $(kr_S/\pi)^{1/2}$, and $L^{-1}e^{ikL}$ approximates to $r^{-1}e^{ikr}e^{ikr_S}$.

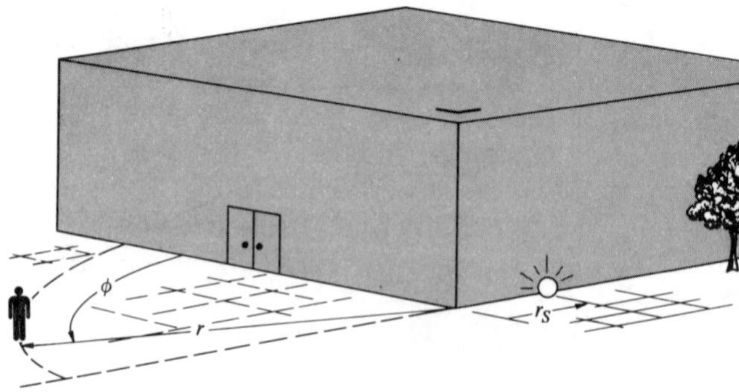

Figure 9-28 Geometry adopted for discussion of diffraction of sound around the corner of a building. The source is on the ground adjacent to the building's side; r is much larger than r_S.

Another factor of 2 emerges because $\cos \nu(\phi \pm \beta) = -\cos \nu\phi$ requires that the two terms associated with $\phi + \phi_S$ and $\phi - \phi_S$ be the same in Eq. (9-8.10). In the direct-wave term (present only if $\phi > \pi/2$), a factor of 2 is included because the reflected wave coincides with it; another factor 2 accounts for ground reflection. Also, since $r \gg r_S$, the factor $R^{-1}e^{ikR}$ approximates to $e^{ikr_S \sin\phi}\, r^{-1}e^{ikr}$. Consequently, the far field becomes

$$\hat{p} = \hat{S}\,\hat{F}(\phi)\,\frac{e^{ikr}}{r} \tag{9-9.9}$$

with

$$\hat{F}(\phi) = 4e^{ikr_S \sin\phi} H\left(\phi - \frac{\pi}{2}\right) - \frac{6^{1/2}e^{i\pi/4}e^{ikr_S}A_D(X)}{[1 - \frac{1}{2}\cos(2\phi/3)]^{1/2}} \tag{9-9.10}$$

$$X = \frac{3}{2}\left(\frac{kr_S}{\pi}\right)^{1/2} \frac{\frac{1}{2} - \cos(2\phi/3)}{[1 - \frac{1}{2}\cos(2\phi/3)]^{1/2}} \tag{9-9.11}$$

Our interest here is in values of ϕ between 0 and, say, $3\pi/4$ (135°); the above result neglects diffraction from all but one corner of the building, so it may not be applicable near $\phi = 0$ when r extends beyond the front of the building. Neither may it be applicable near $\phi = 3\pi/2$ (270°) when r extends behind the rear of the building.

The description in Eq. (9) is that of the spherical spreading in the far field of a directional sound source. Its form dispels any misconception that diffracted waves always spread cylindrically (amplitude proportional to $r^{-1/2}$), although such may be a good approximation in the other limit when $r \ll r_S$.

The quantity $|\hat{F}(\phi)|^2$ describes the source's far-field radiation pattern. Its value is 4 for $\phi = \pi/2$ and, given kr_S moderately large compared with 1, it approaches 16 at larger $\phi - \pi/2$. On the shadow side ($\phi < \pi/2$), the asymptotic limit of $A_D(X)$ yields for $\phi - \pi/2$ negative and not small

$$|\hat{F}(\phi)|^2 \to \frac{8/(3\pi kr_S)}{[\cos(2\phi/3) - \frac{1}{2}]^2} \tag{9-9.12}$$

where the limiting expression is bounded from below by $32/(3\pi kr_S)$, occurring when $\phi = 0$. Thus the far-field intensity ultimately decreases with distance r_S from the corner as $1/kr_S$ for any fixed angle ϕ less than $\pi/2$.

Near $\phi = \pi/2$, we can set $\sin\phi \approx 1 - (\Delta\phi)^2/2$, $1 - \frac{1}{2}\cos(2\phi/3) \approx \frac{3}{4}$, and $\cos(2\phi/3) - \frac{1}{2} \approx -\Delta\phi/\sqrt{3}$, where $\Delta\phi = \phi - \pi/2$, such that Eq. (10) yields for the radiation pattern in the transition region

$$|\hat{F}(\phi)|^2 = 16\left|H(\bar{X}) - \frac{e^{i\pi/4}}{2^{1/2}}A_D(\bar{X})e^{i(\pi/2)\bar{X}^2}\right|^2 \tag{9-9.13}$$

with

$$\bar{X} = \left(\frac{kr_S}{\pi}\right)^{1/2}\Delta\phi \tag{9-9.14}$$

Thus the characteristic single-edge diffraction pattern, plotted in Fig. 5-13, emerges once again.

Backscattering from an Edge

Anomalous echoes of higher-frequency sound can often be explained in terms of diffraction by edges. The analysis in Sec. 9-8 applies both when the source is in the interior of a wedge-shaped region and when it is exterior to a wedge-shaped obstacle. In either case, the echo from the edge is predicted by Eq. (9-8.10) with $r = r_S$, $z = z_S$, $\phi = \phi_S$, such that $L = 2r$ and $\Gamma = (kr/2\pi)^{1/2} = (r/\lambda)^{1/2}$. This yields

$$\hat{p}_{\text{diffr}} = \hat{S}\,\frac{e^{i2kr}}{2r}\,e^{i\pi/4}\left[-\cos\frac{\nu\pi}{2}\,A_D\!\left(\left(\frac{2r}{\lambda}\right)^{1/2}\nu^{-1}\sin\frac{\nu\pi}{2}\right) + \frac{2^{-1/2}\sin\nu\pi}{(1-\cos\nu\pi\cos 2\nu\phi)^{1/2}}\,A_D\!\left(\left(\frac{r}{\lambda}\right)^{1/2} M_\nu(2\phi)\right)\right] \qquad (9\text{-}9.15)$$

for the backscattered echo.

Among the particular cases for which the above result simplifies is that when $\phi = 0$ (or equivalently $\phi = \beta$), which yields

$$\hat{p}_{\text{diffr}} = -\hat{S}\,\frac{e^{i2kr}}{r}\,e^{i\pi/4}\cos\frac{\nu\pi}{2}\,A_D\!\left(\left(\frac{2r}{\lambda}\right)^{1/2}\nu^{-1}\sin\frac{\nu\pi}{2}\right) \approx -\frac{\hat{S}}{\beta}\cot\left[\frac{\nu\pi}{2}\right]\left(\frac{\lambda}{2}\right)^{1/2}\frac{e^{i(2kr+\pi/4)}}{r^{3/2}} \qquad (9\text{-}9.16)$$

The latter results when the asymptotic limit, $1/\pi X$ for $A_D(X)$, applies and is valid for moderately large r/λ provided $\sin(\nu\pi/2)$ is not inordinately small.

Insight into how "strong" such an echo would appear to be can be obtained by comparing the above with the reflection from a wall making 90° with the surface $\phi = 0$, this wall being also at distance r. The latter would give an echo $\hat{p}_{\text{refl}} = 2\hat{S}(2r)^{-1}e^{i2kr}$, where the extra factor of 2 is because the source rests on a

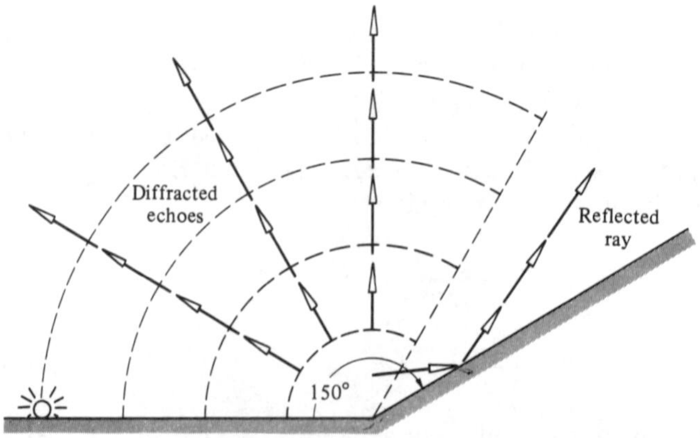

Figure 9-29 Echoes generated by a source on an interior surface of a 150° wedge. The diffracted echoes radiate from the intersection of the two planes.

rigid surface. The relative weakness of the diffracted echo is accordingly

$$\frac{|\hat{p}_{\text{diffr}}|}{|\hat{p}_{\text{refl}}|} = \left|\cot \frac{\nu\pi}{2}\right| \frac{1}{\beta} \left(\frac{\lambda}{2r}\right)^{1/2} \tag{9-9.17}$$

Thus for a source (see Fig. 9-29) on an interior surface of a 150° wedge ($\beta = 5\pi/6$, $\nu = \frac{6}{5}$), one obtains an amplitude ratio of $|\cot(3\pi/5)|(\lambda/2r)^{1/2}6/5\pi = 0.0878(\lambda/r)^{1/2}$.

Alternatively, one can characterize the edge-diffracted echo by how much farther removed a perfectly reflecting surface that returns an echo of the same amplitude would be. If the latter is at distance r^*, then $r/r^* = |\hat{p}_{\text{diffr}}|/|\hat{p}_{\text{refl}}|$. For example, for the 150° wedge example mentioned above, the edge-diffracted wave from an edge 10 wavelengths away appears to come from a reflector at a distance of $(10)^{3/2}\,\lambda/0.0878 = 360$ wavelengths.

PROBLEMS

9-1 Derive formulas for the target strength of a fixed rigid sphere of radius a appropriate to the limiting cases of (*a*) $ka \ll 1$ and (*b*) $ka \gg 1$.

(*c*) What effect does a doubling of frequency have on target strength in these two limits?

9-2 A harmonic plane wave impinges obliquely on a circular disk of radius a centered at the origin. The disk's faces are parallel to the xy plane, and the incident wave has propagation direction $\mathbf{n}_k = \mathbf{e}_x \sin\theta_k + \mathbf{e}_z \cos\theta_k$.

(*a*) Determine an expression for the differential cross section of the disk in the limit $ka \ll 1$. Use the spherical coordinates θ and ϕ.

(*b*) What is the backscattering cross section for the disk under the same circumstances?

(*c*) What is the target strength?

(*d*) Explain in simple terms whatever results when θ_k is set to $\pi/2$ in your answers to parts (*a*) to (*c*).

9-3 Prove that the tensor $\mathbf{M}_{\text{eff}}$ whose cartesian elements are given by Eq. (9-1.25*b*) is symmetric.

9-4 Derive the expression (9-1.41) for the scattering of sound by a Helmholtz resonator in an open space when the incident sound's frequency is close to the resonance frequency.

9-5 A solid sphere of radius a and mass M can move back and forth along the z axis about its equilibrium position at $z = 0$ under the influence of a spring with spring constant k_{sp} N/m. A plane wave of angular frequency ω and acoustic-pressure amplitude P propagating in the $+z$ direction impinges on the sphere, causes it to vibrate, and gives rise to a scattered wave. Consider M and k_{sp} to be such that a resonance scattering occurs at an incident frequency for which $ka \ll 1$.

(*a*) At what ω does the resonance scattering occur?

(*b*) Show that the scattered wave is predominantly dipole.

(*c*) Give an expression for the scattered field at frequencies near the resonance frequency.

(*d*) What is the total scattering cross section at the resonance frequency?

(*e*) How does the result in part (*d*) compare with the upper limit of λ^2/π that results (see Sec. 9-1) for monopole resonance scattering?

9-6 A fluid contains a large number of similar discrete scattering centers, each of which is small compared with the average distance between scatterers. Given that multiple scattering can be neglected and that the scatterers are randomly dispersed, give a heuristic argument or else refute the hypothesis that when the scattering volume is sufficiently large, the scattering from individual scatterers can be regarded as incoherent.

9-7 A narrow-beam but broadband sound wave whose pressure variation has spectral density $p_f^2(f)$ is incident on a bubble with radius a, resonance frequency f_{res}, and acoustic resistance R_A.

(*a*) Estimate in the limit of small R_A the total energy scattered per unit time out of the incident beam by the bubble.

(*b*) Suppose that there are N bubbles per unit volume and that each such bubble has a slightly different resonance frequency but the numbers a, f_{res}, R_A are roughly representative of all the bubbles. Discuss how the spectral density of the acoustic pressure decreases with increasing propagation distance along the axis of the incident beam.

9-8 Sound is propagating along a rigid-walled narrow tube under circumstances for which the Webster horn equation (7-8.5*a*) is applicable. Consider $|(A')^2 - 2AA''|$ to be much smaller than $4k^2A^2$ and use the Born approximation to predict the echo returned back to $x = 0$ when a narrow-band pulse $A^{1/2}p = f(t - x/c)$ is propagating down the axis of the tube. Discuss the feasibility of deriving the x dependence of the tube's cross-sectional area $A(x)$ from the results of pulse-echo soundings.

9-9 A narrow-beam reciprocal transducer whose far field is as described by Eq. (9-2.7) transmits a pulse of nearly constant frequency along the z axis. The ambient medium is nearly homogeneous except for a weak planar discontinuity at $z = h$, where ρ and c change by small increments $\delta\rho$ and δc. The echo from this discontinuity is subsequently received by the same transducer when it is in its reception mode.

(*a*) What is the apparent mean squared pressure received by the transducer during the duration of the echo?

(*b*) What is the apparent backscattering cross-section?

(*c*) What is the apparent target strength?

9-10 Answer the questions in Prob. 9-9 when the planar surface of discontinuity is tilted so that its unit normal makes an angle ϕ with respect to the z axis. The discontinuity plane continues to pass through the point $(0, 0, h)$. Let the beam pattern of the transducer be described by $|\hat{F}_{tr}|^2 = e^{-\alpha\theta^2}$, where α is somewhat larger than 1, and discuss what variations result in the answers when ϕ is small but nonzero.

9-11 The transmitter and receiver in a bistatic echo-sounding configuration both have narrow beam patterns described by $|\hat{F}_{tr}(\theta, \phi)|^2 = e^{-\alpha\theta^2}$, where α is substantially larger than 1. Both transmitter and receiver beams make a 45° angle with the ground and lie in a common vertical plane. The two beams intersect at height $L/2$, where L is the transmitter-receiver separation distance. Determine, to lowest nonvanishing order in $1/\alpha$ and $c\tau/L$, a simple expression (or a numerical value) for the aspect factor $\mathscr{A}$ that appears in Eq. (9-2.29).

9-12 A moving sound source of nominal angular frequency ω_0 moves at speed $V = c/3$ along the x axis past a listener at $x = 0$ and at cylindrical radial distance r.

(*a*) Determine an expression for ω/ω_0 in terms of ct/r, where $t = 0$ is the time the source passes the origin. Here ω is the angular frequency perceived by the listener.

(*b*) Give a sketch of $(\omega - \omega_0)/\omega_0$ versus ct/r. Explain any asymmetries between the $+t$ and $-t$ portions of the curve.

9-13 Two vehicles, one from the north and the other from the east, approach an intersection in such a way that they are likely to collide at the origin at time $t = 0$. Both vehicles have speed $c/10$. The southward-moving vehicle sounds a warning device of frequency f_0 Hz. What is the frequency detected by passengers in the westward-moving vehicle, and how does it vary with time? Assume that the two vehicles barely miss each other at the intersection and that the warning device continues to sound past the intersection.

9-14 A spherical inhomogeneity of mass m and radius a, where m is slightly larger than the displaced fluid mass m_d, is drifting along with the flow at height h in a medium where the ambient velocity varies with height z as $\mathbf{v}_0 = \mathbf{e}_x Vz/h$. A nearly sinusoidal pulse of angular frequency ω_0 is transmitted by a point source resting on a rigid ground at the origin. The source has monopole amplitude $\hat{S}$, and the transmission pulse-excitation time is such that the pulse impinges on the moving inhomogeneity when it is at $x = L$, $y = 0$, $z = h$. Consider the sound speed c and ambient density ρ to be constant and L to be substantially larger than h but less than $(2ch^2/V)^{1/2}$. Use geometrical acoustics and the approximation in which ray paths resemble arcs of circles to determine the incident wave impinging on the inhomogeneity and to trace the evolution of the scattered pulse back to the transmitter (which also functions as a receiver).

(*a*) What is the delay time (to first order in V/c) before reception of the backscattered pulse?

(*b*) From what direction does the echo appear to come?

(*c*) What Doppler shift is evidenced by the echo's frequency?

(*d*) What is the rms amplitude of the acoustic pressure in the echo pulse returned to the transducer?

9-15 The incoming portion of the acoustic pressure in a *conically converging wave* of wave number k has complex amplitude approximately described in cylindrical coordinates by

$$\hat{p}_i \approx \frac{K}{w^{1/2}} e^{-ikw \sin\bar{\theta}} e^{ikz \cos\bar{\theta}}$$

at larger kw, with specified constants K and $\bar{\theta}$. Develop a theory analogous to that given in Sec. 9-4 to explain (*a*) the amplitude of the resulting overall disturbance near $w = 0$ and (*b*) the phase shift associated with ray passage past the focus of a conically converging-diverging ray tube. [The solution requires the use of a Bessel function and of its asymptotic limiting expression. See, for example, J. N. Brune, J. E. Nafe, and L. E. Alsop, *Bull. Seismol. Soc. Am.*, **51**:247–257 (1961).]

9-16 In an atmosphere whose temperature decreases with height near a rigid ground ($z = 0$), the sound field near a point (0, 0, 0) on the inner border of a zone of abnormal audibility (see Sec. 8-4) has the following ray structure. Each ray is an arc of a circle of radius R and moves parallel to the xz plane, bending upward with increasing x. A caustic surface described by the plane, $z = -(\tan\alpha)x$, intersects the ground at the origin with a grazing angle α, so that no rays pass through the region $x < 0$, $z < -(\tan\alpha)x$. Devise an applicable expression for the complex acoustic-pressure amplitude along the ground near $x = 0$. Choose the normalization to be such that $\hat{p}(0, 0, 0)$ is P. Sketch $|\hat{p}/P|^2$ versus x/R for $kR = 100$ and $\alpha = 15°$. Here k is ω/c, with ω equaling the angular frequency and c equaling the sound speed at the ground.

9-17 (*a*) Derive the equation corresponding to (9-5.13) that gives the asymptotic behavior of $w_1(\tau - \eta)$ at large positive η.

(*b*) Show that the function Φ that represents the phase of $e^{ik_0 x} e^{i\xi\tau} w_1(\tau - \eta)$ with $\xi = (k_0 R/2)^{1/3} x/R$ and $\eta = (2k_0^2 R^2)^{1/3} z/R$, is an approximate solution of the eikonal equation

$$\left(\frac{\partial\Phi}{\partial x}\right)^2 + \left(\frac{\partial\Phi}{\partial z}\right)^2 = \frac{k_0^2}{(1 - z/R)^2}$$

when $z \ll R$ and $w_1(\tau - \eta)$ is replaced by the large η asymptotic limit.

(*c*) Verify that the corresponding ray paths are such that $dx/dz > 0$ and that they are propagating obliquely upward when dx/dt is positive.

9-18 (*a*) Show that the acoustic energy shed per unit time and area by a creeping wave propagating along a surface of finite impedance is approximately given by either of the two expressions

$$\frac{(p_{\text{cw}}^2)_{\text{av},0}}{\rho c} \frac{1}{4\pi} \frac{(2/kR)^{1/3}}{|\text{Ai}(b_1)|^2} \quad \text{or} \quad \rho c (v_{z,\text{cw}}^2)_{\text{av},0} \frac{1}{4\pi} \frac{(kR/2)^{1/3}}{|\text{Ai}'(b_1)|^2}$$

where $(p_{\text{cw}}^2)_{\text{av},0}$ is the mean squared pressure of the creeping wave at the surface and $(v_{z,\text{cw}}^2)_{\text{av},0}$ is the corresponding normal component of the fluid velocity. What expression and numerical coefficient would you use for the limiting case of (*b*) a rigid surface and (*c*) a pressure-release surface?

9-19 A creeping wave propagating along a surface of finite impedance loses energy because of ray shedding and because of absorption at the surface. Show that the ratio of the absorption loss to the ray-shedding loss is

$$4\pi \,\text{Re}[\text{Ai}'(b_1)\text{Ai}^*(b_1)e^{i\pi/6}]$$

What limiting expressions, proportional to Re $(1/Z_S)$ or Re Z_S, are applicable when $|Z_S|$ is much greater or much less than ρckl?

9-20 Develop a heuristic argument supporting the conclusion that the energy per unit surface area associated with a creeping wave is $\dot{e}_{av}/2\alpha c$. Here $\dot{e}_{av}$ is the energy lost per unit area and time due to ray shedding and surface absorption and α is the exponential decay rate (nepers per meter) associated with the creeping wave. Show that this result, in conjunction with that in Prob. 9–18, leads to $l/3.2$ for the approximate boundary-layer thickness of a creeping wave propagating along a rigid surface.

9-21 A point harmonic source is adjacent ($\theta = 0$) to a large ($kR \gg 1$) rigid sphere of radius R in an otherwise unbounded homogeneous medium.

(*a*) Use the earth-flattening approximation and the results in Sec. 9-5 to argue that the acoustic pressure near the sphere (but θ not near 0 or π) has complex amplitude approximately given by

$$\hat{p} = \frac{\hat{S}e^{ikR\theta}}{R\theta^{1/2}(\sin\theta)^{1/2}} V\left(\frac{R\theta}{2kl^2}, 0, \frac{r-R}{l}, 0\right)$$

where $l = (R/2k^2)^{1/3}$.

(*b*) Show that the leading term in the residue series for V leads to a creeping-wave description for the field.

(*c*) Why is the denominator factor $\theta^{1/2}(\sin\theta)^{1/2}$ given above in (*a*) a better choice than simply θ?

(*d*) Give a numerical value for $R\hat{p}/\hat{S}$ when $kR = 100$, $r = R$, and $\theta = \pi/2$.

9-22 (*a*) For the circumstances described in Prob. 9-21, show that the field in the shadow zone at points near neither the sphere's surface nor the shadow-zone boundary is approximately

$$\hat{p} = \frac{\hat{S}e^{i(\omega/v_{ph})R\Delta\theta_0}e^{-\alpha R\Delta\theta_0}e^{i\omega\tau_{TR}}e^{-i\pi/12}}{[2krl^2\sin\theta]^{1/2}[c\tau_{TR}/(2Rl)^{1/2}]^{1/2}[-a_1'\mathrm{Ai}(a_1')]}$$

where $\Delta\theta_0$ and τ_{TR} refer to the path of least travel time that connects source and reception point. The path follows the surface through angle $\Delta\theta_0$, then traverses a distance $c\tau_{TR}$ along a straight line that is tangential to the sphere. Assume that αR is substantially larger than 1. *Hint:* Match a geometrical acoustics field to the large $(r - R)/l$ limit of the result from Prob. 9-21.

(*b*) Show that the result in (*a*) reduces for $r \gg kR^2$ and $\theta > \pi/2$ to

$$\hat{p} \to \hat{S}\frac{e^{ikr}}{r}\frac{e^{i(\omega/v_{ph})R(\theta-\pi/2)}e^{-\alpha R(\theta-\pi/2)}e^{-i\pi/12}}{(2krl^2\sin\theta)^{1/2}(2Rl)^{-1/4}[-a_1'\mathrm{Ai}(a_1')]}$$

providing θ is not close to π. Sketch the resulting radiation pattern and discuss its dependence on frequency.

9-23 Apply the concepts implied by the statements in Prob. 9-15 to extend the solution of part (*b*) of Prob. 9-22 to points near and including those where $\theta = \pi$.

9-24 The principle of reciprocity can transform the results in Probs. 9-22 and 9-23 to the solution for the acoustic pressure on the shadow side of the surface of a large rigid sphere when a plane wave is incident.

(*a*) Explain why this is so and summarize the desired solution.

(*b*) Interpret the solution from part (*a*) in terms of creeping waves.

9-25 The analogy between sound penetration into a shadow zone caused by upward refraction of rays in a stratified medium and sound diffraction around a curved surface is demonstrated by the following two exercises.

(*a*) Show that the function $\xi^{-1/2}V(\xi, \eta_0, \eta, q)$ in Eq. (9-5.17) is a solution of the *parabolic equation*

$$\left(\frac{\partial^2}{\partial\eta^2} + \eta + \frac{i\partial}{\partial\xi}\right)\xi^{-1/2}V = 0$$

with the boundary condition

$$\left(\frac{\partial}{\partial\eta} + q\right)\xi^{-1/2}V = 0 \qquad \text{at} \qquad \eta = 0$$

(*b*) Show that if a plane wave impinges at normal incidence (toward the $+x$ direction) on a very wide barrier with a cylindrical locally reacting top, the acoustic-pressure amplitude near the barrier top is approximately ($\varepsilon \ll 1$)

$$\hat{p} = e^{i2\xi/\varepsilon^2} e^{-i(2/3)\eta^{3/2}} F(\xi, \eta)$$

where $F(\xi, \eta)$ satisfies the parabolic equation and $\varepsilon = (2/kR)^{1/3}$. Here R is the radius of the top and ξ and η are related to cartesian coordinates x and y (see the figure) by the transformation

$$x = R\varepsilon\xi - \tfrac{1}{6}R\varepsilon^3(\xi^3 - 3\xi\eta + 2\eta^{3/2}) \qquad y \approx \tfrac{1}{2}R\varepsilon^2(\xi^2 - \eta)$$

The surface $\eta = 0$ corresponds to the barrier top. (See the paper by V. A. Fock and L. A. Weinstein, reprinted in Fock, *Electromagnetic Diffraction and Propagation Problems*, pp. 171–187.)

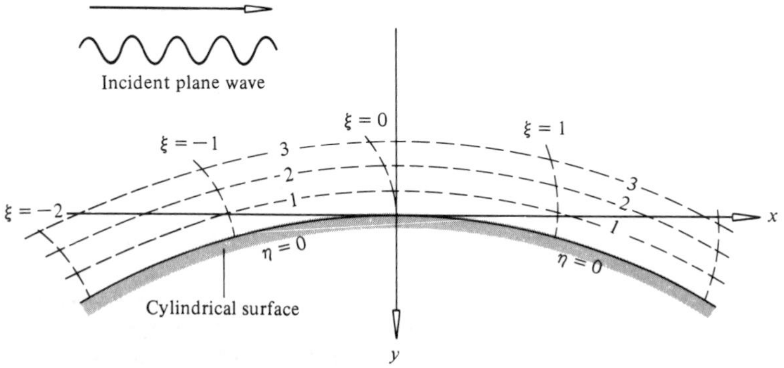

Problem 9-25

9-26 A point source is at distance R from an exterior corner (a point where three edges meet) of a large rectangular rigid box. Given that P is the pressure amplitude that would be measured at the same point if the box were not present, what is the pressure amplitude at the corner?

9-27 The source and the listener are adjacent but on opposite sides ($z = z_S$, $r = r_S$, $\phi = 0$, $\phi_S = \beta$) of a thin rigid screen ($\beta = 2\pi$). Given that the source has monopole amplitude $\hat{S}$ and that $kr \gg 1$, what is the acoustic-pressure amplitude at the listener location?

9-28 Verify that the Sommerfeld solution (page 495*n*.) for plane-wave diffraction by a thin screen ($\beta = 2\pi$) reduces to

$$\hat{p} = \hat{p}_{\text{inc}}\left[1 - 2(1 - i)\left(\frac{kr}{\pi}\right)^{1/2} \cos\frac{\phi}{2}\cos\frac{\phi_s}{2}\right]$$

in the limit $kr \ll 1$. Is this consistent with Eq. (9-7.12)? What does this imply concerning the fluid velocity near the edge? Show that $r^{1/2}\cos\phi/2$ is a solution of Laplace's equation and discuss the significance of this fact.

9-29 A heuristic simplified method for prediction of barrier insertion loss proposed by R. S. Redfearn, *Phil. Mag.*, (7)**30**:223–236 (1940), leads to an insertion loss that is a function of h/λ and ϕ when $z = z_S$, where h and ϕ are the quantities indicated in the figure.

(*a*) Show that such an assumption is consistent for small ϕ with the Fresnel number approximation and that, in such a limit, N_F is approximately $(h/\lambda)\phi$.

(*b*) Show that an alternative substitution for the Fresnel number is $(h^2/\lambda)(r^{-1} + r_S^{-1})$ [Z. Maekawa, *Appl. Acoust.*, **1**:157–173 (1968)].

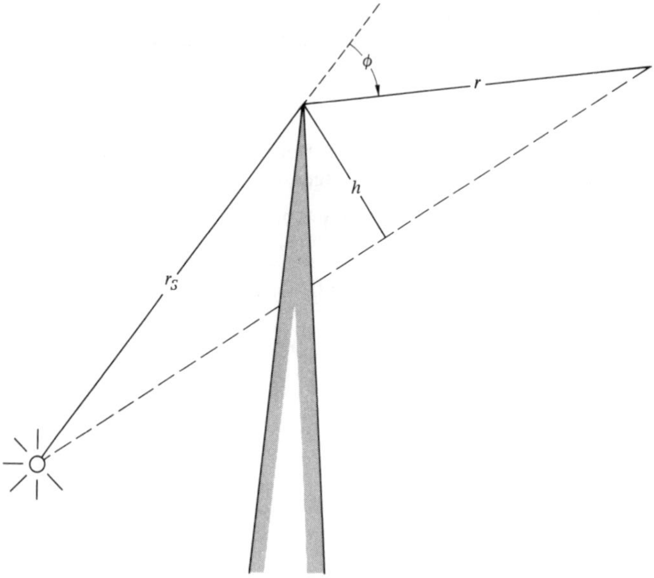

Problem 9-29

9-30 A square thin rigid plate occupies the region $-a < x < a$, $-a < y < a$, of the $z = 0$ plane. A harmonic point source of monopole amplitude $\hat{S}$ is directly in front ($z = 0^+$) of the plate's center. Consider ka as large and consider the field on the $z > 0$ side to be made up of a direct-plus-reflected wave combination plus diffracted waves from each of the four plate edges.

(*a*) Determine an expression for the complex acoustic-pressure amplitude along the $+z$ axis.

(*b*) Describe the locations of any points along the axis where interference from the diffracted waves may cause the acoustic pressure to be inordinately small.

(*c*) Repeat part (*a*) for the $-z$ axis.

9-31 A point source lies on the $\phi_s = \beta$ interior surface of a 120° wedge ($\nu = \frac{3}{2}$).

(*a*) Given that kr_s and kr are both large, express the bistatic reflected field for points near the plane $\phi = 60°$ in terms of single-edge diffraction formulas.

(*b*) What corresponds to a Fresnel number for the circumstances just described?

9-32 A square plate of dimensions a on a side is at sufficient distance R from an acoustic transmitter to be regarded as being in the far field; ka, however, is substantially larger than 1. Use edge-diffraction theory to estimate the target strength of the plate when the incident propagation direction is normal to the plate. Take the transmitter to be omnidirectional and reciprocal and take R to be substantially larger than ka^2.

9-33 The question of whether interior or exterior edges cause the stronger echoes arises in the following example. The terrain is flat and coincides with the $z = 0$ plane for $x < 0$. Between $x = 0$ and $x = 40\lambda$, the terrain slopes upward, rising 3 units for every 4 horizontal units, to a height of 30λ. Beyond $x = 40\lambda$, the terrain is once again level. The transitions from level to sloped and from sloped to level at $x = 0$ and $x = 40\lambda$ are abrupt in terms of a wavelength λ. When an omnidirectional transducer at $x = -100\lambda$, $y = 0$, $z = 0$ transmits a pulse of nearly constant frequency, it subsequently receives two echoes. What is the ratio of the amplitude of the second echo to that of the first echo?

9-34 A simple method for estimating diffraction around thick barriers (*double-edge diffraction*) rests on the following heuristic concepts. When the direct wave from the source strikes the nearest edge, it excites a diffracted wave that travels along the barrier top to the farther edge; there the incident diffracted wave gives rise to a second diffracted wave that travels to the listener on the far side of the barrier. The propagation from source to edge, edge to edge, and from edge to listener is in accord with geometrical-acoustic principles; the generation of diffracted waves by an incident wave at an edge is predicted with Eq. (9-8.16). Apply the method just described when source and listener are on opposite sides of a long rigid rectangular three-sided barrier of width 10λ. A point source of monopole amplitude $\hat{S}$ is adjacent to one side at a distance 10λ from the top and the listener is adjacent to the opposite side ($z = z_S$), also at a distance 10λ from the top. What is the complex pressure amplitude at the listener location?

CHAPTER

TEN

EFFECTS OF VISCOSITY AND OTHER DISSIPATIVE PROCESSES

Phenomena that cannot be explained within the strict confines of the ideal fluid-dynamic equations include attenuation of sound, radiation caused by flow past obstacles, wave structure near a shock front, acoustic streaming, and finite amplitudes of resonating systems. Pertinent physical processes are not necessarily the same for each phenomenon, but the processes commonly entering into consideration involve viscosity, thermal conduction, or relaxation. We here first consider viscosity and thermal conductivity and show how the fluid-dynamic equations are modified when these processes are taken into account. Subsequent sections explore the basic acoustical implications of the resulting equations. Relaxation processes occupy our attention in the final portions of the chapter.

10-1 THE NAVIER-STOKES-FOURIER MODEL

The Stress Tensor

To include viscosity in the basic fluid-dynamic equations, one must first abandon the assumption that the force exerted per unit area by adjacent fluid particles on the surface enclosing a given fluid particle is normal to the surface. Consideration of phenomena involving viscosity, e.g., the drag on a solid body when fluid is flowing past it, requires that this force $\mathbf{f}_S(\mathbf{n}, \mathbf{x})$ per unit area also have a tangential component (see Fig. 10-1). The assumption is made, however, that the molecular interactions between adjacent fluid particles are of such short range that $\mathbf{f}_S(\mathbf{n}, \mathbf{x})$ is independent of the detailed shape and volume of the fluid

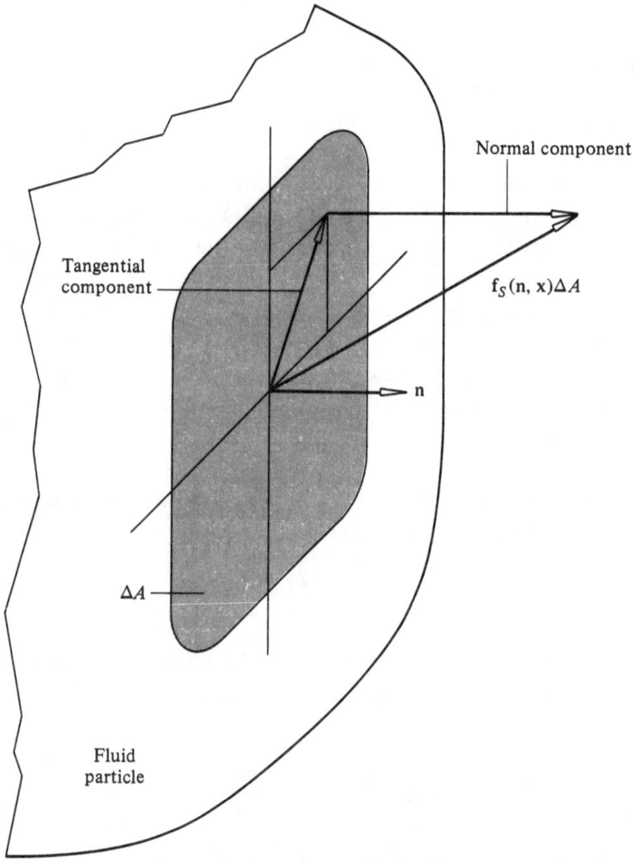

Figure 10-1 Surface force on an area element of an internal surface in a viscous fluid.

particle considered, so that it depends only on the point $\mathbf{x}$ on the surface at which it is applied, on the outward normal $\mathbf{n}$ of the surface at $\mathbf{x}$, and on time t.

Newton's third law applied to neighboring fluid particles requires that $\mathbf{f}_S(-\mathbf{n}, \mathbf{x}) = -\mathbf{f}_S(\mathbf{n}, \mathbf{x})$. Furthermore, the requirement that the net force on a tetrahedron-shaped fluid particle divided by the mass of that particle be finite in the limit as the volume becomes zero leads to the relation†

$$\mathbf{f}_S(\mathbf{n}, \mathbf{x}) = (\mathbf{n} \cdot \mathbf{e}_x)\mathbf{f}_S(\mathbf{e}_x, \mathbf{x}) + (\mathbf{n} \cdot \mathbf{e}_y)\mathbf{f}_S(\mathbf{e}_y, \mathbf{x}) + (\mathbf{n} \cdot \mathbf{e}_z)\mathbf{f}_S(\mathbf{e}_z, \mathbf{x}) \quad (10\text{-}1.1)$$

where $\mathbf{e}_x$, $\mathbf{e}_y$, $\mathbf{e}_z$ are unit vectors in the x, y, z (or x_1, x_2, x_3) directions. The three cartesian components of this vector equation take the form (*Cauchy's stress*

† G. K. Batchelor, *An Introduction to Fluid Dynamics*, Cambridge University Press, 1967, pp. 1–10; Y. C. Fung, *Foundations of Solid Mechanics*, Prentice-Hall, Englewood Cliffs, N.J., 1965, pp. 62–65. The proof is due to A.-L. Cauchy.

relation)

$$\mathbf{e}_i \cdot \mathbf{f}_S(\mathbf{n}, \mathbf{x}) = \sum_{j=1}^{3} \sigma_{ij}(\mathbf{x}) n_j \qquad i = 1, 2, 3 \tag{10-1.2}$$

where
$$\sigma_{ij}(\mathbf{x}) = \mathbf{e}_i \cdot \mathbf{f}_S(\mathbf{e}_j, \mathbf{x}) \qquad i, j = 1, 2, 3 \tag{10-1.3}$$

represents the ith component of the force exerted per unit area at a point $\mathbf{x}$ on the surface of a fluid particle where the outward normal is in direction $\mathbf{e}_j$.

The nine quantities $\sigma_{ij}(\mathbf{x})$ constitute the components of the *stress tensor;* the off-diagonal elements are the *shear stresses*. If the components are known for any one given cartesian coordinate system, the components appropriate to any other choice of axes can be derived from Eqs. (2) and from the geometrical properties of vectors. The stress tensor must be symmetric, $\sigma_{ij}(\mathbf{x}) = \sigma_{ji}(\mathbf{x})$, because the net torque about the center of any fluid particle corresponds to a finite angular acceleration, even in the limit when the particle size becomes vanishingly small.

The expression (2) for the cartesian components of $\mathbf{f}_S(\mathbf{n}, \mathbf{x})$ allows the net surface force on a given fluid particle to be written as

$$\sum_{i=1}^{3} \sum_{j=1}^{3} \mathbf{e}_i \iint [\sigma_{ij}(\mathbf{x})\mathbf{e}_j] \cdot \mathbf{n}\, dS = \sum_{ij} \mathbf{e}_i \iiint \nabla \cdot (\sigma_{ij}\mathbf{e}_j)\, dV$$

where the latter integral is over the volume of the particle. Consequently, the steps in Sec. 1-3, which led there to Euler's equation, lead here instead to the *Cauchy equation of motion*

$$\rho \frac{D\mathbf{v}}{Dt} = \sum_{ij} \mathbf{e}_i \frac{\partial \sigma_{ij}}{\partial x_j} \tag{10-1.4}$$

Without any additional assumptions concerning the stress tensor, this holds equally well for solids and fluids.

The Energy Equation

A basic law of mechanics† is that the net rate of change of energy within a moving fluid particle (occupying time-dependent volume V^*) must be equal to the rate at which work is done on it by the surface forces plus the net rate at which heat energy is flowing into it. Thus we write

$$\frac{d}{dt} \iiint_{V^*} (\tfrac{1}{2}\rho v^2 + \rho u)\, dV = \iint_{S^*} \mathbf{f}_S \cdot \mathbf{v}\, dS - \iint_{S^*} \mathbf{q} \cdot \mathbf{n}\, dS \tag{10-1.5}$$

where u is the *internal energy* per unit mass within the particle and $\mathbf{q}$ is the *heat-flux vector,* defined so that $-\mathbf{q} \cdot \mathbf{n}$ is heat flowing per unit area into the volume at a point on the surface where the outward unit normal is $\mathbf{n}$. The left

† For an extensive discussion and references, see Y. Elkana, *The Discovery of the Conservation of Energy,* Harvard University Press, Cambridge, Mass., 1974.

side of this can be argued, in a manner similar to that in which Eq. (1-3.5) was derived, to be equivalent to the volume integral of $\rho(D/Dt)(\frac{1}{2}v^2 + u)$. Also, with the components of $\mathbf{f}_S$ given by Eq. (2), one has $\mathbf{f}_S \cdot \mathbf{v} = \Sigma(\sigma_{ij}v_i\mathbf{e}_j) \cdot \mathbf{n}$, so Gauss' theorem transforms both of the surface integrals in (5) into volume integrals. The result applies for an arbitrary volume, and thus the equation holds for the integrands themselves; so we obtain the *Fourier-Kirchhoff-Neumann energy equation*†

$$\rho \frac{D}{Dt}(\tfrac{1}{2}v^2 + u) = \sum_{ij} \frac{\partial}{\partial x_j} \sigma_{ij}v_i - \nabla \cdot \mathbf{q} \tag{10-1.6}$$

A simplication in the above results if one subtracts from it the dot product of $\mathbf{v}$ with the momentum equation, this product being

$$\rho\mathbf{v} \cdot \frac{D\mathbf{v}}{Dt} = \rho \frac{D}{Dt}(\tfrac{1}{2}v^2) = \sum_{ij} v_i \frac{\partial \sigma_{ij}}{\partial x_j} = \sum_{ij} \frac{\partial}{\partial x_j} v_i\sigma_{ij} - \sum_{ij} \sigma_{ij} \frac{\partial v_i}{\partial x_j} \tag{10-1.7}$$

Thus, with the subtraction, one has

$$\rho \frac{Du}{Dt} = \sum_{ij} \sigma_{ij} \frac{\partial v_i}{\partial x_j} - \nabla \cdot \mathbf{q} \tag{10-1.8}$$

which replaces the ideal-fluid relation $Ds/Dt = 0$.

Constitutive Relations for a Fluid

Relations between the σ_{ij}, $\mathbf{q}$, and other variables describing the dynamical and thermodynamical state of the fluid are called *constitutive equations*. The Navier-Stokes model adopted here is a generalization of the observation that, for common types of fluids (*newtonian‡ fluids*), the shear stress is proportional to the rate of shear. For a steady unidirectional flow in which $\mathbf{v}$ has only an x component $v_x(y)$, the stress component σ_{xy} is found for such a fluid to equal $\mu\partial v_x/\partial y$, where the *viscosity* μ is independent of v_x and of its spatial variation.

The generalization of the newtonian constitutive relation to an arbitrary state of motion is that any shear-stress component ($i \neq j$) must be a linear combination of the spatial derivatives $\partial v_i/\partial x_j$ and that the shear stresses vanish when all the $\partial v_i/\partial x_j$ are zero. Furthermore, the relation between the σ_{ij} and the $\partial v_i/\partial x_j$ must be independent of the choice of coordinate system. To determine such a relation, it is expedient to first define σ_n as the *average normal component* (one-third of the trace) of the stress tensor. Then the tensor with components $\sigma_{ij} - \sigma_n\delta_{ij}$ is a symmetric tensor with zero trace. The only way this can be linearly related to the $\partial v_i/\partial x_j$ in a form independent of choice of coordinate

† The name is suggested by C. Truesdell, *Continuum Mechanics*, vol. I, *The Mechanical Foundations of Elasticity and Fluid Mechanics*, Gordon and Breach, New York, 1966, p. 40.

‡ The term derives from a statement in F. Cajori, *Newton's Principia: Motte's Translation Revised*, University of California, Berkeley, 1947, p. 385: "The resistance arising from the want of lubricity in the parts of a fluid, is, other things being equal, proportional to the velocity with which the parts of the fluid are separated from one another."

system is for its components to be linear combinations† of the components of whatever tensors can be formed from the $\partial v_i/\partial x_j$ that are also symmetric and also have zero trace. Apart from a multiplicative constant, there is only one such tensor, so

$$\sigma_{ij} - \sigma_n \delta_{ij} = \mu \phi_{ij} \tag{10-1.9}$$

$$\phi_{ij} = \frac{\partial v_i}{\partial x_j} + \frac{\partial v_j}{\partial x_i} - \tfrac{2}{3}\nabla \cdot \mathbf{v}\, \delta_{ij} \tag{10-1.10}$$

The components ϕ_{ij} of the *rate-of-shear tensor* have the desired properties because $\phi_{ij} = \phi_{ji}$ and $\Sigma\phi_{ii} = 0$. That the proportionality factor is the viscosity μ follows from the requirement that (9) must imply $\sigma_{xy} = \mu\ \partial v_x/\partial y$ when $\mathbf{v} = \mathbf{e}_x v_x(y)$.

In regard to the first term on the right side of the energy equation (8), Eqs. (9) and (10) lead to

$$\sum_{ij} \sigma_{ij} \frac{\partial v_i}{\partial x_j} = -\frac{\sigma_n}{\rho}\frac{D\rho}{Dt} + \frac{\mu}{2}\sum_{ij} \phi_{ij}^2 \tag{10-1.11}$$

because $\Sigma\phi_{ii} = 0$ and because the mass-conservation equation (1-2.4) implies $\nabla \cdot \mathbf{v} = -\rho^{-1}\, D\rho/Dt$. Thus (8) becomes

$$\rho\left(\frac{Du}{Dt} - \sigma_n \frac{D\rho^{-1}}{Dt}\right) = \frac{\mu}{2}\sum_{ij} \phi_{ij}^2 - \nabla \cdot \mathbf{q} \tag{10-1.12}$$

For quasi-static processes (disturbances of low frequency and with little spatial variation), the fluid may be regarded as being in local thermodynamic equilibrium. In this limit, particular values of internal energy u per unit mass and specific (per unit mass) volume $1/\rho$ correspond to an entropy $s(u, 1/\rho)$ per unit mass whose differential ds is $(1/T)\ du + (p/T)\ d(1/\rho)$, where p and T are the pressure and temperature corresponding to the equilibrium state associated with the values of u and $1/\rho$. In the equilibrium state, σ_n must also be taken as $-p$. Also, for near-equilibrium states, one expects that $\mathbf{q}$ should be proportional to ∇T but oppositely directed to ∇T because heat flows from high temperature to low temperature, so one would adopt *Fourier's law*,‡ $\mathbf{q} = -\kappa\nabla T$, where κ is the *coefficient of thermal conductivity* (referred to for brevity as the thermal conductivity).

Within the context of the above discussion, the simplest assumptions§ con-

† A proof along such lines follows from the analysis of M. Reiner, "A Mathematical Theory of Dilatancy," *Am. J. Math.*, **67**:350–362 (1945). See, for example, C.-S. Yih, *Fluid Mechanics*, McGraw-Hill, New York, 1969, pp. 26–32. The original derivation of Eq. (9) from continuum-mechanical principles is due to G. G. Stokes, "On the Theories of the Internal Friction of Fluids in Motion, and of the Equilibrium and Motion of Elastic Solids," *Trans. Camb. Phil. Soc.*, **8**:75–102 (1845).

‡ J. Fourier, *The Analytical Theory of Heat*, 1822, trans. by A. Freeman, 1878; reprinted by Dover, New York, 1955, p. 52.

§ Stokes' original derivation gave (in the notation of the present text) $\sigma_n = -p + \mu_B\, \nabla \cdot \mathbf{v}$, where μ_B is the bulk viscosity. The modification to the fluid-dynamic equations caused by the bulk viscosity is discussed in Sec. 10-7; here we proceed as if it were zero.

cerning σ_n and $\mathbf{q}$ are that $\sigma_n = -p$ and $\mathbf{q} = -\kappa \nabla T$, where p and T have the same relation to u and $1/\rho$ as for a fluid in equilibrium. Also, since the equation of state $s = s(u, 1/\rho)$ should be independent of time for any fluid particle, one has

$$T\frac{Ds}{Dt} = \frac{Du}{Dt} + p\frac{D}{Dt}\frac{1}{\rho} \tag{10-1.13}$$

The right side here, with $\sigma_n = -p$, is the quantity in parentheses in Eq. (12).

The assumptions just stated allow us to write (4) as the *Navier-Stokes equation*†

$$\rho\frac{D\mathbf{v}}{Dt} = -\nabla p + \sum_{ij} \mathbf{e}_i \frac{\partial}{\partial x_j}\mu\phi_{ij} \tag{10-1.14}$$

and to write (12) as the *Kirchhoff-Fourier equation*‡

$$\rho T\frac{Ds}{Dt} = \frac{\mu}{2}\sum_{ij}\phi_{ij}^2 + \nabla\cdot(\kappa\nabla T) \tag{10-1.15}$$

where it is understood that the relations between s, ρ, T, and p are the same as for the fluid in equilibrium. Those thermodynamic relations, plus the mass-conservation relation (1-2.4), along with the two equations above and with some specification for κ and μ, constitute what we here call the *Navier-Stokes-Fourier model of a compressible fluid*. The model's chief limitation from the standpoint of acoustics, as discussed in Secs. 10-7 and 10-8, is that it often fails to explain the actual values and the frequency dependence of sound attenuation in extended regions remote from solid boundaries. In other instances, however, it is adequate for understanding phenomena not explicable with the ideal fluid-dynamic equations.

Values of Viscosity and Thermal Conductivity

For gases, μ and κ are functions of temperature T only. For air, in particular, the data and detailed calculations based on the molecular structure of its constituents and on kinetic theory are consistent with the semiempirical formulas§

$$\frac{\mu}{\mu_0} = \left(\frac{T}{T_0}\right)^{3/2}\frac{T_0 + T_S}{T + T_S} \tag{10-1.16a}$$

$$\frac{\kappa}{\kappa_0} = \left(\frac{T}{T_0}\right)^{3/2}\frac{T_0 + T_A e^{-T_B/T_0}}{T + T_A e^{-T_B/T}} \tag{10-1.16b}$$

† The origins of this appear in nineteenth-century works by Navier (1822), Poisson (1829), Saint-Venant (1843), and Stokes (1845), the last work being of greatest influence on the subsequent development of fluid mechanics. For the original references, see Yih, *Fluid Mechanics*, pp. 58–59.

‡ The terminology is somewhat inaccurate since neither Kirchhoff nor Fourier used the concept of entropy in their relevant publications, but it is convenient to refer to Eq. (15) and to the overall model by brief names.

§ J. Hilsenrath et al., *Tables of Thermodynamic and Transport Properties of Air*, etc., Pergamon Press, Oxford, 1960, pp. 7, 10, 11, 26, 57–62. Equation (16*a*) is due to W. Sutherland, "The Viscosity of Gases and Molecular Force," *Phil. Mag.*, (5)**36**:507–531 (1893).

where μ_0 and κ_0 correspond to temperature T_0. If these formulas hold for any given choice of T_0, they also hold for any other choice of T_0. The constants T_S, T_A, and T_B are $T_S = 110.4$ K, $T_A = 245.4$ K, and $T_B = 27.6$ K. If T_0 is 300 K (27°C), then $\mu_0 = 1.846 \times 10^{-5}$ kg/(m · s) and $\kappa_0 = 2.624 \times 10^{-2}$ W/(m · K).

Transport Properties of Water

Typical values for the viscosity and thermal conductivity† of water are $\mu = 1.002 \times 10^{-3}$ kg/(m · s) and $\kappa = 0.597$ W/(m · K) for distilled water at 20°C and atmospheric pressure. Since the corresponding values for seawater are $\mu = 1.081 \times 10^{-3}$ kg/(m · s) and $\kappa = 0.574$ W/(m · K), salinity effects are minor, less than 8 or 4 percent for μ or κ. The variation due to pressure changes at fixed temperature are less than 5 percent up to pressures of the order of 10,000 atm (10^9 Pa) in the case of μ, and it is expected that the pressure dependence of κ is also weak. Thus, one may regard μ and κ as functions only of temperature for most purposes.

The viscosity of pure water decreases with temperature (the opposite from that of air) from 1.787×10^{-3} at 0°C to 0.2818×10^{-3} at 100°C. An approximate expression (accurate to 1 percent between 10 and 30°C) for the dependence near 20°C is

$$\mu = 1.002 \times 10^{-3} e^{-0.0248\,\Delta T} \tag{10-1.17a}$$

where ΔT is the difference between the temperature and 20°C. The temperature dependence of κ is relatively weak; an approximate fit to the data near 20°C is

$$\kappa = 0.597 + 0.0017\Delta T - 7.5 \times 10^{-6}(\Delta T)^2 \tag{10-1.17b}$$

A dimensionless quantity characterizing the relative magnitudes of μ and κ is the *Prandtl number*, $\mathrm{Pr} = \mu c_p/\kappa$, where c_p is the specific heat at constant pressure. For gases, an approximate kinetic-theory analysis‡ suggests that Pr is $4\gamma/(9\gamma - 5)$, where γ is the specific-heat ratio. For a diatomic gas ($\gamma = 1.4$), this gives $\mathrm{Pr} = 0.737$, and for air this value is not markedly different over the temperature range of normal interest from what would be computed from the actual values of μ, c_p, and κ. [With $c_p = \gamma R/(\gamma - 1)$, $R = 287$, $\gamma = 1.4$ (see Sec. 1-9), and with μ and κ as given by Eqs. (16), one finds Pr at 300 K is 0.707.]

For water, the temperature dependence of the Prandtl number is roughly the same as that of the viscosity. The value at 20°C for Pr is 7.0, about 10 times the corresponding value for air.

† The values cited are extracted from the *Handbook of Chemistry and Physics*, 49th ed., Chemical Rubber, Cleveland, 1968, and from R. A. Horne, *Marine Chemistry*, Wiley-Interscience, New York, 1969.

‡ A. Eucken, "On the Thermal Conductivity, the Specific Heat, and the Internal Friction of Gases," *Phys. Z.*, **14**:324–332 (1913). For a commentary and suggested replacements, see S. Chapman and T. G. Cowling, *The Mathematical Theory of Nonuniform Gases*, Cambridge University Press, Cambridge, 1939, 1952, p. 237; M. J. Lighthill, "Viscosity Effects in Sound Waves of Finite Amplitude," in G. K. Batchelor and R. M. Davies (eds.), *Surveys in Mechanics*, Cambridge University Press, London, 1956, pp. 250–351, especially p. 259.

10-2 LINEAR ACOUSTIC EQUATIONS AND ENERGY DISSIPATION

Linear acoustic equations governing small-amplitude disturbances result from the discard of terms of second order in the deviations of p, ρ, $\mathbf{v}$, T, s from their ambient values p_0, ρ_0, $\mathbf{v}_0$, T_0, s_0. For simplicity, we here regard the ambient state as homogeneous and quiescent, such that $\mathbf{v}_0 = 0$ and p_0, ρ_0, T_0, and s_0 are independent of position and time.

Linear Acoustic Equations

The deviations p', ρ', T', s' are related by the thermodynamic equations of state, $\rho = \rho(p, s)$ and $T = T(p, s)$, whose linearized versions give ρ' and T' as linear combinations of p' and s'. With the thermodynamic identities $(\partial\rho/\partial s)_p = -\rho\beta T/c_p$ and $(\partial T/\partial p)_s = T\beta/\rho c_p$, the coefficients can be expressed in terms of $c_p = T(\partial s/\partial T)_p$, $c^2 = (\partial p/\partial\rho)_s$, and $\beta = \rho[\partial(1/\rho)/\partial T]_p$ (representing the specific heat at constant pressure, the sound speed squared, and the coefficient of thermal expansion). One has, in particular,

$$\rho' = \frac{1}{c^2}\,p - \left(\frac{\rho\beta T}{c_p}\right)_0 s \tag{10-2.1a}$$

$$T' = \left(\frac{T\beta}{\rho c_p}\right)_0 p + \left(\frac{T}{c_p}\right)_0 s \tag{10-2.1b}$$

where, for convenience in subsequent writing, the primes on p' and s' have been omitted and the coefficients are understood to be evaluated at the ambient state. (For an ideal gas, $p = \rho RT$ implies $\beta = 1/T$, so βT can be replaced by 1 in the above.)

The remaining linear equations for the model come from the conservation-of-mass relation (1-2.4), the Navier-Stokes equation (10-1.14), and the Kirchhoff-Fourier equation (10-1.15). The quantities ϕ_{ij} and ∇T are automatically first order, so μ and κ need only be taken to zero order and are constants for any given choice of ambient state. Thus, the linear equations reduce to

$$\frac{\partial\rho'}{\partial t} + \rho_0\nabla\cdot\mathbf{v} = 0 \tag{10-2.2a}$$

$$\rho_0\frac{\partial\mathbf{v}}{\partial t} = -\nabla p + \sum_{ij}\mu\mathbf{e}_i\frac{\partial\phi_{ij}}{\partial x_j} \tag{10-2.2b}$$

$$\rho_0 T_0\frac{\partial s}{\partial t} = \kappa\nabla^2 T' \tag{10-2.2c}$$

where we adhere to our previous convention of omitting unnecessary primes. Alternatively, with the definition (10-1.10) for the components of the rate-of-shear tensor, Eq. (2b) can be written as

$$\rho_0\frac{\partial\mathbf{v}}{\partial t} = -\nabla p + \mu[\nabla^2\mathbf{v} + \tfrac{1}{3}\nabla(\nabla\cdot\mathbf{v})] \tag{10-2.2b'}$$

For a given ambient state, the coefficients $1/c^2$, $(\rho\beta T/c_p)_0$, ρ_0, κ, etc., in Eqs. (1) and (2) can be regarded as numerical constants.

The Energy Conservation-Dissipation Corollary

We here examine the changes the model described by Eqs. (1) and (2) above necessitates in the acoustic energy-conservation law (1-11.2). Taking the dot product of Eq. (2*b*) with **v** and adding to it p/ρ_0 times (2*a*) and T'/T_0 times (2*c*) yields

$$\frac{\partial}{\partial t}\left(\tfrac{1}{2}\rho_0 v^2\right) + \frac{p}{\rho_0}\frac{\partial \rho'}{\partial t} + \rho_0 T'\frac{\partial s}{\partial t} = -\nabla\cdot(p\mathbf{v}) + \mu\sum_{ij}\frac{\partial}{\partial x_j}v_i\phi_{ij} - \mu\sum_{ij}\phi_{ij}\frac{\partial v_i}{\partial x_j} + \frac{\kappa}{T_0}\nabla\cdot(T'\nabla T') - \frac{\kappa}{T_0}(\nabla T')^2 \tag{10-2.3}$$

The sum of the second and third terms on the left side reduces, because of Eqs. (1), to

$$\frac{p}{\rho_0}\frac{\partial \rho'}{\partial t} + \rho_0 T'\frac{\partial s}{\partial t} = \frac{\partial}{\partial t}\left[\frac{1}{2}\frac{p^2}{\rho_0 c^2} + \frac{1}{2}\left(\frac{\rho T}{c_p}\right)_0 s^2\right]$$

Also, as in the derivation of Eq. (10-1.11), we can replace the sum over i and j of $\phi_{ij}\,\partial v_i/\partial x_j$ by a similar sum over $\frac{1}{2}\phi_{ij}^2$.

The substitutions just described reduce Eq. (3) to

$$\frac{\partial w}{\partial t} + \nabla\cdot\mathbf{I} = -\mathscr{D} \tag{10-2.4}$$

where

$$w = \tfrac{1}{2}\rho_0 v^2 + \frac{1}{2}\frac{p^2}{\rho_0 c^2} + \frac{1}{2}\left(\frac{\rho T}{c_p}\right)_0 s^2 \tag{10-2.5a}$$

$$\mathbf{I} = p\mathbf{v} - \mu\sum_{ij}\mathbf{e}_j v_i\phi_{ij} - \frac{\kappa}{T_0}T'\,\nabla T' \tag{10-2.5b}$$

$$\mathscr{D} = \tfrac{1}{2}\mu\sum_{ij}\phi_{ij}^2 + \frac{\kappa}{T_0}(\nabla T')^2 \tag{10-2.5c}$$

These equations should be compared with the analogous acoustic-energy-conservation theorem in Sec. 1-11 that results when viscosity and thermal conduction are neglected.

The energy interpretation of Eq. (4) is most apparent when both sides are integrated over some fixed control volume, so that an application of Gauss' theorem yields

$$\frac{d}{dt}\iiint w\,dV + \iint \mathbf{I}\cdot\mathbf{n}\,dS = -\iiint \mathscr{D}\,dV \tag{10-2.6}$$

Here the first term on the left is the time rate of change of disturbance energy in the control volume; the second term is the net rate at which such energy is flowing out through the control volume's surface. Therefore the nonzero term on the right (with the indicated minus sign) must be the negative of the rate at which energy is "unaccountably" being lost. Since what is lost in this context is said to be dissipated, $\mathcal{D}$ is the energy dissipated† per unit volume and time. The two terms in expression (5*c*) for $\mathcal{D}$ are the rates of energy dissipation per unit volume caused by viscosity and thermal conduction, respectively. Their nonnegative values are in accord with the expectation that the net energy associated with any disturbance must always decrease after the cessation of the source excitation.

Our expression for the disturbance energy w per unit volume in Eq. (5*a*) includes an additional term proportional to the square of the entropy deviation s. For disturbances normally classified as sound, this term is negligibly small compared with the other two, but there are other types of disturbances characterized primarily by heat conduction for which this term dominates. As regards the energy-flux vector $\mathbf{I}$, the dot product of the second term in (5*b*) with $\mathbf{n}$ is the power transmitted per unit area by viscous stresses across a surface with unit normal $\mathbf{n}$; its contribution to the surface integral in (6) is the work done per unit time and surface area by the viscous stresses on the environment external to the control volume. The first two terms in (5*b*) combine to give [in accord with Eq. (10-1.9)] $-\Sigma\sigma'_{ij}v_i\mathbf{e}_j$, where σ'_{ij} is the deviation of the corresponding stress tensor component from its ambient value $-p_0\delta_{ij}$. The net contribution of these terms to $\mathbf{I}\cdot\mathbf{n}$ is accordingly $-\mathbf{f}'_S(\mathbf{n}, \mathbf{x}, t)\cdot\mathbf{v}$, where $-\mathbf{f}'_S$ is the deviation of the force per unit area exerted by the control volume on its external environment. The third term in (5*b*) represents the flux of disturbance energy associated with heat conduction, but because of the factor T'/T_0, it cannot be interpreted literally as heat energy flowing per unit area and time.

Attenuation of Plane Sound Waves

A simple application of the energy conservation-dissipation theorem is the calculation of the attenuation of a plane wave propagating in the $+x_1$ direction (specified by unit vector $\mathbf{e}_1 = \mathbf{e}_x$) through a medium with small μ and κ. The relations between the acoustic pressure p, fluid velocity $\mathbf{v}$, temperature deviation T', and their spatial dependences are then nearly the same over any local region as predicted by the idealized model discussed in Chap. 1. Thus $\mathbf{v} \approx \mathbf{e}_1 p/\rho_0 c$ and $T' \approx (T\beta/\rho c_p)_0 p$. Also, since the dependence of these on t and $\mathbf{x}$ is approximately such that they vary only with $t - x_1/c$, one has [with

† Alternatively, $\mathcal{D}$ may be regarded as T_0 times the rate per unit volume at which entropy is being irreversibly generated by the disturbance. See, for example, R. C. Tolman and P. C. Fine, "On the Irreversible Production of Entropy," *Rev. Mod. Phys.*, **20**:51–77 (1948); C. Eckart, "The Thermodynamics of Irreversible Processes," *Phys. Rev.*, **58**:267–269 (1940).

$\nabla f(t - x_1/c) = -(\mathbf{e}_1/c)\partial f/\partial t]$

$$\nabla T' \approx \left(\frac{T\beta}{\rho c_p}\right)_0 \left(\frac{-\mathbf{e}_1}{c}\right) \frac{\partial p}{\partial t} \tag{10-2.7a}$$

$$\nabla \cdot \mathbf{v} = \frac{\partial v_1}{\partial x_1} \approx -\frac{1}{\rho_0 c^2}\frac{\partial p}{\partial t} \tag{10-2.7b}$$

$$\phi_{11} = \frac{4}{3}\frac{\partial v_1}{\partial x_1} \qquad \phi_{22} = \phi_{33} = -\frac{2}{3}\frac{\partial v_1}{\partial x_1} \tag{10-2.7c}$$

so [with the thermodynamic identity $\gamma - 1 = T\beta^2 c^2/c_p$ from Eq. (1-9.9)]

$$(\nabla T')^2 \approx \left(\frac{T}{\rho^2 c_p c^4}\right)_0 (\gamma - 1)\left(\frac{\partial p}{\partial t}\right)^2 \tag{10-2.8}$$

$$\sum_{ij} \phi_{ij}^2 \approx \frac{8}{3}\frac{(\partial p/\partial t)^2}{(\rho_0 c^2)^2} \tag{10-2.9}$$

Thus the dissipation, Eq. (5c), per unit volume and time is approximately

$$\mathscr{D} \approx \left[\tfrac{4}{3}\mu + \frac{(\gamma - 1)\kappa}{c_p}\right] \frac{(\partial p/\partial t)^2}{(\rho_0 c^2)^2} \tag{10-2.10}$$

For a plane wave of constant angular frequency ω, in the absence of viscosity and thermal conductivity, the time average of $(\partial p/\partial t)^2$ is $\omega^2(p^2)_{\text{av}}$ or $\omega^2 \rho_0 c I_{\text{av}}$, where I_{av} is the intensity in the direction of propagation. Consequently, to lowest nonzero order in κ and μ, Eq. (10) implies

$$\mathscr{D}_{\text{av}} \approx 2\alpha_{\text{cl}} I_{\text{av}} \tag{10.2.11}$$

where we use the abbreviations† (cl for classical)

$$\alpha_{\text{cl}} = \frac{\omega^2 \delta_{\text{cl}}}{c^3} \qquad \delta_{\text{cl}} = \frac{\mu}{2\rho_0}\left(\frac{4}{3} + \frac{\gamma - 1}{\text{Pr}}\right) \tag{10-2.12}$$

and Pr is the Prandtl number $\mu c_p/\kappa$.

Since the time average of the energy conservation-dissipation theorem, Eq. (4), requires, for plane waves propagating in the x direction, that $dI_{\text{av}}/dx = -\mathscr{D}_{\text{av}}$, the approximation (11) yields

$$I_{\text{av}} = I_{\text{av},0}\, e^{-2\alpha_{\text{cl}} x} \qquad |\hat{p}| = |\hat{p}|_{x=0} e^{-\alpha_{\text{cl}} x} \tag{10-2.13}$$

The second version follows because I_{av} is proportional to the square of any field amplitude associated with the disturbance. Thus α_{cl} gives the attenuation of the

† The original derivations of α_{cl} proceeded along lines analogous to those described below in the derivations of Eq. (10-3.6) and (10-8.10). The result without thermal conductivity is due to Stokes, "On the Theories of the Internal Friction." The inclusion of thermal conduction was carried through for an ideal gas by G. Kirchhoff, "On the Influence of Heat Conduction in a Gas on Sound Propagation," *Ann. Phys. Chem.*, (5)**134:**177–193 (1868). The generalization to other classes of fluids is due to P. Langevin, whose work was reported by P. Biquard, "On the Absorption of Ultrasonic Waves by Liquids," *Ann. Phys.*, (11)**6:**195–304 (1936).

disturbance in nepers per meter as predicted by the Navier-Stokes-Fourier model to lowest order in μ and κ.

Except for a monatomic gas,† the *classical attenuation coefficient* α_{cl} is generally not in accord with experiment and gives an underestimate. Extended models that remove such discrepancies are discussed in Sec. 10-7; however, the Navier-Stokes-Fourier model is often sufficient when the bulk of the disturbance energy is being dissipated within a wavelength or less from a solid surface.

10-3 VORTICITY, ENTROPY, AND ACOUSTIC MODES

At frequencies of normal interest, any disturbance governed by the linear equations derived in the previous section can be considered as a superposition of vorticity, entropy, and acoustic *modal wave fields*.‡ The individual modal fields satisfy equations considerably simpler than those for the disturbance as a whole and are uncoupled in the linear approximation except at boundaries. To show such a decomposition is possible and to arrive at the appropriate equations for the component fields, we begin with an analysis of plane-wave disturbances.

Dispersion Relations for the Component Modes§

A plane-wave disturbance of angular frequency ω in a homogeneous time-independent medium is one for which each field quantity (ψ_n denoting one of these) varies with t and $\mathbf{x}$ as

$$\psi_n(\mathbf{x}, t) = \mathrm{Re}\ \hat{\psi}_n e^{-i\omega t} e^{i\mathbf{k}\cdot\mathbf{x}} \tag{10-3.1}$$

where the wave-number vector $\mathbf{k}$ is the same for each field quantity. The number $\hat{\psi}_n$ is independent of $\mathbf{x}$ and t and is in general complex, as are the components of $\mathbf{k}$. For an *isotropic medium,* where there is no preferred direction in space (as for the model in the previous section but not when gravity is taken into account), any set of values (k_x, k_y, k_z) yielding an appropriate $k^2 = k_x^2 + k_y^2 + k_z^2$ is possible. However, there are only a small number of k^2 for a given ω for which a nontrivial solution (at least one $\hat{\psi}_n$ not zero) exists. (In the

† M. Greenspan, "Propagation of Sound in Rarefied Helium," *J. Acoust. Soc. Am.*, **22**:568–571 (1951); "Propagation of Sound in Five Monatomic Gases," ibid., **28**:644–648 (1956). Greenspan's data show that the so-called classical theory is valid if $\rho_0 c^2/\omega\mu\gamma$ is greater than 10.

‡ Although this point of view was implicit in Kirchhoff's (1868) solution for sound attenuation in a circular tube, its modern origins began with L. Cremer, "On the Acoustic Boundary Layer outside a Rigid Wall," *Arch. Elektr. Uebertrag.*, **2**:136–139 (1948); P. A. Lagerstrom, J. D. Cole, and L. Trilling, "Problems in the Theory of the Viscous Compressible Fluids," *Calif. Inst. Technol. Guggenheim Aeronaut. Lab. Rep. Off. Nav. Res.*, 1949 (reprinted 1950); and L. S. G. Kovasznay, "Turbulence in Supersonic Flow," *J. Aeronaut. Sci.*, **20**:657–674, 682 (1953).

§ The discussion here is comparable to that developed by E. O. Astrom (1950) and others for electromagnetic disturbances in an ionized gas with an impressed ambient magnetic field. See, for example, T. H. Stix, *The Theory of Plasma Waves*, McGraw-Hill, New York, 1962, pp. 11–13.

present case, there are three such values.) The resulting relations between k^2 and ω are the *dispersion relations* for the possible modes of propagation.

Given that k^2 has one of the allowed values, there is at least one set of $\hat{\psi}_n$'s for which the governing equations are satisfied by the substitution (1). The equations do, however, impose linear relations (generically called *polarization relations*†) between the members of the set. A procedure for finding the possible k^2's and the corresponding polarization relations begins with a formal substitution of expressions like (1) into the governing linear partial-differential equations; all the requisite differentiations are then carried out, and each such equation is written in the form

$$\text{Re}\,[(\text{something})e^{-i\omega t}e^{i\mathbf{k}\cdot\mathbf{x}}] = 0$$

One subsequently argues that this will be true in general only if the "something" is zero. This leads to the prescription that all such amplitude equations emerge from the original partial-differential equations with the replacement of $\partial/\partial x$ by ik_x, $\partial/\partial y$ by ik_y, $\partial/\partial z$ by ik_z, $\partial/\partial t$ by $-i\omega$, and ψ_n by $\hat{\psi}_n$. In this manner, Eqs. (10-2.2) are replaced by

$$\omega\left(\frac{\hat{p}}{c^2} - \frac{\rho\beta T}{c_p}\hat{s}\right) - \rho\mathbf{k}\cdot\hat{\mathbf{v}} = 0 \qquad (10\text{-}3.2a)$$

$$-i\omega\rho\hat{\mathbf{v}} = -i\mathbf{k}\hat{p} - \mu(k^2\hat{\mathbf{v}} + \tfrac{1}{3}\mathbf{k}\,\mathbf{k}\cdot\hat{\mathbf{v}}) \qquad (10\text{-}3.2b)$$

$$i\omega\hat{s} = \frac{\kappa}{\rho c_p}k^2\left(\hat{s} + \frac{\beta}{\rho}\hat{p}\right) \qquad (10\text{-}3.2c)$$

In writing these, we have also used Eqs. (10-2.1) to replace $\hat{\rho}'$ and $\hat{T}'$ by the corresponding expressions in terms of $\hat{p}$ and $\hat{s}$. (Here and in what follows the subscript 0 is omitted on symbols for ambient quantities whenever the risk of misinterpretation is small.)

Taking the cross product and dot products, respectively, of $\mathbf{k}$ with Eq. (2*b*) yields

$$(-i\omega\rho + \mu k^2)(\mathbf{k}\times\hat{\mathbf{v}}) = 0 \qquad (10\text{-}3.3a)$$

$$(\omega\rho + i\tfrac{4}{3}\mu k^2)\mathbf{k}\cdot\hat{\mathbf{v}} = k^2\hat{p} \qquad (10\text{-}3.3b)$$

The first of these allows two possibilities: $\mathbf{k}\times\hat{\mathbf{v}} = 0$ or $k^2 = i\omega\rho/\mu$. The first possibility requires $\hat{\mathbf{v}}$ be parallel to $\mathbf{k}$. The second possibility, with k^2 replaced by $i\omega\rho/\mu$ in Eqs. (3*b*), (2*a*), and (2*c*), requires zero values for $\mathbf{k}\cdot\hat{\mathbf{v}}$, $\hat{p}$, and $\hat{s}$ (providing $\omega \neq 0$). In particular, $\mathbf{k}$ and $\hat{\mathbf{v}}$ must be perpendicular. This gives us one possible plane-wave mode for the fluid: $\mathbf{k}\times\hat{\mathbf{v}} \neq 0, \mathbf{k}\cdot\hat{\mathbf{v}} = 0$, the remaining field quantities, p, ρ', T', and s, all zero; k^2 is $i\omega\rho/\mu$.

Returning to the first possibility ($\mathbf{k}\times\hat{\mathbf{v}} = 0$), we simplify our algebra if we

† C. O. Hines, "Internal Atmospheric Gravity Waves at Ionospheric Heights," *Can. J. Phys.*, **38:**1441–1481 (1960).

abbreviate

$$X = \frac{c^2k^2}{\omega^2} \qquad \varepsilon_\mu = i\,\frac{4}{3}\,\frac{\mu\omega}{\rho c^2} \tag{10-3.4a}$$

$$\varepsilon_\kappa = \frac{i\,\kappa\omega}{\rho c^2 c_p} \tag{10-3.4b}$$

Equations (2c) and (3b), with $\mathbf{k}\cdot\hat{\mathbf{v}}$ taken from (2a), represent two simultaneous linear equations for $\hat{s}$ and $\hat{p}$, which can be written, with the definitions (4) and with the thermodynamic identity $\gamma - 1 = \beta^2 Tc^2/c_p$ [see Eq. (1-9.9)], as

$$\begin{bmatrix} 1 + \varepsilon_\kappa X & \varepsilon_\kappa X \\ -(\gamma - 1)(1 + \varepsilon_\mu X) & 1 + \varepsilon_\mu X - X \end{bmatrix} \begin{bmatrix} \hat{s} \\ \beta\hat{p}/\rho \end{bmatrix} = \begin{bmatrix} 0 \\ 0 \end{bmatrix} \tag{10-3.5}$$

A nontrivial solution of Eq. (5) exists only if the determinant of coefficients vanishes, yielding the following quadratic equation† (*Kirchhoff's dispersion relation*):

$$(-\varepsilon_\kappa + \gamma\varepsilon_\mu\varepsilon_\kappa)X^2 + (\varepsilon_\mu + \gamma\varepsilon_\kappa - 1)X + 1 = 0 \tag{10-3.6}$$

The radical resulting from the exact solution of this is awkward to handle when one considers generalizations to phenomena not describable as plane waves of constant frequency. However, for all conceivable cases of interest, both $|\varepsilon_\mu|$ and $|\varepsilon_\kappa|$ are much less than 1, so the roots can be expressed as truncated power series in ε_μ and ε_κ, causing the following approximate dispersion and polarization relations to result from Eqs. (5) and (6):

$$X \approx 1 + \varepsilon_\mu + (\gamma - 1)\varepsilon_\kappa \qquad \hat{s} \approx -\frac{\varepsilon_\kappa \beta\hat{p}}{\rho} \tag{10-3.7a}$$

$$X \approx -\frac{1}{\varepsilon_\kappa} + (\gamma - 1)\left(1 - \frac{\varepsilon_\mu}{\varepsilon_\kappa}\right) \qquad \frac{\beta\hat{p}}{\rho} \approx (\gamma - 1)(\varepsilon_\kappa - \varepsilon_\mu)\hat{s} \tag{10-3.7b}$$

A Generalization Based on the Superposition Principle

Each of the three dispersion relations derived above can be written

$$k^2 + f(i\omega) = 0 \tag{10-3.8}$$

where $f(i\omega)$ is a power series in $i\omega$ with real coefficients. If a $\psi_n(\mathbf{x}, t)$ described by Eq. (1) has a wave-number vector that conforms to one such dispersion relation, then

$$\text{Re}\,\{[-k^2 - f(i\omega)]\hat{\psi}_n e^{-i\omega t} e^{i\mathbf{k}\cdot\mathbf{x}}\} = 0 \tag{10-3.9}$$

† Given first for an ideal gas by G. Kirchhoff (1868). An extensive discussion of its solutions without the restriction that $|\varepsilon_\mu|$ and $|\varepsilon_\kappa|$ be small and with the bulk viscosity included (such that $4\mu/3$ is replaced by $4\mu/3 + \mu_B$) is given by C. Truesdell, "Precise Theory of the Absorption and Dispersion of Forced Plane Infinitesimal Waves according to the Navier-Stokes Equations," *J. Ration. Mech. Anal.*, **2**:643–730 (1953).

However, in this context $-k^2$ is equivalent to ∇^2 and $i\omega$ is equivalent to $-\partial/\partial t$, so one could alternatively write

$$\left[\nabla^2 - f\left(-\frac{\partial}{\partial t}\right)\right] \psi_n(\mathbf{x}, t) = 0 \tag{10-3.10}$$

Furthermore, this is true for any superposition of plane-wave disturbances that conform to the same dispersion relation. Similarly, the polarization relations associated with each dispersion relation lead to partial-differential equations.†

Vorticity Mode

The dispersion relation $k^2 = i\omega\rho/\mu$ leads to the *diffusion equation*

$$\nabla^2 \mathbf{v}_{\text{vor}} = \frac{\rho}{\mu} \frac{\partial \mathbf{v}_{\text{vor}}}{\partial t} \tag{10-3.11}$$

The corresponding polarization relations, as explained in the sentences following Eq. (3), must be

$$\nabla \cdot \mathbf{v}_{\text{vor}} = 0 \qquad p_{\text{vor}} = s_{\text{vor}} = T'_{\text{vor}} = \rho'_{\text{vor}} = 0 \tag{10-3.12}$$

These correspond to an incompressible flow that does not alter any of the thermodynamic state variables. Since of the three classes of disturbance fields this is the only one for which the vorticity $\nabla \times \mathbf{v}$ is nonzero, we refer to it as the *vorticity-mode field.*

Acoustic Mode

The dispersion relation in Eq. (7*a*), given the definitions (4), leads to the partial-differential equation

$$\nabla^2 p_{\text{ac}} - \frac{1}{c^2}\frac{\partial^2 p_{\text{ac}}}{\partial t^2} + \frac{2}{c^4}\delta_{\text{cl}}\frac{\partial^3 p_{\text{ac}}}{\partial t^3} = 0 \tag{10-3.13}$$

where δ_{cl} is defined by Eq. (10-2.12). This may be regarded as the wave equation for acoustic disturbances with a slight correction for viscosity and thermal conduction. The differential-equation versions of the polarization relations for this mode, with all terms of first or higher order in ε_μ and ε_κ deleted, are

$$\nabla \times \mathbf{v}_{\text{ac}} = 0 \qquad s_{\text{ac}} \approx 0 \qquad \rho\frac{\partial \mathbf{v}_{\text{ac}}}{\partial t} \approx -\nabla p_{\text{ac}}$$

$$T'_{\text{ac}} \approx \left(\frac{T\beta}{\rho c_p}\right)_0 p_{\text{ac}} \qquad \rho'_{\text{ac}} \approx \frac{p_{\text{ac}}}{c^2} \tag{10.3.14}$$

The first of these follows from $\mathbf{k} \times \hat{\mathbf{v}} = 0$, the second from Eq. (7*a*), the third from (2*b*), and the fourth and fifth from Eqs. (10-2.1) with $s_{\text{ac}} \approx 0$.

† L. Trilling, "On Thermally Induced Sound Fields," *J. Acoust. Soc. Am.*, **27**:425–431 (1955).

The Entropy Mode

The dispersion relation in Eq. (7*b*), with the retention of only the leading term, $-1/\varepsilon_\kappa$, on the right side, leads to the *thermal-diffusion equation* of conduction heat transfer:

$$\nabla^2 s_{\text{ent}} = \frac{\rho c_p}{\kappa} \frac{\partial s_{\text{ent}}}{\partial t} \tag{10-3.15}$$

the same equation being satisfied for all components of the field, T'_{ent} in particular. (The development leading to this is the explanation of why c_p rather than c_v should appear in the coefficient of $\partial T/\partial t$ in the thermal-diffusion equation.)

The differential-equation versions of the polarization relations for this mode, with all terms of first or higher order in ε_κ and ε_μ deleted, are

$$p_{\text{ent}} \approx 0 \qquad \mathbf{v}_{\text{ent}} \approx \left(\frac{\beta T \kappa}{\rho c_p^2}\right)_0 \nabla s_{\text{ent}} \qquad \nabla \times \mathbf{v}_{\text{ent}} = 0$$

$$T' \approx \left(\frac{T}{c_p}\right)_0 s_{\text{ent}} \qquad \rho'_{\text{ent}} \approx -\left(\frac{\rho\beta T}{c_p}\right)_0 s_{\text{ent}} \tag{10-3.16}$$

The first follows from the polarization relation in Eq. (7*b*); the third from $\mathbf{k} \times \hat{\mathbf{v}} = 0$; the fourth and fifth from Eqs. (10-2.1) with $p_{\text{ent}} \approx 0$. To develop the equation for $\mathbf{v}_{\text{ent}}$, it is insufficient to set $\hat{p}_{\text{ent}}$ to 0 in Eq. (2*b*) because $|\mathbf{k}|$ is large; instead, use Eq. (3*b*) to eliminate the $\mathbf{k} \cdot \hat{\mathbf{v}}$ term in (2*b*) and thereby obtain $[\omega\rho + i(\frac{4}{3})\mu k^2]\hat{\mathbf{v}}$ for $\mathbf{k}\hat{p}$. Substitution of k^2 and $\hat{p}$ from Eq. (7b) then yields, with some manipulation [involving the definitions (4) and the thermodynamic identity for $\gamma - 1$], the equation $\hat{\mathbf{v}} = -i(\beta T\kappa/\rho c_p^2)_0 \mathbf{k}\hat{s}$, so the second relation in Eq. (16) results. The velocity $\hat{\mathbf{v}}_{\text{ent}}$ is small but not negligible because the dispersion relation $k^2 = i\omega\rho c_p/\kappa$ allows the possibility that $|\nabla s_{\text{ent}}|$ will be much larger than $(\omega/c)|s_{\text{ent}}|$.

Since Eqs. (16) indicate that $\mathbf{v}_{\text{ent}} \approx (\beta\kappa/\rho c_p)_0 \nabla T'_{\text{ent}}$, in this mode (with $\beta > 0$) the fluid flows from colder regions toward hotter regions. Although this might contradict one's intuition, it is dictated by the conservation of mass. At a local temperature maximum, the diffusion equation (15) predicts that the temperature is decreasing with time; thermodynamic considerations (with $p \approx 0$) require the density to be simultaneously increasing with time. The fluid flows toward the temperature maximum to cause this density increase.

The label "entropy mode" applies because entropy fluctuations are a major feature; in contrast, entropy fluctuations are totally absent in the vorticity mode, and they are relatively small compared with those of, say, $\beta p/\rho_0$ in the acoustic mode.

10-4 ACOUSTIC BOUNDARY-LAYER THEORY

Any superposition of vorticity-, acoustic-, and entropy-mode fields will satisfy the linear equations for a fluid with finite viscosity and thermal conductivity.

The converse statement, that any disturbance satisfying those equations can be represented as such a superposition, is, for brevity, not proved here but may be considered a reasonable premise† with which to begin an analysis of any given boundary-value problem. Thus we write, for example,

$$\mathbf{v} = \mathbf{v}_{\text{vor}} + \mathbf{v}_{\text{ac}} + \mathbf{v}_{\text{ent}} \tag{10-4.1}$$

for the acoustic fluid velocity.

For a given ω, the dispersion relations, $k^2 = i\omega\rho/\mu$ and $k^2 \approx i\omega\rho c_p/\kappa$, for the vorticity and entropy modes are such that the imaginary part of k (associated with attenuation) for such modes is much larger than ω/c. This suggests that the vorticity- and entropy-mode fields die out rapidly with increasing distances from boundaries, interfaces, and sources. Consequently, one expects a disturbance in an extended space to be primarily made up of the acoustic-mode field (or else be inordinately small) except near such perturbations.

Measures of how far from a boundary the vorticity- and entropy-mode fields extend are the respective values of $1/|k_I|$. These *boundary-layer thicknesses* l_{vor} and l_{ent} are [with $i^{1/2} = (1 + i)/2^{1/2}$]

$$l_{\text{vor}} = \left(\frac{2\mu}{\omega\rho}\right)^{1/2} \qquad l_{\text{ent}} = \left(\frac{2\kappa}{\omega\rho c_p}\right)^{1/2} = \frac{l_{\text{vor}}}{(\text{Pr})^{1/2}} \tag{10-4.2}$$

While these lengths are not necessarily small (they tend toward ∞ as $\omega \to 0$), they are nevertheless much less than the corresponding acoustic wavelength divided by 2π,

$$\frac{2\pi l_{\text{vor}}}{\lambda_{\text{ac}}} = \left(\frac{\omega}{c}\right) l_{\text{vor}} = \left(\frac{2\omega\mu}{\rho c^2}\right)^{1/2} \ll 1$$

[For example, for 500 Hz in air, with $\mu = 1.85 \times 10^{-5}$, $\rho = 1.2$, $c = 340$ (SI units), l_{vor} is 10^{-4} m, while λ_{ac} is 0.68 m; the ratio above is 10^{-3}.]

In previous chapters it has tacitly been assumed that the physical dimensions of the space and sources are much larger than l_{vor} and l_{ent}. Thus, for example, the analysis in Sec. 7-3 of low-frequency sound in ducts presumes, for a circular duct of radius a, that ω be low enough to ensure that $\omega \ll c/a$ but still high enough to ensure that $l_{\text{vor}} \ll a$. Although this forces ω to lie between $2\mu/\rho a^2$ and c/a, these limits often encompass a wide range. In the present section we continue to assume that l_{vor} and l_{ent} are much smaller than the

† A proof for an ideal gas with a Prandtl number of $\frac{3}{4}$ is given by T. Y. Wu, "Small Perturbations in the Unsteady Flow of a Compressible, Viscous, and Heat-Conducting Fluid," *J. Math. Phys.*, **35:**13–27 (1956). A general proof could be constructed beginning with the proposition that any solution of

$$(\nabla^2 + \lambda_1)(\nabla^2 + \lambda_2)(\nabla^2 + \lambda_3)\psi = 0$$

can be written $\psi_1 + \psi_2 + \psi_3$, where $(\nabla^2 + \lambda_i)\psi_i = 0$ and no two of the λ_i are equal. (The latter premise is not valid in the limit $\omega = 0$.) The one-dimensional version of this is a fundamental theorem for homogeneous linear differential equations of arbitrary order with constant coefficients. See, for example, R. Courant, *Differential and Integral Calculus,* vol. 2, Wiley-Interscience, Glasgow, 1936, pp. 438–442.

physical dimensions, but we recognize the presence of vorticity-mode and entropy-mode boundary layers.

Boundary Conditions at a Solid Surface

Once viscosity is taken into account, the requirement that the normal component of fluid velocity be continuous at an interface is no longer sufficient (along with the other conditions of continuity of pressure and of causality, described in Chap. 3) to guarantee a unique solution of the fluid-dynamic equations. This is so because the Navier-Stokes equation, unlike Euler's equation, is not of first order in the spatial derivatives. An additional condition invariably imposed is that the tangential components of velocity also be continuous, the rationale being that a fluid should not slide any more freely with respect to an interface than it does with itself; this lack of slip is observed when the motion is examined sufficiently close to an interface.†

The surface force per unit area $\mathbf{f}_S(\mathbf{n}, \mathbf{x})$ must also be continuous (in accord with Newton's third law) across any interface with unit normal $\mathbf{n}$. Thus, if $\mathbf{n}$ is in the x_1 direction, Eq. (10-1.2) requires that σ_{11}, $\sigma_{12} = \sigma_{21}$, and $\sigma_{13} = \sigma_{31}$ all be continuous. The other components, σ_{22}, $\sigma_{23} = \sigma_{32}$, and σ_{33}, however, can be discontinuous. Similarly, conservation of energy requires that $\mathbf{q} \cdot \mathbf{n}$, the normal component of the heat flux vector, be continuous at an interface. In addition, the temperature is continuous.

Since solids are generally much better conductors of heat than fluids, the requirements that $\mathbf{q} \cdot \mathbf{n}$ and T be continuous at a solid-fluid interface are often replaced by the simpler requirement that the solid's surface be at ambient temperature, or equivalently that

$$T' = 0 \tag{10-4.3}$$

at the surface. A brief analysis suggests that the criteria for this being a valid replacement are‡

$$(\rho c_p \kappa)_{\text{fluid}} \ll (\rho c_p \kappa)_{\text{solid}} \tag{10-4.4a}$$

$$(\kappa \rho c_p)^{1/2}_{\text{fluid}} \ll \omega^{1/2} (\rho c_p)_{\text{solid}} \left(\frac{\text{volume}}{\text{surface}}\right)_{\text{solid}} \tag{10-4.4b}$$

The premise on which (3) is based is that although an external disturbance may impart heat to the solid, it also periodically extracts heat; the extra energy within the solid at any given time is never sufficient to change the average

† A. H. Shapiro, *Shape and Flow: The Fluid Dynamics of Drag*, Doubleday, Garden City, N.Y., 1961, pp. 59–63.

‡ The first equation results from an analysis of plane-wave reflection at normal incidence from an elastic half space with finite thermal conductivity. The second is based on a computation of the heat flow into the solid that uses the plane-wave result; this energy is assumed to be uniformly distributed within the solid, and the requirement is imposed that the peak temperature rise within the solid be substantially less than the peak temperature rise of the incident wave.

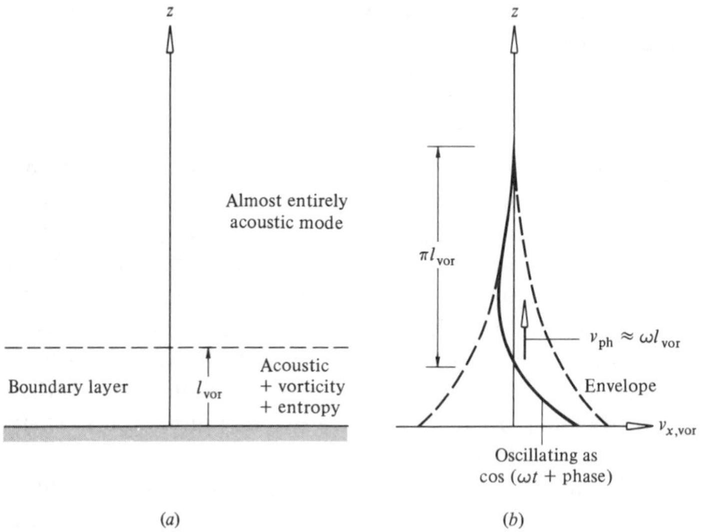

Figure 10-2 (*a*) Concept of an acoustic boundary layer. (*b*) Vorticity-mode portion of oscillating fluid velocity at a surface; $v_{x,\text{vor}}$ is confined within an envelope that dies exponentially as $e^{-z/l_{\text{vor}}}$. The lines of constant phase have apparent upward phase velocity of ωl_{vor}; moving nodal lines are at intervals of πl_{vor}.

temperature within the solid perceptibly and since the body is a good conductor, the average temperature is the same as the surface temperature.

Vorticity- and Entropy-Mode Fields near a Solid Surface†

We consider a solid-fluid interface nominally occupying the xy plane with the z axis pointing into the fluid (see Fig. 10-2). The disturbance is assumed to have constant angular frequency ω, where ω is such that both $\mu\omega/\rho c^2$ and $\kappa\omega/\rho c^2 c_p$ are much less than 1. This allows us to take the polarization relations for the acoustic and entropy modes in the approximate forms Eqs. (10-3.14) and (10-3.16).

Since the boundary conditions discussed above apply to the sum of the three modal fields, rather than to each individually, we first do not consider them explicitly. However, since we are interested in cases when vorticity- and entropy-mode fields are caused by sound of much longer wavelength than l_{vor} or l_{ent}, we assume that these fields vary much more rapidly with the z coordinate than with the x and y coordinates and consequently approximate the operator ∇^2 by $\partial^2/\partial z^2$ in the two diffusion equations. Note also that the solution of the equation

$$\frac{\partial^2}{\partial z^2}\hat{\psi}(x, y, z) = -\frac{2i}{l^2}\hat{\psi}(x, y, z) \tag{10-4.5}$$

† Cremer, "On the Acoustic Boundary Layer . . . ."

is

$$\hat{\psi}(x, y, z) = \hat{\psi}(x, y, 0)e^{-(1-i)z/l} \tag{10-4.6}$$

The sign in the exponent is chosen such that the solution is bounded at large distances from the wall. Applied to Eqs. (10-3.11) and (10-3.15), this approximate result leads to the prediction that the complex spatially dependent amplitudes of the components of the vorticity- and entropy-mode fields should vary with z as in Eq. (6), where $l = l_{\text{vor}}$ for the vorticity mode and $l = l_{\text{ent}}$ for the entropy mode. Thus, in the modal relations (10-3.12) and (10-3.16) between the complex spatially dependent amplitudes, one can replace $\partial/\partial z$ wherever it appears by $-(1 - i)/l_{\text{vor}}$ and $-(1 - i)/l_{\text{ent}}$ for the vorticity- and entropy-mode fields, respectively.

Applying the prescription just described yields the z-independent relations

$$\nabla_T \cdot \hat{\mathbf{v}}_{\text{vor},T} - (1 - i)\hat{\mathbf{v}}_{\text{vor}} \cdot \frac{\mathbf{n}}{l_{\text{vor}}} = 0 \tag{10-4.7a}$$

$$\hat{\mathbf{v}}_{\text{ent},T} = \frac{\beta\kappa}{\rho c_p} \nabla_T \hat{T}'_{\text{ent}} \approx 0 \tag{10-4.7b}$$

$$\hat{\mathbf{v}}_{\text{ent}} \cdot \mathbf{n} = -\frac{\beta\kappa}{\rho c_p}(1 - i)\frac{\hat{T}'_{\text{ent}}}{l_{\text{ent}}} \tag{10-4.7c}$$

where the subscript T (for tangential) denotes the tangential component and $\mathbf{n} = \mathbf{e}_z$ is the unit vector normal to the surface. The approximation $\hat{\mathbf{v}}_{\text{ent},T} \approx 0$ is in accord with the expectation $|\nabla_T \hat{T}'_{\text{ent}}| \ll |\hat{T}'_{\text{ent}}|/l_{\text{ent}}$.

Boundary Condition on the Acoustic-Mode Field

If the surface is oscillating as a rigid body such that every material point on the surface has a velocity with complex amplitude $\hat{\mathbf{v}}_{\text{wall}}$, the no-slip condition requires

$$\hat{\mathbf{v}}_{\text{wall}} = \hat{\mathbf{v}}_{\text{vor}} + \hat{\mathbf{v}}_{\text{ac}} + \hat{\mathbf{v}}_{\text{ent}} \qquad z = 0 \tag{10-4.8}$$

at the surface's nominal location. If, in addition, the solid is highly conducting and has a high "capacity for storing heat," Eq. (3) requires

$$T'_{\text{ent}} + T'_{\text{ac}} = 0 \tag{10-4.9}$$

at the surface.

Taking the horizontal divergence (operating with $\nabla_T\cdot$) of (8) and using Eqs. (7) yields

$$0 = (1 - i)\hat{\mathbf{v}}_{\text{vor}} \cdot \frac{\mathbf{n}}{l_{\text{vor}}} + \nabla_T \cdot \hat{\mathbf{v}}_{\text{ac},T} \tag{10-4.10}$$

Similarly the normal component of (8) gives [with (7c) replacing $\hat{\mathbf{v}}_{\text{ent}} \cdot \mathbf{n}$]

$$\hat{\mathbf{v}}_{\text{wall}} \cdot \mathbf{n} = \hat{\mathbf{v}}_{\text{vor}} \cdot \mathbf{n} + \hat{\mathbf{v}}_{\text{ac}} \cdot \mathbf{n} - \frac{\beta\kappa}{\rho c_p}(1 - i)\frac{\hat{T}'_{\text{ent}}}{l_{\text{ent}}} \tag{10-4.11}$$

With an elimination of $\hat{\mathbf{v}}_{\text{vor}} \cdot \mathbf{n}$ from these, the subsequent replacement of $\hat{T}'_{\text{ent}}$ by $-\hat{T}'_{\text{ac}}$, of $\hat{T}'_{\text{ac}}$ by $(T\beta/\rho c_p)_0 \hat{p}'_{\text{ac}}$ [from (10-3.14)], of κ by $\omega\rho c_p l_{\text{ent}}^2/2$ [from Eq. (2)], and of $\beta^2 T_0$ by $(\gamma - 1)c_p/c^2$ (a thermodynamic identity), one obtains

$$\hat{\mathbf{v}}_{\text{wall}} \cdot \mathbf{n} = \hat{\mathbf{v}}_{\text{ac}} \cdot \mathbf{n} - (1 + i)\frac{l_{\text{vor}}}{2} \nabla_T \cdot \hat{\mathbf{v}}_{\text{ac},T} + (1 - i)(\gamma - 1)\frac{\omega}{c}\frac{l_{\text{ent}}}{2}\frac{\hat{p}_{\text{ac}}}{\rho c} \tag{10-4.12}$$

at the surface ($z = 0$). Because this involves only the acoustic-mode field variables, it represents an approximate boundary condition for that modal field. In the limit $l_{\text{vor}} \to 0$ and $l_{\text{ent}} \to 0$ it reduces to the commonly applied boundary condition $\hat{\mathbf{v}}_{\text{wall}} \cdot \mathbf{n} = \hat{\mathbf{v}}_{\text{ac}} \cdot \mathbf{n}$.

The analysis above also suggests that within the boundary layer the flow field associated with the vorticity mode is approximately described by

$$\hat{\mathbf{v}}_{\text{vor}} \approx (\hat{\mathbf{v}}_{\text{wall},T} - \hat{\mathbf{v}}_{\text{ac},T})_{z=0} e^{-(1-i)z/l_{\text{vor}}} \tag{10-4.13}$$

the vertical component being negligible in comparison. Similarly, the entropy-mode-field temperature is approximately

$$\hat{T}'_{\text{ent}} = -\left(\frac{T\beta}{\rho c_p}\right)_0 (\hat{p}_{\text{ac}})_{z=0}\, e^{-(1-i)z/l_{\text{ent}}} \tag{10-4.14}$$

In this approximation, the acoustic field variables at the surface suffice to determine the vorticity- and entropy-mode fields. Alternatively, since $\hat{p}_{\text{ac}}$ and the tangential velocity $\hat{\mathbf{v}}_{\text{ac},T}$ are expected to vary insignificantly over distance intervals comparable to l_{ent} and l_{vor}, the quantities $\hat{\mathbf{v}}_{\text{ac},T}$ and $\hat{p}_{\text{ac}}$ at $z = 0$ can be interpreted as the total disturbance pressure and tangential fluid velocity just outside the boundary layer, e.g., at $z = 10l_{\text{vor}}$.

Insofar as the two boundary conditions can be adequately approximated by $\mathbf{v}_{\text{wall}} \cdot \mathbf{n} = \mathbf{v}_{\text{ac}} \cdot \mathbf{n}$ at $z = 0$, the ideal acoustic model (with viscosity and thermal conductivity neglected and with slip relative to boundaries allowed) produces accurate predictions except near solid surfaces. If one wants to know the tangential velocity and the temperature near such surfaces, one need only add expressions (13) and (14) to the predictions of the ideal acoustic model.

Energy Loss from the Acoustic Mode at a Boundary

In most instances, one is not interested in the total energy loss per se but in the energy irreversibly lost from the acoustic-mode field, because this loss accounts for the attenuation of sound. Since the acoustic-mode field constitutes a solution of the overall set of equations developed in Sec. 10-2, the energy-conservation-dissipation theorem applies equally to that field by itself. The net power flowing out of that field (into other modal fields) at a boundary per unit area is very nearly $-p_{\text{ac}}\mathbf{v}_{\text{ac}} \cdot \mathbf{n}$ since the other terms contributing to the acoustic intensity are considerably smaller for the acoustic-mode field. The time average of this, with the normal component $\hat{\mathbf{v}}_{\text{ac}} \cdot \mathbf{n}$ taken from the boundary condition

(12) and with a vector identity for $\hat{p}^* \nabla_T \cdot \hat{\mathbf{v}}_{ac,T}$, is

$$-(\mathbf{I}_{ac} \cdot \mathbf{n})_{av} = -\frac{1}{2} \operatorname{Re} (\hat{p}^* \hat{\mathbf{v}}_{wall} \cdot \mathbf{n}) - \frac{1}{2} \frac{l_{vor}}{2} \nabla \cdot \{\operatorname{Re} [(1+i)\hat{p}^* \hat{\mathbf{v}}_{ac,T}]\}$$
$$+ \frac{1}{2} \frac{l_{vor}}{2} \operatorname{Re} [(1+i) \nabla_T \hat{p}^* \cdot \hat{\mathbf{v}}_{ac,T}]$$
$$+ \frac{1}{2} \frac{l_{ent}}{2} (\gamma - 1) \frac{\omega}{c} \frac{|\hat{p}_{ac}|^2}{\rho c} \qquad (10\text{-}4.15)$$

The first term is the negative of the work done per unit time and area by the wall motion against the surface pressure on the fluid; the second term is a total derivative and therefore averages out to zero over a sufficiently large area and is of no consequence as regards the calculation of irreversible energy loss. The third term can be reexpressed with $\nabla_T \hat{p}_{ac} = i\omega\rho\hat{\mathbf{v}}_{ac,T}$, so with l_{vor} and l_{ent} replaced by Eqs. (2) we identify

$$\left(\frac{d^2E}{dA\,dt}\right)_{diss} = \left(\frac{\omega\rho\mu}{2}\right)^{1/2} (v^2_{ac,T})_{av} + (\gamma - 1) \left(\frac{\omega\rho\kappa}{2c_p}\right)^{1/2} \frac{(p^2)_{av}}{(\rho c)^2} \qquad (10\text{-}4.16)$$

as the energy dissipated per unit area and time at the surface.†

Plane-Wave Reflection at a Solid Surface

The boundary condition (12) allows an examination of the effects of viscosity and thermal conduction on the reflection of plane waves. For a plane wave at angle of incidence θ_i (see Fig. 10-3), the trace-velocity matching principle requires that all field quantities vary with t and with tangential coordinates in the combination $t - \mathbf{n}_{i,T} \cdot \mathbf{x}$, so $c\ \nabla_T p$ is $-\mathbf{n}_{i,T}\ \partial p/\partial t$. The component $\mathbf{n}_{i,T}$ of the unit vector in the direction of incidence is such that $\mathbf{n}_{i,T} \cdot \mathbf{n}_{i,T} = \sin^2 \theta_i$. Consequently, an application of the divergence operator to the tangential portion $\mathbf{v}_{ac,T} = \mathbf{n}_{i,T} p/\rho c$ of the plane-wave relation (which holds for sum of incident and reflected waves) yields

$$\nabla_T \cdot \mathbf{v}_{ac,T} = -\frac{\sin^2 \theta_i}{\rho c^2} \frac{\partial p}{\partial t} \qquad (10\text{-}4.17)$$

Subsequent insertion of the above into Eq. (12), with the wall assumed motionless, leads to

$$\frac{1}{Z} = -\frac{\hat{\mathbf{v}}_{ac} \cdot \mathbf{n}_{wall}}{\hat{p}} = \tfrac{1}{2}(1 - i) \frac{\omega}{\rho c^2} [l_{vor} \sin^2 \theta_i + (\gamma - 1) l_{ent}]$$
$$= \frac{e^{-i\pi/4}}{\rho c} \eta_\mu(\omega) \left[\sin^2 \theta_i + \frac{\gamma - 1}{(\mathrm{Pr})^{1/2}}\right] \qquad (10\text{-}4.18)$$

as the apparent specific admittance (reciprocal of specific impedance) of the

† R. F. Lambert, "Wall Viscosity and Heat Conduction Losses in Rigid Tubes," *J. Acoust. Soc. Am.*, **23**:480–481 (1951).

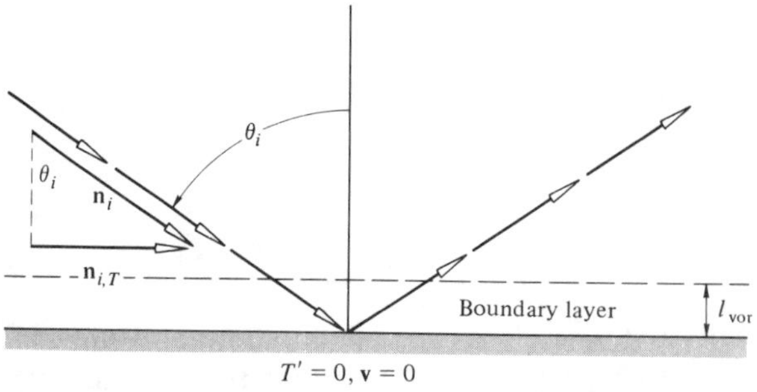

Figure 10-3 Definitions of symbols used in discussion of the reflection of a plane acoustic wave at a rigid wall when viscosity and thermal conduction are taken into account.

surface. Here and in what follows, we abbreviate

$$\eta_\mu(\omega) = \left(\frac{\omega\mu}{\rho c^2}\right)^{1/2} \qquad \eta_\kappa(\omega) = (\gamma - 1)\left(\frac{\omega\kappa}{\rho c^2 c_p}\right)^{1/2} \tag{10-4.19}$$

such that $\eta_\kappa/\eta_\mu = (\gamma - 1)/(\mathrm{Pr})^{1/2}$ (approximately 0.48 for air). Because Z depends on θ_i, the surface cannot be regarded as locally reacting.

Insertion of the above expression for Z into Eq. (3-3.4) yields the reflection coefficient for the acoustic pressure. The absorption coefficient $1 - |\mathscr{R}|^2$ is subsequently found to be

$$\alpha(\theta_i) = \frac{4\bar{\eta}\sqrt{2}\cos\theta_i}{(\sqrt{2}\cos\theta_i + \bar{\eta})^2 + \bar{\eta}^2} \tag{10-4.20}$$

where $\bar{\eta}$ is used as an abbreviation for $\eta_\mu \sin^2\theta_i + \eta_\kappa$. When θ_i is 0, this has the approximate value (since $\eta_\kappa \ll 1$)

$$\alpha(0) = 2\sqrt{2}\,\eta_\kappa \tag{10-4.21}$$

With increasing θ_i, the absorption coefficient rises to a maximum (see Fig. 10-4) and then drops to zero at grazing incidence, $\theta_i = \pi/2$. Because $\bar{\eta}$ is generally small compared with 1, the maximum occurs when θ_i is close to $\pi/2$, so its location can be determined by setting $\bar{\eta}$ equal to $\eta_\mu + \eta_\kappa$ in Eq. (20). Doing this and setting the derivative to zero yields $\theta_i = \cos^{-1}(\eta_\mu + \eta_\kappa)$, which in turn is approximately $\pi/2 - \eta_\mu - \eta_\kappa$. The corresponding maximum value is

$$\alpha_{\max} = \frac{4\sqrt{2}}{(\sqrt{2} + 1)^2 + 1} = 0.828 \tag{10-4.22}$$

Such a large value, however, is not representative for typical choices of θ_i. At $\theta_i = 45°$, for example, Eq. (20) yields approximately $4\bar{\eta}$ when η_κ and η_μ are small compared with 1; so the mirror-reflection model is usually an excellent first approximation.

The specific impedance in Eq. (18) also implies that the phase ϕ_R of the reflection coefficient $|\mathscr{R}|e^{i\phi_R}$ increases from 0 to π as θ_i varies from 0 to $\pi/2$.

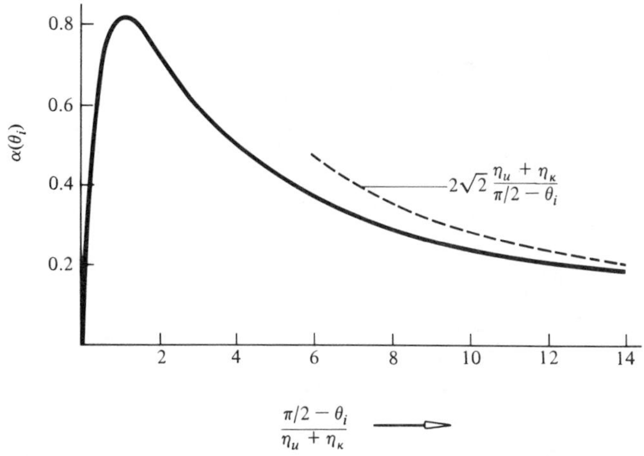

Figure 10-4 Angular dependence of absorption coefficient $\alpha(\theta_i)$ for reflection from a rigid wall with acoustic boundary layer taken into account. The absorption coefficient is largest for angles near grazing incidence and (in such a limit) is a function only of $(\pi/2 - \theta_i)/(\eta_\mu + \eta_\kappa)$, where η_μ is $(\omega\mu/\rho c^2)^{1/2}$ and η_κ/η_μ is $(\gamma - 1)/(\mathrm{Pr})^{1/2}$.

However, when η_μ and η_κ are small, ϕ_R remains close to 0 until θ_i approaches grazing incidence. The value $\pi/2$ for ϕ_R is obtained when θ_i has that value for which $\alpha = \alpha_{\max}$.

The absorption coefficient $\alpha(\theta_i)$ in Eq. (20) is compatible with the expression, Eq. (16), for the rate at which energy is absorbed by the surface because $(p^2)_{\mathrm{av}}$ is $|1 + \mathscr{R}|^2(p_i^2)_{\mathrm{av}}$ and because $(v^2_{\mathrm{ac},T})_{\mathrm{av}}$ is $|1 + \mathscr{R}|^2(p_i^2)_{\mathrm{av}}(\sin^2\theta_i)/(\rho c)^2$. Since the incident energy per unit area and time is $(p_i^2)_{\mathrm{av}}(\cos\theta_i)/\rho c$, Eqs. (16) and (18) lead to

$$\alpha = \frac{|1 + \mathscr{R}|^2\bar{\eta}}{\sqrt{2}\cos\theta_i} = \frac{4\,\mathrm{Re}\,(\rho c/Z\cos\theta_i)}{|1 + (\rho c/Z\cos\theta_i)|^2} \tag{10-4.23}$$

which is the same as $1 - |\mathscr{R}|^2$.

10-5 ATTENUATION AND DISPERSION IN DUCTS AND THIN TUBES

The effects of viscosity and thermal conduction on sound in ducts† are much greater than for propagation in free space because of the boundary conditions imposed by the duct walls. Here we consider the walls to be rigid and always at

† H. Helmholtz, "On the Influence of Friction in the Air on Sound Motion," *Verhandl. Naturhist. Med. Ver. Heidelberg*, **3**:16–20 (1863), reprinted in *Wissenschaftliche Abhandlungen*, vol. 1, Barth, Leipzig, 1882, pp. 383–387; Kirchhoff, "On the Influence of Heat Conduction"; D. E. Weston, "The Theory of the Propagation of Plane Sound Waves in Tubes," *Proc. Phys. Soc. (Lond.)*, **B66**:695–709 (1953).

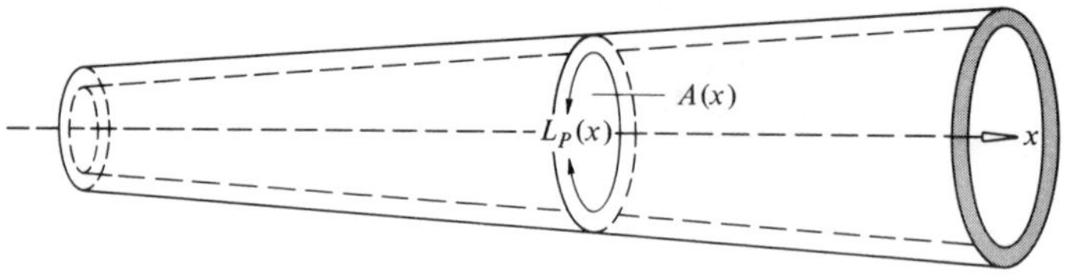

(a)

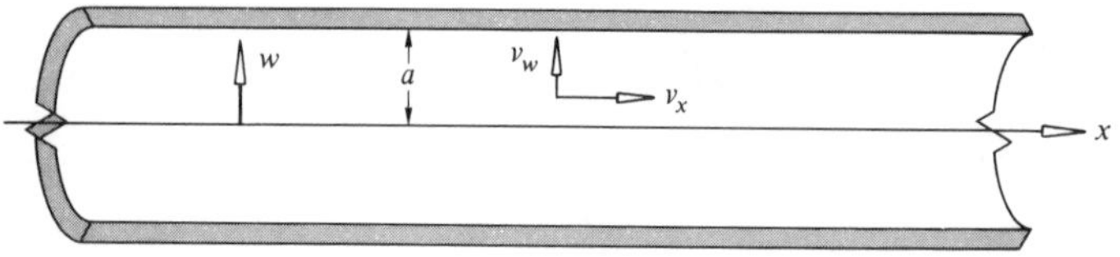

(b)

Figure 10-5 (*a*) Duct of variable cross-sectional area $A(x)$ and perimeter L_P. (*b*) Circular duct of radius a. The indicated geometries are used in the discussion of thermoviscous effects on sound propagation in ducts.

ambient temperature, so $\mathbf{v} = 0$ and $T' = 0$ at the walls. Two limiting cases are of principal interest, i.e., when a representative cross-sectional dimension is (1) much larger and (2) much smaller than the boundary-layer thicknesses l_{vor} and l_{ent}.

Propagation in Wide Ducts

We consider the duct to be large enough for the boundary layers to occupy a very small fraction of the duct's cross-sectional area A. If a nominally plane wave is propagating down the duct, most of the disturbance is associated with the acoustic-mode field and, for the most part, the field quantities vary only with distance x along the axis of the duct (see Fig. 10-5*a*).

An approximate equation for the pressure perturbation can be derived by variational techniques.† Starting with the partial-differential equation (10-3.13), recognizing that $p_{\text{ac}} \approx p$, and letting $p = \text{Re}\,[\hat{p}(x, y, z)e^{-i\omega t}]$ yields

$$\nabla^2\hat{p} + M\hat{p} = 0 \qquad M = \frac{\omega^2}{c^2} + \frac{2i\omega^3\delta_{\text{cl}}}{c^4} \tag{10-5.1}$$

Multiplying Eq. (1) by a small variation $\delta\hat{p}$, recognizing that $\delta\hat{p}\ \nabla^2\hat{p}$ is

† S. H. Crandall, D. C. Karnopp, E. F. Kurtz, Jr., and D. C. Pridmore-Brown, *Dynamics of Mechanical and Electromechanical Systems,* McGraw-Hill, New York, 1968, pp. 336–343, 417–424; P. M. Morse and H. Feshbach, *Methods of Theoretical Physics,* vol. 1, McGraw-Hill, 1953, pp. 301–318.

$\nabla \cdot (\delta\hat{p}\,\nabla\hat{p}) - \delta[\frac{1}{2}(\nabla\hat{p})^2]$ and $\delta\hat{p}\,\hat{p}$ is $\delta(\frac{1}{2}\hat{p}^2)$ to first order, subsequently integrating over a slice of the duct between x_1 and x_2, applying Gauss' theorem, and requiring that $\delta\hat{p} = 0$ at x_1 and x_2 gives

$$\delta \int_{x_1}^{x_2} \iint [\tfrac{1}{2}M(\hat{p})^2 - \tfrac{1}{2}(\nabla\hat{p})^2]\, dA\, dx + \int_{x_1}^{x_2} \oint \delta\hat{p}\, \nabla\hat{p} \cdot \mathbf{n}_{\text{wall}}\, dl\, dx = 0 \qquad (10\text{-}5.2)$$

where l denotes distance around the perimeter of the duct.

The disturbance resembles a plane wave, so to lowest nonvanishing order in κ and μ, Eq. (10-4.18) gives the boundary condition

$$\nabla\hat{p} \cdot \mathbf{n}_{\text{wall}} = \frac{i\omega\rho}{Z}\hat{p} \qquad \frac{\rho c}{Z} = e^{-i\pi/4}[\eta_\mu(\omega) + \eta_\kappa(\omega)] \qquad (10\text{-}5.3)$$

where the apparent specific impedance is evaluated with $\theta_i = \pi/2$ (grazing incidence). Thus our variational indicator becomes

$$\delta \int_{x_1}^{x_2} \left\{ \iint [\tfrac{1}{2}M\hat{p}^2 - \tfrac{1}{2}(\nabla\hat{p})^2]\, dA + \frac{i\omega\rho}{2Z} \oint \hat{p}^2\, dl \right\} dx = 0 \qquad (10\text{-}5.4)$$

If we restrict our set of trial functions to those which vary with x only (which approximates the actual case), then the "best choice" for $\hat{p}(x)$ is such that

$$\delta \int_{x_1}^{x_2} \left\{ \left[\tfrac{1}{2}M\hat{p}^2 - \frac{1}{2}\left(\frac{\partial\hat{p}}{\partial x}\right)^2 \right] A + \frac{i\omega\rho}{2Z}\hat{p}^2 L_P \right\} dx = 0 \qquad (10\text{-}5.5)$$

where L_P is the perimeter of the duct cross section.

Upon taking the variation of the above integral, using

$$\frac{1}{2}\delta\left(\frac{\partial\hat{p}}{\partial x}\right)^2 = \frac{\partial\hat{p}}{\partial x}\frac{\partial(\delta\hat{p})}{\partial x}$$

then integrating by parts, and invoking the requirement that $\delta\hat{p}$ vanish at x_1 and x_2, we obtain an expression of the form

$$\int_{x_1}^{x_2} (\text{something})\ \delta\hat{p}\, dx = 0$$

But the factor (something) must be zero because of the arbitrariness in x_1 and x_2, so

$$\frac{d}{dx}\left(A\frac{d\hat{p}}{dx}\right) + \left(MA + \frac{i\omega\rho}{Z}L_P\right)\hat{p} = 0 \qquad (10\text{-}5.6)$$

is the appropriate partial-differential equation for $\hat{p}(x)$.

The above equation when A and L_P vary with x is the generalization of the Webster horn equation (Sec. 7-8) that includes dissipation effects. The interest here is in the uniform-duct case where A and L_P are independent of x, such that Eq. (6) has solutions of the form $\hat{p}e^{ikx}$, where k^2A is the coefficient of $\hat{p}$ in Eq. (6). With M and Z taken from Eqs. (1) and (3), the square of the complex wave number becomes

$$k^2 = \frac{\omega^2}{c^2} + \frac{2i\omega}{c}[\alpha_{\text{cl}} + (1 - i)\alpha_{\text{walls}}] \tag{10-5.7}$$

where

$$\alpha_{\text{walls}} = 2^{-3/2}\eta_\mu(\omega)\left[1 + \frac{\gamma - 1}{(\text{Pr})^{1/2}}\right]\frac{L_P}{A} \tag{10-5.8}$$

The quantity η_μ is $(\omega\mu/\rho c^2)^{1/2}$, as defined in Eq. (10-4.19).

For the frequencies of interest, α_{cl} and α_{walls} are much less than ω/c. (The latter assertion stems from the restriction that l_{vor} and l_{ent} be much smaller than A/L_P.) Consequently, the square root of (7) is approximately $\omega/c + i\alpha_{\text{cl}} + (1 + i)\alpha_{\text{walls}}$. The frequencies are nevertheless assumed sufficiently low to ensure that $\alpha_{\text{cl}} \ll \alpha_{\text{walls}}$, so we discard the $i\alpha_{\text{cl}}$ term. This implies that the dissipation within the interior of the duct is much less than that within the boundary layer. The two assumptions $\alpha_{\text{walls}} \ll \omega/c$ and $\alpha_{\text{walls}} \gg \alpha_{\text{cl}}$ restrict ω to the range

$$\left(\frac{L_P}{A}\right)^2 \frac{\mu}{8\rho} \ll \omega \ll \left[\frac{9}{32}\left(\frac{L_P}{A}\right)^2 \frac{\rho c^4}{\mu}\right]^{1/3} \tag{10-5.9}$$

Because $\mu/\rho c$ is of the order of 5×10^{-8} and 7×10^{-10} m for air and water, respectively, such a range exists for any macroscopic value of A/L_P.

The approximations just described lead to the dispersion relation†

$$k = \frac{\omega}{c} + (1 + i)\alpha_{\text{walls}} \tag{10-5.10}$$

for the propagation of sound waves in a duct.

The attenuation coefficient α_{walls}, given by the imaginary part of the above expression, varies with ω as $\omega^{1/2}$ and thus has a relatively strong dependence on frequency at lower frequencies. Another feature is that the real part of k is not identically ω/c but is shifted. Thus, a traveling wave $p = \text{Re}\, Pe^{-i\omega t}e^{ikx}$ is of the form (taking the constant P as real) $Pe^{-\alpha x}\cos(\omega t - k_R x)$, where the apparent phase velocity $v_{\text{ph}} = \omega/k_R$ is

$$v_{\text{ph}} = \frac{\omega}{\omega/c + \Delta k_R} \approx c - \frac{c^2}{\omega}\Delta k_R \approx c - \frac{c^2\alpha_{\text{walls}}}{\omega} \tag{10-5.11}$$

This is lower than the speed of sound in an open space by an increment that varies as $\omega^{-1/2}$ and becomes larger the smaller the frequency. Thus, sound in pipes travels slower than sound in open air. [A pulse of sound of nearly constant angular frequency travels with a group velocity‡ v_g of the order of $1/(dk_R/d\omega)$ or $1/(c^{-1} + \frac{1}{2}\Delta k_R/\omega)$ since Δk_R varies with ω as $\omega^{1/2}$. This would give $c - v_g \approx \frac{1}{2}(c - v_{\text{ph}})$ so the group would travel with only half the reduction in speed of a point of constant phase. However, v_g is still less than c.]

† This was experimentally verified by W. P. Mason, "The Propagation Characteristics of Sound Tubes and Acoustic Filters," *Phys. Rev.*, **31**:283–295 (1928).

‡ P. S. H. Henry, "The Tube Effect in Sound-Velocity Measurements," *Proc. Phys. Soc.*, **43**:340–361 (1931).

Propagation in Narrow Tubes

In the other limit, when the cross-sectional dimensions of the tube are very small or when the frequency is sufficiently low, the boundary layer encompasses the entire duct and the theory of Sec. 10-4 is no longer applicable. For simplicity, we here limit our consideration to a circular cylinder (Fig. 10-5*b*) whose radius a is sufficiently small for the criteria $\omega\rho a^2/\mu \ll 1$ and $\omega\rho a^2 c_p/\kappa \ll 1$ to be satisfied. The analysis can be carried out for arbitrary radius a with some exactitude in terms of Bessel functions of complex argument, but here we confine ourselves to a brief heuristic derivation for the small a case† that leads to the same results as the exact solution in the same limit.

Our starting point is the x component of the linearized version (10-2.2*b*′) of the Navier-Stokes equation

$$\rho_0 \frac{\partial v_x}{\partial t} = -\frac{\partial p}{\partial x} + \mu\left[\nabla^2 v_x + \frac{1}{3}\frac{\partial}{\partial x}\nabla\cdot\mathbf{v}\right] \tag{10-5.12}$$

Given that $\omega\rho \ll \mu/a^2$ (as assumed above) and presupposing that $\nabla^2 v_x$ is of the order of v_x/a^2, we discard the inertial term on the left side at the outset. The fluid is flowing for the most part in the $+x$ direction and this, in conjunction with the requirement $v_x = 0$ at $r = a$, suggests that the radial velocity's contribution to the right side is minor. Also, we anticipate that the r dependence of v_x will be much greater than its x dependence (as is so for a steady flow), so we discard all terms involving x derivatives of v_x. This leaves us with

$$\frac{1}{r}\frac{\partial}{\partial r}\left(r\frac{\partial v_x}{\partial r}\right) = \frac{1}{\mu}\frac{\partial p}{\partial x} \tag{10-5.13}$$

The no-slip requirement, $v_x = 0$ at $r = a$, implies that v_x should vary relatively strongly with r, but we anticipate that the r dependence of $\partial p/\partial x$ will be minor, so we integrate the above treating $\partial p/\partial x$ as being independent of r. One constant of integration is obtained from the requirement that v_x be finite at $r = 0$, the other from $v_x = 0$ at $r = a$, so the result‡ is

$$v_x = -\frac{1}{4\mu}\frac{\partial p}{\partial x}(a^2 - r^2) \tag{10-5.14}$$

Another assumption, compatible with the restriction $\omega\rho a^2 c_p/\kappa \ll 1$, is that the implication of the linearized version of the Kirchhoff-Fourier equation and the boundary condition $T' \approx 0$ at $r = a$ is that $T' \approx 0$ throughout the interior of the tube; i.e., the flow is isothermal. This would then require, from Eqs. (10-

† J. W. S. Rayleigh, "On Porous Bodies in Relation to Sound," *Phil. Mag.*, (5)**16:**181–186 (1883).

‡ This is the fundamental result for *Poiseuille flow*, steady flow of an incompressible viscous fluid in a circular tube. The term stems from Poiseuille's experimental discovery (1840–1841, 1846) that the mass flowing per unit time through a tube is proportional to $-a^4\, dp/dx$. See H. Lamb, *Hydrodynamics*, 6th ed., reprinted by Dover, New York, 1945, pp. 585–586.

2.1), that

$$\rho' \approx \left(\frac{1}{c^2} + \frac{\beta^2 T}{c_p}\right) p = \frac{1}{c_T^2} p \tag{10-5.15}$$

where $c_T = c/\gamma^{1/2}$ is the isothermal sound speed (see Sec. 1-10).

The conservation-of-mass equation (10-2.2a) with the above substitution for ρ' becomes

$$\frac{1}{c_T^2}\frac{\partial p}{\partial t} + \rho\left(\frac{\partial v_x}{\partial x} + \frac{1}{r}\frac{\partial}{\partial r} r v_r\right) = 0 \tag{10-5.16}$$

and a subsequent integration over the cross-sectional area of the tube, with the boundary condition $v_r = 0$ at $r = a$, yields

$$\frac{1}{c_T^2}\frac{\partial}{\partial t}\iint p\, dA + \rho_0 \frac{\partial}{\partial x}\iint v_x\, dA = 0 \tag{10-5.17}$$

But since v_x is approximately given by Eq. (14), and since p is nearly independent of r, this approximates to

$$\frac{\partial^2 p}{\partial x^2} = \frac{8\mu}{\rho c_T^2 a^2}\frac{\partial p}{\partial t} \tag{10-5.18}$$

which is a diffusion equation.

The volume velocity U_x through the tube, defined by the integral of v_x over a cross-sectional area, satisfies the same differential equation and is related to p by what results† from integrating both sides of (14) over a cross-sectional area:

$$U_x = -\frac{\pi a^4}{8\mu}\frac{\partial p}{\partial x} \tag{10-5.19}$$

Also, in terms of U_x, Eq. (17) leads to

$$\pi a^2 \frac{\partial p}{\partial t} = -\rho c_T^2 \frac{\partial U_x}{\partial x} \tag{10-5.20}$$

The last two equations have the energy corollary

$$\frac{\partial}{\partial t}\left(\frac{\pi a^2}{2\rho c_T^2} p^2\right) + \frac{\partial}{\partial x} p U_x = -\frac{8\mu}{\pi a^4} U_x^2 \tag{10-5.21}$$

with the identification of pU_x as power transported in the $+x$ direction and of $(8\mu/\pi a^4)U_x^2$ as energy dissipated per unit time and per unit distance along the tube axis. (We have no kinetic-energy term because we discarded the inertial term in the Navier-Stokes equation. For the type of flow considered, the time

† For a tube of other than circular cross section, Eqs. (19) and (20) remain valid providing πa^2 is replaced by the tube cross-sectional area and $8\mu/a^2$ is replaced by a coefficient of resistance R that is proportional to μ and depends on the size and shape of the cross section. See H. Lamb, *The Dynamical Theory of Sound,* 2d ed., 1925, reprinted by Dover, New York, 1960, pp. 197–199. For an elliptical cross section, R is $4\mu(a^2 + b^2)/a^2b^2$, where a, b are semiaxes, this result being due to Boussinesq (1868).

rate of change of kinetic energy is always much less than the rate at which energy is being dissipated by viscosity at the walls.)

The differential equation (18) predicts that a constant-frequency disturbance traveling in the $+x$ direction in a tube of infinite length will be such that the complex wave number $\omega/v_{ph} + i\alpha$ is $e^{i\pi/4}(8\mu\omega/\rho c_T^2 a^2)^{1/2}$, so

$$v_{ph} = c_T\left(\frac{\rho\omega a^2}{4\mu}\right)^{1/2} \qquad \text{and} \qquad \alpha = \left(\frac{4\mu\omega}{\rho c_T^2 a^2}\right)^{1/2} \tag{10-5.22}$$

describe the phase velocity and attenuation coefficient. For the considered range of frequencies, one has $v_{ph} \ll c_T$, $\alpha \gg \omega/c$, and so the disturbance is traveling slowly with a high attenuation.

Slab with Circular Pores

A rudimentary model of a porous material† consists of a thick rigid slab (see Fig. 10-6) with many long cylindrical holes bored perpendicular to its face. If the number of such holes per unit area is N, and if each has radius a (such that the *porosity* is $N\pi a^2$), what is the absorption coefficient of the slab?

If $\hat{p}$ is the complex pressure amplitude just outside the slab, the volume velocity flowing into the pores per unit area of slab is $\hat{U}/A = N\hat{p}/Z_{A,h}$, where $Z_{A,h}$ is the acoustic impedance of a single hole. The ratio $\hat{p}/(\hat{U}/A)$, however, is the apparent specific impedance Z_s of the slab. Equation (20) gives

$$Z_{A,h} = \frac{\rho c_T^2}{\pi a^2}\frac{k}{\omega} = \left(\frac{8\mu\rho c_T^2}{\pi^2\omega a^6}\right)^{1/2} e^{i\pi/4} \tag{10-5.23}$$

with the wave number k identified as $e^{i\pi/4}(8\mu\omega/\rho c_T^2 a^2)^{1/2}$. Consequently,

$$\frac{Z_s}{\rho c} = \frac{1}{N\pi a^2}\left(\frac{8\mu}{\rho\omega\gamma a^2}\right)^{1/2} e^{i\pi/4} \tag{10-5.24}$$

and the corresponding absorption coefficient results when this replaces $Z/\rho c$ in the second version of Eq. (10-4.23).

For normal incidence and in the low-frequency limit, the absorption coefficient is

$$\alpha(0) = N\pi a^2\left(\frac{\rho\omega\gamma a^2}{\mu}\right)^{1/2} \tag{10-5.25}$$

and increases with ω as $\omega^{1/2}$ and with pore radius a, for fixed N, as a^3. However, the larger a is the thicker the slab must be to permit the assumption that reflections from the far ends of the pores have negligible effect. The analysis

† The modern theory of sound propagation in porous materials involves the porosity, the apparent compressibility of the fluid, the flow resistivity, and a structure factor, equal to the ratio of apparent to actual densities of the fluid in the pores: C. Zwikker and C. W. Kosten, *Sound Absorbing Materials,* Elsevier, Amsterdam, 1949; L. L. Beranek, *Acoustic Measurements,* Wiley, New York, 1949, pp. 844–860; P. M. Morse and K. U. Ingard, *Theoretical Acoustics,* McGraw-Hill, New York, 1968, pp. 252–255.

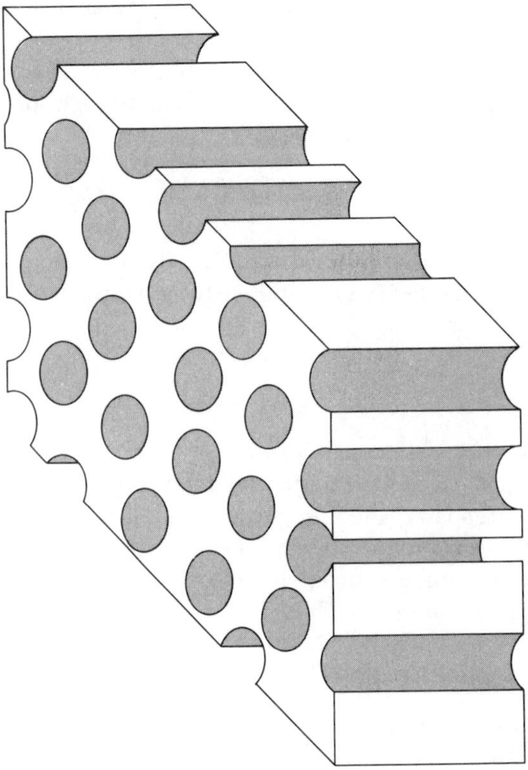

Figure 10-6 Rudimentary model of a porous material: a thick slab with many circular holes drilled perpendicular to the face.

here presumes that the thickness is somewhat larger than the reciprocal of the attenuation coefficient α in Eq. (22).

10-6 VISCOSITY EFFECTS ON SOUND RADIATION

The coupling of vorticity-mode and acoustic-mode fields at a surface affects the radiation of sound from that surface. To see how this is possible, we extend the analysis of sound generation, developed in Chap. 4, to include viscous effects.

Revision of the Kirchhoff-Helmholtz Theorem

A general result, expressing pressure external to a surface in terms of field quantities on the surface, can be derived in a manner similar to that described in Sec. 4-6. For simplicity, we ignore thermal conduction and take (from Sec. 10-3) the governing equations for a field of constant angular frequency $\omega = ck$ to be

$$\mathbf{v} = \mathbf{v}_{\mathrm{ac}} + \mathbf{v}_{\mathrm{vor}} \qquad \nabla \times \hat{\mathbf{v}}_{\mathrm{ac}} = 0 \qquad \nabla \cdot \hat{\mathbf{v}}_{\mathrm{vor}} = 0 \tag{10-6.1}$$

$$-i\omega\rho\hat{\mathbf{v}}_{\mathrm{vor}} = \mu\,\nabla^2\hat{\mathbf{v}}_{\mathrm{vor}} \qquad -i\omega\rho\hat{\mathbf{v}}_{\mathrm{ac}} = -\nabla\hat{p} \tag{10-6.2}$$

$$-i\omega\hat{p} + \rho c^2\,\nabla \cdot \hat{\mathbf{v}}_{\mathrm{ac}} = 0 \tag{10-6.3}$$

Here the far-field viscous attenuation of the acoustic-mode field is neglected.

From the above equations, it follows with some vector identities† that, for any function G,

$$\nabla \cdot (i\omega\rho G\hat{\mathbf{v}}_{ac} - \hat{p}\nabla G) = -\hat{p}(\nabla^2 + k^2)G \tag{10-6.4}$$

$$\nabla \cdot [i\omega\rho G\hat{\mathbf{v}}_{vor} - \mu(\nabla \times \hat{\mathbf{v}}_{vor}) \times \nabla G] = 0 \tag{10-6.5}$$

The sum of these two relations in turn implies

$$\nabla \cdot [i\omega\rho G\hat{\mathbf{v}} - \hat{p}\nabla G - \mu(\nabla \times \hat{\mathbf{v}}) \times \nabla G] = -\hat{p}(\nabla^2 + k^2)G \tag{10-6.6}$$

The derivation now proceeds as in Sec. 4-6 with the integration of Eq. (6) over the volume external to a closed surface S and with G taken as the free-space Green's function. The Kirchhoff-Helmholtz theorem of Eq. (4-6.6) is consequently replaced‡ by

$$\begin{aligned} p(\mathbf{x}, t) = {} & \frac{\rho}{4\pi}\iint \frac{\dot{v}_n(\mathbf{x}_S, t - R/c)}{R}\,dS \\ & + \frac{1}{4\pi c}\iint \mathbf{e}_R \cdot \mathbf{n}_S \left(\frac{\partial}{\partial t} + \frac{c}{R}\right)\frac{p(\mathbf{x}_S, t - R/c)}{R}\,dS \\ & - \frac{\mu}{4\pi c}\iint \mathbf{n}_S \cdot \left(\frac{\partial}{\partial t} + \frac{c}{R}\right)\frac{\mathbf{e}_R \times \mathbf{\Omega}(\mathbf{x}_S, t - R/c)}{R}\,dS \end{aligned} \tag{10-6.7}$$

where $\mathbf{\Omega} = \nabla \times \mathbf{v}$ is the *vorticity*. The assumptions adopted in the derivation are the same as in Sec. 4-6, except that here the existence of the vorticity mode is taken into account. It is required in addition that the vorticity-mode field vanish sufficiently rapidly at great distances from the source that the integral over the outer sphere can be discarded.

The multipole expansion of Eqs. (4-6.8) and (4-6.9) is similarly modified; retention of only the monopole and dipole terms yields

$$p = S\left(t - \frac{r}{c}\right) - \nabla \cdot \frac{\mathbf{D}(t - r/c)}{r} \tag{10-6.8}$$

with

$$S(t) = \frac{\rho}{4\pi}\iint \dot{\mathbf{v}} \cdot \mathbf{n}_S\,dS \tag{10-6.9a}$$

$$\mathbf{D}(t) = \frac{1}{4\pi}\iint (\rho\mathbf{x}_S\dot{\mathbf{v}} \cdot \mathbf{n}_S + \mathbf{n}_S p + \mu\mathbf{n}_S \times \mathbf{\Omega})\,dS \tag{10-6.9b}$$

† Note that (with $\mathbf{v}_{vor}$ replaced by $\mathbf{A}$)

$$\begin{aligned} \nabla \cdot [(\nabla \times \mathbf{A}) \times \nabla G] &= (\nabla G) \cdot [\nabla \times (\nabla \times \mathbf{A})] - (\nabla \times \mathbf{A}) \cdot (\nabla \times \nabla G) \\ &= (\nabla G) \cdot [\nabla(\nabla \cdot \mathbf{A}) - \nabla^2\mathbf{A}] = -(\nabla G) \cdot (\nabla^2\mathbf{A}) \qquad \text{if } \nabla \cdot \mathbf{A} = 0 \end{aligned}$$

‡ This is similar to, and can be regarded as a special case of, the fundamental aeroacoustic theorem derived and extended in the following papers: N. Curle, "The Influence of Solid Boundaries on Aerodynamic Sound," *Proc. R. Soc. Lond.*, **A286**:559–572 (1965); W. F. Möhring, E.-A. Müller, and F. F. Obermeier, "Sound Generation by Unsteady Flow as a Singular Perturbation Problem," *Acustica*, **21**:184–188 (1969); J. E. Ffowcs-Williams and D. L. Hawkings, "Sound Generation by Turbulence and Surfaces in Arbitrary Motion," *Phil. Trans. R. Soc. Lond.*, **A264**:321–342 (1969).

The distinction from the inviscid case is the term $\mu\mathbf{n}_S \times \mathbf{\Omega}$ in the integrand of (9b).

Transversely Oscillating Rigid Bodies

For a transversely oscillating rigid body, the quantity $S(t)$ is zero and $\dot{\mathbf{v}} = \dot{\mathbf{v}}_C$ is constant along the surface, so the operator ∇ can be regarded as having only an $\mathbf{n}_S$ component in the evaluation of $\mathbf{\Omega}$ at the surface. Consequently

$$\mathbf{n}_S \times (\nabla \times \mathbf{v}) = \mathbf{n}_S \times [\mathbf{n}_S \times (\mathbf{n}_S \cdot \nabla)\mathbf{v}] = -[(\mathbf{n}_S \cdot \nabla)\mathbf{v}]_T \qquad (10\text{-}6.10)$$

where the subscript T denotes the component tangential to the surface. Since the tangential derivative of any cartesian component of $\mathbf{v}$ is zero at the surface, one can rewrite this as

$$\mathbf{n}_S \times \mathbf{\Omega} = -\sum_{ij} \mathbf{n}_S \cdot \mathbf{e}_i \left(\frac{\partial v_i}{\partial x_j} + \frac{\partial v_j}{\partial x_i}\right) \mathbf{e}_{j,T} \approx -\sum_{ij} \mathbf{n}_S \cdot \mathbf{e}_i \phi_{ij} \mathbf{e}_j \qquad (10\text{-}6.11)$$

where ϕ_{ij} is the rate of shear tensor. [The indicated approximation makes negligible change in Eq. (9b), provided $|\hat{p}| \gg \mu|\nabla \cdot \hat{\mathbf{v}}|$. Moreover, it is con-

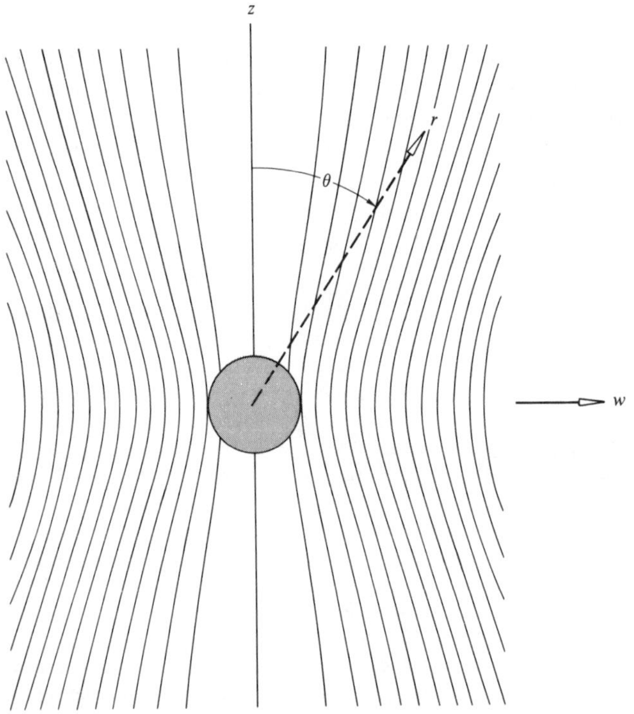

Figure 10-7 Streamlines about a transversely oscillating sphere in the Stokes' flow limit. Each streamline is a line along which $\sin^2\theta\,(3r/a - a/r)$ is constant. *(After H. Lamb, Hydrodynamics, 6th ed., Dover Publications, New York, 1945, p. 599.)*

sistent with the neglect of the viscous term in Eq. (2*b*).] Consequently, the second and third terms in Eq. (9*b*) combine to give

$$\mathbf{n}_S p + \mu \mathbf{n}_S \times \mathbf{\Omega} = -\sum_{ij} (\mathbf{n}_S \cdot \mathbf{e}_i)(-p\delta_{ij} + \mu\phi_{ij})\mathbf{e}_j = -\mathbf{f}_S(\mathbf{n}_S, \mathbf{x}_S) \qquad (10\text{-}6.12)$$

where $\mathbf{f}_S(\mathbf{n}_S, \mathbf{x}_S)$ is the force per unit area exerted on the surface by the external fluid.

With the substitution (12) and with the surface integral of $\mathbf{x}_S \dot{\mathbf{v}} \cdot \mathbf{n}_S$ replaced by $\dot{\mathbf{v}}_C \rho^{-1} m_d$, as in Eq. (4-7.11), the function $\mathbf{D}(t)$ appropriate to the dipole field reduces to

$$\mathbf{D}(t) = \frac{1}{4\pi} [m_d \dot{\mathbf{v}}_C(t) + \mathbf{F}(t)] \qquad (10\text{-}6.13)$$

where m_d is the displaced mass and where $\mathbf{F}(t)$ is the force exerted on the fluid by the body [opposite in sense to $\mathbf{f}_S(\mathbf{n}_S, \mathbf{x}_S)$]. This is exactly the same as results when viscosity is ignored; here, however, $\mathbf{F}(t)$ can include a force caused by shear stresses as well as a force caused by surface pressures.

Stokes Flow Limit

An example that can be analyzed in some detail† is that of a transversely oscillating sphere of radius a. If the oscillation is very slow, such that $(\omega\rho/\mu)^{1/2} a \ll 1$, then the force $\mathbf{F}(t)$ is the same as for low-Reynolds-number incompressible flow (see Fig. 10-7) past a sphere; the governing equations are what results when inertial terms and nonlinear terms are neglected in the Navier-Stokes equation. The solution, due to Stokes,‡ gives

$$\mathbf{F}(t) = 6\pi a \mu \mathbf{v}_C(t) \qquad (10\text{-}6.14)$$

In the same limit the inertial term in $\mathbf{D}(t)$ is negligible, so the far-field acoustic pressure in Eq. (8) reduces to

$$p = -\tfrac{3}{2} a \mu \nabla \cdot \left(\frac{\mathbf{v}_C(t - r/c)}{r} \right) \qquad (10\text{-}6.15)$$

Thin-Boundary-Layer Approximation

If the frequency is high enough (for the transversely oscillating sphere example just discussed) to ensure that $(\omega\rho/\mu)^{1/2} a \gg 1$, the boundary-layer model of Sec. 10-4 is applicable. The boundary condition perceived at the surface of the sphere (see Fig. 10-8) by the acoustic-mode field is identified from Eq. (10-4.12)

† Lamb, *Hydrodynamics*, 6th ed., pp. 654–657.

‡ G. G. Stokes, "On the Effect of the Internal Friction of Fluids on the Motion of Pendulums," *Trans. Camb. Phil. Soc.*, vol. 9 (1851), reprinted in *Mathematical and Physical Papers*, vol. 3, Johnson Reprint, New York, 1966, pp. 1–141; G. K. Batchelor, *An Introduction to Fluid Dynamics*, Cambridge University Press, London, 1967, pp. 230–234.

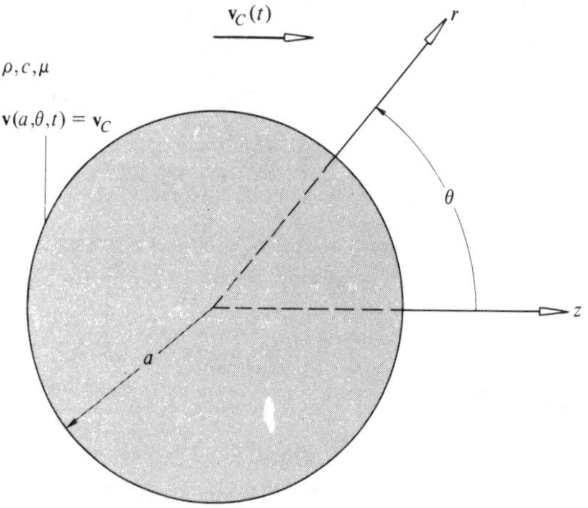

Figure 10-8 Boundary conditions and geometry for discussion of radiation from a transversely oscillating sphere in a viscous fluid.

as (at $r = a$)

$$\hat{v}_C \cos\theta = \hat{v}_{\mathrm{ac},r} - (1+i)\frac{l_{\mathrm{vor}}}{2}\frac{1}{a\sin\theta}\frac{\partial}{\partial\theta}(\hat{v}_{\mathrm{ac},\theta}\sin\theta) \tag{10-6.16}$$

The above boundary condition is satisfied if we take the solution of the Helmholtz equation in a form analogous to that adopted in Sec. 4-2:

$$\hat{p} = i\omega\rho\hat{v}_C a^3 B\cos\theta\,\frac{\partial}{\partial r}\frac{e^{ik(r-a)}}{r} \tag{10-6.17}$$

The second of Eqs. (2) then requires that

$$\hat{v}_{\mathrm{ac},r} = a^3\hat{v}_C B\left(\frac{\cos\theta}{r^3}\right)(2 - 2ikr - k^2r^2)e^{ik(r-a)} \tag{10-6.18a}$$

$$\hat{v}_{\mathrm{ac},\theta} = a^3\hat{v}_C B\left(\frac{\sin\theta}{r^2}\right)(1 - ikr)e^{ik(r-a)} \tag{10-6.18b}$$

The constant B is therefore identified from (16) as being such that

$$1 = (2 - 2ika - k^2a^2)B - (1+i)\frac{l_{\mathrm{vor}}}{a}(1 - ika)B \tag{10-6.19}$$

This implies that in the limit of small ka the effect of viscosity on the pressure amplitude is to multiply it by a factor

$$\frac{\hat{p}_{\mathrm{with}\,\mu}}{\hat{p}_{\mathrm{no}\,\mu}} = \frac{1}{1 - (1+i)(l_{\mathrm{vor}}/2a)} \tag{10-6.20}$$

For the thin-boundary-layer case, $(l_{\mathrm{vor}}/a) \ll 1$, the magnitude of the above factor is greater than 1, so viscosity increases the sound radiation, given that the amplitude of oscillation remains constant. This is consistent with the Stokes-flow-limit result (15), which predicts the amplitude to increase linearly with μ. An increase in viscosity increases the force that the fluid exerts on the

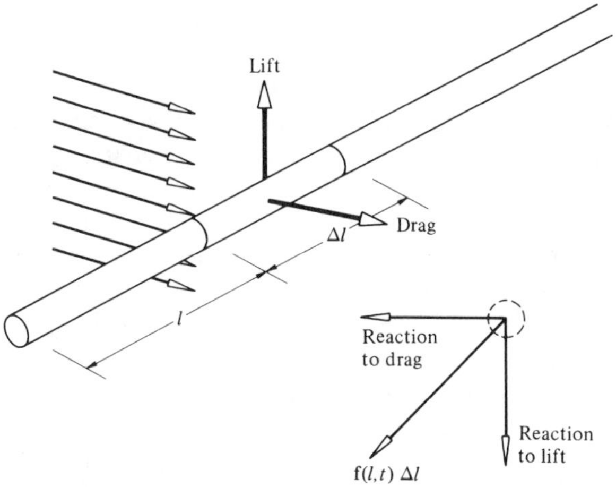

Figure 10-9 Concepts applicable to the generation of aeolian tones by flow past a cylinder. The acoustic field can be regarded as being caused by the fluctuating portions of the forces (reactions to lift and drag) exerted by the cylinder on the fluid.

oscillating sphere; the reaction to this force, equal and opposite, generates the sound; more force, more sound.

Gutin's Principle

A principle† implied by Eqs. (8) and (13) is that forces exerted on a surface generate sound regardless of how such forces originate. Thus, if a flow past a cylinder‡ (Fig. 10-9) generates sound (*aeolian tones*), one can regard it as being caused by the reactions to the fluctuations of the forces, e.g., lift and drag, exerted by the unsteady flow on the cylinder. Superposition of such forces yields

$$p(\mathbf{x}, t) = -\nabla \cdot \left\{ \frac{1}{4\pi} \int \left[\rho \pi a^2 \dot{\mathbf{v}}_C \left(l, t - \frac{R}{c} \right) + \mathbf{f} \left(l, t - \frac{R}{c} \right) \right] \frac{1}{R} \, dl \right\} \tag{10-6.21}$$

where

l = distance along cylinder

$R = |\mathbf{x} - \mathbf{x}_C(l)|$ = distance of listener from contributing element of cylinder

$\mathbf{f}(l, t)$ = force that element exerts per unit cylinder length on surrounding fluid

† L. Gutin, "On the Sound Field of a Rotating Airscrew," *Phys. Z. Sowjetunion*, **9**:57–71 (1936).

‡ P. Leehey and C. E. Hanson, "Aeolian Tones Associated with Resonant Vibration," *J. Sound Vib.*, **13**:465–483 (1970); O. M. Phillips, "The Intensity of Aeolian Tones," *J. Fluid Mech.*, **1**:607–624 (1956).

If the cylinder is constrained not to move, one is left with just the force contribution.

The principle just described reduces the problem of determining the sound field to the problem of determining the force. The latter, however, may be nearly independent of the compressibility of the fluid, such that its analysis can be guided by a model of incompressible flow. Even if the force is unsteady and random, similitude considerations can yield gross predictions. For example, aeolian tones of a nonmoving cylinder are usually of nearly constant frequency (for Reynolds number between 50 and 10^4). The frequency should be a function of the nominal steady-flow velocity U past the cylinder, of the fluid density ρ, of the viscosity μ, of the cylinder diameter $d = 2a$, and of nothing else. Dimensional considerations† then require that the *Strouhal number*

$$\mathrm{S} = \frac{fd}{U} \qquad (10\text{-}6.22)$$

depend only on the *Reynolds number* $\mathrm{Re} = U\rho d/\mu$. Experiments indicate that $\mathrm{S} = 0.13$ when $\mathrm{Re} = 50$, it increases to 0.2 at $\mathrm{Re} = 300$, and thereafter remains nearly constant up to $\mathrm{Re} \approx 10^4$. Thereafter the sound is not narrow-band, so the identification of a unique Strouhal number becomes difficult. The force is associated with the alternate shedding of oppositely rotating vortices from the top and the bottom of the cylinder. These vortices move downstream from the cylinder in an array called the *von Kármán vortex street.*

Helicopter Rotor Noise

The classical application of Gutin's principle is to sound radiation by a rotating helicopter rotor (see Fig. 10-10). The simplest model‡ considers the blades to be infinitesimally thin and the lift and drag forces (caused by viscosity) on the blades to be time-independent. The force exerted on the air, however, is fluctuating because the blades are rotating. Thus, when the helicopter is hovering, the force per unit area of rotor acting on the air due to blade n (defined such that its integral over an annular segment of area $w\,\Delta w\,\Delta\phi$ is the force on that segment) is

$$w^{-1}[-f_L(w)\mathbf{e}_z + f_D(w)\mathbf{e}_\phi]\,\delta^{(2\pi)}(\phi - \phi_n - \omega_R t) \qquad (10\text{-}6.23)$$

† J. W. S. Rayleigh, *The Theory of Sound,* vol. 2, 2d ed., 1896, reprinted by Dover, New York, 1945, pp. 412–414; V. Strouhal, "On a Special Type of Tone Excitation," *Ann. Phys.*, n.s., **5**:216–251 (1878); L. S. G. Kovásznay, "Hot-Wire Investigation of the Wake behind Cylinders at Low Reynolds Numbers," *Proc. R. Soc. Lond.*, **A198**: 174–190 (1949); T. von Kármán, "On the Resistance Mechanism, Which a Moving Body in a Fluid Experiences," *Nachr. K. Ges. Wiss. Goettingen, Math. Phys. Kl.*, **1912**:547–556 (1912).

‡ For later work and improved models, see I. E. Garrick and E. W. Watkins, "A Theoretical Study of the Effect of Forward Speed on the Free Space Sound Pressure Field around Helicopters," *NACA* TR1198, 1954; M. V. Lowson and J. B. Ollerhead, "A Theoretical Study of Helicopter Rotor Noise," *J. Sound Vib.*, **9**:197–222 (1969); J. W. Leverton and F. W. Taylor, "Helicopter Blade Slap," ibid., **4**:345–357 (1966); A. R. George, "Helicopter Noise: State of the Art," *J. Aircraft,* **15**:707–715 (1978).

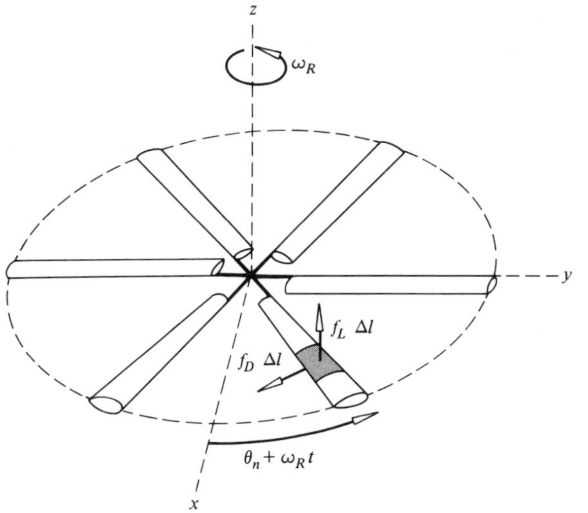

Figure 10-10 Geometry and parameters adopted for discussion of sound radiation by a helicopter rotor.

where $f_L(w)$ and $f_D(w)$ are the lift and drag forces per unit blade length at radial distance w from the hub. The function $\delta^{(2\pi)}(\phi)$ is defined so that it behaves like a delta function near wherever its argument is an integer multiple of 2π. Thus it is described formally by the Fourier series, as in Eq. (2-7.1),

$$\delta^{(2\pi)}(\phi) = \frac{1}{2\pi} \sum_{\nu=-\infty}^{\infty} e^{i\nu\phi} \tag{10-6.24}$$

With the argument taken as $\phi - \phi_n - \omega_n t$, the singularities occur at the angular position, $\phi_n + \omega_R t$, mod 2π, of the nth blade, where ω_R is the angular velocity of the rotor.

With the forces on the air as described above, the superposition principle, in conjunction with Eqs. (8) and (13), then leads to

$$p = -\frac{1}{4\pi} \nabla \cdot \left[\int_0^{2\pi} \int_0^L \mathbf{f}\left(l, \phi', t - \frac{R}{c}\right) \frac{1}{R} l\, dl\, d\phi' \right] \tag{10-6.25}$$

where

$$\mathbf{f}(w, \phi, t) = \sum_{n=1}^{N_B} w^{-1}[-f_L \mathbf{e}_z - (\sin \phi) f_D \mathbf{e}_x + (\cos \phi) f_D \mathbf{e}_y] \delta^{(2\pi)}(\phi - \phi_n - \omega_R t) \tag{10-6.26}$$

represents the force per unit area of rotor plane exerted on the fluid by the N_B blades. Here L is the length of a blade and R is distance from the integration point.

If the blades are symmetrically spaced, so that $\phi_n = 2\pi n/N_B$, the sum over n of the $\delta^{(2\pi)}(\phi - \phi_n - \omega_R t)$ becomes, from Eq. (24),

$$\frac{N_B}{2\pi} \sum_{\nu=-\infty}^{\infty} \exp\left[i\nu \left(\phi - \omega_R t + \frac{\omega_R}{c} R \right) \right] \frac{I_\nu}{N_B}$$

where
$$\frac{I_\nu}{N_B} = N_B^{-1} \sum_{n=1}^{N_B} e^{-i2\pi\nu n/N_B}$$
$$= \begin{cases} 1 & \nu = \text{integer multiple of } N_B \\ 0 & \text{otherwise} \end{cases}$$

Consequently, Eq. (25) reduces to

$$p = \sum_{m=0}^{\infty} \varepsilon_m \operatorname{Re} \hat{p}_m e^{-i\omega_m t} \tag{10-6.27}$$

$$\hat{p}_m = -\frac{1}{4\pi} \nabla \cdot \left[\int_0^{2\pi} \int_0^L \hat{\mathbf{f}}_m(l, \phi') R^{-1} e^{ik_m R}\, l\, dl\, d\phi' \right] \tag{10-6.28}$$

$$\hat{\mathbf{f}}_m(w, \phi) = \frac{N_B}{2\pi} \frac{1}{w} [-f_L \mathbf{e}_z - (\sin\phi) f_D \mathbf{e}_x + (\cos\phi) f_D \mathbf{e}_y] e^{imN_B\phi} \tag{10-6.29}$$

Here $\varepsilon_m = 1$ for $m = 0$ and $\varepsilon_m = 2$ for $m \geqslant 1$; the quantity $\omega_m = mN_B\omega_R$ is the m-th harmonic of the *blade-passage frequency*, while k_m is ω_m/c.

The far-field approximation results when R is replaced by r in the denominator and by $r - l \sin\theta \cos(\phi - \phi')$ in the exponential in Eq. (28). With subsequent discard of terms smaller than $1/r$ resulting after the divergence operation, one obtains

$$\hat{p}_m = -\frac{e^{ik_m r}}{4\pi r} ik_m \mathbf{e}_r \cdot \left[\int_0^{2\pi} \int_0^L \hat{\mathbf{f}}_m(l, \phi') e^{-ik_m l \sin\theta \cos(\phi - \phi')}\, l\, dl\, d\phi' \right] \tag{10-6.30}$$

where $\mathbf{e}_r$ is the unit vector in the radial direction. The insertion, into the above, of the expression in Eq. (29) subsequently yields

$$\hat{p}_m = \frac{i(-i)^N k_m N_B e^{ik_m r} e^{iN\phi}}{4\pi r} \left[\cos\theta \int_0^L f_L(l) J_N(k_m l \sin\theta)\, dl - \frac{N}{k_m} \int_0^L l^{-1} f_D(l) J_N(k_m l \sin\theta)\, dl \right] \tag{10-6.31}$$

Here we abbreviate $N_B m$ by N and recognize† that

$$\frac{i^N}{2\pi} \int_0^{2\pi} e^{iN\phi'} e^{-iX\cos\phi'}\, d\phi' = J_N(X) \tag{10-6.32}$$

is the Bessel function of Nth order.

† G. N. Watson, *A Treatise on the Theory of Bessel Functions*, 2d ed., Cambridge University Press, London, 1944, pp. 17, 20. Watson's eq. (3), p. 20, leads to

$$J_N(X) = \frac{1}{2\pi} \int_{\pi/2}^{5\pi/2} e^{i(N\theta - X\sin\theta)}\, d\theta$$

which, with θ replaced by $\phi' + \pi/2$, yields Eq. (32) above. Note also that (32) implies, with $\sin\phi'$ replaced by $(e^{i\phi'} - e^{-i\phi'})/2i$,

$$\frac{i^N}{2\pi} \int_0^{2\pi} (\sin\phi') e^{iN\phi'} e^{-iX\cos\phi'}\, d\phi' = \frac{1}{2i} \left[\frac{1}{i} J_{N+1}(X) - iJ_{N-1}(X) \right] = -\frac{1}{2} \frac{2N}{X} J_N(X)$$

where the second equality follows from Watson's eq. (1), p. 17.

where s_{fr} is the entropy that would result were the vibrational degrees of freedom frozen and s_ν is the entropy associated with the internal vibrations of the νth species of molecules. The former satisfies

$$T\,ds_{\text{fr}} = d(u_{\text{tr}} + u_{\text{rot}}) + p\,d\rho^{-1}$$

which, with $u_{\text{tr}} + u_{\text{rot}} = c_{v,\text{fr}}T$ and $p = \rho RT$, integrates to

$$s_{\text{fr}} = c_{v,\text{fr}} \ln\,(u - \sum_\nu u_\nu) + R \ln \rho^{-1} + \text{const} \tag{10-7.13}$$

Here $c_{v,\text{fr}}$, identified as $R/(\gamma - 1)$ from Eq. (5), is the coefficient of specific heat at constant volume when the vibrational degrees of freedom are frozen. The s_ν are such that $T_\nu\,ds_\nu = du_\nu$, so Eq. (4), with $T_\nu^* \gg T_\nu$, yields $s_\nu \approx u_\nu/T_\nu$. The affinity A_ν is consequently identified, from Eqs. (11) to (13), as

$$A_\nu = T\left(\frac{-c_{v,\text{fr}}}{u_{\text{tr}} + u_{\text{rot}}}\frac{du_\nu}{dT_\nu} + \frac{1}{T_\nu}\frac{du_\nu}{dT_\nu}\right) = \left(\frac{T}{T_\nu} - 1\right)c_{v\nu} \tag{10-7.14}$$

where

$$c_{v\nu} = \frac{du_\nu}{dT_\nu} = \frac{n_\nu}{n} R \left(\frac{T_\nu^*}{T_\nu}\right)^2 e^{-T_\nu^*/T_\nu} \tag{10-7.15}$$

is the specific heat associated with the internal vibrations of the ν-type molecules.

Fluid-Dynamic Equations with Relaxation Included†

In regard to the energy equation (10-1.12), Eqs. (10) and (11) allow us to write

$$\frac{Du}{Dt} - \sigma_n \frac{D\rho^{-1}}{Dt} = T\frac{Ds}{Dt} - \sum_\nu A_\nu \frac{DT_\nu}{Dt} - \mu_B \nabla \cdot \mathbf{v} \frac{D\rho^{-1}}{Dt} \tag{10-7.16}$$

Since $D\rho^{-1}/Dt$ is $\rho^{-1}\nabla \cdot \mathbf{v}$ and since $\nabla \cdot \mathbf{q}$ is $T\ \nabla \cdot (\mathbf{q}/T) + (\mathbf{q}/T) \cdot \nabla T$, the above transforms Eq. (10-1.12) into the *entropy-balance equation*

$$\rho \frac{Ds}{Dt} + \nabla \cdot \frac{\mathbf{q}}{T} = \sigma_s \tag{10-7.17}$$

where

$$T\sigma_s = \mu_B(\nabla \cdot \mathbf{v})^2 + \tfrac{1}{2}\mu \sum_{ij} \phi_{ij}^2 + \frac{\kappa}{T}(\nabla T)^2 + \rho \sum_\nu A_\nu \frac{DT_\nu}{Dt} \tag{10-7.18}$$

Similarly, the Navier-Stokes equation (10-1.14), with the introduction of the bulk viscosity, becomes

$$\rho \frac{D\mathbf{v}}{Dt} = -\nabla p + \nabla(\mu_B \nabla \cdot \mathbf{v}) + \mu \sum_{ij} \mathbf{e}_i \frac{\partial \phi_{ij}}{\partial x_j} \tag{10-7.19}$$

† Equations (16) to (19) apply to other fluids (including seawater) if the T_ν are replaced by appropriate "internal variables" n_ν. For freshwater, no internal variables are needed if a bulk viscosity is included in the formulation.

An alternate version of Eq. (17) resulting with the substitution,† $s_{fr} + \Sigma s_\nu$ for s, is

$$\rho \frac{Ds_{fr}}{Dt} + \sum_\nu \frac{\rho}{T_\nu} c_{v\nu} \frac{DT_\nu}{Dt} - \nabla \cdot \left(\frac{\kappa}{T} \nabla T\right) = \sigma_s \tag{10-7.20}$$

Note that if s_{fr} is regarded as a function of any two of the variables p, ρ, or T, then it is independent of the T_ν. The thermodynamic identities relating s_{fr}, p, T, and ρ are the same as when there are no molecular vibrations.

The Relaxation Equations

The fluid-dynamic model must be supplemented by one additional equation for each vibrational temperature T_ν included as a thermodynamic variable. (For air, detailed experiments and calculations‡ based on molecular kinetics suggest it is

† For liquids, an appropriate decomposition is

$$s(u, \rho^{-1}, n_\nu) \approx s_{eq}(p, \rho^{-1}) + \Delta s$$

where $s_{eq}(p, \rho^{-1})$ is the equilibrium value that corresponds to the local instantaneous value of p [as defined by Eq. (11) with T_ν replaced by n_ν]; the quantity Δs is of first order in the A_ν. For seawater, where the relaxation processes are chemical, a simplified model takes $\Delta s = \Delta s_1 + \Delta s_2$, with

$$\Delta s_\nu = \frac{c_p(\Delta K_T^{-1})_\nu}{\beta T} \Delta\xi_\nu \qquad \rho A_\nu \frac{Dn_\nu}{Dt} = (\Delta K_T^{-1})_\nu \frac{(\Delta\xi_\nu)^2}{\tau_\nu} \qquad \Delta\xi_\nu = \frac{n_\nu - n_\nu^e(p, T)}{\partial n_\nu^e(p, T)/\partial p}$$

where the $\Delta\xi_\nu$ satisfy the relaxation equations

$$\left(\frac{D}{Dt} + \frac{1}{\tau_\nu}\right) \Delta\xi_\nu = -\frac{Dp}{Dt}$$

The quantity n_1 is the number of dissolved $B(OH)_3$ (boric acid) molecules per unit mass of water as a whole that are in the fully associated state (rather than being broken into two spatially separated ions); n_2 is the analogous number of dissolved $MgSO_4$ (magnesium sulfate) molecules; the superscript e denotes the equilibrium value. In the above relations, β is the coefficient of volume expansion; $(\Delta K_T^{-1})_\nu$ is the contribution of the dissolved molecules of species ν to the isothermal compressibility (reciprocal of bulk modulus). The theory that a pressure-dependent chemical reaction can cause absorption and dispersion of sound is due to L. N. Liebermann, "Sound Propagation in Chemically Active Media," *Phys. Rev.*, **76**:1520–1524 (1949) and was further developed by M. Eigen and K. Tamm, "Sound Absorption in Electrolyte Solutions as a Sequence of Chemical Reactions," *Z. Elektrochem.*, **66**:93–121 (1962). The identifications of $MgSO_4$ and $B(OH)_3$ as the principal contributors to relaxation processes in seawater are due to O. B. Wilson, Jr., and R. W. Leonard, "Sound Absorption in Aqueous Solutions of Magnesium Sulfate and in Sea Water," *J. Acoust. Soc. Am.*, **23**:624A (1951) and to E. Yeager, F. Fisher, J. Miceli, and R. Bressel, "Origin of the Low-Frequency Sound Absorption in Sea Water," ibid., **53**:1705–1707 (1973).

‡ J. E. Piercy, "Noise Propagation in the Open Atmosphere," pap. presented at *84th Meet. Acoust. Soc. Am., Miami Beach, Fl., November 1972;* and L. C. Sutherland, J. E. Piercy, H. E. Bass, and L. B. Evans, "A Method for Calculating the Absorption of Sound in the Atmosphere," pap. presented at *88th Meet., St. Louis, Mo., November 1974,* rev. November 1975. The analysis in these papers constitutes part of the background for the absorption calculation procedure in ANSI Standard S1.26/ASA23-1978, American National Standard Method for the Calculation of the Absorption of Sound by the Atmosphere. More extensive models are described by L. B.

The relaxation times τ_1 and τ_2 are sensitive to the fraction h of air molecules that are H_2O molecules; an O_2 molecule or an N_2 molecule colliding with a H_2O molecule is much more likely to experience a change in vibrational energy than when colliding with another O_2 or N_2 molecule. Experimental data and calculations carried out for a CO_2 fraction of 3.1×10^{-5} (representative of normal air) yield the semiempirical formulas† (see Fig. 10-11)

$$\frac{p_{\text{ref}}}{p}\frac{1}{2\pi\tau_1} = 24 + 4.41 \times 10^6\, h\, \frac{0.05 + 100h}{0.391 + 100h} \tag{10-7.24a}$$

$$\frac{p_{\text{ref}}}{p}\frac{1}{2\pi\tau_2} = \left(\frac{T_{\text{ref}}}{T}\right)^{1/2} (9 + 3.5 \times 10^4\, h\, e^{-F}) \tag{10-7.24b}$$

$$F = 6.142\left[\left(\frac{T_{\text{ref}}}{T}\right)^{1/3} - 1\right] \tag{10-7.24c}$$

where $p_{\text{ref}} = 1.013 \times 10^5$ Pa and $T_{\text{ref}} = 293.16$ K. (These equations should be accurate to within 10 percent between 0 and 40°C.) The relative humidity RH (expressed as percentage) is defined such that

$$h = \frac{10^{-2}(\text{RH})p_{\text{vp}}(T)}{p} \tag{10.7.25}$$

where $p_{\text{vp}}(T)$ is the vapor pressure of water at temperature T. Representative values of $p_{\text{vp}}(T)$ are

T,°C	5	10	15	20	30	40
$p_{\text{vp}}(T)$, Pa	872	1228	1705	2338	4243	7376

10-8 ABSORPTION OF SOUND

Linear Acoustic Equations for Air

For a homogeneous quiescent medium, the equations developed in the previous section, with the neglect of nonlinear terms, yield‡

$$\frac{\partial \rho'}{\partial t} + \rho_0 \nabla \cdot \mathbf{v} = 0 \tag{10-8.1a}$$

$$\rho_0 \frac{\partial \mathbf{v}}{\partial t} = -\nabla p + \mu_B \nabla(\nabla \cdot \mathbf{v}) + \mu \sum_{ij} \mathbf{e}_i \frac{\partial \phi_{ij}}{\partial x_j} \tag{10-8.1b}$$

† The first relation is due to J. E. Piercy, "Comparison of Standard Methods of Calculating the Attenuation of Sound in Air with Laboratory Measurements," presented orally to *82nd Meet. Acoust. Soc. Am., Denver, Co., October 1971;* the second is due to Sutherland, Piercy, Bass, and Evans. "A Method for Calculating the Absorption of Sound" and is based in major part on experimental data of C. M. Harris and W. Tempest (1965).

‡ For seawater, the corresponding versions of (1*c*) and (*d*), resulting from the relations on p. 552*n*., are

$$\rho_0 \frac{\partial s_{\text{fr}}}{\partial t} + \sum_{\nu} \left(\frac{\rho}{T}\right)_0 c_{v\nu} \frac{\partial T_\nu}{\partial t} - \frac{\kappa}{T_0} \nabla^2 T' = 0 \tag{10-8.1c}$$

$$\frac{\partial T_\nu}{\partial t} = \frac{1}{\tau_\nu}(T' - T_\nu) \tag{10-8.1d}$$

$$\rho' = \frac{1}{c^2} p - \left(\frac{\rho \beta T}{c_p}\right)_0 s_{\text{fr}} \tag{10-8.1e}$$

$$T' = \left(\frac{T\beta}{\rho c_p}\right)_0 p + \left(\frac{T}{c_p}\right)_0 s_{\text{fr}} \tag{10-8.1f}$$

$$\phi_{ij} = \frac{\partial v_i}{\partial x_j} + \frac{\partial v_j}{\partial x_i} - \tfrac{2}{3}\nabla \cdot \mathbf{v}\, \delta_{ij} \tag{10-8.1g}$$

Here Eq. ($1a$) is a restatement of the linearized version of the conservation-of-mass equation; Eqs. ($1b$) to ($1d$) are the linearized versions of Eqs. (10-7.19) to (10-7.21); Eqs. ($1e$) to ($1g$) are restatements of Eqs. (10-2.1a), (10-2.1b), and (10-1.10). The primes on v', p', s'_{fr}, and T'_ν have been deleted, so s_{fr}, for example, here represents the deviation from its ambient value of the entropy for the gas when molecular vibrations are frozen. The thermodynamic coefficients in Eqs. ($1e$) and ($1f$) are those appropriate to such a frozen state, although the deviations from the values appropriate to a gas in thermodynamic equilibrium are slight. For a gas, β is $1/T_0$, c_p is $\gamma R/(\gamma - 1)$, c^2 is $\gamma R T_0$, and γ is (dof + 2)/dof.

Energy Corollary

An energy-conservation-dissipation theorem,

$$\frac{\partial w}{\partial t} + \nabla \cdot \mathbf{I} = -\mathscr{D} \tag{10-8.2}$$

also holds for the model represented by Eqs. (1). A derivation similar to that described in Sec. 10-2 leads to the identifications

$$w = \tfrac{1}{2}\rho_0 v^2 + \frac{1}{2}\frac{p^2}{\rho_0 c^2} + \frac{1}{2}\left(\frac{\rho T}{c_p}\right)_0 s_{\text{fr}}^2 + \sum_{\nu} \frac{1}{2}\left(\frac{\rho c_{v\nu}}{T}\right)_0 T_\nu^2 \tag{10-8.3a}$$

$$\mathbf{I} = p\mathbf{v} - \mu_B \mathbf{v}(\nabla \cdot \mathbf{v}) - \mu \sum_{ij} v_i \phi_{ij} \mathbf{e}_j - \kappa T_0^{-1} T' \,\nabla T' \tag{10-8.3b}$$

$$\mathscr{D} = \mu_B (\nabla \cdot \mathbf{v})^2 + \tfrac{1}{2}\mu \sum_{ij} \phi_{ij}^2 + \kappa T_0^{-1} (\nabla T)^2 + \sum_{\nu} \left(\frac{\rho_0 c_{v\nu}}{T \tau_\nu}\right)_0 (T' - T_\nu)^2 \tag{10-8.3c}$$

$$\rho_0 \frac{\partial s'_{\text{eq}}}{\partial t} + \sum_{\nu} \left[\frac{\rho c_p (\Delta K_T^{-1})_\nu}{\beta T}\right]_0 \frac{\partial(\Delta \xi_\nu)}{\partial t} = \frac{\kappa}{T_0} \nabla^2 T' \tag{i}$$

$$\left(\frac{\partial}{\partial t} + \frac{1}{\tau_\nu}\right) \Delta \xi_\nu = -\frac{\partial p}{\partial t} \tag{ii}$$

Equations ($1e$) and ($1f$) remain unchanged except that s_{fr} should be replaced by s'_{eq}. Whether the thermodynamic coefficients are evaluated at the equilibrium state or the frozen state makes little quantitative difference in the predictions.

$d(\ln \alpha)/d(\ln f)$, equal to 2 over the frequency intervals of $0 < f < f_2/2$, $2f_2 < f < f_1/2$, and $2f_1 < f$. As one moves upward in frequency, the successive line segments will be displaced downward, although α will increase monotonically with frequency f. From the highest-frequency segment, one identifies the coefficient α'_{cl}/f^2. Then a plot of $(\alpha - \alpha'_{\text{cl}})\lambda$ versus f should have a peak value of $(\alpha_1\lambda)_m$ at a frequency f_1. [This presumes that $(\alpha_2\lambda)_m f_2$ is substantially less than $(\alpha_1\lambda)_m f_1$.] Then, to determine $(\alpha_2\lambda)_m$ and f_2 one plots $(\alpha - \alpha'_{\text{cl}} - \alpha_1)\lambda$ versus f, where α_1 is taken from Eq. (12).

In the low-frequency limit, the absorption due to a relaxation process is indistinguishable from that due to an additional increment

$$\Delta\mu_B \doteq \frac{2\rho_0 c^2 \tau_\nu}{\pi} (\alpha_\nu \lambda)_m \tag{10-8.15}$$

being added to the bulk viscosity.† Consequently, the apparent bulk viscosity within a given frequency range is composed of contributions from all relaxation processes whose relaxation frequencies are higher than the upper limit of that frequency range. In the model described in the preceding section for air, μ_B was ascribed to rotational relaxation; since the rotational relaxation frequency is much higher than any acoustical frequency of interest, the inclusion of this process with the bulk viscosity is appropriate.

Phase-Velocity Changes due to Relaxation Processes

A fundamental property of a relaxation process is that different frequencies propagate with different phase velocities, so the propagation is dispersive. Taking the real part k_R of the k in Eq. (9b) and neglecting second-order terms in $(k_R - \omega/c)/(\omega/c)$, we find

$$\frac{\omega}{k_R} = v_{\text{ph}} = c_0 + \frac{c}{\pi} \sum_\nu \frac{(\alpha_\nu \lambda)_m \omega^2 \tau_\nu^2}{1 + \omega^2 \tau_\nu^2} \tag{10-8.16}$$

$$= c - \frac{c}{\pi} \sum_\nu \frac{(\alpha_\nu \lambda)_m}{1 + \omega^2 \tau_\nu^2} \tag{10-8.16a}$$

where c_0 and c, as noted previously, are the low- and high-frequency limits of the phase velocity. Thus, with increasing frequency, the phase velocity increases monotonically from c_0 to c. Over any frequency decade centered at an isolated relaxation frequency, the phase velocity increases (see Fig. 10-12) by an increment Δc_ν equal to $(c/\pi)(\alpha_\nu\lambda)_m$. In air at 20°C, the corresponding increments are 0.11 and 0.023 m/s for the O_2 and N_2 vibrational relaxation processes, respectively.

For gases, the two limiting sound speeds are associated with the values γ_0 and γ_{fr} of the specific-heat ratio γ appropriate to the equilibrium and frozen

† L. Tisza, "Supersonic Absorption and Stokes's Viscosity Relation," *Phys. Rev.*, **61**:531–536 (1942); J. Meixner, "Flows of Fluid Media with Internal Transformations and Bulk Viscosity," *Z. Phys.*, **131**:456–469 (1952).

states, given by

$$\gamma_0 = \frac{c_{p,\text{fr}} + \Sigma c_{v\nu}}{c_{v,\text{fr}} + \Sigma c_{v\nu}} \qquad \gamma_{\text{fr}} = \frac{c_{p,\text{fr}}}{c_{v,\text{fr}}} \tag{10-8.17}$$

where $c_{p,\text{fr}}$ and $c_{v,\text{fr}}$ are the specific heats that result when the molecular vibrations are frozen. The corresponding sound speeds, c_0 and c_{fr}, are $(\gamma_0 RT)^{1/2}$ and $(\gamma_{\text{fr}} RT)^{1/2}$. [The latter is what is denoted by c in Eqs. (9) and (10).] From this point of view,† the phase velocity in Eqs. (16) can be regarded as the sound speed in a gas whose apparent specific-heat ratio increases monotonically from γ_0 to γ_{fr} as ω ranges from 0 to ∞.

PROBLEMS

10-1 Suppose Eqs. (10-1.16) with particular choices of μ_0, κ_0, and T_0 yield values of μ_0' and κ_0' at temperature T_0'. Prove that the predicted values of μ and κ at any third temperature T are unchanged when μ_0, κ_0, and T_0 are replaced by μ_0', κ_0', and T_0'.

10-2 What fractional error (order of magnitude) would result in the plane-wave attenuation coefficient of water if thermal conduction is neglected at the outset?

10-3 Show that the components of the viscous portion of the stress tensor are given in spherical coordinates by

$$\sigma_{rr} = 2\mu\left(\frac{\partial v_r}{\partial r} - \tfrac{1}{3}\nabla\cdot\mathbf{v}\right) \qquad \sigma_{\theta\theta} = 2\mu\left(\frac{1}{r}\frac{\partial v_\theta}{\partial \theta} + \frac{v_r}{r} - \tfrac{1}{3}\nabla\cdot\mathbf{v}\right)$$

$$\sigma_{\phi\phi} = 2\mu\left(\frac{1}{r\sin\theta}\frac{\partial v_\phi}{\partial\phi} + \frac{v_r}{r} + \frac{v_\theta\cot\theta}{r} - \tfrac{1}{3}\nabla\cdot\mathbf{v}\right)$$

$$\sigma_{r\theta} = \mu\left(r\frac{\partial}{\partial r}\frac{v_\theta}{r} + \frac{1}{r}\frac{\partial v_r}{\partial\theta}\right) \qquad \sigma_{r\phi} = \mu\left(\frac{1}{r\sin\theta}\frac{\partial v_r}{\partial\phi} + r\frac{\partial}{\partial r}\frac{v_\phi}{r}\right)$$

$$\sigma_{\phi\theta} = \mu\left(\frac{\sin\theta}{r}\frac{\partial}{\partial\theta}\frac{v_\phi}{\sin\theta} + \frac{1}{r\sin\theta}\frac{\partial v_\theta}{\partial\phi}\right)$$

10-4 A model for explaining sonically induced rises in ambient temperature takes the ambient temperature to satisfy

$$\rho c_p \frac{\partial T}{\partial t} - \nabla\cdot(\kappa\nabla T) = \mathcal{D}_{\text{ac}}$$

where $\mathcal{D}_{\text{ac}}$ is the acoustic energy dissipated per unit time and volume. Discuss a possible rationale for this equation, starting from the Navier-Stokes-Kirchhoff fluid dynamic model of Sec. 10-1. Suppose plane waves are at normal incidence from a first medium, with negligible attenuation and thermal conductivity, onto a second medium, within which the attenuation is α Np/m and the thermal conductivity is κ. If the intensity of the incident wave is I (time-averaged), and if ρc for both media are the same, show that the temperature in medium 2 tends asymptotically to increase with time as $t^{1/2}$. (Assume $\alpha\lambda \ll 1$ and that the temperature measurement is at a point many wavelengths from the interface.)

10-5 (*a*) Show that the acoustic pressure in a plane wave propagating in the $+x$ direction through a medium for which Eqs. (10-2.1) and (10-2.2) are applicable approximately satisfied either

$$\frac{\partial p}{\partial x} + \frac{1}{c}\frac{\partial p}{\partial t} - \frac{\delta}{c^3}\frac{\partial^2 p}{\partial t^2} = 0 \quad \text{or} \quad \frac{\partial p}{\partial x} + \frac{1}{c}\frac{\partial p}{\partial t} - \frac{\delta}{c}\frac{\partial^2 p}{\partial x^2} = 0$$

† H. O. Kneser, "The Dispersion Theory of Sound," *Ann. Phys.*, (5) **20**:761–776(1931); P. S. H. Henry, "The Energy Exchanges between Molecules," *Proc. Camb. Phil. Soc.*, **28**:249–255(1932).

(*b*) Hence show that if p is given by $f(x)$ at $t = 0$, then

$$p(x,t) = \int_{-\infty}^{\infty} G(x - ct - \xi,\ 4t\delta_{\text{cl}})f(\xi)d\xi$$

where
$$G(x, y^2) = (\pi y^2)^{-1/2}e^{-(x/y)^2}$$

(*c*) Suppose that $f(x)$ is $P\sin(\pi x/L)$ for x between $-L$ and L and is 0 for other values of x. Take L to be $10\delta_{\text{cl}}/c$. Determine the pulse's waveform versus x at a time such that $(4t\delta_{\text{cl}})^{1/2}$ is $3L$. Make any approximations that seem appropriate and if necessary evaluate the integral numerically. Sketch your result.

10-6 The superposition principle requires that the energy-conservation-dissipation theorem, Eq. (10-2.4), hold for the acoustic-, vorticity-, and entropy-mode fields separately. Show that this is so and give the appropriate expressions for w, $\mathbf{I}$, and $\mathcal{D}$ for each of the mode fields in as simple a form as possible that is consistent with the approximations entailed in the tabulations in Sec. 10-3.

10-7 A large flat immovable surface of a solid with high thermal conductivity is adjacent to a fluid with thermal conductivity κ, sound speed c, ambient density ρ, specific-heat ratio γ, coefficient of volume expansion β, and coefficient c_p of specific heat at constant pressure. The surface temperature of the solid is made to oscillate about ambient temperature T_0 with a deviation $(\Delta T)_S \cos \omega t$. Determine the resulting acoustic disturbance within the fluid at large distances from the surface to lowest nonvanishing order in κ and $(\Delta T)_S$.

10-8 A plane wave of constant frequency is incident on a rigid immovable sphere for which $ka \ll 1$ but $a/l_{\text{vor}} \gg 1$. Estimate, to lowest nonvanishing order in μ and ka, how much energy is dissipated per unit time in the viscous boundary layer when the incident wave's time-averaged intensity is I. (The ratio of these two quantities is the *absorption cross section*.)

10-9 Estimate the attenuation in nepers per meter of the higher modes in a rectangular duct. Assume that the attenuation is due solely to thermal- and viscous-energy dissipation within the boundary layers at the duct walls. Use Eq. (10-5.2) as a starting point and replace $\nabla\hat{p}\cdot\mathbf{n}_{\text{wall}}$ by $i\omega\rho\hat{\mathbf{v}}_{\text{ac}}\cdot\mathbf{n}_{\text{wall}}$, where the latter is identified from Eq. (10-4.12). Take

$$\hat{p} = \hat{\psi}(x)\cos\frac{n_y\pi y}{L_y}\cos\frac{n_z\pi z}{L_z}$$

and derive from the variational principle a differential equation for $\hat{\psi}(x)$ whose solution of the form e^{ikx} determines $\alpha = \operatorname{Im} k$.

10-10 A Helmholtz resonator resembling a bottle with a long neck has a resonance frequency of 250 Hz and a neck 5 cm long with a 1-cm inner diameter. Use one of the models discussed in Sec. 10-5 for sound waves in tubes to estimate the resistive part of the acoustic impedance of the resonator. Assuming that the resonator is in air at 27°C, determine which is dominant: loss of energy through viscous friction or through radiation out the mouth. What is the Q of the resonator?

10-11 How should the absorption coefficient in Eq. (10-5.25) for reflection from a thick slab with cylindrical holes be modified when the angle of incidence θ_i is not zero?

10-12 Modify the model leading to Eq. (10-5.25) to account for the finite thickness h of the slab. Determine an expression for the transmission loss of the slab.

10-13 (*a*) If a flat rigid surface of extensive area is oscillating tangential to itself with displacement $\operatorname{Re}\hat{\xi}e^{-i\omega t}$, show that the force exerted on the adjacent fluid per unit area of surface is

$$\mathbf{f} = -\frac{\omega\mu}{l_{\text{vor}}}\operatorname{Re}[(1+i)\hat{\xi}e^{-i\omega t}]$$

where $l_{\text{vor}} = (2\mu/\omega\rho)^{1/2}$.

(*b*) Estimate the force exerted on the adjoining fluid by a thin circular disk of radius a that is oscillating in such a manner in the limit $a \gg l_{\text{vor}}$. (Assume a laminar boundary layer.)

(*c*) Given that $ka \ll 1$, what would be the far-field acoustic pressure and the time-averaged radiated acoustic power for the circumstances of (*b*)?

10-14 Given that the force amplitude on a cylinder immersed in a nominally steady flow is nearly independent of viscosity over a wide range of Reynolds number, how should the radiated acoustic power associated with the aeolian tone vary with the nominal velocity of the flow past the cylinder?

10-15 A steady flow past an obstacle of characteristic dimension a causes a radiation of sound. Assume that Gutin's principle applies and that the frequencies of interest are such that $ka \ll 1$. If the Reynolds number of the incoming flow is held constant, how would you expect the radiated acoustic power to vary with the velocity (much slower than the sound speed) of the flow? Devise a similitude theory for the sound radiation that expresses the spectral density of the far-field acoustic pressure in terms of dimensionless parameters, including the Strouhal number fa/U and the Reynolds number $U\rho a/\mu$.

10-16 The $m = 0$ term in the far-field acoustic pressure of a helicopter rotor is responsible for the transmission of the helicopter weight W to the ground. A helicopter's rotor has six blades, each 6 m long, and is rotating at 200 r/min; it slowly flies at 100 m altitude over the ground. If the helicopter mass is 3000 kg, what would you estimate as the nonoscillating part of the pressure increment at the ground caused by the helicopter's passage?

10-17 Some additional simplification in the helicopter-noise model discussed in Sec. 10-6 results when $f_L(l)$ and $f_D(l)$ are assumed to be concentrated at radial distance L_{eff} such that $f_L(l) = (W/N_B)\delta(l - L_{\text{eff}})$, $f_D(l) = f_L(l)/R_{L/D}$, where $R_{L/D} \approx 0.2$ is the lift-to-drag-ratio. For such circumstances determine and sketch the radiation patterns for the fundamental and first two harmonics ($m = 1, 2, 3$) when $(N_B\omega_R/c)L_{\text{eff}}$ is 1.0. (Approximate the Bessel functions by the leading term in their power-series expansions.)

10-18 A Bessel function of large order is well approximated over the range of arguments where the function has its largest values by the expression (M. Abramowitz and I. Stegun (eds.), *Handbook of Mathematical Functions,* Dover, New York, 1965, p. 367)

$$J_N(z) \approx \left(\frac{2}{N}\right)^{1/3} \text{Ai}\left[-\left(\frac{2}{N}\right)^{1/3}(z - N)\right]$$

where $\text{Ai}(\eta)$ is the Airy function. Using this approximation and the properties of the Airy function, discuss the radiation of helicopter noise by the higher harmonics of the blade-passage frequency for the circumstances described in Prob. 10-17 but with $N_B(\omega_R/c)L_{\text{eff}}$ not fixed. For what threshold value of the latter parameter does the radiated power begin to rise abruptly? Within what range of angle θ does the sound appear to be concentrated when $N_B(\omega_R/c)L_{\text{eff}}$ has a specified value that exceeds this threshold?

10-19 Use the energy-conservation-dissipation theorem, Eq. (10-8.2), to derive the absorption coefficient for constant-frequency plane-wave sound propagation in air, following a procedure analogous to that in the derivation of Eq. (10-2.11).

10-20 Carry through the steps leading to Eq. (10-8.9*b*) for a gas with μ, μ_B, and κ set to zero and with only one relaxation process. Show that the resulting dispersion relation can be written

$$k = \frac{\omega}{c_0} + \omega(c_0^{-1} - c_\infty^{-1})\frac{i\omega\tau_\nu}{1 - i\omega\tau_\nu}$$

where c_∞ is the sound speed in the high-frequency limit. If $p(x, t)$ describes a transient plane wave propagating in the $+x$ direction, each Fourier component $\hat{p}e^{-i\omega t}e^{ikx}$ of which satisfies this dispersion relation, what partial-differential equation would be appropriate for $p(x, t)$?

10-21 The Sabine-Franklin reverberation-time formula (6-1.12) can be modified to take into account absorption within the interior of a room if one assumes the field is a superposition of a large number of plane waves, each of which is attenuated by α_{pl} Np per unit propagation distance.

(*a*) What is the resulting modified version of the formula for T_{60}?

(*b*) If the room has dimensions 8 by 8 by 4 m and a nominal reverberation time of 6 s, what must α_{pl} be to cause a 10 percent reduction in the reverberation time?

(*c*) Discuss possible circumstances for which α_{pl} might have a value of this magnitude.

10-22 Sound of frequency 2000 Hz is propagating in the plane-wave mode in a square duct of dimensions a on a side. The air temperature is 20°C, the relative humidity is such that the relaxation frequency for O_2 vibrations is 2000 Hz. How large must the dimension a be before the dissipation by molecular relaxation within the interior of the duct exceeds that within the thermoviscous boundary layer?

10-23 A sound wave of 5000 Hz frequency is propagating through air at 20°C with ambient pressure of 10^5 Pa. Plot the absorption coefficient in nepers per meter versus relative humidity. At what relative humidity is α a maximum? What is the corresponding value of α?

10-24 With as little mathematical detail as possible explain why the contributions to the attenuation coefficient from different mechanisms are usually assumed to be additive.

10-25 Carry through the derivation of the dispersion relation (10-8.9*b*), taking the linear acoustic equations for seawater as a starting point.

10-26 Determine the magnitudes of the contributions from the different mechanisms (viscosity, thermal conduction, bulk viscosity, O_2 vibrational relaxation, and N_2 vibrational relaxation) to the plane-wave attenuation for 50-Hz sound in air at 10°C. Carry out the calculation for relative humidities of 0, 50, and 100 percent. Repeat the calculation for a frequency of 5000 Hz. What inferences would you draw concerning the relative importances of the various mechanisms over the range of audible frequencies?

10-27 An airplane flying at 3000 m causes a sound-pressure level of 90 dB on the ground for the octave band centered at 500 Hz. The humidity is not measured, but the ambient temperature is 20°C. To estimate an upper limit for the sound-pressure level to be expected under similar circumstances, a noise-control consultant assumes that the number 90 dB applies when the humidity is such that the attenuation from airplane to the ground is a maximum but the upper-limit number applies when the humidity causes the attenuation to be minimal. What is the calculated upper limit? What would it be were the plane to fly at 6000 m instead?

CHAPTER
ELEVEN
NONLINEAR EFFECTS IN SOUND PROPAGATION

Acoustics is ordinarily concerned only with small-amplitude disturbances, so nonlinear effects are typically of minor significance. There are instances, however, when a small nonlinear term in the fluid-dynamic equations can lead to novel and substantial phenonema. In some instances, e.g., shock waves, the predominant behavior develops because of a long-term accumulation of small nonlinear perturbations. In other instances, e.g., radiation pressure, nonlinear effects cause a small but nonzero magnitude to be associated with a physical entity, the existence of which the linear model precludes.

The present chapter is concerned primarily with instances of the first type and in particular with how cumulative nonlinear effects distort acoustic waveforms propagating through fluids.

11-1 NONLINEAR STEEPENING

To study nonlinear aspects of sound propagation, we begin with the ideal fluid-dynamic equations with the neglect of viscous and other dissipative terms. The restriction of our attention to one-dimensional flow allows us to recast the basic model in the form

$$\frac{\partial \rho}{\partial t} + \frac{\partial}{\partial x} \rho v = 0 \qquad (11\text{-}1.1a)$$

$$\rho \left(\frac{\partial v}{\partial t} + v \frac{\partial v}{\partial x} \right) = - \frac{\partial p}{\partial x} \qquad (11\text{-}1.1b)$$

$$\rho = \rho(p, s) \qquad (11\text{-}1.1c)$$

$$s = \text{const} \qquad (11\text{-}1.1d)$$

Here, in our initial discussions, the specific entropy s is considered initially constant so that it is always constant. This enables us to regard ρ and $c = (\partial\rho/\partial p)^{-1/2}$ as functions of the total pressure p.

Plane Waves in Homogeneous Media

Particular solutions† analogous to plane waves traveling in the $+x$ or $-x$ directions ($v' \approx \pm p'/\rho_0 c^2$) result from the stipulation that v be a single-valued function of p, so that $\partial v/\partial t = (dv/dp)\,\partial p/\partial t$, etc. This assumption inserted into the mass conservation equation and into Euler's equation yields

$$\frac{d\rho}{dp}\frac{\partial p}{\partial t} + \frac{d(\rho v)}{dp}\frac{\partial p}{\partial x} = 0 \qquad (11\text{-}1.2a)$$

$$\rho\frac{dv}{dp}\frac{\partial p}{\partial t} + \left(\rho v\frac{dv}{dp} + 1\right)\frac{\partial p}{\partial x} = 0 \qquad (11\text{-}1.2b)$$

These will be equivalent if the determinant of coefficients vanishes; such a condition, with $d\rho/dp = 1/c^2$, leads to $dv/dp = \pm 1/\rho c$. The choice of the plus sign corresponds to propagation in the $+x$ direction and reduces either (2*a*) or (2*b*) to the nonlinear partial-differential equation

$$\frac{\partial p}{\partial t} + (v + c)\frac{\partial p}{\partial x} = 0 \qquad (11\text{-}1.3)$$

The implication of Eq. (3) is that if $p(x_{\text{obs}}(t), t)$ represents the pressure at a moving observation point $x_{\text{obs}}(t)$, then p will appear constant in time if $dx_{\text{obs}}/dt = v + c$. This time invariance follows from a comparison of the equation $dp(x_{\text{obs}}, t)/dt = 0$ with (3). Since $v + c$ is a function of p, and since p appears constant to someone moving with speed $v + c$, each point with fixed pressure amplitude p appears to move with constant (time-independent) velocity, although two points of different amplitudes move with different velocities (see Fig. 11-1).

† An alternate approach defines

$$\lambda(\rho) = \int_{\rho_0}^{\rho}\frac{c(\rho)}{\rho}\,d\rho$$

so that (subscripts denoting partial derivatives) $\rho_t = (\rho/c)\lambda_t$, $p_x = \rho c\lambda_x$, etc., and Eqs. (1) reduce to

$$\lambda_t + v\lambda_x + cv_x = 0 \qquad v_t + vv_x + c\lambda_x = 0 \qquad \text{(i)}$$

or

$$(\lambda + v)_t + (v + c)(\lambda + v)_x = 0 \qquad (\lambda - v)_t + (v - c)(\lambda - v)_x = 0 \qquad \text{(ii)}$$

A particular solution (*simple wave*) results with $v = \lambda$, yielding

$$v_t + (v + c)v_x = 0 \qquad p_t + (v + c)p_x = 0 \qquad \text{(iii)}$$

which is the same as Eq. (3). [B. Riemann, "On the Propagation of Plane Air Waves of Finite Amplitude," *Abhandl. Ges. Wiss. Goettingen* (1860), reprinted in *The Collected Works of Bernhard Riemann,* Dover, New York, 1953, pp. 156–175.]

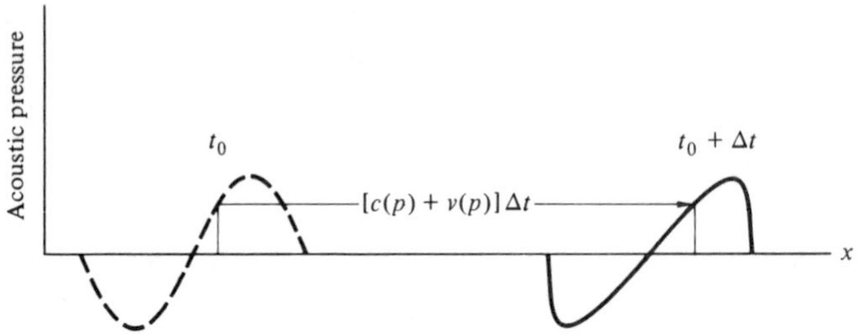

Figure 11-1 Evolution of an acoustic-pressure waveform in a plane traveling wave. Each amplitude portion travels with a characteristic amplitude-dependent speed $c(p) + v(p)$.

A parametric description of the solution results with the specification of $p(x, t)$ at time $t = 0$. Setting $p = p_0 + p'(x, t)$ and $p'(x, 0) = f(x)$ yields†

$$p'(x, t) = f(\phi) \qquad x = \phi + (v + c)t \tag{11.1.4}$$

$$p' = f(x - (v + c)t) \tag{11.1.4a}$$

where v and c are evaluated at $p_0 + f(\phi)$; at time t, the point at which p' equals $f(\phi)$ is displaced a distance $(v + c)t$ beyond where x is ϕ.

For given t, a plot of $p'(x, t)$ versus x results from letting ϕ run through all values for which $f(\phi)$ is nonzero, simultaneously tabulating p' and x from Eqs. (4). A possibility ignored at this point but discussed further below is that the resulting graph of p' versus x may not be single-valued.

For small-amplitude acoustic waves, the relations $dv/dp = 1/\rho c$ and $c = c(p)$ yield

$$v \approx \frac{p'}{\rho_0 c_0} \qquad c \approx c_0 + \left(\frac{\partial c}{\partial p}\right)_0 p' \tag{11-1.5}$$

where the ambient fluid velocity v_0 is presumed zero. The derivative $(\partial c/\partial p)_0$ (at constant entropy) is evaluated at the ambient state and is therefore constant.

The two expressions in Eq. (5) combine into

$$c + v \approx c_0 + \frac{\beta_0 p'}{\rho_0 c_0} \approx c_0 + \beta_0 v \tag{11-1.6}$$

where the constant β_0 (which should not be confused with the coefficient of volume expansion) is

$$\beta_0 = 1 + \left(\rho c \frac{\partial c}{\partial p}\right)_0 = \frac{1}{2}\left(\rho^3 c^4 \frac{\partial^2 \rho^{-1}}{\partial p^2}\right)_0 \tag{11-1.7}$$

(The second version here follows from $\partial\rho^{-1}/\partial p = -\rho^{-2}c^{-2}$.) Alternatively, if p is regarded as a function of s and ρ, then $\partial c^2/\partial p$ is $(\partial c^2/\partial\rho)/(\partial p/\partial\rho)$ or

† S. Earnshaw, "On the Mathematical Theory of Sound," *Phil. Trans. R. Soc. Lond.*, **150:**133–148 (1859). A similar result for a gas in which p is directly proportional to ρ had been obtained somewhat earlier by S. D. Poisson, "Memoir on the Theory of Sound," *J. Ec. Polytech.*, **7:**319–392 (1808).

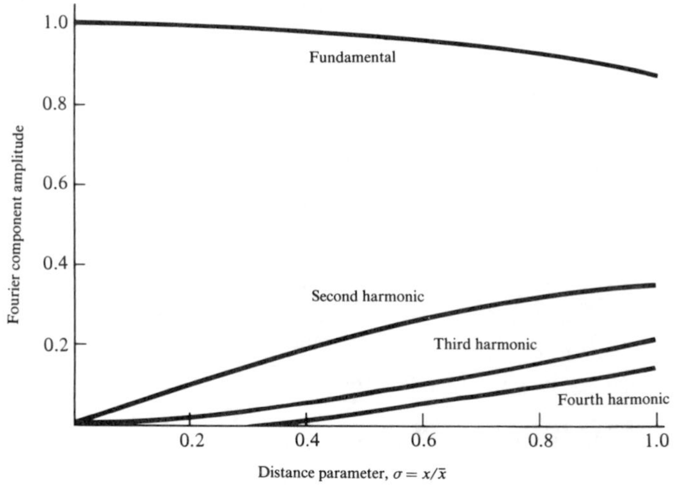

Figure 11-3 Amplitudes of harmonics (units of P_0) versus distance x in units of $\bar{x}$ for a plane wave that is sinusoidal at $x = 0$; depicted curves without energy dissipation.

and therefore varies linearly with x; higher harmonics grow more slowly (see Fig. 11-3).

Conservation of Energy

The growth of the overtones must be at the expense of the fundamental. Since the model represented by Eq. (2) incorporates no dissipation mechanisms, one expects the energy per cycle to be independent of σ:

$$\frac{d}{d\sigma}\int_0^{2\pi} p^2(\theta,\sigma)\,d\theta = 0 \qquad (11\text{-}2.11)$$

To show that this follows from (2), change the integration variable to $\xi = \omega\psi$ so that the left side above becomes

$$\frac{d}{d\sigma}\int_0^{2\pi} P_0^2 \sin^2\xi\,(1 - \sigma\cos\xi)\,d\xi = -P_0^2\int_0^{2\pi} \sin^2\xi \cos\xi\,d\xi$$

which integrates to zero.

Parseval's theorem (see Sec. 2-7) consequently requires that

$$\sum_{n=1}^{\infty} p_{n,\text{pk}}^2(\sigma) = P_0^2 \qquad (11\text{-}2.12)$$

be independent of σ (providing $\sigma < 1$). This deduction is consistent with Eqs. (9) and (10); the decrease of $p_{1,\text{pk}}^2$ for small σ is compensated by the growth of $p_{2,\text{pk}}^2$.

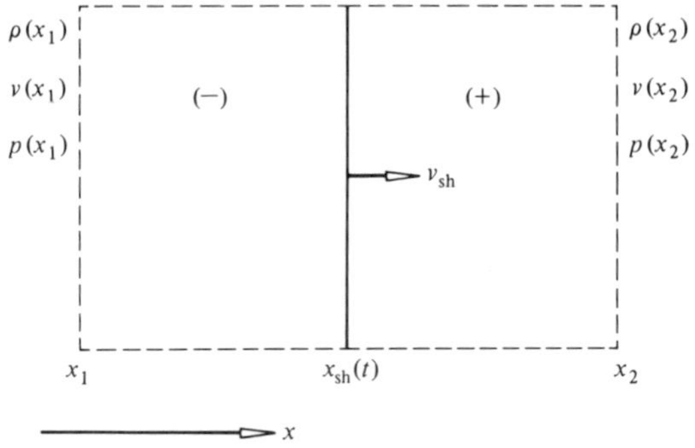

Figure 11-4 Fixed control volume containing a moving surface of discontinuity.

11-3 WEAK-SHOCK THEORY

The Rankine-Hugoniot Relations

The resolution of the multivalued waveform dilemma,† which was a major unsolved problem during most of the nineteenth century, came with the discovery and physical understanding of *shock waves*. The governing partial differential equations require the assumption (see Secs. 1-2 and 1-3) that ρ, v, and p are continuous. If they are not, then one must back up to the original integral equations. For a fixed control volume of unit cross section and with fixed endpoints x_1 and x_2, conservation of mass requires (see Fig. 11-4)

$$\frac{d}{dt}\int_{x_1}^{x_2} \rho\, dx = (\rho v)_{x_1} - (\rho v)_{x_2} \tag{11-3.1a}$$

or that the time rate of change of mass in the volume be the difference of the rate at which mass is flowing in at x_1 minus that at which it is flowing out at x_2. (Here the subscript denotes the point at which the indicated quantity is evaluated.)

Similarly, the time rate of change of momentum in the volume is equal to the rate (per unit area) $(\rho v)_{x_1}(v)_{x_1}$ momentum is flowing in minus that rate $(\rho v)_{x_2}(v)_{x_2}$ at which it is flowing out plus the net force (per unit area) $p_{x_1} - p_{x_2}$ exerted on the control volume:

$$\frac{d}{dt}\int_{x_1}^{x_2} \rho v\, dx = (\rho v^2 + p)_{x_1} - (\rho v^2 + p)_{x_2} \tag{11-3.1b}$$

A third relation comes from the consideration of the time rate of change of energy (energy density equal to $\frac{1}{2}\rho v^2$ plus ρu, where $\frac{1}{2}\rho v^2$ represents

† J. Challis, "On the Velocity of Sound," *Phil. Mag.*, (3)**32:**494–499 (1848); G. G. Stokes, "On a Difficulty in the Theory of Sound," ibid., **33:**349–356 (1848); G. B. Airy, "The Astronomer Royal on a Difficulty in the Problem of Sound," ibid., **34:**401–405 (1849).

kinetic energy per unit volume and u represents internal energy per unit mass). For the control volume, this should equal the rate $(\frac{1}{2}\rho v^2 + \rho u)_{x_1} v_{x_1}$ energy is being convected in by the flow minus the rate at which it is convected out plus the rate $(pv)_{x_1} - (pv)_{x_2}$ at which work is being done on the control volume by external pressures, or

$$\frac{d}{dt}\int_{x_1}^{x_2} \rho(\tfrac{1}{2}v^2 + u)\, dx = [(\tfrac{1}{2}\rho v^2 + \rho u + p)v]_{x_1} - [(\tfrac{1}{2}\rho v^2 + \rho u + p)v]_{x_2} \tag{11-3.1c}$$

If one considers x_1 and x_2 as arbitrary and all quantities as continuous and differentiable, the first two of these lead to the one-dimensional partial-differential equations displayed in Secs. 1-2 and 1-3. To derive $Ds/Dt = 0$, one uses the second law of thermodynamics in the form (1-4.4) with ds replaced by Ds/Dt, etc., and eliminates Dv^2/Dt from the differential-equation version of (1c) by using what results from the product of v with Euler's equation. If discontinuities are present, however, these steps cannot be carried through.

To see what results when a discontinuity is present, one postulates a moving point $x_{\text{sh}}(t)$ (eventually identified as the location of a shock) between x_1 and x_2, at which p, v, ρ, u are discontinuous. Each of the integrals over x in Eqs. (1) can be split into integrals from x_1 to x_{sh} and from x_{sh} to x_2. Then, standard rules for differentiation yield, for example,

$$\frac{d}{dt}\int_{x_1}^{x_2} \rho\, dx = (\rho_- - \rho_+)v_{\text{sh}} + \int_{x_1}^{x_{\text{sh}-}} \frac{\partial \rho}{\partial t}\, dx + \int_{x_{\text{sh}+}}^{x_2} \frac{\partial \rho}{\partial t}\, dx$$

where ρ_- and ρ_+ represent the values of ρ on the $-x$ and $+x$ sides of the discontinuity and $v_{\text{sh}} = dx_{\text{sh}}/dt$ is the velocity of the discontinuity surface. In the limit in which x_1 and x_2 are arbitrarily close to x_{sh}, the integrals on the right become negligible and $(\rho v)_{x_1} \to (\rho v)_-$, $(\rho v)_{x_2} \to (\rho v)_+$, so Eq. (1$a$) yields

$$[\rho(v - v_{\text{sh}})]_+ = [\rho(v - v_{\text{sh}})]_- \tag{11-3.2a}$$

In a similar manner, Eqs. (1b) and (1c) imply

$$[\rho v(v - v_{\text{sh}}) + p]_+ = [\rho v(v - v_{\text{sh}}) + p]_- \tag{11-3.2b}$$

$$[\rho(\tfrac{1}{2}v^2 + u)(v - v_{\text{sh}}) + pv]_+ = [\rho(\tfrac{1}{2}v^2 + u)(v - v_{\text{sh}}) + pv]_- \tag{11-3.2c}$$

Equations (2) are the *Rankine-Hugoniot relations.*†

An equivalent way of writing the second relation above is to subtract from

† W. J. M. Rankine, "On the Thermodynamic Theory of Waves of Finite Longitudinal Disturbance," *Phil. Trans. R. Soc. Lond.*, **160:**277–288 (1870); H. Hugoniot, "On the Propagation of Movement through a Body and Especially through an Ideal Gas," *J. Ec. Polytech.*, **58:**1–125 (1889); G. I. Taylor, "The Conditions Necessary for Discontinuous Motion in Gases," *Proc. R. Soc. Lond.*, **A84:**371–377 (1910). When the flow is not perpendicular to the shock front, the above still hold with v_+ and v_- interpreted at the normal components of $\mathbf{v}_+$ and $\mathbf{v}_-$. The tangential component of the velocity must be continuous across the shock surface. See, for example, L. D. Landau and E. M. Lifshitz, *Fluid Mechanics*, Pergamon, London, 1959, pp. 317–319.

it v_{sh} times the first, so that

$$[\rho(v - v_{sh})^2 + p]_+ = [\rho(v - v_{sh})^2 + p]_- \tag{11-3.2b'}$$

Similarly Eq. (2c) minus v_{sh} times (2b) plus $v_{sh}^2/2$ times (2a) all divided by (2a) yields

$$[h + \tfrac{1}{2}(v - v_{sh})^2]_+ = [h + \tfrac{1}{2}(v - v_{sh})^2]_- \tag{11-3.2c'}$$

where we abbreviate $h = u + p/\rho$ for the *enthalpy* per unit mass. In dividing by $[\rho(v - v_{sh})]_+$ we have ruled out *contact discontinuities* from consideration (for which $v_+ = v_- = v_{sh}$, $p_+ = p_-$, $\rho_+ \neq \rho_-$). The analysis here applies to *shock waves*, for which $v_+ \neq v_{sh}$.

With the abbreviations $\Delta v = v_- - v_+$, $\Delta h = h_- - h_+$, $v_{av} = (v_+ + v_-)/2$, etc., Eqs. (2a), (2b'), and (2c') yield, after some algebraic manipulations,†

$$\Delta p = \frac{\Delta h}{(1/\rho)_{av}} \qquad \Delta v = (v_{sh} - v_{av})\frac{\Delta\rho}{\rho_{av}}$$

$$(v_{sh} - v_{av})^2 = -\frac{(\rho^{-1})_{av}^2\,\Delta p}{\Delta(\rho^{-1})} \qquad (\Delta v)^2 = -\Delta\rho^{-1}\,\Delta p \tag{11-3.3}$$

It follows from these that Δp, $\Delta\rho$, Δh, and $\Delta v/(v_{sh} - v_{av})$ must all have the same sign.

One further restriction comes from the inequality version of the second law of thermodynamics. If the shock is advancing in the $+x$ direction relative to the fluid, so that $v_{sh} - v_{av} > 0$, then $s_- \geq s_+$; a fluid particle's entropy cannot be decreased by passage of the shock, so Δs and $v_{sh} - v_{av}$ have the same sign. (Below it is demonstrated that Δs and Δp must have the same sign, so Δp and $v_{sh} - v_{av}$ have the same sign.)

Weak Shocks

If $|\Delta\rho|/\rho_{av} \ll 1$, the consequences of the first of Eqs. (3) can be explored by expanding $h(p, s)$ in a Taylor series in $\delta p = p - p_{av}$ and $\delta s = s - s_{av}$, the various coefficients being denoted by h^0, h_p^0, h_s^0, h_{pp}^0, h_{ps}^0, etc., such that, for example, h_{ps}^0 is $\partial^2 h/(\partial p\,\partial s)$ evaluated at p_{av} and s_{av}. To obtain h_+, one sets $\delta p = -\Delta p/2$, $\delta s = -\Delta s/2$ in this expansion; to obtain h_-, one sets $\delta p = \Delta p/2$, $\delta s = \Delta s/2$. An expansion for ρ^{-1} follows from the thermodynamic identity $\rho^{-1} = \partial h/\partial p$. (Note that $dh = T\,ds + \rho^{-1}\,dp$ follows from $T\,ds = du + p\,d\rho^{-1}$ and $h = u + p/\rho$.) The so-derived expansions for $(\rho^{-1})_+$ and $(\rho^{-1})_-$ in terms of Δs and Δp lead in turn to

$$\Delta h - (\rho^{-1})_{av}\,\Delta p = h_s^0\,\Delta s - \tfrac{1}{12}h_{ppp}^0(\Delta p)^3 - \tfrac{1}{8}h_{pps}^0(\Delta p)^2\,\Delta s - \cdots \tag{11-3.4}$$

† W. D. Hayes, "The Basic Theory of Gasdynamic Discontinuities," in H. W. Emmons (ed.), *Fundamentals of Gas Dynamics*, Princeton University Press, Princeton, N.J., 1958, pp. 416–481. The first of Eqs. (3) is the *Hugoniot equation*; the corresponding plot of p_- versus $1/\rho_-$ for fixed p_+ and $1/\rho_+$ is a *Hugoniot diagram*.

length $2L(t)$ increases, but the overpressure decreases; the product $L(t)P(t) = L_0P_0$ remains constant.

Dissipation of Acoustic Energy

In the absence of shocks, nonlinear effects do not change the net acoustic energy associated with a pulse; they merely cause a rearrangement of the frequency distribution of the energy. The demonstration of this is similar to that of Eq. (11-2.11); the energy density is $p^2/\rho c^2$ for a traveling wave because $v \approx p/\rho c$. The net energy per unit area transverse to propagation direction for a pulse of finite duration is then

$$E(t) = \frac{1}{\rho c^2} \int_{-\infty}^{\infty} p^2\, dx \tag{11-4.8}$$

If Eqs. (11-1.9) are valid for a single-valued description of the pulse, this can alternatively be written

$$E(t) = \frac{1}{\rho c^2} \int_{-\infty}^{\infty} f^2(\phi) \frac{\partial x}{\partial \phi}\, d\phi = \frac{1}{\rho c^2} \int_{-\infty}^{\infty} f^2(\phi) \left(1 + \frac{\beta f'(\phi) t}{\rho c}\right) d\phi \tag{11-4.9}$$

The second term, however, integrates to zero since $f^3(\phi) \to 0$ as $\phi \to \pm\infty$. Consequently, $E(t)$ is independent of time.

On the other hand, if a shock is present, the integral must be broken into integrals from $-\infty$ to $\phi_-(t)$ and $\phi_+(t)$ to ∞. The time derivative of $E(t)$ consequently yields, with some algebraic manipulation, the relation

$$\begin{aligned} \rho c^2 \frac{dE(t)}{dt} &= f^2(\phi_-) \frac{d}{dt}\left[\phi_- + \frac{\beta f(\phi_-) t}{\rho c}\right] \\ &\quad - f^2(\phi_+) \frac{d}{dt}\left[\phi_+ + \frac{\beta f(\phi_+) t}{\rho c}\right] - \frac{2\beta}{3\rho c}[f^3(\phi_-) - f^3(\phi_+)] \end{aligned} \tag{11-4.10}$$

The first two quantities here in brackets [see Eq. (11-1.9)] are $x_{sh} - ct$, so their time derivatives [see Eq. (11-3.8)] are both $\frac{1}{2}[f(\phi_+) + f(\phi_-)]\beta/\rho c$. Then, with additional manipulations Eq. (10) yields†

$$\frac{dE}{dt} = -\frac{\beta}{6\rho^2 c^3} [f(\phi_-) - f(\phi_+)]^3 = -\rho c T_0\, \Delta s \tag{11-4.11}$$

where the latter version follows from Eq. (11-3.7). Since $\Delta s > 0$, the presence of the shock causes the energy in the wave to decrease with time.

The validity of Eq. (11) is substantiated by our N-wave example, for which $E(t) = \frac{2}{3}P^2L/\rho c^2$ decreases with t as $1/(1 + t/\tau_N)^{1/2}$. With $P(t)$, $L(t)$, and τ_N taken from Eqs. (6), (5), and (3), we find

$$\frac{dE}{dt} = -\frac{1}{3\rho c^2} \frac{P_0^2 L_0/\tau_N}{(1 + t/\tau_N)^{3/2}} = -\frac{\beta P^3(t)}{3\rho^2 c^3} \tag{11-4.12}$$

† I. Rudnick, "On the Attenuation of a Repeated Sawtooth Shock Wave," *J. Acoust. Soc. Am.*, **25**:1012–1013 (1953).

The extra factor of 2 detected from a comparison of this with Eq. (11) is because the N wave has two shocks.

Why do we find a dissipation of acoustic energy when no dissipation mechanisms are explicitly taken into account? An explanation proceeds from the observation that if the model were modified to include a typical dissipation mechanism such as viscosity, the resulting solutions would never be discontinuous. However, if the coefficient characterizing the dissipation were gradually reduced in magnitude, regions of steep gradients would become evident. In the limit as the coefficient approaches zero, these steep gradients approach discontinuities with all the properties of the shocks predicted by the ideal-fluid model. The dissipation rate per unit area transverse to the propagation direction approaches a limit independent of the magnitude of the dissipation coefficient. Thus, one can regard the dissipation at a shock as caused by *some* physical mechanism, but given that the real dissipation mechanisms are weak, it is a fortunate occurrence† that the magnitude of the dissipation is nearly independent of the nature and strength of the mechanism.

11-5 EVOLUTION OF SAWTOOTH WAVEFORMS

A plane wave with sufficient amplitude and generated by a transducer oscillating at constant frequency approaches a sawtooth shape at large distances.‡ To investigate the transition, we let $p(0, t) = P_0 \sin \omega t$ be the acoustic pressure at the face of the transducer. With the neglect of ambient flow, the pressure amplitude $P_0 \sin \omega t_0$ created at time t_0 will be at a point

$$x = \left[c + \frac{\beta P_0}{\rho c} \sin \omega t_0\right](t - t_0) \tag{11-5.1}$$

at time t; the quantity in brackets is the speed of the wave portion with amplitude $P_0 \sin \omega t_0$. The above, along with $p = P_0 \sin \omega t_0$, yields a parametric description of the distorted waveform, with which one can construct p versus x for any given t. (The description is equivalent to that in Sec. 11-2 given that $\beta P_0 \ll \rho c^2$.)

A single cycle of the waveform (see Fig. 11-7) generated at times t_0 between $-\pi/\omega$ and π/ω nominally lies between $x = ct - c\pi/\omega$ and $x = ct + c\pi/\omega$. The portion generated between $t_0 = -\pi/\omega$ and $t_0 = -\pi/2\omega$ is such that $\omega t_0 = -\sin^{-1}(p/P_0) - \pi$, with the arc sine understood to be between $-\pi/2$ and $\pi/2$. Similarly, the portion generated between $t_0 = -\pi/2\omega$ and $t_0 = \pi/2\omega$ is such that $\omega t_0 = \sin^{-1}(p/P_0)$, while the portion generated between $t_0 = \pi/2\omega$ and $t_0 = \pi/\omega$ is such that $\omega t_0 = \pi - \sin^{-1}(p/P_0)$. These expressions for t_0,

† See, for example, the comments by W. Heisenberg, "Nonlinear Problems in Physics," *Phys. Today,* **20**(5):27–33 (May 1967).

‡ Whitham, "The Flow Pattern of a Supersonic Projectile"; Blackstock, "Connection between the Fay and Fubini Solutions."

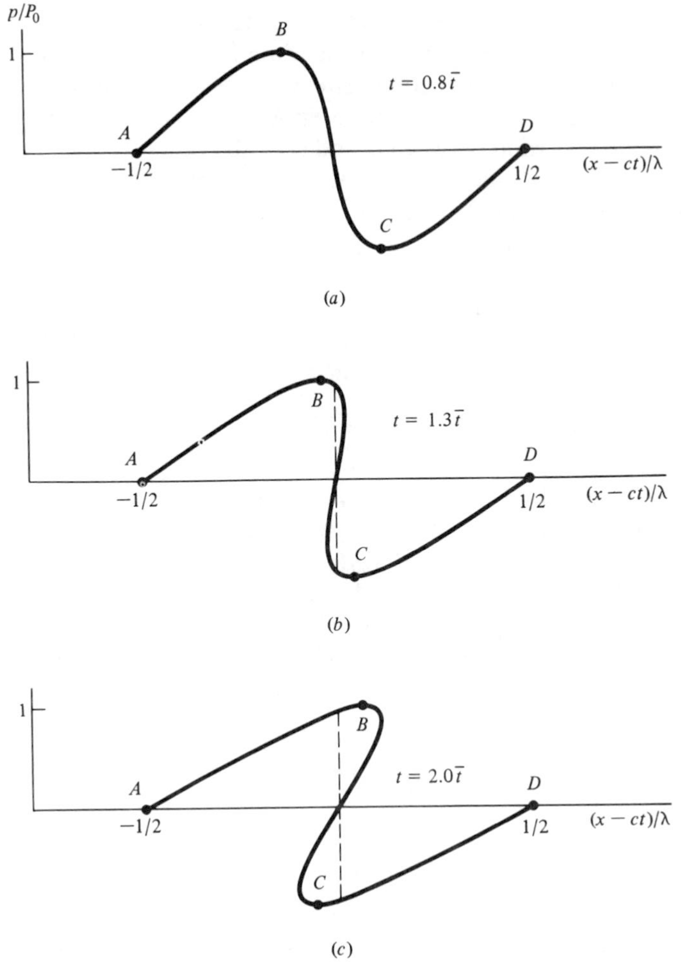

Figure 11-7 Waveform segment of a plane traveling wave generated by a single cycle of an oscillating transducer. (*a*) Segment before first formation of shock; (*b*) after shock formation but before peak (at *B*) overtakes trough (at *C*); (*c*) after peak overtakes trough. Transitions occur at $\bar{t}$ and at $(\pi/2)\bar{t}$. Vertical coordinate is ratio of acoustic pressure p to peak amplitude P_0 that waveform has before peak overtakes trough.

when inserted into Eq. (1), give three relations for x in terms of p, which (for low-amplitude acoustic waves where $\beta p/\rho c^2 \ll 1$ but t may be large) approximate to

$$x = \begin{cases} ct + \dfrac{\beta p t}{\rho c} + \dfrac{c}{\omega}\left(\pi + \sin^{-1}\dfrac{p}{P_0}\right) & -\dfrac{\pi}{\omega} < t_0 < -\dfrac{\pi}{2\omega} \quad (11\text{-}5.2a) \\ ct + \dfrac{\beta p t}{\rho c} - \dfrac{c}{\omega}\sin^{-1}\dfrac{p}{P_0} & -\dfrac{\pi}{2\omega} < t_0 < \dfrac{\pi}{2\omega} \quad (11\text{-}5.2b) \\ ct + \dfrac{\beta p t}{\rho c} - \dfrac{c}{\omega}\left(\pi - \sin^{-1}\dfrac{p}{P_0}\right) & \dfrac{\pi}{2\omega} < t_0 < \dfrac{\pi}{\omega} \quad (11\text{-}5.2c) \end{cases}$$

The corresponding ranges of p are 0 to $-P_0$, $-P_0$ to P_0, and P_0 to 0.

If the above three curves for x versus p are each plotted with t fixed, taking p as varying over the ranges specified, the composite curve will be of one of the forms sketched in Fig. 11-7. The tail portion between A and B corresponds to the first equation, the middle portion between B and C to the second equation, and the leading portion between C and D to the third equation. The curve so constructed is always such that it is symmetric under inversions ($x - ct \rightarrow ct - x$, $p \rightarrow -p$) about the point $x = ct$, $p = 0$. Consequently, the equal-area rule requires that if a shock is present it must be at $x = ct$, so the shock is moving at the ambient sound speed c.

The earliest time a shock forms for the waveform segment considered above is when Eq. (11-1.12) predicts $\partial x/\partial p = 0$ at $p = 0$. This is when

$$t = \bar{t} = \frac{\rho c^2}{P_0}\frac{1}{\beta\omega} = \frac{\bar{x}}{c} \tag{11-5.3}$$

where $\bar{x}$ is the same as defined by Eq. (11-2.3). The shock at $x = ct$ for $t > \bar{t}$ continues to grow up until point B reaches $x = ct$. This occurs when the x predicted by Eq. (2c) is ct at $p = P_0$, such that $\beta P_0 t/\rho c = (c/\omega)\pi/2$, or when $t = (\pi/2)\bar{t}$. Up until this time the peak amplitude of the waveform is still P_0.

After time $t = (\pi/2)\bar{t}$, the shock at $x = ct$ erodes the wave peak, and the waveform resembles a sawtooth. The peak amplitude $p_{\max}$ at times $t > (\pi/2)\bar{t}$ for the waveform segment considered is found by setting the x of Eq. (2c) equal to ct, giving

$$\frac{t}{\bar{t}} = \frac{\pi - \sin^{-1}(p_{\max}/P_0)}{p_{\max}/P_0} \tag{11-5.4}$$

so the following tabulation results:

$p_{\max}/P_0$	1	0.9	0.8	0.7	0.6	0.5	0.4	0.3	0.2	0.1
$t/\bar{t}$	1.57	2.25	2.77	3.38	4.16	5.24	6.83	9.46	14.70	30.41

For $t/\bar{t} > 3$, a good approximation (to within 4 percent) results from setting $\sin^{-1}(p_{\max}/P_0) \approx p_{\max}/P_0$, such that (4) yields

$$p_{\max} = \frac{\pi P_0}{1 + t/\bar{t}} \tag{11-5.5}$$

In the same limit, Eq. (2c) reduces to

$$\frac{p}{p_{\max}} = 1 + \frac{\omega}{c\pi}(x - ct) \tag{11-5.6}$$

for the description of the positive phase behind the shock ($x - ct$ between $-c\pi/\omega$ and 0). Similar considerations hold for the peak underpressure and the negative phase before the shock; the waveform remains symmetric under inversions about $x = ct$, $p = 0$. The net discontinuity in pressure at the shock is $2p_{\max}$.

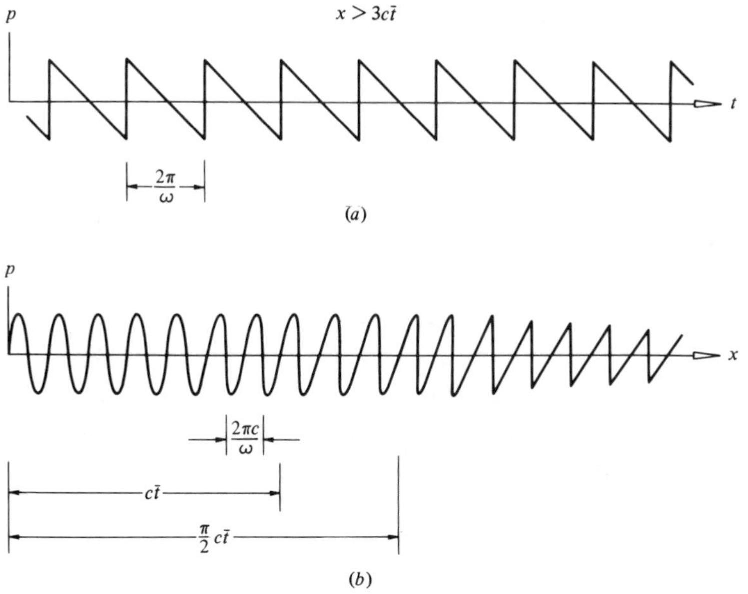

Figure 11-8 (*a*) Sketch of acoustic pressure versus time at a point sufficiently distant from an oscillating transducer for a sawtooth profile to have formed. (*b*) Sketch of acoustic pressure versus x for a particular instant of time, showing the evolution of the sawtooth profile.

Because the transducer oscillations are periodic, the foregoing analysis applies to any cycle of the overall waveform; each cycle has the same history, given an appropriate shift in time origin. Thus at a given point where $x > \bar{x}$ the disturbance passing by will have shocks at time intervals of $\Delta t = 2\pi/\omega$. The received signal will have the same period. At x greater than approximately $3\bar{x}$, the waveform will be nearly sawtooth in shape, and the peak overpressure of each cycle will be given by Eq. (5) with t replaced by x/c. After each shock, each with net discontinuity $2p_{\text{max}}$, the pressure decreases linearly with increasing time until p reaches $-p_{\text{max}}$; then another shock arrives, and the cycle repeats itself (see Fig. 11-8*a*).

When considered as a function of x for fixed t, the disturbance is not periodic, although zero crossings are equally spaced at intervals of $\pi c/\omega$. Beyond $x = \bar{x}$, there are shocks at intervals of $\Delta x = 2\pi c/\omega$. Beyond $x = (\pi/2)\bar{x}$, the successive peak amplitudes are smaller and smaller, all smaller than P_0 (see Fig. 11-8*b*).

The expression describing the waveform in the sawtooth limit when $\sigma = x/\bar{x}$ is larger than, say, 3 can be taken as

$$p = \frac{\pi P_0}{1+\sigma} f_{\text{ST}}(\omega t') \qquad t' = t - \frac{x}{c} \tag{11-5.7}$$

where the sawtooth wave function is

$$f_{\text{ST}}(\omega t') = 1 - \frac{\omega t'}{\pi} \qquad 0 < \omega t' < 2\pi \tag{11-5.8}$$

and is periodic in $\omega t'$ with period 2π. The pressure has the equivalent Fourier-series representation

$$p = \sum_{n=1}^{\infty} \frac{2P_0/n}{1+\sigma} \sin\left[n\omega\left(t - \frac{x}{c}\right)\right] \qquad (11\text{-}5.9)$$

which differs from the $\sigma < 1$ version, Eq. (11-2.7), in that $1/(1+\sigma)$ replaces $J_n(n\sigma)/\sigma$.

At large distances, such that $1 + x/\bar{x} \approx x/\bar{x}$, the peak overpressure at fixed x becomes

$$p_{\max}(x) \approx \frac{P_0 \pi \bar{x}}{x} = \frac{\pi \rho c^3}{\beta \omega x} \qquad (11\text{-}5.10)$$

which is independent of P_0 and which decreases inversely with x. The only feature characteristic of the excitation that remains is the driving frequency.

The phenomenon described by Eq. (10) leads to the concept of *saturation.*† At a fixed far-field value of x, the received peak pressure varies with P_0 as

$$p_{\max} = P_0 \qquad P_0 < \frac{\pi \rho c^3}{2\beta x \omega} \qquad (11\text{-}5.11a)$$

$$p_{\max} \approx \frac{\pi P_0}{1 + (x\beta\omega/\rho c^3)P_0} \qquad P_0 > \frac{3\rho c^3}{\beta x \omega} \qquad (11\text{-}5.11b)$$

so one infers that $p_{\max}$ increases monotonically with P_0; but regardless of how high P_0 is raised, $p_{\max}$ cannot exceed the saturation value in Eq. (10), which gives the theoretical upper limit to what can be received at a distance x from a transducer oscillating at angular frequency ω. The amplitude is within 90° of the upper limit when P_0 is greater than $9\rho c^3/\beta\omega x$.

The above discussion, based on the weak-shock theory, presumes that P_0 is somewhat less than ρc^2 (say, less than $0.1\rho c^2$). The neglect of dissipative mechanisms requires, moreover, that $p_{\max}$ be greater than $3\alpha\rho c^3/\omega\beta$ [see Eq. (11-3.10)]. Consequently, Eq. (10) implies that, for the analysis to be applicable, x should be less than $\pi/3\alpha \approx 1/\alpha$, where α is the plane-wave attenuation coefficient. The sawtooth region therefore extends from $x \approx (\pi/2)\rho c^3/\beta\omega P_0$ to $x \approx 1/\alpha$. As discussed further in Sec. 11-7, beyond the upper distance (the "old-age" region), the wave resembles a sinusoidal wave whose peak amplitude decreases as $e^{-\alpha x}$.

† The possibility that finite-amplitude effects may limit the acoustic efficiency of a sound source was suggested by L. V. King, "On the Propagation of Sound in the Free Atmosphere and the Acoustic Efficiency of Fog-Signal Machinery: An Account of Experiments Carried Out at Father Point, Quebec, September, 1913," *Phil. Trans. R. Soc. Lond.*, **A218:**211–293 (1919). The first correctly interpreted observation of saturation is due to C. H. Allen, "Finite Amplitude Distortion in a Spherically Diverging Sound Wave in Air," Ph.D. thesis, Pennsylvania State University, 1950. For recent reviews, see J. A. Shooter, T. G. Muir, and D. T. Blackstock, "Acoustic Saturation of Spherical Waves in Water," *J. Acoust. Soc. Am.*, **55:**54–62 (1974); D. A. Webster and D. T. Blackstock, "Finite-Amplitude Saturation of Plane Sound Waves in Air," ibid., **62:**518–523 (1977).

The thickness parameter l has the property

$$l = \frac{v_{\text{sh}}}{(-dv/d\xi)_{v=\frac{1}{2}v_{\text{sh}}}} \qquad (11\text{-}6.14)$$

so a straight line tangent to the waveform at its half-peak point ($\xi = 0$) reaches from the line $v = v_{\text{sh}}$ to the line $v = 0$ over a distance interval of l. This accordingly allows us to regard l as the *shock thickness*. From an analogous point of view, l/c is the *shock rise time*. Both are inversely proportional to the shock overpressure.

Relaxation Effects on Shock Structure

To examine how a relaxation process affects the propagation of a weak shock, we consider a medium with only one such process. The application of the operator $(1 + \tau\partial/\partial t)$ in such a case to both sides of Eq. (5*a*) then yields,† with the help of Eq. (3*b*),

$$\left(1 + \tau\frac{\partial}{\partial t}\right)[v_t + (c + \Delta c + \beta v)v_x - \delta\, v_{xx}] = (\Delta c)v_x \qquad (11\text{-}6.15)$$

As in the preceding discussion, we assume that v is of the form $v(\xi)$, where $\xi = x - Vt$, and that $v \to v_{\text{sh}}$ as $\xi \to -\infty$. The latter requirement leads to $V = c + \frac{1}{2}\beta v_{\text{sh}}$, so the following ordinary differential equation for v results:

$$\left(1 - V\tau\frac{d}{d\xi}\right)\left[v(v - v_{\text{sh}}) - \frac{v_{\text{sh}}l}{4}v_\xi\right] = V\tau\phi v_{\text{sh}}v_\xi \qquad (11\text{-}6.16)$$

where l is the characteristic length given by Eq. (13) and where

$$\phi = \frac{2\Delta c}{\beta v_{\text{sh}}} = 2\rho c\frac{\Delta c}{\beta p_{\text{sh}}} \qquad (11\text{-}6.17)$$

is a dimensionless quantity that measures the relative strength of the relaxation process and of the nonlinearity. The limit $\phi \to 0$ yields the same differential equation (9) as neglect of relaxation processes.

If $l + 4V\tau\phi$ is substantially larger than $V\tau$, the differential equation (16) can be approximated by setting the operator $1 - V\tau\, d/d\xi$ equal to 1 on the left side, so that Eq. (12) results but with l replaced by an augmented shock thickness

$$l^* = l + 4V\tau\phi \approx l + 4c\tau\phi \qquad (11\text{-}6.18)$$

The conclusion is the same as that from a low-frequency approximation to the relaxation process's contribution to the dispersion relation, so that the bulk viscosity is augmented by an amount $\Delta\mu_B$ given by Eq. (10-8.15). This applies in particular for shocks sufficiently weak to ensure that $\phi \gg 1$. Any semblance of

† A. L. Polyakova, S. I. Soluyan, and R. V. Khokhlov, "Propagation of Finite Disturbances in a Relaxing Medium," *Sov. Phys. Accoust.*, **8**(1):78–82 (1962); Ockendon and Spence, "Nonlinear Wave Propagation"; O. V. Rudenko and S. I. Soluyan, *Theoretical Foundations of Nonlinear Acoustics*, Consultants Bureau, New York, 1977, pp. 88–96.

a shock in the waveform will be lost if l^* is larger than one-fourth a representative wavelength. Under such circumstances, the weak-shock theory discussed in Sec. 10-3 loses applicability, even as a gross approximation.

The differential equation (16) is difficult to solve in general, but some insight results from setting l to zero at the outset, so that

$$\frac{1}{V\tau v_\xi} = \frac{d}{dv}\frac{\xi}{V\tau} = \frac{(\phi - 1)v_{sh} + 2v}{v(v - v_{sh})} = \frac{1 + \phi}{v - v_{sh}} - \frac{\phi - 1}{v} \qquad (11\text{-}6.19)$$

which integrates to

$$e^{\xi/V\tau} = (\text{const})(v_{sh} - v)^{1+\phi}v^{1-\phi} \qquad (11\text{-}6.20)$$

At this point, one must distinguish between the cases $\phi < 1$ and $\phi > 1$. If $\phi < 1$, the frozen sound speed $c + \Delta c$ is less than the wave speed $c + \frac{1}{2}\beta v_{sh}$, so the only possibility for a wave advancing into an undisturbed medium is for the waveform to begin with a discontinuity of net overpressure $\rho c v_f$ (f for front), where $v_f < v_{sh}$. This discontinuity must move with a speed $c + \Delta c + (\beta/2)v_f$; one uses the frozen sound speed here rather than the equilibrium sound speed because the fluid just behind the discontinuity must behave as if the internal degrees of freedom were frozen over any time interval small compared with the relaxation time τ. The speed $c + \Delta c + (\beta/2)v_f$ must be the same as V, however, so we identify $v_f = (1 - \phi)v_{sh}$. If $\xi = 0$ locates the discontinuous beginning of the waveform, the constant of integration in (20) must be such that $v = (1 - \phi)v_{sh}$ when $\xi = 0$. Thus, we have ($\phi < 1$)

$$v = 0 \qquad \xi > 0 \qquad (11\text{-}6.21a)$$

$$e^{\xi/V\tau} = \left(\frac{1 - v/v_{sh}}{\phi}\right)^{1+\phi}\left(\frac{v/v_{sh}}{1 - \phi}\right)^{1-\phi} \qquad 1 - \phi < \frac{v}{v_{sh}} < 1 \qquad (11\text{-}6.21b)$$

In the other case, when $\phi > 1$, the waveform is continuous and has a precursor (which arises because the frozen sound speed exceeds the nominal shock velocity). To pinpoint the region of transition near $\xi = 0$, we choose the constant of integration to be such that $v = v_{sh}/2$ when $\xi = 0$. Thus, Eq. (20) yields ($\phi > 1$)

$$e^{\xi/V\tau} = \frac{(2 - 2v/v_{sh})^{\phi+1}}{(2v/v_{sh})^{\phi-1}} \qquad 0 < v < v_{sh} \qquad (11\text{-}6.22)$$

which in turn leads to

$$4V\tau\phi = \frac{v_{sh}}{\left(-\dfrac{dv}{d\xi}\right)_{v=\frac{1}{2}v_{sh}}} \qquad (11\text{-}6.23)$$

so $4V\tau\phi \approx 4c\tau\phi$ can be regarded as the apparent shock thickness for the case $\phi > 1$. However, the waveform is not symmetric about the half-peak crossing, except in the limit when $\phi \gg 1$. In the latter case, Eq. (22) reduces to Eq. (12) but with l replaced by $4V\tau\phi$.

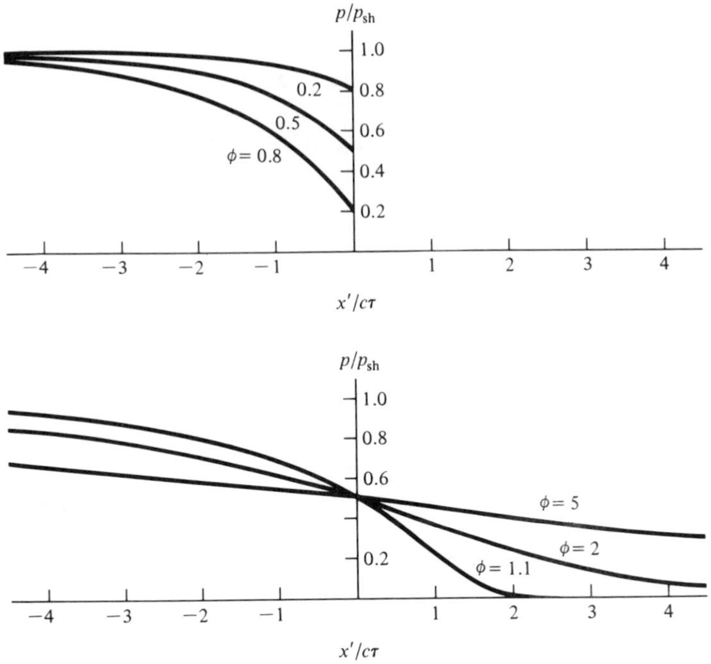

Figure 11-10 Profiles of the leading portions of shock waves of various amplitude in a medium with a single relaxation process and with viscosity and thermal conduction neglected. The parameter ϕ is $2\rho c\ \Delta c/\beta p_{sh}$, where Δc is difference between frozen sound speed and equilibrium sound speed, p_{sh} is shock overpressure, and β is $1 + \frac{1}{2}B/A$. (*a*) If $\phi < 1$, the asymptotic waveform begins with a discontinuity, while (*b*) if $\phi > 1$, it has a precursor and no discontinuity.

Plots of $v/v_{sh} = p/p_{sh}$ versus $\xi/V\tau$, derived from Eqs. (21) and (22), are given in Fig. 11-10 for various values of ϕ. From the inspection of such plots and from the analysis above, one concludes that the relaxation process has relatively little effect on the transition region if the overpressure amplitude is such that $\phi < 0.2$. In the other limit, when the overpressure is sufficiently low for ϕ to be greater than, say, 4, the effect of a relaxation process can be formally taken into account by the thermoviscous model, i.e., that leading to Burgers' equation, with a suitable augmentation of the bulk viscosity. This presumes that $l + 4V\tau\phi$ is substantially smaller than one-fourth of a representative wavelength of the disturbance. In the opposite circumstance, the nonlinear effects would be of negligible importance compared with dissipation and dispersion.

For sound in air, the value of Δc at 20°C is 0.11 m/s for O_2 vibrational relaxation and is 0.023 m/s for the N_2 vibrational relaxation. The corresponding values of $\rho c\ \Delta c/\beta$ are 39.0 and 7.8 Pa, respectively. Consequently, these relaxation processes have minor influence on shock structure if $p_{sh} > 200$ Pa. If $p_{sh} < 2$ Pa, the presence of relaxation processes is accounted for by an appropriate augmentation of the bulk viscosity, providing the shock duration is somewhat longer than what the model would predict for the rise time.

11-7 TRANSITION TO OLD AGE

The gradual rounding of the shocks in a sawtooth waveform results ultimately in a sinusoidal waveform. We here complete the discussion of the example begun in Sec. 11-2 with an analysis of the corresponding solution of Mendousse's version, Eq. (11-6.7), of Burgers' equation.

Reduction to the Linear Diffusion Equation

The insertion† of

$$v(x, t') = a \frac{F_{t'}(x, t')}{F(x, t')} \tag{11-7.1}$$

into Eq. (11-6.7), where a is a constant, yields the differential equation

$$F^2(F_{t'x} - \delta c^{-3} F_{t't't'}) - FF_{t'}[F_x - (3\delta c^{-3} - \beta a c^{-2})F_{t't'}] + (F_{t'})^3(\beta a c^{-2} - 2\delta c^{-3}) = 0 \tag{11-7.2}$$

Consequently, Eq. (1) satisfies the Mendousse-Burgers equation if

$$a = \frac{2\delta}{\beta c} \tag{11-7.3}$$

$$F_x = \delta\, c^{-3} F_{t't'} \tag{11-7.4}$$

Thus, the problem of solving the nonlinear partial-differential equation is reduced to that of solving the linear diffusion equation.

Solution of the Boundary-Value Problem

If the acoustic pressure at $x = 0$ is $P_0 \sin \omega t$, as in Sec. 11-2, the function F should satisfy the boundary condition

$$\frac{P_0}{\rho c} \sin \omega t = \frac{2\delta}{\beta c} \frac{\partial}{\partial t}(\ln F) \qquad x = 0$$

or

$$F = \exp\left(-\frac{\Gamma}{2} \cos \omega t\right) \qquad x = 0 \tag{11-7.5}$$

$$\Gamma = \frac{\beta P_0}{\rho \omega \delta} = \frac{c^3/\delta}{\omega^2 \bar{x}} \tag{11-7.6}$$

The diffusion equation (4) is separable and has particular solutions

$$e^{-n^2\omega^2(\delta/c^3)x} \cos n\,\omega t' = e^{-n^2 \alpha x} \cos n \omega t'$$

† Cole, "On a Quasi-linear Parabolic Equation"; E. Hopf, "The Partial Differential Equation $u_t + uu_x = \mu u_{xx}$," *Commun. Pure Appl. Math.*, 3:201–230 (1950); the adaption to Eq. (11-6.7) of this technique for solving quasi-linear partial-differential equations is included in Mendousse, "Nonlinear Dissipative Distortion."

The analysis, given the F function appropriate to the ray path connecting airplane flight track and observation point, is then along the lines summarized here. Taking into account the variation of atmospheric properties with height proceeds in the manner outlined in Sec. 11-8.

PROBLEMS

11-1 Use data for water summarized in Sec. 1-9 to derive the parameter of nonlinearity B/A for pure water at 10°C.

11-2 For a simple wave, not necessarily of low amplitude, advancing in the $+x$ direction through a gas of original ambient pressure p_0, density ρ_0, and sound speed c_0, give explicit expressions for fluid velocity $v(p)$ and sound speed $c(p)$ as functions of total pressure p.

11-3 Tabulations of Bessel functions indicate (see Prob. 10-18) that

$$J_n(n\sigma) \rightarrow \frac{0.447}{n^{1/3}} + 0.411(\sigma - 1)n^{1/3}$$

in the limit of small $\sigma - 1$ and large n. Use this result to show that the Fubini-Ghiron solution is convergent at $\sigma = 1$ but its derivative with respect to t' diverges at $\sigma = 1$ for some value of $\omega t'$. What does the latter imply is occurring in the waveform as $\sigma = 1$ is approached?

11-4 Prove that the tangential component of the fluid velocity must be continuous across a shock.

11-5 The weak-shock model predicts that a shock of overpressure p_{sh} advances with speed $c + \frac{1}{2}\beta p_{sh}/\rho c$ into a medium at rest with ambient sound speed c and ambient density. Derive an expression for the lowest nonvanishing-order (in $p_{sh}/\rho c^2$) correction to this, assuming that the fluid is an ideal gas.

11-6 The signature (acoustic pressure versus time) of a wave recorded at a point $x = 0$ is shown in the figure. The wave propagates in a homogeneous medium without ambient flow in the $+x$ direction.

(*a*) To what distance x_{onset} must the wave propagate before a shock is first formed?

(*b*) Sketch the waveform giving expressions for peak overpressure and positive-phase duration for $x = x_{onset}$.

(*c*) Describe the evolution in the signature for $x > x_{onset}$.

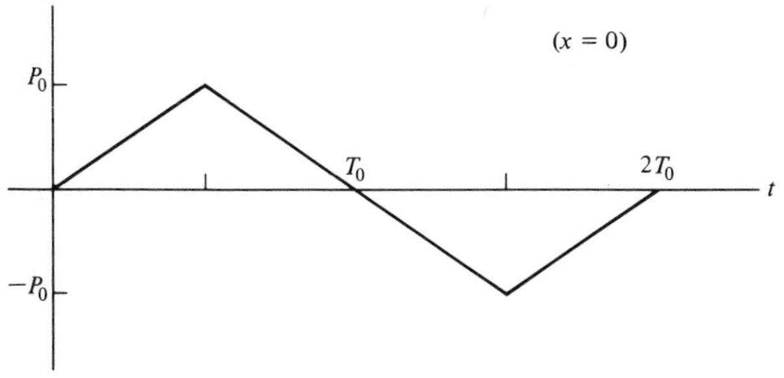

Problem 11-6

11-7 A microphone at $x = 0$ records a transient waveform whose early portion is shown in the figure. The disturbance is a plane wave propagating in the $+x$ direction.

(*a*) How far must the wave propagate beyond $x = 0$ before the second shock overtakes the first?

(*b*) Sketch the waveform's early portion for x less than and for x greater than the value determined in (*a*) and give expressions for all times and overpressures that characterize the waveform.

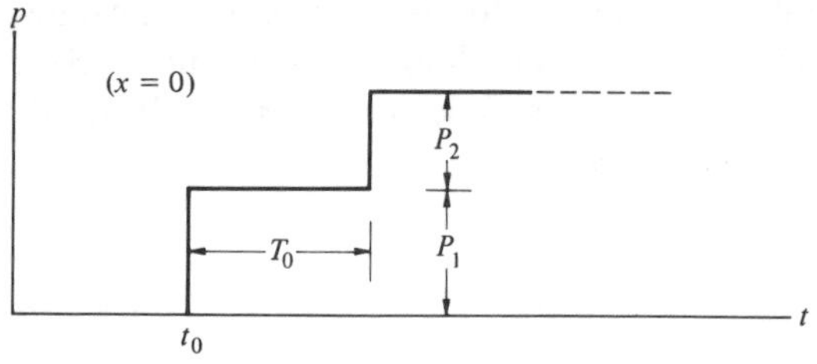

Problem 11-7

11-8 Neglecting thermal conduction, determine an expression for the ambient-temperature rise in a fluid after the passage of an N wave of overpressure P and positive-phase duration T.

11-9 (*a*) Show that the solution of the Mendousse version (11-6.7) of Burgers' equation in the limit of small-amplitude disturbances is

$$v = \frac{B}{x^{1/2}} \int_{-\infty}^{t'} v(0, \tau) e^{-K(t'-\tau)^2/x} \, d\tau$$

where $v(0, \tau)$ is the value of v at $x = 0$ at time τ.

(*b*) What are appropriate identifications for the constants K and B?

(*c*) Explain whether this result is consistent with the particular solution (11-7.12).

11-10 Show that there is a logarithmic derivative substitution analogous to that in Eq. (11-7.1) which reduces the solution of Burgers' equation (11-6.6) to the solution of the linear diffusion equation. Explain how this technique might yield a solution of Burgers' equation when v is specified versus x at $t = 0$.

11-11 (*a*) Show that the approximate dispersion relation derived in Sec. 10-5 for quasi-planar waves in a duct leads in the same spirit of approximation for a transient pulse to the integrodifferential equation

$$\frac{\partial p}{\partial t} + c \frac{\partial p}{\partial x} = -\delta_D \frac{\partial}{\partial t} \int_{-\infty}^{t} \frac{p(x,t_0)}{(t-t_0)^{1/2}} \, dt_0$$

(*b*) What is the appropriate identification for the parameter δ_D?

(*c*) What would be a simple modification of this equation that takes nonlinear effects into account?

11-12 (*a*) Show that Burgers' equation (11-6.6) has the energy-conservation-dissipation corollary

$$\tfrac{1}{2}\rho(v^2)_t + \rho[\tfrac{1}{3}\beta v^3 - \delta v v_{x'}]_{x'} = -\delta\rho(v_{x'})^2$$

(*b*) Hence show that the energy dissipated (per unit time and per unit area transverse to propagation direction) by a stepped shock of overpressure p_{sh} is independent of δ if the propagation is governed by Burgers' equation.

(*c*) How does your expression for the energy-dissipation rate compare with the result in Eq. (11-4.11)?

11-13 An N wave measured at 10 cm from an electric spark in air has a half duration of 10 μs and a pressure amplitude of 1600 Pa. What should these two parameters be at 60 cm from the spark? [B. A. Davy and D. T. Blackstock, *J. Acoust. Soc. Am.*, **49**:732–737 (1971).]

11-14 The waveform described in Prob. 11-6 is a cylindrically symmetric wave radiating outward from the z axis and corresponds to the radial distance r_0, where $r_0 \gg cT_0$.

(*a*) At what value of r would a shock first be formed?

(*b*) Determine peak overpressure and positive-phase duration as functions of r.

11-15 (*a*) Determine an expression for the age variable for a cylindrically diverging wave.

(*b*) What rules apply for extrapolation from values of shock overpressure and positive-phase duration of an N wave received at radius r_1 to values appropriate to radius r_2?

11-16 A pulse propagating radially outward has the form $P_0 \sin \omega t$ for $0 < \omega t < 2\pi$ and is otherwise 0 at the radius r_0.

(*a*) Determine expressions for the asymptotic r dependence of the resulting N-wave overpressure $P(r)$ and positive-phase duration $T(r)$. Assume $P_0 \ll \rho c^2$ and make whatever approximations are appropriate to the model of a weak shock.

(*b*) What are the corresponding values of the constants r^* and K that appear in Eqs. (11-9.8)?

(*c*) Give numerical values appropriate for $P_0 = 10^4$ Pa, $2\pi/\omega = 10\ \mu$s, $r_0 = 5$ cm, the medium being air at a pressure of 10^5 Pa and at a temperature of 20°C. The far-field prediction is desired for a radius of 10 m.

11-17 The sound wave passing into the throat of an exponential horn, throat radius r_t and flare constant m, has pressure amplitude P_0 and angular frequency ω. Determine an approximate expression for the fraction of the radiated power that goes into the higher harmonics. Ignore dissipation and assume that the parameters are such that no shocks are formed within the horn. Assume also that $k^2 \gg m^2$ and that the horn can be regarded as a ray tube.

11-18 A plane wave is propagating obliquely downward so that its wavefront normal makes an angle of θ with the $-z$ axis. The ambient medium is idealized as an isothermal atmosphere whose density decreases exponentially with height, so that $d\rho/dz = -\rho/H$, where H is a constant. At height h_f the acoustic pressure is given by $\varepsilon \sin (2\pi t/T)$ for $0 < t < T$ and is zero otherwise. For given $\rho_0(h_f)$, H, θ, T, and c, there is some value of ε below which a shock can never be formed, regardless of how far the wave propagates. Determine this critical value of ε.

11-19 For the circumstances described in Prob. 11-18 and for $\theta = 0$, determine the asymptotic form of the waveform at heights h many multiples of H below h_f given that ε has one-half the critical value determined in Prob. 11-18.

11-20 A typical sonic boom received on the ground below a supersonic airliner (flying at Mach 2 and 13 km altitude) has an overpressure of 100 Pa and a positive-phase duration of 0.1 s. If the air is at 20°C and has a relative humidity of 50 percent, which of the following processes should have the greatest effect on the shape of the waveform near the shock front: viscosity, O_2 vibrational relaxation, or N_2 vibrational relaxation? (The prevalent view is that atmospheric turbulence is more important for explaining waveform shape alterations than any dissipative mechanism.)

11-21 A pressure pulse at $x = 0$ has the form $p = K/\Delta$ for $-\Delta/2 < t < \Delta/2$ at $x = 0$ and is otherwise 0. Discuss the nonlinear plane-wave propagation of this pulse in the limit of large distance x. How does the asymptotic result evolve when Δ is allowed to become vanishingly small?

11-22 A theory of sonic-boom generation caused by lift proceeds from the model of a distribution of forces moving at supersonic speed through the air. Suppose that the forces are such that Euler's equation in the linear approximation becomes

$$\rho \frac{\partial \mathbf{v}}{\partial t} + \nabla p = -\mathbf{f}(Vt - x)\, \delta(y)\, \delta(z) \quad \text{where} \quad \mathbf{f}(\xi) = \mathbf{e}_z \frac{\pi F_L}{2L} \sin \frac{\pi \xi}{L}$$

for $0 < \xi < L$ and is otherwise zero. Here F_L is the total lift force, and $\mathbf{f}(\xi)$ is the lift force per unit length.

(*a*) Determine the linearized acoustics solution for the resulting sound field at a large distance $|z|$ below ($y = 0$) the flight trajectory.

(*b*) What is the appropriate identification for the Whitham F function?

(*c*) Determine the asymptotic form of the pulse below the source when accumulative nonlinear propagation effects are taken into account.

11-23 How should Burgers' equation be modified to apply to a spherically spreading wave?

11-24 (*a*) Determine analytical expressions for the Whitham F function of the body depicted in Fig. 11-16.

(*b*) What is the corresponding value for the constant K that appears in Eq. (11-10.19)?

11-25 Determine asymptotic expressions for the far-field pressure waveform generated by the supersonic motion of the body of revolution depicted in Fig. 11-18.

APPENDIX
ANSWERS AND HINTS TO PROBLEMS

ANSWERS AND HINTS TO PROBLEMS

1-1. $d = 3; \quad \gamma = 5/3; \quad c = (\gamma R_0 T/M)^{1/2}$

1-2. $\dfrac{d}{dt}\iiint_{V*} dV = \iint_{S*} v_n\, dS$

1-3. $\dfrac{d}{dt}\iiint_{V*} \rho\, dV = \iiint_{V*} \dfrac{\partial \rho}{\partial t}\, dV + \iint_{S*} \rho v_n\, dS$

1-4. $\mathbf{f}_B = -\rho g \mathbf{e}_z$

1-5. a) $T ds = c_v dT - (RT/\rho) d\rho; \quad u = c_v T$

b) $p = K\rho^{\gamma}; \quad K = p_0 \rho_0^{-\gamma} e^{(s - s_0)/c_v}$

1-6. a) $\mathbf{v}_0 = 0; \quad \partial p_0/\partial t = 0; \quad \partial \rho_0/\partial t = 0$

b) $\partial p/\partial t = c^2[\partial \rho'/\partial t + \mathbf{v}\cdot\nabla \rho_0]$

c) $\nabla\cdot\left(\dfrac{1}{\rho_0}\nabla p\right) = -\dfrac{\partial}{\partial t}(\nabla\cdot\mathbf{v})$

1-7. a) $p_0(z) = p_0(0) e^{-(\gamma g/c^2)z}; \quad \dfrac{d\rho_0}{dz} = -\dfrac{\gamma g}{c^2}\rho_0$

b) $\partial \rho'/\partial t + \partial(\rho_0 v_z)/\partial z = 0; \quad \partial(\rho_0 v_z)/\partial t = -\partial p'/\partial z - g\rho'$

c) $\dfrac{\partial}{\partial z}(p' - c^2\rho') = (\gamma - 1) g \rho'$

$$\frac{\partial^2 p'}{\partial t^2} + (\gamma - 1) g \left[-\frac{\partial p'}{\partial z} - g\rho' \right] = c^2 \frac{\partial^2 \rho'}{\partial t^2}$$

1-8. $h = 0.0339; \qquad c = 351$ m/s

2-17. Applicable intermediate result for cited special case is

$$\mathscr{L}\{p(t)\} = \mathrm{Re}\left\{\frac{Ae^{-i\omega t}}{2\pi}\int_0^t h(\xi)e^{i\omega\xi}\,d\xi\right\}$$

2-18. The decibel loss (with $Q^2 = 2\beta x f_0^2$) is

$$-10\log\left\{\frac{\sqrt{2}}{Q}\int_{Q/\sqrt{2}}^{Q\sqrt{2}} e^{-y^2}dy\right\}$$

2-19. $\mathscr{D}_p(\tau) = S_0\Delta f\dfrac{\sin(\pi\tau\Delta f)}{\pi\tau\Delta f}\cos(\pi[f_1+f_2]\tau)$, where $\Delta f = f_2 - f_1$

2-20. Variance in $(p_b^2)_{\mathrm{est}}$ is $\langle p_b^2\rangle^2/[T\Delta f]$ in both cases

2-21. $L_{125} = 10\log[10^{(L_C+0.6)/10} - 10^{(L_A+5.4)/10}]$

2-22. a) $p_f^2(f) = (2\times10^{-3})(f/10^3)^4e^{-2(f/10^3)^2}$

b) $L_A \approx 87.7$ dB

2-23. Occasional pass-by's of noisy vehicles, firing of different cylinders on same engine, atmospheric turbulence, rush hour traffic, pavement roughness and irregularities, aerodynamic noise of flow around moving vehicles.

2-24. 0.63 m

2-25. 3 dB per doubling of distance

2-26. Ratio is $1/[T\Delta f]$

2-27. $N > 100$

2-28. $I = \dfrac{\pi}{12\sqrt{3}}$

2-29. 15, 19, or 22 keys per octave

2-30. To carry through heuristic derivation involving interchange of integration order, insert convergence guarantor $e^{-\epsilon\tau}$ and recognize a Dirac delta function in limit $\epsilon\to0$

2-31. Proper assumptions imply n-th peak of running time average is $1/T$ times total time integral of $p_{n,F}^2$ where $p_{n,F}$ is acoustic pressure, after filtering, of n-th pulse. Use Parseval's theorem.

2-32. $1-(6/\pi^2) = 0.392$

2-33. Insert a factor of $e^{-\epsilon t^2}$ on left side before inserting Fourier transform relations and interchanging order of integration.

2-34. $p_f^2(f) \approx \frac{8\pi^2}{100} \int_{100/T}^{200/T} |\hat{g}(2\pi f)|^2 df$

2-35. $\hat{p}(\omega) = \frac{ip_{pk}}{2\pi\omega}$; $p_F(t) = \frac{p_{pk}}{\pi} \int_{\omega_0/\sqrt{2}}^{\omega_0\sqrt{2}} \frac{\sin[\omega(t-\tau)]}{\omega} d\omega$

fraction $= 1 - \frac{1}{2^{3/2}\pi^2} = 0.964$

2-36. a) $v_f^2(f) = \frac{\omega^2 F_f^2(f)}{(k-m\omega^2)^2 + \omega^2 b^2}$

b) $(v^2)_{av} = \frac{F_f^2(f_r)}{4mb}$, where $2\pi f_r = (k/m)^{1/2}$

2-37. If L is measured in nepers, then $L_1 \oplus L_2 = L_1 + \frac{1}{2}\ln[1 + e^{-2(L_1 - L_2)}]$.

2-38. a) Admissible. b) Admissible. c) Admissible only if $b < 2a$.

2-39. $L_E = 10 \log \left\{ \frac{p_{pk}^2/p_{ref}^2}{2\sqrt{2}\pi^2 f_0 t_{ref}} \right\}$

which decreases by 3 dB when f_0 doubles.

2-40. a) Derive $\hat{p}(\omega) = \frac{i}{2\pi\omega} \int \frac{dp}{dt} e^{i\omega t} dt$ and let $\frac{dp}{dt}$ equal $(\Delta p)\delta(t - t_0)$ plus a bounded quantity. The contribution from the latter goes to zero at large ω at least as fast as ω^{-2}.

b) $\hat{p}(\omega) = -\frac{1}{2\pi\omega^2} \int \frac{d^2p}{dt^2} e^{i\omega t} dt$ where $\frac{d^2p}{dt^2}$ is $(\Delta \dot{p})\delta(t - t_0)$ plus a bounded quantity.

2-41. a) Integrate by parts and use $(d/d\tau)h_F(t-\tau) = -(d/dt)h_F(t-\tau)$.

b) Prove that filtering operation commutes with time and spatial differentiations.

3-1. $v_r = -\omega b \sin\theta \sin(\omega t - \phi)$ at $r = a$

3-2. Applicable intermediate result is the ratio of the octave band contribution to the mean squared pressure, when reflection is included, to that due to incident wave alone, this ratio being

$$2 + 2\left\{ \frac{\sin(\Psi f_2) - \sin(\Psi f_1)}{(f_2 - f_1)\Psi} \right\}$$

where $\Psi = (4\pi y/c)\cos\theta_I$. Required minimum distance for y is 1.62 m.

3-3. Let $\eta(t-[(x/c)\sin\theta_I])$ be the displacement of the interface, such that

$$(v_y)_0^{(+)} = \left(\frac{\partial}{\partial t} + v_0 \frac{\partial}{\partial x}\right)\eta$$

3-4. $\dfrac{Z}{\rho c} = \dfrac{3\sqrt{2}}{5} - i\dfrac{4\sqrt{2}}{5}$; $\alpha = 0.723$

3-5. a) $\theta_I = 85.4°$

3-6. With $Z = \rho c(\zeta_R + i\zeta_I)$, one finds

$$\alpha(\theta_I) = \frac{4\zeta_R \cos\theta_I}{(\zeta_R \cos\theta_I + 1)^2 + (\zeta_I \cos\theta_I)^2}$$

Values of α for any two angles of incidence allow ζ_R to be uniquely determined, but one can only determine the absolute magnitude of ζ_I.

3-7. $Z = \left(\dfrac{\rho c}{\cos\theta_I}\right)\dfrac{Be^{i\psi}}{2A - Be^{i\psi}}$

3-8. $\tau = \dfrac{(2\rho c\omega\pi a^2)^2}{(2\rho c\omega\pi a^2)^2 + (k_{\text{eff}} - \omega^2 m_{\text{eff}})^2}$

3-9. $\frac{1}{4}k(x_p^0)^2$; nonoscillatory if $k_{\text{eff}}\, m_{\text{eff}} < (\rho c\pi a^2)^2$

3-10. $\Delta f = -3.8$ Hz; $Q = 39.3$

3-11. Right side of equation for fraction reflected,

$$|\mathscr{R}_{\mathrm{I,II}}|^2 = \left[\frac{(\rho c)_{\mathrm{II}}\sec\theta_{\mathrm{II}} - (\rho c)_{\mathrm{I}}\sec\theta_{\mathrm{I}}}{(\rho c)_{\mathrm{II}}\sec\theta_{\mathrm{II}} + (\rho c)_{\mathrm{I}}\sec\theta_{\mathrm{I}}}\right]^2$$

and Snell's law equation are unchanged when the wave comes from medium II at angle of incidence θ_{II}

3-12. $c_{\mathrm{II}} = 5596$ m/s; $L = 0.070$ m; 100% transmitted

3-13. $k \approx \dfrac{\omega}{c}\left[1 + \dfrac{2i}{(500)^2}\right]$; $\alpha \approx \dfrac{\omega}{c}\left[\dfrac{2+i}{500}\right]$

$$I_x \approx \frac{|\hat{p}|^2}{2\rho c}; \quad I_y \approx -\frac{|\hat{p}|^2}{1000\rho c}$$

$$|\hat{p}|^2 = P^2 \exp\left\{-2(\omega/c)y/500\right\}\exp\left\{-4(\omega/c)x/(500)^2\right\}$$

3-14. $Z = R_f + i\rho c \cot kL$; the fraction absorbed is

$$\frac{4R_f\rho c}{(R_f+\rho c)^2 + (\rho c \cot kL)^2}.$$

The maximum value $4R_f\rho c/(R_f+\rho c)^2$ occurs when $L = (2n+1)\pi/2k$, with n integer.

3-15. Energy at time $10L/c$ is $50(\rho AL)V_0^2$.

3-16. No; Yes; No

3-17. At the ground, $v_{pk} = 0.005$ m/s; intensity was 0.005 W/m^2; at the cited ionospheric height, $v_{pk} = 50$ m/s; intensity was 0.005 W/m^2.

3-18. $R_f = 1.5\rho c$; $\alpha = 0.96$; if wall not present then $R_{TL} = 4.9$ dB

3-19. a) 4792 Hz; b) 2727 Hz; c) 5455 Hz; d) Ratio is always 2:1

3-20. $\dfrac{d}{\lambda} = \dfrac{1}{2} - \dfrac{1}{2\pi}\tan^{-1}(X/2)$; $\Delta R_{TL} = 10\log\left[1 + \dfrac{X^4}{4X^2+4}\right]$;

$$X = \frac{\omega m_{pl}}{\rho c}$$

3-21. Elliptical counterclockwise path:

$$(8/9)(\delta x)^2 + (\delta y)^2 = (V_0/\omega)^2\exp\left\{-(32)^{1/2}\omega y/c\right\}$$

Lowest point in trajectory corresponds to surface wave trough.

3-22. a) $\dfrac{\hat{R}_I}{\hat{T}_{III}} = \dfrac{i}{2}\left[\dfrac{(\rho c)_I}{(\rho c)_{II}} - \dfrac{(\rho c)_{II}}{(\rho c)_I}\right]\sin(\omega d/c_{II})$

3-23. k_{II} determined from

$$\frac{\{(B/A)e^{ik\Delta L}\}_b}{\{(B/A)e^{ik\Delta L}\}_a} = \frac{\sin(k_{II}d_b)}{\sin(k_{II}d_a)}$$

Then Z_{II} determined (data from either "a" or "b" experiment) from

$$(B/A)e^{ik\Delta L} = -\frac{i}{2}\left[\frac{(\rho c)_I}{Z_{II}} - \frac{Z_{II}}{(\rho c)_I}\right]\sin k_{II}d$$

3-24. $p = \dfrac{2P}{1+\epsilon^2}\{\epsilon f_{\text{even}}(\tau,D) + f_{\text{odd}}(\tau,D)\}$

$$f_{\text{even}} = -2 + D\Psi - (\tau/2)\ln\Phi; \quad f_{\text{odd}} = -\tau\Psi - (D/2)\ln\Phi$$

$$\Psi = \tan^{-1}\frac{1-\tau}{D} + \tan^{-1}\frac{1+\tau}{D}; \quad \Phi = \frac{D^2+(\tau-1)^2}{D^2+(\tau+1)^2}$$

where $\epsilon = \dfrac{\beta_{\text{II}}\rho_{\text{I}}c_{\text{I}}}{\rho_{\text{II}}c_{\text{II}}\cos\theta_{\text{I}}}$; $D = \dfrac{\beta_{\text{II}}d}{c_{\text{II}}T}$;

$$\tau = t/T; \quad \beta_{\text{II}}^2 = \left(\frac{c_{\text{II}}}{c_{\text{I}}}\right)^2 \sin^2\theta_{\text{I}} - 1$$

3-25. $\hat{p} = A\sin(n\pi x/L)$; $f = nc/2L$.

3-26. $f_r = c/4L$; $Q = \pi/4\epsilon$; $\mathscr{P} = \dfrac{\rho c V_0/2\epsilon}{1+4Q^2(\Delta f/f_r)^2}$

3-27. $c_W = (T/\sigma)^{1/2}\Phi(\eta)$; $\eta = \dfrac{2\rho T^{1/2}}{\sigma^{3/2}\omega}$

with $\Phi(\eta)$ determined from numerical solution of $\eta = \Phi^{-3} - \Phi^{-1}$.

3-28. Applicable intermediate results are

$$\frac{d}{dy}\ln\hat{p} = i\omega\rho Z_{\text{local}}^{-1}; \quad \frac{d}{dy}\ln\hat{v}_y = \frac{i\omega}{\rho}(c^{-2} - v_{\text{tr}}^{-2})Z_{\text{local}}$$

Mass law follows from zeroth order approximation to

$$Z_{\text{local}}(0) - Z_{\text{local}}(d) = -i\omega\int_0^d \rho\, dy + i\omega\int_0^d [c^{-2} - v_{\text{tr}}^{-2}]\rho^{-1}Z_{\text{local}}^2\, dy$$

where $\rho = m_{\text{pl}}/d$ and d is plate thickness.

4-1. $p = (\rho c V_0)\left(\dfrac{a}{r}\right)e^{-(c/a)(t-c^{-1}r)}$ if $t > r/c$

Half of the energy stays in near field

4-2. The quantity $e^{ct/a}\psi$ satisfies inhomogeneous ordinary differential equation for a harmonic oscillator under influence of a transient force. Green's function $G(t\,|\,\tau)$ is 0 if $t<\tau$ and is $(a/c)\sin[(c/a)(t-\tau)]$ if $t>\tau$.

4-3. $p = (\Delta p)e^{-s}[\cos s - \sin s]H(s)$; where $s = \dfrac{c}{a}\left(t - \dfrac{r}{c} + \dfrac{a}{c}\right)$

and $\Delta p = \rho c v_C \dfrac{a}{r}\cos\theta$ is pressure jump at $r \gg a$

4-4. $\mathscr{P} = \dfrac{4}{3}\pi\dfrac{\rho c^3(\Omega a/c)^6 b^2}{4+(\Omega a/c)^4}$

4-5. $E_K - E_P \to \frac{1}{4} m_d (v_C^2)_{av} \frac{1 + \frac{1}{2}(ka)^2}{1 + \frac{1}{4}(ka)^4}$

4-6. $F_n = \sum_{u=2-n}^{1} F_{n,u} (a/r)^u; \quad F_{n,u} = \frac{i^{n+1}}{(1-u)!} \sum_{t=0}^{n-2+u} \frac{(-1)^t}{t!}$

$G_n = \sum_{u=-1}^{\infty} G_{n,u} (kr)^u; \quad G_{n,u} = \frac{i^{u+1+n}}{(u+1)!} \sum_{t=0}^{n-2} \frac{(-1)^t}{t!}$

Method of matched asymptotic expansions requires $G_{n+u,-u} = F_{n,u}$

4-7. $p_f^2(f) = \frac{\rho^2 a^4}{r^2[1 + (ka)^2]}; \quad L_{b+1} - L_b \approx 3$ dB

4-8. $L_{b+1} - L_b \approx 9$ dB

4-9. $\mathscr{P} = 2\mathscr{P}_1 \left[1 - \frac{\sin kd}{kd}\right]$

where $\mathscr{P}_1$ is the power when only one source is active

4-10. Write Helmholtz equation, surface boundary condition, radiation condition, and Eqs. (4-6.9) in dimensionless form using a as a length scale and $\hat{v}_{typ}$ as a velocity scale; conclude that $\hat{p}/\rho c\hat{v}_{typ}$ is function of ka and $\mathbf{x}/a$.

4-11. Power proportional to $pM^{7/2}/\gamma^{5/2}$, equal to 0.01 $\mathscr{P}_{av,0}$ and 7500 $\mathscr{P}_{av,0}$ for second and third cases.

4-12. $\mathscr{P} = 2\pi\rho c \left[\frac{k^2 a^4}{1 + (ka)^2} |\hat{v}_S|^2 + \frac{k^4 a^6}{4 + (ka)^4} \frac{|\hat{v}_C|^2}{3}\right]$

$|\hat{v}_C|/|\hat{v}_S| = 34.5$ for equal contributions when $ka = 0.1$

4-13. Ratio $= \frac{3[27 + 6(ka)^2 + (ka)^4] + i3(ka)^5}{81 + 9(ka)^2 - 2(ka)^4 + (ka)^6}$;

The real part is less than 1.25 up to $ka = 1.278$; the imaginary part is less than one-fourth of the real part up to $ka = 1.666$.

4-14. An applicable intermediate result ($kr \gg 1$) is

$$\hat{p} \approx -\frac{i\omega\hat{v}_S\rho}{4\pi r} e^{ikr} \int_0^{\pi} e^{-ika\cos\theta_S} \left[1 + \frac{ka}{i + ka} \cos\theta_S\right] 2\pi a^2 \sin\theta_S \, d\theta_S$$

4-15. In the limit of large r, the integral for $\hat{p}$ reduces to

$$-\frac{i\omega\rho\hat{v}_C}{4\pi} \frac{e^{ikr}}{r} \iint e^{\beta \mathbf{n}_S \cdot \mathbf{e}_r} [1 + D\mathbf{e}_r \cdot \mathbf{n}_S] \mathbf{n}_S \cdot \mathbf{e}_r dS$$

where $D = \frac{\beta^2 + \beta}{2 + \beta^2 + 2\beta}$ with $\beta = -ika$

4-16. $|\delta\phi| < 0.57$ degrees if $kr = 0.1$; $|\delta\phi| < 5.7$ degrees if $kr = 1.0$

4-17. Total power $= \dfrac{2\pi}{\rho c} Q_{11}^2 k^4$; for one alone it is $(1/5)$-th of this value.

4-18. For $r > a$, $t > 0$, the acoustic pressure is nonzero only if $r - a < ct < r + a$ and then has value $[(\Delta p)/2][1 - (ct/r)]$

4-19. $$|\hat{p}|^2 \approx \frac{4p_{10}^2 10^6}{[(97)^2 + (30)^2]} \frac{1}{k^2 r^2} \sin^4\theta \cos^2\phi \sin^2\phi$$

$$\mathcal{P}_{\text{av}} = 163 \frac{c p_{10}^2}{\omega^2 \rho}$$

p_{10} varies with ω as ω^5, increases by factor of 32; $\mathcal{P}_{\text{av}}$ varies as ω^8, increases by factor of 256.

4-20. $$\mathcal{P}_{\text{av}} = \frac{2\pi}{3\rho c} \frac{A^2 k^2 a^4}{1 + (ka)^2}$$

4-21. $$t = (\ln 10) \frac{a}{c} \frac{4 + (ka)^4}{(ka)^4} \frac{M}{m_d}; \quad \text{where } k = \frac{1}{c}\left(\frac{k_{\text{sp}}}{M}\right)^{1/2}$$

4-22. If all four in phase, power increases by factor of 16. For the other stated phasing, one has two perpendicular dipoles, 90° out of phase; radiation is predominantly horizontal with intensity proportional to $\sin^2\theta$.

4-23. $L_A = 106$ dB

4-24. $\mathcal{P} = 12.6$ W

4-25. $$\mathcal{P}_{\text{av}} = \frac{2\pi K^2 k^2}{3\rho c}$$

4-26. $$p_{\text{pk}} = \frac{2^{1/2} 10^{-3} K_1 \cos\theta}{kr}\left[1 + \left(\frac{1}{kr}\right)^2\right]^{1/2}; \quad k = \frac{\omega_1}{c_1}$$

4-27. $p_{\text{rms}} = 0.5$ Pa

4-28. Applicable intermediate results are $\Phi_{\text{in}} = \dfrac{2\Omega a^2}{3\pi} \cos\eta \sin\eta \sin\phi F_2^1(\xi)$

$$F_2^1(\xi) \to -\frac{16}{5}\left(\frac{a}{2r}\right)^3; \quad \Phi_{\text{in}} \to -\frac{4\Omega a^5}{45\pi} \frac{\partial^2}{\partial y \partial z}\left(\frac{1}{r}\right)$$

4-29. Superimpose solution represented by Eqs. (4-8.8), (4-8.10), and (4-8.11) with result from problem 4-28. Let $v_C = -\Omega\Delta$ and use $p = -\rho\partial\Phi/\partial t$.

4-30. A simple example is two closed loops with a common segment. Each loop should have a voltage source and other circuit elements. Let $\hat{e}_1$ be the voltage of the left loop's voltage source and let $\hat{i}_1$ be the corresponding current. You must prove that $\hat{i}_1/\hat{e}_2$ when $\hat{e}_1 = 0$ equals $\hat{i}_2/\hat{e}_1$ when $\hat{e}_2 = 0$.

4-31. Start with Eq. (4-9.7) with surface S consisting of spheres S_1 and S_2 enclosing points $\mathbf{x}_1$ and $\mathbf{x}_2$, respectively. When S_1 and S_2 become small, $\hat{\mathbf{v}}_b$ and $\hat{p}_b$ are regarded as constant over S_1, etc. One must also prove that $\iint \mathbf{n}_1 \hat{p}_a \, dS_1 \to 0$ in the limit of vanishing sphere radius.

5-1. 103 dB

5-2. $G = R^{-1}e^{ikR} - R_I^{-1}e^{ikR_I} \to -2d\dfrac{d}{dz}(r^{-1}e^{ikr})$

where $\{R^2, R_I^2\} = x^2 + y^2 + (z \mp d)^2$

5-3. $\dfrac{\mathcal{P}}{\mathcal{P}_{\text{av,ff}}} = 1 + 3\dfrac{\sin 2kd}{2kd} + 3\dfrac{\sin 2\sqrt{2}kd}{2\sqrt{2}kd} + \dfrac{\sin 2\sqrt{3}kd}{2\sqrt{3}kd}$

$kd > 23$ is necessary criterion

5-4. b) $F(k\mathbf{x}, \mathbf{e}_i) = 8\cos(kx\mathbf{e}_i \cdot \mathbf{e}_x)\cos(ky\mathbf{e}_i \cdot \mathbf{e}_y)\cos(kz\mathbf{e}_i \cdot \mathbf{e}_z)$

c) $\hat{p} = \hat{p}_i(0,0,0)F(k\mathbf{x}, \mathbf{e}_i)$

5-5. a) Method of images gives combination of four free-field Green's functions, with appropriate signs.

b) $r^2|G_k^2| = 16\cos^2(kx_S \sin\theta\cos\phi)\sin^2(kz_S\cos\theta)$

c) $\mathcal{P} = \mathcal{P}_{\text{ff}}\left[1 + \dfrac{\sin 2kx_S}{2kx_S} - \dfrac{\sin 2kz_S}{2kz_S} - \dfrac{\sin 2k(x_S^2 + z_S^2)^{1/2}}{2k(x_S^2 + z_S^2)^{1/2}}\right]$

5-6. $\mathcal{P} = \dfrac{\rho c k^2}{2\pi}|\hat{v}_n|^2 A^2$

5-7. a) $\mathcal{P}_{\text{av}} = 4\rho c(ka)^2\pi a^2|\hat{v}_n|^2$

5-8. a) $|\hat{v}_n| = 0.32\text{m/s}$ b) $|\hat{F}| = 15.5\text{N}$

5-9. a) $I = \dfrac{\rho c|\hat{v}_n|^2 k^2 a^4}{8\pi^2 r^2}\left[\dfrac{\sin((1/2)ka\sin\theta\cos\phi)}{(1/2)ka\sin\theta\cos\phi}\right]^2$

$\times\left[\dfrac{\sin((1/2)ka\sin\theta\sin\phi)}{(1/2)ka\sin\theta\sin\phi}\right]^2$

c) $ka = 2\pi$

5-10. $\eta = 0.003$

7-4. a) $\mathscr{P} = \dfrac{6\pi}{k^2 A}\mathscr{P}_{\text{ff}}$; b) 0

7-5. a) $\hat{p} = -\dfrac{i}{kA}\,[2\pi\rho c\mathscr{P}_{\text{ff}}]^{1/2}e^{ikx}$

b) Answer doubles when $x_0 = \lambda/2$. (Cancellation occurs if $x_0 = \lambda/4$.)

7-6. Start with Eq. (4-9.7) and use relations such as

$$\left(\iint \hat{\mathbf{v}}_a \cdot \mathbf{n}_{\text{in}}\, dS\right)_2 = -D_{21}\hat{p}_1$$

where the indicated integral is over side 2.

7-7. b) Continuous-pressure two-port

c) Circuit should have capacitances C_1 and C_2 in series, and these should be in parallel with capacitance C_3.

7-8. a) $Z_{\text{right}} = Z_{\text{left}} = i(\rho c/A)\cot(kL/2)$; $Z_{\text{mid}} = -i(\rho c/A)\sin(kL)$

b) π-network, two acoustic compliances, $C_A = V/(2\rho c^2)$, and an acoustic inertance, $M_A = \rho L/A$

c) Mass between two springs

7-9. $\dfrac{4A_1A_2}{(A_1 + A_2 + A_3)^2}$

7-10. $|\mathscr{T}|^2 = \dfrac{(a^2\omega\rho c/4T)^2}{1 + (a^2\omega\rho c/4T)^2}$

7-11. $\text{IL} = 20 \log\left[1 + \dfrac{A_b}{2A}\right]$

7-12. The fraction into the branch is $\dfrac{(4A/\rho c)|Z_L|^2}{|1 + (Z_L A/\rho c)|^2}\,\text{Re}\left(\dfrac{1}{Z_L} - \dfrac{1}{Z_R}\right)$

7-13. Equations imply $(\pi x/a) + \text{sign}(x)\ln[(\alpha^{-1} + \alpha)/2] \to \Phi/2B$ as $|x| \to \infty$, so Φ has apparent discontinuity at $x = 0$ of $4B\ln[\csc(\pi b/2a)]$. Criterion for ignoring constriction is $ka\ln[\csc(\pi b/2a)] \ll \pi/2$.

7-14. $\dfrac{4X}{(2+X)^2}$ absorbed; $\dfrac{X^2}{(2+X)^2}$ reflected; $\dfrac{4}{(2+X)^2}$ transmitted;

where $X = 0.01\,\rho c A/b$

7-15. $\omega_r = (4c^2 a/V)^{1/2}$

7-16. a) $Z_A = -i(\omega\rho l'/A)[1 - (\omega_r^2/\omega^2)]$

b) $\omega_r = (\rho l'/A)^{-1/2}[(V/\rho c^2) + G]^{-1/2}$

c) $G \ll V/\rho c^2$

7-17. a) $M_A = \dfrac{1}{(V/\rho c^2)(2\pi f_r)^2}$

b) $l' = \dfrac{Ac^2}{(V)(2\pi f_r)^2}$

c) $|\hat{p}_{\text{in}}/\hat{p}_{\text{ext}}| = \dfrac{2\pi c^3}{(V)(2\pi f_r)^3}$

7-18. a) $R_A = 1.18\times 10^4$ kg/(s·m^4); $C_A = 3.6\times 10^{-9}$ m^{4}s^2/kg;
$M_A = 1.12\times 10^2$ kg/m^4

b) $Q = 15$

7-19. b) $Z_A = \dfrac{\omega^4 C_A^2 M_A^2 - 3\omega^2 C_A M_A + 1}{-i\omega C_A[2-\omega^2 C_A M_A]}$

c) $(M_A C_A)^{1/2}\omega_r$ equal to 0.6180 or 1.6180

d) 180° out of phase at higher resonance

7-20. $Z_{\text{HR}} = \pm \dfrac{i\rho c}{2A}\dfrac{\alpha_T}{(1-\alpha_T^2)^{1/2}}$

7-21. a) $V = 0.0325$ m^3; b) Fraction is 0.9944; c) Fraction is 0.9946

7-22. The excess kinetic energy is the limit as $L_- \to \infty$ and $L_+ \to \infty$ of

$$\iiint' \frac{1}{2}\rho(\nabla\Phi)^2 dV - (\rho U_{12}^2 L_+/2A_+) - (\rho U_{12}^2 L_-/2A_-)$$

where the volume integration extends over the region $-L_- < x < L_+$. The integration is accomplished with aid of $\nabla\cdot(\Phi\nabla\Phi) = (\nabla\Phi)^2$ and inner region outer boundary conditions such as

$$\Phi \to \Phi_\infty^+(t) + (U/A_+)x \quad \text{as } x \to \infty$$

7-23. a) 1.11×10^{-10} W; b) $R_{\text{TL}} = 64$ dB

7-24. $\mathscr{P}_{\text{sc}} = \dfrac{32a^2}{\pi} I_{\text{av}}$

7-25. $\Delta L = 10 \log\left[1 + \left(\dfrac{\omega w^2}{ac}\right)^2\right]$

7-26. Power dissipated $\approx \dfrac{(R_f/2\pi a^2)|\hat{p}_{\text{ext}}|^2}{(\omega M_A)^2 + (R_f/\pi a^2)^2}$

Power transmitted $\approx \dfrac{(\omega a/2\pi c)(\omega M_A)|\hat{p}_{\text{ext}}|^2}{(\omega M_A)^2 + (R_f/\pi a^2)^2}$

where $M_A = \rho/2a$ and $|\hat{p}_{\text{ext}}|^2 = 8\rho c I_{i,\text{av}}$

7-27. Use a symmetrical conically converging-diverging flow over a region of length L on each side of orifice. Then vary L. Principle of minimum acoustic inertance yields

$$M_A \leqslant \frac{\rho 2^{3/2}}{\pi a}\,[1-(a/b)]^{3/2}$$

If $a/b \ll 1$, actual M_A should be $\rho/(2a)$

7-28. Result for $b/a \ll 1$ should be same as for open end of duct with infinite flange. King's exact answer is $0.261\rho/b$. Karal's approximate answer in the $b/a \ll 1$ limit is $0.270\rho/b$.

7-29. Fraction of incident power that is radiated is approximately $2(ka)^2$

7-30. a) $l = 0.310$ m; b) $\mathscr{P} = 2.963$ W; c) $Q = 58.9$; d) 750 Hz

7-31. $A_M/A = 8.6$ and $L = 0.085$ m

7-32. The fraction transmitted is $\dfrac{4A_1A_4A_3^2}{(A_1A_4 + A_3^2)^2}$

7-33. $$\frac{\partial}{\partial t}\left(\frac{\rho U^2}{2A} + \frac{Ap^2}{2\rho c^2}\right) + \frac{\partial}{\partial x}(pU) = 0$$

$(pU)_{\text{av}}$ is independent of x.

7-34. Intermediate result is Bessel's equation ($n = 0$ and $\xi = kx$)

$$\frac{d^2\hat{p}}{d\xi^2} + \frac{1}{\xi}\frac{d\hat{p}}{d\xi} + \hat{p} = 0$$

7-35. See discussion on pp. 363–365. Applicable intermediate result is $\hat{U}_{\text{dia}}/\hat{U}_{\text{th}} = 1 - \omega^2 M_A C_A - i\omega C_A Z_{\text{th}}$

7-36. $\dfrac{d^2\hat{p}}{dx^2} + \dfrac{1}{c^2}(\omega^2 - \omega_c^2)\hat{p} = 0$, where $\omega_c = c(2an/\pi b^2)^{1/2}$ is the cutoff frequency. If $b = 0.05$ m, $a = 0.002$ m, and n is such that 10% of the area is holes, then $f_c = 2000$ Hz.

7-37. $$10^{\text{IL}/10} = \frac{1}{(V+U)^2}[V^2 + U^2 + (1+e)^2 + (1+e)^{-2}V^2U^2]$$

where $e = \dfrac{A_{\text{out}}}{A_{\text{pipe}}}$ and $(\beta L)^2 = (kL)^2 - (400/3)$;

$U = \tan(kL/2) + (ek/\beta)\tan(\beta L/2)$;

$V = \cot(kL/2) + (ek/\beta)\cot(\beta L/2)$

8-1. Make use of relations such as

$$\frac{d\mathbf{n}}{dt} = [(c\mathbf{n} + \mathbf{v})\cdot\nabla]\mathbf{n}; \quad \mathbf{n}\cdot\mathbf{n} = 1; \quad (\mathbf{n}\cdot\nabla)\mathbf{n} = -\mathbf{n}\times(\nabla\times\mathbf{n});$$

$$\nabla\times\{\mathbf{n}/(c + \mathbf{v}\cdot\mathbf{n})\} = 0$$

8-2. $\dfrac{d\mathbf{x}}{dt} = c^2\mathbf{s}; \quad \dfrac{d\mathbf{s}}{dt} = -\dfrac{1}{c}\nabla c; \quad \dfrac{d}{dt} = c\dfrac{d}{dl}$

8-3. Applicable intermediate results are

$$T_{AB} = \int_{q_A}^{q_B} \frac{[(dx/dq)^2 + (dy/dq)^2 + (dz/dq)^2]^{1/2}}{c(x,y,z)}\, dq$$

$$\frac{1}{|d\mathbf{x}/dq|}\frac{d}{dq}\left(\frac{1}{c|d\mathbf{x}/dq|}\frac{d\mathbf{x}}{dq}\right) = \nabla\frac{1}{c}$$

8-4. Start with $F = (\omega - \mathbf{v}\cdot\mathbf{k})^2 - c^2k^2$ such that $\partial F/\partial\omega = 2(\omega - \mathbf{v}\cdot\mathbf{k})$, etc. Set $k_i = \omega s_i$ and recognize that $\Omega = 1 - \mathbf{v}\cdot\mathbf{s}$ and $\Omega^2 = c^2s^2$.

8-5. Applicable intermediate results are

$$(\mathbf{v}\cdot\nabla)\mathbf{v} = \nabla(v^2/2) \text{ and } \frac{1}{\rho}\nabla p = \frac{1}{\gamma - 1}\nabla c^2$$

8-6. Applicable intermediate results are $\dfrac{d}{dt}\left(\dfrac{n_\phi}{c}\right) = -\dfrac{n_\phi n_w}{w}$ and $cn_w = \dfrac{dw}{dt}$

8-7. Plane containing path is formed by origin, initial point, and initial ray direction

8-8. $s_\phi r$, s_z, and $n_\phi r/(c + un_\phi)$ are constant along a ray path

8-9. Applicable intermediate results (with $\mathbf{x}_q = d\mathbf{x}/dq$) are

$$c\mathbf{n}\cdot\mathbf{v}_{\text{ray}}\,\partial L/\partial\mathbf{x}_q = 2\mathbf{x}'c\mathbf{n}\cdot\mathbf{x}' - \mathbf{v} - \mathbf{x}'v_{\text{ray}} + 2\mathbf{x}'\mathbf{v}\cdot\mathbf{x}' = c\mathbf{n}$$

$$\frac{\partial L}{\partial\mathbf{x}} = -\frac{dl/dq}{v_{\text{ray}}}\left[(\mathbf{n}\cdot\mathbf{v}_{\text{ray}})^{-1}\nabla c + \sum_{k=1}^{3} s_k\nabla v_k\right]$$

8-10. Start with $ct = (h^2 + w_1^2)^{1/2} + [z^2 + (w - w_1)^2]^{1/2}$ and recognize that $\partial t/\partial w_1 = 0$ implies $\sin\theta_I = \sin\theta_R$, where $w_1/h = \tan\theta_I$.

8-11. Equate 0 to the derivative with respect to w_1 of

$$t = \frac{(h^2 + w_1^2)^{1/2}}{c_{\text{I}}} + \frac{[d^2 + (w - w_1)^2]^{1/2}}{c_{\text{II}}}$$

8-12. $ct = 2(L^2 - R^2)^{1/2} + 2R\sin^{-1}(R/L)$

8-13. Applicable approximations (when $x/R \ll 1$ and $|R - ct| \ll R$) are

$$x \approx \alpha[1 - (ct/R)] + 20.5ct(\alpha/R)^3$$

$$z - R \approx (ct - R)[1 - (\alpha^2/2R^2)] + (\alpha/R)^4 10.375R$$

8-14. a) Both $\frac{\partial}{\partial\alpha} x(\alpha,t) = 0$ and $\frac{\partial}{\partial\alpha} z(\alpha,t) = 0$ yield the same equation for t in terms of α.

b) Substitute for t into equations for $x(\alpha,t)$ and $z(\alpha,t)$.

c) Caustic begins with a cusp and asymptotically approaches the lines $\pm x/R = 0.1027(z/R) - 0.1826$

8-15. $R(\theta_0) = 2H \tan\theta_0 + 20H \cot\theta_0$

Minimum $R_{\min} = 12.65H$ obtained when $\tan^2\theta_0 = 10$

8-16. a) With appropriate definition of angle ϕ, a ray has circle radius $R = (H - h)/\cos\phi$; the ray that grazes ground has radius $R = H$ and touches ground at $w = [2Hh - h^2]^{1/2}$.

c) $c_0 t = H \ln\left(\frac{H + (2Hh - h^2)^{1/2}}{H - h}\right) + [w - (2Hh - h^2)^{1/2}]$

8-17. $\sin(x/H) = (\tan\theta_0)\sinh(z/H)$

8-18. b) $x = d/2$; additional roots (possible when $d > 2b$) are

$x = (d/2) \pm [(d/2)^2 - b^2]^{1/2}$

8-19. a) Ray leaving surface at angle θ_0 with respect to z-axis has circle radius $(c_0/\alpha)\csc\theta_0$. Rotating radius vector makes angle θ with horizontal, such that $\theta = \pi/2$ at trajectory's lowest point.

b) Caustic condition is $(2n + 1)\cos\theta = \cos\theta_0$

8-20. $(p^2)_{\text{av}} = \frac{\rho c \mathscr{P}}{4\pi x^2} \frac{1}{1 + (x/2H)^2}$

8-21. No. The caustic condition $(\partial w/\partial\theta_0)(\partial z/\partial\theta) - (\partial w/\partial\theta)(\partial z/\partial\theta_0) = 0$ is satisfied only at the source point.

8-22. a) Applicable intermediate results are

$\tanh(c_0\tau/2H) = \cos\theta_0 + (z/w)\sin\theta_0$;

$\cot\theta_0 = (w/2H) - (z/w) + (z^2/2Hw)$

b) $w^2 + \{z + 2H\sinh^2(c_0\tau/2H)\}^2 = H^2\sinh^2(c_0\tau/H)$

8-23. a) $v = \dfrac{2e^5 f(t - c_w^{-1}d - 10c_a^{-1}H)}{[\rho_w c_w + \rho_{a,0}c_{a,0}][10(c_a/c_w)H + d]}$

b) The ratio of intensities, source above ground and source below ground, observed at height $10H$ is

$$\frac{2(\rho_w c_w + \rho_{a,0}c_{a,0})^2}{\rho_w c_w \rho_{a,0} c_{a,0}}\left(\frac{c_a}{c_w}\right)^2\left[1 + \frac{(c_w/c_a)d}{10H}\right]^2$$

8-24. A ray initially making small angle ϵ (radians) with z-axis has path $w \approx \frac{\epsilon}{c_0}\int_0^z c\,dz$. The ray tube area is πw^2 and the power passing through ray tube is $(\mathscr{P}/4\pi)\pi\epsilon^2$.

8-25. One must prove that

$$\frac{k_{z,\mathrm{I}}}{\rho_\mathrm{I}\omega^2} = \frac{\mathscr{R}^2 k_{z,\mathrm{I}}}{\rho_\mathrm{I}\omega^2} + \frac{\mathscr{T}^2 k_{z,\mathrm{II}}}{\rho_\mathrm{II}(\omega - k_x v_\mathrm{II})^2}$$

8-26. a) Applicable intermediate results are $\nabla p_0 = c^2\nabla\rho_0$ and $(\mathbf{v}_0\cdot\nabla)\mathbf{v}' + (\mathbf{v}'\cdot\nabla)\mathbf{v}_0 = \nabla(\mathbf{v}'\cdot\mathbf{v}_0) - \mathbf{v}_0\times(\nabla\times\mathbf{v}')$

b) Take dot product of first displayed equation with $\rho_0\mathbf{v}' + \mathbf{v}_0 p'/c^2$; multiply second displayed equation by $\mathbf{v}'\cdot\mathbf{v}_0 + p'/\rho_0$.

c) $\mathscr{W} = w + \mathbf{I}\cdot\mathbf{v}_0/c^2$ and $\mathbf{I} \approx wc\mathbf{n}$ yield $\mathscr{W} \approx w/\Omega$ where $\Omega = c/(c + \mathbf{n}\cdot\mathbf{v}_0)$

8-27. b)
$$\frac{\partial}{\partial t}\left\{A\left[\frac{1}{2}\rho_0(v')^2 + \frac{(p')^2}{2\rho_0 c^2} + \frac{p'v'v_0}{c^2}\right]\right\} + \frac{\partial}{\partial x}\left\{A\,[p' + \rho_0 v_0 v']\left[v' + \frac{p'v_0}{\rho_0 c^2}\right]\right\} = 0$$

d)
$$\frac{P^2 A(v_0 + c)^2}{\rho_0 c^3} = \text{constant}$$

8-28. If the lens surface is taken as flat on the source side, with thickness h_0 at $r = 0$, and if d is distance from source side of lens to focal point, then

$$h(r) = h_0 + 0.634(d - h_0) - \{[0.634(d - h_0)]^2 - (1.224r)^2\}^{1/2}$$

which is the equation of an ellipse.

8-29. $\hat{p} = Pe^{-ikz}\{1 + (-2z)^{-1}R_0^{1/2}(w - R_0)^{1/2}\exp[ik(w - R_0)^2/(-2z)]\}$

8-30. a)
$$\frac{I_\text{with}}{I_\text{without}} = \frac{8}{5} + \frac{2\sqrt{3}}{\sqrt{5}}\cos(4\pi\cos\theta)$$

b) Radiation pattern given in parametric form (θ_i ranging from 0 to $\pi/2$) by

$$\theta = 2\theta_i - \sin^{-1}([3/4]\sin\theta_i);$$

$$\frac{I_\text{with}}{I_\text{without}} = 1 + \delta + 2\delta^{1/2}\cos(2kR_i\cos^2\theta_i)$$

$$R_i = (1/2)(\delta^{-1} - 1)R_C\cos\theta_i;$$

$$\delta^{-1} + 1 = (8/3)\sec\theta_i[1 - (3/4)^2\sin^2\theta_i]^{1/2}$$

8-31. $(p^2)_{av} = \left(\dfrac{10 \cos\phi}{5\cos\phi + 1}\right)^2 \dfrac{\sin^3\theta_0}{\sin\phi} \dfrac{\rho_0 c_0 \mathscr{P}}{4\pi w^2}$

where $\cot\phi = (2wH)^{-1}(0.19H^2 - w^2)$; $\cot\theta_0 = \dfrac{5(0.19H^2 + w^2)}{9wH}$

8-32. With the abbreviations, $\zeta = z/R_0$ and $u = w/R_0$, the caustic is described by $\zeta + (1/2) = (1/2)[1 - (1 - u^{2/3})^{1/2}] - u^{2/3}(1 - u^{2/3})^{1/2} \approx -(3/4)u^{2/3}$

9-1. a) $\text{TS} = 10\log(\sigma_{\text{back}}/4\pi R_{\text{ref}}^2)$; $\sigma_{\text{back}} = (25/9)\pi a^2(ka)^4$

b) $\sigma_{\text{back}} = \pi a^2$

c) Increases TS by 12 dB and 0 dB, respectively.

9-2. a) $\dfrac{d\sigma}{d\Omega} = (4/9\pi^2)a^2(ka)^4\cos^2\theta\cos^2\theta_k$

b) $\sigma_{\text{back}} = (16/9\pi)a^2(ka)^4\cos^4\theta_k$

c) $\text{TS} = 10\log(\sigma_{\text{back}}/4\pi R_{\text{ref}}^2)$

d) The flow velocity is parallel to the disk's faces, so the disk does not disturb the flow.

9-3. An intermediate result, obtained by the use of Gauss's theorem, is

$$\iiint\left[(\Phi_\nu - x_\nu)\frac{\partial\Delta_2}{\partial x_\mu} - (\Phi_\mu - x_\mu)\frac{\partial\Delta_2}{\partial x_\nu}\right]dV = 0$$

9-4. Applicable equations are $\hat{U}_{\text{into}} = 4\pi\hat{S}/i\omega\rho$;

$\hat{p}_{\text{out}} = B + ik\hat{S}$; $\hat{p}_{\text{out}} = Z_{\text{HR}}\hat{U}_{\text{into}}$

9-5. a) $\omega_r^2 = k_{\text{sp}}/(M + \frac{1}{2}M_d)$, where $M_d = \frac{4}{3}\pi a^3\rho$ is the displaced mass

c) $\hat{D} \approx \dfrac{(i/4)k_r a^3 M_d B}{(M + \frac{1}{2}M_d)[1 - (\omega_r/\omega)^2] + (i/6)(k_r a)^3 M_d}$

9-6. Relative phases, associated with travel time differences, must be randomly distributed over a range of at least 2π for the assumption to hold. Dimension of the scattering volume in the direction $\mathbf{e}_i - \mathbf{e}_{\text{sc}}$ must be at least $\lambda/[2\sin(\theta/2)]$.

9-7. a) Energy scatter per unit time is approximately $\pi^2(a/c)f_{\text{res}}^2 p_f^2(f_{\text{res}})/R_A$

b) Attenuation in nepers per unit propagation distance is

$$\alpha(f) = \frac{4\pi a^2 N}{[1 - (f/f_{\text{res}})^2]^2 + (2aR_A/\rho f_{\text{res}})^2}$$

With increasing x, the spectral density loses a narrow notch of frequencies centered at f_{res}.

9-8. One must solve (numerically) the integral equation

$$(A^{1/2}p_{\text{echo}})_{x=0} = \int_0^\infty J(x_0) f(t - 2x_0/c)\,dx_0$$

and then determine $A(x)$ by solving the ordinary differential equation $4A^2 dJ/dx = (A')^2 = 2AA''$

9-9. a) $(p^2_{\text{sc,ap}})_{\text{av}} \approx \left(\dfrac{\Delta\Omega_{\text{tr}}}{4\pi}\right)^2 (kh)^2 \left[\dfrac{\delta(\rho c)}{\rho c}\right]^2 (p_i^2)_{\text{av}}$

b) $\sigma_{\text{back}} = \dfrac{k^2 h^4 (\Delta\Omega_{\text{tr}})^2}{4\pi} \left[\dfrac{\delta(\rho c)}{\rho c}\right]^2$

9-10. An approximate analysis suggests the replacement

$$\Delta\Omega_{\text{tr}} \to \int_0^{2\pi} \int_0^{\pi} e^{-\alpha\theta^2} \exp\{2ikh \sin\theta \cos\mu \tan\phi\} \sin\theta\, d\theta\, d\mu$$

which approximates to $(\pi/\alpha)\exp[-(kh\phi/\alpha)^2]$.

9-11. $\mathscr{A} = 1$

9-12. a) $\dfrac{\omega - \omega_0}{\omega_0} = \dfrac{1}{8} - \dfrac{(3/8)(ct/r)}{[8 + (ct/r)^2]^{1/2}}$

b) At time $t = 0$ one is still hearing sound that left the source when x was negative.

9-13. $f/f_0 = 1.1526$ if $t < 0$ and $f/f_0 = 0.8676$ if $t > 0$.

9-14. a) $T \approx \dfrac{2}{c}(h^2 + L^2)^{1/2} - VL/c^2$

b) $\dfrac{dz}{dx} = \dfrac{h}{L} - \dfrac{VL}{2ch}$

c) $\dfrac{\omega_{\text{rec}} - \omega_{\text{tr}}}{\omega_{\text{tr}}} \approx -\dfrac{2V}{c} \dfrac{L}{[h^2 + L^2]^{1/2}}$

d) $\sigma_{\text{back}} = \dfrac{k^4}{4\pi}\left(\dfrac{4}{3}\pi a^3\right)^2 \left(\dfrac{3(m - m_d)}{2m + m_d}\right)^2$

$$(p^2)_{\text{av,echo}} = \frac{2|\hat{S}|^2}{\pi} \frac{\sigma_{\text{back}}}{[h^2 + L^2]^2}$$

9-15. a) $\hat{p} = (2\pi k \sin\bar{\theta}) K e^{-i\pi/4} e^{ikz\cos\bar{\theta}}$

b) $\mathscr{R} = e^{-i\pi/2}$

10-13. d) $\text{power} = \dfrac{\omega^5 \pi \mu a^4}{6c^3} |\hat{\xi}|^2$

10-14. $(p^2)_{\text{av}}$ and the power both vary with U as U^6.

10-15. $p_f^2(f) \approx \dfrac{\rho^2 U^5 a^3 Q}{c^2 r^2}$, where the dimensionless quantity Q is a function of the Strouhal number, the Reynolds number, and angular coordinates.

10-16. $p = \dfrac{W}{2\pi h} = 0.47$ Pa

10-17. If one takes $N_B = 6$, the $\hat{p}_m$ depend on m and θ through the factor

$$(R_{L/D}\cos\theta - 6)\,\frac{m[(m/2)\sin\theta]^{6m}}{(6m)!}$$

10-18. A marked increase is expected when $\omega_R L_{\text{eff}}/c$ goes from below unity to above unity. If one requires the amplitude of the Airy function to exceed 1/2 of its peak value, then $[\omega_R L_{\text{eff}}/c]\theta$ lies between the limits, $1 - 0.28(\omega_R/\omega)^{2/3}$ and $1 + 16(\omega_R/\omega)^{2/3}$.

10-19. An applicable intermediate result is

$$((T' - T_\nu)^2)_{\text{av}} = \frac{(\omega\tau_\nu)^2}{1 + (\omega\tau_\nu)^2}((T')^2)_{\text{av}}.$$

Use the approximation $T' \approx (T\beta/\rho c_p)_0 p$ and the thermodynamic relation

$$\beta^2 = \frac{(\gamma - 1)c_p}{c^2 T}$$

10-20.

$$\left(1 + \tau_\nu \frac{\partial}{\partial t}\right)\left(\frac{\partial}{\partial x} + \frac{1}{c_0}\frac{\partial}{\partial t}\right)p = (c_0^{-1} - c_\infty^{-1})\tau_\nu \frac{\partial^2 p}{\partial t^2}$$

10-21. a) $T_{60} = T_{60,n}\left[1 + \dfrac{2cT_{60,n}\alpha_{\text{pl}}}{6\ln 10}\right]^{-1}$

where the nominal reverberation time $T_{60,n}$ corresponds to $\alpha_{\text{pl}} = 0$.

b) $\alpha_{\text{pl}} = 3.8 \times 10^{-4}$ Np/m.

c) It can occur at any frequency above 117 Hz if the humidity is right, and at almost any frequency if the frequency is greater than 5000 Hz.

10-22. $a > 2.3$ m.

10-23. Maximum of 0.0155 Np/m occurs when RH$\approx$10.5%.

10-24. Expand the complex wave number $k(\omega,\mu,\mu_B,\kappa,c_{v1},c_{v2})$ in a power series in μ, κ, etc., and keep only up through the first order terms. The coefficient of any such term should be independent of the parameters that are associated with dissipation.

10-25. Applicable first order intermediate result is

$$i\omega\rho_0\hat{s}_{\text{eq}}/\hat{p} = \frac{2c_p}{\pi c^2\beta T_0}\sum_{\nu}\frac{(\alpha_\nu\lambda)_m\omega^2\tau_\nu}{1-i\omega\tau_\nu} + k^2\kappa(\beta/\rho c_p)_0$$

10-26. At 50 Hz: $\alpha_\mu = 2.5\times10^{-8}$, $\alpha_{\mu_B} = 1.1\times10^{-8}$, $\alpha_\kappa = 1.0\times10^{-8}$, and with relative humidities of 0, 50, and 100%, $\alpha_{\nu1} = 1.0\times10^{-4}$, 7.2×10^{-7}, and 3.0×10^{-7}, while $\alpha_{\nu2} = 7.4\times10^{-6}$, 9.3×10^{-6}, and 4.9×10^{-6}. At 5000 Hz: $\alpha_\mu = 2.5\times10^{-4}$, $\alpha_{\mu_B} = 1.1\times10^{-4}$, $\alpha_\kappa = 1.0\times10^{-4}$, and with relative humidities of 0, 50, and 100%, $\alpha_{\nu1} = 1.2\times10^{-4}$, 6.7×10^{-3}, and 3.0×10^{-3}, while $\alpha_{\nu2} = 7.7\times10^{-6}$, 1.8×10^{-4}, and 3.5×10^{-4}.

10-27. If the plane were flying at 3000m, the calculated upper limit would be 102.2 dB; at 6000m, it would be 114.4 dB.

11-1. $B/A = 2\rho c\left[\left(\frac{\partial c}{\partial p}\right)_T + \frac{\beta T}{\rho c_p}\left(\frac{\partial c}{\partial T}\right)_p\right]$ yields 4.7 for fresh water and 5.0 for sea water.

11-2. $c = c_0\left(\frac{p}{p_0}\right)^{(\gamma-1)/2\gamma}$; $\quad v = \frac{2c_0}{\gamma-1}\left[\left(\frac{p}{p_0}\right)^{(\gamma-1)/2\gamma} - 1\right]$

11-3. $\sum_{n=N}^{\infty}\frac{1}{n^{4/3}}$ converges, but $\sum_{n=N}^{\infty}\frac{1}{n^{1/3}}$ diverges

11-4. Integral form of y-th component of Euler's equation for a stationary control volume is

$$\frac{d}{dt}\int\int\int\rho v_y dV + \int\int[\rho v_y \mathbf{v}\cdot\mathbf{n} + pn_y]dS = 0$$

A derivation similar to that of Section (11-3) yields

$$[\rho v_y(v_x - v_{\text{sh}})]_+ = [\rho v_y(v_x - v_{\text{sh}})]_-$$

11-5. $v_{\text{sh}} \approx c + \frac{1}{2}\beta\frac{\Delta p}{\rho c} - \frac{1}{8}\beta^2\frac{(\Delta p)^2}{\rho^2 c^3}$

NAME INDEX

SUBJECT INDEX